Texts and
Monographs
in Physics

Texts and Monographs in Physics

R. Bass: **Nuclear Reactions with Heavy Ions** (1980).

A. Bohm: **Quantum Mechanics: Foundations and Applications,** Second Edition (1986).

O. Bratteli and D.W. Robinson: **Operator Algebras and Quantum Statistical Mechanics.** Volume I: C*- and W*-Algebras. Symmetry Groups. Decomposition of States (1979). Volume II: Equilibrium States. Models in Quantum Statistical Mechanics (1981).

K. Chadan and P.C. Sabatier: **Inverse Problems in Quantum Scattering Theory** (1977).

M. Chaichian and N.F. Nelipa: **Introduction to Gauge Field Theories** (1984).

G. Gallavotti: **The Elements of Mechanics** (1983).

W. Glöckle: **The Quantum Mechanical Few-Body Problem** (1983).

W. Greiner, B. Müller, and J. Rafelski: **Quantum Electrodynamics of Strong Fields** (1985).

J.M. Jauch and F. Rohrlich: **The Theory of Photons and Electrons: The Relativistic Quantum Field Theory of Charged Particles with Spin One-half,** Second Expanded Edition (1980).

J. Kessler: **Polarized Electrons (1976).** Out of print. (Second Edition available as Springer Series in Atoms and Plasmas, Vol. 1.)

G. Ludwig: **Foundations of Quantum Mechanics I** (1983).

G. Ludwig: **Foundations of Quantum Mechanics II** (1985).

R.G. Newton: **Scattering Theory of Waves and Particles,** Second Edition (1982).

A. Perelomov: **Generalized Coherent States and Their Applications** (1986).

H. Pilkuhn: **Relativistic Particle Physics** (1979).

R.D. Richtmyer: **Principles of Advanced Mathematical Physics.** Volume I (1978). Volume II (1981).

W. Rindler: **Essential Relativity: Special, General, and Cosmological,** Revised Second Edition (1980).

P. Ring and P. Schuck: **The Nuclear Many-Body Problem** (1980).

R.M. Santilli: **Foundations of Theoretical Mechanics.** Volume I: The Inverse Problem in Newtonian Mechanics (1978). Volume II: Birkhoffian Generalization of Hamiltonian Mechanics (1983).

M.D. Scadron: **Advanced Quantum Theory and Its Applications Through Feynman Diagrams** (1979).

N. Straumann: **General Relativity and Relativistic Astrophysics** (1984).

C. Truesdell and S. Bharatha: **The Concepts and Logic of Classical Thermodynamics as a Theory of Heat Engines: Rigourously Constructed upon the Foundation Laid by S. Carnot and F. Reech** (1977).

F.J. Ynduráin: **Quantum Chromodynamics: An Introduction to the Theory of Quarks and Gluons** (1983).

Ola Bratteli
Derek W. Robinson

Operator Algebras and Quantum Statistical Mechanics 1

C^*- and W^*-Algebras
Symmetry Groups
Decomposition of States

Second Edition

Springer-Verlag
New York Berlin Heidelberg
London Paris Tokyo

Ola Bratteli
Institute of Mathematics
University of Trondheim
N-7034 Trondheim NTH
Norway

Derek W. Robinson
Department of Mathematics
Institute of Advanced Studies
Australian National University
Canberra, A.C.T.
Australia

Series Editors

Wolf Beiglböck
Institut für Angewandte Mathematik
Universität Heidelberg
D-6900 Heidelberg 1
Federal Republic of Germany

Joseph L. Birman
Department of Physics
The City College of the
City University of New York
New York, NY 10031
U.S.A.

Robert P. Geroch
Department of Physics
University of Chicago
Chicago, IL 60632
U.S.A.

Elliott H. Lieb
Department of Physics
Joseph Henry Laboratories
Princeton University
Princeton, NJ 08540
U.S.A.

Tullio Regge
Istituto de Fisica Teorica
University di Torino
I-10125 Torino
Italy

Walter Thirring
Institut für Theoretische Physik
der Universität Wien
A-1090 Wien
Austria

Library of Congress Cataloging in Publication Data
Bratteli, Ola.
Operator algebras and quantum statistical mechanics.
(Texts and monographs in physics)
Bibliography: v. 1, p.
Includes index.
Contents: v. 1. C^*- and W^*-algebras, symmetry groups, decomposition of states.
1. Operator algebras. 2. Statistical mechanics. 3. Quantum statistics. I. Robinson, Derek W. II. Title. III. Series.
QA326.B74 1987 512′.55 86-27877

With 1 Illustration

Printed and bound by R. R. Donnelley & Sons, Harrisonburg, Virginia.
Printed in the United States of America.

9 8 7 6 5 4 3 2 1

ISBN 0-387-17093-6 Springer-Verlag New York Berlin Heidelberg
ISBN 3-540-17093-6 Springer-Verlag Berlin Heidelberg New York

Preface to the Second Edition

The second edition of this book differs from the original in three respects. First, we have eliminated a large number of typographical errors. Second, we have corrected a small number of mathematical oversights. Third, we have rewritten several subsections in order to incorporate new or improved results. The principal changes occur in Chapters 3 and 4.

In Chapter 3, Section 3.1.2 now contains a more comprehensive discussion of dissipative operators and analytic elements. Additions and changes have also been made in Sections 3.1.3, 3.1.4, and 3.1.5. Further improvements occur in Section 3.2.4. In Chapter 4 the only substantial changes are to Sections 4.2.1 and 4.2.2. At the time of writing the first edition it was an open question whether maximal orthogonal probability measures on the state space of a C^*-algebra were automatically maximal among all the probability measures on the space. This question was resolved positively in 1979 and the rewritten sections now incorporate the result.

All these changes are nevertheless revisionary in nature and do not change the scope of the original edition. In particular, we have resisted the temptation to describe the developments of the last seven years in the theory of derivations, and dissipations, associated with C^*-dynamical systems. The current state of this theory is summarized in [[Bra 1]] published in Springer-Verlag's Lecture Notes in Mathematics series.

Canberra and Trondheim, 1986

Ola Bratteli
Derek W. Robinson

Preface to the First Edition

In this book we describe the elementary theory of operator algebras and parts of the advanced theory which are of relevance, or potentially of relevance, to mathematical physics. Subsequently we describe various applications to quantum statistical mechanics. At the outset of this project we intended to cover this material in one volume but in the course of development it was realized that this would entail the omission of various interesting topics or details. Consequently the book was split into two volumes, the first devoted to the general theory of operator algebras and the second to the applications.

This splitting into theory and applications is conventional but somewhat arbitrary. In the last 15–20 years mathematical physicists have realized the importance of operator algebras and their states and automorphisms for problems of field theory and statistical mechanics. But the theory of 20 years ago was largely developed for the analysis of group representations and it was inadequate for many physical applications. Thus after a short honeymoon period in which the new found tools of the extant theory were applied to the most amenable problems a longer and more interesting period ensued in which mathematical physicists were forced to redevelop the theory in relevant directions. New concepts were introduced, e.g. asymptotic abelianness and KMS states, new techniques applied, e.g. the Choquet theory of barycentric decomposition for states, and new structural results obtained, e.g. the existence of a continuum of nonisomorphic type-three factors. The results of this period had a substantial impact on the subsequent development of the theory of operator algebras and led to a continuing period of fruitful collaboration between mathematicians and physicists. They also led to an intertwining of the theory and applications in which the applications often forced the formation of the theory. Thus in this context the division of this book has a certain arbitrariness.

The two volumes of the book contain six chapters, four in this first volume and two in the second. The chapters of the second volume are numbered consecutively with those of the first and the references are cumulative. Chapter 1 is a brief historical introduction and it is the five subsequent chapters that form the main body of material. We have encountered various difficulties in our attempts to synthesize this material into one coherent book. Firstly there are broad variations in the nature and difficulty of the different chapters. This is partly because the subject matter lies between the main-streams of pure mathematics and theoretical physics and partly because it is a mixture of standard theory and research work which has not previously appeared in book form. We have tried to introduce a uniformity and structure and we hope the reader will find our attempts are successful. Secondly the range of topics relevant to quantum statistical mechanics is certainly more extensive than our coverage. For example we have completely omitted discussion of open systems, irreversibility, and semi-groups of completely positive maps because these topics have been treated in other recent monographs [[Dav 1]] [[Eva 1]].

This book was written between September 1976 and July 1979. Most of Chapters 1–5 were written whilst the authors were in Marseille at the Université d'Aix-Marseille II, Luminy, and the Centre de Physique Théorique CNRS. During a substantial part of this period O. Bratteli was supported by the Norwegian Research Council for Science and Humanities and during the complementary period by a post of "Professeur Associé" at Luminy. Chapter 6 was partially written at the University of New South Wales and partially in Marseille and at the University of Oslo.

Chapters 2, 3, 4 and half of Chapter 5 were typed at the Centre de Physique Théorique, CNRS, Marseille. Most of the remainder was typed at the Department of Pure Mathematics, University of New South Wales. It is a pleasure to thank Mlle. Maryse Cohen-Solal, Mme. Dolly Roche, and Mrs. Mayda Shahinian for their work.

We have profited from discussions with many colleagues throughout the preparation of the manuscript. We are grateful to Gavin Brown, Ed Effros, George Elliott, Uffe Haagerup, Richard Herman, Daniel Kastler, Akitaka Kishimoto, John Roberts, Ray Streater and André Verbeure for helpful comments and corrections to earlier versions.

We are particularly indebted to Adam Majewski for reading the final manuscript and locating numerous errors.

Oslo and Sydney, 1979

Ola Bratteli
Derek W. Robinson

Contents (Volume 1)

Contents (Volume 2)

Introduction

The theory of algebra of operators on Hilbert space began in the 1930s with a series of papers by von Neumann, and Murray and von Neumann. The principal motivations of these authors were the theory of unitary group representations and certain aspects of the quantum mechanical formalism. They analyzed in great detail the structure of a family of algebras which are referred to nowadays as von Neumann algebras, or W^*-algebras. These algebras have the distinctive property of being closed in the weak operator topology and it was not until 1943 that Gelfand and Naimark characterized and partially analyzed uniformly closed operator algebras, the so-called C^*-algebras. Despite Murray and von Neumann's announced motivations the theory of operator algebras had no significant application to group representations for more than fifteen years and its relevance to quantum mechanical theory was not fully appreciated for more than twenty years. Despite this lapse there has been a subsequent fruitful period of interplay between mathematics and physics which has instigated both interesting structural analysis of operator algebras and significant physical applications, notably to quantum statistical mechanics. We intend to describe this theory and these applications. Although these results have also stimulated further important applications of algebraic theory to group representations and relativistic field theory we will only consider these aspects peripherally.

In order to understand the significance of operator algebras for mathematical physics, and also to appreciate the development of the theory, it is of interest to retrace the history of the subject in a little more detail.

In late 1924, and early 1925, Heisenberg and Schrödinger independently proposed equivalent, although seemingly disparate, explanations for the empirical quantization rules of Bohr and Sommerfeld. These rules had been developed as an aid to the classification of experimental data accumulated in the previous two decades which indicated an atomic and subatomic structure that failed to conform to the accepted rules of classical Newtonian mechanics. These explanations, which were originally known as matrix mechanics and wave mechanics, were almost immediately synthesized into the present theory of atomic structure known as quantum mechanics. This theory differs radically from all previous mechanical theories insofar that it has a probabilistic interpretation and thus replaces classical determinism by a philosophy of indeterminism.

Heisenberg's formalism identified the coordinates of particle momentum and position with operators p_i and q_j satisfying the canonical commutation relations

$$p_i p_j - p_j p_i = 0 = q_i q_j - q_j q_i,$$
$$p_i q_j - q_j p_i = -i\hbar\delta_{ij},$$

and the equation determining the change of any such operator A with the time t was specified as

$$\frac{\partial A_t}{\partial t} = \frac{i(HA_t - A_t H)}{\hbar}.$$

In these equations $\hbar$ is Planck's constant and H denotes the Hamiltonian operator, which is conventionally a function of the particle position and momenta, e.g.,

$$H = \sum_{i=1}^{n} \frac{p_i^2}{2m} + V(q_1, q_2, \ldots, q_n),$$

where the first term corresponds to the kinetic energy of particles with mass m and the second to the energy of interaction between the particles. Although Heisenberg's formalism was tentatively proposed in terms of matrix operators a simple calculation with the commutation relations shows that at least one of each p_i and q_i cannot be a bounded operator. Thus the operators were assumed to act on an infinite-dimensional Hilbert space $\mathfrak{H}$. Physically, each vector $\psi \in \mathfrak{H}$ corresponds to a state of the system and for ψ normalized the values $(\psi, A_t\psi)$ correspond to the values of the observables A at the time t. Schrödinger's wave mechanics, on the other hand, was directly formulated in terms of a function ψ of n variables, the particle coordinates. The function ψ represents the state of the system and the dynamics of particles of mass m with mutual interaction V are determined by the Schrödinger equation

$$i\hbar \frac{\partial \psi_t}{\partial t}(x_1, \ldots, x_n) = \left(-\frac{\hbar^2}{2m} \sum_{i=1}^{n} \frac{\partial^2}{\partial x_i^2} + V(x_1, \ldots, x_n) \right) \psi_t(x_1, \ldots, x_n).$$

Physically, the distribution

$$\rho_t(x_1, \ldots, x_n) = |\psi_t(x_1, \ldots, x_n)|^2 \, dx_1 \cdots dx_n$$

corresponds to the probability distribution that the particle coordinates should assume the values $x_1, \ldots, x_n$ at time t when the system is in the state ψ. In particular, this presumes that ψ_t is a normalized vector of the Hilbert space $L^2(\mathbb{R}^n)$.

The relation between the two formalisms, and their equivalence, follows from the identifications $\mathfrak{H} = L^2(\mathbb{R}^n)$ and

$$(p_i\psi)(x_1, \ldots, x_n) = -i\hbar \frac{\partial \psi}{\partial x_i}(x_1, \ldots, x_n),$$

$$(q_i\psi)(x_1, \ldots, x_n) = x_i\psi(x_1, \ldots, x_n),$$

$$(H\psi)(x_1, \ldots, x_n) = \left(\left(\sum_{i=1}^{n} \frac{p_i^2}{2m} + V(q_1, \ldots, q_n) \right) \psi \right)(x_1, \ldots, x_n).$$

The complementarity between the dynamical algorithms then follows by the transposition law

$$(\psi, A_t\psi) = (\psi_t, A\psi_t),$$

where A and ψ correspond to A_t and ψ_t at time $t = 0$.

The work of Stone and von Neumann in the late 1920s and early 1930s clarified the connection between the above formalisms, provided a mathematically coherent description of quantum mechanics, and proved that the theory was essentially unique. Firstly, these authors extended Hilbert's spectral theorem to general selfadjoint operators on Hilbert space. This theorem assigns a projection-valued measure on the real line $E_A(\lambda)$ to each such operator A and thus for every unit vector ψ, $\lambda \in \mathbb{R} \mapsto (\psi, E_A(\lambda)\psi)$ is a probability measure and in von Neumann's interpretation of quantum mechanics this measure determines the distribution of values obtained when the observable corresponding to A is measured while the system is in the state ψ. In this theory functions of observables still have an interpretation as observables, i.e., if f is a real Borel function then the function f of the observable represented by A is the observable with probability distribution given by $B \mapsto (\psi, E_A(f^{-1}(B))\psi)$ in the state ψ, where B ranges over the Borel sets. The latter observable is then represented by the operator $f(A)$. In particular, A^2, $A^3, \ldots$ have observable significance. Secondly, in 1930, Stone showed that if $t \mapsto U_t$ is a continuous unitary representation of the real line then there exists a unique selfadjoint H such that

$$\frac{dU_t}{dt} = iU_tH$$

on the domain of H and, conversely, if H is selfadjoint then this equation determines a unique continuous unitary representation of $\mathbb{R}$. The connection between U and H corresponds to an exponential relation $U_t = \exp\{itH\}$ and Stone's theorem shows that the Schrödinger equation has a unique solution ψ_t, satisfying the relations $\|\psi_t\| = \|\psi\|$ of probability conservation, if, and only if, the Hamiltonian H is selfadjoint. This solution is given by $\psi_t = U_{-t}\psi$, where we have chosen units such that $\hbar = 1$. The corresponding solution of Heisenberg's equation of motion is then given by

$$A_t = U_tAU_{-t}$$

and equivalence of the two dynamics results because

$$(\psi_t, A\psi_t) = (U_{-t}\psi, AU_{-t}\psi) = (\psi, A_t\psi).$$

Finally, Stone announced a uniqueness result for operators satisfying Heisenberg's commutation relations $p_iq_j - q_jp_i = -i\delta_{ij}$, etc., and von Neumann provided a detailed proof in 1931. As the p_i, or q_i, are necessarily unbounded it is best to formulate this result in terms of the unitary groups

$U_i(t) = \exp\{ip_i t\}$, $V_j(t) = \exp\{iq_j t\}$ associated with the selfadjoint operators p_i, q_j. These groups satisfy the Weyl form of the commutation relations,

$$U_i(s)V_j(t) = V_j(t)U_i(s)e^{ist\delta_{ij}},$$
$$U_i(s)U_j(t) - U_j(t)U_i(s) = 0 = V_i(s)V_j(t) - V_j(t)V_i(s),$$

and the Stone–von Neumann uniqueness theorem states that the only representation of these relations by continuous unitary groups on Hilbert space are sums of copies of the Schrödinger representation. (In Chapter 5 we derive this result from a more general theorem concerning the C^*-algebra generated by an arbitrary number of unitary groups satisfying the Weyl relations.)

Thus by the early 1930s the theory of quantum mechanics was firmly founded and the basic rules could be summarized as follows:

(1) an observable is a selfadjoint operator A on a Hilbert space $\mathfrak{H}$;
(2) a (pure) state is a vector in $\mathfrak{H}$;
(3) the expected value of A in the state ψ is given by $(\psi, A\psi)$;
(4) the dynamical evolution of the system is determined by specification of a selfadjoint operator H through either of the algorithms

$$A \mapsto A_t = e^{itH}Ae^{-itH}, \quad \text{or} \quad \psi \mapsto \psi_t = e^{-itH}\psi.$$

Detailed models are then given by specific identification of A, H, etc., as above.

A slight extension of this formalism is necessary for applications to statistical mechanics. In order to allow the extra uncertainty inherent in the statistical description one needs a broader notion of state. A mixed state ω is defined as a functional over the observables of the form

$$\omega(A) = \sum_i \lambda_i(\psi_i, A\psi_i),$$

where $\lambda_i \geq 0$, $\sum_i \lambda_i = 1$, and $\|\psi_i\| = 1$. If all bounded selfadjoint operators on $\mathfrak{H}$ represent observables then these mixed states are automatically of the form

$$\omega(A) = \mathrm{Tr}(\rho A),$$

where ρ is a positive trace-class operator with trace equal to one.

Subsequently, various algebraic reformulations of this quantum mechanical formalism were proposed. In 1932 von Neumann observed that the product

$$A \circ B = \frac{AB + BA}{2}$$

of two observables A, B can again be interpreted as an observable in his theory because

$$A \circ B = \frac{(A + B)^2 - A^2 - B^2}{2}.$$

This product satisfies the distributive laws,

$$A \circ (B + C) = A \circ B + A \circ C, \qquad (B + C) \circ A = B \circ A + C \circ A,$$
$$\lambda(A \circ B) = (\lambda A) \circ B = A \circ (\lambda B),$$

and is commutative, $A \circ B = B \circ A$, but the product is not associative in general, i.e., $(A \circ B) \circ C$ may be different from $A \circ (B \circ C)$. On the other hand, the following rule, which is weaker than associativity, is satisfied:

$$((A \circ A) \circ B) \circ A = (A \circ A) \circ (B \circ A).$$

In 1933 Jordan suggested that the quantum observables should be characterized by this algebraic structure. Algebras over the reals satisfying these axioms are now commonly known as Jordan algebras. In the mid-1930s Jordan, von Neumann, and Wigner classified the finite-dimensional Jordan algebras over the reals with the additional reality property

$$A_1 \circ A_1 + A_2 \circ A_2 + \cdots + A_n \circ A_n = 0 \Rightarrow A_1 = A_2 = \cdots = A_n = 0.$$

These algebras are direct sums of simple ones, and, with one exception, a simple, finite-dimensional Jordan algebra is an algebra of selfadjoint operators on a Hilbert space with Jordan product defined by the anticommutator. The exceptional algebra is the algebra of hermitian 3×3 matrices over the Cayley numbers, and is denoted by $M_3{}^8$.

Despite these investigations and subsequent analysis of infinite-dimensional Jordan algebras by von Neumann, it appeared that the most fruitful algebraic reformulation of quantum mechanics was in terms of the W^*-algebras of Murray and von Neumann. Thus the quantum observables were identified with the selfadjoint elements of a weakly closed *-algebra of operators $\mathfrak{M}$ on a Hilbert space $\mathfrak{H}$ and the states as the mixed states described above. These mixed states are linear functionals over $\mathfrak{M}$ which assume positive values on positive elements and equal one on the identity. It is now conventional to call all normalized, positive, linear functionals states. The mixed states are usually called normal states and can be characterized by various algebraic, or analytic, properties among the set of all states. (The general structure of topological algebras and their states is discussed in Chapter 2.)

In 1947, following the characterization of C^*-algebras by Gelfand and Naimark, Segal argued that the uniform convergence of observables has a direct physical interpretation, while weak convergence has only analytical significance (a viewpoint which is disputable). Thus he proposed that the observables should be identified as elements of a uniformly closed Jordan algebra and demonstrated that this was sufficient for spectral theory and hence for the quantum mechanical interpretation. Nevertheless the lack of structural classification of the Jordan algebras compelled the stronger assumption that the observables formed the selfadjoint part of a C^*-algebra $\mathfrak{A}$ with identity and the physical states a subset of the states over $\mathfrak{A}$. Subsequently, the structure of Jordan algebras has been developed, and this latter assumption appears well founded. (Albert and Paige, in 1959, showed that the exceptional algebra $M_3{}^8$ has no Hilbert space representation, but Alfsen, Schultz, and Størmer, in 1974, developed Segal's theory for Jordan algebras $\mathfrak{A}$ with identity which are Banach spaces in a norm $\|\cdot\|$ having the three properties

(1) $\|A \circ B\| \leq \|A\| \, \|B\|$,
(2) $\|A \circ A\| = \|A\|^2$,
(3) $\|A \circ A\| \leq \|A \circ A + B \circ B\|$

for all A, B in $\mathfrak{A}$. An equivalent order-theoretic definition states that $(\mathfrak{A}, \mathbb{1})$ is a complete order unit space such that $A \circ A \geq 0$ for all $A \in \mathfrak{A}$, and $-\mathbb{1} \leq A \leq \mathbb{1} \Rightarrow A \circ A \leq \mathbb{1}$. Alfsen, Schultz, and Størmer proved that in this case $\mathfrak{A}$ contains a Jordan ideal $\mathfrak{J}$ such that $\mathfrak{A}/\mathfrak{J}$ has a faithful, isometric representation as a Jordan algebra of selfadjoint operators on a Hilbert space, and each "irreducible" Jordan representation of $\mathfrak{A}$ not annihilating $\mathfrak{J}$ is onto $M_3{}^8$.)

Segal also developed the mutual correspondence between states and representations of a C^*-algebra $\mathfrak{A}$, which is of importance both mathematically and physically, and subsequently reinterpreted the Stone–von Neumann uniqueness theorem as a result concerning states. If $\mathfrak{A}$ and $\mathfrak{M}$ are the C^*-algebra and W^*-algebra, respectively, generated by the Weyl operators $\{U_i(s), V_j(t); s, t \in \mathbb{R}, i, j = 1, 2, \ldots, n\}$ in the Schrödinger representation then Segal defined a state ω over $\mathfrak{A}$ to be regular if $\omega(U_i(s)V_j(t))$ is jointly continuous in s and t, for all i and j. This regularity property is directly related to the existence of position and momentum operators and the rephrased uniqueness theorem states that ω is regular if, and only if, it is the restriction to $\mathfrak{A}$ of a normal state over $\mathfrak{M}$. The representations corresponding to such states are sums of copies of the Schrödinger representation.

The latter form of the uniqueness theorem indicates that the distinction between the W^*-algebra and C^*-algebra description of quantum observables for systems with a finite number of particles is at most a matter of technical convenience. Nevertheless, it is vital in the broader context of systems with an infinite number of degrees of freedom, e.g., systems with an infinite number of particles.

There are several different ways of extending the formalism of Heisenberg to an infinite number of operators p_i, q_j satisfying the canonical commutation relations. Either one can construct a direct analogue of the Schrödinger representation by use of functional integration techniques or one can use a unitarily equivalent reformulation of this representation which is meaningful even for an infinite number of variables. The oldest and most common version of the representation which allows this generalization was proposed by Fock in 1932 and rigorously formulated by Cook some twenty years later (see Chapter 5). Once one has such a formulation one can construct an infinite family of unitary Weyl operators and a C^*-algebra $\mathfrak{A}$ and a W^*-algebra $\mathfrak{M}$ generated by these operators. But now the uniqueness theorem is no longer valid. There exist regular states over $\mathfrak{A}$ which are not obtained by restriction of normal states over $\mathfrak{M}$ and the representations corresponding to these states are no longer sums of copies of the Fock–Cook or Schrödinger representation. This lack of uniqueness was not generally recognized until the 1950s when Segal, Friederichs, and others gave examples of inequivalent regular representations. In 1955 Haag proved a theorem which essentially showed that two pure ground states are either equal or disjoint, i.e., generate inequivalent representations (see Corollary 5.3.41 in Chapter 5). Thus distinct dynamics appeared to determine distinct representations. The significance of the inequivalent representations was also partially clarified by Garding and Wightman in 1954 and completely explained by work of Chaiken, Dell' Antonio, Doplicher, and Ruelle, in the mid-1960s. The regular states over $\mathfrak{A}$ which are restrictions of normal states over $\mathfrak{M}$ are exactly those states in which a finite number of particles occur with probability one. (This result, which was indicated by model calculations of Haag, van Hove, Araki and others, in the 1950s, governs our approach to the Stone–von Neumann uniqueness theorem in Chapter 5). Thus the Schrödinger representation suffices for the description of a finite number of particles, but other representations of the C^*-algebra, or W^*-algebra, are essential for systems with an infinite number of particles. The realization of this distinction was the starting point for most of the subsequent applications of *-algebras to mathematical physics.

It is not, of course, immediately evident that the examination of systems composed of an infinite number of particles is relevant to physics and in the context of statistical mechanics this was often hotly contested in the 1950s and 1960s. To understand the relevance it is necessary to examine the type of idealization which is inherent in physical theories even with a finite number of particles.

As a simple example let us first envisage the scattering of a particle from a fixed target. The experiment consists of repeatedly shooting the particle at the target with a predetermined velocity and then measuring the velocity after scattering. A typical theoretical problem is to calculate the force between the target and the incident particle from the scattering data. The first commonplace theoretical idealization is to reduce the problem to the discussion of two bodies in isolation. This is harmless if the experimental setup has been carefully arranged and the forces exercised by the surroundings are negligible, and this can of course be checked by rearrangement of the apparatus. Such controls are a part of good experimental procedure. In order to have a reliable, reproducible, experiment it should also be ensured that the scattering data is insensitive to small changes in the initial and final measurement processes. For example, if the forces are strong and the particle and target are initially close then a small change in initial velocity would have a large effect on the final velocity. Similarly, if the final velocity is measured too soon after the particle has passed the target then its velocity will not have stabilized and a slight lapse in the measurement procedure would produce a significant change in the data. In a well-executed experiment various checks and counter-checks should be made to ensure the invariance of the data under such perturbations. If this is the case then the theoretician can interpret the results as *asymptotic data*, e.g., the experimentally measured outgoing velocity is considered to represent the velocity the particle would obtain after an arbitrarily long passage of time. Thus while the experimentalist might collect all his data between breakfast and lunch in a small cluttered laboratory his theoretical colleagues interpret the results in terms of isolated systems moving eternally in infinitely extended space. The validity of appropriate idealizations of this type is the heart and soul of theoretical physics and has the same fundamental significance as the reproducibility of experimental data.

The description of thermodynamic systems by statistical mechanics is another source of idealization of a finite physical system by an infinite theoretical model. The measurement of the heat capacity of a liquid is a typical thermal experiment and is accomplished by heating a sufficiently substantial sample of the liquid in a calorimeter. After the appropriate measurements the experimentalist divides the heat capacity by the mass of the sample and quotes the result as a specific heat in calories/gram °K, or other suitable units. This data is presented in this form because the experimentalist is confident that, under the conditions of the experiment and within the accuracy of the measurement, the heat capacity is proportional to the mass but otherwise independent of the size of the sample, the make of the calorimeter, etc. Of course, this can be checked by repetition of the experiment with various volumes of liquid and different calorimeters and again such checks are a concomitant of good experimental procedure. The theoretical quantity to compare with the measured result is then the portion of the calculated heat capacity strictly proportional to the size of the system for a large system, and a convenient mathematical procedure for obtaining this is to divide the heat capacity by the volume and then take the limit as the volume tends to infinity. This is usually called the thermodynamic limit. Thus one again introduces an idealization of a finite system by an infinite system and this idealization is justified if the foregoing tenets of experimental procedure are valid, and if the theoretical model used for the calculation is reliable. But if matter is composed of atomic particles then the idealized infinite liquid, being at nonzero density, contains an infinite number of particles, i.e., the theoretical model of the liquid involves an infinity of particles.

Before explaining the use of algebraic methods for the description of such infinite systems let us emphasize that in thermodynamics not all significant quantities are stable under perturbations. In fact, the most interesting phenomena, phase transitions, involve instabilities. Typically, if one confines a liquid at fixed temperature and reduces the pressure then at a certain critical value the liquid vaporizes and slight variations of the pressure around this critical value produce quite distinct thermodynamic states, or phases. Similarly, if the pressure is held fixed and the temperature raised then at a critical temperature a liquid–vapor phase transition takes place and small variations of the temperature around this critical value produce large changes in quantities such as the density. At the critical pressure, or temperature, various mixtures of vapor and liquid can coexist and hence the state of equilibrium is not unique. Thus the density per unit volume, the specific heat per gram °K, etc., while well defined at most pressures and temperatures appear to have no precise meaning for certain critical values of the thermodynamic parameters. One therefore expects the theoretical counterparts of these quantities to vary sharply with the temperature, etc., at these critical points. Here again the idealization of the thermodynamic limit is useful because the rapid variation of the quantities calculated at finite volume appears as a sharp discontinuity after the infinite volume limit. In fact this appearance of discontinuities is sometimes cited as a justification of the limit process.

Let us now examine the role of C^*-algebras and W^*-algebras in the description of infinite particle systems.

If one first considers a finite system in a subset Λ of space then the algebraic rephrasing of quantum mechanics sketched above identifies the corresponding observables as the selfadjoint elements of a C^*-algebra $\mathfrak{A}_\Lambda$. Therefore the observables corresponding to an arbitrarily large system would be determined by the union of the $\mathfrak{A}_\Lambda$. In specific models of point particles $\mathfrak{A}_\Lambda$ could correspond to the C^*-algebra generated by Weyl operators acting on the Fock–Cook representation space $\mathfrak{H}_\Lambda$, or to all bounded operators on $\mathfrak{H}_\Lambda$. But in any case if $\Lambda_1 \subset \Lambda_2$ then the algebras should satisfy $\mathfrak{A}_{\Lambda_1} \subset \mathfrak{A}_{\Lambda_2}$. It can be shown (Theorem 2.2.5 and Proposition 2.2.7) that if $A \in \mathfrak{A}_{\Lambda_1} \cap \mathfrak{A}_{\Lambda_2}$ then the norm of A as an element in $\mathfrak{A}_{\Lambda_i}$ is the same for $i = 1$ and $i = 2$. It follows that the union of the $\mathfrak{A}_\Lambda$ has a unique norm completion which is a C^*-algebra. The algebra $\mathfrak{A}$ is constructed without reference to a particular state, or representation, of the system and can be understood as the C^*-algebra of observables of the infinite system. Algebras built in this manner from a family of subalgebras $\mathfrak{A}_\Lambda$ are usually referred to as quasi-local algebras and the $\mathfrak{A}_\Lambda$ are called local algebras. The 1934 paper of Murray and von Neumann mentions the interrelationships of algebras corresponding to observables situated in separate parts of a system as a motivation for the examination of W^*-algebras in the context of quantum mechanics. Nevertheless, it was not until 1957 that Haag emphasized the importance of the quasi-local structure in field-theoretic models and in the 1960s that it was applied to quantum statistical mechanics. In this latter context the algebraic structure has provided a useful framework for the analysis of equilibrium states.

The rules of quantum statistical mechanics provide various algorithms, the Gibbs ensembles, for the construction of the equilibrium state $\omega_{\Lambda,\alpha}$, of a system in Λ, as a state over $\mathfrak{A}_\Lambda$. The suffix α denotes the thermodynamic parameters, e.g., α could represent the temperature and density, or the temperature and chemical potential, and the relevant algorithm depends upon the selection of this parametrization. In conformity with the above discussion one attempts to take the thermodynamic limit of the $\omega_{\Lambda,\alpha}$. Thus for each $A \in \mathfrak{A}_\Lambda$ and each Λ one tries to calculate

$$\omega_\alpha(A) = \lim_{\Lambda' \to \infty} \omega_{\Lambda',\alpha}(A),$$

where the limit indicates that Λ' increases to eventually contain any compact subset. The set of values $\omega_\alpha(A)$ then represents equilibrium data which is independent of size and shape. It is to be expected that the limit exists except for certain critical values of α. At these exceptional values various limit points should exist and each set of limit data then describes a possible equilibrium situation, i.e., several independent thermodynamic phases exist. But the limit data determine a state ω_α over $\mathfrak{A}$ and thus the equilibrium states of the system correspond to a subset of the states over $\mathfrak{A}$ and this construction gives a parametrization $\alpha \mapsto \omega_\alpha$ of these states.

Most applications of the algebraic formalism to statistical mechanics have concentrated on equilibrium phenomena and have attempted to justify, and amplify, this interpretation of equilibrium states as states over a C^*-algebra $\mathfrak{A}$. There have been two different types of analysis which are partially related but whose emphases are distinct. The first approach follows the course outlined in the above discussion while the second is less direct and aims at the characterization of the equilibrium states without reference to the thermodynamic limiting process. Nevertheless, both approaches have the same subsequent goal, the derivation of physically significant properties of the set of states designated as equilibrium states, e.g., smoothness properties of the thermodynamic parametrization of these states. Let us next sketch these two approaches in slightly more detail.

In the first type of analysis one begins with a specific Hamiltonian operator H_Λ which incorporates a description of the interactions and boundary conditions for particles in a finite region Λ and then tries to construct the Gibbs equilibrium state of the system. Schematically, these states are of the form

$$\omega_{\Lambda,\beta}(A) = \frac{\mathrm{Tr}(e^{-\beta H_\Lambda}A)}{\mathrm{Tr}(e^{-\beta H_\Lambda})},$$

where β is the inverse temperature in suitable units. Thus for this construction it is necessary to prove that H_Λ is selfadjoint and $\exp\{-\beta H_\Lambda\}$ is of trace-class. Next one examines the existence, or nonexistence, of the limit ω_β of the states $\omega_{\Lambda,\beta}$ as $\Lambda \to \infty$. Discussion of these questions involves a whole range of techniques of functional analysis, e.g., functional integration, convexity and subadditivity inequalities, integral equations, etc. In the simplest models of classical mechanics the results give a rather detailed justification of the *a priori* discussion. Unfortunately, the present results for more realistic models of classical and quantum mechanics are only partial and give little information concerning critical phenomena.

In the second type of analysis one begins with a general prescription for the dynamics of the idealized infinite system and the simplest, and strongest, assumption is that the time development $t \mapsto \tau_t(A)$ of the observables A is given by a continuous one-parameter group τ of *-automorphisms of the C^*-algebra $\mathfrak{A}$ of all observables. One then specifies criteria for a state ω over $\mathfrak{A}$ to be an equilibrium state in terms of properties of ω relative to τ, e.g., the obvious requirement of stationarity with time of ω corresponds to the invariance condition

$$\omega(\tau_t(A)) = \omega(A)$$

for all $A \in \mathfrak{A}$ and $t \in \mathbb{R}$. Several equivalent sets of criteria have been developed and justified either by arguments from first principle, e.g., stationarity, stability, and ergodicity, or by analogy with the Gibbs equilibrium formalism for finite systems. For example, one has the formal identity

$$\mathrm{Tr}(e^{-\beta H_\Lambda}(e^{itH_\Lambda}Ae^{-itH_\Lambda})B) = \mathrm{Tr}(e^{-\beta H_\Lambda}B(e^{i(t+i\beta)H_\Lambda}Ae^{-i(t+i\beta)H_\Lambda})).$$

Hence using the definition of the Gibbs state $\omega_{\Lambda,\beta}$ introduced in the previous paragraph and tentatively identifying $\tau_t(A)$ as the limit of

$$\exp\{itH_\Lambda\}A\exp\{-itH_\Lambda\}$$

as $\Lambda \to \infty$ one would expect the thermodynamic limit ω_β of the $\omega_{\Lambda,\beta}$ to satisfy the relation

$$\omega_\beta(\tau_t(A)B) = \omega_\beta(B\tau_{t+i\beta}(A))$$

for all $A, B \in \mathfrak{A}$ and $t \in \mathbb{R}$. This identity was first noted by Kubo in 1957, and subsequently by Martin and Schwinger in 1959, for the finite-volume Gibbs states. It is now commonly referred to as the KMS condition and was proposed as a criterion for equilibrium by Haag, Hugenholtz, and Winnink, in 1967. The condition presupposes that the function $t \mapsto \omega(B\tau_t(A))$ is analytic in the strip $0 < \mathrm{Im}\, t < \beta$ and then expresses an approximate commutation of observables within the state ω. The general analysis of equilibrium states then proceeds by the examination of the states satisfying the given criterion of equilibrium such as the KMS condition. One attempts to prove existence and settle questions of uniqueness and nonuniqueness, etc.

The KMS condition has played an important role in the synthesis of the mathematical and physical theories largely because an almost identical relation occurred in Tomitas analysis of von Neumann algebras. In the mid-1960s Tomita assigned to each "faithful" normal state ω over a W^*-algebra $\mathfrak{M}$ a canonical one-parameter group of *-automorphisms τ^ω. Tomita was solely motivated by questions of structural analysis but in 1970 Takesaki demonstrated that the state ω satisfied the KMS condition with respect to the group τ^ω with the slight difference that $t \mapsto \tau_t{}^\omega(A)$, $A \in \mathfrak{M}$, is not necessarily continuous in norm. The Tomita–Takesaki theory is developed in detail in Chapter 2 and analysis of the KMS condition occurs in Chapter 5.

Although the above approaches to equilibrium statistical mechanics have different points of departure they both lead to a designation of a class of states as equilibrium states and the analysis of these states is aided by the infinite-volume idealization. For example, this idealization allows properties such as homogeneity of a physical sample to be expressed by exact symmetry properties of the theoretical model. The group of space translations acts as *-automorphisms of the C^*-algebra $\mathfrak{A}$ of all observables and homogeneity is reflected by invariance of the equilibrium state under this group. Analysis of invariant states over a C^*-algebra leads immediately to a noncommutative analogue of ergodic theory which itself developed from classical statistical mechanical. Ergodic theory analyzes a dynamical system (X, μ, T) consisting of a measure space X, a probability measure μ, and a one-parameter group T of measure-preserving transformations of X. The direct algebraic analogue $(\mathfrak{A}, \omega, \tau)$ consists of a C^*-algebra $\mathfrak{A}$, a state ω, and a one-parameter group of *-automorphisms τ which leaves ω invariant although it is also of interest to examine more complicated groups. The general analysis of equilibrium states, in the mid-1960s, established that many results of classical ergodic theory could be extended to the non-commutative case if $(\mathfrak{A}, \omega, \tau)$ satisfies a suitable property of asymptotic abelianness, e.g., a property of the kind

$$\lim_{|t|\to\infty} \|\tau_t(A)B - B\tau_t(A)\| = 0.$$

Typically, this condition allows one to deduce that each invariant ω has a unique decomposition in terms of invariant states which are analogues of ergodic measures. These states are characterized by properties of indecomposability, or mixing, and the decomposition appears related to the separation of thermodynamic phases. These developments, which are described in detail in Chapter 4, provide another useful and interesting synthesis of mathematics and physics.

In the foregoing approach to the second type of analysis of equilibrium states we made the idealization that the time development is given by a continuous one-parameter group τ of *-automorphisms of a C^*-algebra $\mathfrak{A}$. This assumption, however, is only satisfied for very simple models. It is false even for the noninteracting Bose gas. Thus for a more realistic theory it is necessary to weaken this assumption and various possibilities are indicated by extension of the first form of analysis. One attempts to construct the equilibrium state ω and the time development τ simultaneously. In the "next best" situations τ is then given as an automorphism group of the weak closure $\pi_\omega(\mathfrak{A})''$ of $\mathfrak{A}$ in the cyclic representation associated with ω. If one adopts the viewpoint that mixtures of vector states in this particular representation constitute all "physically interesting" states or if one includes all selfadjoint operators in $\pi_\omega(\mathfrak{A})''$ among the observables this latter description is acceptable. Thus in this wider context one refines the algebraic notions of physical states and observables.

In 1963 Haag and Kastler made a number of interesting suggestions concerning states and physical equivalence which to a large extent justify the viewpoint mentioned above. They argued that if observables $A_1, \ldots, A_n$ are measured in a state ω one obtains numbers $\omega(A_1) = \lambda_1, \ldots, \omega(A_n) = \lambda_n$ but the inherent imprecision of the measurement process means that these observed values are only determined to lie within small intervals $(\lambda_i - \varepsilon, \lambda_i + \varepsilon)$. Thus ω is physically equivalent to any state ω' which satisfies

$$|\omega(A_i) - \omega'(A_i)| < \varepsilon, \qquad i = 1, \ldots, n,$$

i.e., physical equivalence is determined by neighborhoods in the weak* topology. But mixtures of vector states of any faithful representation $\pi_\omega(\mathfrak{A})$ are weakly* dense in the set of all states and hence this line of reasoning suggests that these states are a sufficiently large set for a full physical description. Nonetheless, if one studies systems which are physically dissimilar, e.g., systems at different temperatures, or densities, then it is necessary to study states which are not mixtures of vector states of one or other representation and it is practical to examine the dynamical evolution in each state separately.

Notes and Remarks

The papers by Murray and von Neumann are conveniently collected in Volume 3 of the collected works of von Neumann [[Neu 1]]. The Gelfand–Naimark characterization of C^*-algebras occurs in [Gel 1].

The appendix to Mackey's Chicago lectures [[Mac 1]] describes the influence of von Neumann algebra theory for the development of group representations. This appendix also discusses much of the same material presented in this chapter.

Many of the early papers on quantum mechanics are accessible in the book edited by van der Waerden [[Wae 1]] and von Neumann's axiomatization of the theory occurs in its earliest form in [[Neu 2]].

The work of Stone and von Neumann can be found in Volume 2 of von Neumann's collected works [[Neu 1]] and the monograph of Stone [[Sto 1]].

Segal's formulation of quantum mechanics in terms of C^*-algebras is summarized in [Seg 1].

The structure theory of Jordan algebras is reviewed by Størmer in [[Str 1]].

In this chapter we have given an historical account of the development of quantum theories, and also mentioned some of the attempts to develop the theory axiomatically, originating with von Neumann's theory. We have, however, avoided developing a detailed axiomatization of quantum statistical mechanics, because at present there exists no set of axioms which covers all models. We mentioned, for example, at the end of the introduction the difficulties associated with the assumption that the dynamics is given by a strongly continuous one-parameter group of $*$-automorphisms of the C^*-algebra associated with the system. For a more complete review of the development of axiomatic quantum theories up to 1974, the reader could consult Wightman's article on Hilbert's sixth problem [Wig 2].

The remainder of the topics discussed in this introductory chapter are described in detail in the following chapters. References can be found in the Notes and Remarks to the relevant chapters.

C^*-Algebras and von Neumann Algebras

2.1. C^*-Algebras

2.1.1. Basic Definitions and Structure

C^*-algebra theory is an abstraction of the structure of certain algebras of bounded operators acting on a Hilbert space and is simultaneously a special case of the theory of Banach algebras. Consequently, the theory can be developed in two different ways. Either one can begin with an abstract description suited to the general analysis of Banach algebras or one may start with a specific representation of the algebra on a Hilbert space. We will follow the first of these approaches.

Let $\mathfrak{A}$ be a vector space with coefficient field $\mathbb{C}$, the field of complex numbers $\alpha, \beta, \ldots$. The space $\mathfrak{A}$ is called an algebra if it is equipped with a multiplication law which associates the product AB to each pair $A, B \in \mathfrak{A}$. The product is assumed to be associative and distributive. Explicitly, one assumes

(1) $A(BC) = (AB)C$,
(2) $A(B + C) = AB + AC$,
(3) $\alpha\beta(AB) = (\alpha A)(\beta B)$.

A subspace $\mathfrak{B}$ of $\mathfrak{A}$ which is also an algebra with respect to the operations of $\mathfrak{A}$ is called a subalgebra. The algebra $\mathfrak{A}$ is commutative, or abelian, if the product is commutative, i.e., if

$$AB = BA.$$

A mapping $A \in \mathfrak{A} \to A^* \in \mathfrak{A}$ is called an involution, or adjoint operation, of the algebra $\mathfrak{A}$ if it has the following properties:

(1) $A^{**} = A$;
(2) $(AB)^* = B^*A^*$;
(3) $(\alpha A + \beta B)^* = \bar{\alpha}A^* + \bar{\beta}B^*$.

($\bar{\alpha}$ is the complex conjugate of α). An algebra with an involution is called a $*$-algebra and a subset $\mathfrak{B}$ of $\mathfrak{A}$ is called selfadjoint if $A \in \mathfrak{B}$ implies that $A^* \in \mathfrak{B}$.

The algebra $\mathfrak{A}$ is a normed algebra if to each element $A \in \mathfrak{A}$ there is associated a real number $\|A\|$, the norm of A, satisfying the requirements

(1) $\|A\| \geq 0$ and $\|A\| = 0$ if, and only if, $A = 0$,
(2) $\|\alpha A\| = |\alpha| \, \|A\|$,
(3) $\|A + B\| \leq \|A\| + \|B\|$,
(4) $\|AB\| \leq \|A\| \, \|B\|$.

The third of these conditions is called the triangle inequality and the fourth is the product inequality.

The norm defines a metric topology on $\mathfrak{A}$ which is referred to as the uniform topology. The neighborhoods of an element $A \in \mathfrak{A}$ in this topology are given by

$$\mathscr{U}(A; \varepsilon) = \{B; B \in \mathfrak{A}, \|B - A\| < \varepsilon\},$$

where $\varepsilon > 0$. If $\mathfrak{A}$ is complete with respect to the uniform topology then it is called a Banach algebra. A normed algebra with involution which is complete and has the property $\|A\| = \|A^*\|$ is called a Banach *-algebra.

Our principal definition is the following:

Definition 2.1.1. A *C^*-algebra* is a Banach *-algebra $\mathfrak{A}$ with the property

$$\|A^*A\| = \|A\|^2$$

for all $A \in \mathfrak{A}$.

The norm property which characterizes a C^*-algebra is a relic of an underlying Hilbert space structure. Note that this property combined with the product inequality yields $\|A^*\| = \|A\|$ automatically, because

$$\|A\|^2 = \|A^*A\| \leq \|A^*\| \, \|A\|$$

and hence $\|A\| \leq \|A^*\|$. Interchanging the roles of A and A^* one concludes that

$$\|A\| = \|A^*\|$$

for all $A \in \mathfrak{A}$.

The following examples illustrate how the C^*-norm condition arises. Note that here and throughout the sequel, we use the term Hilbert space to mean a complex Hilbert space.

EXAMPLE 2.1.2. Let $\mathfrak{H}$ be a Hilbert space and denote by $\mathscr{L}(\mathfrak{H})$ the set of all bounded operators on $\mathfrak{H}$. Define sums and products of elements of $\mathscr{L}(\mathfrak{H})$ in the standard manner and equip this set with the operator norm

$$\|A\| = \sup\{\|A\psi\|; \psi \in \mathfrak{H}, \|\psi\| = 1\}.$$

The Hilbert space adjoint operation defines an involution on $\mathscr{L}(\mathfrak{H})$ and with respect to these operations and this norm $\mathscr{L}(\mathfrak{H})$ is a C^*-algebra. In particular, the C^*-norm property follows from observing that

$$\begin{aligned}\|A\|^2 &= \sup\{(A\psi, A\psi); \psi \in \mathfrak{H}, \|\psi\| = 1\}\\ &= \sup\{(\psi, A^*A\psi); \psi \in \mathfrak{H}, \|\psi\| = 1\}\\ &\leq \sup\{\|A^*A\psi\|; \psi \in \mathfrak{H}, \|\psi\| = 1\}\\ &= \|A^*A\|\\ &\leq \|A^*\|\,\|A\|\\ &= \|A\|^2.\end{aligned}$$

Note that any uniformly closed subalgebra $\mathfrak{A}$ of $\mathscr{L}(\mathfrak{H})$ which is selfadjoint is also a C^*-algebra.

EXAMPLE 2.1.3. Let $\mathscr{LC}(\mathfrak{H})$ denote the algebra of compact operators acting on the Hilbert space $\mathfrak{H}$. It follows that $\mathscr{LC}(\mathfrak{H})$ is a C^*-algebra. Firstly, $\mathscr{LC}(\mathfrak{H})$ is a selfadjoint subalgebra of $\mathscr{L}(\mathfrak{H})$ and, secondly, it is uniformly closed because the uniform limit of a set of compact operators on $\mathfrak{H}$ is automatically compact.

Function algebras provide other examples of C^*-algebras which appear at first sight to be of a slightly different nature to the foregoing.

EXAMPLE 2.1.4. Let X be a locally compact space and $C_0(X)$ the continuous functions over X which vanish at infinity. By this we mean that for each $f \in C_0(X)$ and $\varepsilon > 0$ there is a compact $K \subseteq X$ such that $|f(x)| < \varepsilon$ for all $x \in X \backslash K$, the complement of K in X. Define the algebraic operations by $(f + g)(x) = f(x) + g(x)$, $(\alpha f)(x) = \alpha f(x)$, $(fg)(x) = f(x)g(x)$, and an involution by $f^*(x) = \overline{f(x)}$. Finally, introduce a norm by

$$\|f\| = \sup\{|f(x)|; x \in X\}.$$

It follows that $C_0(X)$ is a commutative C^*-algebra. In particular, the norm identity is valid because

$$\|ff^*\| = \sup\{|f(x)|^2; x \in X\} = \|f\|^2.$$

Note that if μ is a measure on X and $\mathfrak{H} = L^2(X; \mu)$ is the Hilbert space of μ-square integrable functions over X then $C_0(X)$ may be interpreted as an algebra of multiplication operators on $\mathfrak{H}$. Thus $C_0(X)$ is a sub-C^*-algebra of $\mathscr{L}(\mathfrak{H})$ and is analogous to the previous examples.

An identity $\mathbb{1}$ of a C^*-algebra $\mathfrak{A}$ is an element of $\mathfrak{A}$ such that

$$A = \mathbb{1}A = A\mathbb{1}$$

for all $A \in \mathfrak{A}$. It follows by involution that $\mathbb{1}^*$ is also an identity. But $\mathfrak{A}$ can have at most one identity because a second such element $\mathbb{1}'$ would satisfy

$$\mathbb{1}' = \mathbb{1}\mathbb{1}' = \mathbb{1}.$$

Thus $\mathbb{1} = \mathbb{1}'$. Note also that the relations

$$\|\mathbb{1}\| = \|\mathbb{1}^*\mathbb{1}\| = \|\mathbb{1}\|^2,$$

$$\|A\| = \|\mathbb{1}A\| \leq \|\mathbb{1}\| \, \|A\|$$

imply that $\|\mathbb{1}\| = 0$ or 1. But if $\|\mathbb{1}\| = 0$ then one must have $\|A\| = 0$ for all $A \in \mathfrak{A}$ and the algebra is identically zero. We will systematically ignore this trivial case and assume that $\|\mathbb{1}\| = 1$.

Although a C^*-algebra can have at most one identity element it is not automatic that it possesses an identity. For example, the algebra $\mathscr{L}\mathscr{C}(\mathfrak{H})$ (cf. Example 2.1.3) possesses an identity if, and only if, $\mathfrak{H}$ is finite-dimensional while the algebra $C_0(X)$ (cf. Example 2.1.4) has an identity if, and only if, X is compact. The absence of an identity can complicate the structural analysis of $\mathfrak{A}$ but these complications can to a large extent be avoided by embedding $\mathfrak{A}$ in a larger algebra $\tilde{\mathfrak{A}}$ which has an identity. The construction of this larger algebra is accomplished as follows:

Proposition 2.1.5. *Let $\mathfrak{A}$ be a C^*-algebra without identity and let $\tilde{\mathfrak{A}}$ denote the algebra of pairs $\{(\alpha, A); \alpha \in \mathbb{C}, A \in \mathfrak{A}\}$ with operations $(\alpha, A) + (\beta, B) = (\alpha + \beta, A + B)$, $(\alpha, A)(\beta, B) = (\alpha\beta, \alpha B + \beta A + AB)$, $(\alpha, A^*) = (\bar{\alpha}, A^*)$. It follows that the definition*

$$\|(\alpha, A)\| = \sup\{\|\alpha B + AB\|, B \in \mathfrak{A}, \|B\| = 1\}$$

yields a norm on $\tilde{\mathfrak{A}}$ with respect to which $\tilde{\mathfrak{A}}$ is a C^*-algebra. The algebra $\mathfrak{A}$ is identifiable as the C^*-subalgebra of $\tilde{\mathfrak{A}}$ formed by the pairs $(0, A)$.

PROOF. The triangle and product inequalities for $\|(\alpha, A)\|$ are easily verified. We next show that $\|(\alpha, A)\| = 0$ implies $\alpha = 0$, $A = 0$. But $\|(0, A)\| = \|A\|$ and hence $\|(0, A)\| = 0$ implies $A = 0$. Thus we can assume $\alpha \neq 0$ and by scalar multiplication we can even take $\alpha = 1$. But

$$\|B - AB\| \leq \|B\| \, \|(1, -A)\|$$

and hence $\|(1, -A)\| = 0$ implies $B = AB$ for all $B \in \mathfrak{A}$. By involution one then has $B = BA^*$ for all $B \in \mathfrak{A}$. In particular, $A^* = AA^* = A$,

$$B = AB = BA,$$

and A is an identity, which is a contradiction.

The C^*-norm property follows by noting that

$$\begin{aligned}\|(\alpha, A)\|^2 &= \sup\{\|\alpha B + AB\|^2; B \in \mathfrak{A}, \|B\| = 1\} \\ &= \sup\{\|B^*(\bar{\alpha}\alpha B + \bar{\alpha}AB + \alpha A^*B + A^*AB)\|; B \in \mathfrak{A}, \|B\| = 1\} \\ &\leq \|(\alpha, A)^*(\alpha, A)\| \\ &\leq \|(\alpha, A)^*\| \, \|(\alpha, A)\|.\end{aligned}$$

Therefore

$$\|(\alpha, A)\| \leq \|(\alpha, A)^*\|.$$

But the reverse inequality follows by replacing (α, A) with $(\alpha, A)^*$. Hence

$$\|(\alpha, A)\|^2 \leq \|(\alpha, A)^*(\alpha, A)\| \leq \|(\alpha, A)\|^2$$

and the desired conclusion is established.

The completeness of $\tilde{\mathfrak{A}}$ follows straightforwardly from the completeness of $\mathbb{C}$ and $\mathfrak{A}$.

Definition 2.1.6. Let $\mathfrak{A}$ be a C^*-algebra without identity. The *C^*-algebra $\tilde{\mathfrak{A}}$ obtained by adjoining an identity $\mathbb{1}$ to $\mathfrak{A}$* is defined as the algebra of pairs (α, A) described in Proposition 2.1.5. We use the notation $\alpha\mathbb{1} + A$ for the pair (α, A) and write $\tilde{\mathfrak{A}} = \mathbb{C}\mathbb{1} + \mathfrak{A}$.

Note that if $\mathfrak{A}$ is a C^*-algebra with identity $\mathbb{1}$ then it is quite possible to have a C^*-subalgebra $\mathfrak{B}$ which has no identity. In this situation the algebra $\tilde{\mathfrak{B}}$ obtained by adjoining an identity is identifiable as the smallest C^*-subalgebra of $\mathfrak{A}$ which contains both $\mathfrak{B}$ and $\mathbb{1}$.

Although the above construction provides a powerful tool for investigating C^*-algebras without identity it does not solve all problems related to the absence of an identity. An alternative approximation procedure which allows the construction of an "approximate identity" will be discussed in Section 2.2.3.

Next we introduce a number of additional concepts which constantly reoccur in algebraic theory.

A subspace $\mathfrak{B}$ of an algebra $\mathfrak{A}$ is called a left ideal if $A \in \mathfrak{A}$ and $B \in \mathfrak{B}$ imply that $AB \in \mathfrak{B}$. Alternatively, $\mathfrak{B}$ is a right ideal if $A \in \mathfrak{A}$ and $B \in \mathfrak{B}$ imply that $BA \in \mathfrak{B}$. If $\mathfrak{B}$ is both a left ideal and a right ideal, then it is called a two-sided ideal. Note that each ideal is automatically an algebra. For example, if $\mathfrak{B}$ is a left ideal and $B_1, B_2 \in \mathfrak{B}$ then automatically $B_1 \in \mathfrak{A}$, $B_2 \in \mathfrak{B}$, and $B_1B_2 \in \mathfrak{B}$. Further note that if $\mathfrak{B}$ is a left (or right) ideal of an algebra $\mathfrak{A}$ with involution and if $\mathfrak{B}$ is also selfadjoint then $\mathfrak{B}$ is automatically a two-sided ideal. Explicitly, if $B \in \mathfrak{B}$ then $AB \in \mathfrak{B}$ for all $A \in \mathfrak{A}$. But by selfadjointness $B^* \in \mathfrak{B}$ and hence $A^*B^* \in \mathfrak{B}$ for all $A \in \mathfrak{A}$. Again by selfadjointness $BA = (A^*B^*)^* \in \mathfrak{B}$. Thus $\mathfrak{B}$ is two-sided.

If $\mathfrak{A}$ is a Banach $*$-algebra and $\mathfrak{J} \subseteq \mathfrak{A}$ is a closed two-sided $*$-ideal then the quotient space $\mathfrak{A}/\mathfrak{J}$ can also be regarded as a Banach $*$-algebra. Thus an element $\hat{A} \in \mathfrak{A}/\mathfrak{J}$ is a subset of elements defined for each $A \in \mathfrak{A}$ by

$$\hat{A} = \{A + I; I \in \mathfrak{J}\}$$

and multiplication, addition, and involution are defined by $\hat{A}\hat{B} = \widehat{AB}$, $\hat{A} + \hat{B} = \widehat{A + B}$, and $\hat{A}^* = \widehat{A^*}$. The requirement that $\mathfrak{J}$ is a two-sided ideal guarantees that these operations are well defined, i.e., independent of the choice of representative $A + I_1$, $B + I_2$, of $\hat{A}$, $\hat{B}$. For example,

$$(A + I_1)(B + I_2) = AB + I_3,$$

where

$$I_3 = I_1B + AI_2 + I_1I_2 \in \mathfrak{J}.$$

The quotient space $\mathfrak{A}/\mathfrak{I}$ becomes a Banach *-algebra if we also introduce the norm by the definition

$$\|\hat{A}\| = \inf\{\|A + I\|; I \in \mathfrak{I}\}.$$

It is straightforward to check that this definition yields a norm and that $\mathfrak{A}/\mathfrak{I}$ is complete with respect to this norm. It is less evident that if $\mathfrak{A}$ is a C^*-algebra then $\mathfrak{A}/\mathfrak{I}$ equipped with the foregoing structure is also a C^*-algebra. We will give a demonstration of this fact in Section 2.2.3.

EXAMPLE 2.1.7. Let $\mathfrak{A} = \mathscr{L}(\mathfrak{H})$ be the C^*-algebra of all bounded operators on the complex Hilbert space $\mathfrak{H}$. Choose a vector $\Omega \in \mathfrak{H}$ and define $\mathfrak{I}_\Omega$ by

$$\mathfrak{I}_\Omega = \{A; A \in \mathfrak{A}, A\Omega = 0\}.$$

The set $\mathfrak{I}_\Omega$ is a left ideal of $\mathfrak{A}$.

EXAMPLE 2.1.8. Let $\mathfrak{A} = \mathscr{L}(\mathfrak{H})$ and $\mathfrak{B} = \mathscr{L}\mathscr{C}(\mathfrak{H})$, the algebra of compact operators on $\mathfrak{H}$. $\mathfrak{B}$ is a two-sided ideal of $\mathfrak{A}$ because the product of a bounded operator and a compact operator is a compact operator.

EXAMPLE 2.1.9. Let $\mathfrak{A} = C_0(X)$, the commutative C^*-algebra of Example 2.1.4. If F is a closed subset of X, and $\mathfrak{B}$ consists of the elements in $\mathfrak{A}$ which are zero on F then $\mathfrak{B}$ is a closed two-sided ideal of $\mathfrak{A}$, and the quotient algebra $\mathfrak{A}/\mathfrak{B}$ is identifiable as $C_0(F)$. Using the Stone–Weierstrass theorem one can show that each closed, two-sided ideal in $\mathfrak{A}$ has this form.

A C^*-algebra $\mathfrak{A}$ is called simple if it has no nontrivial closed two-sided ideals, i.e., if the only closed two-sided ideals are $\{0\}$ and $\mathfrak{A}$. If $\mathfrak{A}$ has an identity this amounts to saying that $\mathfrak{A}$ has no two-sided ideals at all, closed or not. Simple C^*-algebras play a fundamental role in applications to mathematical physics.

We conclude this introductory section, by stating the basic structure theorems for C^*-algebras. In Examples 2.1.2–2.1.4 we saw that uniformly closed selfadjoint subalgebras of bounded operators on a Hilbert space are C^*-algebras and, furthermore, the function algebras $C_0(X)$ yield examples of commutative C^*-algebras. The structure theorems state that these particular cases in fact describe the general situation.

Theorem 2.1.10. *Let $\mathfrak{A}$ be a C^*-algebra. It follows that $\mathfrak{A}$ is isomorphic to a norm-closed selfadjoint algebra of bounded operators on a Hilbert space.*

Theorem 2.1.11. *Let $\mathfrak{A}$ be a commutative C^*-algebra. It follows that $\mathfrak{A}$ is isomorphic to the algebra $C_0(X)$ of continuous functions, over a locally compact Hausdorff space X, which vanish at infinity.*

The proofs of these theorems will be given in Sections 2.3.4 and 2.3.5.

2.2. Functional and Spectral Analysis

2.2.1. Resolvents, Spectra, and Spectral Radius

In real and complex analysis two of the most important elementary functions are the inverse and the exponential. The function $z \in \mathscr{C} \mapsto (\lambda - z)^{-1} \in \mathbb{C}$ is the crucial element in analysis with Cauchy transforms and the function $x \in \mathbb{R} \mapsto \exp\{i\lambda x\} \in \mathbb{C}$ is the starting point for Fourier analysis. Both these functions are of paramount importance in generalizing functional analysis to other algebraic structures. Study of the inverse function immediately leads to the notions of resolvent and spectrum and we will next analyze these concepts for elements of a C^*-algebra.

If $\mathfrak{A}$ is an algebra with identity $\mathbb{1}$ then an element $A \in \mathfrak{A}$ is said to be invertible if there exists an element $A^{-1} \in \mathfrak{A}$, the inverse of A, such that

$$AA^{-1} = \mathbb{1} = A^{-1}A.$$

There are a number of elementary conclusions which follow directly from this definition. If A is invertible then it has a unique inverse and this inverse is invertible with $(A^{-1})^{-1} = A$; if A and B are invertible then AB is invertible and $(AB)^{-1} = B^{-1}A^{-1}$; if $\mathfrak{A}$ is a $*$-algebra and A is invertible then A^* is invertible and $(A^*)^{-1} = (A^{-1})^*$.

Definition 2.2.1. Let $\mathfrak{A}$ be an algebra with identity $\mathbb{1}$. The *resolvent set* $r_{\mathfrak{A}}(A)$ of an element $A \in \mathfrak{A}$ is defined as the set of $\lambda \in \mathbb{C}$ such that $\lambda\mathbb{1} - A$ is invertible and the *spectrum* $\sigma_{\mathfrak{A}}(A)$ of A is defined as the complement of $r_{\mathfrak{A}}(A)$ in $\mathbb{C}$. The inverse $(\lambda\mathbb{1} - A)^{-1}$, where $\lambda \in r_{\mathfrak{A}}(A)$, is called the *resolvent of* A *at* λ.

The spectrum of an element of a general algebra can be quite arbitrary but in a Banach algebra, and in particular in a C^*-algebra, the situation is quite simple, as we will see.

There are various techniques for analyzing resolvents and spectra, and one of the simplest is by series expansion and analytic continuation. If, for example, $\lambda \in \mathbb{C}$ and $|\lambda| > \|A\|$ then the series

$$\lambda^{-1} \sum_{m \geq 0} \left(\frac{A}{\lambda}\right)^m$$

is Cauchy in the uniform topology. But, by completeness, the series must define an element of $\mathfrak{A}$ and one immediately verifies that this element is the inverse of $\lambda\mathbb{1} - A$. In particular, $\lambda \in r_{\mathfrak{A}}(A)$ and the spectrum $\sigma_{\mathfrak{A}}(A)$ is bounded, $\sigma_{\mathfrak{A}}(A) \subseteq \{\lambda; \lambda \in \mathbb{C}, |\lambda| \leq \|A\|\}$. More generally, if $\lambda_0 \in r_{\mathfrak{A}}(A)$ and $|\lambda - \lambda_0| < \|(\lambda_0\mathbb{1} - A)^{-1}\|$ then the Neumann series

$$\sum_{m \geq 0} (\lambda_0 - \lambda)^m(\lambda_0\mathbb{1} - A)^{-m-1}$$

defines an element of $\mathfrak{A}$ and by explicit calculation this element is $(\lambda\mathbb{1} - A)^{-1}$. Thus $\lambda \in r_{\mathfrak{A}}(A)$. This latter argument also establishes that $r_{\mathfrak{A}}(A)$ is open and $\lambda \mapsto (\lambda\mathbb{1} - A)^{-1}$ is continuous on $r_{\mathfrak{A}}(A)$. As $\sigma_{\mathfrak{A}}(A)$ is the complement of $r_{\mathfrak{A}}(A)$ it is automatically closed and hence compact. One can also show that the spectrum is nonempty.

Proposition 2.2.2. *Let A be an element of a Banach algebra with identity and define the spectral radius $\rho(A)$ of A by*

$$\rho(A) = \sup\{|\lambda|, \lambda \in \sigma_{\mathfrak{A}}(A)\}.$$

It follows that

$$\rho(A) = \lim_{n\to\infty} \|A^n\|^{1/n} = \inf_n \|A^n\|^{1/n} \leq \|A\|.$$

In particular, the limit exists. Thus the spectrum of A is a nonempty compact set.

PROOF. Let $|\lambda|^n > \|A^n\|$ for some $n > 0$. As each $m \in \mathbb{Z}$ can be decomposed as $m = pn + q$ with $p, q \in \mathbb{Z}$ and $0 \leq q < n$ one again establishes that the series

$$\lambda^{-1} \sum_{m \geq 0} \left(\frac{A}{\lambda}\right)^m$$

is Cauchy in the uniform topology and defines $(\lambda\mathbb{1} - A)^{-1}$. Therefore

$$\rho(A) \leq \|A^n\|^{1/n}$$

for all $n > 0$, and consequently

$$\rho(A) \leq \inf_n \|A^n\|^{1/n} \leq \liminf_{n\to\infty} \|A^n\|^{1/n}.$$

Thus to complete the proof it suffices to establish that $\rho(A) \geq r_A$, where

$$r_A = \limsup_{n\to\infty} \|A^n\|^{1/n}.$$

There are two cases.

Firstly, assume $0 \in r_{\mathfrak{A}}(A)$, i.e., A is invertible. Then $1 = \|A^n A^{-n}\| \leq \|A^n\| \, \|A^{-n}\|$ and hence $1 \leq r_A r_{A^{-1}}$. This implies $r_A > 0$. Consequently, if $r_A = 0$ one must have $0 \in \sigma_{\mathfrak{A}}(A)$ and $\rho(A) \geq r_A$.

Secondly, we may assume $r_A > 0$. We will need the following simple observation. If A_n is any sequence of elements such that $R_n = (\mathbb{1} - A_n)^{-1}$ exists then $\mathbb{1} - R_n =$

$-A_n(\mathbb{1} - A_n)^{-1}$ and $A_n = -(\mathbb{1} - R_n)(\mathbb{1} - (\mathbb{1} - R_n))^{-1}$. Therefore $\|\mathbb{1} - R_n\| \to 0$ is equivalent to $\|A_n\| \to 0$ by power series expansion.

Define $S_A = \{\lambda; \lambda \in \mathbb{C}, |\lambda| \geq r_A\}$. We assume that $S_A \subseteq r_{\mathfrak{A}}(A)$ and obtain a contradiction. Let ω be a primitive nth root of unity. By assumption

$$R_n(A;\lambda) = n^{-1} \sum_{k=1}^{n} \left(\mathbb{1} - \frac{\omega^k A}{\lambda}\right)^{-1}$$

is well defined for all $\lambda \in S_A$. But an elementary calculation shows that

$$R_n(A;\lambda) = \left(\mathbb{1} - \frac{A^n}{\lambda^n}\right)^{-1}.$$

Next one has the continuity estimate

$$\left\|\left(\mathbb{1} - \frac{\omega^k A}{r_A}\right)^{-1} - \left(\mathbb{1} - \frac{\omega^k A}{\lambda}\right)^{-1}\right\|$$

$$= \left\|\left(\mathbb{1} - \frac{\omega^k A}{r_A}\right)^{-1} \omega^k A\left(\frac{1}{\lambda} - \frac{1}{r_A}\right)\left(\mathbb{1} - \frac{\omega^k A}{\lambda}\right)^{-1}\right\|$$

$$\leq |\lambda - r_A| \, \|A\| \sup_{\gamma \in S_A} \|(\gamma\mathbb{1} - A)^{-1}\|^2,$$

which is uniform in k. (The supremum is finite since $\lambda \mapsto \|(\lambda\mathbb{1} - A)^{-1}\|$ is continuous on $r_{\mathfrak{A}}(A)$ and for $|\lambda| > \|A\|$ one has

$$\|(\lambda\mathbb{1} - A)^{-1}\| \leq |\lambda|^{-1} \sum_{n \geq 0} \|A\|^n/|\lambda|^n = (|\lambda| - \|A\|)^{-1}.)$$

It immediately follows that for each $\varepsilon > 0$ there is a $\lambda > r_A$ such that

$$\left\|\left(\mathbb{1} - \frac{A^n}{r_A{}^n}\right)^{-1} - \left(\mathbb{1} - \frac{A^n}{\lambda^n}\right)^{-1}\right\| < \varepsilon$$

uniformly in n. But $\|A^n\|/\lambda^n \to 0$ and by the above observation $\|(\mathbb{1} - A^n/\lambda^n)^{-1} - \mathbb{1}\| \to 0$. This implies that $\|(\mathbb{1} - A^n/r_A{}^n)^{-1} - \mathbb{1}\| \to 0$ and $\|A^n\|/r_A{}^n \to 0$ by another application of the same observation. This last statement contradicts, however, the definition of r_A and hence the proof is complete.

A second useful technique for analyzing resolvents, etc., is by transformation. For example, the identity

$$(\lambda^n\mathbb{1} - A^n) = (\lambda\mathbb{1} - A)(\lambda^{n-1}\mathbb{1} + \lambda^{n-2}A + \cdots + A^{n-1})$$

demonstrates that if $\lambda^n \in r_{\mathfrak{A}}(A^n)$ then $\lambda \in r_{\mathfrak{A}}(A)$. Therefore, by negation, $\sigma_{\mathfrak{A}}(A)^n \subseteq \sigma_{\mathfrak{A}}(A^n)$. Other examples of relations which follow from simple transformations are contained in the next proposition.

Proposition 2.2.3. *Let $\mathfrak{A}$ be a *-algebra with identity. For $A \in \mathfrak{A}$ and $\lambda \in \mathbb{C}$*

$$\sigma_{\mathfrak{A}}(\lambda\mathbb{1} - A) = \lambda - \sigma_{\mathfrak{A}}(A),$$

$$\sigma_{\mathfrak{A}}(A^*) = \overline{\sigma_{\mathfrak{A}}(A)},$$

and if A is invertible

$$\sigma_{\mathfrak{A}}(A^{-1}) = \sigma_{\mathfrak{A}}(A)^{-1}$$

Moreover, for each pair $A, B \in \mathfrak{A}$ one has

$$\sigma_{\mathfrak{A}}(AB) \cup \{0\} = \sigma_{\mathfrak{A}}(BA) \cup \{0\}.$$

PROOF. The first property is evident; the second follows from the relation

$$(\lambda\mathbb{1} - A^*) = (\bar{\lambda}\mathbb{1} - A)^*.$$

The third statement is a consequence of

$$(\lambda\mathbb{1} - A) = \lambda A(A^{-1} - \lambda^{-1}\mathbb{1})$$

and

$$(\lambda^{-1}\mathbb{1} - A^{-1}) = \lambda^{-1}A^{-1}(A - \lambda\mathbb{1}).$$

Explicitly, one argues that if $\lambda \neq 0$ and $\lambda\mathbb{1} - A$ is not invertible then the first relation shows that $\lambda^{-1}\mathbb{1} - A^{-1}$ is not invertible. The second relation establishes the converse. The exceptional point $\lambda = 0$ is dealt with by noting that the invertibility implies that $\{0\} \notin \sigma_{\mathfrak{A}}(A)$ and

$$\sigma_{\mathfrak{A}}(A^{-1}) \subseteq \{\lambda; |\lambda| \leq \|A^{-1}\| < +\infty\}.$$

Finally, if $\lambda \in r_{\mathfrak{A}}(BA)$ then one calculates that

$$(\lambda\mathbb{1} - AB)(\mathbb{1} + A(\lambda\mathbb{1} - BA)^{-1}B) = \lambda\mathbb{1}.$$

This demonstrates that $\lambda\mathbb{1} - AB$ is invertible with the possible exception of $\lambda = 0$. Therefore $\sigma_{\mathfrak{A}}(BA) \cup \{0\} \supseteq \sigma_{\mathfrak{A}}(AB) \cup \{0\}$. Interchanging A and B gives the reverse inclusion and hence equality.

Further examples of spectral relations which arise from simple transformations occur in the subsequent discussion of elements of C^*-algebras. But first we must adopt a convention for defining the spectrum if the algebra does not contain an identity. The simplest procedure is to adjoin an identity.

Definition 2.2.4. Let $\mathfrak{A}$ be a $*$-algebra without identity and $\tilde{\mathfrak{A}} = \mathbb{C}\mathbb{1} + \mathfrak{A}$ the $*$-algebra obtained by adjoining an identity. If $A \in \mathfrak{A}$ then the *resolvent set* $r_{\mathfrak{A}}(A)$ and the *spectrum* $\sigma_{\mathfrak{A}}(\mathfrak{A})$ *of* A are defined, respectively, by

$$r_{\mathfrak{A}}(A) = r_{\tilde{\mathfrak{A}}}(A), \qquad \sigma_{\mathfrak{A}}(A) = \sigma_{\tilde{\mathfrak{A}}}(A).$$

Next, we partially characterize the spectra of special classes of elements of a C^*-algebra $\mathfrak{A}$. The most important elements are the normal, selfadjoint, isometric, and unitary elements.

An element $A \in \mathfrak{A}$ is defined to be normal if

$$AA^* = A^*A$$

and selfadjoint if

$$A = A^*.$$

If $\mathfrak{A}$ has an identity $\mathbb{1}$ then A is called an isometry whenever

$$A^*A = \mathbb{1}$$

and A is unitary if

$$A^*A = \mathbb{1} = AA^*.$$

Note that a general element $A \in \mathfrak{A}$ has a unique decomposition in terms of selfadjoint elements A_1, A_2 of the form

$$A = A_1 + iA_2.$$

The real and imaginary parts A_1, A_2 of A are given, respectively, by $A_1 = (A + A^*)/2$ and $A_2 = (A - A^*)/2i$.

Our convention concerning the spectrum in Definition 2.2.4 essentially allows us to assume the existence of an identity in discussing the C^*-algebra situation. One has the following:

Theorem 2.2.5. *Let $\mathfrak{A}$ be a C^*-algebra with identity.*

(a) *If $A \in \mathfrak{A}$ is normal or selfadjoint then the spectral radius $\rho(A)$ of A is given by*

$$\rho(A) = \|A\|.$$

(b) *If $A \in \mathfrak{A}$ is isometric, or unitary then*

$$\rho(A) = 1.$$

(c) *If $A \in \mathfrak{A}$ is unitary then*

$$\sigma_{\mathfrak{A}}(A) \subseteq \{\lambda; \lambda \in \mathbb{C}, |\lambda| = 1\}.$$

(d) *If A is selfadjoint*

$$\sigma_{\mathfrak{A}}(A) \subseteq [-\|A\|, \|A\|], \qquad \sigma_{\mathfrak{A}}(A^2) \subseteq [0, \|A\|^2].$$

(e) *For general $A \in \mathfrak{A}$ and each polynomial P*

$$\sigma_{\mathfrak{A}}(P(A)) = P(\sigma_{\mathfrak{A}}(A)).$$

PROOF. (a) The normality of A and the C^*-norm identity imply that

$$\begin{aligned}
\|A^{2^n}\|^2 &= \|(A^*)^{2^n}(A)^{2^n}\| \\
&= \|(A^*A)^{2^n}\| \\
&= \|(A^*A)^{2^{n-1}}\|^2 \\
&= \cdots = \|A^*A\|^{2^n} = \|A\|^{2^{n+1}}.
\end{aligned}$$

Therefore

$$\rho(A) = \lim_{n\to\infty} \|A^{2^n}\|^{2^{-n}} = \|A\|.$$

(b) The proof is similar to (a). One has

$$\begin{aligned}\|A^n\|^2 &= \|(A^*)^n A^n\| \\ &= \|(A^*)^{n-1} A^{n-1}\| \\ &= \|\mathbb{1}\| = 1.\end{aligned}$$

(c) As each unitary element is isometric $\sigma_{\mathfrak{A}}(A)$ is contained in the unit disc by part (b). But from Proposition 2.2.3 one has

$$\sigma_{\mathfrak{A}}(A) = \overline{\sigma_{\mathfrak{A}}(A^*)} = \overline{\sigma_{\mathfrak{A}}(A^{-1})} = (\overline{\sigma_{\mathfrak{A}}(A)})^{-1}.$$

It follows immediately from these two observations that $\sigma_{\mathfrak{A}}(A)$ is contained in the unit circle.

(d) Each selfadjoint A is automatically normal and hence $\rho(A) = \|A\|$. Thus if $|\lambda^{-1}| > \|A\|$ one has $\lambda^{-1} \in r_{\mathfrak{A}}(A)$ and $\mathbb{1} + i|\lambda|A$ is invertible. Define $U \in \mathfrak{A}$ by

$$U = (\mathbb{1} - i|\lambda|A)(\mathbb{1} + i|\lambda|A)^{-1}.$$

It is straightforward to check that U is unitary. Hence $\mathbb{1}(1 - i|\lambda|\alpha)(1 + i|\lambda|\alpha)^{-1} - U$ is invertible for all $\alpha \in \mathbb{C}$ with $\operatorname{Im}\alpha \neq 0$ by part (c). But

$$\mathbb{1}(1 - i|\lambda|\alpha)(1 + i|\lambda|\alpha)^{-1} - U = 2i|\lambda|(1 + i|\lambda|\alpha)^{-1}(A - \alpha\mathbb{1})(\mathbb{1} + i|\lambda|A)^{-1}$$

and hence $A - \alpha\mathbb{1}$ is invertible for all α such that $\operatorname{Im}\alpha \neq 0$. Therefore $\sigma_{\mathfrak{A}}(A) \subseteq \mathbb{R} \cap \{\lambda; |\lambda| \leq \|A\|\} = [-\|A\|, \|A\|]$. The statement concerning $\sigma_{\mathfrak{A}}(A^2)$ then follows from part (e).

(e) First note that if

$$B = \prod_{i=1}^{n} A_i,$$

where $A_i \in \mathfrak{A}$ and $A_i A_j = A_j A_i$ for $i, j = 1, \ldots, n$ then B is invertible if, and only if, each A_i is invertible. This follows because invertibility of the A_i together with commutativity of the A_i, A_j implies $A_j^{-1}A_i^{-1} = A_i^{-1}A_j^{-1}$ and hence

$$B^{-1} = \prod_{i=1}^{n} A_i^{-1},$$

where the order of the factors is irrelevant.

Conversely, if B is invertible

$$A_i^{-1} = B^{-1} \prod_{j \neq i} A_j.$$

Now choose $\alpha_i, \alpha \in \mathbb{C}$ such that

$$P(x) - \lambda = \alpha \prod_{i=1}^{n} (x - \alpha_i),$$

and then

$$P(A) - \lambda\mathbb{1} = \alpha \prod_{i=1}^{n} (A - \alpha_i \mathbb{1}).$$

Hence $\lambda \in \sigma_{\mathfrak{A}}(P(A))$ if, and only if, $\alpha_i \in \sigma_{\mathfrak{A}}(A)$ for some $i = 1, \ldots, n$. But $P(\alpha_i) = \lambda$ and hence $\lambda \in \sigma_{\mathfrak{A}}(P(A))$ is equivalent to $\lambda \in P(\sigma_{\mathfrak{A}}(A))$.

Remark. If A is normal the last statement may be extended to the equality $\sigma_{\mathfrak{A}}(f(A)) = f(\sigma_{\mathfrak{A}}(A))$ for all continuous functions f. This result is known as the spectral mapping theorem.

The spectral radius formula $\rho(A) = \|A\|$ for A selfadjoint or A normal is of fundamental significance and will be repeatedly used without comment.

Corollary 2.2.6. *If $\mathfrak{A}$ is a *-algebra and there exists a norm of $\mathfrak{A}$ with the C^*-norm property and with respect to which $\mathfrak{A}$ is closed then this norm is unique.*

PROOF. If $A \in \mathfrak{A}$ then $\sigma_{\mathfrak{A}}(A)$ depends only on the algebraic structure of $\mathfrak{A}$. Thus if $A = A^*$, $\|A\| = \rho(A)$ is uniquely determined. For general A

$$\|A\| = \|A^*A\|^{1/2} = \rho(A^*A)^{1/2}.$$

The foregoing result on the spectra of selfadjoint elements can now be used to remove an ambiguity in the definition of the spectrum. If $\mathfrak{B}$ is a subalgebra of $\mathfrak{A}$ and $A \in \mathfrak{B}$ then there are two possible spectra, $\sigma_{\mathfrak{A}}(A)$ and $\sigma_{\mathfrak{B}}(A)$. In general these spectra are distinct although the inclusion $\mathfrak{B} \subseteq \mathfrak{A}$ does imply that $\sigma_{\mathfrak{A}}(A) \subseteq \sigma_{\mathfrak{B}}(A)$. The situation for C^*-algebras is, however, simple.

Proposition 2.2.7. *Let $\mathfrak{B}$ be a C^*-subalgebra of the C^*-algebra $\mathfrak{A}$. If $A \in \mathfrak{B}$ then*

$$\sigma_{\mathfrak{A}}(A) = \sigma_{\mathfrak{B}}(A).$$

PROOF. We may assume that $\mathfrak{A}$ and $\mathfrak{B}$ have a common identity element. We must show that if $\lambda\mathbb{1} - A$ is invertible in $\mathfrak{A}$ then it is invertible in $\mathfrak{B}$. In fact, we will show that it is invertible in the C^*-subalgebra $\mathfrak{C}$ generated by $\mathbb{1}$, A, and A^*. This will then give

$$\sigma_{\mathfrak{C}}(A) = \sigma_{\mathfrak{B}}(A) = \sigma_{\mathfrak{A}}(A).$$

Thus we need to establish that if $A \in \mathfrak{A}$ is invertible then $A^{-1} \in \mathfrak{C}$. Suppose first that A is selfadjoint; then $\sigma_{\mathfrak{B}}(A) \subseteq \mathbb{R}$ by Proposition 2.2.5. Now we will obtain A^{-1} by analytically continuing $(A - \lambda\mathbb{1})^{-1}$ along the imaginary axis from $\lambda = \lambda_0 = 2i\|A\|$. First note that $(A - \lambda_0\mathbb{1})^{-1}$ is determined by a uniformly convergent series

$$(A - \lambda_0\mathbb{1})^{-1} = -\sum_{n\geq 0} \lambda_0^{-1}\left(\frac{A}{\lambda_0}\right)^n$$

each of whose terms is contained in $\mathfrak{C}$. Thus $(A - \lambda_0\mathbb{1})^{-1} \in \mathfrak{C}$. Secondly, for $\lambda \notin \sigma_{\mathfrak{A}}(A)$ the resolvent $R(\lambda) = (A - \lambda\mathbb{1})^{-1}$ is a normal operator and $\sigma_{\mathfrak{A}}(R(\lambda)) = \sigma_{\mathfrak{A}}(A - \lambda\mathbb{1})^{-1}$ $= (\sigma_{\mathfrak{A}}(A) - \lambda)^{-1}$ by Proposition 2.2.3. If $d(\lambda)$ denotes the distance from λ to $\sigma_{\mathfrak{A}}(A)$ it follows from Theorem 2.2.5 that $\|R(\lambda)\| = d(\lambda)^{-1}$ and this estimate implies that the series

$$R(\lambda) = \sum_{n\geq 0} (\lambda - \lambda_0)^n R(\lambda_0)^{n+1}$$

converges in a sphere of radius $d(\lambda_0) = \|R(\lambda_0)\|^{-1}$ around λ_0. This ensures the validity of the analytic continuation argument because $d(\lambda_0) > |\lambda_0|$ for λ_0 pure imaginary, the invertibility of A implies that $\sigma_{\mathfrak{A}}(A) \in \mathbb{R} \backslash [-\varepsilon, \varepsilon]$ for some $\varepsilon > 0$, and hence

$$A^{-1} = \sum_{n \geq 0} (-\lambda_0)^n R(\lambda_0)^{n+1}$$

for some λ_0 pure imaginary and $|\lambda_0| > \|A\|$.

Next let A be invertible but not necessarily selfadjoint. It follows that A^*A is invertible and by the foregoing argument $(A^*A)^{-1}$ is contained in the C^*-subalgebra of $\mathfrak{C}$ generated by $\mathbb{1}$ and A^*A. Finally, define

$$X = (A^*A)^{-1}A^*.$$

One has $X \in \mathfrak{C}$ but the relation $XA = \mathbb{1}$ implies that $X = A^{-1}$ and hence A is invertible in $\mathfrak{C}$.

As the spectrum $\sigma_{\mathfrak{A}}(A)$ of each element A of the C^*-algebra $\mathfrak{A}$ has the above independence property we will simplify our notation by dropping the suffix $\mathfrak{A}$. Thus in the sequel we denote the spectrum of an element A of a C^*-algebra by $\sigma(A)$.

2.2.2. Positive Elements

Probably the most important class of elements of a C^*-algebra is the class of positive elements because the notion of positivity allows the introduction of an order relation between various elements of the algebra and gives a method of making quantitative comparisons.

There are various equivalent characterizations of positivity but the most convenient definition appears to be in terms of the spectrum.

Definition 2.2.8. An element A of a $*$-algebra $\mathfrak{A}$ is defined to be *positive* if it is selfadjoint and its spectrum $\sigma(A)$ is a subset of the positive half-line. The set of all positive elements of $\mathfrak{A}$ is denoted by $\mathfrak{A}_+$.

We begin the analysis of positive elements by examining their square roots. It is worthwhile emphasizing that the square root operation plays a distinguished role in complex function analysis. This operation, together with the elementary algebraic operations, allows the easy construction of absolute values of a function, e.g., $|f| = \sqrt{\bar{f}f}$, and this in turn provides the starting point for decomposition of a real function into positive and negative parts, $f_\pm = (|f| \pm f)/2$, etc. Thus for the generalization of function analysis it is convenient to have an algebraic algorithm for forming the square root. We next examine such algorithms.

As a preliminary we deduce the following simple characterization of positive elements.

Lemma 2.2.9. *Let $\mathfrak{A}$ be a C^*-algebra with identity $\mathbb{1}$. A selfadjoint element $A \in \mathfrak{A}$ is positive if, and only if, $\|\mathbb{1} - A/\|A\|\,\| \leq 1$. If A is selfadjoint, and $\|\mathbb{1} - A\| \leq 1$ then A is positive, and $\|A\| \leq 2$.*

PROOF. If A is positive then $\sigma(A) \subseteq [0, \|A\|]$ by Theorem 2.2.5. Thus $\sigma(\mathbb{1} - A/\|A\|) \subseteq [0, 1]$ and $\|\mathbb{1} - A/\|A\|\,\| \leq 1$. Conversely, $\|\mathbb{1} - A/\|A\|\,\| \leq 1$ implies $\sigma(\mathbb{1} - A/\|A\|) \subseteq [-1, 1]$, or $\sigma(A) \subseteq [0, 2\|A\|]$, and hence A is positive. The proof of the second statement is identical.

Theorem 2.2.10. *Let $\mathfrak{A}$ be a C^*-algebra. A selfadjoint element $A \in \mathfrak{A}$ is positive if, and only if, $A = B^2$ for some selfadjoint $B \in \mathfrak{A}$. Moreover, if A is positive there exists a unique positive B such that $A = B^2$ and this B lies in the abelian C^*-subalgebra of $\mathfrak{A}$ generated by A.*

PROOF. If B is selfadjoint then B^2 is selfadjoint and $\sigma(B^2) \subseteq [0, \|B\|^2]$ by Theorem 2.2.5(d). Thus B^2 is positive. To prove the converse we construct a positive B such that $B^2 = A$.

If $\mathfrak{A}$ does not have an identity then we first adjoin one. Next for $\lambda > 0$ and A positive we note that $\lambda\mathbb{1} + A$ is invertible and

$$A(\lambda\mathbb{1} + A)^{-1} = \mathbb{1} - \lambda(\lambda\mathbb{1} + A)^{-1}.$$

It now follows easily from Proposition 2.2.3 that

$$\sigma(A(\lambda\mathbb{1} + A)^{-1}) \subseteq [0, \|A\|(\lambda + \|A\|)^{-1}]$$

and hence

$$\|A(\lambda\mathbb{1} + A)^{-1}\| \leq \|A\|(\lambda + \|A\|)^{-1}.$$

This estimate allows us to define $B \in \mathfrak{A}$ by a Riemann integral

$$B = \frac{1}{\pi}\int_0^\infty \frac{d\lambda}{\lambda^{1/2}}\, A(\lambda\mathbb{1} + A)^{-1}.$$

Convergence of the integral is measured with respect to the algebraic norm and it follows from the foregoing estimate that the integral is well defined at both zero and infinity. It can now be verified by explicit calculation that $A = B^2$. One writes B^2 as a double integral, separates the integrand into partial fractions, and then integrates with respect to one variable. We omit the details of this verification (see Notes and Remarks at the end of this chapter for detailed references). But we will demonstrate that B is positive. Since

$$1 = \frac{1}{\pi}\int_0^\infty \frac{d\lambda}{\lambda^{1/2}}\,\frac{1}{\lambda + 1}$$

one has

$$\begin{aligned}\|\mathbb{1} - B\| &\leq \frac{1}{\pi}\int_0^\infty \frac{d\lambda}{\lambda^{1/2}}\,\|(\lambda + 1)^{-1}\mathbb{1} - A(\lambda\mathbb{1} + A)^{-1}\| \\ &\leq \frac{1}{\pi}\int_0^\infty \frac{d\lambda}{\lambda^{1/2}}\,\frac{1}{\lambda + 1}\,\|\lambda(\lambda\mathbb{1} + A)^{-1}\|\cdot\|\mathbb{1} - A\|.\end{aligned}$$

One has, however, $\|\lambda(\lambda\mathbb{1} + A)^{-1}\| \leq 1$ for $\lambda > 0$ and $\|\mathbb{1} - A\| \leq 1$ for A positive with $\|A\| = 1$ (Lemma 2.2.9). Therefore one concludes that $\|\mathbb{1} - B\| \leq 1$. This implies that B is positive by a second application of Lemma 2.2.9.

Next let $\mathfrak{A}_A$ denote the abelian C^*-algebra generated by A. If $\lambda > 0$ one has $(\lambda\mathbb{1} + A)^{-1} \in \tilde{\mathfrak{A}}_A = \mathbb{C}\mathbb{1} + \mathfrak{A}_A$ by Proposition 2.2.7. Hence $A(\lambda\mathbb{1} + A)^{-1} \in \mathfrak{A}_A$ and $B \in \mathfrak{A}_A$.

Finally, we must prove that B is the unique positive element with the property $A = B^2$. As a preliminary first note that one may repeat the above construction to find a positive C such that $B = C^2$. Furthermore, C is in the algebra generated by B which is, of course, equal to $\mathfrak{A}_A$. Thus A, B, and C mutually commute. Next assume a second positive element B' such that $B'^2 = A$, and a positive C' such that $C'^2 = B'$. Clearly C' commutes with B' and then $C'A = C'B'^2 = B'^2C' = AC'$ implies that C' commutes with A. But then C' commutes with B and C because they are in $\mathfrak{A}_A$. In this manner one deduces that A, B, B', C, and C' all commute. Now note that

$$\begin{aligned} 0 &= (B^2 - B'^2)(B - B') \\ &= (B - B')B(B - B') + (B - B')B'(B - B') \\ &= ((B - B')C)^2 + ((B - B')C')^2. \end{aligned}$$

Both elements in this final expression are positive and as their sum is zero both elements must be zero (if $X = ((B - B')C)^2$ one has $X \in \mathfrak{A}_+$ and $-X \in \mathfrak{A}_+$ which imply $\sigma(X) = 0$, or $X = 0$). Taking the difference of the two elements then gives $(B - B')^3 = 0$ and hence $\|(B - B')^3\| = 0$. Applying Proposition 2.2.2 and Theorem 2.2.5 one then concludes $\rho(B - B') = 0$ and $\|B - B'\| = 0$, i.e., $B = B'$.

This result allows us to define the square root of a positive element A of a C^*-algebra $\mathfrak{A}$ as the unique positive element B, of $\mathfrak{A}$, such that $B^2 = A$. The square root is denoted by $\sqrt{A}$, or $A^{1/2}$. If A is selfadjoint then the modulus of A can also be defined as $\sqrt{A^2}$. The modulus, or absolute value, is denoted by $|A|$.

Remark. The square root of A was constructed through an integral algorithm whose main utility was to prove that $A^{1/2}$ is in the algebra generated by A. Once this, and the uniqueness of $A^{1/2}$, have been established one may use other, easier, algorithms. For example,

$$A^{1/2} = \|A\|^{1/2}\left[\mathbb{1} - \sum_{n\geq 1} c_n\left(\mathbb{1} - \frac{A}{\|A\|}\right)^n\right],$$

where the c_n are the coefficients of the Taylor series for the function $x \in [0, 1] \mapsto \sqrt{1 - x} \in [0, 1]$. Convergence of the series is assured by Lemma 2.2.9.

Next we examine properties of the set of positive elements and the decomposition of selfadjoint elements into positive and negative parts.

Proposition 2.2.11. *The set $\mathfrak{A}_+$ of positive elements of the C^*-algebra $\mathfrak{A}$ is a uniformly closed convex cone with the property*

$$\mathfrak{A}_+ \cap (-\mathfrak{A}_+) = \{0\}.$$

If A is a selfadjoint element of $\mathfrak{A}$ and one defines $A_{\pm} = (|A| \pm A)/2$ it follows that

(1) $A_{\pm} \in \mathfrak{A}_{+}$,
(2) $A = A_{+} - A_{-}$,
(3) $A_{+}A_{-} = 0$.

Moreover, $A_{\pm}$ are the unique elements with these properties.

PROOF. It clearly suffices to prove the proposition in the case that $\mathfrak{A}$ has an identity. If $A \in \mathfrak{A}_{+}$ and $\lambda \geq 0$ then $\lambda A \in \mathfrak{A}_{+}$ by Lemma 2.2.9. Next we show that if $A, B \in \mathfrak{A}_{+}$ then $(A + B)/2 \in \mathfrak{A}_{+}$. It is sufficient to consider the case $\|A\| = 1$, $\|B\| = 1$ but then $\|(A + B)/2\| \leq 1$ and

$$\left\| \mathbb{1} - \frac{A + B}{2} \right\| \leq \frac{\|\mathbb{1} - A\|}{2} + \frac{\|\mathbb{1} - B\|}{2} \leq 1,$$

where the last estimate uses the first statement of Lemma 2.2.9. The desired result then follows from the second statement of this lemma. Now, as we have already noted, $A \in \mathfrak{A}_{+} \cap (-\mathfrak{A}_{+})$ implies $\sigma(A) = 0$ and hence by the selfadjointness of A, $\|A\| = 0$, or $A = 0$. To deduce that $\mathfrak{A}_{+}$ is closed consider $A_n \in \mathfrak{A}_{+}$ such that $\|A_n - A\| \to 0$. Then $\|A_n\| - \|A\| \to 0$. But $A_n \in \mathfrak{A}_{+}$ is equivalent to $\| \|A_n\| - A_n\| \leq \|A_n\|$ and in the limit one has $\| \|A\| - A\| \leq \|A\|$, which is equivalent to $A \in \mathfrak{A}_{+}$.

Now consider the decomposition statement. It is evident that $A = A_{+} - A_{-}$ but

$$4A_{+}A_{-} = A^2 - |A|A + A|A| - A^2 = 0$$

because A commutes with $|A| = \sqrt{A^2}$ as this latter element is in the abelian algebra generated by A^2. We next prove that A_{+} is positive. The proof for A_{-} is identical. First define A_n by

$$A_n = n(\mathbb{1} + nA_{+}^2)^{-1}A_{+}^2$$

and note that

$$A_n|A| = A_nA_{+}.$$

Next we estimate

$$\begin{aligned} \|A_n|A| - A_{+}\|^2 &= \|n(\mathbb{1} + nA_{+}^2)^{-1}A_{+}^3 - A_{+}\|^2 \\ &= \|(\mathbb{1} + nA_{+}^2)^{-1}A_{+}\|^2 \\ &= \|(\mathbb{1} + nA_{+}^2)^{-2}A_{+}^2\| \\ &\leq \|(\mathbb{1} + nA_{+}^2)^{-1}A_{+}^2\| \, \|(\mathbb{1} + nA_{+}^2)^{-1}\| \\ &= \frac{1}{n}\|\mathbb{1} - (\mathbb{1} + nA_{+}^2)^{-1}\| \, \|(\mathbb{1} + nA_{+}^2)^{-1}\|. \end{aligned}$$

But $\mathbb{1} + nA_{+}^2$ has spectrum in $[1, \infty\rangle$, and hence $(\mathbb{1} + nA_{+}^2)^{-1}$ has spectrum in $[0, 1]$, by Proposition 2.2.3. It immediately follows that

$$\|A_n|A| - A_{+}\| \leq n^{-1/2}$$

Thus A_+ is the uniform limit of $A_n|A|$. But $|A|$, $A_+{}^2$, etc., are positive and commute. Therefore

$$A_n|A| = \left(|A|^{1/4}|A_+|^{1/2}\left(\frac{1}{n} + A_+{}^2\right)^{-1/2}|A_+|^{1/2}|A|^{1/4}\right)^2 \in \mathfrak{A}_+.$$

The positivity of A_+ then follows because $\mathfrak{A}_+$ is closed.

Finally, if $A_1, A_2 \in \mathfrak{A}_+$, $A = A_1 - A_2$, and $A_1A_2 = 0$ then $A^2 = A_1{}^2 + A_2{}^2 = (A_1 + A_2)^2$. Therefore $|A| = A_1 + A_2$ by the uniqueness of the positive square root and $A_+ = (|A| + A)/2 = A_1$. Similarly $A_- = A_2$.

The decomposition $A = A_+ - A_-$ described in Proposition 2.2.11 is often referred to as the orthogonal decomposition of A. Its existence is useful in deducing the final, and most important, characterization of positive elements.

Theorem 2.2.12. *Let $\mathfrak{A}$ be a C^*-algebra. The following conditions of $A \in \mathfrak{A}$ are equivalent:*

(1) *A is positive;*
(2) *$A = B^*B$ for some $B \in \mathfrak{A}$.*

PROOF. (1) $\Rightarrow$ (2) is already contained in Theorem 2.2.10.

(2) $\Rightarrow$ (1) Denote the orthogonal decomposition of B^*B by

$$B^*B = C - D.$$

Thus, $C, D \in \mathfrak{A}_+$ and $CD = 0 = DC$. We must show that $D = 0$. But first we have

$$(BD)^*(BD) = D(C - D)D = -D^3 \in -\mathfrak{A}_+.$$

Next remark that

$$BD = S + iT$$

with S and T selfadjoint, and one then calculates that

$$(BD)(BD)^* = -(BD)^*(BD) + 2(S^2 + T^2) \in \mathfrak{A}_+,$$

where we have used the fact that $\mathfrak{A}_+$ is a convex cone. Thus $\sigma((BD)(BD)^*) \subseteq [0, \|B\|^2\|D\|^2]$ and by Proposition 2.2.3 $\sigma((BD)^*(BD)) \subseteq [0, \|B\|^2\|D\|^2]$. But we already concluded that $(BD)^*(BD) \in -\mathfrak{A}_+$ and therefore $\sigma(D^3) = \{0\}$. The spectral radius formula then gives $\|D^3\| = 0 = \|D\|^3$ and hence $D = 0$.

Let us now examine some of the implications of the foregoing characterizations of positive elements. As $\mathfrak{A}_+$ is a convex cone with $\mathfrak{A}_+ \cap (-\mathfrak{A}_+) = \{0\}$ one can introduce an order relation $A - B \geq 0$ between selfadjoint elements. The relation $A - B \geq 0$ is interpreted to mean that $A - B \in \mathfrak{A}_+$ and we also write $A \geq B$, or $B \leq A$. If $A \geq B$ and $A \neq B$ one of course writes $A > B$.

This order relation has the two properties

(1) $A \geq 0$ and $A \leq 0$ imply $A = 0$,
(2) $A \geq B$ and $B \geq C$ imply $A \geq C$,

but the special properties of positive elements of a C^*-algebra yield some less obvious orderings.

Proposition 2.2.13. *Let A, B, C be elements of a C^*-algebra $\mathfrak{A}$. The following implications are valid:*

(a) *if* $A \geq B \geq 0$ *then* $\|A\| \geq \|B\|$;
(b) *if* $A \geq 0$ *then* $A\|A\| \geq A^2$;
(c) *if* $A \geq B \geq 0$ *then*

$$C^*AC \geq C^*BC \geq 0$$

for all $C \in \mathfrak{A}$;
(d) *if* $\mathfrak{A}$ *possesses an identity,* $A \geq B \geq 0$, *and* $\lambda > 0$ *then*

$$(B + \lambda \mathbb{1})^{-1} \geq (A + \lambda \mathbb{1})^{-1}.$$

PROOF. (a) We adjoin an identity $\mathbb{1}$ to $\mathfrak{A}$ if necessary. The spectral radius formula of Theorem 2.2.5 then gives $A \leq \|A\|\mathbb{1}$ and hence $0 \leq B \leq \|A\|\mathbb{1}$. But this implies that $\|B\| \leq \|A\|$ by a second application of the same formula.

(b) One has $\sigma(A - \|A\|\mathbb{1}/2) \subseteq [-\|A\|/2, \|A\|/2]$ and hence $\sigma((A - \|A\|\mathbb{1}/2)^2) \subseteq [0, \|A\|^2/4]$ by Theorem 2.2.5(d). Thus

$$0 \leq \left(A - \frac{\|A\|\mathbb{1}}{2}\right)^2 \leq \frac{\mathbb{1}\|A\|^2}{4},$$

which is equivalent to $0 \leq A^2 \leq \|A\|A$.

(c) As $A - B \in \mathfrak{A}_+$ one has $A - B = D^*D$ for some $D \in \mathfrak{A}$ by Theorem 2.2.12. But then

$$C^*AC - C^*BC = (DC)^*(DC) \in \mathfrak{A}_+$$

by the same theorem.

(d) One has

$$A + \lambda\mathbb{1} \geq B + \lambda\mathbb{1} \geq \lambda\mathbb{1}$$

and both $A + \lambda\mathbb{1}$ and $B + \lambda\mathbb{1}$ are positive invertible. Therefore by part (c)

$$(B + \lambda\mathbb{1})^{-1/2}(A + \lambda\mathbb{1})(B + \lambda\mathbb{1})^{-1/2} \geq \mathbb{1}.$$

If, however, $X = X^*$ and $X \geq \mathbb{1}$ then $\sigma(X) \subseteq [1, \infty\rangle$ and $\sigma(X^{-1}) \subseteq [0, 1]$ by Proposition 2.2.3. Thus $X^{-1} \leq \mathbb{1}$. This gives

$$(B + \lambda\mathbb{1})^{1/2}(A + \lambda\mathbb{1})^{-1}(B + \lambda\mathbb{1})^{1/2} \leq \mathbb{1}.$$

Finally, multiplying each side by $(B + \lambda\mathbb{1})^{-1/2}$ and invoking part (c), one finds

$$(A + \lambda\mathbb{1})^{-1} \leq (B + \lambda\mathbb{1})^{-1}.$$

There are many other interesting inequalities which may be deduced from Proposition 2.2.13(d) by integration with suitable functions of λ. For example, if $A \geq B \geq 0$ one has

$$\begin{aligned} A^{1/2} &= \frac{1}{\pi}\int_0^\infty \frac{d\lambda}{\lambda^{1/2}}(\mathbb{1} - \lambda\mathbb{1}(\lambda\mathbb{1} + A)^{-1}) \\ &\geq \frac{1}{\pi}\int_0^\infty \frac{d\lambda}{\lambda^{1/2}}(\mathbb{1} - \lambda\mathbb{1}(\lambda\mathbb{1} + B)^{-1}) \\ &= B^{1/2}, \end{aligned}$$

i.e., $A^{1/2} \geq B^{1/2} \geq 0$. By use of similar transforms one can deal with other fractional powers and deduce that $A \geq B \geq 0$ implies $A^\alpha \geq B^\alpha \geq 0$ for all $0 \leq \alpha \leq 1$. But this is not necessarily true for $\alpha > 1$.

The following decomposition lemma is often useful and is another application of the structure of positive elements.

Lemma 2.2.14. *Let $\mathfrak{A}$ be a C^*-algebra with identity. Every element $A \in \mathfrak{A}$ has a decomposition of the form*

$$A = a_1 U_1 + a_2 U_2 + a_3 U_3 + a_4 U_4$$

where the U_i are unitary elements of $\mathfrak{A}$ and the $a_i \in \mathbb{C}$ satisfy $|a_i| \leq \|A\|/2$.

PROOF. It suffices to consider the case $\|A\| = 1$. But then $A = A_1 + iA_2$ with $A_1 = (A + A^*)/2$ and $A_2 = (A - A^*)/2i$ selfadjoint, $\|A_1\| \leq 1$, $\|A_2\| \leq 1$. A general selfadjoint element B with $\|B\| \leq 1$ can, however, be decomposed into two unitary elements $B = (U_+ + U_-)/2$ by the explicit construction $U_\pm = B \pm i\sqrt{\mathbb{1} - B^2}$.

As a final application of the properties of positive elements we consider another type of decomposition. First let us extend our definition of the modulus. If $\mathfrak{A}$ is a C^*-algebra then A^*A is positive for all $A \in \mathfrak{A}$ by Theorem 2.2.11. The modulus of $A \in \mathfrak{A}$ is then defined by $|A| = \sqrt{A^*A}$. If A is selfadjoint this coincides with the previous definition. Now note that if $\mathfrak{A}$ contains an identity and A is invertible then A^*A is invertible and its inverse is positive. It follows that $|A|$ is invertible and $|A|^{-1} = \sqrt{(A^*A)^{-1}}$. But one then has

$$A = U|A|.$$

where $U = A|A|^{-1}$. Moreover, $U^*U = \mathbb{1}$ and U is invertible ($U^{-1} = |A|A^{-1}$). Therefore U is a unitary element of $\mathfrak{A}$ and in fact lies in the C^*-subalgebra generated by A and A^*. This decomposition of A is a special case of the so-called polar decomposition. The general polar decomposition concerns operators on a Hilbert space and represents each closed, densely defined operator A as a product $A = V(A^*A)^{1/2}$ of a partial isometry V and a positive selfadjoint operator $|A| = (A^*A)^{1/2}$. We illustrate this and other Hilbert space properties in the following:

EXAMPLE 2.2.15. Let $\mathcal{L}(\mathfrak{H})$ denote the algebra of all bounded operators on the complex Hilbert space $\mathfrak{H}$; then $\mathcal{L}(\mathfrak{H})$ is a C^*- algebra by Example 2.1.2. If $A \in \mathcal{L}(\mathfrak{H})$ then the abstract definition of positivity is equivalent to $A = B^*B$ for some $B \in \mathcal{L}(\mathfrak{H})$ and this implies that $(\psi, A\psi) = \|B\psi\|^2 \geq 0$ for all $\psi \in \mathfrak{H}$. In Hilbert space theory this last property is usually taken as the definition of positivity but it is equivalent to the abstract definition by the following reasoning. If the values of $(\psi, A\psi)$ are positive then they are, in particular, real and $(\psi, A\psi) = (A\psi, \psi)$. Therefore the polarization identity

$$(\psi, A\varphi) = \frac{1}{4} \sum_{k=0}^{3} i^{-k}((\psi + i^k\varphi), A(\psi + i^k\varphi))$$

demonstrates that $(\psi, A\varphi) = (A\psi, \varphi)$ for all $\psi, \varphi \in \mathfrak{H}$, i.e., A is selfadjoint. But if $\lambda < 0$ then

$$\begin{aligned}\|(A - \lambda\mathbb{1})\psi\|^2 &= \|A\psi\|^2 + 2|\lambda|(\psi, A\psi) + \lambda^2\|\psi\|^2 \\ &\geq \lambda^2\|\psi\|^2\end{aligned}$$

and $A - \lambda\mathbb{1}$ is invertible. Consequently $\sigma(A) \in [0, \|A\|]$ and A is positive in the general sense.

EXAMPLE 2.2.16. Let $A \in \mathscr{L}(\mathfrak{H})$ and $|A| = (A^*A)^{1/2}$. Now define an operator V on all vectors of the form $|A|\psi$ by the action

$$V|A|\psi = A\psi.$$

This is a consistent definition of a linear operator because $|A|\psi = 0$ is equivalent to $0 = \||A|\psi\| = \|A\psi\|$ and hence $A\psi = 0$. Moreover, V is isometric because $\|V|A|\psi\| = \|A\psi\| = \||A|\psi\|$. We may extend V to a partial isometry on $\mathfrak{H}$ by setting it equal to zero on the orthogonal complement of the set $\{|A|\psi;\, \psi \in \mathfrak{H}\}$ and extending by linearity. This yields the polar decomposition of A, i.e., $A = V|A|$. This decomposition is unique in the sense that if $A = UB$ with $B \geq 0$ and U a partial isometry such that $U\varphi = 0$ just for φ orthogonal to the range of B then $U = V$ and $B = |A|$. This follows because $A^*A = BU^*UB = B^2$ and hence B is equal to the unique positive square root $|A|$ of A^*A. But then $U|A| = V|A|$ and both U and V are equal to zero on the orthogonal complement of the range of $|A|$. In general, V will not be an element of the C^*-algebra $\mathfrak{A}_A$ generated by A and A^*, although we have seen that this is the case whenever A has a bounded inverse. Nevertheless, in Section 2.4 we will see that V is an element of the algebra obtained by adding to $\mathfrak{A}_A$ all strong or weak limit points of nets of elements of $\mathfrak{A}_A$.

2.2.3. Approximate Identities and Quotient Algebras

In Section 2.2.1 we gave examples of C^*-algebras which failed to have an identity element and demonstrated that it is always possible to adjoin such an element. Nevertheless, situations often occur in which the absence of an identity is fundamental and it is therefore useful to introduce the notion of an approximate identity.

Definition 2.2.17. If $\mathfrak{J}$ is a right ideal of a C^*-algebra $\mathfrak{A}$ then an *approximate identity of* $\mathfrak{J}$ is defined to be a net[1] $\{E_\alpha\}$ of positive elements $E_\alpha \in \mathfrak{J}$ such that

(1) $\|E_\alpha\| \leq 1$,
(2) $\alpha \leq \beta$ implies $E_\alpha \leq E_\beta$,
(3) $\lim_\alpha \|E_\alpha A - A\| = 0$ for all $A \in \mathfrak{J}$.

[1] A set $\mathscr{U}$ is said to be directed when there exists an order relation, $\alpha \leq \beta$, between certain pairs of elements $\alpha, \beta \in \mathscr{U}$ which is reflexive ($\alpha \leq \alpha$), transitive ($\alpha \leq \beta$ and $\beta \leq \gamma$ imply $\alpha \leq \gamma$), antisymmetric ($\alpha \leq \beta$ and $\beta \leq \alpha$ imply $\beta = \alpha$) and when for each pair $\alpha, \beta \in \mathscr{U}$ there exists a γ such that $\alpha \leq \gamma$ and $\beta \leq \gamma$. A net is a family of elements, of a general set M, which is indexed by a directed set $\mathscr{U}$.

The definition of an approximate identity of a left ideal is similar but condition (3) is replaced by

(3′) $\lim_\alpha \|AE_\alpha - A\| = 0$ for all $A \in \mathfrak{J}$.

It is necessary to prove the existence of approximate identities.

Proposition 2.2.18. *Let $\mathfrak{J}$ be a right ideal of a C^*-algebra $\mathfrak{A}$. $\mathfrak{J}$ possesses an approximate identity.*

PROOF. First adjoin an identity to $\mathfrak{A}$, if necessary. Next let $\mathcal{U}$ denote the set of finite families of $\mathfrak{J}$. The set $\mathcal{U}$ can be ordered by inclusion, i.e., if $\alpha = \{A_1, \ldots, A_m\}$ and $\beta = \{B_1, \ldots, B_n\}$ then $\alpha \geq \beta$ is equivalent to β being a subfamily of α. Now for the foregoing choice of α define $F_\alpha \in \mathfrak{A}$ by

$$F_\alpha = \sum_{i=1}^{m} A_i A_i^*$$

and introduce E_α by

$$E_\alpha = mF_\alpha(\mathbb{1} + mF_\alpha)^{-1}.$$

As each $A_i \in \mathfrak{J}$ one has $E_\alpha, F_\alpha \in \mathfrak{J}$. Furthermore $\|E_\alpha\| \leq 1$ and

$$\begin{aligned}(E_\alpha A_i - A_i)(E_\alpha A_i - A_i)^* &\leq \sum_{i=1}^{m} (E_\alpha - \mathbb{1})A_i A_i^*(E_\alpha - \mathbb{1}) \\ &= (\mathbb{1} + mF_\alpha)^{-1}F_\alpha(\mathbb{1} + mF_\alpha)^{-1} \\ &= F_\alpha^{1/2}(\mathbb{1} + mF_\alpha)^{-2}F_\alpha^{1/2} \\ &\leq F_\alpha^{1/2}(\mathbb{1} + mF_\alpha)^{-1}F_\alpha^{1/2} \\ &= \frac{1}{m}(\mathbb{1} - (\mathbb{1} + mF_\alpha)^{-1}) \\ &\leq \frac{1}{m}\mathbb{1}.\end{aligned}$$

Here we have used $0 \leq (\mathbb{1} + mF_\alpha)^{-1} \leq \mathbb{1}$ and Proposition 2.2.13(c). Therefore by part (a) of the same proposition

$$\|E_\alpha A_i - A_i\|^2 \leq \frac{1}{m}$$

and consequently $\|E_\alpha A - A\| \underset{\alpha}{\to} 0$ for all $A \in \mathfrak{J}$. Finally, note that

$$E_\alpha - E_\beta = (\mathbb{1} + nF_\beta)^{-1} - (\mathbb{1} + mF_\alpha)^{-1}$$

but $\alpha \geq \beta$ implies $mF_\alpha \geq nF_\beta$ and hence $E_\alpha \geq E_\beta$ by Proposition 2.2.13(d). Therefore the E_α form an approximate identity.

The existence of an approximate identity allows us to complete the discussion of quotient algebras which we began in Section 2.1. The principal result is the following:

Proposition 2.2.19. *Let $\mathfrak{I}$ be a closed two-sided ideal of a C^* algebra $\mathfrak{A}$. It follows that $\mathfrak{I}$ is selfadjoint and the quotient algebra $\mathfrak{A}/\mathfrak{I}$ defined in Section 2.1.1 is a C^*-algebra.*

PROOF. Let $\{E_\alpha\}$ be an approximate identity of $\mathfrak{I}$. If $A \in \mathfrak{I}$ then $\|A^*E_\alpha - A^*\| = \|E_\alpha A - A\| \to 0$. But $A^*E_\alpha \in \mathfrak{I}$ and hence $A^* \in \mathfrak{I}$ because $\mathfrak{I}$ is closed. This proves that $\mathfrak{I}$ is selfadjoint.

To complete both the discussion of the quotient algebra given in Section 2.1.1 and the proof of the proposition we must show that the norm on the quotient algebra,

$$\|\hat{A}\| = \inf\{\|A + I\|;\, I \in \mathfrak{I}\},$$

has the C^*-norm property. To prove this we first establish that

$$\|\hat{A}\| = \lim_\alpha \|A - E_\alpha A\|.$$

This follows by adjoining, if necessary, an identity to $\mathfrak{A}$, noting that for $I \in \mathfrak{I}$, $\|E_\alpha I - I\| \to 0$, and

$$\begin{aligned}\limsup_\alpha \|A - E_\alpha A\| &= \limsup_\alpha \|(\mathbb{1} - E_\alpha)(A + I)\| \\ &\leq \|A + I\|.\end{aligned}$$

The inequality follows because $\sigma(E_\alpha) \in [0, 1]$. Therefore $\sigma(\mathbb{1} - E_\alpha) \in [0, 1]$ and $\|\mathbb{1} - E_\alpha\| \leq 1$. But then one concludes that

$$\begin{aligned}\|\hat{A}\| &\geq \limsup_\alpha \|A - E_\alpha A\| \\ &\geq \liminf_\alpha \|A - E_\alpha A\| \\ &\geq \inf\{\|A + I\|;\, I \in \mathfrak{I}\} = \|\hat{A}\|.\end{aligned}$$

The C^*-norm property is then a consequence of the following calculation:

$$\begin{aligned}\|\hat{A}\|^2 &= \lim_\alpha \|A - E_\alpha A\|^2 \\ &= \lim_\alpha \|(A - E_\alpha A)(A - E_\alpha A)^*\| \\ &= \lim_\alpha \|(\mathbb{1} - E_\alpha)(AA^* + I)(\mathbb{1} - E_\alpha)\| \\ &\leq \|AA^* + I\|,\end{aligned}$$

where I is an arbitrary element of $\mathfrak{I}$. Thus

$$\|\hat{A}\|^2 \leq \|\hat{A}\hat{A}^*\| \leq \|\hat{A}\|\,\|\hat{A}^*\|,$$

which implies firstly that $\|\hat{A}\| = \|\hat{A}^*\|$ and, secondly, that

$$\|\hat{A}\|^2 = \|\hat{A}\hat{A}^*\|.$$

2.3. Representations and States

2.3.1. Representations

In the previous sections we partially described the abstract theory of C^*-algebras and illustrated the general theory by examples of C^*-algebras of operators acting on a Hilbert space. Next we discuss representation theory and develop the connection between the abstract description and the operator examples. The two key concepts in this development are the concepts of representation and state. The states of $\mathfrak{A}$ are a class of linear functionals which take positive values on the positive elements of $\mathfrak{A}$ and they are of fundamental importance for the construction of representations. We precede the discussion of these states by giving the precise definition of a representation and by developing some general properties of representations.

First let us define a *-morphism between two *-algebras $\mathfrak{A}$ and $\mathfrak{B}$ as a mapping π; $A \in \mathfrak{A} \mapsto \pi(A) \in \mathfrak{B}$, defined for all $A \in \mathfrak{A}$ and such that

(1) $\pi(\alpha A + \beta B) = \alpha\pi(A) + \beta\pi(B)$,
(2) $\pi(AB) = \pi(A)\pi(B)$,
(3) $\pi(A^*) = \pi(A)^*$

for all A, $B \in \mathfrak{A}$ and $\alpha \in \mathbb{C}$. The name morphism is usually reserved for mappings which only have properties (1) and (2). As all morphisms we consider are *-morphisms we occasionally drop the * symbol.

Now each *-morphism π between C^*-algebras $\mathfrak{A}$ and $\mathfrak{B}$ is positive because if $A \geq 0$ then $A = B^*B$ for some $B \in \mathfrak{A}$ by Theorem 2.2.12 and hence

$$\pi(A) = \pi(B^*B) = \pi(B)^*\pi(B) \geq 0.$$

It is less evident that π is automatically continuous.

Proposition 2.3.1. *Let $\mathfrak{A}$ be a Banach *-algebra with identity, $\mathfrak{B}$ a C^*-algebra, and π a *-morphism of $\mathfrak{A}$ into $\mathfrak{B}$. Then π is continuous and*

$$\|\pi(A)\| \leq \|A\|$$

for all $A \in \mathfrak{A}$. Moreover, if $\mathfrak{A}$ is a C^*-algebra then the range $\mathfrak{B}_\pi = \{\pi(A);$ $A \in \mathfrak{A}\}$ of π is a C^*-subalgebra of $\mathfrak{B}$.

PROOF. First assume $A = A^*$. Then since $\mathfrak{B}$ is a C^*-algebra and $\pi(A) \in \mathfrak{B}$, one has

$$\|\pi(A)\| = \sup\{|\lambda|; \lambda \in \sigma(\pi(A))\}$$

by Theorem 2.2.5(a). Next define $P = \pi(\mathbb{1}_{\mathfrak{A}})$ where $\mathbb{1}_{\mathfrak{A}}$ denotes the identity of $\mathfrak{A}$. It follows from the definition of π that P is a projection in $\mathfrak{B}$. Hence replacing $\mathfrak{B}$ by the C^*-algebra $P\mathfrak{B}P$ the projection P becomes the identity $\mathbb{1}_{\mathfrak{B}}$ of the new algebra $\mathfrak{B}$. Moreover, $\pi(\mathfrak{A}) \subseteq \mathfrak{B}$. Now it follows from the definitions of a morphism and of the spectrum that $\sigma_{\mathfrak{B}}(\pi(A)) \subseteq \sigma_{\mathfrak{A}}(A)$. Therefore

$$\|\pi(A)\| \leq \sup\{|\lambda|; \lambda \in \sigma_{\mathfrak{A}}(A)\} \leq \|A\|$$

by Proposition 2.2.2. Finally, if A is not selfadjoint one can combine this inequality with the C^*-norm property and the product inequality to deduce that

$$\|\pi(A)\|^2 = \|\pi(A^*A)\| \leq \|A^*A\| \leq \|A\|^2.$$

Thus $\|\pi(A)\| \leq \|A\|$ for all $A \in \mathfrak{A}$ and π is continuous.

The range $\mathfrak{B}_\pi$ is a *-subalgebra of $\mathfrak{B}$ by definition and to deduce that it is a C^*-subalgebra we must prove that it is closed, under the assumption that $\mathfrak{A}$ is a C^*-algebra.

Now introduce the kernel $\ker \pi$ of π by

$$\ker \pi = \{A \in \mathfrak{A}; \pi(A) = 0\}$$

then $\ker \pi$ is a closed two-sided *-ideal. For example if $A \in \mathfrak{A}$ and $B \in \ker \pi$ then $\pi(AB) = \pi(A)\pi(B) = 0$, $\pi(BA) = \pi(B)\pi(A) = 0$, and $\pi(B^*) = \pi(B) = 0$. The closedness follows from the estimate $\|\pi(A)\| \leq \|A\|$. Thus we can form the quotient algebra $\mathfrak{A}_\pi = \mathfrak{A}/\ker \pi$ and $\mathfrak{A}_\pi$ is a C^*-algebra by Proposition 2.2.19. The elements of $\mathfrak{A}_\pi$ are the classes $\hat{A} = \{A + I; I \in \ker \pi\}$ and the morphism π induces a morphism $\hat{\pi}$ from $\mathfrak{A}_\pi$ onto $\mathfrak{B}_\pi$ by the definition $\hat{\pi}(\hat{A}) = \pi(A)$. The kernel of $\hat{\pi}$ is zero by construction and hence $\hat{\pi}$ is an isomorphism between $\mathfrak{A}_\pi$ and $\mathfrak{B}_\pi$. Thus we can define a morphism $\hat{\pi}^{-1}$ from the *-algebra $\mathfrak{B}_\pi$ onto the C^*-algebra $\mathfrak{A}_\pi$ by $\hat{\pi}^{-1}(\hat{\pi}(\hat{A})) = \hat{A}$ and then applying the first statement of the proposition to $\hat{\pi}^{-1}$ and $\hat{\pi}$ successively one obtains

$$\|\hat{A}\| = \|\hat{\pi}^{-1}(\hat{\pi}(\hat{A}))\| \leq \|\hat{\pi}(\hat{A})\| \leq \|\hat{A}\|.$$

Thus $\|\hat{A}\| = \|\hat{\pi}(\hat{A})\| = \|\pi(A)\|$. Consequently, if $\pi(A_n)$ converges uniformly in $\mathfrak{B}$ to an element A_π then $\hat{A}_n$ converges in $\mathfrak{A}_\pi$ to an element $\hat{A}$ and $A_\pi = \hat{\pi}(\hat{A}) = \pi(A)$ where A is any element of the equivalence class $\hat{A}$. Thus $A_\pi \in \mathfrak{B}_\pi$ and $\mathfrak{B}_\pi$ is closed.

Next we define the concept of *-isomorphism between C^*-algebras.

A *-morphism π of $\mathfrak{A}$ to $\mathfrak{B}$ is a *-isomorphism if it is one-to-one and onto, i.e., if the range of π is equal to $\mathfrak{B}$ and each element of $\mathfrak{B}$ is the image of a unique element of $\mathfrak{A}$. Thus a *-morphism π of the C^*-algebra $\mathfrak{A}$ onto a C^*-algebra $\mathfrak{B}$ is a *-isomorphism if, and only if, $\ker \pi = 0$.

Now we can introduce the basic definition of representation theory.

Definition 2.3.2. A *representation of a C^*-algebra* $\mathfrak{A}$ is defined to be a pair $(\mathfrak{H}, \pi)$, where $\mathfrak{H}$ is a complex Hilbert space and π is a *-morphism of $\mathfrak{A}$ into $\mathscr{L}(\mathfrak{H})$. The representation $(\mathfrak{H}, \pi)$ is said to be *faithful* if, and only if, π is a *-isomorphism between $\mathfrak{A}$ and $\pi(\mathfrak{A})$, i.e., if, and only if, $\ker \pi = \{0\}$.

There is a variety of rather obvious terminology associated with this definition. The space $\mathfrak{H}$ is called the representation space, the operators $\pi(A)$ are called the representatives of $\mathfrak{A}$ and, by implicit identification of π and the set of representatives, one also says that π is a representation of $\mathfrak{A}$ on $\mathfrak{H}$.

The discussion preceding Definition 2.3.2 established that each representation $(\mathfrak{H}, \pi)$ of a C^*-algebra $\mathfrak{A}$ defines a faithful representation of the quotient algebra $\mathfrak{A}_\pi = \mathfrak{A}/\ker \pi$. In particular, every representation of a simple C^*-algebra is faithful. Naturally, the most important representations are the faithful ones and it is useful to have criteria for faithfulness.

Proposition 2.3.3. *Let $(\mathfrak{H}, \pi)$ be a representation of the C^*-algebra $\mathfrak{A}$. The representation is faithful if, and only if, it satisfies each of the following equivalent conditions:*

(1) $\ker \pi = \{0\}$;
(2) $\|\pi(A)\| = \|A\|$ *for all* $A \in \mathfrak{A}$;
(3) $\pi(A) > 0$ *for all* $A > 0$.

PROOF. The equivalence of condition (1) and faithfulness is by definition. We now prove (1) ⇒ (2) ⇒ (3) ⇒ (1).

(1) ⇒ (2) As $\ker \pi = \{0\}$ we can define a morphism π^{-1} from the range of π into $\mathfrak{A}$ by $\pi^{-1}(\pi(A)) = A$ and then applying Proposition 2.3.1 to π^{-1} and π successively one has

$$\|A\| = \|\pi^{-1}(\pi(A))\| \leq \|\pi(A)\| \leq \|A\|.$$

(2) ⇒ (3) If $A > 0$ then $\|A\| > 0$ and hence $\|\pi(A)\| > 0$, or $\pi(A) \neq 0$. But $\pi(A) \geq 0$ by Proposition 2.3.1 and therefore $\pi(A) > 0$.

(3) ⇒ (1) If condition (1) is false then there is a $B \in \ker \pi$ with $B \neq 0$ and $\pi(B^*B) = 0$. But $\|B^*B\| \geq 0$ and as $\|B^*B\| = \|B\|^2$ one has $B^*B > 0$. Thus condition (3) is false.

A *-automorphism τ of a C^*-algebra $\mathfrak{A}$ is defined to be a *-isomorphism of $\mathfrak{A}$ into itself, i.e., τ is a *-morphism of $\mathfrak{A}$ with range equal to $\mathfrak{A}$ and kernel equal to zero.

The foregoing argument utilizing the invertibility of τ implies the following:

Corollary 2.3.4. *Each *-automorphism τ of a C^*-algebra $\mathfrak{A}$ is norm preserving, i.e., $\|\tau(A)\| = \|A\|$ for all $A \in \mathfrak{A}$.*

Now we turn our attention to various kinds of representation and methods of composing or decomposing representations.

First we introduce the notion of a subrepresentation. If $(\mathfrak{H}, \pi)$ is a representation of the C^*-algebra $\mathfrak{A}$ and $\mathfrak{H}_1$ is a subspace of $\mathfrak{H}$ then $\mathfrak{H}_1$ is said to be invariant, or stable, under π if $\pi(A)\mathfrak{H}_1 \subseteq \mathfrak{H}_1$ for all $A \in \mathfrak{A}$. If $\mathfrak{H}_1$ is a closed subspace of $\mathfrak{H}$ and $P_{\mathfrak{H}_1}$ the orthogonal projector with range $\mathfrak{H}_1$ then the invariance of $\mathfrak{H}_1$ under π implies that

$$P_{\mathfrak{H}_1}\pi(A)P_{\mathfrak{H}_1} = \pi(A)P_{\mathfrak{H}_1}$$

for all $A \in \mathfrak{A}$. Hence

$$\begin{aligned}\pi(A)P_{\mathfrak{H}_1} &= (P_{\mathfrak{H}_1}\pi(A^*)P_{\mathfrak{H}_1})^* \\ &= (\pi(A^*)P_{\mathfrak{H}_1})^* \\ &= P_{\mathfrak{H}_1}\pi(A)\end{aligned}$$

for all $A \in \mathfrak{A}$, i.e., the projector $P_{\mathfrak{H}_1}$ commutes with each of the representatives $\pi(A)$. Conversely, this commutation property implies that $\mathfrak{H}_1$ is invariant under π. Hence one deduces that $\mathfrak{H}_1$ is invariant under π if, and only if,

$$\pi(A)P_{\mathfrak{H}_1} = P_{\mathfrak{H}_1}\pi(A)$$

for all $A \in \mathfrak{A}$. Furthermore, we may conclude that if $\mathfrak{H}_1$ is invariant under π and if π_1 is defined by

$$\pi_1(A) = P_{\mathfrak{H}_1}\pi(A)P_{\mathfrak{H}_1}$$

then $(\mathfrak{H}_1, \pi_1)$ is a representation of $\mathfrak{A}$, e.g.,

$$\begin{aligned}\pi_1(A)\pi_1(B) &= (P_{\mathfrak{H}_1}\pi(A))(\pi(B)P_{\mathfrak{H}_1}) \\ &= P_{\mathfrak{H}_1}\pi(AB)P_{\mathfrak{H}_1} = \pi_1(AB).\end{aligned}$$

A representation constructed in this manner is called a subrepresentation of $(\mathfrak{H}, \pi)$.

Note that the foregoing method of passing to a subrepresentation gives a decomposition of π in the following sense. If $\mathfrak{H}_1$ is invariant under π then its orthogonal complement $\mathfrak{H}_1{}^\perp$ is also invariant. Setting $\mathfrak{H}_2 = \mathfrak{H}_1{}^\perp$ one can define a second subrepresentation $(\mathfrak{H}_2, \pi_2)$ by $\pi_2(A) = P_{\mathfrak{H}_2}\pi(A)P_{\mathfrak{H}_2}$. But $\mathfrak{H}$ has a direct sum decomposition, $\mathfrak{H} = \mathfrak{H}_1 \oplus \mathfrak{H}_2$, and each operator $\pi(A)$ then decomposes as a direct sum $\pi(A) = \pi_1(A) \oplus \pi_2(A)$. Thus we write $\pi = \pi_1 \oplus \pi_2$ and $(\mathfrak{H}, \pi) = (\mathfrak{H}_1, \pi_1) \oplus (\mathfrak{H}_2, \pi_2)$.

A particularly trivial type of representation of a C^*-algebra is given by $\pi = 0$, i.e., $\pi(A) = 0$ for all $A \in \mathfrak{A}$. A representation might be nontrivial but nevertheless have a trivial part. Thus if $\mathfrak{H}_0$ is defined by

$$\mathfrak{H}_0 = \{\psi; \psi \in \mathfrak{H}, \pi(A)\psi = 0 \text{ for all } A \in \mathfrak{A}\}$$

then $\mathfrak{H}_0$ is invariant under π and the corresponding subrepresentation $\pi_0 = P_{\mathfrak{H}_0}\pi P_{\mathfrak{H}_0}$ is zero. With this notation a representation $(\mathfrak{H}, \pi)$ is said to be nondegenerate if $\mathfrak{H}_0 = \{0\}$. Alternatively, one says that a set $\mathfrak{M}$ of bounded operators acts nondegenerately on $\mathfrak{H}$ if

$$\{\psi; A\psi = 0 \text{ for all } A \in \mathfrak{M}\} = \{0\}.$$

An important class of nondegenerate representations is the class of cyclic representations. To introduce these representations we first define a vector Ω in a Hilbert space $\mathfrak{H}$ to be cyclic for a set of bounded operators $\mathfrak{M}$ if the set $\{A\Omega; A \in \mathfrak{M}\}$ is dense in $\mathfrak{H}$. Then we have the following:

Definition 2.3.5. A *cyclic representation of a C^*-algebra* $\mathfrak{A}$ is defined to be a triple $(\mathfrak{H}, \pi, \Omega)$, where $(\mathfrak{H}, \pi)$ is a representation of $\mathfrak{A}$ and Ω is a vector in $\mathfrak{H}$ which is cyclic for π, in $\mathfrak{H}$.

In the sequel, if there is no possible ambiguity we will often abbreviate the terminology and say that Ω is a cyclic vector, or Ω is cyclic for π. There is a more general concept than a cyclic vector which is also often useful. If $\mathfrak{K}$ is a closed subspace of $\mathfrak{H}$ then $\mathfrak{K}$ is called a cyclic subspace for $\mathfrak{H}$ whenever the set

$$\left\{\sum_i \pi(A_i)\psi_i;\ A_i \in \mathfrak{A},\ \psi_i \in \mathfrak{K}\right\}$$

is dense in $\mathfrak{H}$. The orthogonal projector $P_{\mathfrak{K}}$, whose range is $\mathfrak{K}$, is also called a cyclic projector.

It is evident from these definitions that every cyclic representation is nondegenerate but there is a form of converse to this statement. To describe this converse we need the general notion of a direct sum of representations.

Let $(\mathfrak{H}_\alpha, \pi_\alpha)_{\alpha \in I}$ be a family of representations of the C^*-algebra $\mathfrak{A}$ where the index set I can be countable or noncountable. The direct sum

$$\mathfrak{H} = \bigoplus_{\alpha \in I} \mathfrak{H}_\alpha$$

of the representation spaces $\mathfrak{H}_\alpha$ is defined in the usual manner[2] and one defines the direct sum representatives

$$\pi = \bigoplus_{\alpha \in I} \pi_\alpha$$

by setting $\pi(A)$ equal to the operator $\pi_\alpha(A)$ on the component subspace $\mathfrak{H}_\alpha$. This definition yields bounded operators $\pi(A)$ on $\mathfrak{H}$ because $\|\pi_\alpha(A)\| \leq \|A\|$, for all $\alpha \in I$, by Proposition 2.3.1. It is easily checked that $(\mathfrak{H}, \pi)$ is a representation and it is called the direct sum of the representations $(\mathfrak{H}_\alpha, \pi_\alpha)_{\alpha \in I}$. One has the following result.

Proposition 2.3.6. *Let $(\mathfrak{H}, \pi)$ be a nondegenerate representation of the C^*-algebra $\mathfrak{A}$. It follows that π is the direct sum of a family of cyclic subrepresentations.*

PROOF. Let $\{\Omega_\alpha\}_{\alpha \in I}$ denote a maximal family of nonzero vectors in $\mathfrak{H}$ such that

$$(\pi(A)\Omega_\alpha, \pi(B)\Omega_\beta) = 0$$

for all $A, B \in \mathfrak{A}$, whenever $\alpha \neq \beta$. The existence of such a family can be deduced with the aid of Zorn's lemma. Next define $\mathfrak{H}_\alpha$ as the Hilbert subspace formed by closing the linear subspace $\{\pi(A)\Omega_\alpha, A \in \mathfrak{A}\}$. This is an invariant subspace so we can introduce π_α by $\pi_\alpha(A) = P_{\mathfrak{H}_\alpha}\pi(A)P_{\mathfrak{H}_\alpha}$ and it follows that each $(\mathfrak{H}_\alpha, \pi_\alpha, \Omega_\alpha)$ is a cyclic

[2] The finite subsets F of the index set I form a directed set when ordered by inclusion and $\mathfrak{H}$ consists of those families $\psi = \{\psi_\alpha\}$, $\varphi = \{\varphi_\alpha\}$ of vectors such that $\varphi_\alpha, \psi_\alpha \in \mathfrak{H}_\alpha$ and

$$\lim_F \sum_{\alpha \in F} \|\psi_\alpha\|^2_{\mathfrak{H}_\alpha} < +\infty, \qquad \lim_F \sum_{\alpha \in F} \|\varphi_\alpha\|^2 < +\infty.$$

The scalar product on $\mathfrak{H}$ is then defined by

$$(\varphi, \psi) = \sum_\alpha (\varphi_\alpha, \psi_\alpha)_{\mathfrak{H}_\alpha} = \lim_F \sum_{\alpha \in F} (\varphi_\alpha, \psi_\alpha)_{\mathfrak{H}_\alpha}.$$

representation of $\mathfrak{A}$. But the maximality of the $\{\Omega_\alpha\}_{\alpha \in I}$ and the nondegeneracy of π imply that there is no nonzero Ω which is orthogonal to each subspace $\mathfrak{H}_\alpha$ and hence

$$\mathfrak{H} = \bigoplus_{\alpha \in I} \mathfrak{H}_\alpha, \qquad \pi = \bigoplus_{\alpha \in I} \pi_\alpha.$$

The foregoing proposition essentially reduces the discussion of general representations to that of cyclic representations. This is of importance because there is a canonical manner of constructing cyclic representations which we will discuss in detail in Section 2.3.3. The type of decomposition used to reduce the general situation to the cyclic situation depends upon the existence of nontrivial invariant subspaces. No further reduction is possible in the absence of such subspaces and this motivates the next definition.

Definition 2.3.7. A set $\mathfrak{M}$ of bounded operators on the Hilbert space $\mathfrak{H}$ is defined to be *irreducible* if the only closed subspaces of $\mathfrak{H}$ which are invariant under the action of $\mathfrak{M}$ are the trivial subspaces $\{0\}$ and $\mathfrak{H}$. A representation $(\mathfrak{H}, \pi)$ of a C^*-algebra $\mathfrak{A}$ is defined to be *irreducible* if the set $\pi(\mathfrak{A})$ is irreducible on $\mathfrak{H}$.

The term topologically irreducible is sometimes used in place of irreducible. The term irreducible is defined by the demand that the only invariant subspaces, closed or not, are $\{0\}$ and $\mathfrak{H}$. Actually, the two notions coincide for representations of a C^*-algebra but we will not prove this equivalence.

There are two standard criteria for irreducibility.

Proposition 2.3.8. *Let $\mathfrak{M}$ be a selfadjoint set of bounded operators on the Hilbert space $\mathfrak{H}$. The following conditions are equivalent:*

(1) *$\mathfrak{M}$ is irreducible;*
(2) *the commutant $\mathfrak{M}'$ of $\mathfrak{M}$, i.e., the set of all bounded operators on $\mathfrak{H}$ which commute with each $A \in \mathfrak{M}$, consists of multiples of the identity operator;*
(3) *every nonzero vector $\psi \in \mathfrak{H}$ is cyclic for $\mathfrak{M}$ in $\mathfrak{H}$, or $\mathfrak{M} = 0$ and $\mathfrak{H} = \mathbb{C}$.*

Proof. (1) $\Rightarrow$ (3) Assume there is a nonzero ψ such that the set $\{A\psi;\, A \in \mathfrak{M}\}$ is not dense in $\mathfrak{H}$. The orthogonal complement of this set then contains at least one nonzero vector and is invariant under $\mathfrak{M}$ (unless $\mathfrak{M} = \{0\}$ and $\mathfrak{H} = \mathbb{C}$), and this contradicts condition (1).

(3) $\Rightarrow$ (2) If $T \in \mathfrak{M}'$ then $T^* \in \mathfrak{M}'$ and, furthermore, $T + T^* \in \mathfrak{M}'$ and $(T - T^*)/i \in \mathfrak{M}'$. Thus if $\mathfrak{M}' \neq \mathbb{C}\mathbb{1}$ then there is a selfadjoint operator $S \in \mathfrak{M}'$ such that $S \neq \lambda\mathbb{1}$ for any $\lambda \in \mathbb{C}$. As all bounded functions of S must also be in the commutant one deduces that the spectral projectors of S also commute with $\mathfrak{M}$. But if E is any such projector and ψ a vector in the range of E then $\psi = E\psi$ cannot be cyclic and condition (3) is false.

(2) $\Rightarrow$ (1) If condition (1) is false then there exists a closed subspace $\mathfrak{K}$ of $\mathfrak{H}$ which is invariant under $\mathfrak{M}$. But then $P_{\mathfrak{K}} \in \mathfrak{M}'$ and condition (2) is false.

We conclude this survey of the basic properties of representations by remarking that if one has a representation $(\mathfrak{H}, \pi)$ of a C^*-algebra then it is easy to construct other representations. For example if U is a unitary operator on $\mathfrak{H}$ and we introduce π_U by $\pi_U(A) = U\pi(A)U^*$ then $(\mathfrak{H}, \pi_U)$ is a second representation. This type of distinction is, however, not important so we define two representations $(\mathfrak{H}_1, \pi_1)$ and $(\mathfrak{H}_2, \pi_2)$ to be equivalent, or unitarily equivalent, if there exists a unitary operator U from $\mathfrak{H}_1$ to $\mathfrak{H}_2$ such that

$$\pi_1(A) = U\pi_2(A)U^*$$

for all $A \in \mathfrak{A}$. Equivalence of π_1 and π_2 is denoted by $\pi_1 \simeq \pi_2$.

2.3.2. States

Although we have derived various properties of representations of a C^*-algebra $\mathfrak{A}$ we have not, as yet, demonstrated their existence. The positive linear forms, or functionals, over $\mathfrak{A}$ play an important role both in this existence proof and in the construction of particular representations. We next investigate the properties of such forms. We denote the dual of $\mathfrak{A}$ by $\mathfrak{A}^*$, i.e., $\mathfrak{A}^*$ is the space of continuous, linear functionals over $\mathfrak{A}$, and we define the norm of any functional f over $\mathfrak{A}$ by

$$\|f\| = \sup\{|f(A)|; \|A\| = 1\}.$$

The functionals of particular interest are defined as follows:

Definition 2.3.9. A linear functional ω over the *-algebra $\mathfrak{A}$ is defined to be *positive* if

$$\omega(A^*A) \geq 0$$

for all $A \in \mathfrak{A}$. A positive linear functional ω over a C^*-algebra $\mathfrak{A}$ with $\|\omega\| = 1$ is called a *state*.

Notice that we have not demanded that the positive forms be continuous. For a C^*-algebra continuity is in fact a consequence of positivity, as we will see in Proposition 2.3.11. Note also that every positive element of a C^*-algebra is of the form A^*A and hence positivity of ω is equivalent to ω being positive on positive elements.

The origin and relevance of the notion of state is best illustrated by first assuming that one has a representation $(\mathfrak{H}, \pi)$ of the C^*-algebra $\mathfrak{A}$. Now let $\Omega \in \mathfrak{H}$ be any nonzero vector and define ω_Ω by

$$\omega_\Omega(A) = (\Omega, \pi(A)\Omega)$$

for all $A \in \mathfrak{A}$. It follows that ω_Ω is a linear function over $\mathfrak{A}$ but it is also positive because

$$\omega_\Omega(A^*A) = \|\pi(A)\Omega\|^2 \geq 0.$$

It can be checked, e.g., from Proposition 2.3.11 and Corollary 2.3.13 below, that $\|\omega_\Omega\| = 1$ whenever $\|\Omega\| = 1$ and π is nondegenerate. Thus in this case

ω_Ω is a state. States of this type are usually called vector states of the representation $(\mathfrak{H}, \pi)$. Although this example of a state appears very special we will eventually see that it describes the general situation. Every state over a C^*-algebra is a vector state in a suitable representation. As a preliminary to further examination of the connection between states and representations we derive some general properties of states.

The basic tool for exploitation of the positivity of states is the general Cauchy–Schwarz inequality.

Lemma 2.3.10 (Cauchy–Schwarz inequality). *Let ω be a positive linear functional over the *-algebra $\mathfrak{A}$. It follows that*

(a) $\omega(A^*B) = \overline{\omega(B^*A)}$,
(b) $|\omega(A^*B)|^2 \leq \omega(A^*A)\omega(B^*B)$ *for all pairs* $A, B \in \mathfrak{A}$.

PROOF. For $A, B \in \mathfrak{A}$ and $\lambda \in \mathbb{C}$ positivity of ω implies that

$$\omega((\lambda A + B)^*(\lambda A + B)) \geq 0.$$

By linearity this becomes

$$|\lambda|^2\omega(A^*A) + \bar{\lambda}\omega(A^*B) + \lambda\omega(B^*A) + \omega(B^*B) \geq 0.$$

The necessary, and sufficient, conditions for the positivity of this quadratic form in λ are exactly the two conditions of the lemma.

As a first application of this result we derive the following interrelationships between positivity, continuity, and normalization for functionals over a C^*-algebra.

Proposition 2.3.11. *Let ω be a linear functional over a C^*-algebra $\mathfrak{A}$. The following conditions are equivalent:*

(1) *ω is positive;*
(2) *ω is continuous, and*

$$\|\omega\| = \lim_\alpha \omega(E_\alpha{}^2)$$

for some approximate identity $\{E_\alpha\}$ of $\mathfrak{A}$.
If these conditions are fulfilled, i.e., if ω is positive, then

(a) $\omega(A^*) = \overline{\omega(A)}$,
(b) $|\omega(A)|^2 \leq \omega(A^*A)\|\omega\|$,
(c) $|\omega(A^*BA)| \leq \omega(A^*A)\|B\|$,
(d) $\|\omega\| = \sup\{\omega(A^*A), \|A\| = 1\}$

for all $A, B \in \mathfrak{A}$, and

$$\|\omega\| = \lim_\alpha \omega(E_\alpha),$$

where $\{E_\alpha\}$ is any approximate identity of $\mathfrak{A}$.

PROOF. (1) ⇒ (2) Let $A_1, A_2, \ldots$ be a sequence of positive elements with $\|A_i\| \leq 1$. Now if $\lambda_i \geq 0$ and $\sum_i \lambda_i < +\infty$ then $\sum_i \lambda_i A_i$ converges uniformly, and, monotonically, to some positive A and hence, by linearity and positivity

$$\sum_i \lambda_i \omega(A_i) \leq \omega(A) < +\infty.$$

Since this is true for any such sequence λ_i the $\omega(A_i)$ must be uniformly bounded. Thus

$$M_+ = \sup\{\omega(A);\, A \geq 0,\, \|A\| \leq 1\} < +\infty.$$

But it follows easily from Proposition 2.2.11 that each $A \in \mathfrak{A}$ has a decomposition

$$A = \sum_{n=0}^{3} i^n A_n$$

with $A_n \geq 0$ and $\|A_n\| \leq 1$. Hence $\|\omega\| \leq 4M_+ < +\infty$, i.e., ω is continuous.

Next let us apply the Cauchy–Schwarz inequality of Lemma 2.3.10 to obtain

$$|\omega(AE_\alpha)|^2 \leq \omega(A^*A)\omega(E_\alpha^2) \leq M_+\|A\|^2\omega(E_\alpha^2).$$

Taking the limit over α one finds

$$|\omega(A)|^2 \leq M_+ M\|A\|^2$$

where $M = \sup_\alpha \omega(E_\alpha^2)$. Thus $\|\omega\|^2 \leq M_+ M$. But $M \leq \|\omega\|$, because $\|E_\alpha\| \leq 1$, and $M_+ \leq \|\omega\|$. Therefore $\|\omega\| = M_+ = M = \lim_\alpha \omega(E_\alpha^2)$. Incidentally, because $E_\alpha^2 \leq E_\alpha$, one also has $\|\omega\| \leq \lim_\alpha \omega(E_\alpha) \leq \|\omega\|$. Thus $\|\omega\| = \lim_\alpha \omega(E_\alpha)$ and the last statement of the proposition is established.

(2) ⇒ (1) We may assume $\|\omega\| = 1$. If $\mathfrak{A}$ has an identity $\mathbb{1}$ then

$$\|\mathbb{1} - E_\alpha{}^2\| \leq \|\mathbb{1} - E_\alpha\| + \|\mathbb{1} - E_\alpha\|\,\|E_\alpha\|$$

and we have $\lim_\alpha E_\alpha{}^2 = \mathbb{1}$. Hence $\omega(\mathbb{1}) = 1$. If $\mathfrak{A}$ does not have an identity we adjoin one and extend ω to a functional $\tilde{\omega}$ on $\tilde{\mathfrak{A}} = \mathbb{C}\mathbb{1} + \mathfrak{A}$ by

$$\tilde{\omega}(\lambda\mathbb{1} + A) = \lambda + \omega(A).$$

Because $A - AE_\alpha{}^2 = (A - AE_\alpha) + (A - AE_\alpha)E_\alpha$ we have $\lim_\alpha AE_\alpha{}^2 = A$. Using the definition of the norm on $\tilde{\mathfrak{A}}$, Proposition 2.1.5, we tnen have

$$|\tilde{\omega}(\lambda\mathbb{1} + A)| = |\lambda + \omega(A)| = \lim_\alpha |\lambda\omega(E_\alpha{}^2) + \omega(AE_\alpha{}^2)|$$

$$\leq \limsup_\alpha \|\lambda E_\alpha{}^2 + AE_\alpha{}^2\| \leq \|\lambda\mathbb{1} + A\|.$$

Thus in any case we may assume that $\mathfrak{A}$ has an identity and

$$\omega(\mathbb{1}) = 1 = \|\omega\|.$$

Next we show that $A = A^*$ implies that $\omega(A)$ is real. Set

$$\omega(A) = \alpha + i\beta, \qquad \alpha, \beta \in \mathbb{R}.$$

For any real γ we then have

$$\omega(A + i\gamma\mathbb{1}) = \alpha + i(\beta + \gamma).$$

But $A + i\gamma\mathbb{1}$ is normal with spectrum in

$$\sigma(A) + i\gamma \subseteq [-\|A\|, \|A\|] + i\gamma.$$

Hence

$$\|A + i\gamma\mathbb{1}\| = \rho(A + i\gamma\mathbb{1}) = \sqrt{\|A\|^2 + \gamma^2}.$$

Since $|\omega(A + i\gamma\mathbb{1})| \geq |\beta + \gamma|$ we obtain

$$|\beta + \gamma| \leq \sqrt{\|A\|^2 + \gamma^2}$$

for any $\gamma \in \mathbb{R}$. This implies that $\beta = 0$, i.e., $\omega(A)$ is real.

Finally,

$$\left\| \mathbb{1} - \frac{A^*A}{\|A\|^2} \right\| \leq 1$$

for any $A \in \mathfrak{A}$ by Lemma 2.2.9. Hence

$$\left| \omega(\mathbb{1}) - \frac{\omega(A^*A)}{\|A\|^2} \right| \leq 1$$

But $\omega(\mathbb{1}) = 1$ and $\omega(A^*A)$ is real and it is necessary that

$$\omega(A^*A) \geq 0.$$

Thus ω is positive.

Finally, note that (a) and (b) follow by applying Lemma 2.3.10 to A and E_α, and then taking a limit over α. The same lemma implies that

$$|\omega(A^*BA)|^2 \leq \omega(A^*A)\omega(A^*B^*BA)$$

and inequality (c) follows by remarking that

$$A^*B^*BA \leq \|B\|^2 A^*A$$

and hence

$$\omega(A^*B^*BA) \leq \|B\|^2 \omega(A^*A).$$

Property (d) follows from (b).

Corollary 2.3.12. *Let ω_1 and ω_2 be positive linear functionals over the C^*-algebra $\mathfrak{A}$. It follows that $\omega_1 + \omega_2$ is a positive linear functional and*

$$\|\omega_1 + \omega_2\| = \|\omega_1\| + \|\omega_2\|.$$

In particular, the states over $\mathfrak{A}$ form a convex subset of the dual of $\mathfrak{A}$.

PROOF. The positivity of $\omega_1 + \omega_2$ is evident and

$$\begin{aligned}\|\omega_1 + \omega_2\| &= \lim_\alpha(\omega_1(E_\alpha^2) + \omega_2(E_\alpha^2)) \\ &= \lim_\alpha \omega_1(E_\alpha^2) + \lim_\alpha \omega_2(E_\alpha^2) = \|\omega_1\| + \|\omega_2\|.\end{aligned}$$

Finally, if ω_1 and ω_2 are states then $\omega = \lambda\omega_1 + (1 - \lambda)\omega_2$ is positive for $0 \leq \lambda \leq 1$ and $\|\omega\| = \lambda\|\omega_1\| + (1 - \lambda)\|\omega_2\| = 1$. Thus ω is a state.

Next remark that if $\mathfrak{A}$ is a C^*-algebra without identity element and $\tilde{\mathfrak{A}} = \mathbb{C}\mathbb{1} + \mathfrak{A}$ is the algebra obtained by adjoining an identity then every $\omega \in \mathfrak{A}^*$ has an extension $\tilde{\omega} \in \tilde{\mathfrak{A}}^*$ defined by $\tilde{\omega}(\lambda\mathbb{1} + A) = \lambda\|\omega\| + \omega(A)$. This extension $\tilde{\omega}$ is usually called the canonical extension of ω and it is a state extension.

Corollary 2.3.13. *Let $\mathfrak{A}$ be a C^*-algebra without identity and $\tilde{\mathfrak{A}}$ the C^*-algebra obtained by adjoining an identity. Further, let ω be a positive functional over $\mathfrak{A}$ and $\tilde{\omega}$ its canonical extension to $\tilde{\mathfrak{A}}$. It follows that $\tilde{\omega}$ is positive and $\|\tilde{\omega}\| = \|\omega\|$. Moreover, if ω_1, ω_2 are two positive forms and $\tilde{\omega}_1, \tilde{\omega}_2$ their canonical extensions then*

$$\widetilde{\omega_1 + \omega_2} = \tilde{\omega}_1 + \tilde{\omega}_2 .$$

PROOF. Applying Proposition 2.3.11(b) one estimates that

$$\begin{aligned}\tilde{\omega}((\lambda\mathbb{1} + A)^*(\lambda\mathbb{1} + A)) &= |\lambda|^2\|\omega\| + \bar{\lambda}\omega(A) + \lambda\omega(A^*) + \omega(A^*A)\\ &\geq (|\lambda|\,\|\omega\|^{1/2} - \omega(A^*A)^{1/2})^2 \geq 0\end{aligned}$$

and hence $\tilde{\omega}$ is positive. But, as $\tilde{\mathfrak{A}}$ contains the identity, $\|\tilde{\omega}\| = \tilde{\omega}(\mathbb{1}) = \|\omega\|$ by Proposition 2.3.11. Finally,

$$\tilde{\omega}_1(\lambda\mathbb{1} + A) + \tilde{\omega}_2(\lambda\mathbb{1} + A) = \lambda(\|\omega_1\| + \|\omega_2\|) + \omega_1(A) + \omega_2(A),$$

and

$$\|\omega_1\| + \|\omega_2\| = \|\omega_1 + \omega_2\|,$$

which yields the last statement of the corollary.

The property of positivity introduces a natural ordering of functionals. If ω_1 and ω_2 are positive linear functionals we write $\omega_1 \geq \omega_2$, or $\omega_1 - \omega_2 \geq 0$, whenever $\omega_1 - \omega_2$ is positive and we say that ω_1 majorizes ω_2. The properties of states with respect to this ordering will be of great significance throughout the sequel.

If ω_1, ω_2, are states over $\mathfrak{A}$ and $0 < \lambda < 1$ then $\omega = \lambda\omega_1 + (1 - \lambda)\omega_2$ is a state with the property that $\omega \geq \lambda\omega_1$ and $\omega \geq (1 - \lambda)\omega_2$.

Thus if ω is a convex combination of two distinct states then it majorizes multiples of both states. It is natural to call a state pure whenever it cannot be written as a convex combination of other states and the foregoing remark on majorization motivates the following definition:

Definition 2.3.14. A state ω over a C^*-algebra is defined to be *pure* if the only positive linear functionals majorized by ω are of the form $\lambda\omega$ with $0 \leq \lambda \leq 1$. The set of all states is denoted by $E_{\mathfrak{A}}$ and the set of pure states by $P_{\mathfrak{A}}$.

To conclude this section we derive some elementary properties of the sets of states $E_{\mathfrak{A}}$ and $P_{\mathfrak{A}}$. As these sets are subsets of the dual $\mathfrak{A}^*$ of $\mathfrak{A}$ they can be topologized through restriction of any of the topologies of $\mathfrak{A}^*$. There are two obvious such topologies. The norm, or uniform, topology is determined by specifying the neighborhoods of ω to be

$$\mathscr{U}(\omega; \varepsilon) = \{\omega'; \omega' \in \mathfrak{A}^*, \|\omega - \omega'\| < \varepsilon\},$$

where $\varepsilon > 0$. In the weak* topology the neighborhoods of ω are indexed by finite sets of elements, $A_1, A_2, \ldots, A_n \in \mathfrak{A}$, and $\varepsilon > 0$. One has

$$\mathscr{U}(\omega; A_1, \ldots, A_n; \varepsilon) = \{\omega'; \omega' \in \mathfrak{A}^*, |\omega'(A_i) - \omega(A_i)| < \varepsilon, \quad i = 1, 2, \ldots, n\}$$

In practice it appears that the weak* topology is of greatest use although we will later have recourse to the uniform topology.

Theorem 2.3.15. *Let $\mathfrak{A}$ be a C^*-algebra and let $B_{\mathfrak{A}}$ denote the positive linear functionals over $\mathfrak{A}$ with norm less than or equal to one. It follows that $B_{\mathfrak{A}}$ is a convex, weakly* compact subset of the dual $\mathfrak{A}^*$ whose extremal points are 0 and the pure states $P_{\mathfrak{A}}$. Moreover, $B_{\mathfrak{A}}$ is the weak* closure of the convex envelope of its extremal points.*

The set of states $E_{\mathfrak{A}}$ is convex but it is weakly compact if, and only if, $\mathfrak{A}$ contains an identity. In this latter case the extremal points of $E_{\mathfrak{A}}$ are the pure states $P_{\mathfrak{A}}$ and $E_{\mathfrak{A}}$ is the weak* closure of the convex envelope of $P_{\mathfrak{A}}$.*

PROOF. $B_{\mathfrak{A}}$ is a convex, weakly* closed subset of the unit ball $\mathfrak{A}_1{}^*$ of $\mathfrak{A}^*$, i.e., $\mathfrak{A}_1{}^* = \{\omega; \omega \in \mathfrak{A}^*, \|\omega\| \leq 1\}$. But $\mathfrak{A}_1{}^*$ is weakly* compact by the Alaoglu–Banach theorem.

Now 0 is an extreme point of $B_{\mathfrak{A}}$ because if $\omega \in B_{\mathfrak{A}}$, and $-\omega \in B_{\mathfrak{A}}$ then $\omega(A^*A) = 0$ for all $A \in \mathfrak{A}$ and Proposition 2.3.11(b) gives $\omega(A) = 0$ for all $A \in \mathfrak{A}$, i.e., $\omega = 0$.

Next suppose $\omega \in P_{\mathfrak{A}}$ and $\omega = \lambda\omega_1 + (1 - \lambda)\omega_2$ with $0 < \lambda < 1$ and $\omega_1, \omega_2 \in B_{\mathfrak{A}}$. It follows that $\omega \geq \lambda\omega_1$ and hence $\lambda\omega_1 = \mu\omega$ for some $0 \leq \mu \leq 1$ by purity. But $1 = \|\omega\| = \lambda\|\omega_1\| + (1 - \lambda)\|\omega_2\|$ and one must have $\|\omega_1\| = 1 = \|\omega_2\|$. Therefore $\lambda = \mu$ and $\omega = \omega_1$. Similarly, $\omega = \omega_2$ and hence ω is an extremal point of $B_{\mathfrak{A}}$.

Suppose now that ω is an extremal point of $B_{\mathfrak{A}}$ and $\omega \neq 0$. One must have $\|\omega\| = 1$. Thus ω is a state and we must deduce that it is pure. Suppose the contrary; then there is a state $\omega_1 \neq \omega$ and a λ with $0 < \lambda < 1$ such that $\omega \geq \lambda\omega_1$. Define ω_2 by $\omega_2 = (\omega - \lambda\omega_1)/(1 - \lambda)$; then $\|\omega_2\| = (\|\omega\| - \lambda\|\omega_1\|)/(1 - \lambda) = 1$ and ω_2 is also a state. But $\omega = \lambda\omega_1 + (1 - \lambda)\omega_2$ and ω is not extremal, which is a contradiction.

The set $B_{\mathfrak{A}}$ is the closed convex hull of its extremal points by the Krein–Milman theorem. This theorem asserts in particular the existence of such extremal points, which is not at all evident *a priori*.

Finally, if $\mathfrak{A}$ contains an identity $\mathbb{1}$ then $E_{\mathfrak{A}}$ is the intersection of $B_{\mathfrak{A}}$ with the hyperplane $\omega(\mathbb{1}) = 1$. Thus the convexity, weak* compactness, and the generation properties of $E_{\mathfrak{A}}$ follow from the similar properties of $B_{\mathfrak{A}}$. It remains to prove that $E_{\mathfrak{A}}$ is not weakly* compact if $\mathbb{1} \notin \mathfrak{A}$ and this will be deduced in Section 2.3.4.

2.3.3. Construction of Representations

If $(\mathfrak{H}, \pi)$ is a nondegenerate representation of a C^*-algebra $\mathfrak{A}$ and Ω is a vector in $\mathfrak{H}$ with $\|\Omega\| = 1$ then we have deduced in the previous section that the linear functional

$$\omega_\Omega(A) = (\Omega, \pi(A)\Omega)$$

is a state over $\mathfrak{A}$. This type of state is called a vector state. Now we want to prove the converse. Every state is a vector state for some nondegenerate representation. Thus starting from a state ω we must construct a representation $(\mathfrak{H}_\omega, \pi_\omega)$ of $\mathfrak{A}$ and a vector $\Omega_\omega \in \mathfrak{H}_\omega$ such that ω is identified as the vector state ω_{Ω_ω}, i.e., such that

$$\omega(A) = (\Omega_\omega, \pi_\omega(A)\Omega_\omega)$$

for all $A \in \mathfrak{A}$.

The idea behind this construction is very simple. First consider the definition of the representation space $\mathfrak{H}_\omega$. The algebra $\mathfrak{A}$ is a Banach space and with the aid of the state ω it may be converted into a pre-Hilbert space by introduction of the positive semidefinite scalar product

$$\langle A, B\rangle = \omega(A^*B).$$

Next define $\mathfrak{J}_\omega$ by

$$\mathfrak{J}_\omega = \{A; A \in \mathfrak{A}, \omega(A^*A) = 0\},$$

The set $\mathfrak{J}_\omega$ is a left ideal of $\mathfrak{A}$ because $I \in \mathfrak{J}_\omega$ and $A \in \mathfrak{A}$ implies that

$$0 \le \omega((AI)^*AI) \le \|A\|^2\omega(I^*I) = 0$$

by Proposition 2.3.11, i.e., $AI \in \mathfrak{J}_\omega$.

Now define equivalence classes ψ_A, ψ_B by

$$\psi_A = \{\hat{A}; \hat{A} = A + I, I \in \mathfrak{J}_\omega\}$$

and remark that these equivalence classes also form a complex vector space when equipped with the operations inherited from $\mathfrak{A}$; $\psi_A + \psi_B = \psi_{A+B}$, $\alpha\psi_A = \psi_{\alpha A}$. Furthermore, this latter space is a strict pre-Hilbert space with respect to the scalar product

$$(\psi_A, \psi_B) = \langle A, B\rangle = \omega(A^*B).$$

It must, of course, be checked that this is a coherent and correct definition but this is easily verified with the aid of Proposition 2.3.11. For example, (ψ_A, ψ_B) is independent of the particular class representative used in its definition because

$$\begin{aligned}\omega((A + I_1)^*(B + I_2)) &= \omega(A^*B) + \overline{\omega(B^*I_1)} + \omega(A^*I_2) + \omega(I_1{}^*I_2)\\ &= \omega(A^*B)\end{aligned}$$

whenever $I_1, I_2 \in \mathfrak{I}_\omega$. It is well known that a strict pre-Hilbert space may be completed, i.e., linearly embedded as a dense subspace of a Hilbert space in a manner which preserves the scalar product, and the completion of this space is defined as the representation space $\mathfrak{H}_\omega$.

Next let us consider the definition of the representatives $\pi_\omega(A)$. First we specify their action on the dense subspace of $\mathfrak{H}_\omega$ formed by the vectors ψ_B, $B \in \mathfrak{A}$, by the definition

$$\pi_\omega(A)\psi_B = \psi_{AB}.$$

Note that this relation is again independent of the representative used for the class ψ_B because

$$\pi_\omega(A)\psi_{B+I} = \psi_{AB+AI} = \psi_{AB} = \pi_\omega(A)\psi_B$$

for $I \in \mathfrak{I}_\omega$. Moreover, each $\pi_\omega(A)$ is a linear operator because

$$\begin{aligned}\pi_\omega(A)(\lambda\psi_B + \psi_C) &= \pi_\omega(A)\psi_{\lambda B+C} = \psi_{\lambda AB+AC} \\ &= \lambda\psi_{AB} + \psi_{AC} \\ &= \lambda\pi_\omega(A)\psi_B + \pi_\omega(A)\psi_C.\end{aligned}$$

Finally, by Proposition 2.3.11(c) one finds

$$\begin{aligned}\|\pi_\omega(A)\psi_B\|^2 &= (\psi_{AB}, \psi_{AB}) \\ &= \omega(B^*A^*AB) \\ &\leq \|A\|^2\omega(B^*B) \\ &= \|A\|^2\|\psi_B\|^2\end{aligned}$$

and hence $\pi_\omega(A)$ has a bounded closure, which we also denote by $\pi_\omega(A)$. The algebraic properties of the π_ω follow easily, e.g.,

$$\pi_\omega(A_1)\pi_\omega(A_2)\psi_B = \psi_{A_1A_2B} = \pi_\omega(A_1A_2)\psi_B$$

and hence $\pi_\omega(A_1)\pi_\omega(A_2) = \pi_\omega(A_1A_2)$. Thus we have now constructed the representation $(\mathfrak{H}_\omega, \pi_\omega)$.

It remains to specify the vector Ω_ω.

If $\mathfrak{A}$ contains the identity we define Ω_ω by

$$\Omega_\omega = \psi_{\mathbb{1}}$$

and this gives the correct identification of ω:

$$\begin{aligned}(\Omega_\omega, \pi_\omega(A)\Omega_\omega) &= (\psi_{\mathbb{1}}, \psi_A) \\ &= \omega(A).\end{aligned}$$

Note further that the set $\{\pi_\omega(A)\Omega_\omega; A \in \mathfrak{A}\}$ is exactly the dense set of equivalence classes $\{\psi_A; A \in \mathfrak{A}\}$ and hence Ω_ω is cyclic for $(\mathfrak{H}_\omega, \pi_\omega)$.

If $\mathfrak{A}$ does not contain the identity then we can adjoin it and repeat the above construction for $\tilde{\mathfrak{A}}$. Now, however, it needs an auxiliary argument to prove that Ω_ω is cyclic for the set $\pi_\omega(\mathfrak{A})$. By construction the set $\pi_\omega(\tilde{\mathfrak{A}})\Omega_\omega = \pi_\omega(\mathbb{C}\mathbb{1} + \mathfrak{A})\Omega_\omega$ is dense and thus the cyclicity of Ω_ω, for $\pi_\omega(\mathfrak{A})$, follows if Ω_ω

is in the closure of the set $\pi_\omega(\mathfrak{A})\Omega_\omega$. Let $\{E_\alpha\}$ be an approximate identity of $\mathfrak{A}$; then

$$\begin{aligned}\|\pi_\omega(E_\alpha)\Omega_\omega - \Omega_\omega\|^2 &= \|\Omega_\omega\|^2 + \|\pi_\omega(E_\alpha)\Omega_\omega\|^2 - 2(\Omega_\omega, \pi_\omega(E_\alpha)\Omega_\omega)\\ &= 1 + \omega(E_\alpha{}^2) - 2\omega(E_\alpha).\end{aligned}$$

Therefore

$$\lim_\alpha \|\pi_\omega(E_\alpha)\Omega_\omega - \Omega_\omega\| = 0$$

by Proposition 2.3.11 and the desired result is established.

We have now established the principal part of the following theorem.

Theorem 2.3.16. *Let ω be a state over the C^*-algebra $\mathfrak{A}$. It follows that there exists a cyclic representation $(\mathfrak{H}_\omega, \pi_\omega, \Omega_\omega)$ of $\mathfrak{A}$ such that*

$$\omega(A) = (\Omega_\omega, \pi_\omega(A)\Omega_\omega)$$

for all $A \in \mathfrak{A}$ and, consequently, $\|\Omega_\omega\|^2 = \|\omega\| = 1$. Moreover, the representation is unique up to unitary equivalence.

PROOF. The only statement that we have not as yet proved is the uniqueness. By this we mean that if $(\mathfrak{H}_\omega', \pi_\omega', \Omega_\omega')$ is a second cyclic representation such that

$$\omega(A) = (\Omega_\omega', \pi_\omega'(A)\Omega_\omega')$$

for all $A \in \mathfrak{A}$ then there exists a unitary operator from $\mathfrak{H}_\omega$ onto $\mathfrak{H}_\omega'$ such that

$$U^{-1}\pi_\omega'(A)U = \pi_\omega(A)$$

for all $A \in \mathfrak{A}$, and

$$U\Omega_\omega = \Omega_\omega'.$$

This is, however, established by defining U through

$$U\pi_\omega(A)\Omega_\omega = \pi_\omega'(A)\Omega_\omega'$$

and noting that

$$\begin{aligned}(U\pi_\omega(A)\Omega_\omega, U\pi_\omega(B)\Omega_\omega) &= (\pi_\omega'(A)\Omega_\omega', \pi_\omega'(B)\Omega_\omega')\\ &= \omega(A^*B) = (\pi_\omega(A)\Omega_\omega, \pi_\omega(B)\Omega_\omega).\end{aligned}$$

Thus U preserves the scalar product and is consequently well defined. It easily follows that the closure of U is unitary and has all the desired algebraic properties. We omit the details.

Corollary 2.3.17. *Let ω be a state over the C^*-algebra $\mathfrak{A}$ and τ a *-automorphism of $\mathfrak{A}$ which leaves ω invariant, i.e.,*

$$\omega(\tau(A)) = \omega(A)$$

for all $A \in \mathfrak{A}$. It follows that there exists a uniquely determined unitary operator U_ω, on the space of the cyclic representation $(\mathfrak{H}_\omega, \pi_\omega, \Omega_\omega)$ constructed from ω, such that

$$U_\omega\pi_\omega(A)U_\omega^{-1} = \pi_\omega(\tau(A))$$

for all $A \in \mathfrak{A}$, and

$$U_\omega \Omega_\omega = \Omega_\omega.$$

PROOF. The result follows by applying the uniqueness statement of Theorem 2.3.16 to the cyclic representation $(\mathfrak{H}_\omega, \pi_\omega \circ \tau, \Omega_\omega)$, where $\pi_\omega \circ \tau(A) = \pi_\omega(\tau(A))$.

Definition 2.3.18. The cyclic representation $(\mathfrak{H}_\omega, \pi_\omega, \Omega_\omega)$, constructed from the state ω over the C^*-algebra $\mathfrak{A}$, is defined as the *canonical cyclic representation of $\mathfrak{A}$ associated with ω.*

Next we demonstrate that the notions of purity of a state ω and irreducibility of the representation associated with ω are intimately related.

Theorem 2.3.19. *Let ω be a state over the C^*-algebra $\mathfrak{A}$ and $(\mathfrak{H}_\omega, \pi_\omega, \Omega_\omega)$ the associated cyclic representation. The following conditions are equivalent:*

(1) *$(\mathfrak{H}_\omega, \pi_\omega)$ is irreducible;*
(2) *ω is pure;*
(3) *ω is an extremal point of the set $E_\mathfrak{A}$ of states over $\mathfrak{A}$.*

Furthermore, there is a one-to-one correspondence

$$\omega_T(A) = (T\Omega_\omega, \pi_\omega(A)\Omega_\omega)$$

between positive functionals ω_T, over $\mathfrak{A}$, majorized by ω and positive operators T in the commutant π_ω', of π_ω, *with $\|T\| \leq 1$.*

PROOF. (1) $\Rightarrow$ (2) Assume that (2) is false. Thus there exists a positive functional ρ such that $\rho(A^*A) \leq \omega(A^*A)$ for all $A \in \mathfrak{A}$. But applying the Cauchy–Schwarz inequality one then has

$$\begin{aligned}|\rho(B^*A)|^2 &\leq \rho(B^*B)\rho(A^*A)\\ &\leq \omega(B^*B)\omega(A^*A)\\ &= \|\pi_\omega(B)\Omega_\omega\|^2\|\pi_\omega(A)\Omega_\omega\|^2.\end{aligned}$$

Thus $\pi_\omega(B)\Omega_\omega \times \pi_\omega(A)\Omega_\omega \mapsto \rho(B^*A)$ is a densely defined, bounded, sesquilinear functional, over $\mathfrak{H}_\omega \times \mathfrak{H}_\omega$, and there exists a unique bounded operator T, on $\mathfrak{H}_\omega$, such that

$$(\pi_\omega(B)\Omega_\omega, T\pi_\omega(A)\Omega_\omega) = \rho(B^*A).$$

As ρ is not a multiple of ω the operator T is not a multiple of the identity. Moreover,

$$\begin{aligned}0 &\leq \rho(A^*A)\\ &= (\pi_\omega(A)\Omega_\omega, T\pi_\omega(A)\Omega_\omega)\\ &\leq \omega(A^*A) = (\pi_\omega(A)\Omega_\omega, \pi_\omega(A)\Omega_\omega)\end{aligned}$$

and hence $0 \leq T \leq \mathbb{1}$. But

$$\begin{aligned}(\pi_\omega(B)\Omega_\omega, T\pi_\omega(C)\pi_\omega(A)\Omega_\omega) &= \rho(B^*CA)\\ &= \rho((C^*B)^*A) = (\pi_\omega(B)\Omega_\omega, \pi_\omega(C)T\pi_\omega(A)\Omega_\omega)\end{aligned}$$

and therefore $T \in \pi_\omega'$. Thus condition (1) is false.

(2) ⇒ (1) Assume that (1) is false. If $T \in \pi_\omega'$ then $T^* \in \pi_\omega'$ and $T + T^*, (T - T^*)/i$ are also elements of the commutant. Thus there exists a selfadjoint element S of π_ω' which is not a multiple of the identity. Therefore there exists a spectral projector P of S such that $0 < P < \mathbb{1}$ and $P \in \pi_\omega'$. Consider the functional

$$\rho(A) = (P\Omega_\omega, \pi_\omega(A)\Omega_\omega).$$

This is certainly positive because

$$\rho(A^*A) = (P\pi_\omega(A)\Omega_\omega, P\pi_\omega(A)\Omega_\omega) \geq 0.$$

Moreover,

$$\begin{aligned} \omega(A^*A) - \rho(A^*A) &= (\pi_\omega(A)\Omega_\omega, (\mathbb{1} - P)\pi_\omega(A)\Omega_\omega) \\ &\geq 0. \end{aligned}$$

Thus ω majorizes ρ. It is easily checked that ρ is not a multiple of ω and hence (2) is false.

This proves the equivalence of the first two conditions stated in the theorem and simultaneously establishes the correspondence described by the last statement.

The equivalence of conditions (2) and (3) is already contained in Theorem 2.3.15.

This characterization of pure states has two easy, and useful, consequences.

Corollary 2.3.20. *Let ω be a state over a C^*-algebra $\mathfrak{A}$ without identity and let $\tilde{\omega}$ denote its canonical extension to $\tilde{\mathfrak{A}} = \mathbb{C}\mathbb{1} + \mathfrak{A}$. It follows that ω is a pure state over $\mathfrak{A}$ if, and only if, $\tilde{\omega}$ is a pure state over $\tilde{\mathfrak{A}}$.*

PROOF. If $(\mathfrak{H}_\omega, \pi_\omega, \Omega_\omega)$ is the cyclic representation associated with ω and $\mathbb{1}_\omega$ is the identity operator on $\mathfrak{H}_\omega$ then one can readily identify the representation associated with $\tilde{\omega}$ by $\mathfrak{H}_{\tilde{\omega}} = \mathfrak{H}_\omega$, $\Omega_{\tilde{\omega}} = \Omega_\omega$, and $\pi_{\tilde{\omega}}(\lambda\mathbb{1} + A) = \lambda\mathbb{1}_\omega + \pi_\omega(A)$. The two representations are simultaneously irreducible and hence the two states are simultaneously pure.

Corollary 2.3.21. *Let ω be a state over an abelian C^*-algebra. It follows that ω is a pure state if, and only if, $\omega(AB) = \omega(A)\omega(B)$ for all $A, B \in \mathfrak{A}$.*

PROOF. The state ω is pure if, and only if, the associated representation is irreducible. But $\pi_\omega(\mathfrak{A}) \subseteq \pi_\omega(\mathfrak{A})'$, because $\mathfrak{A}$ is abelian, and hence $(\mathfrak{H}_\omega, \pi_\omega)$ is irreducible if, and only if, $\mathfrak{H}_\omega$ is one-dimensional. This is true if, and only if, the state factors as in condition (2).

2.3.4. Existence of Representations

In continuation of our discussion of representations we next establish the existence of nontrivial representations and prove the basic structure theorem, Theorem 2.1.10, announced in Section 2.1. The proof partially depends upon the properties of convexity, compactness, etc., that we have already established for the states of a C^*-algebra but the principal new ingredient is the Hahn–Banach theorem.

Theorem 2.3.22A (Hahn–Banach). *Let Y be a subspace of a normed linear space X and f a bounded linear functional on Y. It follows that f has a bounded linear extension F, on X, such that $\|F\| = \|f\|$.*

In Section 2.4, and subsequent chapters, we need a generalization of this theorem to spaces with locally convex topologies determined by families of semi-norms. The existence of states of a C^*-algebra follows, however, from the foregoing simple version.

We begin with a result concerning states with specific properties.

Lemma 2.3.23. *Let A be an arbitrary element of the C^*-algebra $\mathfrak{A}$. There exists a pure state ω over $\mathfrak{A}$ such that*

$$\omega(A^*A) = \|A\|^2$$

and hence there exists an irreducible representation $(\mathfrak{H}, \pi, \Omega)$ of $\mathfrak{A}$ such that

$$\|\pi(A)\| = \|A\|.$$

PROOF. First adjoin an identity, if necessary. Next consider the subspace $\mathfrak{B}$, of $\mathfrak{A}$, given by

$$\mathfrak{B} = \{\alpha\mathbb{1} + \beta A^*A;\, \alpha, \beta \in \mathbb{C}\},$$

and define a linear functional f, on $\mathfrak{B}$, by

$$f(\alpha\mathbb{1} + \beta A^*A) = \alpha + \beta\|A\|^2.$$

One has

$$\begin{aligned}|\alpha + \beta\|A\|^2| &\leq \sup\{|\alpha + \beta\lambda|;\, \lambda \in \sigma(A^*A)\}\\ &= \|\alpha\mathbb{1} + \beta A^*A\|\end{aligned}$$

by the spectral radius formula applied to the normal element $\alpha\mathbb{1} + \beta A^*A$. Thus $\|f\| \leq 1$. But $f(\mathbb{1}) = 1$ and hence $\|f\| = 1 = f(\mathbb{1})$. Now apply Theorem 2.3.22A with the identification $X = \mathfrak{A}$, $Y = \mathfrak{B}$. It results that there exists a bounded linear extension ω, of f, with $\|\omega\| = 1 = f(\mathbb{1})$ and hence ω is a state by Proposition 2.3.11. But $\omega(A^*A) = f(A^*A) = \|A\|^2$. Now let E_A denote the set of all states with the property $\omega(A^*A) = \|A\|^2$. This set is a nonempty, convex, weakly* closed, and hence weakly* compact, subset of the set $E_{\mathfrak{A}}$ of all states over $\mathfrak{A}$. Hence E_A possesses extreme points by the Krein–Milman theorem. Let $\hat{\omega}$ be such an extreme point and suppose that $\hat{\omega} = \lambda\omega_1 + (1 - \lambda)\omega_2$ for some pair of states ω_1, ω_2 and some $0 < \lambda < 1$. But $|\omega_i(A^*A)| \leq \|A\|^2$ and $\|A\|^2 = \lambda\omega_1(A^*A) + (1 - \lambda)\omega_2(A^*A)$. This is only possible if $\omega_1(A^*A) = \omega_2(A^*A) = \|A\|^2$ and hence $\omega_1, \omega_2 \in E_A$. But as $\hat{\omega}$ is an extreme point of E_A one concludes that $\omega_1 = \omega_2 = \hat{\omega}$. Thus $\hat{\omega}$ is an extremal point of $E_{\mathfrak{A}}$, and this implies that $\hat{\omega}$ is pure by Theorem 2.3.19. This completes the proof of the first statement of the lemma. The second follows from the relations

$$\begin{aligned}\|A\|^2 = \omega(A^*A) &= \|\pi_\omega(A)\Omega_\omega\|^2\\ &\leq \|\pi_\omega(A)\|^2 \leq \|A\|^2,\end{aligned}$$

where $(\mathfrak{H}_\omega, \pi_\omega, \Omega_\omega)$ is the cyclic representation associated with ω and the last inequality is an application of Proposition 2.3.1.

Now we are in a position to prove the basic structure theorem. Let us first recall its statement.

Theorem 2.1.10. *Let $\mathfrak{A}$ be a C^*-algebra. It follows that $\mathfrak{A}$ is isomorphic to a norm-closed selfadjoint algebra of bounded operators on a Hilbert space.*

PROOF. For each state ω of $\mathfrak{A}$ construct the associated cyclic representation $(\mathfrak{H}_\omega, \pi_\omega, \Omega_\omega)$ and then form the direct sum representation $(\mathfrak{H}, \pi)$,

$$\mathfrak{H} = \bigoplus_{\omega \in E_{\mathfrak{A}}} \mathfrak{H}_\omega, \qquad \pi = \bigoplus_{\omega \in E_{\mathfrak{A}}} \pi_\omega.$$

For each $A \in \mathfrak{A}$ there is an ω_A such that $\|\pi_{\omega_A}(A)\| = \|A\|$ by Lemma 2.3.23. But $\|\pi(A)\| \geq \|\pi_{\omega_A}(A)\| = \|A\|$. Thus $\|\pi(A)\| = \|A\|$ by reapplying Proposition 2.3.1 and π is faithful.

We can also complete the proof of Theorem 2.3.15.

It remains to prove that the states $E_{\mathfrak{A}}$ of the C^*-algebra $\mathfrak{A}$ are not weakly* compact if $\mathfrak{A}$ does not contain the identity. For this it suffices to show that each weak* neighborhood of zero contains a state. Now each element of $\mathfrak{A}$ can be decomposed as a linear combination of four positive elements and it suffices to consider the neighborhoods indexed by $A_1, A_2, \ldots, A_n \in \mathfrak{A}_+$ and $\varepsilon > 0$. Introducing $A = A_1 + \cdots + A_n \in \mathfrak{A}_+$ it is sufficient to find an $\omega \in E_{\mathfrak{A}}$ such that $\omega(A) < \varepsilon$. Let $(\mathfrak{H}, \pi)$ be a faithful nondegenerate representation of $\mathfrak{A}$. As A is not invertible in $\mathfrak{A}$ it is not invertible in $\mathbb{C}\mathbb{1} + \mathfrak{A}$ because if A^{-1} is an inverse in the latter algebra then AA^{-2} is an inverse in $\mathfrak{A}$. Therefore $\pi(A)$ is not invertible in $\mathscr{L}(\mathfrak{H})$ and there must exist a unit vector $\psi \in \mathfrak{H}$ such that $\omega_\psi(A) = (\psi, A\psi) < \varepsilon$. Thus ω_ψ is a state with the desired property.

To conclude this section we give another consequence of the Hahn–Banach theorem which is often useful.

Proposition 2.3.24. *Let $\mathfrak{A}$ be a C^*-algebra, $\mathfrak{B}$ a sub-C^*-algebra of $\mathfrak{A}$, and ω a state over $\mathfrak{B}$. It follows that there exists a state $\hat{\omega}$, over $\mathfrak{A}$, which extends ω. If ω is a pure state of $\mathfrak{B}$ then $\hat{\omega}$ may be chosen to be a pure state of $\mathfrak{A}$.*

PROOF. First remark that we may assume $\mathfrak{A}$ and $\mathfrak{B}$ to have a common identity. The general situation can be reduced to the special situation by adjoining an identity and considering the canonical extension $\tilde{\omega}$ of ω to $\mathbb{C}\mathbb{1} + \mathfrak{B}$. For the second statement of the theorem it is then essential to note that purity of a state and purity of its canonical extension are equivalent (Corollary 2.3.20).

Next ω has a bounded linear extension $\hat{\omega}$ with $\|\hat{\omega}\| = \|\omega\| = 1$ by the Hahn–Banach theorem. But $\hat{\omega}(\mathbb{1}) = \omega(\mathbb{1}) = 1$ and hence $\hat{\omega}$ is positive by Proposition 2.3.11. Thus $\hat{\omega}$ is a state.

Finally, let E_ω denote the set of all states over $\mathfrak{A}$ which extend ω. The set is a nonempty convex subset of the set of all states $E_{\mathfrak{A}}$. Moreover, E_ω is weakly* closed, and hence weakly* compact. By the Krein–Milman theorem, E_ω has at least one extremal point $\hat{\omega}$. We now argue that if ω is pure then $\hat{\omega}$ is pure. Assume that $\hat{\omega} = \lambda\hat{\omega}_1 + (1 - \lambda)\hat{\omega}_2$ for some $0 < \lambda < 1$, where $\hat{\omega}_1$ and $\hat{\omega}_2$ are states over $\mathfrak{A}$. The restrictions ω_1 and ω_2, of $\hat{\omega}_1$ and $\hat{\omega}_2$, are states over $\mathfrak{B}$ and hence $\omega = \lambda\omega_1 + (1 - \lambda)\omega_2$. But ω is assumed to be a pure state over $\mathfrak{B}$ and hence $\omega_1 = \omega_2 = \omega$. Therefore $\hat{\omega}_1, \hat{\omega}_2 \in E_\omega$ and, since $\hat{\omega}$ is an extremal point of E_ω, one has $\hat{\omega}_1 = \hat{\omega}_2 = \hat{\omega}$. This demonstrates that $\hat{\omega}$ is an extremal point of $E_{\mathfrak{A}}$, i.e., $\hat{\omega}$ is a pure state over $\mathfrak{A}$ by Theorem 2.3.19.

2.3.5. Commutative C^*-Algebras

To conclude the discussion of representations we prove the structure theorem for abelian C^*-algebras, Theorem 2.1.11. In fact we are now in a position to prove a more precise form of this theorem in which the topological space X is explicitly identified. The space X is defined in terms of characters, whose formal definition is the following:

Definition 2.3.25. Let $\mathfrak{A}$ be an abelian C^*-algebra. A *character* ω, of $\mathfrak{A}$, is a nonzero linear map, $\omega; A \in \mathfrak{A} \mapsto \omega(A) \in \mathbb{C}$, of $\mathfrak{A}$ into the complex numbers $\mathbb{C}$ such that

$$\omega(AB) = \omega(A)\omega(B)$$

for all $A, B \in \mathfrak{A}$. The *spectrum* $\sigma(\mathfrak{A})$, of $\mathfrak{A}$, is defined to be the set of all characters on $\mathfrak{A}$.

The introduction of characters is quite conventional but, in fact, characters are nothing other than pure states. To establish this we need, however, the following simple result.

Lemma 2.3.26. *If ω is a character of an abelian C^*-algebra $\mathfrak{A}$ then $\omega(A) \in \sigma(A)$, the spectrum of A, for all $A \in \mathfrak{A}$. Hence $|\omega(A)| \leq \|A\|$ and $\omega(A^*A) \geq 0$.*

PROOF. First adjoin an identity, if necessary, and define $\tilde{\omega}$ on $\tilde{\mathfrak{A}} = \mathbb{C}1 + \mathfrak{A}$ by $\tilde{\omega}(\lambda 1 + A) = \lambda + \omega(A)$; then $\tilde{\omega}$ is still a character because

$$\begin{aligned}\tilde{\omega}((\lambda 1 + A)(\mu 1 + B)) &= \lambda\mu + \mu\omega(A) + \lambda\omega(B) + \omega(A)\omega(B)\\ &= \tilde{\omega}(\lambda 1 + A)\tilde{\omega}(\mu 1 + B).\end{aligned}$$

Thus we may assume $1 \in \mathfrak{A}$ and because $\omega(A) = \omega(A1) = \omega(A)\omega(1)$, and $\omega \neq 0$, one must have $\omega(1) = 1$. Next assume $\lambda \notin \sigma(A)$. Then there exists a B such that $(\lambda 1 - A)B = 1$ and consequently

$$\omega(\lambda 1 - A)\omega(B) = \omega(1) = 1.$$

But then $(\lambda - \omega(A))\omega(B) = 1$ and $\lambda \neq \omega(A)$. This proves that $\omega(A) \in \sigma(A)$ and then $|\omega(A)| \leq \rho(A) = \|A\|$ by the spectral radius formula. Finally $\omega(A^*A) \in \sigma(A^*A) \geq 0$.

Proposition 2.3.27. *Let ω be a nonzero linear functional over the abelian C^*-algebra $\mathfrak{A}$. The following conditions are equivalent:*

(1) *ω is a pure state;*
(2) *ω is a character.*

Hence the spectrum $\sigma(\mathfrak{A})$, of $\mathfrak{A}$, is a subset of the dual $\mathfrak{A}^*$, of $\mathfrak{A}$.

PROOF. (1) ⇒ (2) This has already been proved (Corollary 2.3.21).

(2) ⇒ (1) Lemma 2.3.26 demonstrates that ω is a continuous positive form with $\|\omega\| \leq 1$. But if E_α is an approximate identity then

$$\omega(A) = \lim_\alpha \omega(AE_\alpha) = \omega(A) \lim_\alpha \omega(E_\alpha)$$

and hence

$$1 = \lim_\alpha \omega(E_\alpha).$$

Therefore $\|\omega\| = 1$ and ω is a state. Finally, it is a pure state, because it is multiplicative, by another application of Corollary 2.3.21.

Now we may demonstrate a more precise version of Theorem 2.1.11.

Theorem 2.1.11A. *Let $\mathfrak{A}$ be an abelian C^*-algebra and X the set of characters of $\mathfrak{A}$ equipped with the weak* topology inherited from the dual $\mathfrak{A}^*$, of $\mathfrak{A}$. It follows that X is a locally compact Hausdorff space which is compact if, and only if, $\mathfrak{A}$ contains the identity. Moreover, $\mathfrak{A}$ is isomorphic to the algebra $C_0(X)$ of continuous functions over X which vanish at infinity.*

PROOF. First let us prove that X is locally compact. If $\omega_0 \in X$ then we may choose $A \in \mathfrak{A}_+$ such that $\omega_0(A) > 0$ and hence, by scaling, we may assume that $\omega_0(A) > 1$. Thus the set

$$K = \{\omega;\, \omega \in X,\, \omega(A) > 1\}$$

is an open neighborhood of ω_0 whose closure $\bar{K}$ satisfies

$$\bar{K} \subseteq \{\omega;\, \omega \in X,\, \omega(A) \geq 1\}.$$

We now argue that the last set is compact. Clearly, a weak* limit ω, of characters ω_α, has the multiplicative property $\omega(BC) = \omega(B)\omega(C)$. But

$$\omega(A) = \lim_\alpha \omega_\alpha(A) \geq 1$$

and hence ω is nonzero. Thus the set $\{\omega;\, \omega \in X,\, \omega(A) \geq 1\}$ is a closed subset of the weakly* compact unit ball of $\mathfrak{A}^*$ and hence it is itself compact. Note that if $\mathfrak{A}$ has an identity then the set of all characters is closed by the same argument applied to $A = 2\mathbb{1}$ and hence is a weakly* compact subset of the unit ball of $\mathfrak{A}^*$.

Next if $A \in \mathfrak{A}$ we define its representative $\hat{A}$ by $\hat{A}(\omega) = \omega(A)$. It follows immediately that $\hat{A}$ is a complex-valued, continuous, function and, moreover, the map $A \to \hat{A}$ is a morphism, e.g., $\widehat{AB}(\omega) = \omega(AB) = \omega(A)\omega(B) = \hat{A}(\omega)\hat{B}(\omega)$. But the basic existence lemma (Lemma 2.3.23) also proves that

$$\|\hat{A}\|^2 = \sup_{\omega \in X} |\hat{A}(\omega)|^2 = \sup_{\omega \in X} |\widehat{A^*A}(\omega)| = \|A\|^2.$$

Thus $A \mapsto \hat{A}$ is an isomorphism. Next we prove that $\hat{A} \in C_0(X)$. For this it suffices to show that for each $\varepsilon > 0$ the set

$$K_\varepsilon = \{\omega; \omega \in X, |\omega(A)| \geq \varepsilon\}$$

is weakly* compact. But this follows by an argument identical to that used in the previous paragraph.

Finally, note that the functions $\hat{A}$ separate points of X in the sense that if $\omega_1 \neq \omega_2$ then there is an $\hat{A}$ such that $\hat{A}(\omega_1) \neq \hat{A}(\omega_2)$. Indeed, this is the definition of $\omega_1 \neq \omega_2$. Thus the set of $\hat{A}$ gives the whole of $C_0(X)$ by the Stone–Weierstrass theorem. If X is compact then $C_0(X)$ contains the constant functions and $\mathfrak{A}$ must contain the identity.

The transformation $A \mapsto \hat{A}$ is usually called the Gelfand transform. In specific cases the structure theorem can be made more precise.

Theorem 2.1.11B. *If $\mathfrak{A}$ is an abelian C^*-algebra which is generated by one element A (and its adjoint A^*) then $\mathfrak{A}$ is isomorphic to the C^*-algebra of continuous functions on the spectrum $\sigma(A)$, of A, which vanish at* 0.

PROOF. First consider the case that $\mathfrak{A}$ has an identity. Hence A is invertible, i.e., $0 \notin \sigma(A)$, and $X = \sigma(\mathfrak{A})$ is weakly* compact. Next define a mapping φ by

$$\varphi(\omega) = \omega(A)$$

for all $\omega \in X$. Then φ maps X into $\sigma(A)$ by Lemma 2.3.26. If $\omega_1, \omega_2 \in X$ then $\omega_1 = \omega_2$ is equivalent to $\omega_1(A) = \omega_2(A)$ because the ω_i are multiplicative and A generates $\mathfrak{A}$. Thus φ is one-to-one and we next argue that it is a homeomorphism. If $\lambda \in \sigma(A)$ then the closure of the set $\{(\lambda\mathbb{1} - A)B; B \in \mathfrak{A}\}$ is a closed two-sided ideal in $\mathfrak{A}$ which does not contain the ball $\{C; C \in \mathfrak{A}, \|\mathbb{1} - C\| < 1\}$. This follows by noting that all elements of this ball are invertible, $C^{-1} = \sum_{n\geq 0} (\mathbb{1} - C)^n$, and this would contradict $\lambda \in \sigma(A)$. Thus Lemma 2.3.23 implies the existence of a pure state ω with $\lambda\mathbb{1} - A \in \ker \omega$, i.e., $\omega(A) = \lambda$. The map φ is clearly continuous and it is a homeomorphism since X and $\sigma(A)$ are compact.

If $\mathfrak{A}$ does not contain an identity, consider the algebra $\tilde{\mathfrak{A}} = \mathbb{C}\mathbb{1} + \mathfrak{A}$ obtained by adjoining an identity to $\mathfrak{A}$. Then $\tilde{\mathfrak{A}}$ is generated by $\mathbb{1}$ and A. By Corollary 2.3.20 each character ω on $\mathfrak{A}$ has a unique extension to a character $\tilde{\omega}$ on $\tilde{\mathfrak{A}}$ by setting $\tilde{\omega}(\lambda\mathbb{1} + A) = \lambda + \omega(A)$. Conversely, if $\tilde{\omega}$ is a character of $\tilde{\mathfrak{A}}$, then $\tilde{\omega}|_{\mathfrak{A}}$ is a character of $\mathfrak{A}$ unless $\tilde{\omega}|_{\mathfrak{A}} = 0$, in which case $\tilde{\omega}$ is the unique character $\tilde{\omega}_\infty$ of $\tilde{\mathfrak{A}}$ defined by $\tilde{\omega}_\infty(\mathbb{1}) = 1, \tilde{\omega}_\infty|_{\mathfrak{A}} = 0$. Thus $\sigma(\tilde{\mathfrak{A}}) = \sigma(\mathfrak{A}) \cup \{\tilde{\omega}_\infty\}$ as sets.

Now, define a map $\varphi; \sigma(\tilde{\mathfrak{A}}) \to \mathbb{C}$ by

$$\varphi(\tilde{\omega}) = \tilde{\omega}(A).$$

Then φ maps $\sigma(\tilde{\mathfrak{A}})$ into $\sigma(A)$ by Lemma 2.3.26. If $\varphi(\tilde{\omega}_1) = \varphi(\tilde{\omega}_2)$, then $\tilde{\omega}_1|_{\mathfrak{A}} = \tilde{\omega}_2|_{\mathfrak{A}}$ while $\tilde{\omega}_1(\mathbb{1}) = 1 = \tilde{\omega}_2(\mathbb{1})$; thus $\tilde{\omega}_1 = \tilde{\omega}_2$, so φ is one-to-one. The same argument as above shows that $\varphi; \sigma(\tilde{\mathfrak{A}}) \mapsto \sigma(A)$ is onto, and in fact a homeomorphism. If $B \mapsto \hat{B}$

is the Gelfand isomorphism $\tilde{\mathfrak{A}} \mapsto C(\sigma(\tilde{\mathfrak{A}}))$, we thus obtain an isomorphism $B \mapsto \hat{\hat{B}}$; $\mathfrak{A} \mapsto C(\sigma(A))$ by setting $\hat{\hat{B}}(\lambda) = \hat{B}(\varphi^{-1}(\lambda))$. Then, by the definition of φ, we have

$$\hat{\hat{A}}(\lambda) = \lambda, \qquad \lambda \in \sigma(A).$$

Thus $\mathfrak{A} \subseteq \tilde{\mathfrak{A}}$ is isomorphic with the subalgebra of $C(\sigma(A))$ generated by the identity function $\lambda \mapsto \lambda$. By the Stone–Weierstrass theorem this algebra is just the continuous functions on $\sigma(A)$ vanishing at 0.

2.4. von Neumann Algebras

2.4.1. Topologies on $\mathscr{L}(\mathfrak{H})$

Each C^*-algebra can be represented by an algebra of bounded operators acting on a Hilbert space $\mathfrak{H}$. In general, there are many inequivalent methods of representation but in any fixed representation the algebra is closed in the uniform operator topology. Detailed analysis of the representation structure entails study of the action of the algebra on vectors and subspaces of the Hilbert space $\mathfrak{H}$. In this analysis it is natural and interesting to consider all operators which approximate the C^*-algebra representatives on all finite-dimensional subspaces. Thus one is motivated to complete the operator algebra in some topology which is weaker than the uniform topology but which, nevertheless, has some form of uniformity on the finite-dimensional subspaces. There is a large variety of such topologies but it turns out that the closure of the C^*-algebra is independent of the particular choice of topology. The enlarged algebra obtained by this closure procedure is an example of a von Neumann algebra.

Our immediate aim is to study von Neumann algebras but it is first necessary to review the various operator topologies associated with $\mathscr{L}(\mathfrak{H})$.

All the topologies we consider are locally convex topologies respecting the vector space structure of $\mathscr{L}(\mathfrak{H})$. The topologies will be defined by a set of seminorms $\{p\}$. One obtains a basis for the neighborhoods of zero in these topologies by considering, for each finite subsequence $p_1, \ldots, p_n$, of the seminorms, the sets of $A \in \mathscr{L}(\mathfrak{H})$ such that $p_i(A) < 1$, $i = 1, \ldots, n$.

Much of the subsequent analysis relies heavily on the general Hahn–Banach theorem for real, or complex, vector spaces. In Section 2.3.4 we introduced the theorem for normed spaces, Theorem 2.3.22A, and the generalization essentially consists of a restatement with the norm replaced by a seminorm, or any other homogeneous subadditive function.

Theorem 2.3.22B (Hahn–Banach). *Let X be a real vector space and p a real-valued function on X satisfying*

(1) $p(\omega_1 + \omega_2) \leq p(\omega_1) + p(\omega_2)$, $\omega_1, \omega_2 \in X$,
(2) $p(\lambda\omega) = \lambda p(\omega)$, $\lambda \geq 0$, $\omega \in X$.

Further, let Y be a real subspace of X and f a real linear functional on Y satisfying

$$f(\omega) \leq p(\omega), \qquad \omega \in Y.$$

It follows that f has a real linear extension F to X such that

$$F(\omega) \leq p(\omega), \qquad \omega \in X.$$

If X is a normed space and one chooses $p(\omega) = \|\omega\| \, \|f\|$ then this theorem reduces to Theorem 2.3.22A. If, however, X is a locally convex topological Hausdorff space and p is chosen to be one of the seminorms defining the topology then the theorem establishes the existence of continuous linear extensions of continuous functionals over subspaces.

Both versions of the Hahn–Banach theorem that we have presented are statements concerning the existence of extensions of linear functionals. The theorem can, however, be rephrased as a geometric result involving separation properties. A version of this nature is the following:

Theorem 2.3.22C (Hahn–Banach). *Let K be a closed convex subset of a real locally convex topological Hausdorff vector space. If $\omega_0 \notin K$ then there exists a continuous affine functional f such that $f(\omega_0) > 1$ and $f(\omega) \leq 1$ for all $\omega \in K$.*

This third version can be deduced from the second as follows. Fix $\omega' \in K$ and define L by

$$L = \{\omega; \omega = \omega'' - \omega', \omega'' \in K\}.$$

Next introduce p_L by

$$p_L(\omega) = \inf\{\lambda; \lambda \geq 0, \lambda^{-1}\omega \in L\}.$$

One may check that $p_L(\omega_1 + \omega_2) \leq p_L(\omega_1) + p_L(\omega_2)$, $p_L(\lambda\omega) = \lambda p_L(\omega)$ for $\lambda \geq 0$, and $p_L(\omega) \leq 1$ if and only if $\omega \in L$. For this one uses the convexity and closedness of L and the fact that $0 \in L$. But one also has $\omega_0 - \omega' \notin L$ and hence $p_L(\omega_0 - \omega') > 1$. Now define g on the subspace $\{\lambda(\omega_0 - \omega'); \lambda \in \mathbb{R}\}$ by $g(\lambda(\omega_0 - \omega')) = \lambda p_L(\omega_0 - \omega')$. Theorem 2.3.22B then implies that g has a continuous linear extension to X such that $g(\omega) \leq p_L(\omega)$. The function $f(\omega) = g(\omega - \omega')$ has the desired properties.

This argument shows that the third version of the Hahn–Banach theorem is a consequence of the second but the converse is also true. As this is not directly relevant to the sequel we omit the proof.

Now we examine the operator topologies of $\mathscr{L}(\mathfrak{H})$.

The strong and the σ-strong topologies. If $\xi \in \mathfrak{H}$ then $A \mapsto \|A\xi\|$ is a seminorm on $\mathscr{L}(\mathfrak{H})$. The strong topology is the locally convex topology on $\mathscr{L}(\mathfrak{H})$ defined by these seminorms.

The related σ-strong topology is obtained by considering all sequences $\{\xi_n\}$ in $\mathfrak{H}$ such that $\sum_n \|\xi_n\|^2 < \infty$. Then for $A \in \mathscr{L}(\mathfrak{H})$

$$\sum_n \|A\xi_n\|^2 \leq \|A\|^2 \sum_n \|\xi_n\|^2 < \infty.$$

Hence $A \mapsto [\sum_n \|A\xi_n\|^2]^{1/2}$ is a seminorm on $\mathscr{L}(\mathfrak{H})$. The set of these seminorms defines the σ-strong topology.

Proposition 2.4.1. *The σ-strong topology is finer than the strong topology, but the two topologies coincide on the unit sphere $\mathscr{L}_1(\mathfrak{H})$ of $\mathscr{L}(\mathfrak{H})$. $\mathscr{L}_1(\mathfrak{H})$ is complete in the uniform structure defined by these topologies. Multiplication $(A, B) \mapsto AB$ is continuous on $\mathscr{L}_1(\mathfrak{H}) \times \mathscr{L}(\mathfrak{H}) \mapsto \mathscr{L}(\mathfrak{H})$ in these topologies. If $\mathfrak{H}$ is infinite-dimensional multiplication is not jointly continuous on all of $\mathscr{L}(\mathfrak{H})$, and the mapping $A \mapsto A^*$ is not continuous.*

PROOF. Since the seminorms defining the σ-strong topology are uniform limits of strongly continuous seminorms, the first statement follows immediately. The completeness of $\mathscr{L}_1(\mathfrak{H})$ follows from the completeness of $\mathfrak{H}$. The continuity of multiplication on $\mathscr{L}_1(\mathfrak{H}) \times \mathscr{L}(\mathfrak{H})$ is a consequence of the relation

$$AB\xi - A_0B_0\xi = A(B - B_0)\xi + (A - A_0)B_0\xi.$$

The discontinuity of $A \mapsto A^*$, for $\mathfrak{H}$ infinite-dimensional, is illustrated by the following example. Let $\{\xi_n\}$ be an orthonormal basis for $\mathfrak{H}$, and consider the elements $A_n \in \mathscr{L}(\mathfrak{H})$ defined by $A_n\xi = (\xi_n, \xi)\xi_1$. Then $A_n \to 0$ σ-strongly, but $(A_n{}^*\xi_1, \xi) = (\xi_1, A_n\xi) = (\xi_1, \xi_1)(\xi_n, \xi)$, i.e., $A_n{}^*\xi_1 = \xi_n$, so $A_n{}^*\xi_1$ does not tend to zero.

The weak and the σ-weak topologies. If $\xi, \eta \in \mathfrak{H}$, then $A \mapsto |(\xi, A\eta)|$ is a seminorm on $\mathscr{L}(\mathfrak{H})$. The locally convex topology on $\mathscr{L}(\mathfrak{H})$ defined by these seminorms is called the weak topology. The seminorms defined by the vector states $A \mapsto |(\xi, A\xi)|$ suffice to define this topology because $\mathfrak{H}$ is complex and one has the polarization identity

$$4(\xi, A\eta) = \sum_{n=0}^{3} i^{-n}(\xi + i^n\eta, A(\xi + i^n\eta)).$$

Let $\{\xi_n\}$, $\{\eta_n\}$ be two sequences from $\mathfrak{H}$ such that

$$\sum_n \|\xi_n\|^2 < \infty, \qquad \sum_n \|\eta_n\|^2 < \infty.$$

Then for $A \in \mathscr{L}(\mathfrak{H})$

$$\begin{aligned}\sum_n |(\xi_n, A\eta_n)| &\leq \sum_n \|\xi_n\| \, \|A\| \, \|\eta_n\| \\ &\leq \|A\| \left[\sum_n \|\xi_n\|^2\right]^{1/2} \left[\sum_n \|\eta_n\|^2\right]^{1/2} \\ &< \infty.\end{aligned}$$

Hence $A \mapsto \sum_n |(\xi_n, A\eta_n)|$ is a seminorm on $\mathscr{L}(\mathfrak{H})$. The locally convex topology on $\mathscr{L}(\mathfrak{H})$ induced by these seminorms is called the σ-weak topology.

Proposition 2.4.2. *The σ-weak topology is finer than the weak topology, but the two topologies coincide on the unit sphere $\mathscr{L}_1(\mathfrak{H})$ of $\mathscr{L}(\mathfrak{H})$. $\mathscr{L}_1(\mathfrak{H})$ is compact in this topology. The mappings $A \mapsto AB$, $A \mapsto BA$, and $A \mapsto A^*$ are continuous in this topology, but multiplication is not jointly continuous if $\mathfrak{H}$ is infinite-dimensional.*

PROOF. Since the seminorms, defining the σ-weak topology are uniform limits of weakly continuous seminorms, the first statement is immediate. The separate continuity of multiplication is evident, while the continuity of $A \mapsto A^*$ follows from the relation $|(\xi, A^*\eta)| = |(\eta, A\xi)|$.

The compactness of $\mathscr{L}_1(\mathfrak{H})$ in the weak topology follows from the next proposition and the Alaoglu–Bourbaki theorem.

Proposition 2.4.3. *Let* Tr *be the usual trace on $\mathscr{L}(\mathfrak{H})$, and let $\mathscr{T}(\mathfrak{H})$ be the Banach space of trace-class operators on $\mathfrak{H}$ equipped with the trace norm $T \mapsto \mathrm{Tr}(|T|) = \|T\|_{\mathrm{Tr}}$. Then it follows that $\mathscr{L}(\mathfrak{H})$ is the dual $\mathscr{T}(\mathfrak{H})^*$ of $\mathscr{T}(\mathfrak{H})$ by the duality*

$$A \times T \in \mathscr{L}(\mathfrak{H}) \times \mathscr{T}(\mathfrak{H}) \mapsto \mathrm{Tr}(AT).$$

The weak topology on $\mathscr{L}(\mathfrak{H})$ arising from this duality is just the σ-weak topology.*

PROOF. Because of the inequality $|\mathrm{Tr}(AT)| \leq \|A\| \, \|T\|_{\mathrm{Tr}}$, $\mathscr{L}(\mathfrak{H})$ is a subspace of $\mathscr{T}(\mathfrak{H})^*$ by the duality described in the proposition. Conversely, assume $\omega \in \mathscr{T}(\mathfrak{H})^*$ and consider a rank one operator $E_{\varphi,\psi}$ defined for $\varphi, \psi \in \mathfrak{H}$ by

$$E_{\varphi,\psi}\chi = \varphi(\psi, \chi).$$

One has $E_{\varphi,\psi}^* = E_{\psi,\varphi}$ and $E_{\varphi,\psi}E_{\psi,\varphi} = \|\psi\|^2 E_{\varphi,\varphi}$. Hence

$$\|E_{\varphi,\psi}\|_{\mathrm{Tr}} = \|\psi\| \mathrm{Tr}(E_{\varphi,\varphi})^{1/2} = \|\psi\| \, \|\varphi\|.$$

It follows that

$$|\omega(E_{\varphi,\psi})| \leq \|\omega\| \, \|\varphi\| \, \|\psi\|.$$

Hence there exists, by the Riesz representation theorem, an $A \in \mathscr{L}(\mathfrak{H})$ with $\|A\| \leq \|\omega\|$ such that

$$\omega(E_{\varphi,\psi}) = (\psi, A\varphi).$$

Consider $\omega_0 \in \mathscr{T}(\mathfrak{H})^*$ defined by

$$\omega_0(T) = \mathrm{Tr}(AT);$$

then

$$\begin{aligned}\omega_0(E_{\varphi,\psi}) &= \mathrm{Tr}(AE_{\varphi,\psi}) \\ &= (\psi, A\varphi) \\ &= \omega(E_{\varphi,\psi}).\end{aligned}$$

Now for any $T \in \mathscr{T}(\mathfrak{H})$ there exist bounded sequences $\{\psi_n\}$ and $\{\varphi_n\}$ and a sequence $\{\alpha_n\}$ of complex numbers such that

$$\sum_n |\alpha_n| < \infty$$

and

$$T = \sum_n \alpha_n E_{\varphi_n, \psi_n}.$$

The latter series converges with respect to the trace norm and hence

$$\begin{aligned}\omega(T) &= \sum_n \alpha_n \omega(E_{\varphi_n, \psi_n}) \\ &= \sum_n \alpha_n \omega_0(E_{\varphi_n, \psi_n}) = \omega_0(T) = \mathrm{Tr}(AT).\end{aligned}$$

Thus $\mathscr{L}(\mathfrak{H})$ is just the dual of $\mathscr{T}(\mathfrak{H})$.

The weak* topology on $\mathscr{L}(\mathfrak{H})$ arising from this duality is given by the seminorms

$$A \in \mathscr{L}(\mathfrak{H}) \mapsto |\mathrm{Tr}(AT)|.$$

Now for

$$T = \sum_n \alpha_n E_{\varphi_n, \psi_n}$$

one has

$$\begin{aligned}\mathrm{Tr}(AT) &= \sum_n \alpha_n \mathrm{Tr}(E_{\varphi_n, \psi_n} A) \\ &= \sum_n \alpha_n(\psi_n, A\varphi_n).\end{aligned}$$

Thus the seminorms are equivalent to the seminorms defining the σ-weak topology.

Definition 2.4.4. The space of σ-weakly continuous linear functionals on $\mathscr{L}(\mathfrak{H})$ is called the *predual of* $\mathscr{L}(\mathfrak{H})$ and is denoted by $\mathscr{L}_*(\mathfrak{H})$.

As noted in Proposition 2.4.3, $\mathscr{L}_*(\mathfrak{H})$ can be canonically identified with $\mathscr{T}(\mathfrak{H})$, and $\mathscr{L}(\mathfrak{H}) = \mathscr{L}_*(\mathfrak{H})^*$.

The strong and the σ-strong* topologies.* These topologies are defined by seminorms of the form

$$A \mapsto \|A\xi\| + \|A^*\xi\|$$

and

$$A \mapsto \left[\sum_n \|A\xi_n\|^2 + \sum_n \|A^*\xi_n\|^2\right]^{1/2},$$

respectively, where $\sum_n \|\xi_n\|^2 < \infty$. The main difference between the strong* and the strong topology is that $A \mapsto A^*$ is continuous in the former topology but not the latter. Otherwise the following proposition is proved as in the strong case.

Proposition 2.4.5. *The σ-strong* topology is finer than the strong* topology, but the two topologies coincide on the unit sphere $\mathscr{L}_1(\mathfrak{H})$ of $\mathscr{L}(\mathfrak{H})$, Multiplication $(A, B) \mapsto AB$; $\mathscr{L}_1(\mathfrak{H}) \times \mathscr{L}_1(\mathfrak{H}) \mapsto \mathscr{L}(\mathfrak{H})$ is continuous, and $A \mapsto A^*$ is continuous in these topologies, but multiplication $\mathscr{L}(\mathfrak{H}) \times \mathscr{L}(\mathfrak{H}) \mapsto \mathscr{L}(\mathfrak{H})$ is discontinuous if $\mathfrak{H}$ is infinite-dimensional.*

The relation between the various topologies on $\mathscr{L}(\mathfrak{H})$ is as follows:

$$\begin{array}{ccccccc} \text{uniform} < & \sigma\text{-strong*} & < & \sigma\text{-strong} & < & \sigma\text{-weak} \\ & \wedge & & \wedge & & \wedge \\ & \text{strong*} & < & \text{strong} & < & \text{weak} \end{array}$$

Here "$<$" means "finer than," and if $\mathfrak{H}$ is infinite-dimensional then "$<$" can be taken to mean "strictly finer than."

It is interesting to note that the σ-strong*, σ-strong, and σ-weak topologies allow just the same continuous linear functionals. The same is true when the σ- is removed. As the proof of both assertions are the same we prove the former.

Proposition 2.4.6. *Every σ-strongly* continuous linear functional ω on $\mathscr{L}(\mathfrak{H})$ is σ-weakly continuous, hence is in $\mathscr{L}_*(\mathfrak{H})$ and has the form $\omega(A) = \sum_n (\xi_n, A\eta_n)$, where $\sum_n \|\xi_n\|^2 < \infty$, $\sum_n \|\eta_n\|^2 < \infty$.*

PROOF. Suppose that ω is a σ-strongly* continuous linear functional on $\mathscr{L}(\mathfrak{H})$. Then there exists a sequence $\{\xi_n\}$ in $\mathfrak{H}$ such that $\sum_{n\geq 1} \|\xi_n\|^2 < \infty$ and

$$|\omega(A)| \leq \left[\sum_{n\geq 1} (\|A\xi_n\|^2 + \|A^*\xi_n\|^2)\right]^{1/2}.$$

Let $\tilde{\mathfrak{H}} = \bigoplus_{n=-\infty}^{\infty} \mathfrak{H}_n$, where $\mathfrak{H}_n = \mathfrak{H}$ for $n = 1, 2, \ldots$, and $\mathfrak{H}_n = \bar{\mathfrak{H}}$, the conjugate Hilbert space[3] of $\mathfrak{H}$, for $n = -1, -2, \ldots$. Note that $\tilde{\xi} = \{\cdots, \bar{\xi}_2, \bar{\xi}_1, \xi_1, \xi_2, \ldots\}$ is an element in $\tilde{\mathfrak{H}}$. For each $\tilde{\eta} = \{\bar{\eta}_{-m}, \eta_m; m = 1, 2, \ldots\} \in \tilde{\mathfrak{H}}$ and $A \in \mathscr{L}(\mathfrak{H})$ define

$$\tilde{A}\tilde{\eta} = \{\overline{A^*\eta_{-m}}, A\eta_m; m = 1, 2, \ldots\}.$$

Then $\tilde{A} \in \mathscr{L}(\tilde{\mathfrak{H}}), \|\tilde{A}\| = \|A\|$, and the map $A \mapsto \tilde{A}$ is linear. By the inequality on ω, the map $\tilde{A}\tilde{\xi} \mapsto \omega(A)$ is a bounded linear functional on the space $\{\tilde{A}\tilde{\xi}; A \in \mathscr{L}(\mathfrak{H})\}$. The Riesz representation theorem asserts the existence of an $\tilde{\eta} \in \tilde{\mathfrak{H}}$ such that

$$\begin{aligned} \omega(A) &= (\tilde{\eta}, \tilde{A}\tilde{\xi}) \\ &= \sum_{n=1}^{\infty} [(\eta_n, A\xi_n) + (\xi_{-n}, A\eta_{-n})]. \end{aligned}$$

Hence ω is σ-weakly continuous.

An immediate consequence of Proposition 2.4.6 is the equality of closures of convex subsets of $\mathscr{L}(\mathfrak{H})$ in the various topologies.

[3] $\bar{\mathfrak{H}} = \mathfrak{H}$ as a set. If $\xi \in \mathfrak{H}$, let $\bar{\xi}$ denote the corresponding element in $\bar{\mathfrak{H}}$. The Hilbert space structure on $\bar{\mathfrak{H}}$ is then defined by

$$\begin{aligned} \bar{\xi} + \bar{\eta} &= \overline{\xi + \eta}, \\ \lambda\bar{\xi} &= \overline{\bar{\lambda}\xi}, \\ (\bar{\xi}, \bar{\eta}) &= (\eta, \xi). \end{aligned}$$

Theorem 2.4.7. *Let $\mathfrak{K}$ be a convex subset of $\mathscr{L}(\mathfrak{H})$ and $\mathscr{L}_r(\mathfrak{H})$ the ball of radius r in $\mathscr{L}(\mathfrak{H})$. The following conditions are equivalent:*

(1) *$\mathfrak{K}$ is σ-weakly closed;*
(2) *$\mathfrak{K}$ is σ-strongly closed;*
(3) *$\mathfrak{K}$ is σ-strongly* closed;*
(4) *$\mathfrak{K} \cap \mathscr{L}_r(\mathfrak{H})$ is weakly (therefore σ-weakly) closed for all $r > 0$;*
(5) *$\mathfrak{K} \cap \mathscr{L}_r(\mathfrak{H})$ is strongly (therefore σ-strongly) closed for all $r > 0$;*
(6) *$\mathfrak{K} \cap \mathscr{L}_r(\mathfrak{H})$ is strongly* (therefore σ-strongly*) closed for all $r > 0$.*

PROOF. The equivalence $(1) \Leftrightarrow (4)$ follows from the fact that $\mathscr{L}(\mathfrak{H})$ is the dual of $\mathscr{L}_*(\mathfrak{H})$ and a theorem of Banach (see Notes and Remarks). The implications $(4) \Rightarrow (5) \Rightarrow (6)$ and $(1) \Rightarrow (2) \Rightarrow (3)$ are trivial. Since $\mathfrak{K} \cap \mathscr{L}_r(\mathfrak{H})$ is convex for all r, the implications $(3) \Rightarrow (1)$ and $(6) \Rightarrow (4)$ follow from Proposition 2.4.6 and the fact that a closed convex set containing zero is its own bipolar.

2.4.2. Definition and Elementary Properties of von Neumann Algebras

Let $\mathfrak{H}$ be a Hilbert space. For any subset $\mathfrak{M}$ of $\mathscr{L}(\mathfrak{H})$ we again let $\mathfrak{M}'$ denote its commutant, i.e. the set of all bounded operators on $\mathfrak{H}$ commuting with every operator in $\mathfrak{M}$. Clearly $\mathfrak{M}'$ is a Banach algebra of operators containing the identity $\mathbb{1}$. If $\mathfrak{M}$ is selfadjoint then $\mathfrak{M}'$ is a C^*-algebra of operators on $\mathfrak{H}$, which is closed under all the locally convex topologies defined in the preceding section. One has

$$\mathfrak{M} \subseteq \mathfrak{M}'' = \mathfrak{M}^{(\mathrm{iv})} = \mathfrak{M}^{(\mathrm{vi})} = \cdots,$$

$$\mathfrak{M}' = \mathfrak{M}''' = \mathfrak{M}^{(\mathrm{v})} = \mathfrak{M}^{(\mathrm{vii})} = \cdots.$$

Definition 2.4.8. A *von Neumann algebra on* $\mathfrak{H}$ is a *-subalgebra $\mathfrak{M}$ of $\mathscr{L}(\mathfrak{H})$ such that

$$\mathfrak{M} = \mathfrak{M}''.$$

The *center* $\mathfrak{Z}(\mathfrak{M})$ *of a von Neumann algebra* is defined by

$$\mathfrak{Z}(\mathfrak{M}) = \mathfrak{M} \cap \mathfrak{M}'.$$

A von Neumann algebra is called a *factor* if it has a trivial center, i.e., if $\mathfrak{Z}(\mathfrak{M}) = \mathbb{C}\mathbb{1}$.

We next note some elementary facts about von Neumann algebras. Let A be a selfadjoint element in a von Neumann algebra $\mathfrak{M}$. If some operator commutes with A then it also commutes with all the spectral projections of A, hence these spectral projections lie in $\mathfrak{M}$. Since A can be approximated in norm by linear combination of spectral projections and any element in $\mathfrak{M}$ is a linear combination of two selfadjoint operators, $A = (A + A^*)/2 + i(A - A^*)/2i$, the projections in $\mathfrak{M}$ span a norm-dense subspace of $\mathfrak{M}$.

Since any element in a C^*-algebra with identity is a linear combination of four unitary elements (Lemma 2.2.14), it follows that an element $A \in \mathscr{L}(\mathfrak{H})$ lies in $\mathfrak{M}$ if, and only if, $VAV^* = A$ for all unitary elements $V \in \mathfrak{M}'$. Hence if $A = U|A|$ is the polar decomposition (Example 2.2.16) of an element $A \in \mathfrak{M}$, then for any unitary element $V \in \mathfrak{M}'$

$$VU\,V^*V|A|V^* = VU|A|V^* = VAV^* = A = U|A|.$$

Thus by the uniqueness of the polar decomposition

$$VUV^* = U, \qquad V|A|V^* = |A|.$$

Hence $U \in \mathfrak{M}$, $|A| \in \mathfrak{M}$.

Similarly, if $\{A_\alpha\}$ is an increasing net of positive operators from $\mathfrak{M}$ with least upper bound $A \in \mathscr{L}(\mathfrak{H})_+$ then for any unitary element $V \in \mathfrak{M}'$, $\{VA_\alpha V^*\} = \{A_\alpha\}$ has least upper bound VAV^*. Hence $A = VAV^*$ and $A \in \mathfrak{M}$.

EXAMPLE 2.4.9. $\mathscr{L}(\mathfrak{H})$ is a von Neumann algebra and even a factor since $\mathscr{L}(\mathfrak{H})' = \mathbb{C}\mathbb{1}$. $\mathscr{L}\mathscr{C}(\mathfrak{H})$ is not a von Neumann algebra, since $\mathscr{L}\mathscr{C}(\mathfrak{H})' = \mathbb{C}\mathbb{1}$ and hence $\mathscr{L}\mathbb{C}(\mathfrak{H})'' = \mathscr{L}(\mathfrak{H})$. Note that in this case one can easily approximate any operator in $\mathscr{L}(\mathfrak{H})$ by finite-rank operators in any of the locally convex topologies considered in Section 1, i.e., the closure of $\mathscr{L}\mathscr{C}(\mathfrak{H})$ in any of these topologies is $\mathscr{L}(\mathfrak{H})$. This is a special case of a fundamental fact known as the von Neumann density theorem, or the bicommutant theorem, which we prove below (Theorem 2.4.11).

Definition 2.4.10. If $\mathfrak{M}$ is a subset of $\mathscr{L}(\mathfrak{H})$ and $\mathfrak{K}$ is a subset of $\mathfrak{H}$, let $[\mathfrak{M}\mathfrak{K}]$ denote the closure of the linear span of elements of the form $A\xi$, where $A \in \mathfrak{M}$, $\xi \in \mathfrak{K}$. Let $[\mathfrak{M}\mathfrak{K}]$ also denote the orthogonal projection onto $[\mathfrak{M}\mathfrak{K}]$.

Recall that a $*$-subalgebra $\mathfrak{A} \subseteq \mathscr{L}(\mathfrak{H})$ is said to be nondegenerate if $[\mathfrak{A}\mathfrak{H}] = \mathfrak{H}$ (see Section 2.3.1).

If $\mathfrak{A} \subseteq \mathscr{L}(\mathfrak{H})$ contains the identity then it is automatically nondegenerate.

Theorem 2.4.11 (Bicommutant theorem). *Let $\mathfrak{A}$ be a nondegenerate $*$-algebra of operators on $\mathfrak{H}$. Then the following conditions are equivalent:*

(1) $\mathfrak{A}'' = \mathfrak{A}$;
(2) *(resp.* (2a)) $\mathfrak{A}$ *(resp.* $\mathfrak{A}_1$) *is weakly closed;*
(3) *(resp.* (3a)) $\mathfrak{A}$ *(resp.* $\mathfrak{A}_1$) *is strongly closed;*
(4) *(resp.* (4a)) $\mathfrak{A}$ *(resp.* $\mathfrak{A}_1$) *is strongly* closed;*
(5) *(resp.* (5a)) $\mathfrak{A}$ *(resp.* $\mathfrak{A}_1$) *is σ-weakly closed;*
(6) *(resp.* (6a)) $\mathfrak{A}$ *(resp.* $\mathfrak{A}_1$) *is σ-strongly closed;*
(7) *(resp.* (7a)) $\mathfrak{A}$ *(resp.* $\mathfrak{A}_1$) *is σ-strongly* closed.*

PROOF. The equivalence of (2a), (3a), (4a), (5), (5a), (6), (6a), (7), (7a) follows from Theorem 2.4.7. Clearly (1) implies all the other conditions and (2) $\Rightarrow$ (3) $\Rightarrow$ (4) $\Rightarrow$ (7). Hence it remains to show, for example, that (6) $\Rightarrow$ (1). To do this consider a countably infinite sum of replicas of $\mathfrak{H}$: $\tilde{\mathfrak{H}} = \bigoplus_{n=1}^{\infty} \mathfrak{H}_n$, where $\mathfrak{H}_n = \mathfrak{H}$ for all n.

If $A \in \mathscr{L}(\mathfrak{H})$, define $\pi(A) \in \mathscr{L}(\tilde{\mathfrak{H}})$ by

$$\pi(A)\left(\bigoplus_n \xi_n\right) = \bigoplus_n (A\xi_n).$$

π is clearly a *-automorphism of $\mathscr{L}(\mathfrak{H})$ into a subalgebra of $\mathscr{L}(\tilde{\mathfrak{H}})$.

Lemma 2.4.12. *One has the relation $\pi(\mathfrak{A}'') = \pi(\mathfrak{A})''$.*

PROOF. Let E_n be the orthogonal projection from $\tilde{\mathfrak{H}} = \bigoplus_{n=1}^{\infty} \mathfrak{H}_n$ onto $\mathfrak{H}_n = \mathfrak{H}$. Clearly, $B \in \mathscr{L}(\tilde{\mathfrak{H}})$ lies in $\pi(\mathfrak{A})'$ if, and only if, $E_n B E_m \in \mathfrak{A}'$ for all n and m. Hence $C \in \mathscr{L}(\tilde{\mathfrak{H}})$ lies in $\pi(\mathfrak{A})''$ if, and only if, C commutes with all the E_n, and $E_n C E_n$ is a fixed element of $\mathfrak{A}''$, i.e., if, and only if, $C \in \pi(\mathfrak{A}'')$.

Lemma 2.4.13. *If $\mathfrak{M}$ is a nondegenerate *-algebra of operators on a Hilbert space $\mathfrak{H}$, then ξ belongs to $[\mathfrak{M}\xi]$ for every $\xi \in \mathfrak{H}$.*

PROOF. Let $P = [\mathfrak{M}\xi]$; then

$$MP = PMP$$

for all $M \in \mathfrak{M}$. By conjugation one has

$$PM^* = PM^*P$$

for all $M^* \in \mathfrak{M}$. As $\mathfrak{M}$ is a selfadjoint set one deduces that

$$MP = PM = PMP$$

for all $M \in \mathfrak{M}$, i.e., $P \in \mathfrak{M}'$. Now if $\xi' = P\xi$ and $\xi'' = (\mathbb{1} - P)\xi$ one has $\xi = \xi' + \xi''$. But then the relation

$$A\xi' + A\xi'' = A\xi \in [\mathfrak{M}\xi]$$

implies that $A\xi'' = 0$ for all $A \in \mathfrak{M}$. Thus, for an arbitrary $\eta \in \mathfrak{H}$ and $A \in \mathfrak{M}$

$$(\xi'', A\eta) = (A^*\xi'', \eta) = 0.$$

Thus ξ'' is in the orthogonal complement of $[\mathfrak{M}\mathfrak{H}] = \mathfrak{H}$. Hence $\xi'' = 0$ and $\xi \in [\mathfrak{M}\xi]$.

Lemma 2.4.14. *Let $\mathfrak{M}$ be a nondegenerate *-algebra of operators on a Hilbert space $\mathfrak{H}$. Then for any $\xi \in \mathfrak{H}$, $A \in \mathfrak{M}''$, and $\varepsilon > 0$ there exists an element $B \in \mathfrak{M}$ such that*

$$\|(A - B)\xi\| < \varepsilon.$$

PROOF. We must show that $[\mathfrak{M}\xi] = [\mathfrak{M}''\xi]$. Let P be the orthogonal projection onto $[\mathfrak{M}\xi]$. Since $\mathfrak{M}[\mathfrak{M}\xi] \subseteq [\mathfrak{M}\xi]$, $P \in \mathfrak{M}'$. Hence P commutes with $\mathfrak{M}''$, so $\mathfrak{M}''[\mathfrak{M}\xi] \subseteq [\mathfrak{M}\xi]$. By Lemma 2.4.13, $\xi \in [\mathfrak{M}\xi]$, hence $\mathfrak{M}''\xi \subseteq [\mathfrak{M}\xi]$, and $[\mathfrak{M}''\xi] \subseteq [\mathfrak{M}\xi]$.

END OF THE PROOF OF THEOREM 2.4.11. Assume (6); let $A \in \mathfrak{A}''$ and let $\{\xi_n\}$ be a sequence in $\mathfrak{H}$ such that $\sum_n \|\xi_n\|^2 < \infty$. Then $\bigoplus_n \xi_n \in \tilde{\mathfrak{H}}$. Since $\mathfrak{A}$ is nondegenerate, $\pi(\mathfrak{A})$ is nondegenerate. Also $\pi(A) \in \pi(\mathfrak{A})''$ by Lemma 2.4.12. Hence Lemma 2.4.14

applies with A replaced by $\pi(A)$, $\mathfrak{M} = \pi(\mathfrak{A})$ and $\xi = \bigoplus_n \xi_n$. Thus there exists a $B \in \mathfrak{A}$ such that

$$\begin{aligned} \varepsilon &> \|(\pi(A) - \pi(B))\xi\| \\ &= \left[\sum_{n=1}^{\infty} \|(A - B)\xi_n\|^2 \right]^{1/2} \end{aligned}$$

and A must be in the σ-strong closure of $\mathfrak{A}$. Hence $A \in \mathfrak{A}$ and $\mathfrak{A}'' \subseteq \mathfrak{A}$.

Corollary 2.4.15 (von Neumann density theorem). *Let $\mathfrak{A}$ be a non-degenerate *-algebra of operators acting on a Hilbert space $\mathfrak{H}$. It follows that $\mathfrak{A}$ is dense in $\mathfrak{A}''$ in the weak, strong, strong*, σ-weak, σ-strong, and σ-strong* topologies.*

PROOF. If $\bar{\mathfrak{A}}$ is the closure of $\mathfrak{A}$ in any of the topologies above, then $\bar{\mathfrak{A}}' = \mathfrak{A}'$, hence $\bar{\mathfrak{A}}'' = \mathfrak{A}''$. But $\bar{\mathfrak{A}} = \bar{\mathfrak{A}}''$ by Theorem 2.4.11.

Next we prove a useful theorem which immediately implies a stronger version of Corollary 2.4.15.

Theorem 2.4.16 (Kaplansky's density theorem). *If $\mathfrak{A}$ is a *-algebra of operators on a Hilbert space then the unit ball of $\mathfrak{A}$ is σ-strongly* dense in the unit ball of the weak closure of $\mathfrak{A}$.*

PROOF. Let $\mathfrak{M}$ be the weak closure of $\mathfrak{A}$, $\mathfrak{A}_1$ and $\mathfrak{M}_1$ the unit balls of $\mathfrak{A}$ and $\mathfrak{M}$, respectively, and for any subset $\mathfrak{N}$ of $\mathscr{L}(\mathfrak{H})$ let $\mathfrak{N}_{sa}$ be the selfadjoint elements in $\mathfrak{N}$. It is evident that $\mathfrak{A}_1$ is norm dense in the unit ball of the norm closure of $\mathfrak{A}$, so we may assume that $\mathfrak{A}$ is a C^*-algebra.

By Theorem 2.4.11, $\mathfrak{A}$ is σ-strongly* dense in $\mathfrak{M}$, hence $\mathfrak{A}_{sa}$ is σ-strongly dense in $\mathfrak{M}_{sa}$. (Theorem 2.4.11 is applied to the subspace $[\mathfrak{A}\mathfrak{H}]$ of $\mathfrak{H}$.) The real function $t \mapsto 2t(1 + t^2)^{-1}$ increases strictly from -1 to 1 on the interval $[-1, 1]$ and has range in $[-1, 1]$. Hence if one defines a function $f\colon \mathscr{L}(\mathfrak{H})_{sa} \mapsto \mathscr{L}(\mathfrak{H})_{sa}$ by $f(A) = 2A(\mathbb{1} + A^2)^{-1}$ then for any C^*-subalgebra $\mathfrak{B} \subseteq \mathscr{L}(\mathfrak{H})$ f maps $\mathfrak{B}_{sa}$ into $\mathfrak{B}_{1\,sa}$. Moreover, f maps $\mathfrak{B}_{1\,sa}$ in a one-to-one fashion onto itself. Hence if f is continuous in the σ-strong topology it will follow that $\mathfrak{A}_{1\,sa} = f(\mathfrak{A}_{sa})$ is σ-strongly dense in $\mathfrak{M}_{1\,sa} = f(\mathfrak{M}_{sa})$.

For $A, B \in \mathscr{L}(\mathfrak{H})_{sa}$, we estimate

$$\begin{aligned} \tfrac{1}{2}(f(A) - f(B)) &= (\mathbb{1} + A^2)^{-1}[A(\mathbb{1} + B^2) - (\mathbb{1} + A^2)B](\mathbb{1} + B^2)^{-1} \\ &= (\mathbb{1} + A^2)^{-1}(A - B)(\mathbb{1} + B^2)^{-1} \\ &\quad + (\mathbb{1} + A^2)^{-1}A(B - A)B(\mathbb{1} + B^2)^{-1} \\ &= (\mathbb{1} + A^2)^{-1}(A - B)(\mathbb{1} + B^2)^{-1} \\ &\quad + \tfrac{1}{4}f(A)(B - A)f(B). \end{aligned}$$

Hence, f is σ-strongly continuous by Proposition 2.4.1.

To complete the proof, consider the Hilbert space $\tilde{\mathfrak{H}} = \mathfrak{H} \oplus \mathfrak{H}$. Each operator $A \in \mathscr{L}(\tilde{\mathfrak{H}})$ is represented by a 2×2 matrix (A_{ij}), $i, j = 1, 2$. Let $\tilde{\mathfrak{A}}$ (resp. $\tilde{\mathfrak{M}}$) be the operators in $\mathscr{L}(\tilde{\mathfrak{H}})$ such that $A_{ij} \in \mathfrak{A}$, $i, j = 1, 2$ (resp. $A_{ij} \in \mathfrak{M}$). Clearly, $\tilde{\mathfrak{A}}$ and $\tilde{\mathfrak{M}}$

are *-algebras on $\tilde{\mathfrak{H}}$, and $\tilde{\mathfrak{A}}$ is weakly dense in $\tilde{\mathfrak{M}}$. Now pick $B \in \mathfrak{M}$ with $\|B\| \leq 1$ and define $\tilde{B} \in \tilde{\mathfrak{M}}$ by

$$\tilde{B} = \begin{pmatrix} 0 & B \\ B^* & 0 \end{pmatrix}.$$

Then $\tilde{B}^* = \tilde{B}$ and $\|\tilde{B}\| \leq 1$. By the first part of the proof there exist operators

$$\tilde{A} = \begin{pmatrix} A_{11} & A_{12} \\ A_{21} & A_{22} \end{pmatrix} \in \tilde{\mathfrak{A}}_1$$

with $A_{12} = A_{21}^*$ such that $\tilde{A}$ converges σ-strongly to $\tilde{B}$. Then A_{12} converges σ-strongly to B and $A_{12}^* = A_{21}$ converges σ-strongly to B^*. Thus A_{12} converges σ-strongly* to B. But one also has $\|A_{12}\| \leq \|\tilde{A}\| \leq 1$.

2.4.3. Normal States and the Predual

If μ is a σ-finite measure, then $L^\infty(d\mu)$ forms a von Neumann algebra of multiplication operators on the Hilbert space $L^2(d\mu)$. $L^\infty(d\mu)$ is the dual of $L^1(d\mu)$; $L^1(d\mu)$, however, is only a norm-closed subspace of the dual of $L^\infty(d\mu)$. In this section we single out an analogous subset of the dual of a von Neumann algebra $\mathfrak{M}$, called the predual, and study its properties.

Definition 2.4.17. The *predual of a von Neumann algebra* $\mathfrak{M}$ is the space of all σ-weakly continuous linear functionals on $\mathfrak{M}$. It is denoted by $\mathfrak{M}_*$.

Note that we have already introduced this definition in the special case that $\mathfrak{M} = \mathscr{L}(\mathfrak{H})$. If ω is any functional on $\mathfrak{M}$ which is continuous with respect to any locally convex topology induced by $\mathscr{L}(\mathfrak{H})$, then ω extends to a continuous linear functional on $\mathscr{L}(\mathfrak{H})$ by the Hahn–Banach theorem (Theorem 2.3.22B). Thus, by Proposition 2.4.6 one may replace σ-weakly in Definition 2.4.17 by σ-strongly*, and all elements $\omega \in \mathfrak{M}_*$ have the form

$$\omega(A) = \sum_n (\xi_n, A\eta_n),$$

where $\sum_n \|\xi_n\|^2 < \infty$ and $\sum_n \|\eta_n\|^2 < \infty$.

Proposition 2.4.18. *The predual $\mathfrak{M}_*$ of a von Neumann algebra $\mathfrak{M}$ is a Banach space in the norm of $\mathfrak{M}^*$, and $\mathfrak{M}$ is the dual of $\mathfrak{M}_*$ in the duality*

$$(A, \omega) \in \mathfrak{M} \times \mathfrak{M}_* \mapsto \omega(A).$$

PROOF. Let $\mathfrak{M}^\perp$ be the elements in $\mathscr{L}_*(\mathfrak{H})$ which are orthogonal to $\mathfrak{M}$. Since $\mathfrak{M}$ is a σ-weakly closed subspace of $\mathscr{L}(\mathfrak{H})$, $\mathfrak{M} = \mathfrak{M}^{\perp\perp}$. But the Hahn–Banach theorem ensures that any element in $\mathfrak{M}_*$ extends to $\mathscr{L}_*(\mathfrak{H})$. Moreover, any element in $\mathscr{L}_*(\mathfrak{H})$ defines an element in $\mathfrak{M}_*$ by restriction. Thus $\mathfrak{M}_*$ is canonically identifiable with the Banach space $\mathscr{L}_*/\mathfrak{M}^\perp$. Hence $\mathfrak{M}$ is the dual of this space, because $\mathscr{L}(\mathfrak{H})$ is the dual of $\mathscr{L}_*(\mathfrak{H})$ (Proposition 2.4.3).

Proposition 2.4.18 states that each von Neumann algebra is the dual of a Banach space. It is interesting to note that this may serve as an abstract definition of a von Neumann algebra:

Theorem (Sakai). *A C^*-algebra $\mathfrak{A}$ is *-isomorphic with a von Neumann algebra if, and only if, $\mathfrak{A}$ is the dual of a Banach space.*

We will not give a proof of this result since it is not needed in the sequel (see Notes and Remarks).

We turn next to a characterization of the positive functionals in $\mathfrak{M}_*$.

Lemma 2.4.19. *Let $\{A_\alpha\}$ be an increasing net in $\mathscr{L}(\mathfrak{H})_+$ with an upper bound in $\mathscr{L}(\mathfrak{H})_+$. Then $\{A_\alpha\}$ has a least upper bound (l.u.b.) A, and the net converges σ-strongly to A.*

PROOF. Let $\mathfrak{K}_\alpha$ be the weak closure of the set of A_β with $\beta > \alpha$. Since $\mathscr{L}(\mathfrak{H})_1$ is weakly compact, there exists an element A in $\bigcap_\alpha \mathfrak{K}_\alpha$. For all A_α the set of $B \in \mathscr{L}(\mathfrak{H})_+$ such that $B \geq A_\alpha$ is σ-weakly closed and contains $\mathfrak{K}_\alpha$, hence $A \geq A_\alpha$. Hence A majorizes $\{A_\alpha\}$ and lies in the weak closure of $\{A_\alpha\}$. If B is another operator majorizing $\{A_\alpha\}$, then it majorizes its weak closure; thus $B \geq A$ and A is the least upper bound of $\{A_\alpha\}$. Finally, if $\xi \in \mathfrak{H}$ then

$$\begin{aligned}\|(A - A_\alpha)\xi\|^2 &\leq \|A - A_\alpha\| \, \|(A - A_\alpha)^{1/2}\xi\|^2 \\ &\leq \|A\| (\xi, (A - A_\alpha)\xi) \\ &\underset{\alpha}{\to} 0.\end{aligned}$$

Since the strong and σ-strong topology coincide on $\mathscr{L}(\mathfrak{H})_1$, this ends the proof.

Definition 2.4.20. Let $\mathfrak{M}$ be a von Neumann algebra, and ω a positive linear functional on $\mathfrak{M}$. If $\omega(\text{l.u.b.}_\alpha A_\alpha) = \text{l.u.b.}_\alpha \, \omega(A_\alpha)$ for all increasing nets $\{A_\alpha\}$ in $\mathfrak{M}_+$ with an upper bound then ω is defined to be *normal*.

Theorem 2.4.21. *Let ω be a state on a von Neumann algebra $\mathfrak{M}$ acting on a Hilbert space $\mathfrak{H}$. The following conditions are equivalent:*

(1) *ω is normal;*
(2) *ω is σ-weakly continuous;*
(3) *there exists a density matrix ρ, i.e., a positive trace-class operator ρ on $\mathfrak{H}$ with $\mathrm{Tr}(\rho) = 1$, such that*

$$\omega(A) = \mathrm{Tr}(\rho A).$$

PROOF. (3) $\Rightarrow$ (2) follows from Proposition 2.4.3 and (2) $\Rightarrow$ (1) from Lemma 2.4.19. Next we show (2) $\Rightarrow$ (3). If ω is σ-weakly continuous there exist sequences $\{\xi_n\}$, $\{\eta_n\}$ of vectors such that $\sum_n \|\xi_n\|^2 < \infty$, $\sum_n \|\eta_n\|^2 < \infty$, and $\omega(A) = \sum_n (\xi_n, A\eta_n)$. Define $\tilde{\mathfrak{H}} = \bigoplus_{n=1}^\infty \mathfrak{H}$ and introduce a representation π of $\mathfrak{M}$ on $\tilde{\mathfrak{H}}$ by

$\pi(A)(\bigoplus_n \psi_n) = \bigoplus_n (A\psi_n)$. Let $\xi = \bigoplus_n \xi_n$, $\eta = \bigoplus_n \eta_n$ and then $\omega(A) = (\xi, \pi(A)\eta)$. Since $\omega(A)$ is real for $A \in \mathfrak{M}_+$ we have

$$\begin{aligned} 4\omega(A) &= 2(\xi, \pi(A)\eta) + 2(\xi, \pi(A^*)\eta) \\ &= 2(\xi, \pi(A)\eta) + 2(\eta, \pi(A)\xi) \\ &= (\xi + \eta, \pi(A)(\xi + \eta)) - (\xi - \eta, \pi(A)(\xi - \eta)) \\ &\leq (\xi + \eta, \pi(A)(\xi + \eta)). \end{aligned}$$

Hence, by Theorem 2.3.19 there exists a positive $T \in \pi(\mathfrak{M})'$ with $0 \leq T \leq \mathbb{1}/2$ such that

$$\begin{aligned} (\xi, \pi(A)\eta) &= (T(\xi + \eta), \pi(A)T(\xi + \eta)) \\ &= (\psi, \pi(A)\psi). \end{aligned}$$

Now $\psi \in \mathfrak{H}$ has the form $\psi = \bigoplus_n \psi_n$, and therefore

$$\omega(A) = \sum_n (\psi_n, A\psi_n).$$

The right side of this relation can be used to extend ω to a σ-weakly continuous positive linear functional $\tilde{\omega}$ on $\mathscr{L}(\mathfrak{H})$. Since $\tilde{\omega}(\mathbb{1}) = 1$, it is a state. Thus, by Proposition 2.4.3 there exists a trace-class operator ρ with $\mathrm{Tr}(\rho) = 1$ such that

$$\tilde{\omega}(A) = \mathrm{Tr}(\rho A).$$

Let P be the rank one projector with range ξ; then

$$(\xi, \rho\xi) = \mathrm{Tr}(P\rho P) = \mathrm{Tr}(\rho P) = \tilde{\omega}(P) \geq 0.$$

Thus ρ is positive.

We now turn to the proof of (1) $\Rightarrow$ (2). Assume that ω is a normal state on $\mathfrak{M}$. Let $\{B_\alpha\}$ be an increasing net of elements in $\mathfrak{M}_+$ such that $\|B_\alpha\| \leq 1$ for all α and such that $A \mapsto \omega(AB_\alpha)$ is σ-strongly continuous for all α. We can use Lemma 2.4.19 to define B by

$$B = \underset{\alpha}{\mathrm{l.u.b.}}\, B_\alpha = \sigma\text{-strong} \lim_\alpha B_\alpha.$$

Then $0 \leq B \leq \mathbb{1}$ and $B \in \mathfrak{M}$. But for all $A \in \mathfrak{M}$ we have

$$\begin{aligned} |\omega(AB - AB_\alpha)|^2 &= |\omega(A(B - B_\alpha)^{1/2}(B - B_\alpha)^{1/2})|^2 \\ &\leq \omega(A(B - B_\alpha)A^*)\omega(B - B_\alpha) \\ &\leq \|A\|^2\omega(B - B_\alpha). \end{aligned}$$

Hence

$$\|\omega(\cdot B) - \omega(\cdot B_\alpha)\| \leq (\omega(B - B_\alpha))^{1/2}.$$

But ω is normal. Therefore $\omega(B - B_\alpha) \to 0$ and $\omega(\cdot B_\alpha)$ tends to $\omega(\cdot B)$ in norm. As $\mathfrak{M}_*$ is a Banach space, $\omega(\cdot B) \in \mathfrak{M}_*$. Now, applying Zorn's lemma, we can find a maximal element $P \in \mathfrak{M}_+ \cap \mathfrak{M}_1$ such that $A \mapsto \omega(AP)$ is σ-strongly continuous. If $P = \mathbb{1}$ the theorem is proved. So assume *ad absurdum* that $P \neq \mathbb{1}$. Put $P' = \mathbb{1} - P$ and choose an $\xi \in \mathfrak{H}$ such that $\omega(P') < (\xi, P'\xi)$. If $\{B_\alpha\}$ is an increasing net in $\mathfrak{M}_+$ such that $B_\alpha \leq P'$, $\omega(B_\alpha) \geq (\xi, B_\alpha\xi)$, and $B = \mathrm{l.u.b.}_\alpha\, B_\alpha = \sigma\text{-strong} \lim_\alpha B_\alpha$, then $B \in \mathfrak{M}_+$, $B \leq P'$, and $\omega(B) = \sup \omega(B_\alpha) \geq \sup(\xi, B_\alpha\xi) = (\xi, B\xi)$. Hence, by Zorn's lemma, there exists a maximal $B \in \mathfrak{M}_+$ such that $B \leq P'$ and $\omega(B) \geq (\xi, B\xi)$. Put $Q = P' - B$. Then $Q \in \mathfrak{M}_+$, $Q \neq 0$ (since $\omega(P') < (\xi, P'\xi)$), and if $A \in \mathfrak{M}_+$, $A \leq Q$, $A \neq 0$, then $\omega(A) < (\xi, A\xi)$ by the maximality of B.

For any $A \in \mathfrak{M}$ we have

$$QA^*AQ \leq \|A\|^2 Q^2 \leq \|A\|^2 \|Q\| Q.$$

Hence $(QA^*AQ)/\|A\|^2\|Q\| \leq Q$ and $\omega(QA^*AQ) < (\xi, QA^*AQ\xi)$. Combining this with the Cauchy–Schwarz inequality one finds

$$\begin{aligned} |\omega(AQ)|^2 &\leq \omega(\mathbb{1})\omega(QA^*AQ) \\ &< (\xi, QA^*AQ\xi) = \|AQ\xi\|^2. \end{aligned}$$

Thus both $A \mapsto \omega(AQ)$ and $A \mapsto \omega(A(P + Q))$ are σ-strongly continuous. Since $P + Q \leq \mathbb{1}$, this contradicts the maximality of P.

We note in passing that the notion of normal state can be used to give another abstract characterization of von Neumann algebras.

Theorem (Kadison). *If $\mathfrak{A}$ is a C^*-algebra the following two conditions are equivalent:*

(1) *$\mathfrak{A}$ is $*$-isomorphic with a von Neumann algebra;*
(2) *any bounded, increasing net of operators in $\mathfrak{A}$ has a least upper bound and for any positive nonzero element $A \in \mathfrak{A}$ there exists a normal state ω, over $\mathfrak{A}$, such that $\omega(A) \neq 0$.*

Any σ-weakly continuous linear functional is a linear combination of four σ-weakly continuous states by the polarization identity. Thus, Theorem 2.4.21 implies that the σ-weak topology is only dependent on the order structure on a von Neumann algebra and not on the particular Hilbert space representation. This implies that isomorphisms and homomorphisms between von Neumann algebras are automatically continuous in the σ-weak topology. Before stating the formal result, we need a characterization of the σ-weakly closed ideals of a von Neumann algebra.

Proposition 2.4.22. *Let $\mathfrak{M}$ be a von Neumann algebra and $\mathfrak{I}$ a σ-weakly closed two-sided ideal in $\mathfrak{M}$. It follows that there exists a projection $E \in \mathfrak{M} \cap \mathfrak{M}'$ such that $\mathfrak{I} = \mathfrak{M}E$.*

PROOF. Note first that $\mathfrak{I}$ is selfadjoint by the following argument. If $A \in \mathfrak{I}$ has polar decomposition $A = U|A|$ then $A^*A \in \mathfrak{I}$, and $|A| = (A^*A)^{1/2} \in \mathfrak{I}$. Thus $A^* = |A|U^* \in \mathfrak{I}$. Next, by Lemma 2.4.19, there exists a largest projection $E \in \mathfrak{I}$. If $\{E_\alpha\}$ is an approximate identity we may take $E = \sigma\text{-strong} \lim_\alpha E_\alpha$. It is then clear that E is an identity for $\mathfrak{I}$. Hence for $A \in \mathfrak{M}$ one has

$$AE = (AE)E = E(AE) = (EA)E = E(EA) = EA.$$

Thus $E \in \mathfrak{M}'$ and so $E \in \mathfrak{M} \cap \mathfrak{M}'$.

Theorem 2.4.23. *Let $\mathfrak{M}$ and $\mathfrak{N}$ be two von Neumann algebras and τ a $*$-homomorphism from $\mathfrak{M}$ onto $\mathfrak{N}$. It follows that τ is σ-weakly and σ-strongly continuous.*

PROOF. Let $\{A_\alpha\}$ be an increasing net in $\mathfrak{M}_+$ and define A by $A = \text{l.u.b.}_\alpha A_\alpha =$ σ-weak $\lim_\alpha A_\alpha$. Then because τ preserves positivity and is onto $\tau(A) = \text{l.u.b.}_\alpha \tau(A_\alpha)$ $= \sigma$-weak $\lim_\alpha \tau(A_\alpha)$. Hence if ω is a normal state on $\mathfrak{N}$ then $\omega \circ \tau$ is a normal state on $\mathfrak{M}$. Now any σ-weakly continuous functional is a linear combination of σ-weakly continuous states. It then follows from Theorem 2.4.21 that if ω is a σ-weakly continuous functional on $\mathfrak{N}$ then $\omega \circ \tau$ is σ-weakly continuous on $\mathfrak{M}$. Hence τ is σ-weakly continuous.

Next, if A_α converges σ-strongly to 0 then $A_\alpha{}^*A_\alpha$ converges σ-weakly to 0. Hence $\tau(A_\alpha)^*\tau(A_\alpha) = \tau(A_\alpha{}^*A_\alpha)$ converges σ-weakly to 0 and $\tau(A_\alpha)$ converges σ-strongly to 0.

Theorem 2.4.24. *Let $\mathfrak{M}$ be a von Neumann algebra, ω a normal state on $\mathfrak{M}$, and let $(\mathfrak{H}, \pi, \Omega)$ be the associated cyclic representation. It follows that $\pi(\mathfrak{M})$ is a von Neumann algebra and π is normal in the sense that $\pi(\text{l.u.b.}_\alpha A_\alpha) = \text{l.u.b.}_\alpha \pi(A_\alpha)$ for any bounded, increasing net $\{A_\alpha\}$ in $\mathfrak{M}_+$.*

PROOF. If $A_\alpha \nearrow A$ in $\mathfrak{M}$ then $\pi(A_\alpha)$ is increasing and $\pi(A_\alpha) \leq \pi(A)$ for all α. But since ω is normal, we have for any $B \in \mathfrak{M}$

$$\begin{aligned}
(\pi(B)\Omega, \pi(A)\pi(B)\Omega) &= \omega(B^*AB) \\
&= \omega\left(\text{l.u.b.}_\alpha\, B^*A_\alpha B\right) \\
&= \text{l.u.b.}_\alpha\, \omega(B^*A_\alpha B) \\
&= \text{l.u.b.}_\alpha\, (\pi(B)\Omega, \pi(A_\alpha)\pi(B)\Omega)
\end{aligned}$$

for all $B \in \mathfrak{M}$. But the set $\pi(\mathfrak{M})\Omega$ is norm dense in $\mathfrak{H}$ and so π is normal.

Proceeding as in the proof of Theorem 2.4.23 it follows that π is σ-weakly continuous as a map from $\mathfrak{M}$ into $\mathscr{L}(\mathfrak{H})$. Hence the kernel $\mathfrak{J}$ of π is a σ-weakly closed ideal in $\mathfrak{M}$. By Proposition 2.4.22 there exists a projection $E \in \mathfrak{M} \cap \mathfrak{M}'$ such that $\mathfrak{J} = \mathfrak{M}E$.

Hence π lifts to a faithful representation of the von Neumann algebra $\mathfrak{M}(\mathbb{1} - E)$ by $\pi(A(\mathbb{1} - E)) = \pi(A)$, and we may assume that π is faithful. Then by Proposition 2.3.3, π is isometric. Thus π maps $\mathfrak{M}_1$ onto $\pi(\mathfrak{M})_1$. But $\mathfrak{M}_1$ is σ-weakly compact and π is σ-weakly continuous. Thus $\pi(\mathfrak{M})_1$ is σ-weakly compact and, in particular, σ-weakly closed. By Theorem 2.4.11, $\pi(\mathfrak{M})$ is a von Neumann algebra.

2.4.4. Quasi-Equivalence of Representations

Earlier, at the end of Section 2.3.1, we introduced the concept of unitary equivalence of two representations of a C^*-algebra $\mathfrak{A}$. It follows from Theorem 2.3.16 that $(\mathfrak{H}_1, \pi_1)$ and $(\mathfrak{H}_2, \pi_2)$ are unitarily equivalent if, and only if, the unit vectors of $\mathfrak{H}_1$ and the unit vectors of $\mathfrak{H}_2$ define the same set of states of $\mathfrak{A}$. A slightly weaker but more natural concept of equivalence where physical applications are concerned is the concept of quasi-equivalence of two representations

Definition 2.4.25. If π is a representation of a C^*-algebra $\mathfrak{A}$, then a state ω of $\mathfrak{A}$ is said to be *π-normal* if there exists a normal state ρ of $\pi(\mathfrak{A})''$ such that

$$\omega(A) = \rho(\pi(A))$$

for all $A \in \mathfrak{A}$.

Two representations π_1 and π_2 of a C^*-algebra $\mathfrak{A}$ are said to be *quasi-equivalent*, written $\pi_1 \approx \pi_2$, if each π_1-normal state is π_2-normal and conversely.

If $(\mathfrak{H}, \pi)$ is a representation of a C^*-algebra $\mathfrak{A}$, and n is a cardinal, let $n\pi$ denote the representation of $\mathfrak{A}$ on $n\mathfrak{H} = \bigoplus_{k=1}^n \mathfrak{H}$ defined by

$$n\pi(A)\left(\bigoplus_{k=1}^{n} \xi_k\right) = \bigoplus_{k=1}^{n} (\pi(A)\xi_k).$$

We have already proved in Lemma 2.4.12 that $(n\pi(\mathfrak{A}))''$ is isomorphic to $\pi(\mathfrak{A})''$ by an isomorphism which extends $n\pi(A) \mapsto \pi(A)$, $A \in \mathfrak{A}$.

The next theorem shows among other things that quasi-equivalence is the same as unitary equivalence up to multiplicity.

Theorem 2.4.26. *Let $\mathfrak{A}$ be a C^*-algebra and let $(\mathfrak{H}_1, \pi_1)$ and $(\mathfrak{H}_2, \pi_2)$ be nondegenerate representations of $\mathfrak{A}$. The following conditions are equivalent:*

(1) *there exists an isomorphism $\tau : \pi_1(\mathfrak{A})'' \mapsto \pi_2(\mathfrak{A})''$ such that $\tau(\pi_1(A)) = \pi_2(A)$ for all $A \in \mathfrak{A}$;*

(2) *$\pi_1 \approx \pi_2$, i.e., the π_1-normal and the π_2-normal states are the same;*

(3) *there exist cardinals n, m, projections $E_1' \in n\pi_1(\mathfrak{A})'$, $E_2' \in m\pi_2(\mathfrak{A})'$ and unitary elements $U_1 : \mathfrak{H}_1 \mapsto E_2'(m\mathfrak{H}_2)$, $U_2 : \mathfrak{H}_2 \mapsto E_1'(n\mathfrak{H}_1)$ such that*

$$U_1\pi_1(A)U_1{}^* = m\pi_2(A)E_2',$$
$$U_2\pi_2(A)U_2{}^* = n\pi_1(A)E_1'$$

for all $A \in \mathfrak{A}$;

(4) *There exists a cardinal n such that $n\pi_1 \simeq n\pi_2$, i.e., π_1 and π_2 are unitarily equivalent up to multiplicity.*

Remark. This theorem really contains two distinct ideas; one is contained in the equivalence (1) $\Leftrightarrow$ (2) and concerns the representations π_1 and π_2, while the other is contained in (1) $\Leftrightarrow$ (3) $\Leftrightarrow$ (4) and concerns the structure of isomorphisms between von Neumann algebras, in particular the question of when isomorphisms are unitarily implemented. For example, if π_1 and π_2 are irreducible and quasi-equivalent, the isomorphism τ is unitarily implemented and π_1 and π_2 are unitarily equivalent. Analogously, if both $\pi_1(\mathfrak{A})''$ and $\pi_2(\mathfrak{A})''$ have a separating and cyclic vector, the isomorphism τ is unitarily implemented by Corollary 2.5.32, and again π_1 and π_2 are unitarily equivalent.

PROOF OF THEOREM 2.4.26. (1) ⇒ (2) is an immediate consequence of Theorem 2.4.23.

(2) ⇒ (3) By Proposition 2.3.6, π_1 is a direct sum of cyclic representations, i.e., we can find a set $\{\xi_\alpha\}$ of unit vectors in $\mathfrak{H}$ such that $[\pi_1(\mathfrak{A})\xi_\alpha]$ are mutually orthogonal, and $\sum_\alpha [\pi_1(\mathfrak{A})\xi_\alpha] = \mathbb{1}$. Let

$$\omega_\alpha(A) = (\xi_\alpha, \pi_1(A)\xi_\alpha), \qquad A \in \mathfrak{A},$$

be the states corresponding to ξ_α. By assumption, ω_α is π_2-normal for each α. Hence, by Theorem 2.4.21, there exists a sequence $\{\eta_{\alpha,n}\}$ in $\mathfrak{H}_2$ such that $\sum_n \|\eta_{\alpha,n}\|^2 = 1$ and

$$\omega_\alpha(A) = \sum_n (\eta_{\alpha,n}, \pi_2(A)\eta_{\alpha,n})$$

$$= (\eta_\alpha, \aleph_0\pi_2(A)\eta_\alpha)$$

for $A \in \mathfrak{A}$, where $\eta_\alpha = \bigoplus_n \eta_{\alpha,n} \in \aleph_0\mathfrak{H}_2$. Thus, by Theorem 2.3.16, there exists a unitary $U_\alpha\colon [\pi_1(\mathfrak{A})\xi_\alpha] \mapsto [\aleph_0\pi_2(\mathfrak{A})\eta_\alpha]$ such that

$$U_\alpha\pi_1(A)U_\alpha{}^* = \aleph_0\pi_2(A)[\aleph_0\pi_2(\mathfrak{A})\eta_\alpha].$$

If k is the cardinality of $\{\alpha\}$, we obtain, by addition, an isometry $U_1 = \sum_\alpha U_\alpha$ from $\mathfrak{H}_1 = \bigoplus_\alpha [\pi_1(\mathfrak{A})\xi_\alpha]$ to $k\aleph_0\mathfrak{H}_2 = \bigoplus_\alpha (\aleph_0\mathfrak{H}_2)$ with range

$$\bigoplus_\alpha [\aleph_0\pi_2(\mathfrak{A})\eta_\alpha] = E_2{}'\left(\bigoplus_\alpha (\aleph_0\mathfrak{H}_2)\right)$$

such that

$$U_1\pi_1(A)U_1{}^* = k\aleph_0\pi_2(A)E_2{}'.$$

This establishes the first half of (3); the second half follows by interchanging the roles of π_1 and π_2.

(3) ⇒ (4) We may choose n and m in (3) to be infinite, i.e., setting $k = \sup\{n, m\}$ we have that $kn = km = k$. Then it follows from (3) that $k\pi_1$ is unitarily equivalent to a subrepresentation of $mk\pi_2 = k\pi_2$ and $k\pi_2$ is unitarily equivalent to a subrepresentation of $nk\pi_1 = k\pi_1$. A Cantor–Bernstein argument then implies that $k\pi_1$ and $k\pi_2$ are unitarily equivalent.

(4) ⇒ (1) Follows immediately from Lemma 2.4.12.

A state ω of a C^*-algebra $\mathfrak{A}$ is called a *primary state*, or a *factor state*, if $\pi_\omega(\mathfrak{A})''$ is a factor, where π_ω is the associated cyclic representation. Two states ω_1 and ω_2 of $\mathfrak{A}$ are said to be *quasi-equivalent* if π_{ω_1} and π_{ω_2} are quasi-equivalent (in the abelian case this is the same as equivalence of the probability measures corresponding to ω_1 and ω_2). The next proposition is useful in applications to quasi-local algebras (see Section 2.6).

Proposition 2.4.27. *Let ω_1 and ω_2 be factor states of a C^*-algebra $\mathfrak{A}$. It follows that ω_1 and ω_2 are quasi-equivalent if, and only if, $\frac{1}{2}(\omega_1 + \omega_2)$ is a factor state.*

PROOF. Let $(\mathfrak{H}_i, \pi_i, \Omega_i)$ be the cyclic representations defined by ω_i; put $\mathfrak{H} = \mathfrak{H}_1 \oplus \mathfrak{H}_2, \Omega = (1/\sqrt{2})(\Omega_1 \oplus \Omega_2), \pi = \pi_1 \oplus \pi_2$, and $\omega = \frac{1}{2}(\omega_1 + \omega_2)$. It follows that $\omega(A) = (\Omega, \pi(A)\Omega)$ for $A \in \mathfrak{A}$, i.e., $(\mathfrak{H}_\omega, \pi_\omega, \Omega_\omega)$ identifies with the subrepresentation of $(\mathfrak{H}, \pi)$ determined by the projection $E' = [\pi(\mathfrak{A})\Omega] \in \pi(\mathfrak{A})'$.

Assume first that π_1 and π_2 are quasi-equivalent. By Theorem 2.4.26 there exists an isomorphism τ; $\pi_1(\mathfrak{A})'' \mapsto \pi_2(\mathfrak{A})''$ such that $\tau(\pi_1(A)) = \pi_2(A)$. Since any element in $\pi(\mathfrak{A})''$ is a σ-weak limit of elements of the form $\pi_1(A) \oplus \pi_2(A) = \pi_1(A) \oplus \tau(\pi_1(A))$, and τ is σ-weakly continuous, it follows that $\pi_1(\mathfrak{A})''$ is isomorphic to $\pi(\mathfrak{A})''$ by the isomorphism $A \mapsto A \oplus \tau(A)$. Hence $\pi(\mathfrak{A})''$ is a factor. But $A \mapsto AE'$ is a σ-weakly* continuous homomorphism from $\pi(\mathfrak{A})''$ onto $\pi_\omega(\mathfrak{A})''$, and $\pi(\mathfrak{A})''$ has no nontrivial σ-weakly closed ideals, by Proposition 2.4.22. It follows that $\pi_\omega(\mathfrak{A})''$ is isomorphic to $\pi(\mathfrak{A})''$, hence $\pi_\omega(\mathfrak{A})''$ is a factor.

Conversely, assume that π_1 and π_2 are not quasi-equivalent and let $C \in \pi(\mathfrak{A})'$. Let E be the orthogonal projection from $\mathfrak{H}_1 \oplus \mathfrak{H}_2$ onto $\mathfrak{H}_1 \oplus \{0\}$. We will show that $CE = EC$. Since $\pi(\mathfrak{A})'$ is a von Neumann algebra, it is enough to show that $(\mathbb{1} - E)CE = 0$. Assume this is not the case, and let U be the partial isometry in the polar decomposition of $(\mathbb{1} - E)CE$. Then $U \in \pi(\mathfrak{A})'$ since $E, C \in \pi(\mathfrak{A})'$ and $UE = (\mathbb{1} - E)U = U$. Hence $U\pi_1(A) = \pi_2(A)U$, i.e., U establishes a unitary equivalence between a subrepresentation of π_1 and a subrepresentation of π_2.

Since any subrepresentation of a factor representation is quasi-equivalent with the representation itself (by Theorem 2.4.26(1) and Proposition 2.4.22) it follows that $\pi_1 \approx \pi_2$, which is a contradiction. Hence $CE = EC$ for any $C \in \pi(\mathfrak{A})'$, and thus $E \in \pi(\mathfrak{A})''$ i.e., $E \in \pi(\mathfrak{A})'' \cap \pi(\mathfrak{A})'$. It follows that EE' is a nontrivial element in the center of $\pi_\omega(\mathfrak{A})'' = \pi(\mathfrak{A})''E'$ (nontrivial since $E'(\mathfrak{H}_1 \oplus \{0\}) = [\pi_1(\mathfrak{A})\Omega_1] \oplus \{0\} \neq \{0\}$ and $E'(\{0\} \oplus \mathfrak{H}_2) \neq \{0\}$, thus $0 \neq EE' \neq E'$). Hence $\omega = \frac{1}{2}(\omega_1 + \omega_2)$ is not a factor state.

2.5. Tomita–Takesaki Modular Theory and Standard Forms of von Neumann Algebras

Theorem 2.1.11(a) established that an abelian von Neumann algebra $\mathfrak{M}$ has the form $C(X)$ for a compact Hausdorff space X. If ω is a normal state on $\mathfrak{M}$ then the Riesz representation theorem implies the existence of a probability measure μ on X such that $\omega(A) = \int A(x)\, d\mu(x)$ for $A \in \mathfrak{M}$. It follows immediately that if $(\mathfrak{H}, \pi, \Omega)$ is the cyclic representation associated with ω then $\mathfrak{H}$ is identifiable with $L^2(X, \mu)$, Ω with the function which is identically equal to 1, and $\pi(\mathfrak{M})$ with $L^\infty(X, \mu)$ acting as multiplication operators on $L^2(X, \mu)$. Suppose that the support of μ equals X. The predual $\mathfrak{M}_*$ is then identified as $L^1(X, \mu)$. In particular, a positive functional $\rho \in \mathfrak{M}_*$ is represented by a unique positive function in L^1, which is again a square of a unique positive function in L^2. This establishes a one-to-one correspondence between the positive normal states and the positive functions $L^2{}_+$ in L^2 $(=\mathfrak{H})$. This correspondence can then be used to define another one-to-one correspondence between the automorphisms of $\mathfrak{M}$ and the unitary operators on L^2 which map the positive functions onto the positive functions.

Let us briefly examine the structure of the positive elements $L^2{}_+$ of L^2. These elements form a closed convex cone which is self-dual in the sense that the inequality

$$\int d\mu\; fg \geq 0$$

is valid for all $f \in L^2{}_+$ if, and only if, $g \in L^2{}_+$. An abstract description of this cone is given by remarking that the algebra $\mathfrak{M}$ is a C^*-algebra and its positive elements form a uniformly closed convex cone (Proposition 2.2.11). Each element of this cone is of the form A^*A, with $A \in \mathfrak{M}$, the L^2-representative of $\pi(A^*A)\Omega$ is positive, and $L^2{}_+$ is the weak closure of such vectors. The self-duality then arises because

$$\begin{aligned}(\pi(A^*A)\Omega, \pi(B^*B)\Omega) &= \omega(A^*AB^*B)\\ &= \omega((AB)^*AB) \geq 0.\end{aligned}$$

The second step uses the commutativity of $\mathfrak{M}$.

In this section we assign a similar structure to a general von Neumann algebra $\mathfrak{M}$ with a faithful (see Definition 2.5.4) normal state ω. Namely, we construct a Hilbert space $\mathfrak{H}$ with a "positive self-dual cone" $\mathscr{P}$ such that the positive elements in $\mathfrak{M}_*$ correspond to vectors in $\mathscr{P}$ and automorphisms of $\mathfrak{M}$ correspond to unitary elements in $\mathfrak{H}$ which leave $\mathscr{P}$ invariant. Although the general description of $\mathscr{P}$ is similar to the abstract description for abelian $\mathfrak{M}$ there is an essential difference which arises from noncommutativity. Let $\mathfrak{M}$ act on $\mathfrak{H}$ and assume Ω is cyclic for $\mathfrak{M}$. One can form the convex cone $A^*A\Omega$ of vectors in $\mathfrak{H}$ but if $\mathfrak{M}$ is not abelian this cone does not necessarily have the property of self-duality. There is no reason why the associated state $\omega(A) = (\Omega, A\Omega)$ should satisfy $\omega(A^*AB^*B) \geq 0$. This property is, however, valid if ω is a trace, i.e., if $\omega(AB) = \omega(BA)$ for all $A, B \in \mathfrak{M}$, and it is worth comparing this latter situation with the abelian case.

Define the conjugation operator J, on $\mathfrak{H}$, by

$$JA\Omega = A^*\Omega.$$

The trace property of ω gives

$$\|A\Omega\|^2 = \omega(A^*A) = \omega(AA^*) = \|A^*\Omega\|^2$$

and hence J extends to a well-defined antiunitary operator. Moreover,

$$JAJB\Omega = JAB^*\Omega = BA^*\Omega.$$

For $\mathfrak{M}$ abelian this calculation shows that J implements the *-conjugation, i.e., $A^* = j(A)$, where we have defined j by $j(A) = JAJ$. In the trace situation the action of j is more complex. One has, for example,

$$\begin{aligned}(B_1\Omega, A_1 j(A_2)B_2\Omega) &= \omega(B_1{}^*A_1B_2A_2{}^*)\\ &= (B_1\Omega, j(A_2)A_1B_2\Omega)\end{aligned}$$

and this demonstrates that $j(A) \in \mathfrak{M}'$, a property which is of course shared by the abelian case.

The example of a trace indicates that the general self-dual cone should be constructed by modification of the *-conjugation in the set $AA^*\Omega$. The A^* should be replaced by an alternative conjugate element $j(A)$ and the conjugation j should be expected to provide a map from $\mathfrak{M}$ to $\mathfrak{M}'$. Examination of the map $A\Omega \mapsto A^*\Omega$ is the starting point of the Tomita–Takesaki theory which we consider in Section 2.5.2. Prior to this we introduce, in Section 2.5.1, the class of algebras that are analysed in the sequel and which are of importance in applications.

2.5.1. σ-Finite von Neumann Algebras

All the von Neumann algebras encountered in quantum statistical mechanics and quantum field theory fall in the following class.

Definition 2.5.1. A von Neumann algebra $\mathfrak{M}$ is *σ-finite* if all collections of mutually orthogonal projections have at most a countable cardinality.

Note in particular that a von Neumann algebra on a separable Hilbert space is σ-finite. The converse is not, however, valid, i.e., not all σ-finite von Neumann algebras can be represented on a separable Hilbert space.

Definition 2.5.2. Let $\mathfrak{M}$ be a von Neumann algebra on a Hilbert space $\mathfrak{H}$. A subset $\mathfrak{K} \subseteq \mathfrak{H}$ is *separating for* $\mathfrak{M}$ if for any $A \in \mathfrak{M}$, $A\xi = 0$ for all $\xi \in \mathfrak{K}$ implies $A = 0$.

Recall that a subset $\mathfrak{K} \subseteq \mathfrak{H}$ is cyclic for $\mathfrak{M}$ if $[\mathfrak{M}\mathfrak{K}] = \mathfrak{H}$. There is a dual relation between the properties of cyclic for the algebra and separating for the commutant.

Proposition 2.5.3. *Let $\mathfrak{M}$ be a von Neumann algebra on $\mathfrak{H}$ and $\mathfrak{K} \subseteq \mathfrak{H}$ a subset. The following conditions are equivalent:*

(1) *$\mathfrak{K}$ is cyclic for $\mathfrak{M}$;*
(2) *$\mathfrak{K}$ is separating for $\mathfrak{M}'$.*

PROOF. (1) $\Rightarrow$ (2) Assume that $\mathfrak{K}$ is cyclic for $\mathfrak{M}$ and choose $A' \in \mathfrak{M}'$ such that $A'\mathfrak{K} = \{0\}$. Then for any $B \in \mathfrak{M}$ and $\xi \in \mathfrak{K}$, $A'B\xi = BA'\xi = 0$, hence $A'[\mathfrak{M}\mathfrak{K}] = 0$ and $A' = 0$.

(2) $\Rightarrow$ (1) Suppose that $\mathfrak{K}$ is separating for $\mathfrak{M}'$ and set $P' = [\mathfrak{M}\mathfrak{K}]$. P' is then a projection in $\mathfrak{M}'$ and $(\mathbb{1} - P')\mathfrak{K} = \{0\}$. Hence $\mathbb{1} - P' = 0$ and $[\mathfrak{M}\mathfrak{K}] = \mathfrak{H}$.

Definition 2.5.4. A state ω on a von Neumann algebra $\mathfrak{M}$ is *faithful* if $\omega(A) > 0$ for all nonzero $A \in \mathfrak{M}_+$.

EXAMPLE 2.5.5. Let $\mathfrak{M} = \mathscr{L}(\mathfrak{H})$ with $\mathfrak{H}$ separable. Every normal state ω over $\mathfrak{M}$ is of the form

$$\omega(A) = \mathrm{Tr}(\rho A),$$

where ρ is a density matrix. If ω is faithful then $\omega(E) > 0$ for each rank one projector, i.e., $\|\rho^{1/2}\psi\| > 0$ for each $\psi \in \mathfrak{H}\backslash\{0\}$. Thus ρ is invertible (in the densely defined self-adjoint operators on $\mathfrak{H}$). Conversely, if ω is not faithful then $\omega(A^*A) = 0$ for some nonzero A and hence $\|\rho^{1/2}A^*\psi\| = 0$ for all $\psi \in \mathfrak{H}$, i.e., ρ is not invertible. This establishes that ω is faithful if, and only if, ρ is invertible. Remark that if $\mathfrak{H}$ is non-separable then ρ can have at most a countable number of nonzero eigenvalues and hence $\omega(A)$ must vanish for some positive A, i.e., ω is not faithful. Thus $\mathscr{L}(\mathfrak{H})$ is σ-finite, i.e., $\mathfrak{H}$ is separable if, and only if, $\mathscr{L}(\mathfrak{H})$ has a faithful normal state.

The next proposition gives a characterization of σ-finite von Neumann algebras.

Proposition 2.5.6. *Let $\mathfrak{M}$ be a von Neumann algebra on a Hilbert space $\mathfrak{H}$. Then the following four conditions are equivalent:*

(1) *$\mathfrak{M}$ is σ-finite;*
(2) *there exists a countable subset of $\mathfrak{H}$ which is separating for $\mathfrak{M}$;*
(3) *there exists a faithful normal state on $\mathfrak{M}$;*
(4) *$\mathfrak{M}$ is isomorphic with a von Neumann algebra $\pi(\mathfrak{M})$ which admits a separating and cyclic vector.*

PROOF. (1) $\Rightarrow$ (2) Let $\{\xi_\alpha\}$ be a maximal family of vectors in $\mathfrak{H}$ such that $[\mathfrak{M}'\xi_\alpha]$ and $[\mathfrak{M}'\xi_{\alpha'}]$ are orthogonal whenever $\alpha \neq \alpha'$. Since $[\mathfrak{M}'\xi_\alpha]$ is a projection in $\mathfrak{M}$ (in fact the smallest projection in $\mathfrak{M}$ containing ξ_α), $\{\xi_\alpha\}$ is countable. But by the maximality,

$$\sum_\alpha [\mathfrak{M}'\xi_\alpha] = \mathbb{1}.$$

Thus $\{\xi_\alpha\}$ is cyclic for $\mathfrak{M}'$. Hence $\{\xi_\alpha\}$ is separating for $\mathfrak{M}$ by Proposition 2.5.3.

(2) $\Rightarrow$ (3) Choose a sequence ξ_n such that the set $\{\xi_n\}$ is separating for $\mathfrak{M}$ and such that $\sum_n \|\xi_n\|^2 = 1$. Define ω by

$$\omega(A) = \sum_n (\xi_n, A\xi_n).$$

ω is σ-weakly continuous, hence normal (Theorem 2.4.21). If $\omega(A^*A) = 0$ then $0 = (\xi_n, A^*A\xi_n) = \|A\xi_n\|^2$ for all n, hence $A = 0$.

(3) $\Rightarrow$ (4) Let ω be a faithful normal state on $\mathfrak{M}$ and $(\mathfrak{H}, \pi, \Omega)$ the corresponding cyclic representation. By Theorem 2.4.24, $\pi(\mathfrak{M})$ is a von Neumann algebra. If $\pi(A)\Omega = 0$ for an $A \in \mathfrak{M}$ then $\omega(A^*A) = \|\pi(A)\Omega\|^2 = 0$, hence $A^*A = 0$ and $A = 0$. This proves that π is faithful and that Ω is separating for $\pi(\mathfrak{M})$.

(4) $\Rightarrow$ (1) Let Ω be the separating (and cyclic) vector for $\pi(\mathfrak{M})$, and let $\{E_\alpha\}$ be a family of mutually orthogonal projections in $\mathfrak{M}$. Set $E = \sum_\alpha E_\alpha$. Then

$$\begin{aligned} \|\pi(E)\Omega\|^2 &= (\pi(E)\Omega, \pi(E)\Omega) \\ &= \sum_{\alpha\alpha'} (\pi(E_\alpha)\Omega, \pi(E_{\alpha'})\Omega) \\ &= \sum_\alpha \|\pi(E_\alpha)\Omega\|^2 \end{aligned}$$

by Lemma 2.4.19. Since $\sum_\alpha \|\pi(E_\alpha)\Omega\|^2 < +\infty$, only a countable number of the $\pi(E_\alpha)\Omega$ is nonzero, and thus the same is true for the E_α.

2.5.2. The Modular Group

If $\mathfrak{M}$ is a σ-finite von Neumann algebra we may assume, by Proposition 2.5.6, that $\mathfrak{M}$ has a separating and cyclic vector Ω. The mapping $A \in \mathfrak{M} \mapsto A\Omega \in \mathfrak{H}$ then establishes a one-to-one linear correspondence between $\mathfrak{M}$ and a dense subspace $\mathfrak{M}\Omega$ of $\mathfrak{H}$. This correspondence may be used to transfer algebraic operations on $\mathfrak{M}$ to operations on $\mathfrak{M}\Omega$. In this section we study the antilinear operator S_0 on $\mathfrak{M}\Omega$ which comes from the *operation on $\mathfrak{M}$ and various operators associated with S_0.

Before starting with the proper subject of this section we need a definition and a lemma.

Definition 2.5.7. Let $\mathfrak{M}$ be a von Neumann algebra on a Hilbert space $\mathfrak{H}$. A closed operator A on $\mathfrak{H}$ is said to be *affiliated with* $\mathfrak{M}$, written $A\eta\mathfrak{M}$, if $\mathfrak{M}'D(A) \subseteq D(A)$ and $AA' \supseteq A'A$ for all $A' \in \mathfrak{M}'$.

A relation between elements of the algebra and unbounded affiliated operators is provided by the following:

Lemma 2.5.8. *Assume that A is a closed operator affiliated with a von Neumann algebra $\mathfrak{M}$. If $A = U|A|$ is the polar decomposition of A, then U and the spectral projections of $|A|$ lie in $\mathfrak{M}$.*

PROOF. Let U' be a unitary element in $\mathfrak{M}'$. Then

$$U'UU'^*U'|A|U'^* = U'U|A|U'^* = U|A|,$$

so by the uniqueness of the polar decomposition

$$U'UU'^* = U$$

and

$$U'|A|U'^* = |A|.$$

Hence $U \in \mathfrak{M}$. If

$$|A| = \int_0^\infty \lambda \, dE(\lambda)$$

is the spectral decomposition of $|A|$, then by the second relation above

$$\int_0^\infty \lambda U' \, dE(\lambda) \, U'^* = U' \int_0^\infty \lambda \, dE(\lambda) \, U'^* = \int_0^\infty \lambda \, dE(\lambda).$$

By the uniqueness of the spectral decomposition of $|A|$ it follows that

$$U'E(\lambda)U'^* = E(\lambda),$$

i.e., $E(\lambda) \in \mathfrak{M}$ for all $\lambda \geq 0$.

Now we return to the study of the antilinear operator S_0.

In the introduction we examined the special example of a trace state. In the trace case the operator J corresponding to S_0 was antiunitary and $J\mathfrak{M}J \subseteq \mathfrak{M}'$. In particular, $J\mathfrak{M}\Omega \subseteq \mathfrak{M}'\Omega$. But one also has $\mathfrak{M}\Omega = J(J\mathfrak{M}\Omega) \subseteq J\mathfrak{M}'\Omega$. Thus in examination of J, or S_0, it is natural to define a supplementary operator on $\mathfrak{M}'\Omega$ which should correspond to inversion of the conjugation. Thus we actually study two antilinear operators, S_0 and F_0, one associated with $\mathfrak{M}\Omega$ and the other with $\mathfrak{M}'\Omega$.

Proposition 2.5.3 established that if Ω is cyclic and separating for $\mathfrak{M}$ then it is also cyclic and separating for $\mathfrak{M}'$. Therefore the two antilinear operators S_0 and F_0, given by

$$S_0 A\Omega = A^*\Omega$$

for $A \in \mathfrak{M}$ and

$$F_0 A'\Omega = A'^*\Omega$$

for $A' \in \mathfrak{M}'$, are both well defined on the dense domains $D(S_0) = \mathfrak{M}\Omega$ and $D(F_0) = \mathfrak{M}'\Omega$.

Proposition 2.5.9. *Adopt the foregoing definitions. It follows that S_0 and F_0 are closable and*

$$S_0{}^* = \bar{F}_0, \qquad F_0{}^* = \bar{S}_0,$$

where the bar denotes closure. Moreover, for any $\psi \in D(\bar{S}_0)$ there exists a closed operator Q, on $\mathfrak{H}$, such that

$$Q\Omega = \psi, \qquad Q^*\Omega = \bar{S}_0\psi,$$

and Q is affiliated with $\mathfrak{M}$. The corresponding result is also true for $\bar{F}_0$.

PROOF. For $A \in \mathfrak{M}$, $A' \in \mathfrak{M}'$ we have

$$\begin{aligned}(A'\Omega, S_0 A\Omega) &= (A'\Omega, A^*\Omega)\\ &= (A\Omega, A'^*\Omega)\\ &= (A\Omega, F_0 A'\Omega).\end{aligned}$$

Thus $F_0 \subseteq S_0{}^*$. Hence $S_0{}^*$ is densely defined and S_0 is closable. Analogously, $S_0 \subseteq F_0{}^*$.

To show that $S_0{}^*$ is actually the closure of F_0 we first pick a $\xi \in D(S_0{}^*)$ and set $\psi = S_0{}^*\xi$. Then for $A \in \mathfrak{M}$ one has

$$\begin{aligned}(A\Omega, \psi) &= (A\Omega, S_0{}^*\xi)\\ &= (\xi, S_0 A\Omega) = (\xi, A^*\Omega).\end{aligned}$$

Next define operators Q_0 and Q_0^+ by

$$Q_0 \colon A\Omega \mapsto A\xi,$$

$$Q_0^+ : A\Omega \mapsto A\psi$$

for all $A \in \mathfrak{M}$. The foregoing relation then establishes that

$$\begin{aligned}(B\Omega, Q_0 A\Omega) &= (B\Omega, A\xi) = (A^*B\Omega, \xi)\\ &= (\psi, B^*A\Omega) = (B\psi, A\Omega) = (Q_0{}^+ B\Omega, A\Omega)\end{aligned}$$

for all $A, B \in \mathfrak{M}$. Hence $Q_0{}^+ \subseteq Q_0{}^*$ and Q_0 is closable. Let $Q' = \overline{Q_0}$.

If $A, B \in \mathfrak{M}$ then

$$Q_0 AB\Omega = AB\xi = AQ_0 B\Omega.$$

Hence by closure

$$Q'A \supseteq AQ',$$

i.e., A maps $D(Q')$ into $D(Q')$ and commutes with Q' on $D(Q')$. Hence if $Q' = U'|Q'|$ is the polar decomposition of Q' then $U' \in \mathfrak{M}'$ and all the spectral projections of $|Q'|$ lie in $\mathfrak{M}'$ (Lemma 2.5.8). Let $E_n' \in \mathfrak{M}'$ be the spectral projection of $|Q'|$ corresponding to the interval $[0, n]$ and set

$$Q_n' = U'E_n'|Q'|.$$

It follows that $Q_n' \in \mathfrak{M}'$ and

$$\begin{aligned} Q_n'\Omega &= U'E_n'|Q'|\Omega = U'E_n'U'^*U'|Q'|\Omega \\ &= U'E_n'U'^*Q_0\Omega = U'E_n'U'^*\xi. \end{aligned}$$

Moreover,

$$Q_n'^*\Omega = E_n'|Q'|U'^*\Omega = E_n'Q_0^+\Omega = E_n'\psi.$$

Hence $U'E_n'U'^*\xi \in D(F_0)$ and

$$F_0(U'E_n'U'^*\xi) = E_n'\psi.$$

Now, E_n' converges strongly to the identity $\mathbb{1}$ and $U'U'^*$ is the projector with range equal to the range of Q'. This set contains $\xi = \mathbb{1}\xi$. Hence $\xi \in D(\bar{F}_0)$ and $\bar{F}_0\xi = \psi = S_0^*\xi$. Thus $S_0^* \subseteq \bar{F}_0 \subseteq S_0^*$ and $\bar{F}_0 = S_0^*$. Interchanging S_0 and F_0 in this argument yields $\bar{S}_0 = F_0^*$.

Definition 2.5.10. Define S and F as the closures of S_0 and F_0, respectively, i.e.,

$$S = \bar{S}_0, \qquad F = \bar{F}_0.$$

Let Δ be the unique, positive, selfadjoint operator and J the unique anti-unitary operator occurring in the polar decomposition

$$S = J\Delta^{1/2}$$

of S. Δ is called the *modular operator associated with the pair* $\{\mathfrak{M}, \Omega\}$ and J is called the *modular conjugation.*

There are various relatively straightforward connections between the operators S, F, Δ, and J.

Proposition 2.5.11. *The following relations are valid:*

$$\Delta = FS, \qquad \Delta^{-1} = SF,$$

$$S = J\Delta^{1/2}, \qquad F = J\Delta^{-1/2},$$

$$J = J^*, \qquad J^2 = \mathbb{1},$$

$$\Delta^{-1/2} = J\Delta^{1/2}J.$$

PROOF. $\Delta = S^*S = FS$, and $S = J\Delta^{1/2}$ by Definition 2.5.10 and Proposition 2.5.9. Since $S_0 = S_0^{-1}$ it follows by closure that $S = S^{-1}$, and hence

$$J\Delta^{1/2} = S = S^{-1} = \Delta^{-1/2}J^*,$$

so that $J^2\Delta^{1/2} = J\Delta^{-1/2}J^*$. Clearly, $J\Delta^{-1/2}J^*$ is a positive operator, and by the uniqueness of the polar decomposition one deduces that

$$J^2 = \mathbb{1}$$

and then

$$J^* = J, \qquad \Delta^{-1/2} = J\Delta^{1/2}J.$$

But this implies that

$$F = S^* = (\Delta^{-1/2}J)^* = J\Delta^{-1/2}$$

and

$$SF = \Delta^{-1/2}JJ\Delta^{-1/2} = \Delta^{-1}.$$

Before proceeding let us again consider the abelian and trace examples discussed in the Introduction. In both these examples Ω is separating because cyclicity and the trace condition $\omega(B^*A^*AB) = \omega(BB^*A^*A)$ establish that $A\Omega = 0$ is equivalent to $A = 0$. One checks that $\Delta = \mathbb{1}$, $S = F = J$ and, moreover,

$$J\mathfrak{M}J \subseteq \mathfrak{M}', \qquad J\mathfrak{M}'J \subseteq \mathfrak{M},$$

thus

$$J\mathfrak{M}J = \mathfrak{M}'.$$

Thus the modular operator Δ reflects in some manner the nontracial character of ω. Although these examples provide no direct insight into the possible properties of Δ, one can infer the structural characteristics of Δ from the following discussion.

Consider the action of the operator SAS, with $A \in \mathfrak{M}$. For each pair B, $C \in \mathfrak{M}$ one has

$$(SAS)BC\Omega = SAC^*B^*\Omega = BCA^*\Omega$$

and

$$B(SAS)C\Omega = BSAC^*\Omega = BCA^*\Omega.$$

Thus SAS is affiliated with $\mathfrak{M}'$.

Now suppose that Δ is bounded, and thus $\Delta^{-1} = J\Delta J$, S, and F are bounded. Then by the above reasoning

$$S\mathfrak{M}S \subseteq \mathfrak{M}', \qquad F\mathfrak{M}'F \subseteq \mathfrak{M}.$$

Thus

$$\begin{aligned}\Delta\mathfrak{M}\Delta^{-1} &= \Delta^{1/2}JJ\Delta^{1/2}\mathfrak{M}\Delta^{-1/2}JJ\Delta^{-1/2} \\ &= FS\mathfrak{M}SF \subseteq F\mathfrak{M}'F \subseteq \mathfrak{M},\end{aligned}$$

and by iteration

$$\Delta^n\mathfrak{M}\Delta^{-n} \subseteq \mathfrak{M}$$

for $n = 0, 1, 2, \ldots$. Now for $A \in \mathfrak{M}$, $A' \in \mathfrak{M}'$, $\varphi, \psi \in \mathfrak{H}$, consider the entire analytic function

$$f(z) = \|\Delta\|^{-2z}(\varphi, [\Delta^z A \Delta^{-z}, A']\psi).$$

Then $f(z) = 0$ for $z = 0, 1, 2, 3, \ldots$, and we have the estimate (using $\|\Delta^{-1}\| = \|J\Delta J\| = \|\Delta\|$)

$$|f(z)| = O(\|\Delta\|^{-2\,\mathrm{Re}\,z}(\|\Delta\|^{|\mathrm{Re}\,z|})^2) = O(1)$$

for $\mathrm{Re}\,z \geq 0$. By Carlson's theorem it follows that $f(z) = 0$ for all $z \in \mathbb{C}$. Hence

$$\Delta^z \mathfrak{M} \Delta^{-z} \subseteq \mathfrak{M}'' = \mathfrak{M}$$

for all $z \in \mathbb{C}$. Since $\mathfrak{M} = \Delta^z(\Delta^{-z}\mathfrak{M}\Delta^z)\Delta^{-z} \subseteq \Delta^z\mathfrak{M}\Delta^{-z}$ it follows that

$$\Delta^z \mathfrak{M} \Delta^{-z} = \mathfrak{M}, \qquad z \in \mathbb{C}.$$

Next

$$J\mathfrak{M}J = J\Delta^{1/2}\mathfrak{M}\Delta^{-1/2}J = S\mathfrak{M}S \subseteq \mathfrak{M}'$$

and, analogously,

$$J\mathfrak{M}'J = J\Delta^{-1/2}\mathfrak{M}'\Delta^{1/2}J = F\mathfrak{M}'F \subseteq \mathfrak{M}.$$

Hence

$$J\mathfrak{M}J = \mathfrak{M}'.$$

The principal result of the Tomita–Takesaki theory is that these relations persist in the general case when Δ is not necessarily bounded, i.e.,

$$J\mathfrak{M}J = \mathfrak{M}' \quad \text{and} \quad \Delta^{it}\mathfrak{M}\Delta^{-it} = \mathfrak{M}$$

for all $t \in \mathbb{R}$.

The proof in the general case follows different lines from the proof above. To describe these lines assume in the trace case that we know *a priori* that

$$\mathfrak{M}\Omega = \mathfrak{M}'\Omega.$$

Thus for any $A \in \mathfrak{M}$ there exists an $A' \in \mathfrak{M}'$ such that $A^*\Omega = A'\Omega$. The relation $JAJ = A'$ then follows because

$$JAJB\Omega = JAB^*\Omega = BA^*\Omega = BA'\Omega = A'B\Omega$$

for all $B \in \mathfrak{M}$. Hence $J\mathfrak{M}J \subseteq \mathfrak{M}'$, and by a symmetric argument $J\mathfrak{M}'J \subseteq \mathfrak{M}$, i.e., $J\mathfrak{M}J = \mathfrak{M}'$.

The next lemma demonstrates a result analogous to $\mathfrak{M}\Omega = \mathfrak{M}'\Omega$, but the resolvent of the modular operator intervenes in the general case to modify this equality.

Lemma 2.5.12. *If* $\lambda \in \mathbb{C}$, $-\lambda \notin \mathbb{R}_+$, *and* $A' \in \mathfrak{M}'$ *then there exists an* $A_\lambda \in \mathfrak{M}$ *such that*

$$A_\lambda^*\Omega = (\Delta + \lambda 1)^{-1}A'\Omega.$$

The following estimate is valid for A_λ:

$$\|A_\lambda\| \le (2|\lambda| + \lambda + \bar{\lambda})^{-1/2}\|A'\|.$$

PROOF. The proof is based on the following simple fact.

Observation. *Let $a, b, K \in \mathbb{R}$, and $\lambda \in \mathbb{C}\backslash\{0\}$ be numbers such that*

$$|a + \lambda b| \le K.$$

It follows that

$$(ab)^{1/2} \le (2|\lambda| + \lambda + \bar{\lambda})^{-1/2} \cdot K.$$

Indeed, this observation is a consequence of the calculation:

$$\begin{aligned} K^2 &\ge |a + \lambda b|^2 - (a - |\lambda| b)^2 \\ &= (2|\lambda| + \lambda + \bar{\lambda})ab. \end{aligned}$$

Now, let $\xi = (\Delta + \lambda \mathbb{1})^{-1} A'\Omega$, and let $B' \in \mathfrak{M}'$ be an arbitrary element. Since

$$(\Delta + \bar{\lambda}\mathbb{1})^{-1} B'^* B'\xi \in D(\Delta) \subseteq D(S)$$

it follows from Proposition 2.5.9 that there exists a closed operator B affiliated with $\mathfrak{M}$ such that $\Omega \in D(B) \cap D(B^*)$ and

$$B\Omega = (\Delta + \bar{\lambda}\mathbb{1})^{-1} B'^* B'\xi.$$

Thus

$$B'^* B'\xi = (\Delta + \bar{\lambda}\mathbb{1}) B\Omega.$$

Let $B = U|B|$ be the polar decomposition of B. We have that

$$\begin{aligned} &|\lambda(\Omega, |B|\Omega) + (\Omega, U|B|U^*\Omega)| \\ &\quad = |\lambda(B\Omega, U\Omega) + (U^*\Omega, B^*\Omega)| \\ &\quad = |\lambda(B\Omega, U\Omega) + (\Delta B\Omega, U\Omega)| \\ &\quad = |((\Delta + \bar{\lambda}\mathbb{1})B\Omega, U\Omega)| \\ &\quad = |(B'^* B'\xi, U\Omega)| \\ &\quad = |(B'\xi, UB'\Omega)| \le \|B'\xi\| \, \|B'\Omega\|. \end{aligned}$$

The Observation now implies that

$$\| |B|^{1/2}\Omega\| \cdot \| |B|^{1/2} U^*\Omega\| \le (2|\lambda| + \lambda + \bar{\lambda})^{-1/2} \|B\xi\| \, \|B'\Omega\|.$$

On the other hand

$$\begin{aligned} \|B'\xi\|^2 &= (B'^* B'\xi, \xi) \\ &= ((\Delta + \bar{\lambda}\mathbb{1})B\Omega, (\Delta + \lambda\mathbb{1})^{-1} A'\Omega) \\ &= (B\Omega, A'\Omega) \\ &= (|B|^{1/2}\Omega, A'|B|^{1/2} U^*\Omega) \\ &\le \|A'\| \, \| |B|^{1/2}\Omega\| \, \| |B|^{1/2} U^*\Omega\| \\ &\le (2|\lambda| + \lambda + \bar{\lambda})^{-1/2} \|A'\| \, \|B'\xi\| \, \|B'\Omega\|. \end{aligned}$$

Dividing both sides by $\|B'\xi\| = \|B'(\Delta + \lambda\mathbb{1})^{-1}A'\Omega\|$ we get the estimate

$$\|B'(\Delta + \lambda\mathbb{1})^{-1}A'\Omega\| \leq (2|\lambda| + \lambda + \bar{\lambda})^{-1/2}\|A'\|\,\|B'\Omega\|.$$

This means that the mapping defined on $\mathfrak{M}'\Omega$ by

$$B'\Omega \mapsto B'(\Delta + \bar{\lambda}\mathbb{1})^{-1}A'\Omega$$

is a bounded densely defined operator with norm less than or equal to

$$(2|\lambda| + \lambda + \bar{\lambda})^{-1/2}\|A'\|.$$

Denoting the closure of this operator by $A_\lambda{}^*$ we have $A_\lambda{}^* \in \mathfrak{M}'' = \mathfrak{M}$ and

$$A_\lambda{}^*\Omega = (\Delta + \lambda\mathbb{1})^{-1}A'\Omega.$$

The second fundamental lemma now explicitly relates the elements $A_\lambda \in \mathfrak{M}$, $A' \in \mathfrak{M}'$ of Lemma 2.5.12.

Lemma 2.5.13. *If $\lambda \in \mathbb{C}$, $-\lambda \notin \mathbb{R}_+$, and $A' \in \mathfrak{M}'$, let $A_\lambda \in \mathfrak{M}$ be the element in $\mathfrak{M}$ such that*

$$A_\lambda{}^*\Omega = (\Delta + \lambda\mathbb{1})^{-1}A'\Omega$$

(the existence of A_λ follows from Lemma 2.5.12). It follows that

$$JA'J = \Delta^{-1/2}A_\lambda\Delta^{1/2} + \bar{\lambda}\Delta^{1/2}A_\lambda\Delta^{-1/2}$$

as a relation between bilinear forms on $D(\Delta^{1/2}) \cap D(\Delta^{-1/2})$.

Proof. Let B', C' be arbitrary elements of $\mathfrak{M}'$ and let $B, C \in \mathfrak{M}$ be such that

$$B^*\Omega = (\Delta + \mathbb{1})^{-1}B'\Omega,$$
$$C^*\Omega = (\Delta + \mathbb{1})^{-1}C'\Omega.$$

Since $(\Delta + \lambda\mathbb{1})A_\lambda{}^*\Omega = A'\Omega$ we have that

$$(\Delta A_\lambda{}^*\Omega, B^*C\Omega) + \bar{\lambda}(A_\lambda{}^*\Omega, B^*C\Omega) = (A'\Omega, B^*C\Omega).$$

We now consider the individual terms in this relation. The first term is

$$\begin{aligned}(\Delta A_\lambda{}^*\Omega, B^*C\Omega) &= (FSA_\lambda{}^*\Omega, B^*C\Omega)\\ &= (SB^*C\Omega, SA_\lambda{}^*\Omega) = (C^*B\Omega, A_\lambda\Omega)\\ &= (B\Omega, CA_\lambda\Omega) = (SB^*\Omega, SA_\lambda{}^*C^*\Omega)\\ &= (A_\lambda{}^*C^*\Omega, \Delta B^*\Omega)\\ &= (C'\Omega, (\Delta + \mathbb{1})^{-1}A_\lambda\Delta(\Delta + \mathbb{1})^{-1}B'\Omega).\end{aligned}$$

The second term, divided by $\bar{\lambda}$, is

$$\begin{aligned}(A_\lambda{}^*\Omega, B^*C\Omega) &= (BA_\lambda{}^*\Omega, C\Omega)\\ &= (SA_\lambda B^*\Omega, SC^*\Omega) = (\Delta C^*\Omega, A_\lambda B^*\Omega)\\ &= (C'\Omega, (\Delta + \mathbb{1})^{-1}\Delta A_\lambda(\Delta + \mathbb{1})^{-1}B'\Omega).\end{aligned}$$

The last term is

$$\begin{aligned}(A'\Omega, B^*C\Omega) &= (A'B\Omega, C\Omega)\\ &= (A'SB^*\Omega, SC^*\Omega) = (C^*\Omega, FA'SB^*\Omega)\\ &= (C'\Omega, (\Delta + \mathbb{1})^{-1}\Delta^{1/2}JA'J\Delta^{1/2}(\Delta + \mathbb{1})^{-1}B'\Omega).\end{aligned}$$

Since $\mathfrak{M}'\Omega$ is dense in $\mathfrak{H}$, we find the relation between bounded operators:

$$(\Delta + 1)^{-1}A_\lambda\Delta(\Delta + 1)^{-1} + \bar{\lambda}(\Delta + 1)^{-1}\Delta A_\lambda(\Delta + 1)^{-1}$$
$$= (\Delta + 1)^{-1}\Delta^{1/2}JA'J\Delta^{1/2}(\Delta + 1)^{-1}.$$

Multiplying to the left and right by $(\Delta + 1)\Delta^{-1/2}$ then gives the desired result.

Now we exploit the relation of the preceding lemma to identify $JA'J$ as an element of $\mathfrak{M}$. This relation has the form

$$JA'J = (D^{-1/2} + \bar{\lambda}D^{1/2})(A_\lambda)$$

with $D^{1/2}(B) = \Delta^{1/2}B\Delta^{-1/2}$, and for $\lambda > 0$ a formal application of the Fourier integral

$$\int_{-\infty}^{\infty} dt\, \frac{e^{ipt}}{e^{\pi t} + e^{-\pi t}} = \frac{1}{e^{p/2} + e^{-p/2}}$$

yields the inverse relation

$$A_\lambda = \lambda^{-1/2} \int_{-\infty}^{\infty} dt\, \frac{\lambda^{it}}{e^{\pi t} + e^{-\pi t}}\, D^{it}(JA'J),$$

with $D^{it}(B) = \Delta^{it}B\Delta^{-it}$. The key to the proof of the following theorem is to justify this inversion because Fourier analysis then shows that $A_\lambda \in \mathfrak{M}$ implies that $D^{it}(JA'J) \in \mathfrak{M}$. This outlines the proof of the principal result.

Theorem 2.5.14 (Tomita–Takesaki theorem). *Let $\mathfrak{M}$ be a von Neumann algebra with cyclic and separating vector Ω, and let Δ be the associated modular operator and J the associated modular conjugation. It follows that*

$$J\mathfrak{M}J = \mathfrak{M}'$$

and, moreover,

$$\Delta^{it}\mathfrak{M}\Delta^{-it} = \mathfrak{M}$$

for all $t \in \mathbb{R}$.

Proof. If $B \in \mathscr{L}(\mathfrak{H})$ then, by spectral analysis, the functions $t \in \mathbb{R} \mapsto (\psi, \Delta^{it}B\Delta^{-it}\varphi)$ are continuous and are bounded by $\|B\|\,\|\psi\|\,\|\varphi\|$ for all $\psi, \varphi \in \mathfrak{H}$. Thus by integration of bilinear forms one can define the transform $I_\lambda(B)$, of B, for each $\lambda > 0$ by

$$I_\lambda(B) = \lambda^{-1/2} \int_{-\infty}^{\infty} dt\, \frac{\lambda^{it}}{e^{\pi t} + e^{-\pi t}}\, \Delta^{it}B\Delta^{-it}.$$

Now take $\psi, \varphi \in D(\Delta^{1/2}) \cap D(\Delta^{-1/2})$ and consider the function

$$f(\lambda) = (\Delta^{-1/2}\psi, I_\lambda(B)\Delta^{1/2}\varphi) + \lambda(\Delta^{1/2}\psi, I_\lambda(B)\Delta^{-1/2}\varphi)$$
$$= \int_{-\infty}^{\infty} dt\, \frac{\lambda^{it}}{e^{\pi t} + e^{-\pi t}} \{\lambda^{-1/2}(\Delta^{-(1/2)-it}\psi, B\Delta^{(1/2)-it}\varphi)$$
$$+ \lambda^{1/2}(\Delta^{(1/2)-it}\psi, B\Delta^{-(1/2)-it}\varphi)\}.$$

Using the spectral decomposition of Δ,

$$\Delta = \int dE_\Delta(\mu)\,\mu,$$

one finds

$$f(\lambda) = \int_{-\infty}^{\infty} dt\,\frac{\lambda^{it}}{e^{\pi t} + e^{-\pi t}} \int d^2(E_\Delta(\mu)\psi, BE_\Delta(\rho)\varphi)\left(\frac{\mu}{\rho}\right)^{it}\left\{\left(\frac{\rho}{\mu\lambda}\right)^{1/2} + \left(\frac{\mu\lambda}{\rho}\right)^{1/2}\right\}.$$

But the domain restrictions on φ, ψ allow interchange of the order of integration and one has

$$\begin{aligned} f(\lambda) &= \int d^2(E_\Delta(\mu)\psi, BE_\Delta(\rho)\varphi)\left\{\left(\frac{\rho}{\mu\lambda}\right)^{1/2} + \left(\frac{\mu\lambda}{\rho}\right)^{1/2}\right\}\int_{-\infty}^{\infty}\frac{dt}{e^{\pi t} + e^{-\pi t}}\left(\frac{\mu\lambda}{\rho}\right)^{it} \\ &= \int d^2(E_\Delta(\mu)\psi, BE_\Delta(\rho)\varphi) = (\psi, B\varphi), \end{aligned}$$

where the first step uses the Fourier relation quoted before the theorem. Thus as a relation between bilinear forms on $D(\Delta^{1/2}) \cap D(\Delta^{-1/2})$ one has

$$\Delta^{-1/2}I_\lambda(B)\Delta^{1/2} + \lambda\Delta^{1/2}I_\lambda(B)\Delta^{-1/2} = B.$$

It follows easily from the definitions that the operators $(D^{-1/2} + \lambda D^{1/2})$ and I_λ on $\mathscr{L}(\mathfrak{H})$ commute, hence it follows from the last relation that $I_\lambda = (D^{-1/2} + \lambda D^{1/2})^{-1}$.

Comparison with Lemma 2.5.13 then establishes that the relation between A' and A_λ has the inverse

$$A_\lambda = I_\lambda(JA'J).$$

Now $A_\lambda \in \mathfrak{M}$ and hence if $B' \in \mathfrak{M}'$

$$(\psi, [B', I_\lambda(JA'J)]\varphi) = 0.$$

Setting $\lambda = e^p$ and using the definition of I_λ one deduces that

$$\int_{-\infty}^{\infty} dt\,\frac{e^{ipt}}{e^{\pi t} + e^{-\pi t}}(\psi, [B', \Delta^{it}JA'J\Delta^{-it}]\varphi) = 0$$

for all $p \in \mathbb{R}$. By Fourier transformation one concludes that

$$\Delta^{it}JA'J\Delta^{-it} \in \mathfrak{M}'' = \mathfrak{M}. \tag{*}$$

Setting $t = 0$ yields

$$J\mathfrak{M}'J \subseteq \mathfrak{M}.$$

But by Proposition 2.5.11 the conjugation of the pair $(\mathfrak{M}', \Omega)$ is the same as the conjugation of the pair $(\mathfrak{M}, \Omega)$. Thus, by the same reasoning,

$$J\mathfrak{M}J \subseteq \mathfrak{M}'.$$

Using $J^2 = \mathbb{1}$ we find $\mathfrak{M} \subseteq J\mathfrak{M}'J \subseteq \mathfrak{M}$, which gives the fundamental result

$$J\mathfrak{M}J = \mathfrak{M}'.$$

Now because $J\mathfrak{M}'J = \mathfrak{M}$ one deduces that any element $A \in \mathfrak{M}$ has the form $A = JA'J$, where $A' \in \mathfrak{M}'$. Thus by (*)

$$\Delta^{it}A\Delta^{-it} \in \mathfrak{M}.$$

Remark. We have shown that the elements A_λ and A' of Lemma 2.5.12 are related by $A_\lambda = I_\lambda(JA'J)$ and by Fourier transformation one has

$$\Delta^{it}JA'J\Delta^{-it} = \pi^{-1}\cosh(\pi t)\int_0^\infty d\lambda\, \lambda^{-((1/2)+it)}A_\lambda \in \mathfrak{M},$$

where the integral is in the weak distribution sense.

Definition 2.5.15. Let $\mathfrak{M}$ be a von Neumann algebra, ω a faithful, normal state on $\mathfrak{M}$, $(\mathfrak{H}_\omega, \pi_\omega, \Omega_\omega)$ the corresponding cyclic representation, and Δ the modular operator associated with the pair $(\pi_\omega(\mathfrak{M}), \Omega_\omega)$. Theorem 2.5.14 establishes the existence of a σ-weakly continuous one-parameter group $t \to \sigma_t{}^\omega$ of *-automorphisms of $\mathfrak{M}$ through the definition

$$\sigma_t{}^\omega(A) = \pi_\omega^{-1}(\Delta^{it}\pi_\omega(A)\Delta^{-it}).$$

The group $t \mapsto \sigma_t{}^\omega$ is called the *modular automorphism group associated with the pair* $(\mathfrak{M}, \omega)$.

The modular automorphism group is one of the most useful elements in the further analysis of von Neumann algebras. It is also of paramount importance in applications to quantum statistical mechanics (see Chapters 5 and 6) because the equilibrium dynamics is usually given by a modular group. In this latter context the modular condition

$$\begin{aligned}(\Delta^{1/2}\pi_\omega(A)\Omega_\omega, \Delta^{1/2}\pi_\omega(B)\Omega_\omega) &= (J\pi_\omega(A^*)\Omega_\omega, J\pi_\omega(B^*)\Omega_\omega)\\ &= (\pi_\omega(B^*)\Omega_\omega, \pi_\omega(A^*)\Omega_\omega)\end{aligned}$$

is of utmost importance. Note that in terms of the modular group, evaluated at the imaginary point $t = i/2$, this condition has the form

$$\omega(\sigma_{i/2}(A)\sigma_{-i/2}(B)) = \omega(BA).$$

EXAMPLE 2.5.16 If $\mathfrak{M} = \mathscr{L}(\mathfrak{H})$, with $\mathfrak{H}$ separable, then each normal state ω is of the form

$$\omega(A) = \mathrm{Tr}(\rho A)$$

and ω is faithful if, and only if, ρ is invertible (Example 2.5.5). In this case one may calculate that the modular group is given by $\sigma_t(A) = \rho^{it}A\rho^{-it}$. For example, the modular condition is satisfied because

$$\omega(BA) = \mathrm{Tr}(\rho BA) = \mathrm{Tr}(\rho(\rho^{-1/2}A\rho^{1/2})(\rho^{1/2}B\rho^{-1/2})).$$

(This condition in fact determines σ uniquely, see Theorem 5.3.10.)

2.5.3. Integration and Analytic Elements for One-Parameter Groups of Isometries on Banach Spaces

The construction of the self-dual cone discussed in the introduction to this section proceeds with the aid of the modular group described in the previous subsection. In order to fully exploit this tool we need some general results on one-parameter groups. Such groups will be studied in detail in Chapter 3 and the preliminary results of this subsection will again be of use.

We consider a complex Banach space X and a norm-closed subspace F of the dual X^* of X such that either $F = X^*$ or $X = F^*$. In the latter case we write $F = X_*$.

Let $\sigma(X, F)$ be the locally convex topology on X induced by the functionals in F.

Definition 2.5.17. A one-parameter family $t \in \mathbb{R} \mapsto \tau_t$ of bounded, linear maps of X into itself is called a *$\sigma(X, F)$-continuous group of isometries of X* if

(1) $\tau_{t_1+t_2} = \tau_{t_1}\tau_{t_2}$, $t_1, t_2 \in \mathbb{R}$, and $\tau_0 = \iota$;
(2) $\|\tau_t\| = 1$, $t \in \mathbb{R}$;
(3) $t \mapsto \tau_t(A)$ is $\sigma(X, F)$-continuous for all $A \in X$, i.e., $t \mapsto \eta(\tau_t(A))$ is continuous for all $A \in X$ and $\eta \in F$;
(4) $A \mapsto \tau_t(A)$ is $\sigma(X, F)$–$\sigma(X, F)$ continuous for all $t \in \mathbb{R}$, i.e., $\eta \circ \tau_t \in F$ for $\eta \in F$.

Note that if $F = X^*$, (4) is automatically satisfied. In any case, (4) implies that we can define a one-parameter family $\tau_t{}^*$ of maps of F by $(\tau_t{}^*\eta)(A) = \eta(\tau_t(A))$. It is then easy to verify that $t \mapsto \tau_t{}^*$ is a $\sigma(F, X)$-continuous group of isometries of F. We will see later (Corollary 2.5.23) that if $F = X^*$, then (3) amounts to the requirement that $t \mapsto \tau_t$ be strongly continuous, i.e., $t \mapsto \tau_t(A)$ be continuous in norm for each $A \in X$. If $F = X^*$ we refer to τ_t as a C_0-group. If $F = X_*$ we call τ_t a $C_0{}^*$-group. The principal groups considered in this book fall into one of the following three categories:

(1) strongly continuous unitary groups on Hilbert spaces, i.e., $X = F = \mathfrak{H}$, where $\mathfrak{H} = \mathfrak{H}^*$ is a Hilbert space;
(2) strongly continuous groups of *-automorphisms on C^*-algebras. These are automatically isometries by Corollary 2.3.4;
(3) weakly continuous groups of *-automorphisms of von Neumann algebras, i.e., $X = \mathfrak{M}$, $F = \mathfrak{M}_*$. In this case, (4) is automatically satisfied due to Theorem 2.4.23. Note that if $t \mapsto U_t$ is a strongly continuous group of unitary elements such that $U_t \mathfrak{M} U_t{}^* \subseteq \mathfrak{M}$ for all t, then it is easily established that $t \mapsto \tau_t(A) = U_t A U_t{}^*$ is a weakly continuous group of *-automorphisms of $\mathfrak{M}$.

Proposition 2.5.18. *Let $t \mapsto \tau_t$ be a $\sigma(X, F)$—continuous group of isometries, and let μ be a Borel measure of bounded variation on $\mathbb{R}$. It follows* *that for each $A \in X$ there exists a $B \in X$ such that*

$$\eta(B) = \int \eta(\tau_t(A)) \, d\mu(t)$$

for any $\eta \in F$.

This result allows us to introduce a notation for the averaging process which indicates more clearly its nature.

Definition 2.5.19. If A and B are related as in Proposition 2.5.18 we write

$$B = \int \tau_t(A) \, d\mu(t)$$

PROOF OF PROPOSITION 2.5.18. First note that the convex closure of any $\sigma(X, F)$—precompact subset of X is $\sigma(X, F)$-compact. If $F = X_*$ this follows from Alaoglu's theorem; if $F = X^*$ it follows from the Krein–Smulian theorem.

Because of the estimate

$$\left| \int \eta(\tau_t(A)) \, d\mu(t) \right| \leq \|A\| \, \|\mu\| \, \|\eta\|$$

there exists an $f \in F^*$ such that

$$f(\eta) = \int \eta(\tau_t(A)) \, d\mu(t), \qquad \eta \in F$$

(if $F = X_*$, this ends the proof).

To show the existence of a $B \in X$ such that $f(\eta) = \eta(B)$, it suffices to show that f is $\sigma(F, X)$-continuous. Now by Mackey's theorem the $\sigma(F, X)$-continuous functionals of F are just the $\tau(F, X)$-continuous functionals, where $\tau(F, X)$ is the Mackey topology on F. This last topology is defined by the seminorms

$$\eta \mapsto \sup_{C \in K} |\eta(C)|,$$

with K ranging over all convex, compact, circled subsets of X in the $\sigma(X, F)$-topology. (Circled means that if $A \in K$ and $\lambda \in \mathbb{C}$ with $|\lambda| = 1$, then $\lambda A \in K$.) Now assume first that μ has compact support contained in $[-\lambda, \lambda]$. By continuity of the map $\gamma, t \mapsto \gamma\tau_t(A)$, the set $\{\gamma\tau_t(A); |\gamma| = 1, t \in [-\lambda, \lambda]\}$ is compact in X, and hence its convex closure K is also compact. The estimate

$$|f(\eta)| \leq \|\mu\| \sup_{t \in [-\lambda, \lambda]} |\eta(\tau_t(A))|$$

then implies that

$$|f(\eta)| \leq \|\mu\| \sup_{C \in K} |\eta(C)|.$$

Hence f is $\tau(F, X)$-continuous. Thus f is $\sigma(F, X)$-continuous and the existence of B is established.

If μ does not have compact support, pick an increasing sequence $\{K_n\}$ of compact subsets of $\mathbb{R}$ such that $|\mu|(\mathbb{R} \backslash K_n) \to 0$. But then for each n we can find a $B_n \in X$ such that

$$\eta(B_n) = \int_{K_n} \eta(\tau_t(A)) \, d\mu(t), \qquad \eta \in F.$$

The estimate

$$|\eta(B_n) - f(\eta)| \leq \|\eta\| \|A\| |\mu|(\mathbb{R}\backslash K_n)$$

then implies that $\{B_n\}$ is a norm Cauchy sequence in X and the limit, $B = \lim_n B_n$, satisfies $f(\eta) = \eta(B)$.

Note that Proposition 2.5.18 is valid under more general circumstances than stated, e.g., property (4) in the definition of τ_t is superfluous, and the only properties of F used are:

(i) $\|A\| = \sup\{|\eta(A)|; \eta \in F, \|\eta\| \leq 1\}$;
(ii) the $\sigma(X, F)$-closed convex hull of every $\sigma(X, F)$-compact subset of X is $\sigma(X, F)$-compact.

Definition 2.5.20. Let $t \mapsto \tau_t$ be a $\sigma(X, F)$-continuous group of isometries. An element $A \in X$ is called *analytic* for τ_t if there exists a strip

$$I_\lambda = \{z, |\text{Im } z| < \lambda\}$$

in $\mathbb{C}$, a function $f; I_\lambda \mapsto X$ such that

(i) $f(t) = \tau_t(A)$ for $t \in \mathbb{R}$,
(ii) $z \mapsto \eta(f(z))$ is analytic for all $\eta \in F$.

Under these conditions, we write

$$f(z) = \sigma_z(A), \qquad z \in I_\lambda.$$

We immediately show that the weak analyticity of condition (ii) is equivalent to strong analyticity:

Proposition 2.5.21. *If A is τ_t-analytic on the strip I_λ, then A is strongly analytic on I_λ, i.e. if $f(z) = \sigma_z(A)$ then* (ii′) *is true:*

(ii′) $\lim_{h \to 0} h^{-1}(f(z + h) - f(z))$ *exists in norm for $z \in I_\lambda$.*

PROOF. If $z \in I_\lambda$, let C be a circle around z with radius r such that $C \subseteq I_\lambda$, and let K be the ball around z with radius $r/2$. For any $x \in K$ we have

$$\eta(f(x)) = \frac{1}{2\pi i} \int_C \frac{\eta(f(y))}{y - x} \, dy, \qquad \eta \in F,$$

by Cauchy's formula. Hence if $z + h, z + g \in K$,

$$(h - g)^{-1}\{h^{-1}(\eta(f(z + h)) - \eta(f(z))) - g^{-1}(\eta(f(z + g)) - \eta(f(z)))\}$$

$$= \frac{1}{2\pi i} \int_C \eta(f(y))(y - z - h)^{-1}(y - z - g)^{-1}(y - z)^{-1} \, dy.$$

Now, for fixed η the absolute value of the right-hand side is uniformly bounded in h and g, because K has a positive distance $r/2$ from C. Thus, by the uniform boundedness theorem, there exists a constant γ such that

$$\sup_{|h|\le r,\, |g|\le r} \frac{1}{|h-g|}\left\|\frac{f(z+h)-f(z)}{h} - \frac{f(z+g)-f(z)}{g}\right\| \le \gamma.$$

The completeness of X implies that $(d/dz)f(z)$ exists.

Proposition 2.5.22. *If $t \mapsto \tau_t$ is a $\sigma(X, F)$-continuous group of isometries, and $A \in X$, define*

$$A_n = \sqrt{\frac{n}{\pi}} \int \tau_t(A) e^{-nt^2}\, dt, \qquad n = 1, 2, \ldots.$$

It follows that each A_n is an entire analytic element for τ_t, $\|A_n\| \le \|A\|$ for all n, and $A_n \to A$ in the $\sigma(X, F)$-topology as $n \to \infty$. In particular, the τ_t-analytic elements form a $\sigma(X, F)$-dense subspace of X.

PROOF. Proposition 2.5.18 implies that

$$f_n(z) = \sqrt{\frac{n}{\pi}} \int \tau_t(A) e^{-n(t-z)^2}\, dt$$

is well defined for all $z \in \mathbb{C}$ because $t \mapsto e^{-n(t-z)^2} \in L^1(\mathbb{R})$ for each z.

For $z = s \in \mathbb{R}$ we have

$$\begin{aligned} f_n(s) &= \sqrt{\frac{n}{\pi}} \int \tau_t(A) e^{-n(t-s)^2}\, dt \\ &= \sqrt{\frac{n}{\pi}} \int \tau_{t+s}(A) e^{-nt^2}\, dt \\ &= \tau_s\left(\sqrt{\frac{n}{\pi}} \int \tau_t(A) e^{-nt^2}\, dt\right) \\ &= \tau_s(A_n). \end{aligned}$$

But for $\eta \in F$ we have

$$\eta(f_n(z)) = \sqrt{\frac{n}{\pi}} \int \eta(\tau_t(A)) e^{-n(t-z)^2}\, dt.$$

Since $|\eta(\tau_t(A))| \le \|\eta\|\,\|A\|$ it follows from the Lebesgue dominated convergence theorem that $z \mapsto \eta(f_n(z))$ is analytic. Hence each A_n is analytic for τ_t.

Next, one derives the estimate

$$\|A_n\| \le \sup_t\{\|\tau_t(A)\|\} \sqrt{\frac{n}{\pi}} \int e^{-nt^2}\, dt = \|A\|.$$

Next note that

$$\eta(A_n - A) = \sqrt{\frac{n}{\pi}} \int e^{-nt^2}(\eta(\tau_t(A)) - \eta(A))\, dt$$

for all $\eta \in F$. But for any $\varepsilon > 0$ we may choose $\delta > 0$ such that $|t| < \delta$ implies $|\eta(\tau_t(A)) - \eta(A)| < \varepsilon/2$. Further, we may subsequently choose N large enough that

$$\sqrt{\frac{N}{\pi}} \int_{|t| \geq \delta} e^{-Nt^2}\, dt < \frac{\varepsilon}{4\|\eta\|\, \|A\|}.$$

It follows that if $n \geq N$ we have

$$\begin{aligned} |\eta(A_n - A)| &\leq \sqrt{\frac{n}{\pi}} \int_{|t| \leq \delta} e^{-nt^2} |\eta(\tau_t(A)) - \eta(A)|\, dt \\ &\quad + \sqrt{\frac{n}{\pi}} \int_{|t| \geq \delta} e^{-nt^2} |\eta(\tau_t(A)) - \eta(A)|\, dt \\ &\leq \frac{\varepsilon}{2} \sqrt{\frac{n}{\pi}} \int_{|t| \leq \delta} e^{-nt^2}\, dt \\ &\quad + 2\|\eta\|\, \|A\| \sqrt{\frac{n}{\pi}} \int_{|t| \geq \delta} e^{-nt^2}\, dt \\ &< \frac{\varepsilon}{2} + \frac{\varepsilon}{2} = \varepsilon. \end{aligned}$$

Hence $A_n \to A$ in the $\sigma(X, F)$-topology.

Corollary 2.5.23. *If $t \mapsto \tau_t$ is a $\sigma(X, X^*)$-continuous group of isometries, then $t \mapsto \tau_t$ is strongly continuous, i.e., $t \mapsto \tau_t(A)$ is continuous in norm for each $A \in X$, and X contains a norm-dense set of entire analytic elements for τ_t.*

PROOF. By Proposition 2.5.22 the set of entire elements for τ_t forms a $\sigma(X, X^*)$-dense subset of X. This subset is clearly a subspace and hence it is norm dense in X by a simple argument using the Hahn–Banach theorem. (If the subspace were not norm dense then there would exist a nonzero linear functional which vanishes on the norm closure of the subspace. But this contradicts the $\sigma(X, X^*)$-density.) But if A is an analytic element, then $t \mapsto \tau_t(A)$ is norm differentiable by Proposition 2.5.21 and hence $t \mapsto \tau_t(A)$ is norm continuous. Finally, for a general $A \in X$ we can find a sequence A_n of analytic elements converging to A and estimate

$$\begin{aligned} \|\tau_t(A) - A\| &\leq \|\tau_t(A - A_n)\| + \|\tau_t(A_n) - A_n\| \\ &\quad + \|A_n - A\| \\ &= 2\|A_n - A\| + \|\tau_t(A_n) - A_n\|. \end{aligned}$$

Corollary 2.5.24. *If $t \mapsto \tau_t$ is a $\sigma(X, X_*)$-continuous group of isometries then the set Y of elements A such that $t \mapsto \tau_t(A)$ is norm continuous is a norm-closed, $\sigma(X, X_*)$-dense subspace of X, and Y is the norm closure of the entire analytic elements for τ_t.*

PROOF. If Y_0 is the norm closure of the entire analytic elements, then $\tau|_{Y_0}$ is strongly continuous by the same proof as in Corollary 2.5.23, and Y_0 is $\sigma(X, X_*)$-dense in X by Proposition 2.5.22. Now if A is an entire element for τ one sees easily that

$\tau_z(A)$ is entire for $z \in \mathbb{C}$. Hence the entire elements for $\tau|_{Y_0}$ coincide with the entire elements for τ. Since Y is the norm closure of the entire elements for $\tau|_Y$, by Corollary 2.5.23, it follows that $Y_0 = Y$.

Finally, note that if $\tau_t(A) = \Delta^{it}A\Delta^{-it}$ is a weakly continuous group of *-auto-morphisms of a von Neumann algebra, where Δ is a positive, invertible selfadjoint operator, then

$$\tau_z(A) = \Delta^{iz}A\Delta^{-iz}$$

for each analytic A. Both sides of this equation are viewed as bilinear forms on the entire vectors of $t \mapsto \Delta^{it}$. The equality follows from the fact that a function which is analytic in a strip around the real axis is determined by its restriction to this axis. This simple fact will be frequently used in the next subsection.

2.5.4. Self-Dual Cones and Standard Forms

Throughout this section $\mathfrak{M}$ denotes a von Neumann algebra on a Hilbert space $\mathfrak{H}$ with a cyclic and separating vector Ω. We use Δ and J to denote respectively, the modular automorphism and the modular conjugation associated with the pair $(\mathfrak{M}, \Omega)$. The associated modular automorphism group is denoted by σ_t and $\mathfrak{M}_0$ is the *-algebra of entire analytic elements for σ. Finally, j; $\mathfrak{M} \mapsto \mathfrak{M}'$ is the antilinear *-isomorphism defined by $j(A) = JAJ$.

Definition 2.5.25. The *natural positive cone* $\mathscr{P}$ associated with the pair $(\mathfrak{M}, \Omega)$ is defined as the closure of the set

$$\{Aj(A)\Omega;\ A \in \mathfrak{M}\}.$$

Note that this cone corresponds exactly to the positive cones discussed for the abelian and trace examples in the introduction to this section. The cone is the analogue of the positive L^2-functions in the case of an abelian algebra.

Proposition 2.5.26. *The closed subset $\mathscr{P} \subseteq \mathfrak{H}$ has the following properties*:

(1) $\mathscr{P} = \overline{\Delta^{1/4}\mathfrak{M}_+\Omega} = \overline{\Delta^{-1/4}\mathfrak{M}'_+\Omega} = \overline{\Delta^{1/4}\overline{\mathfrak{M}_+\Omega}} = \overline{\Delta^{-1/4}\overline{\mathfrak{M}'_+\Omega}}$ *and hence* $\mathscr{P}$ *is a convex cone*;
(2) $\Delta^{it}\mathscr{P} = \mathscr{P}$ *for all* $t \in \mathbb{R}$;
(3) *if* f *is a positive-definite function then* $f(\log \Delta)\mathscr{P} \subseteq \mathscr{P}$;
(4) *if* $\xi \in \mathscr{P}$, *then* $J\xi = \xi$;
(5) *if* $A \in \mathfrak{M}$, *then* $Aj(A)\mathscr{P} \subseteq \mathscr{P}$.

PROOF. (1) For $A \in \mathfrak{M}_0$ we have

$$\begin{aligned}\Delta^{1/4} A A^* \Omega &= \sigma_{-i/4}(A)\sigma_{-i/4}(A^*)\Omega \\ &= \sigma_{-i/4}(A)\sigma_{i/4}(A)^*\Omega \\ &= \sigma_{-i/4}(A) J \Delta^{1/2} \sigma_{i/4}(A)\Omega \\ &= \sigma_{-i/4}(A) J \sigma_{-i/4}(A) J \Omega \\ &= B j(B)\Omega,\end{aligned}$$

where $B = \sigma_{-i/4}(A)$. Since $\sigma_{-i/4}(\mathfrak{M}_0) = \mathfrak{M}_0$ and $\mathfrak{M}_0$ is strongly* dense in $\mathfrak{M}$ it follows from this relation, and Kaplansky's density theorem, that

$$Bj(B)\Omega \in \overline{\Delta^{1/4}\mathfrak{M}_+ \Omega} \subseteq \overline{\Delta^{1/4}\overline{\mathfrak{M}_+ \Omega}} \quad \text{for all } B \in \mathfrak{M}.$$

Thus

$$\mathscr{P} \subseteq \overline{\Delta^{1/4}\mathfrak{M}_+ \Omega} \subseteq \overline{\Delta^{1/4}\overline{\mathfrak{M}_+ \Omega}}.$$

Conversely, $\mathfrak{M}_{0+}$ is strongly* dense in $\mathfrak{M}_+$, by Kaplansky's density theorem, hence $\mathfrak{M}_{0+}\Omega$ is dense in $\overline{\mathfrak{M}_+\Omega}$. For $\psi \in \overline{\mathfrak{M}_+\Omega}$ we choose a sequence $A_n \in \mathfrak{M}_{0+}$ such that $A_n\Omega \to \psi$. Then $\Delta^{1/4}A_n\Omega \in \mathscr{P}$ by the first relation in this proof. But

$$J\Delta^{1/2}A_n\Omega = A_n\Omega \to \psi = J\Delta^{1/2}\psi$$

and hence

$$\|\Delta^{1/4}(\psi - A_n\Omega)\|^2 = (\psi - A_n\Omega, \Delta^{1/2}(\psi - A_n\Omega)) \to 0.$$

Thus $\Delta^{1/4}\psi \in \mathscr{P}$ and $\overline{\Delta^{1/4}\overline{\mathfrak{M}_+\Omega}} \subseteq \mathscr{P}$. Combining these two conclusions yields

$$\mathscr{P} = \overline{\Delta^{1/4}\mathfrak{M}_+ \Omega} = \overline{\Delta^{1/4}\overline{\mathfrak{M}_+ \Omega}}.$$

If $\mathscr{P}'$ is the natural cone corresponding to $(\mathfrak{M}', \Omega)$ then $\mathscr{P}'$ is the closure of the elements of the form

$$\begin{aligned}A'j(A')\Omega &= j(j(A'))j(A')\Omega \\ &= j(A)A\Omega \\ &= Aj(A)\Omega,\end{aligned}$$

where $A = j(A') \in \mathfrak{M}$. Hence $\mathscr{P}' = \mathscr{P}$. Since Δ^{-1} is the modular operator corresponding to $(\mathfrak{M}', \Omega)$, it follows from the first part of the proof that

$$\mathscr{P} = \mathscr{P}' = \overline{\Delta^{-1/4}\mathfrak{M}'_+\Omega} = \overline{\Delta^{-1/4}\overline{\mathfrak{M}'_+\Omega}}.$$

This completes the proof of property (1).

(2) Follows from (1) and the computation

$$\begin{aligned}\Delta^{it}\Delta^{1/4}\mathfrak{M}_+\Omega &= \Delta^{1/4}\Delta^{it}\mathfrak{M}_+\Omega \\ &= \Delta^{1/4}\sigma_t(\mathfrak{M}_+)\Omega = \Delta^{1/4}\mathfrak{M}_+\Omega.\end{aligned}$$

(3) Follows from (2) by noting that Bochner's theorem implies that a positive definite function has the form $f(x) = \int e^{itx}\, d\mu(t)$, where μ is a positive finite Borel measure on $\mathbb{R}$. Hence $f(\log \Delta) = \int \Delta^{it}\, d\mu(t)$, and (3) follows because $\mathscr{P}$ is a closed cone.

For (4) note that

$$JAj(A)\Omega = j(A)A\Omega = Aj(A)\Omega,$$

and for (5) that

$$Aj(A)Bj(B)\Omega = ABj(A)j(B)\Omega = ABj(AB)\Omega.$$

Next we prepare the ground for the proof that $\mathscr{P}$ is self-dual.

Proposition 2.5.27. (1) *Let $\eta \in \mathfrak{H}$ and assume that $(\eta, A\Omega) \geq 0$ for $A \in \mathfrak{M}_+$. Then there exists a positive, selfadjoint operator Q' affiliated with $\mathfrak{M}'$ such that $\eta = Q'\Omega$.*

(2) *$\overline{\mathfrak{M}_+\Omega}$ and $\overline{\mathfrak{M}'_+\Omega}$ are dual cones in $\mathfrak{H}$, i.e.,*

$$\overline{\mathfrak{M}_+\Omega} = \{\xi \in \mathfrak{H}; (\xi, \eta) \geq 0 \text{ for all } \eta \in \overline{\mathfrak{M}'_+\Omega}\},$$
$$\overline{\mathfrak{M}'_+\Omega} = \{\xi \in \mathfrak{H}; (\xi, \eta) \geq 0 \text{ for all } \eta \in \overline{\mathfrak{M}_+\Omega}\}.$$

PROOF. (1) Define an operator A' on $D(A') = \mathfrak{M}\Omega$ by

$$A'A\Omega = A\eta, \qquad A \in \mathfrak{M}.$$

For any unitary $U \in \mathfrak{M}$, one has

$$A'UA\Omega = UA\eta = UA'A\Omega,$$

i.e.,

$$UA'U^* = A'.$$

Moreover,

$$(A\Omega, A'A\Omega) = (A\Omega, A\eta) = \overline{(\eta, A^*A\Omega)} \geq 0.$$

Thus A' is a positive, symmetric operator. Let Q' be the Friedrichs extension of A'. Then Q' is a positive selfadjoint operator and, by the uniqueness of the Friedrichs extension,

$$UQ'U^* = Q'$$

for all unitary elements in $\mathfrak{M}$. Hence Q' is affiliated with $\mathfrak{M}'$ and

$$Q'\Omega = A'\Omega = \eta.$$

(2) First for each subset $K \subset \mathfrak{H}$ introduce the notation

$$K^\vee = \{\eta \in \mathfrak{H}; (\xi, \eta) \geq 0 \quad \text{for all } \xi \in K\}.$$

But if $A \in \mathfrak{M}_+$ and $A' \in \mathfrak{M}'_+$ then

$$(A\Omega, A'\Omega) = (\Omega, A^{1/2}A'A^{1/2}\Omega) \geq 0.$$

Thus $\overline{\mathfrak{M}_+\Omega} \subseteq \overline{\mathfrak{M}'_+\Omega}^\vee$ and $\overline{\mathfrak{M}'_+\Omega} \subseteq \overline{\mathfrak{M}_+\Omega}^\vee$. Now if $\eta \in \overline{\mathfrak{M}_+\Omega}^\vee$ it follows from part (1) of the proposition that $\eta = Q'\Omega$ for a positive, selfadjoint Q' affiliated with $\mathfrak{M}'$. But if E_n' is the spectral projection of Q' corresponding to the interval $[0, n]$ then $Q'E_n' \in \mathfrak{M}'_+$ and $\lim_{n\to\infty} Q'E_n'\Omega = Q'\Omega = \eta$. Thus $\eta \in \overline{\mathfrak{M}'_+\Omega}$ and $\overline{\mathfrak{M}_+\Omega}^\vee = \overline{\mathfrak{M}'_+\Omega}$.

We now are able to establish the most important geometric properties of the cone $\mathscr{P}$.

Proposition 2.5.28.

(1) *$\mathscr{P}$ is a self-dual cone, i.e., $\mathscr{P} = \mathscr{P}^\vee$, where*

$$\mathscr{P}^\vee = \{\eta \in \mathfrak{H}; (\xi; \eta) \geq 0 \text{ for all } \xi \in \mathscr{P}\}.$$

(2) *$\mathscr{P}$ is a pointed cone, i.e.,*

$$\mathscr{P} \cap (-\mathscr{P}) = \{0\}.$$

(3) *If $J\xi = \xi$ then ξ has a unique decomposition $\xi = \xi_1 - \xi_2$, where $\xi_1, \xi_2 \in \mathscr{P}$ and $\xi_1 \perp \xi_2$.*

(4) *$\mathfrak{H}$ is linearly spanned by $\mathscr{P}$.*

PROOF. (1) If $A \in \mathfrak{M}_+$, $A' \in \mathfrak{M}'_+$ then

$$\begin{aligned}(\Delta^{1/4}A\Omega, \Delta^{-1/4}A'\Omega) &= (A\Omega, A'\Omega)\\ &= (\Omega, A^{1/2}A'A^{1/2}\Omega) \geq 0\end{aligned}$$

so $\mathscr{P} \subseteq \mathscr{P}^\vee$, by Proposition 2.5.26(1). Conversely, assume $\xi \in \mathscr{P}^\vee$, i.e., $(\xi, \eta) \geq 0$ for all $\eta \in \mathscr{P}$. Put

$$\xi_n = f_n(\log \Delta)\xi,$$

where $f_n(x) = e^{-x^2/2n^2}$. Then $\xi_n \in \bigcap_{\alpha \in \mathbb{C}} D(\Delta^\alpha)$ and $\xi_n \to \xi$.

Let $\eta \in \mathscr{P}$. Since f_n is positive definite $f_n(\log \Delta)\eta \in \mathscr{P}$ by Proposition 2.5.26(3). Thus

$$(\xi_n, \eta) = (\xi, f_n(\log \Delta)\eta) \geq 0, \qquad \eta \in \mathscr{P}.$$

Let $A \in \mathfrak{M}_+$. Then $\Delta^{1/4}A\Omega \in \mathscr{P}$ and consequently

$$(\Delta^{1/4}\xi_n, A\Omega) = (\xi_n, \Delta^{1/4}A\Omega) \geq 0.$$

Thus $\Delta^{1/4}\xi_n \in \overline{\mathfrak{M}_+\Omega}^\vee = \overline{\mathfrak{M}'_+\Omega}$ by Proposition 2.5.27(2). Hence $\xi_n \in \Delta^{-1/4}\overline{\mathfrak{M}'_+\Omega} \subseteq \mathscr{P}$. Since $\mathscr{P}$ is closed, $\xi = \lim_n \xi_n \in \mathscr{P}$ and the self-polarity of $\mathscr{P}$ is established.

Properties (2), (3), and (4) follow from the fact that $\mathscr{P}$ is a self-polar cone alone.

(2) If $\xi \in \mathscr{P} \cap (-\mathscr{P}) = \mathscr{P} \cap (-\mathscr{P}^\vee)$ then $(\xi, -\xi) \geq 0$, hence $\xi = 0$.

(3) Assume $J\xi = \xi$. Since $\mathscr{P}$ is a closed convex set in a Hilbert space there is a unique $\xi_1 \in \mathscr{P}$ such that

$$\|\xi - \xi_1\| = \inf\{\|\xi - \eta\|; \eta \in \mathscr{P}\}.$$

Put $\xi_2 = \xi_1 - \xi$. Let $\eta \in \mathscr{P}$ and $\lambda > 0$. Then $\xi_1 + \lambda\eta \in \mathscr{P}$ and

$$\|\xi_1 - \xi\|^2 \leq \|\xi_1 + \lambda\eta - \xi\|^2,$$

i.e., $\|\xi_2\|^2 \leq \|\xi_2 + \lambda\eta\|^2$. But this is equivalent to

$$\lambda^2\|\eta\|^2 + 2\lambda \operatorname{Re}(\xi_2, \eta) \geq 0$$

for all $\lambda > 0$. Hence one must have $\operatorname{Re}(\xi_2, \eta) \geq 0$. Now $J\xi_2 = \xi_2$ by assumption and $J\eta = \eta$ because $\eta \in \mathscr{P}$. Thus

$$(\xi_2, \eta) = (J\xi_2, J\eta) = \overline{(\xi_2, \eta)}$$

and (ξ_2, η) must be real. Therefore $(\xi_2, \eta) \geq 0$. Since $\mathscr{P} = \mathscr{P}^\vee$ it follows that $\xi_2 \in \mathscr{P}$. Hence $\xi = \xi_1 - \xi_2$, ξ_1, $\xi_2 \in \mathscr{P}$. We now show $\xi_1 \perp \xi_2$. Since $(1 - \lambda)\xi_1 \in \mathscr{P}$ for $0 \leq \lambda \leq 1$ one has

$$\|\xi_1 - \xi\|^2 \leq \|(1 - \lambda)\xi_1 - \xi\|^2,$$

i.e.,

$$\|\xi_2\|^2 \leq \|\xi_2 - \lambda\xi_1\|^2.$$

Again this is equivalent to

$$\lambda^2\|\xi_1\|^2 - 2\lambda(\xi_1, \xi_2) \geq 0$$

and one must then have $(\xi_1, \xi_2) \leq 0$. But both ξ_1 and ξ_2 are in $\mathscr{P}$ and hence $(\xi_1, \xi_2) = 0$. To show uniqueness of the decomposition, consider two decompositions

$$\xi = \xi_1 - \xi_2, \qquad \xi_1, \xi_2 \in \mathscr{P}, \xi_1 \perp \xi_2,$$

and

$$\xi = \eta_1 - \eta_2, \qquad \eta_1, \eta_2 \in \mathscr{P}, \eta_1 \perp \eta_2.$$

Then one has

$$\xi_1 - \eta_1 = \xi_2 - \eta_2$$

and hence

$$\begin{aligned}\|\xi_1 - \eta_1\|^2 &= (\xi_1 - \eta_1, \xi_2 - \eta_2)\\ &= -(\eta_1, \xi_2) - (\xi_1, \eta_2) \leq 0.\end{aligned}$$

Thus $\xi_1 = \eta_1$ and consequently $\xi_2 = \eta_2$, i.e., the decompositions are identical.

(4) If ξ is orthogonal to the linear span of $\mathscr{P}$ then $\xi \in \mathscr{P}^\vee = \mathscr{P}$, hence $(\xi, \xi) = 0$ and $\xi = 0$.

EXAMPLE 2.5.29. Let $\mathfrak{M} = \mathscr{L}(\mathfrak{H})$ with $\mathfrak{H}$ finite-dimensional and consider the normal state ω given by the density matrix ρ, i.e.,

$$\omega(A) = \text{Tr}(\rho A)$$

for all $A \in \mathfrak{M}$. In Example 2.5.5 we demonstrated that ω is faithful if, and only if, ρ is invertible. Using the identification of the modular group, $\sigma_t^{\omega}(A) = \rho^{it}A\rho^{-it}$ given by Example 2.5.16, one finds that

$$\mathscr{P} = \{\psi_A; \psi_A = \pi_\omega(\rho^{1/4}A^*A\rho^{-1/4})\Omega_\omega, A \in \mathfrak{M}\}.$$

The duality condition arises because

$$\begin{aligned}(\psi_B, \psi_A) &= \text{Tr}(\rho(\rho^{-1/4}B^*B\rho^{1/4})(\rho^{1/4}A^*A\rho^{-1/4}))\\ &= \text{Tr}(\rho^{1/2}B^*B\rho^{1/2}A^*A)\\ &= \text{Tr}((B\rho^{1/2}A)^*(B\rho^{1/2}A)) \geq 0.\end{aligned}$$

Proposition 2.5.30 (Universality of the cone $\mathscr{P}$).

(1) *If $\xi \in \mathscr{P}$ then ξ is cyclic for $\mathfrak{M}$ if, and only if, ξ is separating for $\mathfrak{M}$.*

(2) *If $\xi \in \mathscr{P}$ is cyclic, and hence separating, then the modular conjugation J_ξ and the natural positive cone $\mathscr{P}_\xi$ associated with the pair $(\mathfrak{M}, \xi)$ satisfy*

$$J_\xi = J, \qquad \mathscr{P}_\xi = \mathscr{P}.$$

PROOF. (1) If $\xi \in \mathscr{P}$ is cyclic for $\mathfrak{M}$ then $J\xi$ is cyclic for $\mathfrak{M}' = J\mathfrak{M}J$. Hence $\xi = J\xi$ is separating for $\mathfrak{M}$, and conversely.

(2) Let S_ξ be the closure of the map

$$A\xi \mapsto A^*\xi, \qquad A \in \mathfrak{M},$$

and F_ξ the closure of the map

$$A'\xi \mapsto A'^*\xi, \qquad A' \in \mathfrak{M}'.$$

For each $A \in \mathfrak{M}$ one has

$$\begin{aligned} JF_\xi JA\xi &= JF_\xi(JAJ)\xi \\ &= J(JAJ)^*\xi \\ &= A^*\xi = S_\xi A\xi. \end{aligned}$$

Hence $S_\xi \subseteq JF_\xi J$.

By a symmetric argument

$$F_\xi \subseteq JS_\xi J$$

and consequently

$$JS_\xi = F_\xi J.$$

But then

$$(JS_\xi)^* = S_\xi^* J = F_\xi J = JS_\xi$$

and hence JS_ξ is selfadjoint.

We next show that JS_ξ is positive. As JS_ξ is the closure of its restriction to $\mathfrak{M}\xi$, it is enough to show that $(A\xi, JS_\xi A\xi) \geq 0$ for $A \in \mathfrak{M}$. But

$$\begin{aligned} (A\xi, JS_\xi A\xi) &= (A\xi, JA^*\xi) \\ &= (\xi, A^*j(A^*)\xi) \geq 0 \end{aligned}$$

since both ξ and $A^*j(A^*)\xi$ are in $\mathscr{P}$.

Now we have $S_\xi = J_\xi \Delta_\xi^{1/2} = J(JS_\xi)$. Thus by the uniqueness of the polar decomposition it follows that

$$J_\xi = J.$$

To prove the last statement of the theorem we note that $\mathscr{P}_\xi$ is generated by elements of the form

$$Aj_\xi(A)\xi = Aj(A)\xi.$$

But $\xi \in \mathscr{P}$ and hence $Aj(A)\xi \in \mathscr{P}$ by Proposition 2.5.26(5). Consequently

$$\mathscr{P}_\xi \subseteq \mathscr{P}.$$

But

$$\mathscr{P} = \mathscr{P}^\vee \subseteq \mathscr{P}_\xi^\vee = \mathscr{P}_\xi$$

and hence

$$\mathscr{P} = \mathscr{P}_\xi.$$

After this discussion of the geometric properties of the natural positive cone $\mathscr{P}$, we next show that all positive, normal forms on $\mathfrak{M}$ are represented by a unique vector in the cone. As a corollary one deduces that all automorphisms of $\mathfrak{M}$ are implemented by unitary elements which leave the cone invariant.

Theorem 2.5.31. *For each $\xi \in \mathscr{P}$ define the normal positive form $\omega_\xi \in \mathfrak{M}_{*+}$ by*

$$\omega_\xi(A) = (\xi, A\xi), \qquad A \in \mathfrak{M}.$$

It follows that

(a) *for any $\omega \in \mathfrak{M}_{*+}$ there exists a unique $\xi \in \mathscr{P}$ such that $\omega = \omega_\xi$,*
(b) *the mapping $\xi \mapsto \omega_\xi$ is a homeomorphism when both $\mathscr{P}$ and $\mathfrak{M}_{*+}$ are equipped with the norm topology. Moreover, the following estimates are valid:*

$$\|\xi - \eta\|^2 \leq \|\omega_\xi - \omega_\eta\| \leq \|\xi - \eta\| \, \|\xi + \eta\|.$$

Remark. In the theorem we have defined a mapping $\xi \in \mathscr{P} \mapsto \omega_\xi \in \mathfrak{M}_{*+}$. We will denote the inverse mapping by $\omega \mapsto \xi(\omega)$. One can demonstrate that $\omega \mapsto \xi(\omega)$ is monotonously increasing and concave with respect to the natural ordering of the cones $\mathfrak{M}_{*+}$ and $\mathscr{P}$. One can also derive a formula for $\xi(\omega)$ if $\omega \leq C\omega_\Omega$ for some constant C. In that case $\omega(A) = (A'\Omega, AA'\Omega)$ for a unique $A' \in \mathfrak{M}'_+$, by Theorem 2.3.19. Then $\xi(\omega) = |A'\Delta^{-1/2}|\Omega$, where $|A'\Delta^{-1/2}|$ is the positive part of the polar decomposition of $A'\Delta^{-1/2}$. We omit the proofs of these statements.

We next state an important corollary of the theorem:

Corollary 2.5.32. *There exists a unique unitary representation*

$$\alpha \in \mathrm{Aut}(\mathfrak{M}) \mapsto U(\alpha)$$

of the group $\mathrm{Aut}(\mathfrak{M})$ *of all *-automorphisms of $\mathfrak{M}$ on $\mathfrak{H}$ satisfying the following properties:*

(a) $U(\alpha)AU(\alpha)^* = \alpha(A), \qquad A \in \mathfrak{M};$
(b) $U(\alpha)\mathscr{P} \subseteq \mathscr{P}$ *and, moreover,*

$$U(\alpha)\xi(\omega) = \xi(\alpha^{-1*}(\omega)), \qquad \omega \in \mathfrak{M}_{*+},$$

where $(\alpha^*\omega)(A) = \omega(\alpha(A))$;
(c) $[U(\alpha), J] = 0.$

The mapping $\alpha \in \mathrm{Aut}\,\mathfrak{M} \mapsto U(\alpha) \in U(\mathrm{Aut}\,\mathfrak{M})$ *is a homeomorphism when* $\mathrm{Aut}\,\mathfrak{M}$ *and* $U(\mathrm{Aut}\,\mathfrak{M})$ *are equipped with their norm topologies. It is also a homeomorphism when* $U(\mathrm{Aut}\,\mathfrak{M})$ *is equipped with the weak, strong, or strong-* topology (these are equivalent) and* $\mathrm{Aut}\,\mathfrak{M}$ *is equipped with the topology of strong convergence of* $\mathrm{Aut}(\mathfrak{M})^*$ *on* $\mathfrak{M}_*$. *($\alpha \to \beta$ in this topology if, and only if, $\alpha^*(\omega) \mapsto \beta^*(\omega)$ in norm for each $\omega \in \mathfrak{M}_*$.)*

Remark. One can also obtain a partial converse. If U is a unitary element on $\mathfrak{H}$ such that $U\mathscr{P} = \mathscr{P}$, then there exists a projection $E \in \mathfrak{M} \cap \mathfrak{M}'$ such that

$$U\mathfrak{M}U^* = \mathfrak{M}E + \mathfrak{M}'(\mathbb{1} - E).$$

This will be established in Theorem 3.2.15. (Note that all the algebras $\mathfrak{M}E + \mathfrak{M}'(\mathbb{1} - E)$ have the same natural cone $\mathscr{P}$.)

The proof of the theorem and its corollary is rather long. It relies upon several straightforward but tedious calculations combined with an ingenious technique of comparison which involves doubling, or quadrupling, the system. We divide the main burden of the proof into several lemmas which the reader might be well advised to omit on a first reading.

Lemma 2.5.33. *Let ξ_1 and ξ_2 be cyclic and separating for $\mathfrak{M}$ and let $\mathfrak{H}_4$ be a four-dimensional Hilbert space with an orthogonal basis η_{ij}, $i, j = 1, 2$. Next let $\mathfrak{F}$ be the 2×2 matrix algebra generated, on $\mathfrak{H}_4$, by the matrices E_{ij} which are defined by $E_{ij}\eta_{kl} = \delta_{jk}\eta_{il}$. Moreover, let $\mathfrak{F}'$ be the commutant of $\mathfrak{F}$, i.e., the 2×2 matrix algebra generated by those F_{ij} such that $F_{ij}\eta_{kl} = \delta_{jl}\eta_{ki}$. Finally, let $\mathfrak{K} = \mathfrak{H} \otimes \mathfrak{H}_4$, $\Omega_0 = \xi_1 \otimes \eta_{11} + \xi_2 \otimes \eta_{22}$, $\mathfrak{N} = \mathfrak{M} \otimes \mathfrak{F}$, and let U_{ij}; $\mathfrak{H} \mapsto \mathfrak{K}$ be the isometry defined by $U_{ij}\xi = \xi \otimes \eta_{ij}$.*

It follows that Ω_0 is cyclic and separating for $\mathfrak{N}$ on $\mathfrak{K}$. The corresponding involution, S_{Ω_0}, associated with $(\mathfrak{N}, \Omega_0)$ satisfies

$$S_{\Omega_0} = U_{11}S_{\xi_1}U_{11}^* + U_{21}S_{\xi_1,\xi_2}U_{12}^* + U_{12}S_{\xi_2,\xi_1}U_{21}^* + U_{22}S_{\xi_2}U_{22}^*,$$

where S_{ξ_1,ξ_2} denotes the closure of the operator defined on $\mathfrak{M}\xi_2$ by

$$S_{\xi_1,\xi_2}A\xi_2 = A^*\xi_1, \qquad A \in \mathfrak{M}.$$

PROOF. Any element $A \in \mathfrak{N}$ has the form

$$A = \sum_{ij} A_{ij} \otimes E_{ij},$$

where $A_{ij} \in \mathfrak{M}$. Thus

$$A\Omega_0 = A_{11}\xi_1 \otimes \eta_{11} + A_{12}\xi_2 \otimes \eta_{12} + A_{21}\xi_1 \otimes \eta_{21} + A_{22}\xi_2 \otimes \eta_{22}.$$

This shows that Ω_0 is separating and cyclic for $\mathfrak{N}$, so we may define S_{Ω_0} as the closure of the map $A\Omega_0 \mapsto A^*\Omega_0$, $A \in \mathfrak{N}$. Using the above notation

$$A^* = A_{11}^* \otimes E_{11} + A_{21}^* \otimes E_{12} + A_{12}^* \otimes E_{21} + A_{22}^* \otimes E_{22}$$

and hence

$$A^*\Omega_0 = A_{11}^*\xi_1 \otimes \eta_{11} + A_{21}^*\xi_2 \otimes \eta_{12} + A_{12}^*\xi_1 \otimes \eta_{21} + A_{22}^*\xi_2 \otimes \eta_{22}.$$

Taking the closure on both sides one obtains the closability of S_{ξ_1,ξ_2} and

$$S_{\Omega_0} = U_{11}S_{\xi_1}U_{11}^* + U_{21}S_{\xi_1,\xi_2}U_{12}^* + U_{12}S_{\xi_2,\xi_1}U_{21}^* + U_{22}S_{\xi_2}U_{22}^*.$$

Lemma 2.5.34. *Adopt the same notation as in Lemma 2.5.33 and let $S_{\xi_1,\xi_2} = J_{\xi_1,\xi_2}\Delta_{\xi_1,\xi_2}^{1/2}$ be the polar decomposition of S_{ξ_1,ξ_2}. It follows that*

$$\begin{aligned} J_{\Omega_0} = {} & U_{11}J_{\xi_1}U_{11}^* + U_{21}J_{\xi_1,\xi_2}U_{12}^* \\ & + U_{12}J_{\xi_2,\xi_1}U_{21}^* + U_{22}J_{\xi_2}U_{22}^*, \end{aligned} \tag{$*$}$$

and

$$\begin{aligned} \Delta_{\Omega_0} = {} & U_{11}\Delta_{\xi_1}U_{11}^* + U_{21}\Delta_{\xi_2,\xi_1}U_{21}^* \\ & + U_{12}\Delta_{\xi_1,\xi_2}U_{12}^* + U_{22}\Delta_{\xi_2}U_{22}^*. \end{aligned} \tag{$**$}$$

Proof. Viewing S_{Ω_0}, J_{Ω_0}, and Δ_{Ω_0} as 4×4 matrices one sees immediately that the right-hand side of (*) is an isometry of $\mathfrak{K}$ onto $\mathfrak{K}$ and that the right-hand side of (**) is a positive selfadjoint operator. By uniqueness of the polar decomposition it is enough to verify $J_{\Omega_0}\Delta_{\Omega_0}^{1/2} = S_{\Omega_0}$. This is an easy calculation.

Lemma 2.5.35. *Adopt the notation of Lemmas* 2.5.33 *and* 2.5.34. *There exists a unique unitary element* $U' \in \mathfrak{M}'$ *such that* $J_{\Omega_0}(\mathbb{1} \otimes E_{21})J_{\Omega_0} = U' \otimes F_{21}$ *and for this unitary element*

$$J_{\xi_2, \xi_1} = U'J_{\xi_1}, \qquad J_{\xi_1, \xi_2} = J_{\xi_1}U'^*, \qquad J_{\xi_2} = U'J_{\xi_1}U'^*.$$

Proof. Since $J_{\Omega_0}^2 = \mathbb{1}$, it follows from Lemma 2.5.34 that

$$J_{\xi_1, \xi_2}J_{\xi_2, \xi_1} = J_{\xi_2, \xi_1}J_{\xi_1, \xi_2} = \mathbb{1}.$$

For $\xi_{ij} \in \mathfrak{H}$ we then have

$$\begin{aligned} J_{\Omega_0}(\mathbb{1} \otimes E_{11})J_{\Omega_0}\left(\sum_{ij} \xi_{ij} \otimes \eta_{ij}\right) &= \xi_{11} \otimes \eta_{11} + \xi_{21} \otimes \eta_{21} \\ &= (\mathbb{1} \otimes F_{11})\left(\sum_{ij} \xi_{ij} \otimes \eta_{ij}\right). \end{aligned}$$

Therefore

$$J_{\Omega_0}(\mathbb{1} \otimes E_{11})J_{\Omega_0} = \mathbb{1} \otimes F_{11}$$

and, by a similar argument,

$$J_{\Omega_0}(\mathbb{1} \otimes E_{22})J_{\Omega_0} = \mathbb{1} \otimes F_{22}.$$

But $\mathbb{1} \otimes E_{21}$ is a partial isometry with initial projection $\mathbb{1} \otimes E_{11}$ and final projection $\mathbb{1} \otimes E_{22}$. Thus $J_{\Omega_0}(\mathbb{1} \otimes E_{21})J_{\Omega_0}$ must be a partial isometry with initial projection $\mathbb{1} \otimes F_{11}$ and final projection $\mathbb{1} \otimes F_{22}$. Since

$$J_{\Omega_0}(\mathbb{1} \otimes E_{21})J_{\Omega_0} \in \mathfrak{N}' = \mathfrak{M}' \otimes \mathfrak{F}'$$

it follows that there exists a unitary element $U' \in \mathfrak{M}'$ such that

$$J_{\Omega_0}(\mathbb{1} \otimes E_{21})J_{\Omega_0} = U' \otimes F_{21}.$$

Next, for $\xi \in \mathfrak{H}$ we use Lemma 2.5.34 to compute

$$\begin{aligned} (J_{\xi_2, \xi_1}\xi) \otimes \eta_{12} &= J_{\Omega_0}(\xi \otimes \eta_{21}) \\ &= J_{\Omega_0}(\mathbb{1} \otimes E_{21})(\xi \otimes \eta_{11}) \\ &= J_{\Omega_0}(\mathbb{1} \otimes E_{21})J_{\Omega_0}J_{\Omega_0}(\xi \otimes \eta_{11}) \\ &= (U' \otimes F_{21})(J_{\xi_1}\xi \otimes \eta_{11}) \\ &= (U'J_{\xi_1}\xi) \otimes \eta_{12}. \end{aligned}$$

Hence $J_{\xi_2, \xi_1} = U'J_{\xi_1}$. Taking the adjoint we obtain $J_{\xi_1, \xi_2} = J_{\xi_1}U'^*$. Finally, for $\xi \in \mathfrak{H}$

$$\begin{aligned} (J_{\xi_2}\xi) \otimes \eta_{22} &= J_{\Omega_0}(\xi \otimes \eta_{22}) \\ &= J_{\Omega_0}(\mathbb{1} \otimes E_{21})J_{\Omega_0}J_{\Omega_0}(\xi \otimes \eta_{12}) \\ &= (U' \otimes F_{21})(J_{\xi_1}U'^*\xi \otimes \eta_{21}) \\ &= (U'J_{\xi_1}U'^*\xi) \otimes \eta_{22}. \end{aligned}$$

Therefore

$$J_{\xi_2} = U'J_{\xi_1}U'^*.$$

Next, for any pair ξ_1, ξ_2 of vectors which are both cyclic and separating for $\mathfrak{M}$ we let $\theta(\xi_2, \xi_1)$ denote the unitary element $U' \in \mathfrak{M}'$ obtained in Lemma 2.5.35.

Lemma 2.5.36. *If ξ_1, ξ_2 are cyclic and separating for $\mathfrak{M}$ and $U' \in \mathfrak{M}'$ is unitary then it follows that*

$$\theta(U'\xi_2, \xi_1) = U'\theta(\xi_2, \xi_1).$$

PROOF. For $A \in \mathfrak{M}$ one has

$$S_{U'\xi_2, \xi_1}A\xi_1 = A^*U'\xi_2 = U'A^*\xi_2 = U'S_{\xi_2, \xi_1}A\xi_1.$$

Hence

$$S_{U'\xi_2, \xi_1} = U'S_{\xi_2, \xi_1}.$$

Therefore

$$J_{U'\xi_2, \xi_1} = U'J_{\xi_2, \xi_1}.$$

Now the result follows from Lemma 2.5.35.

Note that the last two lemmas imply that θ satisfies a chain rule:

$$\theta(\xi_3, \xi_1) = \theta(\xi_3, \xi_2)\theta(\xi_2, \xi_1).$$

Lemma 2.5.37. *If ξ is a cyclic and separating vector for $\mathfrak{M}$ the following three statements are equivalent:*

(i) $\theta(\xi, \Omega) = \mathbb{1}$;
(ii) $\xi \in \mathscr{P}_\Omega$;
(iii) $\mathscr{P}_\xi = \mathscr{P}_\Omega$.

PROOF. (i) $\Rightarrow$ (ii) By Lemma 2.5.35, if $\theta(\xi, \Omega) = \mathbb{1}$ then $J_\xi = J_\Omega = J_{\xi,\Omega} = J$. Hence for $A \in \mathfrak{M}$

$$\begin{aligned}(\xi, Aj(A)\Omega) &= (A^*\xi, j(A)\Omega)\\ &= (S_{\xi,\Omega}A\Omega, JA\Omega)\\ &= (J\Delta_{\xi,\Omega}^{1/2}A\Omega, JA\Omega)\\ &= (A\Omega, \Delta_{\xi,\Omega}^{1/2}A\Omega) \geq 0.\end{aligned}$$

Thus ξ is in the dual cone of $\mathscr{P}_\Omega$, which is equal to $\mathscr{P}_\Omega$.

(ii) $\Rightarrow$ (iii) This is Proposition 2.5.30(2).

(iii) $\Rightarrow$ (ii) This implication is trivial.

(ii) $\Rightarrow$ (i) If $\xi \in \mathscr{P}_\Omega$ then for all $A \in \mathfrak{M}$

$$\begin{aligned}0 &\leq (\xi, AJ_\Omega A\Omega)\\ &= (A^*\xi, J_\Omega A\Omega)\\ &= (S_{\xi,\Omega}A\Omega, J_\Omega A\Omega) = (A\Omega, J_\Omega S_{\xi,\Omega}A\Omega).\end{aligned}$$

Hence

$$(\eta, J_\Omega S_{\xi,\Omega}\eta) \geq 0, \qquad \eta \in D(S_{\xi,\Omega}).$$

By Lemma 2.5.33 and Proposition 2.5.9, the adjoint of $S_{\xi,\Omega}$ is the closure $F_{\xi,\Omega}$ of the mapping

$$A'\Omega \mapsto A'^*\xi, \qquad A' \in \mathfrak{M}'.$$

Thus for $A \in \mathfrak{M}$

$$\begin{aligned} J_\Omega F_{\xi,\Omega} J_\Omega A\Omega &= J_\Omega F_{\xi,\Omega} J_\Omega A J_\Omega \Omega \\ &= J_\Omega (J_\Omega A J_\Omega)^* \xi \\ &= A^*\xi = S_{\xi,\Omega} A\Omega, \end{aligned}$$

where we have used $J_\Omega \xi = \xi$. Hence

$$S_{\xi,\Omega} \subseteq J_\Omega F_{\xi,\Omega} J_\Omega.$$

But by a symmetric argument

$$F_{\xi,\Omega} \subseteq J_\Omega S_{\xi,\Omega} J_\Omega$$

and consequently

$$J_\Omega S_{\xi,\Omega} = F_{\xi,\Omega} J_\Omega.$$

Thus

$$(J_\Omega S_{\xi,\Omega})^* = S_{\xi,\Omega}^* J_\Omega = F_{\xi,\Omega} J_\Omega = J_\Omega S_{\xi,\Omega}.$$

We have proved that $J_\Omega S_{\xi,\Omega}$ is positive and selfadjoint. By the uniqueness of the polar decomposition it follows that $J_\Omega = J_{\xi,\Omega}$ because $S_{\xi,\Omega} = J_\Omega(J_\Omega S_{\xi,\Omega})$. Hence $\theta(\xi, \Omega) = \mathbb{1}$ by Lemma 2.5.35.

The next lemma will prove Theorem 2.5.31(a) for a dense set of forms in $\mathfrak{M}_{*+}$. The complete theorem will follow from this partial result and the estimate in Theorem 2.5.31(b).

Lemma 2.5.38. *Let $\eta \in \mathfrak{H}$ be a vector which is cyclic and separating for $\mathfrak{M}$. It follows that there exists a unique vector $\xi \in \mathscr{P}$ with the property*

$$(\eta, A\eta) = (\xi, A\xi)$$

for all $A \in \mathfrak{M}$.

PROOF. Let $U' = \theta(\eta, \Omega)$, and set $\xi = U'^*\eta$. Since U' is a unitary element of $\mathfrak{M}'$, one has

$$(\eta, A\eta) = (\xi, A\xi)$$

for all $A \in \mathfrak{M}$. But by Lemma 2.5.36

$$\theta(\xi, \Omega) = \theta(U'^*\eta, \Omega) = U'^*\theta(\eta, \Omega) = \mathbb{1}$$

and by Lemma 2.5.37 one concludes that $\xi \in \mathscr{P}$.

To prove uniqueness, assume $\xi' \in \mathscr{P}$ with $\omega_{\xi'} = \omega_\xi$. But ξ' is separating for $\mathfrak{M}$ and consequently cyclic by Proposition 2.5.30. Thus $\theta(\xi', \Omega)$ is defined. Since $\omega_{\xi'} = \omega_\xi$

there exists a unitary $U' \in \mathfrak{M}'$ such that $\xi' = U'\xi$, by Theorem 2.3.16. Thus by Lemma 2.5.36

$$\theta(\xi', \Omega) = \theta(U'\xi, \Omega) = U'.$$

Since $\xi' \in \mathscr{P}$ it follows from Lemma 2.5.37 that $U' = \mathbb{1}$; thus $\xi' = \xi$.

Lemma 2.5.39. *The set of positive forms ω_η, where η is cyclic and separating for $\mathfrak{M}$, is norm dense in $\mathfrak{M}_{*+}$.*

PROOF. If $\omega \in \mathfrak{M}_{*+}$ then ω has the form $\omega(A) = \sum_n (\xi_n, A\xi_n)$, with $\sum_n \|\xi_n\|^2 < \infty$ by Theorem 2.4.21. Each ξ_n may be approximated by a vector of the form $A_n'\Omega$, where $A_n' \in \mathfrak{M}'$. But for $A \in \mathfrak{M}_+$

$$\begin{aligned}(A_n'\Omega, AA_n'\Omega) &= (A^{1/2}\Omega, A_n'^*A_n'A^{1/2}\Omega)\\ &\leq \|A_n'\|^2(\Omega, A\Omega).\end{aligned}$$

Thus the set of positive forms ω such that

$$\omega(A) \leq \alpha(\Omega, A\Omega), \qquad A \in \mathfrak{M}_+,$$

for some constant α, is norm dense in $\mathfrak{M}_{*+}$. But by Theorem 2.3.19, the latter states have the form

$$\omega(A) = (A'\Omega, AA'\Omega)$$

where $A' \in \mathfrak{M}'_+$. Now each $A' \in \mathfrak{M}'_+$ can be approximated in norm by an invertible $B' \in \mathfrak{M}'_+$. Set $\eta = B'\Omega$. Then η is cyclic and separating for $\mathfrak{M}'$, by the following argument. If $A\eta = 0$ for $A \in \mathfrak{M}$ then $0 = B'^{-1}A\eta = AB'^{-1}\eta = A\Omega$, thus $A = 0$ and η is separating for $\mathfrak{M}$. But if $A'\eta = 0$ for $A' \in \mathfrak{M}'$ then $A'B\Omega = 0$ and $A'B' = 0$. The invertibility of B' implies that $A' = 0$ and η is separating for $\mathfrak{M}'$, and hence cyclic for $\mathfrak{M}$ by Proposition 2.5.3.

Next we turn to the estimate in Theorem 2.5.31. First we need a lemma.

Lemma 2.5.40. *Set $\mathfrak{H}_{sa} = \mathscr{P}-\mathscr{P}$, $\mathfrak{M}_{sa} = \mathfrak{M}_+-\mathfrak{M}_+$. Then the map Φ; $\mathfrak{M}_{sa} \mapsto \mathfrak{H}_{sa}$; $A \mapsto \Delta^{1/4}A\Omega$ is an order isomorphism of $\mathfrak{M}_{sa}$ on the set of $\xi \in \mathfrak{H}_{sa}$ such that*

$$-\alpha\Omega \leq \xi \leq \alpha\Omega$$

for some constant $\alpha > 0$ (the orders are those induced by the cones $\mathfrak{M}_+$ and $\mathscr{P}$).

PROOF. We have $A \in \mathfrak{M}_+ \Rightarrow \Delta^{1/4}A\Omega \in \mathscr{P}$ by Proposition 2.5.26. Conversely, if $A = A^*$ and $\Delta^{1/4}A\Omega \in \mathscr{P}$ then for any $A' \in \mathfrak{M}'$

$$\begin{aligned}(A'\Omega, AA'\Omega) &= (A'^*A'\Omega, A\Omega)\\ &= (\Delta^{-1/4}|A'|^2\Omega, \Delta^{1/4}A\Omega) \geq 0.\end{aligned}$$

Hence $A \geq 0$.

Thus $\Phi: \mathfrak{M}_{sa} \mapsto \Phi(\mathfrak{M}_{sa})$ is an order isomorphism. Next we show that Φ is $\sigma(\mathfrak{M}, \mathfrak{M}_*)$–$\sigma(\mathfrak{H}, \mathfrak{H})$ continuous ($\sigma(\mathfrak{H}, \mathfrak{H})$ is the weak topology on $\mathfrak{H}$). For $A \in \mathfrak{M}$ we have

$$(\mathbb{1} + \Delta^{1/2})A\Omega = A\Omega + JA^*\Omega;$$

thus

$$A\Omega = (\mathbb{1} + \Delta^{1/2})^{-1}(A\Omega + JA^*\Omega)$$

and

$$\Delta^{1/4}A\Omega = (\Delta^{1/4} + \Delta^{-1/4})^{-1}(A\Omega + JA^*\Omega).$$

Consequently, for $\eta \in \mathfrak{H}_{sa}$

$$(\eta, \Delta^{1/4}A\Omega) = ((\Delta^{1/4} + \Delta^{-1/4})^{-1}\eta, A\Omega + JA^*\Omega).$$

Since $\|(\Delta^{1/4} + \Delta^{-1/4})^{-1}\| \leq 1/2$, the continuity statement follows.

Now assume $-\alpha\Omega \leq \xi \leq \alpha\Omega$. By suitable renormalization we may assume $0 \leq \xi \leq \Omega$. Put

$$\xi_n = f_n(\log \Delta)\xi$$

with $f_n(x) = e^{-x^2/2n^2}$. Then $\xi_n \in D(\Delta^\beta)$ for all $\beta \in \mathbb{C}$ and by Proposition 2.5.26(3)

$$0 \leq \xi_n = f_n(\log \Delta)\xi \leq f_n(\log \Delta)\Omega = \Omega.$$

Now for any $A' \in \mathfrak{M}'_+$

$$(\Delta^{-1/4}\xi_n, A'\Omega) = (\xi_n, \Delta^{-1/4}A'\Omega) \geq 0$$

since $\Delta^{-1/4}A'\Omega \in \mathscr{P}$. Hence by Proposition 2.5.27

$$\Delta^{-1/4}\xi_n \in \overline{\mathfrak{M}'_+\Omega}^{\vee} = \overline{\mathfrak{M}_+\Omega}.$$

Likewise, one finds

$$\Omega - \Delta^{-1/4}\xi_n = \Delta^{-1/4}(\Omega - \xi_n) \in \overline{\mathfrak{M}_+\Omega}$$

and thus, for $A' \in \mathfrak{M}'_+$,

$$0 \leq (\Delta^{-1/4}\xi_n, A'\Omega) \leq (\Omega, A'\Omega).$$

Hence by Proposition 2.5.27(1) there exists an operator $A_n \in \mathfrak{M}$ such that $0 \leq A_n \leq \mathbb{1}$ and

$$\Delta^{-1/4}\xi_n = A_n\Omega$$

and therefore $\xi_n \in \mathfrak{B}$, where

$$\mathfrak{B} = \{\Phi(A); A \in \mathfrak{M}_{sa}, 0 \leq A \leq \mathbb{1}\}.$$

Since $\{A; A \in \mathfrak{M}_{sa}, 0 \leq A \leq \mathbb{1}\}$ is a σ-weakly closed, and thus σ-weakly compact, subset of $\mathfrak{M}_1$, and Φ is $\sigma(\mathfrak{M}, \mathfrak{M}_*)$–$\sigma(\mathfrak{H}, \mathfrak{H})$ continuous it follows that $\mathfrak{B}$ is $\sigma(\mathfrak{H}, \mathfrak{H})$-compact. Thus $\xi = \lim_n \xi_n \in \mathfrak{B}$.

Lemma 2.5.41. *For $\xi, \eta \in \mathscr{P}$,*

$$\|\xi - \eta\|^2 \leq \|\omega_\xi - \omega_\eta\| \leq \|\xi - \eta\| \, \|\xi + \eta\|.$$

PROOF. The second inequality is valid for all $\xi, \eta \in \mathfrak{H}$ since

$$(\omega_\xi - \omega_\eta)(A) = \tfrac{1}{2}[(\xi - \eta, A(\xi + \eta)) + (\xi + \eta, A(\xi - \eta))].$$

To prove the other inequality, assume first that $\xi + \eta$ is cyclic and separating. Since $\xi + \eta \in \mathscr{P}$, $\mathscr{P}_{\xi+\eta} = \mathscr{P}$ by Proposition 2.5.30(2). Also

$$-(\xi + \eta) \leq \xi - \eta \leq (\xi + \eta).$$

Thus applying Lemma 2.5.40 with $\Omega = \xi + \eta$ one deduces the existence of an $A = A^* \in \mathfrak{M}$, with

$$-\mathbb{1} \leq A \leq \mathbb{1}$$

and such that

$$\xi - \eta = \Delta_{\xi+\eta}^{1/4} A(\xi + \eta).$$

Hence

$$\begin{aligned}\|\omega_\xi - \omega_\eta\| &\geq (\omega_\xi - \omega_\eta)(A)\\ &= (\xi, A\xi) - (\eta, A\eta)\\ &= \mathrm{Re}(\xi - \eta, A(\xi + \eta))\\ &= (\xi - \eta, \Delta_{\xi+\eta}^{-1/4}(\xi - \eta)).\end{aligned}$$

As $J(\xi - \eta) = \xi - \eta$, and $J\Delta_{\xi+\eta}^{-1/4} = \Delta_{\xi+\eta}^{1/4} J$, one has

$$(\xi - \eta, \Delta_{\xi+\eta}^{-1/4}(\xi - \eta)) = (\xi - \eta, \Delta_{\xi+\eta}^{1/4}(\xi - \eta)).$$

Thus

$$\begin{aligned}\|\omega_\xi - \omega_\eta\| &\geq (\xi - \eta, \tfrac{1}{2}(\Delta_{\xi+\eta}^{1/4} + \Delta_{\xi+\eta}^{-1/4})(\xi - \eta))\\ &\geq \|\xi - \eta\|^2\end{aligned}$$

because $\frac{1}{2}(\Delta_{\xi+\eta}^{1/4} + \Delta_{\xi+\eta}^{-1/4}) \geq \mathbb{1}$.

Now for general ξ and η in $\mathscr{P}$ we can find sequences $A_n', B_n' \in \mathfrak{M}'_+$ of entire analytic elements for σ' such that $\xi_n = \Delta^{-1/4} A_n'\Omega \to \xi$, $\eta_n = \Delta^{-1/4} B_n'\Omega \to \eta$. By adding $\varepsilon_n \mathbb{1}$ to A_n', B_n' we may assume $A_n' \geq \varepsilon_n \mathbb{1} > 0$ and $B_n' \geq \varepsilon_n \mathbb{1} > 0$. But then $A_n' + B_n' \geq 2\varepsilon_n \mathbb{1}$ and so the $A_n' + B_n'$ are invertible. Hence the $\Delta^{-1/4}(A_n' + B_n')\Delta^{1/4} \in \mathfrak{M}'$ are invertible and consequently $\xi_n + \eta_n = \Delta^{-1/4}(A_n' + B_n')\Omega$ is separating and cyclic for $\mathfrak{M}$. Thus

$$\|\omega_{\xi_n} - \omega_{\eta_n}\| \geq \|\xi_n - \eta_n\|^2$$

and the first inequality of the lemma follows from the second by limiting.

END OF PROOF OF THEOREM 2.5.31. Part (b) is simply Lemma 2.5.41, while part (a) follows from part (b), Lemma 2.5.38, and Lemma 2.5.39.

PROOF OF COROLLARY 2.5.32. Let $\alpha ; \mathfrak{M} \mapsto \mathfrak{M}$ be an automorphism, and let $\xi \in \mathscr{P}$ be the vector representing the state

$$A \mapsto (\Omega, \alpha^{-1}(A)\Omega),$$

i.e.,

$$(\xi, A\xi) = (\Omega, \alpha^{-1}(A)\Omega).$$

Then ξ is separating for $\mathfrak{M}$ and hence cyclic by Proposition 2.5.30. Next, define an operator $U = U(\alpha)$ on $\mathfrak{M}\Omega$ by

$$UA\Omega = \alpha(A)\xi.$$

One has

$$\|UA\Omega\|^2 = (\xi, \alpha(A^*A)\xi) = (\Omega, A^*A\Omega) = \|A\Omega\|^2$$

and so U extends by closure to an isometry, also denoted by U. Since ξ is cyclic, the range of U is dense, hence U is unitary and $U^* = U^{-1}$, i.e.,

$$U^*A\xi = \alpha^{-1}(A)\Omega.$$

Now, for $A, B \in \mathfrak{M}$

$$\begin{aligned} UAU^*B\xi &= UA\alpha^{-1}(B)\Omega \\ &= \alpha(A\alpha^{-1}(B))\xi = \alpha(A)B\xi. \end{aligned}$$

Hence

$$\alpha(A) = UAU^*, \qquad A \in \mathfrak{M},$$

which proves (a) with the identification $U(\alpha) = U$. Next, note that

$$\begin{aligned} SU^*A\xi &= S\alpha^{-1}(A)\Omega \\ &= \alpha^{-1}(A)^*\Omega \\ &= \alpha^{-1}(A^*)\Omega \\ &= U^*A^*\xi \\ &= U^*S_\xi A\xi. \end{aligned}$$

Hence by closure

$$J\Delta^{1/2}U^* = U^*J_\xi\Delta_\xi^{1/2} = U^*J\Delta_\xi^{1/2}$$

or

$$UJU^*U\Delta^{1/2}U^* = J\Delta_\xi^{1/2}.$$

By the uniqueness of the polar decomposition $UJU^* = J$ or, equivalently,

$$[U, J] = 0,$$

which proves (c). Now, using (c) and (a) we have for $A \in \mathfrak{M}$

$$UAj(A)\Omega = \alpha(A)j(\alpha(A))\xi.$$

Since $\xi \in \mathscr{P}$, we deduce from Proposition 2.5.26(5) and Proposition 2.5.30(2) that

$$U\mathscr{P} = \mathscr{P}.$$

If $\varphi \in \mathfrak{M}_{*+}$ we then have

$$\begin{aligned} (U\xi(\varphi), AU\xi(\varphi)) &= (\xi(\varphi), U^*AU\xi(\varphi)) \\ &= (\xi(\varphi), \alpha^{-1}(A)\xi(\varphi)) = \varphi(\alpha^{-1}(A)) \\ &= (\alpha^{-1*}(\varphi))(A) = (\xi(\alpha^{-1*}(\varphi)), A\xi(\alpha^{-1*}(\varphi))) \end{aligned}$$

for all $A \in \mathfrak{M}$. Hence, by the uniqueness of the representing vector in $\mathscr{P}$,

$$U(\alpha)\xi(\varphi) = \xi(\alpha^{-1*}(\varphi)).$$

This establishes (b), and also immediately implies that $\alpha \mapsto U(\alpha)$ is a representation and that $U(\alpha)$ is unique.

The continuity of the maps $\alpha \mapsto U(\alpha)$ and $U(\alpha) \mapsto \alpha$ in the various topologies described in the corollary follows from Theorem 2.5.31 (b) and the fact that

$$\|U(\alpha) - U(\beta)\| = \|(U(\alpha) - U(\beta))|_{\mathscr{P}}\|.$$

The last equality results from the fact that each vector $\psi \in \mathfrak{H}$ has a unique decomposition (Proposition 2.5.28):

$$\psi = \psi_1 - \psi_2 + i(\psi_3 - \psi_4),$$

where $\psi_i \in \mathscr{P}$, $\psi_1 \perp \psi_2$, $\psi_3 \perp \psi_4$, and the $U(\alpha)$ respect this decomposition. The equality of the weak, strong, and strong* topology on the unitary group $U(\mathfrak{H})$ on $\mathfrak{H}$ arises from the identity

$$\begin{aligned} \|(V - U)\psi\|^2 &= ((V - U)\psi, (V - U)\psi) \\ &= 2\|\psi\|^2 - (V\psi, U\psi) - (U\psi, V\psi). \end{aligned}$$

Thus if $V \to U$ weakly then $V \to U$ strongly and, analogously, $V^* \to U^*$ strongly.

2.6. Quasi-Local Algebras

2.6.1. Cluster Properties

In the preceding sections of this chapter we described the structure of general C^*-algebras and von Neumann algebras. Now we discuss a specific class of C^*-algebras, quasi-local algebras, and partially analyze a distinguished set of states, the locally normal states, over these algebras.

The distinctive feature of quasi-local algebras is that they are generated by an increasing net of subalgebras, the local algebras, and we are particularly interested in states which have an approximate factorization property on these subalgebras. Such factorizations are commonly referred to as cluster properties and they are closely related to "purity" or "irreducibility" criteria.

Theorem 2.6.1. *Let $\mathfrak{A}$ be a C^*-algebra on a Hilbert space $\mathfrak{H}$ with a cyclic unit vector Ω, and define a state ω on $\mathscr{L}(\mathfrak{H})$ by*

$$\omega(B) = (\Omega, B\Omega), \qquad B \in \mathscr{L}(\mathfrak{H}).$$

Let $\{\mathfrak{M}_\alpha\}$ be a decreasing net of von Neumann algebras and define $\mathfrak{M}$ by

$$\mathfrak{M} = \bigcap_\alpha \mathfrak{M}_\alpha.$$

If $\mathfrak{M} \subseteq \mathfrak{A}'$ or if $\mathfrak{M} \subseteq \mathfrak{A}''$ and Ω is separating for $\mathfrak{A}''$, then the following conditions are equivalent:

(1) *$\mathfrak{M}$ consists of multiples of the identity;*
(2) *given $A \in \mathfrak{A}$ there exists an α such that*

$$|\omega(AM) - \omega(A)\omega(M)| \leq \|M\|$$

for all $M \in \mathfrak{M}_\alpha$;
(3) *given $A \in \mathfrak{A}''$ there exists an α such that*

$$|\omega(AM) - \omega(A)\omega(M)| \leq \{\omega(M^*M) + \omega(MM^*)\}^{1/2}$$

for all $M \in \mathfrak{M}_\alpha$.

PROOF. (3) $\Rightarrow$ (2) First remark that replacement of A by A/ε demonstrates that the inequality of condition (2) could be replaced by

$$|\omega(AM) - \omega(A)\omega(M)| < \varepsilon\|M\|.$$

An analogous remark is valid for condition (3). Therefore (3) $\Rightarrow$ (2) because

$$\{\omega(M^*M) + \omega(MM^*)\}^{1/2} \leq \sqrt{2}\|M\|.$$

(2) $\Rightarrow$ (1) If condition (1) is false then there exist $A, B \in \mathfrak{A}$ and $M \in \mathfrak{M}$ such that

$$(\Omega, AMB\Omega) \neq \omega(M)(\Omega, AB\Omega).$$

Rescaling A and M one deduces that there exist $A, B \in \mathfrak{A}$ and $M \in \mathfrak{M}$ such that

$$|\omega(AMB) - \omega(M)\omega(AB)| > \|M\|.$$

Thus if $\mathfrak{M} \subseteq \mathfrak{A}'$ then (2) is false and (2) $\Rightarrow$ (1). If, alternatively, Ω is separating for $\mathfrak{A}''$, and $\mathfrak{M} \subseteq \mathfrak{A}''$, and if $\mathfrak{M} \neq \mathbb{C}\mathbb{1}$, then there exists an $M \in \mathfrak{M}$ such that $M\Omega \neq \omega(M)\Omega$. Thus there is an $A \in \mathfrak{A}$ such that $|(A^*\Omega, (M - \omega(M)\mathbb{1})\Omega)| > \|M\|$, i.e.,

$$|\omega(AM) - \omega(A)\omega(M)| > \|M\|.$$

It remains to prove that (1) $\Rightarrow$ (3). This depends upon the following result:

Lemma 2.6.2. *Let $\mathfrak{M}_\alpha$ be a decreasing net of von Neumann algebras on a Hilbert space $\mathfrak{H}$ and define $\mathfrak{M}$ by*

$$\mathfrak{M} = \bigcap_\alpha \mathfrak{M}_\alpha.$$

Assume Ω is cyclic for $\mathfrak{M}'$, in $\mathfrak{H}$, and that there exists a net of elements $M_\alpha \in \mathfrak{M}_\alpha$ such that the following weak limits exist:

$$\varphi = \operatorname*{weak\,lim}_\alpha M_\alpha\Omega, \qquad \varphi^* = \operatorname*{weak\,lim}_\alpha M_\alpha^*\Omega.$$

It follows that φ and φ^* belong to the closure $\overline{\mathfrak{M}\Omega}$ of the set $\mathfrak{M}\Omega$.

Proof. Define $N_\alpha = (M_\alpha + M_\alpha^*)/2$ and $\psi = (\varphi + \varphi^*)/2$; then for $A \in \mathfrak{M}_\alpha'$ one has

$$\begin{aligned}(A\psi, \Omega) &= \lim_\beta (AN_\beta\Omega, \Omega) \\ &= \lim_\beta (A\Omega, N_\beta\Omega) \\ &= (A\Omega, \psi).\end{aligned}$$

It follows that the same relation is valid for all $A \in \{\bigcup_\alpha \mathfrak{M}_\alpha'\}''$. But

$$\left\{\bigcup_\alpha \mathfrak{M}_\alpha'\right\}'' = \mathfrak{M}'.$$

Now let $P \in \mathfrak{M}'$ be the projector with range $\overline{\mathfrak{M}\Omega}$. If $A \in \mathfrak{M}'$ one has

$$\begin{aligned}(A\Omega, P\psi) &= (PA\Omega, \psi) = (PA\psi, \Omega) \\ &= (A\psi, P\Omega) = (A\psi, \Omega) = (A\Omega, \psi).\end{aligned}$$

Thus $\psi - P\psi$ is orthogonal to the dense set of vectors $\mathfrak{M}'\Omega$ and

$$\frac{\varphi + \varphi^*}{2} = \psi = P\psi \in \overline{\mathfrak{M}\Omega}.$$

A similar argument with N_α replaced by $N_\alpha = (M_\alpha - M_\alpha^*)/2i$ and ψ replaced by $(\varphi - \varphi^*)/2i$ gives

$$\frac{\varphi - \varphi^*}{2i} \in \overline{\mathfrak{M}\Omega}$$

and we conclude that $\varphi, \varphi^* \in \overline{\mathfrak{M}\Omega}$.

Now let us complete the proof of Theorem 2.6.1, (1) $\Rightarrow$ (3). Assume that (3) is false. Thus there exists an $A \in \mathfrak{A}''$ and, for every α, an $M_\alpha \in \mathfrak{M}_\alpha$ such that

$$|\omega(AM_\alpha) - \omega(A)\omega(M_\alpha)| > \{\omega(M_\alpha^*M_\alpha) + \omega(M_\alpha M_\alpha^*)\}^{1/2}.$$

Now the right-hand side must be nonzero because $\omega(M_\alpha^*M_\alpha) = 0$ implies $M_\alpha\Omega = 0$, and then the left-hand side is zero, which is a contradiction. Next define N_α by

$$N_\alpha = \frac{M_\alpha}{\{\omega(M_\alpha^*M_\alpha) + \omega(M_\alpha M_\alpha^*)\}^{1/2}}$$

and remark that

$$|\omega(AN_\alpha) - \omega(A)\omega(N_\alpha)| > 1$$

and

$$\omega(N_\alpha^*N_\alpha) + \omega(N_\alpha N_\alpha^*) = 1.$$

The last condition implies that $\|N_\alpha\Omega\| \le 1$ and $\|N_\alpha^*\Omega\| \le 1$. Hence, by weak compactness of the unit ball of $\mathfrak{H}$, there exists a subnet $\{N_{\alpha'}\}$ such that the following limits,

$$\varphi = \lim_{\alpha'} N_{\alpha'}\Omega, \qquad \varphi^* = \lim_{\alpha'} N_{\alpha'}^*\Omega,$$

exist. Therefore

$$|(\Omega, A\varphi) - \omega(A)(\Omega, \varphi)| \ge 1.$$

Now we apply Lemma 2.6.2 to $\mathfrak{M}$ and $N_{\alpha'}$, $N_{\alpha'}^*$. There are two cases. Either: $\mathfrak{M} \subseteq \mathfrak{A}'$, hence $\mathfrak{M}' \supseteq \mathfrak{A}''$, Ω is cyclic for $\mathfrak{M}'$, and the lemma is applicable; or $\mathfrak{M} \subseteq \mathfrak{A}''$ and Ω is separating for $\mathfrak{A}''$. But in the latter case Ω is then separating for $\mathfrak{M}$ and cyclic for $\mathfrak{M}'$ by Lemma 2.5.3, and Lemma 2.6.2 is once more applicable. In either case, since φ cannot be a multiple of Ω by the relation above, $\mathfrak{M} \neq \mathbb{C}\mathbb{1}$, and (1) is false.

Next we want to define quasi-local algebras. These algebras are generated by an increasing net $\{\mathfrak{A}_\alpha\}_{\alpha \in I}$ of subalgebras which satisfy a number of structural relations. In order to introduce this structure it is first necessary to specialize the index set I. In applications the index set typically consists of bounded open subsets of a configuration space $\mathbb{R}^\nu$ ordered by inclusion. This set is not only directed but has various other properties which arise from the operations of union and intersection. We will need partial analogues for more general index sets.

The directed set I is said to possess an orthogonality relation if there is a relation $\perp$ between pairs of elements of I such that

(a) if $\alpha \in I$ then there is a $\beta \in I$ with $\alpha \perp \beta$,
(b) if $\alpha \leq \beta$ and $\beta \perp \gamma$ then $\alpha \perp \gamma$,
(c) if $\alpha \perp \beta$ and $\alpha \perp \gamma$ then there exists a $\delta \in I$ such that $\alpha \perp \delta$ and $\delta \geq \beta, \gamma$.

If I is bounded open subsets of $\mathbb{R}^\nu$ then $\alpha \perp \beta$ could correspond to the disjointness of α and β.

Later we need the analogue of the union of two sets. We will assume that each pair α, β in the index set I has a least upper bound $\alpha \vee \beta$. Thus we assume that if $\alpha, \beta \in I$ then there is a $\alpha \vee \beta \in I$ such that

(d) $\alpha \vee \beta \geq \alpha$ and $\alpha \vee \beta \geq \beta$,
(e) if $\gamma \geq \alpha, \gamma \geq \beta$ then $\gamma \geq \alpha \vee \beta$.

Next remark that if σ is an automorphism of a C^*-algebra $\mathfrak{A}$ which satisfies $\sigma^2 = \iota$, i.e., $\sigma(\sigma(A)) = A$ for all $A \in \mathfrak{A}$, then each element $A \in \mathfrak{A}$ has a unique decomposition into odd and even parts with respect to σ. This decomposition is defined by

$$A = A^+ + A^-, \qquad A^\pm = \frac{A \pm \sigma(A)}{2}.$$

It follows that $\sigma(A^\pm) = \pm A^\pm$, the even elements of $\mathfrak{A}$ form a C^*-subalgebra $\mathfrak{A}^e$, of $\mathfrak{A}$, and the odd elements $\mathfrak{A}^o$ form a Banach space.

Now we are in a position to introduce quasi-local algebras.

Definition 2.6.3. A *quasi-local algebra* is a C^*-algebra $\mathfrak{A}$ and a net $\{\mathfrak{A}_\alpha\}_{\alpha \in I}$ of C^*-subalgebras such that the index set I has an orthogonality relation and the following properties are valid:

(1) if $\alpha \geq \beta$ then $\mathfrak{A}_\alpha \supseteq \mathfrak{A}_\beta$;
(2) $\mathfrak{A} = \overline{\bigcup_\alpha \mathfrak{A}_\alpha}$, where the bar denotes the uniform closure;
(3) the algebras $\mathfrak{A}_\alpha$ have a common identity $\mathbb{1}$;
(4) there exists an automorphism σ such that $\sigma^2 = \iota$, $\sigma(\mathfrak{A}_\alpha) = \mathfrak{A}_\alpha$, and $[\mathfrak{A}_\alpha{}^e, \mathfrak{A}_\beta{}^e] = \{0\}, [\mathfrak{A}_\alpha{}^e, \mathfrak{A}_\beta{}^o] = \{0\}, \{\mathfrak{A}_\alpha{}^o, \mathfrak{A}_\beta{}^o\} = \{0\}$ whenever $\alpha \perp \beta$, where $\mathfrak{A}_\alpha{}^o \subseteq \mathfrak{A}_\alpha$ and $\mathfrak{A}_\alpha{}^e \subseteq \mathfrak{A}_\alpha$ are the odd and even elements with respect to σ.

We have used the notation $\{A, B\} = AB + BA$. One case covered by this definition is $\sigma = \iota$ and then $\mathfrak{A}_\alpha{}^e = \mathfrak{A}_\alpha$ and condition (4) simplifies to the condition

$$[\mathfrak{A}_\alpha, \mathfrak{A}_\beta] = \{0\}$$

whenever $\alpha \perp \beta$. In applications to quantum mechanics $\sigma = \iota$ corresponds to Bose statistics but for Fermi statistics $\sigma \neq \iota$.

Assumption (3) could be replaced by the weaker assumption that each $\mathfrak{A}_\alpha$ has an approximate identity for $\mathfrak{A}$, but this generalization leads to notational complications which are inessential. In applications to mathematical physics the subalgebras $\mathfrak{A}_\alpha$ are indexed by bounded open subsets of $\mathbb{R}^\nu$, ordered by inclusion. The algebra $\mathfrak{A}_\alpha$ is interpreted as the algebra of physical observables for a subsystem localized in the region α of the configuration space $\mathbb{R}^\nu$. The quasi-local algebra $\mathfrak{A}$ corresponds to the extended algebra of observables of an infinite system. A state ω over $\mathfrak{A}$ then represents a physical state of the system and the values $\omega(A), \omega(B), \ldots$ yield the values of the observations $A, B, \ldots$. The representation $(\mathfrak{H}_\omega, \pi_\omega, \Omega_\omega)$ then allows a more detailed description of the individual state ω and the von Neumann algebra $\pi_\omega(\mathfrak{A})''$ is interpreted as the algebra of observables of this state. There are several distinguished subalgebras of $\pi_\omega(\mathfrak{A})''$ which play an important role in the analysis of the states over quasi-local algebras. One such subalgebra is the centre $\mathfrak{Z}_\omega = \pi_\omega(\mathfrak{A})'' \cap \pi_\omega(\mathfrak{A})'$ and two other subalgebras are introduced by the following:

Definition 2.6.4. If ω is a state over the quasi-local algebra $\mathfrak{A}$ then we define the *commutant algebra* $\mathfrak{Z}_\omega{}^c$, of the associated representation

$$(\mathfrak{H}_\omega, \pi_\omega, \Omega_\omega),$$

by

$$\mathfrak{Z}_\omega{}^c = \bigcap_{\alpha \in I} (\pi_\omega(\mathfrak{A}_\alpha)' \cap \pi_\omega(\mathfrak{A}))''$$

and the *algebra at infinity* $\mathfrak{Z}_\omega{}^\perp$ by

$$\mathfrak{Z}_\omega{}^\perp = \bigcap_{\alpha \in I} \left(\bigcup_{\beta \perp \alpha} \pi_\omega(\mathfrak{A}_\beta) \right)''.$$

Note that $\mathfrak{Z}_\omega{}^\perp$ is really an algebra by the condition (c) on the index set I. Further, as $\mathfrak{A}$ is quasi-local and $\mathfrak{Z}_\omega = \pi_\omega(\mathfrak{A})' \cap \pi_\omega(\mathfrak{A})''$ one has

$$\mathfrak{Z}_\omega = \bigcap_{\alpha \in I} (\pi_\omega(\mathfrak{A}_\alpha)' \cap \pi_\omega(\mathfrak{A})'').$$

It easily follows that $\mathfrak{Z}_\omega{}^c \subseteq \mathfrak{Z}_\omega$.

Thus it follows immediately from Theorem 2.6.1 and Kaplansky's density theorem (Theorem 2.4.16) that the following conditions are equivalent if π_ω is faithful:

(1) $\mathfrak{Z}_\omega{}^c$ consists of multiples of the identity;
(2) given $A \in \mathfrak{A}$ there exists an α such that

$$|\omega(AB) - \omega(A)\omega(B)| \leq \|\pi_\omega(B)\|$$

for all $B \in \mathfrak{A}_\alpha{}' \cap \mathfrak{A}$;
(3) given $A \in \mathfrak{A}$ there exists an α such that

$$|\omega(AB) - \omega(A)\omega(B)| \leq \{\omega(B^*B) + \omega(BB^*)\}^{1/2}$$

for all $B \in \mathfrak{A}_\alpha{}' \cap \mathfrak{A}$.

Next we consider properties of the algebra $\mathfrak{Z}_\omega^\perp$. If $\sigma \neq \iota$ then the properties of anticommutation lead to many complications and it is not evident that $\mathfrak{Z}_\omega^\perp \subseteq \mathfrak{Z}_\omega$; but this is indeed the case.

Theorem 2.6.5. *Let ω be a state over a quasi-local algebra. The algebra at infinity $\mathfrak{Z}_\omega^\perp$ is contained in the center $\mathfrak{Z}_\omega$ of the associated representation and, more specifically,*

$$\mathfrak{Z}_\omega^\perp = \bigcap_{\alpha \in I} \left(\bigcup_{\beta \perp \alpha} \pi_\omega(\mathfrak{A}_\beta^{e}) \right)'' \subseteq \mathfrak{Z}_\omega \cap \pi_\omega(\mathfrak{A}^e)''.$$

The following conditions are equivalent:

(1) *$\mathfrak{Z}_\omega^\perp$ consists of multiples of the identity;*
(2) *given $A \in \mathfrak{A}$ there exists an α such that*

$$|\omega(AB) - \omega(A)\omega(B)| \leq \|\pi_\omega(B)\|$$

for all $B \in \mathfrak{A}_\beta$ and all $\beta \perp \alpha$;
(3) *given $A \in \mathfrak{A}$ there exists an α such that*

$$|\omega(AB) - \omega(A)\omega(B)| \leq \{\omega(B^*B) + \omega(BB^*)\}^{1/2}$$

for all $B \in \mathfrak{A}_\beta$ and all $\beta \perp \alpha$.

Proof. If $\mathfrak{M}_\alpha^\perp$ is defined by

$$\mathfrak{M}_\alpha^\perp = \left(\bigcup_{\beta \perp \alpha} \pi_\omega(\mathfrak{A}_\beta) \right)''$$

then

$$\mathfrak{Z}_\omega^\perp = \bigcap_\alpha \mathfrak{M}_\alpha^\perp.$$

Once we establish that $\mathfrak{Z}_\omega^\perp \subseteq \mathfrak{Z}_\omega \subseteq \pi_\omega(\mathfrak{A})'$ the equivalence of the three conditions follows from Theorem 2.6.1 and Kaplansky's density theorem (Theorem 2.4.16). Thus we concentrate on the characterization of $\mathfrak{Z}_\omega^\perp$.

We first claim that if $B \in \mathfrak{Z}_\omega^\perp$ then $B \in \pi_\omega(\mathfrak{A}^e)''$. As $B \in \mathfrak{Z}_\omega^\perp$ implies $B^* \in \mathfrak{Z}_\omega^\perp$, it clearly suffices to consider selfadjoint B. Now consider the sets $\beta = (\alpha, \psi_1, \ldots, \psi_n, \varepsilon)$ with α in the index set I, the vectors $\psi_1, \ldots, \psi_n$ in $\mathfrak{H}_\omega$, and $\varepsilon > 0$. One can form a directed set of these sets by defining $\beta_1 \leq \beta_2$ if, and only if, $\alpha_1 \leq \alpha_2$, $\{\psi_i\}_{\beta_1} \subseteq \{\psi_i\}_{\beta_2}$, and $\varepsilon_1 \geq \varepsilon_2$, where we have used the notation $\beta_j = (\alpha_j, \{\psi_i\}_{\beta_j}, \varepsilon_j)$. Next we define a net B_β indexed by this composite direct set. If $\beta = (\alpha, \psi_1, \ldots, \psi_n, \varepsilon)$ then

$$B \in \left(\bigcup_{\gamma \perp \alpha} \pi_\omega(\mathfrak{A}_\gamma) \right)''$$

and it follows from Kaplansky's density theorem that there exists a γ_β with $\gamma_\beta \perp \alpha$ and a $B_\beta \in \pi_\omega(\mathfrak{A}_{\gamma_\beta})$ such that $B_\beta = B_\beta^*$, $\|B_\beta\| \leq \|B\|$, and

$$\|(B_\beta - B)\psi_i\| < \varepsilon$$

for $i = 1, 2, \ldots, n$. Thus B_β converges strongly to B and as $\|\sigma(B_\beta)\| = \|B_\beta\| \leq \|B\|$ there is a subnet such that $\sigma(B_{\beta'})$ converges weakly. This last statement is a consequence of the weak compactness of the unit ball of $\mathscr{L}(\mathfrak{H}_\omega)$, Proposition 2.4.2. Therefore the odd and even parts $B_{\beta'}^{\mp}$ of $B_{\beta'}$ converge weakly. Let C denote the weak limit of $B_{\beta'}^{-}$. One has

$$\|C\psi\|^2 = \lim_{\beta'} \lim_{\beta''} (B_{\beta'}^{-}\psi, B_{\beta''}^{-}\psi),$$

where both limits are over the same subnet. But for β' fixed in this subnet one has $B_{\beta'} \in \pi_\omega(\mathfrak{A}_{\gamma_{\beta'}})$, and for β'' sufficiently large $B_{\beta''} \in \pi_\omega(\mathfrak{A}_{\gamma_{\beta''}})$, with $\gamma_{\beta''} \perp \gamma_{\beta'}$. Thus

$$\begin{aligned}\lim_{\beta''}(\psi, B_{\beta'}^{-} B_{\beta''}^{-}\psi) &= \lim_{\beta''} -(\psi, B_{\beta''}^{-} B_{\beta'}^{-}\psi) \\ &= -(C\psi, B_{\beta'}^{-}\psi)\end{aligned}$$

by the anticommutation of odd elements. Hence

$$\|C\psi\|^2 = -\lim_{\beta'}(C\psi, B_{\beta'}^{-}\psi) = -\|C\psi\|^2$$

and $C = 0$. We conclude that $B_{\beta'}^{+}$ converges weakly to B and we have established that

$$B \in \bigcap_\alpha \left(\bigcup_{\beta \perp \alpha} \pi_\omega(\mathfrak{A}_\beta{}^{\mathrm{e}})\right)''.$$

But this last set is a subset of $\mathfrak{Z}_\omega{}^\perp$ and hence

$$\begin{aligned}\mathfrak{Z}_\omega{}^\perp &= \bigcap_\alpha \left(\bigcup_{\beta \perp \alpha} \pi_\omega(\mathfrak{A}_\beta{}^{\mathrm{e}})\right)'' \\ &\subseteq \pi_\omega(\mathfrak{A})' \cap \pi_\omega(\mathfrak{A}^{\mathrm{e}})''.\end{aligned}$$

The commutation properties of quasi-local algebras ensure that $\mathfrak{Z}_\omega{}^\perp$ is a subalgebra of $\mathfrak{Z}_\omega$. One also has in general that $\mathfrak{Z}_\omega{}^{\mathrm{c}} \subseteq \mathfrak{Z}_\omega$. For algebras with slightly more structure and a special class of states one can actually deduce equality of these various central subalgebras. It will be convenient to examine algebras with an increasing net of von Neumann algebras.

Definition 2.6.6. Let $\mathfrak{A}$, $\{\mathfrak{M}_\alpha\}_{\alpha \in I}$, be a quasi-local algebra whose generating net is formed of von Neumann algebras $\mathfrak{M}_\alpha$. A state ω over $\mathfrak{A}$ is defined to be *locally normal* if ω is normal in restriction to each $\mathfrak{M}_\alpha$.

In applications to statistical mechanics one typically has that $\mathfrak{M}_\alpha$ is isomorphic to $\mathscr{L}(\mathfrak{H}_\alpha)$ for some $\mathfrak{H}_\alpha$. If ω is locally normal then Theorem 2.4.21 implies that ω in restriction to each $\mathfrak{M}_\alpha$ is determined by a density matrix ρ_α on a Hilbert space $\mathfrak{H}_\alpha$:

$$\omega(A) = \mathrm{Tr}(\rho_\alpha A), \qquad A \in \mathfrak{M}_\alpha.$$

Thus ω could be specified by the family of pairs $\{\mathfrak{H}_\alpha, \rho_\alpha\}$. Note that the inclusion relations $\mathfrak{M}_\alpha \subseteq \mathfrak{M}_\beta$, $\alpha \leq \beta$, then imply certain compatibility conditions on the ρ_α. Specification of ω by the pairs $\{\mathfrak{H}_\alpha, \rho_\alpha\}$ is sometimes useful because one encounters quasi-local algebras with subalgebras $\mathfrak{A}_\alpha$ which are not von Neumann algebras but which are isomorphic to irreducible C^*-subalgebras of $\mathscr{L}(\mathfrak{H}_\alpha)$. Note that in this setting each such ω has a canonical extension $\hat{\omega}$ to the algebra $\hat{\mathfrak{A}}$ generated by the net $\{\mathscr{L}(\mathfrak{H}_\alpha)\}_{\alpha\in I}$ of von Neumann algebras by

$$\hat{\omega}(B) = \mathrm{Tr}(\rho_\alpha B), \qquad B \in \mathscr{L}(\mathfrak{H}_\alpha).$$

In applications to field theory the local algebras $\mathfrak{M}_\alpha$ are more general factors and most of the subsequent analysis is only relevant to the statistical mechanical examples.

Lemma 2.6.7. *Let $\mathfrak{A}$, $\{\mathfrak{M}_\alpha\}_{\alpha\in I}$, be a quasi-local algebra whose generating net is formed of von Neumann algebras $\mathfrak{M}_\alpha$, and let ω be a locally normal state with associated cyclic representation $(\mathfrak{H}, \pi, \Omega)$. It follows that $\pi|_{\mathfrak{M}_\alpha}$ is normal for each α.*

PROOF. Assume that A_γ is an increasing net of positive elements of $\mathfrak{M}_\alpha$ which converges to $A \in \mathfrak{M}_\alpha$ and let $B \in \mathfrak{M}_\beta$, where $\beta \geq \alpha$. Then $B^*A_\gamma B$ converges to B^*AB in $\mathfrak{M}_\beta$. Since $\omega|_{\mathfrak{M}_\beta}$ is normal, we then have

$$\begin{aligned}(\pi(B^*)\Omega, \pi(A_\gamma)\pi(B)\Omega) &= \omega(B^*A_\gamma B)\\ \to \omega(B^*AB) &= (\pi(B^*)\Omega, \pi(A)\pi(B)\Omega).\end{aligned}$$

Now, $\bigcup_\beta \pi(\mathfrak{M}_\beta)\Omega$ is dense in $\mathfrak{H}$; thus $\pi(A_\gamma)$ converges to $\pi(A)$ and $\pi|_{\mathfrak{M}_\alpha}$ is normal.

The essential lemma for the analysis of locally normal states is the following:

Lemma 2.6.8. *Let $\mathfrak{A}$ be a C^*-algebra of operators on a Hilbert space and $\mathfrak{M}$ a von Neumann algebra contained in $\mathfrak{A}$. Assume there exists a projection E from $\mathfrak{A}''$ onto $\mathfrak{M}' \cap \mathfrak{A}''$ which is σ-weakly continuous and such that $E(\mathfrak{A}) \subseteq \mathfrak{A}$. It follows that*

$$\mathfrak{M}' \cap \mathfrak{A}'' = (\mathfrak{M}' \cap \mathfrak{A})''.$$

PROOF. First note that as E is a projection onto $\mathfrak{M}' \cap \mathfrak{A}''$ one has $E(A) = A$ for all $A \in \mathfrak{M}' \cap \mathfrak{A}''$. Next remark that $(\mathfrak{M}' \cap \mathfrak{A})'' \subseteq \mathfrak{M}' \cap \mathfrak{A}''$ and it suffices to prove the reverse inclusion. But if $A \in \mathfrak{M}' \cap \mathfrak{A}''$ then there exists a net $\{A_\beta\} \subseteq \mathfrak{A}$ such that

$$A = \lim_\beta A_\beta$$

and hence

$$A = E(A) = \lim_\beta E(A_\beta)$$

because of the first remark and the σ-weak continuity of E. But $E(A_\beta) \in \mathfrak{M}' \cap \mathfrak{A}$ and hence $A \in (\mathfrak{M}' \cap \mathfrak{A})''$.

An immediate application of this lemma is the following result which derives the equality of $\mathfrak{Z}_\omega{}^c$ with the center.

Proposition 2.6.9. *Let $\mathfrak{A}$, $\{\mathfrak{M}_\alpha\}_{\alpha\in I}$ be a quasi-local algebra whose generating net is formed of von Neumann algebras $\mathfrak{M}_\alpha$, and let ω be a locally normal state over $\mathfrak{A}$. Assume $\mathfrak{M}_\alpha$ is isomorphic to $\mathscr{L}(\mathfrak{H}_\alpha)$ for some $\mathfrak{H}_\alpha$, and all α.*

It follows that the center $\mathfrak{Z}_\omega$ and the commutant algebra $\mathfrak{Z}_\omega^c$ of the associated representation coincide, i.e.,

$$\mathfrak{Z}_\omega = \bigcap_\alpha (\pi_\omega(\mathfrak{M}_\alpha)' \cap \pi_\omega(\mathfrak{A}))'' = \mathfrak{Z}_\omega^c.$$

PROOF. Choose a set of matrix units $\{E_{ij}\}$ for $\mathfrak{M}_\alpha$ and define a projection E on $\mathscr{L}(\mathfrak{H}_\omega)$ by

$$E(B) = \sum_i \pi_\omega(E_{i1})B\pi_\omega(E_{1i}).$$

The properties of matrix units ensure that this sum converges strongly. Since $\pi_\omega|_{\mathfrak{M}_\alpha}$ is normal (Lemma 2.6.7), $\sum_i \pi_\omega(E_{ii}) = \pi_\omega(\mathbb{1}) = \mathbb{1}_{\mathfrak{H}_\alpha}$, and

$$E(B) = \sum_i \pi_\omega(E_{i1})B\pi_\omega(E_{1i}) = B$$

for $B \in \pi_\omega(\mathfrak{M}_\alpha)'$, and for any $B \in \mathscr{L}(\mathfrak{H}_\omega)$ one has $E(B) \in \pi_\omega(\mathfrak{M}_\alpha)'$ because

$$E(B)\pi_\omega(E_{ij}) = \pi_\omega(E_{i1})E(B)\pi_\omega(E_{1j}) = \pi_\omega(E_{ij})E(B).$$

Moreover, if $\mathfrak{N}$ is any von Neumann algebra such that $\pi_\omega(\mathfrak{M}_\alpha) \subseteq \mathfrak{N} \subseteq \mathscr{L}(\mathfrak{H}_\omega)$ then $E(\mathfrak{N}) \subseteq \mathfrak{N}$. Hence $E(\pi_\omega(\mathfrak{M}_\beta)) \subseteq \pi_\omega(\mathfrak{M}_\alpha)' \cap \pi_\omega(\mathfrak{M}_\beta)$ for $\beta \geq \alpha$. As $\mathfrak{A}$ is generated uniformly by the $\mathfrak{M}_\beta$ one concludes that $E(\pi_\omega(\mathfrak{A})) \subseteq \pi_\omega(\mathfrak{A})$ and $E(\pi_\omega(\mathfrak{A})'') \subseteq \pi_\omega(\mathfrak{M}_\alpha)' \cap \pi_\omega(\mathfrak{A})''$. Thus Lemma 2.6.8 applies. Therefore

$$\begin{aligned}\mathfrak{Z}_\omega &= \bigcap_\alpha (\pi_\omega(\mathfrak{M}_\alpha)' \cap \pi_\omega(\mathfrak{A})'') \\ &= \bigcap_\alpha (\pi_\omega(\mathfrak{M}_\alpha)' \cap \pi_\omega(\mathfrak{A}))'' = \mathfrak{Z}_\omega^c.\end{aligned}$$

Next we describe a situation in which $\mathfrak{Z}_\omega = \mathfrak{Z}_\omega^\perp$. Equality of these two algebras depends upon identification of the von Neumann algebras generated by $\mathfrak{M}_\alpha' \cap \mathfrak{A}$ and $\bigcup_{\beta \perp \alpha} \mathfrak{M}_\beta$. The first algebra corresponds to the observables invariant under observations in the region indexed by α and the second corresponds to the observables outside of α. Equality of these two algebras is sometimes referred to as duality. It plays an important role in the discussion of statistics in field theory.

The following result establishes $\mathfrak{Z}_\omega = \mathfrak{Z}_\omega^\perp$ for some systems with $\sigma = \iota$. The more intricate case of $\sigma \neq \iota$ can also be handled (see Notes and Remarks).

Theorem 2.6.10. *Let $\mathfrak{A}$, $\{\mathfrak{M}_\alpha\}_{\alpha \in I}$ be a quasi-local algebra, with $\sigma = \iota$, whose generating net is formed of von Neumann algebras $\mathfrak{M}_\alpha$, and let ω be a locally normal state over $\mathfrak{A}$. Assume that $\mathfrak{M}_\alpha$ is isomorphic to $\mathscr{L}(\mathfrak{H}_\alpha)$, and that $\mathfrak{M}_\alpha \cup \mathfrak{M}_\beta$ generate $\mathfrak{M}_{\alpha \vee \beta}$ in the weak operator topology and that for any pair α, β there exists a $\gamma \perp \alpha$ such that $\beta \leq \gamma \vee \alpha$. It follows that the center $\mathfrak{Z}_\omega$ and the algebra at infinity $\mathfrak{Z}_\omega^\perp$ of the associated representation coincide. Moreover, the following conditions are equivalent:*

(1) *$\mathfrak{Z}_\omega (= \mathfrak{Z}_\omega^\perp)$ consists of multiples of the identity, i.e., ω is a factor state;*
(2) *given α and $\varepsilon > 0$, there exists an α' such that*

$$|\omega(AB) - \omega(A)\omega(B)| \leq \varepsilon \|A\| \, \|B\|$$

for all $A \in \mathfrak{M}_\alpha$, all $B \in \mathfrak{M}_\beta$, and all $\beta \perp \alpha'$;

(3) *given* α *and* $\varepsilon > 0$, *there exists an* α' *such that*

$$|\omega(AB) - \omega(A)\omega(B)| \leq \varepsilon\|A\|\{\omega(B^*B) + \omega(BB^*)\}^{1/2}$$

for all $A \in \mathfrak{M}_\alpha$, *all* $B \in \mathfrak{M}_\beta$, *and all* $\beta \perp \alpha'$.

Proof. Let us construct the projection E as in the proof of Proposition 2.6.9. Our assumptions ensure that each $A \in \pi_\omega(\mathfrak{A})''$ is of the form

$$A = \lim_\gamma \left(\sum_{i=1}^{N_\gamma} \pi_\omega(A_{\gamma,i})\pi_\omega(B_{\gamma,i}) \right),$$

where the limit is in the strong operator topology and

$$A_{\gamma,i} \in \mathfrak{M}_\alpha, \qquad B_{\gamma,i} \in \bigcup_{\beta\perp\alpha} \mathfrak{M}_\beta.$$

But if $A \in \pi_\omega(\mathfrak{M}_\alpha)' \cap \pi_\omega(\mathfrak{A})$ then $E(A) = A$ and

$$A = \lim_\gamma \left(\sum_{i=1}^{N_\gamma} E(\pi_\omega(A_{\gamma,i}))\pi_\omega(B_{\gamma,i}) \right) \in \left(\bigcup_{\beta\perp\alpha} \pi_\omega(\mathfrak{M}_\beta) \right)''$$

because $A_{\gamma,i} \in \mathfrak{M}_\alpha$ ensures that $E(\pi_\omega(A_{\gamma,i}))$ is a multiple of the identity. Thus

$$(\pi_\omega(\mathfrak{M}_\alpha)' \cap \pi_\omega(\mathfrak{A}))'' = \left(\bigcup_{\beta\perp\alpha} \pi_\omega(\mathfrak{M}_\beta) \right)''$$

and application of Proposition 2.6.9 gives

$$\mathfrak{Z}_\omega = \bigcap_\alpha \left(\bigcup_{\beta\perp\alpha} \pi_\omega(\mathfrak{M}_\beta) \right)'' = \mathfrak{Z}_\omega{}^\perp.$$

Now consider the three conditions. Clearly (3) $\Rightarrow$ (2) and (2) implies condition (2) of Theorem 2.6.5 because the $\mathfrak{M}_\alpha$ generate $\mathfrak{A}$ in the uniform topology and $\pi_\omega|_{\mathfrak{M}_\alpha}$ is faithful since it is normal and $\mathfrak{M}_\alpha$ is a factor (Proposition 2.4.22). Application of this theorem then establishes that (2) $\Rightarrow$ (1). It remains to prove that (1) $\Rightarrow$ (3). Now ω restricted to $\mathfrak{M}_\alpha$ is determined by a density matrix ρ_α and, for $\delta > 0$, one may choose a finite rank projector $E \in \mathscr{L}(\mathfrak{H}_\alpha)$ such that

$$\mathrm{Tr}(\rho_\alpha(\mathbb{1} - E)) < \delta.$$

Let E also denote the image of this projector in $\pi_\omega(\mathfrak{M}_\alpha)$. If $A \in \mathfrak{M}_\alpha$ and $B \in \mathfrak{M}_\beta$ with $\alpha \perp \beta$ one has

$$\omega(AB) = \omega(EAEB) + \omega((\mathbb{1} - E)AEB) + \omega(BEA(\mathbb{1} - E)) + \omega((\mathbb{1} - E)A(\mathbb{1} - E)B)$$

by straightforward decomposition and use of quasi-locality to commute B and $EA(\mathbb{1} - E)$. Therefore

$$\begin{aligned} |\omega(AB) - \omega(EAEB)| &\leq \omega(\mathbb{1} - E)\|A\|\{2\omega(B^*B)^{1/2} + \omega(BB^*)^{1/2}\} \\ &< 3\delta\|A\|\{\omega(B^*B) + \omega(BB^*)\}^{1/2}. \end{aligned}$$

Similarly,

$$\begin{aligned} |\omega(AB) - \omega(A)\omega(B)| &< |\omega(EAEB) - \omega(EAE)\omega(B)| \\ &\quad + 6\delta\|A\|\{\omega(B^*B) + \omega(BB^*)\}^{1/2}. \end{aligned}$$

But as E has finite rank on $\mathfrak{H}_\alpha$ the algebra $E\mathfrak{M}E$ has a finite basis and thus by Theorem 2.6.5 one may choose α' such that

$$\sup_{\|A\|=1,\, A\in\mathfrak{M}_\alpha} |\omega(EAEB) - \omega(EAE)\omega(B)| < \delta\{\omega(B^*B) + \omega(BB^*)\}^{1/2}$$

for all $B \in \mathfrak{M}_\beta$ with $\beta \perp \alpha'$. Combining these estimates with $\delta = \varepsilon/7$ gives the desired result.

Theorem 2.6.10 gave a criterion ensuring that a state of a quasi-local algebra is a factor state. We next give conditions for quasi-equivalence of two factor states, stating roughly that they are quasi-equivalent if, and only if, they are equal at infinity.

Corollary 2.6.11. *Let $\mathfrak{A}$, $\{\mathfrak{M}_\alpha\}_{\alpha\in I}$ be a quasi-local algebra satisfying all the requirements in Theorem* 2.6.10, *and let ω_1 and ω_2 be factor states of $\mathfrak{A}$. The following conditions are equivalent:*

(1) *ω_1 and ω_2 are quasi-equivalent;*
(2) *given $\varepsilon > 0$ there exists an α such that*

$$|\omega_1(B) - \omega_2(B)| < \varepsilon\|B\|$$

for all $B \in \mathfrak{M}_\beta$ with $\beta \perp \alpha$;
(3) *given $\varepsilon > 0$ there exists an α such that*

$$|\omega_1(B) - \omega_2(B)| \leq \varepsilon\{\omega_1(B^*B) + \omega_2(B^*B) + \omega_1(BB^*) + \omega_2(BB^*)\}^{1/2}$$

for all $B \in \mathfrak{M}_\beta$ with $\beta \perp \alpha$.

PROOF. By Proposition 2.4.27, ω_1 and ω_2 are quasi-equivalent if, and only if, $\omega = (\omega_1 + \omega_2)/2$ is a factor state. For any A, B in $\mathfrak{A}$ we have

$$\begin{aligned}\omega(AB) - \omega(A)\omega(B) &= \tfrac{1}{2}(\omega_1(AB) - \omega_1(A)\omega_1(B)) \\ &\quad + \tfrac{1}{2}(\omega_2(AB) - \omega_2(A)\omega_2(B)) \\ &\quad + \tfrac{1}{4}(\omega_1(A) - \omega_2(A))(\omega_1(B) - \omega_2(B)).\end{aligned}$$

If ω_1 and ω_2 are factor states it is clear from this calculation that conditions (2) and (3) of the corollary suffice for ω to satisfy conditions (2) and (3) of Theorem 2.6.10. Thus ω is a factor state and, ω_1 and ω_2 are quasi-equivalent.

If, conversely, ω is a factor state there are two possibilities: either $\omega_1 = \omega_2$—then (2) and (3) are trivially fulfilled, or $\omega_1 \neq \omega_2$—in this case, fix some $A \in \bigcup_\alpha \mathfrak{M}_\alpha$ such that $\omega_1(A) \neq \omega_2(A)$. The above calculation then gives an expression of $\omega_1(B) - \omega_2(B)$ as a linear combination of $\omega(AB) - \omega(A)\omega(B)$ and $\omega_i(AB) - \omega_i(A)\omega_i(B)$, $i = 1, 2$. Application of the criteria of Theorem 2.6.10 to the three states ω, ω_1, and ω_2 then gives (2) and (3) of the corollary.

This completes our general discussion of cluster properties. We will return to the subject in Chapter 4.

EXAMPLE 2.6.12. Let I be an arbitrary index set and I_f the directed set of finite subsets of I, where the direction is by inclusion. Associate with each $\alpha \in I$ a finite-dimensional Hilbert space $\mathfrak{H}_\alpha$, with each $\Lambda \in I_f$ the tensor product space

$$\mathfrak{H}_\Lambda = \bigotimes_{\alpha \in \Lambda} \mathfrak{H}_\alpha,$$

and define $\mathfrak{A}_\Lambda = \mathscr{L}(\mathfrak{H}_\Lambda)$. (See Section 2.7.2 for the definition of tensor products.) If $\Lambda_1 \cap \Lambda_2 = \varnothing$ then $\mathfrak{H}_{\Lambda_1 \cup \Lambda_2} = \mathfrak{H}_{\Lambda_1} \otimes \mathfrak{H}_{\Lambda_2}$ and $\mathfrak{A}_{\Lambda_1}$ is isomorphic to the C^*-subalgebra $\mathfrak{A}_{\Lambda_1} \otimes \mathbb{1}_{\Lambda_2}$, of $\mathfrak{A}_{\Lambda_1 \cup \Lambda_2}$, where $\mathbb{1}_{\Lambda_2}$ denotes the identity operator on $\mathfrak{H}_{\Lambda_2}$. Implicitly identifying $\mathfrak{A}_{\Lambda_1}$ and $\mathfrak{A}_{\Lambda_1} \otimes \mathbb{1}_{\Lambda_2}$ one deduces that the algebras $\{\mathfrak{A}_\Lambda\}_{\Lambda \in I_f}$ form an increasing family of matrix algebras. The union of these algebras is an incomplete normed algebra with involution. The minimal norm completion of the union is a quasi-local algebra $\mathfrak{A}$. Taking $\Lambda_1 \perp \Lambda_2$ to mean $\Lambda_1 \cap \Lambda_2 = \varnothing$ one deduces that $\mathfrak{A}$ satisfies the quasilocal commutation conditions, e.g.,

$$A_1 A_2 = A_1 \otimes A_2 = A_2 A_1, \qquad A_1 \in \mathfrak{A}_{\Lambda_1}, A_2 \in \mathfrak{A}_{\Lambda_2}.$$

The least upper bound of Λ_1, Λ_2 is $\Lambda_1 \cup \Lambda_2$ and it follows easily that $\mathfrak{M}_{\Lambda_1} \cup \mathfrak{M}_{\Lambda_2}$ generate $\mathfrak{M}_{\Lambda_1 \cup \Lambda_2}$.

Moreover, every state ω over $\mathfrak{A}$ is locally normal because all states of matrix algebras are normal (for a more general result see Proposition 2.6.13 below). Thus Theorem 2.6.10 applies to an arbitrary state of $\mathfrak{A}$ and gives a characterization of those states which generate factors.

Algebras of the type constructed in the foregoing example are usually called UHF (uniformly hyperfinite) algebras when I is countable. They are of importance in the description of quantum statistical mechanical systems.

2.6.2. Topological Properties

We continue our discussion of quasi-local algebras by discussing some topological properties of locally normal states and, in particular, properties of metrizability. The metrizability will be of relevance in Chapter 4, where we discuss the decomposition of states.

First we need some information concerning the states of irreducible subalgebras of $\mathscr{L}(\mathfrak{H})$ and especially the C^*-algebra $\mathscr{LC}(\mathfrak{H})$ of compact operators on $\mathfrak{H}$. For this discussion we define a state ω over any irreducible subalgebra $\mathfrak{A}$ of $\mathscr{L}(\mathfrak{H})$ to be normal if it is determined by a density matrix ρ in the canonical manner:

$$\omega(A) = \operatorname{Tr}(\rho A), \qquad A \in \mathfrak{A}.$$

Proposition 2.6.13. *Let $\mathscr{LC}(\mathfrak{H})$ denote the C^*-algebra of compact operators on the Hilbert space $\mathfrak{H}$, $\mathscr{T}(\mathfrak{H})$ the Banach space of trace class operators on $\mathfrak{H}$ equipped with the trace norm $T \mapsto \|T\|_{\mathrm{Tr}} = \operatorname{Tr}|T|$, and $\mathscr{L}(\mathfrak{H})$*

the von Neumann algebra of all bounded operators on $\mathfrak{H}$. *It follows that* $\mathscr{T}(\mathfrak{H})$ *is the dual* $\mathscr{LC}(\mathfrak{H})^*$ *of* $\mathscr{LC}(\mathfrak{H})$ *by the duality*

$$T \times A \in \mathscr{T}(\mathfrak{H}) \times \mathscr{LC}(\mathfrak{H}) \mapsto \operatorname{Tr}(TA).$$

Hence $\mathscr{L}(\mathfrak{H})$ *is the bidual of* $\mathscr{LC}(\mathfrak{H})$, *and every state* ω *over* $\mathscr{LC}(\mathfrak{H})$ *is normal.*

PROOF. First remark that $\mathscr{T}(\mathfrak{H})$ is a uniformly dense subspace of $\mathscr{LC}(\mathfrak{H})$ and hence $\mathscr{LC}(\mathfrak{H})^* \subseteq \mathscr{T}(\mathfrak{H})^* = \mathscr{L}(\mathfrak{H})$ by Proposition 2.4.3, where $\mathscr{LC}(\mathfrak{H})^*$ is identified as a subspace of $\mathscr{T}(\mathfrak{H})^*$ by restricting functionals. Thus if $\omega \in \mathscr{LC}(\mathfrak{H})^*$ then

$$\omega(A) = \operatorname{Tr}(TA), \qquad A \in \mathscr{T}(\mathfrak{H}),$$

for some suitable $T \in \mathscr{L}(\mathfrak{H})$. Let $T = U|T|$ be the polar decomposition of T; then $A|T| = AU^*T$ and it follows that

$$\begin{aligned} |\operatorname{Tr}(|T|A)| &= |\operatorname{Tr}(TAU^*)| \\ &= |\omega(AU^*)| \le \|\omega\| \, \|A\| \end{aligned}$$

because $A \in \mathscr{T}(\mathfrak{H})$ implies $AU^* \in \mathscr{T}(\mathfrak{H})$. Thus $A \mapsto \operatorname{Tr}(|T|A)$ is an element of $\mathscr{LC}(\mathfrak{H})^*$.
Next, if $\{P_\alpha\}$ is the increasing net of all finite rank projectors,

$$0 \le \sup_\alpha \operatorname{Tr}(|T|P_\alpha) \le \sup_\alpha \|\omega\| \, \|P_\alpha\| = \|\omega\| < +\infty$$

and hence $|T| \in \mathscr{T}(\mathfrak{H})$. Thus $\mathscr{LC}(\mathfrak{H})^* = \mathscr{T}(\mathfrak{H})$. The statement concerning the bidual follows from Proposition 2.4.3 and the identification of the dual allows one to easily conclude normality of all states.

If one considers algebras $\mathfrak{A}$ which contain $\mathscr{LC}(\mathfrak{H})$ as a subalgebra then the following characterization of the normal states over $\mathfrak{A}$ is often useful.

Proposition 2.6.14. *Let* $\mathfrak{A} \supseteq \mathscr{LC}(\mathfrak{H})$ *be a* C^**-algebra of bounded operators on the Hilbert space* $\mathfrak{H}$ *which contains the compact operators as a subalgebra. A state* ω, *over* $\mathfrak{A}$, *is normal if, and only if,*

$$\sup\{|\omega(A)|; \|A\| = 1, A \in \mathscr{LC}(\mathfrak{H})\} = 1.$$

PROOF. If ω is normal then it is evident that there are finite rank projectors P_α such that

$$\sup_\alpha \omega(P_\alpha) = \sup_\alpha \operatorname{Tr}(\rho P_\alpha) = \operatorname{Tr}(\rho) = 1.$$

Conversely, if the norm condition is satisfied then the restriction of ω to $\mathscr{LC}(\mathfrak{H})$ is a state over this latter algebra and there must exist a density matrix ρ such that

$$\omega(C) = \operatorname{Tr}(\rho C)$$

for all $C \in \mathscr{LC}(\mathfrak{H})$. If $\{P_\alpha\}$ is the increasing net of finite-dimensional projections on $\mathfrak{H}$ one deduces that

$$\lim_\alpha \omega(\mathbb{1} - P_\alpha) = 0.$$

If $A \in \mathfrak{A}$ the Cauchy–Schwarz inequality implies

$$|\omega(A - P_\alpha A)| \leq \omega(\mathbb{1} - P_\alpha)^{1/2}\omega(A^*A)^{1/2}.$$

Hence

$$\omega(A) = \lim_\alpha \omega(P_\alpha A) = \lim_\alpha \mathrm{Tr}(\rho P_\alpha A) = \mathrm{Tr}(\rho A).$$

This shows that ω is normal.

In the general discussion of states over a C^*-algebra $\mathfrak{A}$ (Section 2.3.2), we introduced two topologies, the weak* or $\sigma(\mathfrak{A}^*, \mathfrak{A})$-topology, and the uniform topology. The weak* topology is generally coarser than the uniform topology because each neighborhood in the basis defining the weak* topology clearly contains a neighborhood of the uniform topology. The situation is nicer for the normal states whenever $\mathscr{LC}(\mathfrak{H}) \subseteq \mathfrak{A}$.

Proposition 2.6.15. *Let $\mathfrak{A} \subseteq \mathscr{L}(\mathfrak{H})$ be an irreducible C^*-subalgebra of the C^*-algebra $\mathscr{L}(\mathfrak{H})$ of all bounded operators on the Hilbert space $\mathfrak{H}$. The following conditions are equivalent:*

(1) *the weak* and uniform topologies coincide on the set $N_\mathfrak{A}$ of normal states over $\mathfrak{A}$;*
(2) *the algebra $\mathfrak{A}$ contains the C^*-algebra $\mathscr{LC}(\mathfrak{H})$ of all compact operators as subalgebra.*

PROOF. (1) $\Rightarrow$ (2) Assume (2) is false. We sketch the proof that (1) is false. Firstly, if $\mathfrak{A} \cap \mathscr{LC}(\mathfrak{H})$ is nonzero then one can show that $\mathscr{LC}(\mathfrak{H}) \subseteq \mathfrak{A}$. Hence we may assume $\mathfrak{A} \cap \mathscr{LC}(\mathfrak{H}) = \{0\}$. Secondly, it follows from this condition and the irreducibility of $\mathfrak{A}$ that every state over $\mathfrak{A}$ is a weak* limit of vector states (see Notes and Remarks). In particular, each normal state over $\mathfrak{A}$ is a weak* limit of vector states. If the weak* and uniform topologies were to coincide one would then conclude that each density matrix is the limit in trace norm of rank one projectors. This is absurd and hence (1) is false.

(2) $\Rightarrow$ (1) The states $N_\mathfrak{A}$ can be equipped with the weak* topology arising from $\mathscr{LC}(\mathfrak{H})$ and this is coarser than the weak* topology of $\mathfrak{A}$ because $\mathfrak{A} \supseteq \mathscr{LC}(\mathfrak{H})$. Furthermore, $N_\mathfrak{A}$ can be equipped with the uniform topology from $\mathscr{LC}(\mathfrak{H})$, but it is readily checked that this coincides with the uniform topology from $\mathfrak{A}$ because

$$\begin{aligned}\sup\{|\mathrm{Tr}((\rho - \rho')A)|; A \in \mathscr{LC}(\mathfrak{H}), \|A\| = 1\}\\ = \mathrm{Tr}(|\rho - \rho'|) = \sup\{|\mathrm{Tr}((\rho - \rho')A)|; A \in \mathfrak{A}, \|A\| = 1\}.\end{aligned}$$

Thus it suffices to prove that the weak* and uniform topologies of $\mathscr{LC}(\mathfrak{H})$ coincide on $N_\mathfrak{A}$. But if $\omega \in N_\mathfrak{A}$ then there is a density matrix ρ such that

$$\omega(A) = \mathrm{Tr}(\rho A).$$

Hence for $\varepsilon > 0$ there is a finite rank projector $E \in \mathfrak{A}$ such that

$$0 \leq \mathrm{Tr}(\rho(\mathbb{1} - E)) < \varepsilon,$$

e.g., E could be chosen to be a suitable spectral projector of ρ. Next we define a neighborhood of ω by

$$W(\omega; \varepsilon) = \left\{\omega'; \omega' \in N_{\mathfrak{A}}, \sup_{A \in \mathfrak{A}, \|A\| = 1} |(\omega - \omega')(EAE)| < \varepsilon\right\}.$$

The set $W(\omega; \varepsilon)$ contains a weak* neighborhood because E has finite rank. Hence the bounded operators on $E\mathfrak{H}$ have a finite basis of matrix units and the condition

$$\sup_{A \in \mathscr{LC}(\mathfrak{H}), \|A\| = 1} |(\omega - \omega')(EAE)| < \varepsilon$$

is implied by a finite number of conditions of the form

$$|(\omega - \omega')(A_i)| < \varepsilon.$$

Next, for $\omega' \in W(\omega; \varepsilon)$

$$\mathrm{Tr}(\rho'(\mathbb{1} - E)) = \mathrm{Tr}(\rho(\mathbb{1} - E)) + (\omega - \omega')(E)$$

and hence

$$0 \le \mathrm{Tr}(\rho'(\mathbb{1} - E)) \le \mathrm{Tr}(\rho(\mathbb{1} - E)) + |(\omega - \omega')(E)| < 2\varepsilon.$$

Further, for each $A \in \mathscr{LC}(\mathfrak{H})$ we can apply the triangle inequality and Cauchy–Schwarz inequality to obtain

$$|(\omega - \omega')(A)| \le |(\omega - \omega')(EAE)| + 2\|A\|(\sqrt{\mathrm{Tr}(\rho(\mathbb{1} - E))} + \sqrt{\mathrm{Tr}(\rho'(\mathbb{1} - E))}).$$

Hence one concludes that

$$\sup_{A \in \mathfrak{A}, \|A\| = 1} |(\omega - \omega')(A)| < \varepsilon + 2\sqrt{\varepsilon} + 2\sqrt{2\varepsilon},$$

i.e., the uniform neighborhood

$$U(\omega, \delta) = \{\omega'; \omega' \in E_{\mathfrak{A}}, \|\omega - \omega'\| < \delta\}$$

contains the neighborhood $W(\omega, \varepsilon)$ whenever $\varepsilon + 2\sqrt{\varepsilon} + 2\sqrt{2\varepsilon} < \delta$.

The foregoing coincidence of topologies implies, of course, that the weak* topology is a metric topology on $N_{\mathfrak{A}}$ and that the normal states are metrizable in this topology. If $\mathfrak{A}$ is separable then the set of all states $E_{\mathfrak{A}}$, over $\mathfrak{A}$, is always metrizable in the weak* topology but the result for the normal states follows without any separability assumption.

Let us now consider the set of states $E_{\mathfrak{A}}$ over a quasi-local algebra $\mathfrak{A}$ generated by the subalgebras $\{\mathfrak{A}_\alpha\}_{\alpha \in I}$. The set $E_{\mathfrak{A}}$ can be equipped either with the weak* topology of $\mathfrak{A}$ or the uniform topology as before. The special structure of $\mathfrak{A}$ allows us, however, to introduce a third topology on $E_{\mathfrak{A}}$ which is intermediate to these two topologies. This third topology is called the locally uniform topology and is specified by the set of neighborhoods

$$V(\omega; \alpha, \varepsilon) = \left\{\omega'; \omega' \in E_{\mathfrak{A}}, \sup_{\|A\| = 1, A \in \mathfrak{A}_\alpha} |\omega'(A) - \omega(A)| < \varepsilon\right\},$$

where $\omega \in E_{\mathfrak{A}}$, $\alpha \in I$, and $\varepsilon > 0$. Clearly, the locally uniform topology is finer than the weak *-topology and coarser than the uniform topology. If $\mathfrak{A}$

is generated by an increasing sequence of subalgebras $\mathfrak{A}_n$ then the locally uniform topology is a metric topology and one can construct a metric for it by

$$\|\omega_1 - \omega_2\| = \sum_{n \geq 1} \frac{2^{-n}\|\omega_1 - \omega_2\|_n}{1 + \|\omega_1 - \omega_2\|_n},$$

where

$$\|\omega_1 - \omega_2\|_n = \sup\{|\omega_1(A) - \omega_2(A)|;\ \|A\| = 1,\ A \in \mathfrak{A}_n\}.$$

Theorem 2.6.16. *Let $\mathfrak{A}$, $\{\mathfrak{A}_\alpha\}_{\alpha \in I}$ be a quasi-local algebra and assume that each $\mathfrak{A}_\alpha$ is isomorphic to a subalgebra $\pi(\mathfrak{A}_\alpha)$ of $\mathscr{L}(\mathfrak{H}_\alpha)$ such that $\pi(\mathfrak{A}_\alpha) \supseteq \mathscr{L}\mathscr{C}(\mathfrak{H}_\alpha)$, and define a state ω over $\mathfrak{A}$ to be locally normal if ω restricted to each $\mathfrak{A}_\alpha$ is normal. It follows that the weak* topology and the locally uniform topology coincide on the locally normal states. Thus if $\mathfrak{A}$ is generated by an increasing sequence of subalgebras then the set of locally normal states is metrizable in the weak* topology.*

The proof of this result is an easy consequence of the fact that the $\mathfrak{A}_\alpha$ generate $\mathfrak{A}$ and the coincidence of the local topologies given by Proposition 2.6.15 and the assumption that $\pi(\mathfrak{A}_\alpha) \supseteq \mathscr{L}\mathscr{C}(\mathfrak{H}_\alpha)$. We omit the details.

2.6.3. Algebraic Properties

We conclude our discussion of quasi-local algebras $\mathfrak{A}$, $\{\mathfrak{A}_\alpha\}_{\alpha \in I}$ by proving, under some general conditions on the $\mathfrak{A}_\alpha$, that $\mathfrak{A}$ is simple. In the ensuing discussion an ideal in a C^*-algebra will always mean a closed, two-sided ideal. (Note that such ideals are selfadjoint by Proposition 2.2.19.)

The proof of the next proposition makes use only of conditions (1) and (2) in the definition of quasi-local algebras (Definition 2.6.3).

Proposition 2.6.17. *Let $\mathfrak{A}$, $\{\mathfrak{A}_\alpha\}_{\alpha \in I}$, be a quasi-local algebra and let $\mathfrak{I}$ be an ideal in $\mathfrak{A}$. It follows that $\mathfrak{I}_\alpha = \mathfrak{I} \cap \mathfrak{A}_\alpha$ is an ideal in $\mathfrak{A}_\alpha$ for all α, and*

$$\mathfrak{I} = \overline{\bigcup_\alpha \mathfrak{I}_\alpha}.$$

In particular, any representation π of $\mathfrak{A}$ such that $\pi|_{\mathfrak{A}_\alpha}$ is faithful for all α, is faithful on $\mathfrak{A}$.

PROOF. It is clear that $\mathfrak{I}_\alpha$ is an ideal in $\mathfrak{A}_\alpha$ for all α and that

$$\overline{\bigcup_\alpha \mathfrak{I}_\alpha} \subseteq \mathfrak{I}.$$

To prove the converse, let π be a morphism of $\mathfrak{A}$ such that $\ker \pi = \mathfrak{I}$. We will show that if $A \notin \overline{\bigcup_\alpha \mathfrak{I}_\alpha}$ then $A \notin \mathfrak{I}$. Assume $A \notin \overline{\bigcup_\alpha \mathfrak{I}_\alpha}$ and let $\{A_n\} \subseteq \bigcup_\alpha \mathfrak{A}_\alpha$ be a sequence such that $A_n \to A$. Since $A \notin \overline{\bigcup_\alpha \mathfrak{I}_\alpha}$ we have that

$$\inf\left\{\|A - B\|;\ B \in \bigcup_\alpha \mathfrak{I}_\alpha\right\} = \varepsilon > 0.$$

Choose N such that $n \geq N$ implies $\|A_n - A\| < \varepsilon/2$. For $n \geq N$, if $A_n \in \mathfrak{A}_{\alpha_n}$ we have for any $B \in \mathfrak{I}_{\alpha_n}$

$$\|A_n - B\| \geq \|A - B\| - \|A - A_n\| \geq \varepsilon - \frac{\varepsilon}{2} = \frac{\varepsilon}{2}.$$

Now since $\ker(\pi|_{\mathfrak{A}_{\alpha_n}}) = \mathfrak{I} \cap \mathfrak{A}_{\alpha_n} = \mathfrak{I}_{\alpha_n}$, we have for $n \geq N$, by Proposition 2.2.19,

$$\|\pi(A_n)\| = \inf\{\|A_n - B\|; B \in \mathfrak{I}_{\alpha_n}\} \geq \frac{\varepsilon}{2}$$

because the norm on the C^*-algebra $\pi(\mathfrak{A}_{\alpha_n})$ is the same whether it is viewed as a subalgebra of $\pi(\mathfrak{A})$ or as the image of the quotient map $\mathfrak{A}_{\alpha_n} \mapsto \mathfrak{A}_{\alpha_n}/\mathfrak{I}_{\alpha_n}$ (Corollary 2.2.6). Since π is continuous, $\pi(A_n) \to \pi(A)$. In particular,

$$\|\pi(A)\| = \lim_{n \to \infty} \|\pi(A_n)\| \geq \frac{\varepsilon}{2}.$$

It follows that $A \notin \mathfrak{I}$.

The next corollary gives a general criterion for simplicity. The proof only makes use of conditions (1), (2), and (3) in Definition 2.6.3.

Corollary 2.6.18. *Let $\mathfrak{A}$, $\{\mathfrak{A}_\alpha\}_{\alpha \in I}$ be a quasi-local algebra. The following conditions are equivalent:*

(1) *$\mathfrak{A}$ is simple;*
(2) *for any α and any $A \in \mathfrak{A}_\alpha \backslash \{0\}$ there exists a $\beta \geq \alpha$ such that the ideal generated by A in $\mathfrak{A}_\beta$ is equal to $\mathfrak{A}_\beta$.*

PROOF. (2) $\Rightarrow$ (1) Assume (2), and let $\mathfrak{I}$ be a nonzero ideal in $\mathfrak{A}$. By Proposition 2.6.17 there exists an α such that $\mathfrak{I} \cap \mathfrak{A}_\alpha \neq \{0\}$. But by (2) it follows that there exists $\beta \geq \alpha$ such that $\mathfrak{A}_\beta \subseteq \mathfrak{I}$. Hence $\mathfrak{A}_\gamma \subseteq \mathfrak{I}$ for all $\gamma \geq \beta$, and $\mathfrak{I} = \mathfrak{A}$, i.e., $\mathfrak{A}$ is simple.

(1) $\Rightarrow$ (2) Assume (1), and let $A \in \mathfrak{A}_\alpha \backslash \{0\}$. For any $\beta \geq \alpha$ let $\mathfrak{I}_\beta$ be the ideal generated by A in $\mathfrak{A}_\beta$. Then $\{\mathfrak{I}_\beta\}$ is an increasing net of C^*-algebras, and $\mathfrak{I} = \overline{\bigcup_{\beta \geq \alpha} \mathfrak{I}_\beta}$ is a nonzero ideal in $\mathfrak{A}$ since $\mathfrak{I}_{\beta'} \cap \mathfrak{A}_\beta$ is an ideal in $\mathfrak{A}_\beta$ for $\beta' \geq \beta \geq \alpha$. It follows from (1) that $\mathfrak{I} = \mathfrak{A}$. But then there exists some $\beta \geq \alpha$ and an element $B \in \mathfrak{I}_\beta$ such that $\|B - \mathbb{1}\| < 1$. Hence B is invertible and $\mathbb{1} = B^{-1}B \in \mathfrak{I}_\beta$, i.e., $\mathbb{1} \in \mathfrak{I}_\beta$ and $\mathfrak{I}_\beta = \mathfrak{A}_\beta \mathbb{1} = \mathfrak{A}_\beta$.

The following corollary applies, for example, if all the $\mathfrak{A}_\alpha$ are matrix algebras, or general finite factors, or if the $\mathfrak{A}_\alpha$ are type III factors on separable Hilbert spaces (see Definition 2.7.18).

Corollary 2.6.19. *Let $\mathfrak{A}$, $\{\mathfrak{A}_\alpha\}_{\alpha \in I}$ be a quasi-local algebra and assume that each $\mathfrak{A}_\alpha$ is simple. It follows that $\mathfrak{A}$ is simple.*

PROOF. This is an immediate consequence of Corollary 2.6.18.

The next corollary can be generalized to the case when $\mathfrak{M}_\alpha$ are general factors on a separable Hilbert space such that $\mathfrak{M}_\alpha{}^e = \{A \in \mathfrak{M}_\alpha; \sigma(A) = A\}$

is a properly infinite von Neumann algebra for all α. (See Definition 2.7.15.) The proof of the general case is essentially the same as the proof of the case stated here, but needs a bit more technology.

Corollary 2.6.20. *Let $\mathfrak{A}$, $\{\mathfrak{A}_\alpha\}_{\alpha\in I}$ be a quasi-local algebra with $\sigma = \iota$, and $\mathfrak{M}_\alpha$ isomorphic to $\mathscr{L}(\mathfrak{H}_\alpha)$, where $\mathfrak{H}_\alpha$ is a separable, infinite-dimensional Hilbert space. It follows that $\mathfrak{A}$ is simple.*

PROOF. We will prove this by applying Corollary 2.6.18. Let $A \in \mathfrak{A}_\alpha\backslash\{0\}$. Replacing A by A^*A we may assume A positive. Let E be the spectral projection of A corresponding to the spectral interval $[\|A\|/2, \|A\|]$. Then $E \neq 0$. Next choose some γ, β such that $\gamma \perp \alpha$ and $\beta \geq \gamma \vee \alpha$. We will show that E is an infinite-dimensional projection in $\mathfrak{M}_\beta = \mathscr{L}(\mathfrak{H}_\beta)$.

To this end, let $\{E_{ij}\} \subseteq \mathscr{L}(\mathfrak{H}_\beta)$ be a complete set of matrix units for $\mathfrak{M}_\gamma \subseteq \mathscr{L}(\mathfrak{H}_\beta)$. Then all the E_{ij} commute with E. In particular, EE_{ii} are projections for all i. Now

$$EE_{jj} = EE_{ji}E_{ii}E_{ij} = E_{ji}(EE_{ii})E_{ij}.$$

It follows that if $EE_{ii}=0$ for some i then $EE_{jj}=0$ for all j. But this is impossible since $E = E\mathbb{1} = \sum_i EE_{ii}$. Thus $\{EE_{ii}\}$ is an infinite set of mutually orthogonal nonzero projections dominated by E and E is infinite-dimensional. Hence there exists a partial isometry $W \in \mathscr{L}(\mathfrak{H}_\beta)$ such that

$$WEW^* = \mathbb{1}.$$

But since $A \geq (\|A\|/2)E$ it follows that

$$WAW^* \geq \frac{\|A\|}{2}\mathbb{1}.$$

Thus WAW^* is invertible, and the ideal generated by A in $\mathfrak{M}_\beta$ must be $\mathfrak{M}_\beta$ itself.

The simplicity of $\mathfrak{A}$ now follows from Corollary 2.6.18.

2.7. Miscellaneous Results and Structure

In this section we review some further results in the theory of operator algebras which are important either for applications to mathematical physics or for the structure and classification of the operator algebras (and hence potentially important for applications in mathematical physics). We will not give complete proofs for these results, but in some cases we will give some hints and in all cases references to the complete proofs.

2.7.1. Dynamical Systems and Crossed Products

In analyzing symmetries of physical systems in the setting of an operator algebra, the first concept encountered is that of a dynamical system.

Definition 2.7.1. A *C^*-dynamical system* is a triple $\{\mathfrak{A}, G, \alpha\}$, where $\mathfrak{A}$ is a C^*-algebra, G is a locally compact group, and α is a strongly continuous representation of G in the automorphism group of $\mathfrak{A}$, i.e., for each $g \in G$, α_g is an automorphism of $\mathfrak{A}$ and

$$\alpha_e = \iota,$$

$$\alpha_{g_1}\alpha_{g_2} = \alpha_{g_1 g_2};$$

$g \mapsto \alpha_g(A)$ is continuous in norm for each $A \in \mathfrak{A}$ (e is the identity in G, ι is the identity map of $\mathfrak{A}$).

A W^*-dynamical system is a triple $\{\mathfrak{M}, G, \alpha\}$, where $\mathfrak{M}$ is a von Neumann algebra, G is a locally compact group, and α is a weakly continuous representation of G in the automorphism group of $\mathfrak{M}$.

A *covariant representation* of a dynamical system is a triple $(\mathfrak{H}, \pi, U)$, where $\mathfrak{H}$ is a Hilbert space, π is a nondegenerate representation of the algebra on $\mathfrak{H}$ which is normal in the W^*-case, and U is a strongly continuous unitary representation of G on $\mathfrak{H}$ such that

$$\pi(\alpha_g(A)) = U_g \pi(A) U_g^*, \qquad A \in \mathfrak{A} \text{ (resp. } \mathfrak{M}), \; g \in G.$$

Note that in the W^*-case, the continuity requirement on $g \mapsto \alpha_g$ is equivalent to the requirement that $g \mapsto \alpha_{g^{-1}}^*$ is strongly continuous on $\mathfrak{M}_*$. When

$G = \mathbb{R}$, this follows from Corollary 2.5.23, but it is true in general. This can be seen by averaging, in the sense of Definition 2.5.19, with a continuous function.

To each C^*- (resp. W^*-) dynamical system we associate a new C^*-algebra (resp. von Neumann algebra), called the crossed product of $\mathfrak{A}$ with G and denoted by $C^*(\mathfrak{A}, \alpha) \equiv \mathfrak{A} \otimes_\alpha G$ (resp. $W^*(\mathfrak{M}, \alpha) \equiv \mathfrak{M} \otimes_\alpha G$). Before proceeding with the somewhat messy definition of these objects, we indicate the four main motivations for introducing these concepts.

(1) There is a one-one natural correspondence between covariant representations of a dynamical system and nondegenerate representations of the crossed product.

(2) In many cases of interest, the algebra of a dynamical system is the crossed product of the fixed point subalgebra of $\mathfrak{A}$ (resp. $\mathfrak{M}$)

$$\{A \in \mathfrak{A};\, \alpha_g(A) = A, \text{ all } g \in G\}$$

with a "dual object" of G (which is the dual group if G is abelian), such that the action α is obtained naturally as a dual action on the crossed product. Positive results in this direction exist almost exclusively in the W^*-case, and have been used to analyze "field algebras" in terms of "observable algebras" when G is a gauge group.

(3) Crossed products are important to construct examples both in C^*- and von Neumann algebra theory. When the action α is free and ergodic in a certain sense one can prove that the crossed product is a simple C^*-algebra in the C^*-case, and a factor in the W^*-case.

(4) W^*-crossed products play a fundamental role in the classification of factors and have led to an almost complete classification of hyperfinite factors. These are the factors which are the weak closure of a union of an increasing sequence of finite dimensional subalgebras. All the factors encountered in mathematical physics are hyperfinite but the physical significance of the classification is still unclear. We will list in Section 2.7.3 the main results of the classification.

We now turn to the definition of crossed products.

Definition 2.7.2. Let $\{\mathfrak{A}, G, \alpha\}$ be a C^*-dynamical system. Let dg and $\Delta(g)$ denote, respectively, the left Haar measure and the modular function on G. Let $\mathfrak{K}(\mathfrak{A}, G)$ be the continuous functions from G into $\mathfrak{A}$ with compact support. $\mathfrak{K}(\mathfrak{A}, G)$ is a linear space in a natural way, and we equip $\mathfrak{K}(\mathfrak{A}, G)$ with a multiplication, involution, and norm, defined by

$$(xy)(g) = \int_G x(h)\alpha_h(y(h^{-1}g))\, dh$$

$$x^*(g) = \Delta(g)^{-1}\alpha_g(x(g^{-1}))^*$$

$$\|x\|_1 = \int_G \|x(h)\|\, dh, \qquad x, y \in \mathfrak{K}(\mathfrak{A}, G), g \in G.$$

One verifies that $\mathfrak{K}(\mathfrak{A}, G)$ satisfies all the axioms for a Banach *-algebra except the completeness. Thus the completion $L^1(\mathfrak{A}, G)$ of $\mathfrak{K}(\mathfrak{A}, G)$ is a

Banach $*$-algebra when the algebraic operations are extended by continuity. Now, define a new norm on $L^1(\mathfrak{A}, G)$ by

$$\|x\| = \sup_{\pi} \|\pi(x)\|,$$

where π ranges over all the Hilbert space representations of $L^1(\mathfrak{A}, G)$. Then one easily sees that $\| \ \|$ is a C^*-seminorm on $L^1(\mathfrak{A}, G)$, and we have

$$\|x\| \leq \|x\|_1$$

because, for any representation π of $L^1(\mathfrak{A}, G)$,

$$\begin{aligned}\|\pi(x)\| &= \|\pi(x^*x)\|^{1/2} = \rho(\pi(x^*x))^{1/2}\\ &\leq \rho(x^*x)^{1/2} \leq \|x^*x\|_1^{1/2} \leq \|x\|_1.\end{aligned}$$

Furthermore, using a technique from the definition of the W^*-crossed product, one may show that $L^1(\mathfrak{A}, G)$ has a faithful representation. Thus $\| \ \|$ is a norm. The completion of $L^1(\mathfrak{A}, G)$ in this norm is called the *C^*-crossed product of $\mathfrak{A}$ and G*, and is denoted by

$$C^*(\mathfrak{A}, \alpha) \equiv \mathfrak{A} \otimes_\alpha G.$$

We note that the one-to-one correspondence between covariant representations $\{\mathfrak{H}, \pi, U\}$ of $\{\mathfrak{A}, G, \alpha\}$ and representations ρ of $C^*(\mathfrak{A}, \alpha)$ is given by

$$\rho(x) = \int_G \pi(x(h))U(h)\, dh, \qquad x \in \mathfrak{K}(\mathfrak{A}, G).$$

This gives ρ in terms of $\{\pi, U\}$; to construct $\{\pi, U\}$ when ρ is given one notes that $\mathfrak{A}$ and G act by multiplication on $C^*(\mathfrak{A}, \alpha)$, i.e., one defines π and U by

$$\pi(A)\rho(x) = \rho({}_Ax),$$

$$U(g)\rho(x) = \rho({}_gx),$$

where

$${}_Ax(h) = A(x(h)),$$

$${}_gx(h) = x(g^{-1}h).$$

Note that $\rho(C^*(\mathfrak{A}, \alpha))$ and $\{\pi(\mathfrak{A}), U(G)\}$ generate the same von Neumann algebra on $\mathfrak{H}$ [Dop 1], [Tak 1].

In describing the von Neumann crossed product, it is suitable to work in a concrete representation.

Definition 2.7.3. Let $\{\mathfrak{M}, G, \alpha\}$ be a W^*-dynamical system, and assume $\mathfrak{M}$ acts on a Hilbert space $\mathfrak{H}$. Define a new Hilbert space $L^2(\mathfrak{H}, G, dg)$ as the completion of $\mathfrak{K}(\mathfrak{H}, G)$, where $\mathfrak{K}(\mathfrak{H}, G)$ is the set of continuous functions from G, into $\mathfrak{H}$, with compact support equipped with the inner product

$$(\xi, \eta) = \int_G (\xi(g), \eta(g))\, dg.$$

Define a representation π_0 of $\mathfrak{M}$ and λ of G on $L^2(\mathfrak{H}, G)$ by

$$(\pi_0(A)\xi)(h) = \alpha_h^{-1}(A)\xi(h),$$
$$(\lambda(g)\xi)(h) = \xi(g^{-1}h).$$

It is easily seen that π_0 is a normal faithful representation of $\mathfrak{M}$ and λ is a strongly continuous unitary representation of G such that

$$\lambda(g)\pi_0(A)\lambda(g)^* = \pi_0(\alpha_g(A)).$$

The von Neumann algebra on $L^2(\mathfrak{H}, G)$ generated by $\pi_0(\mathfrak{M})$ and $\lambda(G)$ is called the *crossed product of* $\mathfrak{M}$ *by* G and is denoted by $W^*(\mathfrak{M}, \alpha) = \mathfrak{M} \otimes_\alpha G$.

Note that if $x \in \mathfrak{K}(\mathfrak{M}, G)$, we may define an element $\hat{x} \in W^*(\mathfrak{M}, G)$ by

$$\hat{x} = \int_G \pi_0(x(g))\lambda(g)\, dg.$$

One verifies straightforwardly that

$$\hat{x}\hat{y} = \widehat{xy}, \qquad \hat{x}^* = \widehat{x^*},$$

where

$$xy(g) = \int_G x(h)\alpha_h(y(h^{-1}g))\, dh$$

and

$$x^*(g) = \Delta(g)^{-1}\alpha_g(x(g^{-1}))^*.$$

This connects this definition of crossed product with the C^*-definition, and shows that the algebra $L^1(\mathfrak{A}, G)$ used in the definition of the C^*-crossed product has a faithful representation. The W^*-crossed product is the von Neumann algebra generated by the set of $\hat{x}$ with $x \in \mathfrak{K}(\mathfrak{M}, G)$.

In the sequel we assume the group G to be abelian. Most of the results can be generalized to nonabelian G although the generalization is not always simple. For $s, t \in G$ we write $s + t = st$ and $s - t = st^{-1}$. We let $\hat{G}$ be the dual group of G. Thus $\hat{G}$ is the set $\{\gamma\}$ of characters of G, i.e., homomorphisms of G into the circle group $\{z \in \mathbb{C}, |z| = 1\}$, with a group operation defined by

$$\langle \gamma_1\gamma_2, g\rangle = \langle \gamma_1, g\rangle\langle \gamma_2, g\rangle.$$

We equip $\hat{G}$ with the topology of uniform convergence on compact subsets of G and then $\hat{G}$ is a locally compact group. Moreover, $\hat{\hat{G}} = G$ as topological groups, by Pontryagin's duality theorem.

Next let $\{\mathfrak{A}, G, \alpha\}$ be a C^*-dynamical system with G abelian, and define a mapping from $\hat{G}$ into $\mathscr{L}(\mathfrak{K}(\mathfrak{A}, G))$ by $(\hat{\alpha}_\gamma x)(t) = \overline{\langle \gamma, t\rangle} x(t)$. Then $\hat{\alpha}_\gamma$ is a $*$-automorphism of $\mathfrak{K}(\mathfrak{A}, G)$ and extends by continuity to a $*$-automorphism of $C^*(\mathfrak{A}, \alpha)$. $\{C^*(\mathfrak{A}, \alpha), \hat{G}, \hat{\alpha}\}$ is a C^*-dynamical system.

Analogously, let $\{\mathfrak{M}, G, \alpha\}$ be a W^*-dynamical system with G abelian, and define a mapping μ from $\hat{G}$ into the unitary group on $L^2(\mathfrak{H}, G)$ by

$$\mu(\gamma)\xi(t) = \overline{\langle\gamma, t\rangle}\xi(t).$$

Then

$$\mu(\gamma)\pi_0(A)\mu(\gamma)^* = \pi_0(A), \qquad A \in \mathfrak{M},$$
$$\mu(\gamma)\lambda(t)\mu(\gamma)^* = \overline{\langle\gamma, t\rangle}\lambda(t), \qquad t \in G,$$

and therefore

$$\hat{\alpha}_\gamma(A) = \mu(\gamma)A\mu(\gamma)^*, \qquad A \in W^*(\mathfrak{M}, \alpha)$$

is an automorphism of $W^*(\mathfrak{M}, \alpha)$. It follows immediately that $\{W^*(\mathfrak{M}, \alpha), \hat{G}, \hat{\alpha}\}$ is a W^*-dynamical system.

In both the above cases, the action of $\hat{G}$ on the crossed product is called the *dual action* of the original action of G on the algebra.

Using the canonical embedding of $\mathfrak{K}(\mathfrak{M}, G)$ into $W^*(\mathfrak{M}, \alpha)$, it is not difficult to see that the definitions of dual action in the C^*-case and the W^*-case are essentially the same.

We next give a characterization of the W^*-dynamical systems arising by using dual action on the crossed product. An isomorphism between two W^*-dynamical systems $\{\mathfrak{M}, G, \alpha\}$ and $\{\mathfrak{N}, G, \beta\}$ is of course defined to be a $*$-isomorphism γ between $\mathfrak{M}$ and $\mathfrak{N}$ such that $\gamma\alpha_t = \beta_t\gamma$ for all $t \in G$. The most primitive theorem in this direction is then the following:

Theorem 2.7.4 ([Land 1], [Nak 1]). *Let $\{\mathfrak{M}, G, \alpha\}$ be a W^*-dynamical system where G is abelian. Then the following statements are equivalent:*

(1) *there exists a W^*-dynamical system $\{\mathfrak{N}, \hat{G}, \beta\}$ such that*

$$\{W^*(\mathfrak{N}, \beta), G, \hat{\beta}\}$$

is isomorphic to $\{\mathfrak{M}, G, \alpha\}$;

(2) *there exists a strongly continuous unitary representation U of $\hat{G}$ in $\mathfrak{M}$ such that $\alpha_t(U_\gamma) = \overline{\langle\gamma, t\rangle}U_\gamma$.*

In case (1), *$\mathfrak{N}$ is isomorphic to $\mathfrak{M}^\alpha = \{A \in \mathfrak{M}; \alpha_t(A) = A$ for all $g \in G\}$.*

Proof. The rigorous proofs of these statements are long and tedious and do not give much more insight than a formal argument, provided one is acquainted with the fundamental facts of abelian harmonic analysis. We will therefore only give a formal argument.

(1) $\Rightarrow$ (2) Taking $U_\gamma = \lambda(\gamma)$, this follows immediately from the definition of the dual action.

(2) $\Rightarrow$ (1) In this case introduce $\mathfrak{N}$ by

$$\mathfrak{N} = \{A \in \mathfrak{M}; \alpha_t(A) = A \quad \text{for all } t \in G\}$$

and, more generally, $\mathfrak{N}_\gamma$ by

$$\mathfrak{N}_\gamma = \{A \in \mathfrak{M}; \alpha_t(A) = \overline{\langle\gamma, t\rangle}A \quad \text{for all } t \in G\}.$$

It is then clear that $\mathfrak{N}U_\gamma \subseteq \mathfrak{N}_\gamma$ and $\mathfrak{N}_\gamma U_\gamma^* \subseteq \mathfrak{N}$. Thus $\mathfrak{N}_\gamma = \mathfrak{N}U_\gamma$. Also $U_\gamma \mathfrak{N} U_\gamma^* = \mathfrak{N}$ and hence one may define an action β of $\hat{G}$ on $\mathfrak{N}$ by

$$\beta_\gamma(A) = U_\gamma A U_\gamma^*.$$

One next defines the isomorphism η of $\{\mathfrak{M}, G, \alpha\}$ and $\{W^*(\mathfrak{N}, \beta), G, \hat{\beta}\}$ by setting

$$\eta\left(\sum_i A_i U_{\gamma_i}\right) = \sum_i \pi_0(A_i)\lambda(\gamma_i),$$

where $A_i \in \mathfrak{N} = \mathfrak{M}^\alpha$.

Note that the requirement that the U_γ has the representation property $U_{\gamma_1}U_{\gamma_2} = U_{\gamma_1\gamma_2}$ *cannot* be removed from the condition in (2). This is demonstrated by the example $\mathfrak{M} = \mathscr{L}(L^2(\mathbb{R}))$, $G = \mathbb{R}^2$, and

$$\alpha_{t,s}(A) = U(t)V(s)AV(s)^*U(t)^*$$

where $U(t)\xi(h) = \xi(h - t)$, $V(s)\xi(h) = e^{ish}\xi(h)$. One then has $U(t)V(s) = e^{ist}V(s)U(t)$. It follows easily that $(t, s) \in \mathbb{R}^2 \mapsto \alpha_{t,s}$ is a continuous group, and

$$\mathfrak{N} = \mathfrak{M}^\alpha = \mathbb{C}\mathbb{1}.$$

Thus $\mathfrak{N} \bigotimes_\beta \mathbb{R}^2 = L^\infty(\mathbb{R}^2)$, which is not isomorphic to $\mathscr{L}(L^2(\mathbb{R}))$ for any action β on $\mathfrak{N}$. Next, for $u, v \in \mathbb{R}$, define $W(u, v) = U(-v)V(u)$. Then

$$\alpha_{t,s}(W(u, v)) = e^{i(tu+sv)}W(u, v),$$

but the unitary elements $W(u, v)$ do not define a representation of $\widehat{\mathbb{R}^2} = \mathbb{R}^2$. There exists cases where the requirement $U_{\gamma_1}U_{\gamma_2} = U_{\gamma_1\gamma_2}$ can be removed, for example, if $\mathfrak{M}^\alpha$ is properly infinite (see Definition 2.7.15) and G is separable [Tak 2], [Con 2]. One might also generalize the notion of crossed product to that of a "skew-crossed product" and replace the group property of U by some cocycle property [Zel 1], [Robe 1].

The next motivating reason for introducing crossed products is to construct special operator algebras. In physics the algebra of quantum observables is often obtained from the abelian algebra of classical observables by taking something like the crossed product with the group generated by a set of "conjugate" variables p of the classical variables q [Ara 5], [Heg 1].

We now give an analysis of crossed products in a particularly simple case which, nevertheless, reveals many of the general features. Let G be a finite abelian group, and $\mathfrak{A}$ a finite dimensional abelian C^*-algebra, i.e., $\mathfrak{A} = \mathscr{C}(X)$, where X is a finite set with the discrete topology. The action α of G on $\mathfrak{A}$ then induces an action τ of G on X (the characters on $\mathfrak{A}$) by $(\alpha_g f)(x) = f(\tau_g(x))$. One verifies easily that X is divided into disjoint orbits $X_1, \ldots, X_n$ by this action, i.e., each X_i is of the form $\{\tau_g(x_i); g \in G\}$ for some $x_i \in X$. For each i define $G_i = \{g \in G; gx_i = x_i\}$ and then G_i is a subgroup of G, depending only on X_i up to an inner automorphism of G. The action of G on X_i is isomorphic with the action of G on the residue classes gG_i. It is now a useful exercise to show that

$$\begin{aligned} C^*(\mathscr{C}(X), \alpha) &= \bigoplus_{i=1}^n (\mathscr{L}(L^2(X_i)) \otimes L^\infty(G_i)) \\ &= W^*(L^\infty(X), \alpha). \end{aligned}$$

Hence in order that the crossed product be simple or, equivalently, be a factor or a full matrix algebra, we see that X must contain only one orbit, i.e., the action α must be ergodic, and for each $x \in X$ we must have that $gx = x$ implies $g = e$. Thus the action must be free. This motivates the following definition.

Definition 2.7.5. For a C^*- or W^*-dynamical system $(\mathfrak{A}, G, \alpha)$, where $\mathfrak{A}$ is abelian, we define the action α to be *ergodic* if $\mathfrak{A}$ does not contain a nontrivial closed two-sided globally α-invariant ideal. The action is *free* if for any $g \neq e$ and $A > 0$ there exists a B such that $A \geq B > 0$ and $\alpha_g(B) \neq B$.

Note that in the C^*-case ergodicity means that all the orbits of the action induced by α on the spectrum $\sigma(\mathfrak{A})$ are dense in $\sigma(\mathfrak{A})$, and in the W^*-case ergodicity means that $\mathfrak{A}$ does not contain nontrivial α-invariant projections. This follows from Theorem 2.1.11(A) and Proposition 2.4.22.

Of course this definition makes perfect sense for nonabelian $\mathfrak{A}$, but other notions of ergodicity and freedom are often more appropriate in such cases [Tak 1], [Zel 1], [Taka 1], [Gli 3]. The discussion of the finite-dimensional case now indicates the validity of the following theorem:

Theorem 2.7.6 ([Eff 1]). *Let $\{\mathfrak{A}, G, \alpha\}$ be a C^*-dynamical system, where $\mathfrak{A}$ is abelian and separable, G is discrete, amenable, and countable, and the action α is ergodic and free. Then $C^*(\mathfrak{A}, \alpha)$ is simple.*

In the von Neumann case we have for example the following theorem:

Theorem 2.7.7. *Let $\{\mathfrak{M}, G, \alpha\}$ be a W^*-dynamical system, where $\mathfrak{M}$ is abelian, σ-finite, and G is a countable group acting freely and ergodically on $\mathfrak{M}$. It follows that $W^*(\mathfrak{M}, \alpha)$ is a factor.*

Such a factor is called a *Krieger factor* [Kri 1]. Unless $\{\mathfrak{M}, G, \alpha\}$ is isomorphic to $\{L^\infty(G), G, \text{translation}\}$, this factor is not isomorphic to $\mathcal{L}(\mathfrak{H})$. A Krieger factor is known to be hyperfinite and the converse is "almost" true [Con 3]. Constructions related to this played an important role in the early attempts to construct nonisomorphic factors.

2.7.2. Tensor Products of Operator Algebras

In the proofs in the earlier parts of this chapter, we occasionally encountered the following construction. Given a C^*-algebra $\mathfrak{A}$, we considered all $n \times n$ matrices $(A_{ij})_{1 \leq i, j \leq n}$ with $A_{ij} \in \mathfrak{A}$, and multiplication and involution given by

$$(A_{ij})(B_{ij}) = \left(\sum_k A_{ik} B_{kj}\right),$$

$$(A_{ij})^* = (A_{ji}^*).$$

Denote this *-algebra by $\mathfrak{B} = \mathfrak{A} \otimes M_n$. It is not difficult to show that there is a unique C^*-norm on $\mathfrak{B}$ such that

$$\|(\gamma_{ij}A)\| = \|(\gamma_{ij})\| \, \|A\|,$$

where $\|(\gamma_{ij})\|$ is the C^*-norm of the $n \times n$ matrix (γ_{ij}). $\mathfrak{B}$ is called the tensor product of $\mathfrak{A}$ and M_n.

To generalize this construction consider first a finite collection $X_1, X_2, \ldots, X_n$ of vector spaces. Then there exists a unique vector space $\bigodot_{i=1}^n X_i$ with the following three properties:

(i) for each family $\{x_i\}$, where $x_i \in X_i$, there exists an element $\bigotimes_i x_i$ in $\bigodot_i X_i$ depending multilinearly on the x_i, and all elements in $\bigodot_i X_i$ are a finite linear combination of such elements;

(ii) (universal property) for each multilinear mapping π of the product of the X_i into a vector space Y, there exists a unique linear map $\varphi : \bigodot_i X_i \mapsto Y$ such that

$$\varphi\left(\bigotimes_i x_i\right) = \pi(\{x_i\})$$

for all $x_i \in X_i$;

(iii) (associativity) for each partition $\bigcup_k I_k$ of $\{1, \ldots, n\}$ there exists a unique isomorphism from $\bigodot_i X_i$ onto $\bigodot_k (\bigodot_{i \in I_k} X_i)$ transforming $\bigotimes_i x_i$ into $\bigotimes_k (\bigotimes_{i \in I_k} x_i)$.

Now, if $X_i = \mathfrak{H}_i$ are Hilbert spaces, we may define an inner product on $\bigodot_i \mathfrak{H}_i$ by extending the following definition by linearity

$$\left(\bigotimes_i \xi_i, \bigotimes_i \eta_i\right) = \prod_i (\xi_i, \eta_i).$$

The completion of $\bigodot_i \mathfrak{H}_i$ in the associated norm is called the tensor product of the Hilbert spaces $\mathfrak{H}_i$ and is denoted by

$$\bigotimes_{i=1}^n \mathfrak{H}_i .$$

Note that if $\{\xi_k^{(i)}\}_k$ is an orthonormal basis for $\mathfrak{H}_i$ then the

$$\bigotimes_{i=1}^n \xi_{k_i}^{(i)},$$

where the k_i varies independently for each i, form an orthonormal basis for

$$\bigotimes_{i=1}^n \mathfrak{H}_i .$$

If $X_i = \mathfrak{M}_i$ are von Neumann algebras on Hilbert spaces $\mathfrak{H}_i$, then one may define a *-algebraic structure on $\bigodot_i \mathfrak{M}_i$ by

$$\left(\bigotimes_i A_i\right)\left(\bigotimes_i B_i\right) = \bigotimes_i (A_i B_i),$$

$$\left(\bigotimes_i A_i\right)^* = \bigotimes_i (A_i^*).$$

Now define a map $\pi: \bigodot_i \mathfrak{M}_i \mapsto \mathscr{L}(\bigotimes_i \mathfrak{H}_i)$ by

$$\pi\left(\bigotimes_i A_i\right)\left(\bigotimes_i \xi_i\right) = \bigotimes_i A_i \xi_i .$$

It follows that π is a faithful *-representation of $\bigodot_i \mathfrak{M}_i$. The weak closure of $\pi(\bigodot_i \mathfrak{M}_i)$ is called the von Neumann tensor product of $\{\mathfrak{M}_i\}$ and is denoted by $\bigotimes_{i=1}^n \mathfrak{M}_i$. By Theorem 2.4.26 any isomorphism τ between von Neumann algebras $\mathfrak{M}$ and $\mathfrak{N}$ has the form $\tau(A) = U((A \otimes \mathbb{1})E')U^*$, where $\mathbb{1}$ is the identity on some "large" Hilbert space $\mathfrak{H}$, $E' \in \mathfrak{M}' \otimes \mathscr{L}(\mathfrak{H})$, and U is an isometry. Using this one can show that $\bigotimes_{i=1}^n \mathfrak{M}_i$ depends only on the isomorphism classes of $\mathfrak{M}_i$ and not on $\mathfrak{H}_i$.

If $X_i = \mathfrak{A}_i$ are C^*-algebras the situation is more complicated. Again one can make

$$\bigodot_{i=1}^n \mathfrak{A}_i$$

into a *-algebra. In general, there exist, however, more than one norm on $\bigodot_i \mathfrak{A}_i$ with the C^*-property $\|A^*A\| = \|A\|^2$ and the "cross-norm" property $\|\bigotimes_i A_i\| = \prod_i \|A_i\|$. For applications however, the most useful norm on $\bigodot_i \mathfrak{A}_i$ is the C^*-norm. This is defined by taking faithful representations $(\mathfrak{H}_i, \pi_i)$ of $\mathfrak{A}_i$ and defining

$$\left\| \sum_k \bigotimes_i A_i^{(k)} \right\| = \left\| \sum_k \bigotimes_i \pi_i(A_i^{(k)}) \right\| ,$$

where the latter element is viewed as an operator in $\mathscr{L}(\bigotimes_i \mathfrak{H}_i)$. This norm is independent of the particular faithful representations π_i used. The completion of $\bigodot_i \mathfrak{A}_i$ in this norm is called the C^*-tensor product of the $\mathfrak{A}_i$ and is denoted by

$$\bigotimes_{i=1}^n \mathfrak{A}_i .$$

It is known that if all the $\mathfrak{A}_i$ except one are nuclear (see end of Section 2.7.3) then the norm on $\bigodot_{i=1}^n \mathfrak{A}_i$ constructed above is unique as a C^*-cross norm [Lanc 1].

If $\{\mathfrak{H}_\alpha\}$ is an infinite collection of Hilbert spaces, and $\Omega_\alpha \in \mathfrak{H}_\alpha$ are unit vectors we may form the infinite tensor product $\bigotimes_\alpha^{\Omega_\alpha} \mathfrak{H}_\alpha$ by taking all finite linear combinations of elements of the form $\bigotimes_\alpha \xi_\alpha$, where $\xi_\alpha = \Omega_\alpha$ except for a finite number of α, and completing in the norm arising from the inner product

$$\left(\bigotimes_\alpha \xi_\alpha, \bigotimes_\alpha \eta_\alpha\right) = \prod_\alpha (\xi_\alpha, \eta_\alpha).$$

If $\mathfrak{M}_\alpha$ are von Neumann algebras on $\mathfrak{H}_\alpha$, we may form $\bigotimes_\alpha^{\Omega_\alpha} \mathfrak{M}_\alpha$ by taking the weak closure of linear combinations of elements in $\mathscr{L}(\bigotimes_\alpha^{\Omega_\alpha} \mathfrak{H}_\alpha)$ of the form $\bigotimes_\alpha A_\alpha$, where $A_\alpha = \mathbb{1}_{\mathfrak{H}_\alpha}$ except for finitely many α. The resulting von Neumann algebra $\bigotimes_\alpha^{\Omega_\alpha} \mathfrak{M}_\alpha$ is highly dependent on the choice of the sequence $\{\Omega_\alpha\}$ [Ara 6], [Pow 1]. If each $\mathfrak{M}_\alpha$ is a factor then $\bigotimes_\alpha^{\Omega_\alpha} \mathfrak{M}_\alpha$ is a factor. This can be proved by applying a slight generalization of Theorem 2.6.10 on the state

$$\omega\left(\bigotimes_\alpha A_\alpha\right) = \prod_\alpha (\Omega_\alpha, A_\alpha \Omega_\alpha)$$

If $\{\mathfrak{A}_\alpha\}$ is a collection of C^*-algebras we may, in a similar way, form the infinite C^*-tensor product $\bigotimes_\alpha \mathfrak{A}_\alpha$ by taking an inductive limit of the corresponding finite subproducts. In this case $\bigotimes_\alpha \mathfrak{A}_\alpha$ is uniquely defined (up to the non-uniqueness of the finite tensor products discussed earlier). References for this paragraph are [Gui 1], [Lanc 1].

2.7.3. Weights on Operator Algebras; Self-Dual Cones of General von Neumann Algebras; Duality and Classification of Factors; Classification of C^*-Algebras

In this section we list various results in the theory of operator algebras which are important in the structural theory of these algebras, but which have not yet had any deep impact on mathematical physics. Proofs either will not be given or will only be sketched.

We introduce a generalization of the concept of positive linear functional.

Definition 2.7.8. A *weight* on a C^*-algebra $\mathfrak{A}$ is a function $\omega; \mathfrak{A}_+ \mapsto [0, \infty]$ satisfying

$$\omega(A + B) = \omega(A) + \omega(B), \qquad A, B \in \mathfrak{A}_+$$

$$\omega(\alpha A) = \alpha\omega(A), \qquad \alpha \in \mathbb{R}_+, A \in \mathfrak{A}_+$$

(with the convention $0 \cdot \infty = 0$).

A *trace* on $\mathfrak{A}$ is a weight ω satisfying

$$\omega(A^*A) = \omega(AA^*), \qquad A \in \mathfrak{A}.$$

The following proposition is easy to prove:

Proposition 2.7.9. *If ω is a weight on $\mathfrak{A}$, define $\mathfrak{M}_{\omega+}$ and $\mathscr{L}_\omega$ by*

$$\mathfrak{M}_{\omega+} = \{A \in \mathfrak{A}_+; \omega(A) < \infty\},$$

$$\mathscr{L}_\omega = \{A \in \mathfrak{A}; \omega(A^*A) < \infty\}$$

It follows that $\mathfrak{M}_{\omega+}$ is a cone in $\mathfrak{A}_+$ which is hereditary, i.e.,

$$0 \leq A \leq B \in \mathfrak{M}_{\omega+} \Rightarrow A \in \mathfrak{M}_{\omega+}.$$

*The complex linear span $\mathfrak{M}_\omega$ of $\mathfrak{M}_{\omega+}$ is a *-subalgebra of $\mathfrak{A}$. Moreover, $\mathscr{L}_\omega$ is a left ideal in $\mathfrak{A}$ and*

$$\mathfrak{M}_\omega = \mathscr{L}_\omega{}^* \mathscr{L}_\omega.$$

The weight ω extends to a linear functional (also denoted by ω) on $\mathfrak{M}_\omega$.

Just as the case for states in Theorem 2.3.16 one may equip $\mathscr{L}_\omega$ with a pre-Hilbert structure defined by $A, B \in \mathscr{L}_\omega \mapsto \omega(A^*B)$, and prove:

Theorem 2.7.10. *To each weight ω on $\mathfrak{A}$ there is a Hilbert space $\mathfrak{H}$ and two maps, η; $\mathscr{L}_\omega \mapsto \mathfrak{H}$, π; $\mathfrak{A} \mapsto \mathscr{L}(\mathfrak{H})$, such that η is linear with range dense in $\mathfrak{H}$, π is a representation of $\mathfrak{A}$, and*

$$(\eta(B), \pi(A)\eta(C)) = \omega(B^*AC)$$

for $A \in \mathfrak{A}$, $B, C \in \mathscr{L}_\omega$.

We next generalize the concept of normal positive functional on a von Neumann algebra.

Theorem 2.7.11 ([Haa 3], [Ped 2]). *Let ω be a weight on a von Neumann algebra $\mathfrak{M}$. The following conditions are equivalent:*

(1) *if $\{A_i\}$ is a sequence in $\mathfrak{M}_+$ and $\sum_i A_i = A \in \mathfrak{M}_+$ then $\omega(A) = \sum_i \omega(A_i)$;*
(2) *if $\{A_\alpha\}$ is an increasing net in $\mathfrak{M}_+$ and $A = \text{l.u.b.}_\alpha A_\alpha \in \mathfrak{M}_+$ then $\omega(A) = \text{l.u.b.}_\alpha\, \omega(A_\alpha)$;*
(3) *if $\{A_\alpha\}$ is a σ-weakly convergent net in $\mathfrak{M}_+$ with limit $A \in \mathfrak{M}_+$ then $\omega(A) \leq \text{l.u.b.}_\alpha\, \omega(A_\alpha)$;*
(4) *there is a set $\{\omega_\alpha\}$ of positive, normal functionals on $\mathfrak{M}$ such that $\omega(A) = \sup_\alpha \omega_\alpha(A)$ for all $A \in \mathfrak{M}_+$.*
(5) *there is a set $\{\omega_\alpha\}$ of positive, normal functionals on $\mathfrak{M}$ such that $\omega(A) = \sum_\alpha \omega_\alpha(A)$ for all $A \in \mathfrak{M}_+$.*

Definition 2.7.12. A weight ω on a von Neumann algebra is called *normal* if it satisfies any of the equivalent conditions in Theorem 2.7.11. It is called *faithful* if $A \in \mathfrak{M}$ and $\omega(A) = 0$ implies $A = 0$. It is called *semifinite* if $\mathfrak{M}_\omega$ is σ-weakly dense in $\mathfrak{M}$.

Using part (5) of Theorem 2.7.11 it is not difficult to prove:

Proposition 2.7.13. *Every von Neumann algebra admits a normal, faithful, semifinite weight.*

The first part of the following theorem is proved using so-called left Hilbert algebras [Com 1]; the rest is then proved exactly as in the Tomita–Takesaki theory for states in Section 2.5.2.

Theorem 2.7.14. *Let ω be a faithful, normal, semifinite weight on a von Neumann algebra $\mathfrak{M}$, and define $\mathscr{L}_\omega$, $\mathfrak{H}$, η, and π as in Proposition 2.7.9 and Theorem 2.7.10. Then π is a normal *-isomorphism from $\mathfrak{M}$ onto $\pi(\mathfrak{M}) =$*

$\pi(\mathfrak{M})''$. Set $\pi(\mathfrak{M}) = \mathfrak{M}$. There exists a left ideal $\mathscr{L}_\omega' \in \mathfrak{M}'$ such that $\mathfrak{M}_\omega' = \mathscr{L}'^*_\omega \mathscr{L}_\omega'$ is σ-weakly dense in $\mathfrak{M}'$, and a linear mapping η; $\mathscr{L}_\omega' \mapsto \mathfrak{H}$ with dense range such that $\eta(A'B') = A'\eta(B')$, $A' \in \mathfrak{M}'$, $B' \in \mathscr{L}_\omega'$, and the properties:

(1) $A\eta(A') = A'\psi$ *for all* $A' \in \mathscr{L}_\omega'$ *if, and only if,* $A \in \mathscr{L}_\omega$ *and* $\eta(A) = \psi$;
(2) $A'\eta(A) = A\psi$ *for all* $A \in \mathscr{L}_\omega$ *if, and only if,* $A' \in \mathscr{L}_\omega'$ *and* $\eta(A') = \psi$.

Furthermore, $\eta(\mathfrak{M}_\omega)$ and $\eta(\mathfrak{M}_\omega')$ are dense in $\mathfrak{H}$ and the mappings

$$S_0; \eta(\mathfrak{M}_\omega) \mapsto \mathfrak{H}; \eta(A) \mapsto \eta(A^*),$$

$$F_0; \eta(\mathfrak{M}_\omega') \mapsto \mathfrak{H}; \eta(A') \mapsto \eta(A'^*)$$

are closable with closures S and F satisfying $S^* = F$. If $S = J\Delta^{1/2}$ is the polar decomposition of S then J and Δ satisfy all the identities of Proposition 2.5.10 *and*

$$\Delta^{it}\mathfrak{M}\Delta^{-it} = \mathfrak{M}, \qquad t \in \mathbb{R},$$

$$J\mathfrak{M}J = \mathfrak{M}'.$$

One may now define the modular group, etc., for normal, semifinite, faithful weights just as for normal, faithful states.

One may also define a natural positive cone $\mathscr{P}_\omega$ in the Hilbert space associated to a normal semifinite weight ω as the closure of the set of $\pi(A)J(\eta(A))$, where $A \in \mathfrak{M}_\omega$. The cone $\mathscr{P}_\omega$ has all the properties of natural cones derived in section 2.5.4 with obvious modifications (i.e., $\mathscr{P}_\omega = \overline{\Delta^{1/4}\eta(\mathfrak{M}_{\omega+})}$) [Haa 2].

We now turn to the problem of classifying von Neumann algebras.

Definition 2.7.15. Two projections E, F in a von Neumann algebra $\mathfrak{M}$ are said to be *equivalent* (written $E \sim F$) if there exists a $W \in \mathfrak{M}$ such that $E = W^*W$ and $F = WW^*$. A projection E in $\mathfrak{M}$ is said to be *finite* if it is not equivalent to a proper subprojection of itself; otherwise, it is said to be infinite. The algebra $\mathfrak{M}$ is called *semifinite* if any projection in $\mathfrak{M}$ contains a nonzero finite projection (alternatively: there exists an increasing net $\{E_\alpha\}$ of finite projections in $\mathfrak{M}$ such that $E_\alpha \to \mathbb{1}$). $\mathfrak{M}$ is *finite* if $\mathbb{1}$ is finite; otherwise, $\mathfrak{M}$ is *infinite*. $\mathfrak{M}$ is *properly infinite* if all nonzero projections in the center $\mathfrak{M} \cap \mathfrak{M}'$ are infinite; it is *purely infinite* if all nonzero projections in $\mathfrak{M}$ are infinite.

The following theorem, Connes' Radon–Nikodym theorem, is fundamental for the classification of von Neumann algebras.

Theorem 2.7.16 ([Con 4]). *For any pair φ, ψ of faithful, normal, semifinite weights on $\mathfrak{M}$ there exists a continuous one-parameter family $t \mapsto (D\psi : D\varphi)_t$ of unitaries in $\mathfrak{M}$ with the properties:*

(1) $\sigma_t^\psi(A) = (D\psi : D\varphi)_t \sigma_t^\varphi(A)(D\psi : D\varphi)_t^*$;
(2) $(D\psi : D\varphi)_{t+s} = (D\psi : D\varphi)_t \sigma_t^\varphi((D\psi : D\varphi)_s)$;
(3) $(D\psi : D\varphi)_t^* = (D\varphi : D\psi)_t$, $(D\psi : D\varphi)_t(D\varphi : D\omega)_t = (D\psi : D\omega)_t$;
(4) $\psi(A) = \varphi(UAU^*)$ *with U unitary in* $\mathfrak{M} \Leftrightarrow (D\psi : D\varphi)_t = U^*\sigma_t^\varphi(U)$.

The proof makes use of 2×2 matrix methods in much the same way as in the proof of Lemma 2.5.33. (See Theorem 5.3.34)

We now turn to semifinite von Neumann algebras.

Theorem 2.7.17 ([Dix 1]). *For a von Neumann algebra $\mathfrak{M}$ the following statements are equivalent:*

(1) *$\mathfrak{M}$ is semifinite;*
(2) *For any $A \in \mathfrak{M}_+$, $A \neq 0$, there exists a normal, semifinite trace τ on $\mathfrak{M}$ such that $\tau(A) > 0$;*
(3) *there exists a faithful, normal semifinite trace on $\mathfrak{M}$;*
(4) *σ_t^{ω} is an inner group of automorphisms of $\mathfrak{M}$ for any faithful, normal semifinite weight on $\mathfrak{M}$, i.e., for any ω there exists a positive, invertible selfadjoint operator h affiliated with $\mathfrak{M}$ such that $\sigma_t^{\omega}(A) = h^{it}Ah^{-it}$ for $t \in \mathbb{R}$, $A \in \mathfrak{M}$.*

We remark briefly that (3) ⇒ (1) is trivial, because if ω is a faithful trace, E is a projection, and if $\omega(E) < \infty$, then E must be finite. (2) ⇔ (3) is more or less evident; (4) ⇒ (3) follows by showing that $\tau(A) = \omega(h^{-1}A)$ is a trace, while (3) ⇒ (4) uses Theorem 2.7.16. (See [Tak 3].) Finally, to prove (1) ⇒ (2) one builds up a trace by making a comparison theory for projections in much the same way as one builds up the reals from the integers or constructs the Haar measure of a locally compact group.

If $\{E_\alpha\}$ is a set of mutually orthogonal projections in the center $\mathfrak{Z}$ of the von Neumann algebra $\mathfrak{M}$ with $\sum_\alpha E_\alpha = \mathbb{1}$, then $\mathfrak{M}$ splits in a direct sum $\mathfrak{M} = \sum_\alpha \mathfrak{M}E_\alpha$. $\mathfrak{M}E_\alpha$ is a von Neumann algebra with center $\mathfrak{Z}E_\alpha$. Thus, by going to the limit in a way which will be made precise in Chapter 4, the von Neumann algebra splits in a generalized direct sum, or direct integral, of factors (this applies strictly only to σ-finite von Neumann algebras). Hence the question of classifying von Neumann algebras reduces to the classification of factors.

Definition 2.7.18. A factor $\mathfrak{M}$ is said to be *type* I if it has a minimal, nonzero projection. It is *type* II if it is semifinite, but not type I. A type II factor is *type* II_1 if it is finite and *type* II_∞ if it is infinite. $\mathfrak{M}$ is said to be *type* III if it is not semifinite, i.e., if it is purely infinite.

For type I factors, it is easy to give a complete classification.

Proposition 2.7.19. *If $\mathfrak{M}$ is a type* I *factor, then $\mathfrak{M}$ is isomorphic with $\mathscr{L}(\mathfrak{H})$ for some Hilbert space $\mathfrak{H}$. Thus the dimension of $\mathfrak{H}$ is a complete invariant for type* I *factors.*

Definition 2.7.20. A type I factor is said to be of *type* I_n if n is the dimension of the Hilbert space $\mathfrak{H}$ of Proposition 2.7.19.

It remains to classify type II and type III factors. This can be reduced to the classification of certain semifinite von Neumann algebras and their automorphisms (Connes–Takesaki duality theorem).

Theorem 2.7.21 ([Tak 2], [Con 4]). *If $\mathfrak{M}$ is a properly infinite von Neumann algebra then there exists a semifinite von Neumann algebra $\mathfrak{N}$ with a semifinite, normal, faithful trace τ, and a σ-weakly continuous one-parameter group α_t of *-automorphisms of $\mathfrak{N}$ such that*

$$\tau \circ \alpha_t = e^{-t}\tau, \qquad t \in \mathbb{R},$$

and

$$\mathfrak{M} = W^*(\mathfrak{N}, \alpha).$$

$\{\mathfrak{N}^0, \alpha^0\}$ is another pair with these properties if, and only if, there exists an isomorphism $\gamma: \mathfrak{N} \mapsto \mathfrak{N}^0$ and a one-parameter family U_t of unitary elements in $\mathfrak{N}^0$ such that

$$\alpha_t{}^0(A) = U_t(\gamma\alpha_t\gamma^{-1}(A))U_t{}^*.$$

Furthermore, the center of $\mathfrak{M}$ identifies with the fixed-point subalgebra of the center $\mathfrak{N} \cap \mathfrak{N}'$ of $\mathfrak{N}$ under the action of α. The pair $(\mathfrak{N}, \alpha)$ can be obtained from $\mathfrak{M}$ by picking a faithful, semifinite, normal weight ω on $\mathfrak{M}$ and setting

$$\mathfrak{N} = W^*(\mathfrak{M}, \sigma^\omega);$$

$$\alpha_t = \widehat{\sigma_t{}^\omega} = \textit{the dual action of } \mathbb{R} \textit{ on } W^*(\mathfrak{M}, \sigma^\omega).$$

PROOF. The proof follows by noting that the group σ^ω and the weight ω extend in a natural way to an inner group of automorphisms of $W^*(\mathfrak{M}, \sigma^\omega)$ and a weight on $W^*(\mathfrak{M}, \sigma^\omega)$ such that the extended group turns out to be the modular automorphism group of the extended weight. One then applies Theorem 2.7.17 to get a trace on $\mathfrak{N}$, and finally notes that

$$W^*(W^*(\mathfrak{M}, \sigma), \hat{\sigma}) \cong \mathfrak{M} \otimes \mathscr{L}(L^2(\mathbb{R})) \cong \mathfrak{M}$$

for any $C_0{}^*$-group σ of automorphisms of $\mathfrak{M}$.

Definition 2.7.22. If $\mathfrak{M}$ is a factor, define $S(\mathfrak{M}) = \bigcap_\omega \sigma(\Delta_\omega)$, where ω ranges over all normal semifinite weights on $\mathfrak{M}$ and $\sigma(\Delta_\omega)$ is the spectrum of Δ_ω. Define $T(\mathfrak{M}) = \{t \in \mathbb{R};\, \sigma_t^\omega \text{ is inner}\}$ where ω is a normal semifinite weight on $\mathfrak{M}$. (The latter definition is independent of ω in view of Theorem 2.7.16.)

The set $S(\mathfrak{M})$ is a closed subset of $[0, +\infty\rangle$ such that $S(\mathfrak{M}) \cap \langle 0, +\infty\rangle$ is a subgroup of the multiplicative group $\langle 0, +\infty\rangle$. The set $T\{\mathfrak{M})$ is clearly a subgroup of $\mathbb{R}$. If we define the Connes spectrum $\Gamma(\mathfrak{M}) = \Gamma(\sigma^\omega)$ as in the Notes and Remarks to Section 3.2.3, then $\Gamma(\mathfrak{M})$ is a closed subgroup of $\mathbb{R}$, and $\Gamma(\mathfrak{M})$ is independent of ω by a use of Theorem 2.7.16. The following connections hold between these concepts:

Theorem 2.7.23 ([Con 4]). *If $\mathfrak{M}$ is a factor then*

$$\Gamma(\mathfrak{M}) = \log(S(\mathfrak{M})\backslash\{0\})$$

and if $S(\mathfrak{M}) \neq \{0, 1\}$, then

$$T(\mathfrak{M}) = \{t \in \mathbb{R};\ e^{it\lambda} = 1 \textit{ for all } \lambda \in \Gamma(\mathfrak{M})\}.$$

If $S(\mathfrak{M}) = \{0, 1\}$, the group $T(\mathfrak{M})$ is not determined from $S(\mathfrak{M})$, it might be a dense subgroup of $\mathbb{R}$ or a subgroup of the form $\{nT_0;\ n \in \mathbb{Z}\}$.

Since $S(\mathfrak{M})\backslash\{0\}$ is a closed subgroup of the multiplicative group of positive reals, we have the following possibilities:

(1) $S(\mathfrak{M}) = \{1\}$;
(2) $S(\mathfrak{M}) = [0, \infty\rangle$;
(3) $S(\mathfrak{M}) = \{0\} \cup \{\lambda^n;\ n \in \mathbb{Z}\}, \qquad \lambda \in \langle 0, 1\rangle$;
(4) $S(\mathfrak{M}) = \{0, 1\}$.

Theorem 2.7.17 and Theorem 2.7.23 imply that (1) is the case if, and only if, $\mathfrak{M}$ is semifinite, and (2)–(4) are the case if, and only if, $\mathfrak{M}$ is of type III.

Definition 2.7.24. $\mathfrak{M}$ is a *factor of type* III_1 if $S(\mathfrak{M}) = [0, \infty\rangle$, *of type* III_λ if $S(\mathfrak{M}) = \{0\} + \lambda^{\mathbb{Z}}$, where $0 < \lambda < 1$, and *of type* III_0 if $S(\mathfrak{M}) = \{0, 1\}$.

If $\mathfrak{M}$ is a semifinite factor, then the pair $\{\mathfrak{N}, \alpha\}$ of Theorem 2.7.21 has the form

$$\mathfrak{N} = \mathfrak{M} \otimes L^\infty(\mathbb{R}),$$

$$\alpha_t = \iota \otimes \text{translation by } t$$

so the theorem gives no interesting information.

In the type III case, however, we have the following. (Remember that α acts ergodically on $\mathfrak{N} \cap \mathfrak{N}' = \mathfrak{Z}(\mathfrak{N})$ if, and only if, $\mathfrak{M}$ is a factor):

Theorem 2.7.25. *Let $\mathfrak{M}$ be a von Neumann algebra. One has equivalence between the following pairs of conditions*:

(1_λ) *$\mathfrak{M}$ is a factor of type* III_λ, $0 < \lambda < 1$;
(2_λ) *$\{\mathfrak{Z}(\mathfrak{N}), \alpha\}$ is transitive and periodic with period* $-\log \lambda$;

(1_0) *$\mathfrak{M}$ is a factor of type* III_0;
(2_0) *$\{\mathfrak{Z}(\mathfrak{N}), \alpha\}$ is ergodic, but not isomorphic to $\{L^\infty(\mathbb{R})$, translation$\}$*;

(1_1) *$\mathfrak{M}$ is a factor of type* III_1;
(2_1) *$\mathfrak{N}$ is a factor of type* II_∞.

A special class of factors for which a complete classification is known is the class of so-called hyperfinite factors. Recall that a factor, or more generally a von Neumann algebra, is called hyperfinite if it is generated by an ascending sequence of finite-dimensional factors, i.e., matrix algebras. For a factor, this is equivalent to being generated by an ascending sequence of finite-dimensional von Neumann algebras [Ell 4], [Ell 5]. It is known that a hyperfinite von Neumann algebra has the property E of Tomiyama, i.e., if $\mathfrak{M} \subseteq \mathscr{L}(\mathfrak{H})$ then there exists a projection $E\colon \mathscr{L}(\mathfrak{H}) \mapsto \mathfrak{M}$ of norm one, i.e., $E^2 = E$, $E\mathscr{L}(\mathfrak{H}) = \mathfrak{M}$, $\|E\| = 1$. Such a mapping E also is a conditional expectation: $E(ABC) = AE(B)C$, $A, C \in \mathfrak{M}$, $B \in \mathscr{L}(\mathfrak{H})$ [Tom 1]. The converse is known for all factors $\mathfrak{M}$ [Con 5], [Con 8], [Haa 5]. It is known that Krieger factors, i.e., factors which are the crossed product of a σ-finite abelian von Neumann algebra with a single automorphism acting freely and ergodically, are hyperfinite, and conversely hyperfinite factors are Krieger factors [Con 3], [Con 5], [Con 8], [Haa 5].

As for the classification of hyperfinite factors, it is known that there exists just one hyperfinite type II_1 factor and one hyperfinite II_∞ factor up to isomorphism [Con 5]. The automorphisms of these factors, up to the equivalence relation relevant for Theorems 2.7.21 and 2.7.25, have been classified [Con 6] and, as a result, it has been proved that there exists just one hyperfinite III_λ for each λ between 0 and 1, while the hyperfinite factors of type III_0 are completely classified by the flow $\{\mathfrak{Z}(\mathfrak{N}), \alpha\}$. There exists a hyperfinite III_1 factor [Ara 6], and it was recently proved that this is the only one, [Haa 5], [Con 8].

Parallel to this classification of factors, one has a classification of C^*-algebras. A C^*-algebra $\mathfrak{A}$ is said to be type I if all its factor representations is of type I (a representation π of $\mathfrak{A}$ is a factor representation if $\pi(\mathfrak{A})''$ is a factor). It is known that a separable C^*-algebra $\mathfrak{A}$ is type I if, and only if, for any ideal which is the kernel of an irreducible representation, this representation is unique up to unitary equivalence. Also $\mathfrak{A}$ is type I if, and only if, for any irreducible representation π, of $\mathfrak{A}$, one has, $\mathscr{L}\mathscr{C}(\mathfrak{H}) \subseteq \pi(\mathfrak{A})$ [Gli 4]. Thus the classification of separable type I C^*-algebras is to some extent reduced to a study of its ideal structure.

It is known [Gli 4], [Mar 1] that if $\mathfrak{A}$ is a separable C^*-algebra not of type I and $\mathfrak{M}$ is an infinite hyperfinite factor then there exists a representation π of $\mathfrak{A}$ such that $\pi(\mathfrak{A})'' = \mathfrak{M}$. This suggests that the next simplest class of C^*-algebras to study is the class of nuclear C^*-algebras, i.e., the C^*-algebras such that all their factor representations are hyperfinite. Many studies in this direction have been made [Choi 1], [Eff 2], [Ell 6], [Ped 2].

Notes and Remarks

There are several books and lecture notes on operator algebras which cover the material of this chapter. For example, one may consult Arveson [[Arv 1]], Dixmier [[Dix 1]], [[Dix 2]], Guichardet [[Gui 1]], Kadison [Kad 1], Lanford [Lan 1], Naimark [[Nai 1]], Pedersen [[Ped 1]], Sakai [[Sak 1]], and Schwartz [[Sch 1]]. The references [[Arv 1]], [[Gui 1]], [Kad 1], and [Lan 1], are the most elementary and provide the easiest access to the general theory. There are many books on functional analysis and its various aspects, of which the following are a cross section: Bourbaki [[Bou 1]], Kato [[Kat 1]], Reed and Simon [[Ree 1]], Riesz and Nagy [[Rie 1]], Rudin [[Rud 1]], [[Rud 2]], and Yosida [[Yos 1]]. References [[Ree 1]], [[Rud 2]], and [[Rud 1]] are particularly straightforward.

Section 2.1.1

The material of this section is standard. The study of abstract C^*-algebras originated with the work of Gelfand and Naimark [Gel 1] and Segal [Seg 1]. The original definition of a C^*-algebra in [Gel 1] was a Banach *-algebra whose norm satisfied $\|A^*A\| = \|A\|^2$ and was such that each $\mathbb{1} + A^*A$ had an inverse. This second condition was eventually eliminated by Fukamiya [Fuk 1], Kelley and Vaught [Kel 1], and Kaplansky [Kap 1]. Glimm and Kadison [Gli 1] also showed that the norm condition could be replaced by $\|A^*A\| = \|A^*\| \, \|A\|$. More recently, Araki and Elliott [Ara 1] have shown that the condition $\|AB\| \leq \|A\| \, \|B\|$ is actually a consequence of the remaining C^*-algebraic structure. Theorem 2.1.10 originated in [Gel 1] while Theorem 2.1.11 is due to Gelfand [Gel 2]. Some authors distinguish between abstract C^*-algebras and C^*-algebras concretely realized by bounded operators on a Hilbert space. The abstract algebra is called a B^*-algebra and the concrete algebra a C^*-algebra.

Section 2.2.1

The general theory of spectra of elements of Banach *-algebras was also initiated in [Gel 1] but the usual discussion is based upon the properties of commutative algebraic theory [Gel 2]. We have avoided the use of commutative theory by emphasizing properties which follow from simple transformations.

Section 2.2.2

Theorem 2.2.12, which establishes that an element is positive if, and only if, it is of the form A^*A, is due to Kaplansky [Kap 1]. This is the key result which allows one to deduce that $\mathbb{1} + A^*A$ has an inverse and thus eliminate the surplus axiom of [Gel 1]. Our analysis of positive elements by use of the integral algorithm for the square root is unconventional. The standard procedure is to proceed circuitously through a preliminary analysis of positive elements of abelian algebras. Nevertheless, integral algorithms are often used in the discussion of fractional powers of selfadjoint operators on Hilbert space [[Kat 1]] [[Ree 1]] or, more generally, fractional powers of generators of continuous semigroups [[Yos 1]]. The verification that $A = B^2$, in the proof of Theorem 2.2.10, is described in the semigroup context in Chapter IX, Section 12, of [[Yos 1]]. An alternative proof via contour integration is given in [[Kat 1]]. In relation to Proposition 2.2.13 it is amusing to note that if $\mathfrak{A}$ is a C^*-algebra, $\alpha > 1$ and if $A \geq B \geq 0$ always implies that $A^\alpha \geq B^\alpha$ then $\mathfrak{A}$ must be abelian [Oga 1]. A detailed discussion of the polar decomposition is given in [[Kat 1]].

Section 2.2.3

The theory of approximate identities originates with Segal [Seg 1].

Section 2.3.1

The theory of representations of C^*-algebras and representations of topological groups are intimately entangled. Group theory preceded the algebraic theory and to a large extent motivated it. Proposition 2.3.8, which characterizes irreducibility, is a classic result of group theory and the equivalence of conditions (1) and (2) is usually called Schur's lemma.

Section 2.3.2

The notions of state and pure state are due to Segal [Seg 1]. The terminology is adapted from physics. The Krein–Milman theorem [Kre 1] is discussed and proved in [[Yos 1]] or in [[Rud 2]].

Section 2.3.3

The construction of a representation from a state (Theorem 2.3.16), is due to Gelfand and Naimark [Gel 1] and Segal [Seg 1] and is consequently often called the GNS construction. The connection between irreducibility, purity, and extremality (Theorem 2.3.19), is again in [Seg 1].

Section 2.3.4

The Hahn–Banach theorem occurs in every book on functional analysis. A lucid account is given by [[Rud 1]] or [[Rud 2]]. Our proof of the basic structure theorem via Lemma 2.3.23 is a slight variation on the traditional argument.

Section 2.3.5

As mentioned above the theory of commutative algebras originated with Gelfand [Gel 1].

Section 2.4

The theory of von Neumann algebras preceded that of C^*-algebras and was motivated both by group theory and quantum mechanics. The earliest study of these algebras dates from 1929 [Neu 1]. Much of the standard theory was developed by Murray and von Neumann in a series of papers [Mur 1], [Neu 2].

Section 2.4.1

The Alaoglu–Bourbaki theorem can be found in [[Bou 1]], [[Rud 2]], or in Köthe [[Köt 1]]. The theorem of Banach cited in the proof of Theorem 2.4.7 can be found in [[Bou 1]], Chapter 4, §2, Theorem 5. The assertion concerning the bipolar of a closed convex set is also in [[Bou 1]].

Section 2.4.2

The original commutant theorem and the first density theorem were proved by von Neumann in his earliest paper [Neu 1]. Kaplansky's density theorem came much later [Kap 2]. These theorems are of great technical use and it is worth citing the following result, due to Kadison [Kad 1], [Kad 2], which is of a similar nature.

Theorem. *Let $\mathfrak{A}$ be an irreducible C^*-subalgebra of $\mathscr{L}(\mathfrak{H})$ and let*

$$\{\psi_1, \ldots, \psi_n\}, \{\varphi_1, \ldots, \varphi_n\}$$

denote two finite families of vectors of $\mathfrak{H}$. If there exists a $T \in \mathscr{L}(\mathfrak{H})$ such that $T\varphi_i = \psi_i$ for $i = 1, 2, \ldots, n$ then it follows that there exists an $A \in \mathfrak{A}$ with the same property and with $\|A\| = \|T\|$. If T is selfadjoint then A may be chosen selfadjoint, and if T is unitary then A may be chosen unitary.

This theorem has many applications, e.g., the equivalence of algebraic irreducibility and topological irreducibility cited in Section 2.3.1 follows from this result.

Section 2.4.3

The theorem of Sakai which gives an abstract characterization of von Neumann algebras can be found in [[Sak 1, Theorem 1.16.7]]. The terminology W^*-algebra is often used for the abstractly defined algebra and then the name von Neumann algebra is reserved for the operator algebras. The abstract characterization of von Neumann algebras by Kadison occurs in [Kad 3].

Section 2.5.1

Although a general von Neumann algebra $\mathfrak{M}$ is not σ-finite there exists an increasing net of projections $P_\alpha \in \mathfrak{M}$ such that $P_\alpha \to \mathbb{1}$ and the reduced von Neumann algebras $\mathfrak{M}_\alpha = P_\alpha \mathfrak{M} P_\alpha$ are σ-finite, e.g., one can take $P_\alpha = [\mathfrak{M}'\mathfrak{H}_\alpha]$, where $\mathfrak{H}_\alpha$ is a finite-dimensional subspace of $\mathfrak{H}$. Thus in this sense every von Neumann algebra is generated by σ-finite algebras.

Section 2.5.2

The Tomita–Takesaki theorem originated in an unpublished work of Tomita [Tomi 1]. This manuscript states the principal theorem, Theorem 2.5.14, and detailed a proof which contained all the features of the proof that we have given. An exposé of this work with many refinements and several applications was subsequently given by Takesaki [Tak 3]. The proof of the principal theorem has been simplified and shortened by various authors, notably van Daele [Dae 1], van Daele and Rieffel [Dae 2], and Woronowicz [Wor 1]. The present proof combines features of [Dae 1], [Wor 1] together with some personal refinements. The theorem of Carlson used in the case $\|\Delta\| < +\infty$ may be found in [[Tit 1]].

Section 2.5.3

Analytic elements are of use in the study of group representations [Har 1], [Car 1] and have also been used for the study of selfadjointness properties [Lum 1], [Nel 1]. The Krein–Smulian theorem and Alaoglu's theorem are discussed in [[Dun 1, Chapter V]]. A discussion of the Mackey topology can be found in [[Bou 1]]. The equivalence of weak and strong analyticity is due to Dunford [Dun 1].

Section 2.5.4

The theory of self-dual cones and standard forms was developed by Araki, Connes and Haagerup, independently [Ara 2], [Con 1], [Haa 1]. In our partial description we have also followed [Ara 3], [Haa 2]. The trick of

4×4 matrices used in proving Theorem 2.5.31 originates in [Con 1]. Connes also shows that a self-dual cone (in a Hilbert space) is the natural cone of a von Neumann algebra if, and only if, it satisfies two other properties, which he calls orientability and homogeneity.

Section 2.6.1

The importance of the quasi-local structure of the C^*-algebras of mathematical physics was first emphasized by Haag [Haag 1]. Cluster properties of the type described were first derived by Powers [Pow 1] in the context of UHF algebras. In particular, Powers proved Lemma 2.6.7, Proposition 2.6.8, and Theorem 2.6.9 in this simple case. Powers' work has subsequently been discussed and generalized by many authors, e.g., [Ara 4], [Haag 2], [Lan 2], [Rob 1], [Rue 1]. The notion of algebra at infinity was introduced in [Lan 2] and Lemma 2.6.4 is due to Araki and Kishimoto [Ara 4]. We have generally followed the outline of [Rue 1] with improvements suggested by [Ara 4] and [Rob 1]. A more complete discussion of Fermi statistics is given in [Rob 1]. If $\mathfrak{M}_\alpha = \mathscr{L}(\mathfrak{H}_\alpha)$ then the situation is as follows. For each α there exists an $R_\alpha \in \mathscr{L}(\mathfrak{H}_\alpha)$ such that

$$R_\alpha{}^2 = \mathbb{1}, \qquad R_\alpha A R_\alpha = \sigma(A), \qquad A \in \mathfrak{M}_\alpha,$$

and local normality of ω together with the assumption $\{\mathfrak{M}_\alpha \vee \mathfrak{M}_\beta\}'' = \mathfrak{M}_{\alpha \vee \beta}$ allows the conclusion

$$\pi_\omega(\mathfrak{M}_\alpha' \cap \mathfrak{M}_{\alpha \vee \beta}) = (\pi_\omega(R_\alpha \mathfrak{M}_\beta{}^{\mathrm{o}} + \mathfrak{M}_\beta{}^{\mathrm{e}}))''.$$

Further, if σ is weakly continuous in the representation $(\mathfrak{H}_\omega, \pi_\omega)$ then one can use this identification to conclude that $\mathfrak{Z}_\omega \cap \pi_\omega(\mathfrak{A}^{\mathrm{e}})'' = \mathfrak{Z}_\omega{}^{\perp}$. Cluster properties again characterize the case of trivial $\mathfrak{Z}_\omega$.

Section 2.6.2

The discussion of topologies partially follows [Rob 2]. The two results quoted in the proof of Proposition 2.6.13 can be found in [[Dix 2]]. The fact that $\mathfrak{A} \cap \mathscr{LC}(\mathfrak{H}) \neq \{0\}$ implies $\mathfrak{A} \supseteq \mathscr{LC}(\mathfrak{H})$ is Corollaire 4.1.10 and the statement concerning vector states is Lemma 11.2.1 of this reference.

Section 2.6.3

A more detailed discussion of ideal structure and other algebraic structure in quasi-local algebras can be found in [Gli 2], [Bra 1], [Bra 2] and [Ell 1].

Groups, Semigroups, and Generators

3.1. Banach Space Theory

Physical theories consist essentially of two elements, a kinematical structure describing the instantaneous states and observables of the system, and a dynamical rule describing the change of these states and observables with time. In the classical mechanics of point particles a state is represented by a point in a differentiable manifold and the observables by functions over the manifold. In the quantum mechanics of systems with a finite number of degrees of freedom the states are given by rays in a Hilbert space and the observables by operators acting on the space. For particle systems with an infinite number of degrees of freedom we intend to identify the states with states over appropriate algebras of fields, or operators. In each of these examples the dynamical description of the system is given by a flow, a one-parameter group of automorphisms of the underlying kinematical structure, which represents the motion of the system with time. In classical mechanics one has a group of diffeomorphisms, in quantum mechanics a group of unitary operators on the Hilbert space, and for systems with an infinite number of degrees of freedom a group of automorphisms of the algebra of observables. It is also conventional to describe other symmetries of physical systems by groups of automorphisms of the basic kinematic structure and in this chapter, and Chapter 4, we study various aspects of this group-theoretic description. In this chapter we principally consider one-parameter groups and problems related to dynamics.

In the conventional formulations of theories of interacting particles the dynamical flow is introduced in an implicit manner. The natural description of the motion is in terms of the infinitesimal change of the system. The infinitesimal motion is directly described by some form of Hamiltonian formalism which allows the explicit incorporation of the interparticle interaction. In classical mechanics the infinitesimal change is defined by means of a vector field, in quantum mechanics by a selfadjoint Hamiltonian operator, and for systems with an infinite number of degrees of freedom by some form of derivation of the associated algebra. The first basic problem which occurs is the integration of these infinitesimal prescriptions to give the dynamical flow. This problem, which will be the central theme of this chapter, is, of course, only a minor step in the analysis of the physical theory. The interesting characteristics of the theory are described by the further detailed properties of the flow, dispersive properties, ergodic characteristics, etc. The integration problem only involves the characterization of the completeness, or lack of completeness, of the dynamical prescription which reflects the absence, or presence, of catastrophic behavior of the system. These points are, however, poorly understood for systems with an infinite number of degrees of freedom and this problem constitutes a nontrivial first step in the analysis of the dynamics of such systems.

The general problem is to study the differential equation

$$\frac{dA_t}{dt} = SA_t$$

under a variety of circumstances and assumptions. In each instance the A corresponds to an observable, or state, of the system and will be represented by an element of some suitable space X. The function

$$t \in \mathbb{R} \mapsto A_t \in X$$

describes the motion of A and S is an operator on X, which generates the infinitesimal change of A. The dynamics are given by solutions of the differential equation which respect certain supplementary conditions of growth and continuity. The existence of a sensible, noncatastrophic, time development of the system is equivalent to the existence of global solutions of the equation of motion satisfying the physical boundary conditions.

There are three basic questions concerning such solutions, existence, uniqueness, and stability under small perturbations, and we analyze these problems in a variety of settings, but we always assume that X is a Banach space and S a linear operator on X. Firstly, we examine the differential equation in the purely Banach space setting. Subsequently, we assume that X has an algebraic structure and that S is a derivation.

Formally, the solution of the differential equation is $A_t = U_t A$, where $U_t = \exp\{tS\}$ and the problem is to give a meaning to the exponential. Independently of the manner in which this is done one expects U_t to have the property that U_0 is the identity and that $U_t U_s = U_{t+s}$ and so we seek solutions of this nature. There are, however, many different possible types of continuity of $t \mapsto U_t$ and this leads to a structural hierarchy. We examine uniform, strong, and weak* continuity and sometimes use the assumption that $\|U_t\| \leq 1$ where $\| \;\; \|$ indicates the norm on the bounded operators on X, i.e.,

$$\|S\| = \sup\{\|SA\|; A \in X, \|A\| = 1\}.$$

This growth restriction can be interpreted as a law of conservation, or possible dissipation, of probability. Thus the types of solution that we consider fall into the two following classes:

either $U = \{U_t\}_{t \in \mathbb{R}}$ is a one-parameter group of bounded linear operators on X characterized by the first condition of Definition 2.5.17, i.e.,

$$U_{s+t} = U_s U_t; \qquad t, s \in \mathbb{R}, \quad U_0 = I,$$

where I is the identity map;

or $U = \{U_t\}_{t \in \mathbb{R}_+}$ is only defined for $t \geq 0$ and is a semigroup satisfying

$$U_{s+t} = U_s U_t, \qquad t, s \in \mathbb{R}_+, \quad U_0 = I.$$

If $\|U_t A\| = \|A\|$ for all $t \in \mathbb{R}$, or $t \in \mathbb{R}_+$, $A \in X$, then U is a group, or semigroup, of isometries. If, on the other hand, $\|U_t\| \leq 1$ for all $t \in \mathbb{R}_+$, we refer to U_t as a semigroup of contractions. Note that a one-parameter group of contractions is automatically isometric, because

$$\|U_t\| \leq 1, \qquad \|U_t^{-1}\| = \|U_{-t}\| \leq 1.$$

Now we examine groups and semigroups with various types of continuity.

3.1.1. Uniform Continuity

The theory of uniformly continuous groups and semigroups is straightforward and is described as follows.

Proposition 3.1.1. *Let $\{U_t\}_{t \in \mathbb{R}_+}$ be a one-parameter semigroup of bounded linear operators, $U_t \in \mathscr{L}(X)$, on the Banach space X. The following conditions are equivalent*:

(1) *U_t is uniformly continuous at the origin, i.e.,*

$$\lim_{t \to 0} \|U_t - I\| = 0;$$

(2) *U_t is uniformly differentiable at the origin, i.e., there is a bounded operator $S \in \mathscr{L}(X)$ such that*

$$\lim_{t \to 0} \|(U_t - I)/t - S\| = 0;$$

(3) *there is a bounded operator $S \in \mathscr{L}(X)$ such that*

$$U_t = \sum_{n \geq 0} \frac{t^n}{n!} S^n.$$

If these conditions are fulfilled then U_t extends to a uniformly continuous one-parameter group satisfying

$$\|U_t\| \leq \exp\{|t| \, \|S\|\}.$$

PROOF. Clearly, (3) ⇒ (2) ⇒ (1) and hence we must prove that (1) ⇒ (2) ⇒ (3). For this note that if t is sufficiently small then

$$\left\| \frac{1}{t} \int_0^t ds \, U_s - I \right\| < 1,$$

which implies that the integral

$$X_t = \frac{1}{t} \int_0^t ds \, U_s$$

is invertible and has a bounded inverse. Next consider the identity

$$\begin{aligned}\left(\frac{U_h - I}{h}\right)X_t &= \frac{1}{th}\int_0^t ds(U_{s+h} - U_s)\\ &= \frac{1}{th}\int_t^{t+h} ds\, U_s - \frac{1}{th}\int_0^h ds\, U_s\\ &= \left(\frac{U_t - I}{t}\right)X_h.\end{aligned}$$

The right-hand side converges uniformly, as h tends to zero, to $(U_t - I)/t$ and hence

$$\lim_{h\to 0}\left\|\frac{U_h - I}{h} - \left(\frac{U_t - I}{t}\right)X_t^{-1}\right\| = 0.$$

Thus U_t is uniformly differentiable and we have in fact identified the differential S by

$$S = \left(\frac{U_t - I}{t}\right)X_t^{-1}.$$

Finally, the identity yields the integral equation

$$U_t - I = S\int_0^t ds\, U_s,$$

which may be solved by iteration to give

$$U_t = \sum_{h\geq 0}\frac{t^n}{n!}S^n.$$

The bound on $\|U_t\|$ follows from this formula.

Thus the group $\{U_t\}_{t\in\mathbb{R}}$ of bounded operators on the Banach space X is uniformly continuous if, and only if, its generator S is bounded. By generator we mean the uniform derivative of the group at the origin. Note that this definition implies that sums and uniform limits of generators are again generators. Moreover, if U and V are uniformly continuous groups with generators S and T, respectively, one has

$$U_t - V_t = t\int_0^1 d\lambda\, U_{\lambda t}(S - T)V_{(1-\lambda)t}$$

and hence

$$\|U_t - V_t\| \leq |t|\exp\{|t|(\|S\| + \|T\|)\}\|S - T\|.$$

Thus if S converges uniformly to T then U converges uniformly to V and the convergence is uniform for t in any finite interval of $\mathbb{R}$. Conversely, if $\lambda \geq \|S\|$, $\lambda > \|T\|$, one calculates that

$$S - T = (\lambda I - S)\left[\int_0^\infty dt\, e^{-\lambda t}(U_t - V_t)\right](\lambda I - T)$$

and

$$\|S - T\| \leq (\|S\| + \lambda)(\|T\| + \lambda) \int_0^\infty dt\, e^{-\lambda t} \|U_t - V_t\|.$$

Hence if U converges uniformly to V, on finite intervals of $\mathbb{R}$, then S converges uniformly to T.

This describes the case of uniformly continuous one-parameter groups and indicates that these groups will have limited use in the description of dynamics because their generators, i.e., the associated Hamiltonian operators, are necessarily bounded. Nevertheless, these groups are of some interest in the general context of symmetries.

3.1.2. Strong, Weak, and Weak* Continuity

In this section we examine groups and semigroups of bounded operators on the Banach space X which have weaker continuity properties than those examined in the previous section. To state these continuity properties we consider a formalism similar to that described in Section 2.5.3.

Let F be a norm-closed subspace of the dual X^* of X and let $\sigma(X, F)$ be the locally convex topology on X induced by the functionals in F. We will assume that

(a) $\|A\| = \sup\{|\eta(A)|;\, \eta \in F,\ \|\eta\| = 1\}$;
(b) the $\sigma(X, F)$-closed convex hull of every $\sigma(X, F)$-compact set in X is $\sigma(X, F)$-compact;
(c) the $\sigma(F, X)$-closed convex hull of every $\sigma(F, X)$-compact set in F is $\sigma(F, X)$-compact.

These conditions are satisfied in the special cases $F = X^*$ and $F = X_*$ (see Section 2.5.3) and these two choices of F will be of importance in the sequel.

Another dual topology which is often useful is the Mackey topology $\tau(X, F)$. This topology was briefly used in Section 2.5.3 and it is defined by the seminorms

$$A \in X \mapsto p_K(A) = \sup_{\eta \in K} |\eta(A)|$$

where K ranges over the compact subsets of F. Generally, the Mackey topology restricts K to the convex, compact, circled subsets of F, but by assumption (c) this is equivalent to the definition we have given. The $\tau(X, F)$-topology is particularly simple if $F = X^*$. The unit ball of X^* is $\sigma(X^*, X)$-compact, by the Alaoglu–Bourbaki theorem, and hence the $\tau(X, X^*)$-topology is equal to the norm topology. More generally, one can establish that the $\tau(X, F)$-topology on X is the finest locally convex topology such that all $\tau(X, F)$-continuous functionals lie in F. In particular, this implies that any $\sigma(X, F)$-dense convex subset of X is $\tau(X, F)$-dense.

After these topological preliminaries we return to the examination of groups and semigroups. The groups and semigroups that we wish to examine in the remainder of this section are introduced as follows:

Definition 3.1.2. A semigroup (or group) $t \mapsto U_t$ of bounded linear operators on the Banach space X is called a *$\sigma(X, F)$-continuous semigroup* (*or group*) if

(1) $t \mapsto U_t A$ is $\sigma(X, F)$-continuous for all $A \in X$, i.e., $t \mapsto \eta(U_t A)$ is continuous for all $A \in X$ and $\eta \in F$;

(2) $A \mapsto U_t A$ is $\sigma(X, F)$–$\sigma(X, F)$-continuous for all t, i.e., $\eta \circ U_t \in F$ for $\eta \in F$.

If $F = X^*$ then U is said to be *weakly continuous*, or a C_0-semigroup; if $F = X_*$, U is said to be *weakly* continuous*, or a $C_0{}^*$-semigroup.

We shall see later (Corollary 3.1.8) that a weakly continuous semigroup is in fact strongly continuous, i.e., $t \mapsto U_t A$ is continuous in norm for all $A \in X$. A special case of this has already been demonstrated for C_0-groups of isometries (Corollary 2.5.23). To tackle the general case we must first control the growth of $\|U_t\|$.

Proposition 3.1.3. *Let $U = \{U_t\}_{t \geq 0}$ be a $\sigma(X, F)$-continuous semigroup on the Banach space X. There exists an $M \geq 1$ and $\beta \geq \inf_{t>0}(t^{-1} \log\|U_t\|)$ such that*

$$\|U_t\| \leq Me^{\beta t}.$$

PROOF. Because the function $t \in \mathbb{R}_+ \mapsto \eta(U_t A)$ is continuous for all $\eta \in F$ and $A \in X$ it follows, by two applications of the uniform boundedness principle, that there is an $M < +\infty$ such that $\|U_t\| \leq M$ for $t \in [0, 1]$. Now each $t \geq 0$ is of the form $t = n + s$ with n a nonnegative integer and $s \in [0, 1\rangle$. Thus

$$\|U_t\| = \|U_1{}^n U_s\| \leq M^{n+1} \leq Me^{\beta t},$$

where $\beta = \log M$. Furthermore, $M \geq \|U_0\| = 1$ and

$$\beta + t^{-1} \log M \geq t^{-1} \log\|U_t\| \geq \inf_{s>0} s^{-1} \log\|U_s\|,$$

giving the estimates on M and β.

One consequence of this growth condition is that $\{U_t e^{-\beta t}\}_{t \geq 0}$ is a $\sigma(X, F)$-continuous semigroup which is uniformly bounded, and hence the regularization procedure of Proposition 2.5.18 can be readily applied. In the proof of the next lemma we explicitly use the assumptions (a), (b), and (c) on the pair (X, F).

Proposition 3.1.4. *Let $t \mapsto U_t$ be a $\sigma(X, F)$-continuous semigroup on X such that $\|U_t\| \leq Me^{\beta t}$. Let μ be a complex measure on $\mathbb{R}_+$ such that $\int_0^\infty d|\mu|(t)\, e^{\beta t} < \infty$. It follows that*

$$U_\mu(A) = \int_0^\infty d\mu(t)\, U_t(A)$$

defines a bounded linear operator on X. Furthermore, U_μ is $\sigma(X, F)$–$\sigma(X, F)$-continuous.

PROOF. The first statement is essentially Proposition 2.5.18. For the second statement we have to show that $U_\mu{}^*\eta \in F$ for all $\eta \in F$, where $U_\mu{}^*$ is the dual of U_μ on X^*. But $t \mapsto U_t{}^*$ is a $\sigma(F, X)$-continuous semigroup on F by Definition 3.1.2, and we have

$$\begin{aligned} U_\mu{}^*\eta(A) &= \eta(U_\mu(A)) \\ &= \int d\mu(t)\, \eta(U_t(A)) \\ &= \int d\mu(t)(U_t{}^*\eta)(A). \end{aligned}$$

Since the conditions (a), (b), and (c) on the pair (X, F) are symmetric in X and F we may apply Proposition 2.5.18 again to conclude that $U_\mu{}^*\eta \in F$.

Next we introduce the generator of a $\sigma(X, F)$-continuous semigroup.

Definition 3.1.5. Let U be a $\sigma(X, F)$-continuous semigroup on the Banach space X. The *(infinitesimal) generator of* U is defined as the linear operator S on X, whose domain $D(S)$ is composed of those $A \in X$ for which there exists a $B \in X$ with the property that

$$\eta(B) = \lim_{t \to 0} \frac{\eta((U_t - I)A)}{t}$$

for all $\eta \in F$. If $A \in D(S)$ the action of S is defined by $SA = B$.

Note that the semigroup property of U automatically implies $U_t D(S) \subseteq D(S)$ and

$$SU_t A = U_t SA$$

for all $A \in D(S)$. We now analyze miscellaneous properties of generators and their resolvents. Recall that the resolvent set $r(S)$ of an operator S on the Banach space X is the set of $\lambda \in \mathbb{C}$ for which $\lambda I - S$ has a bounded inverse, the spectrum $\sigma(S)$ of S is the complement of $r(S)$ in $\mathbb{C}$, and if $\lambda \in r(S)$ then $(\lambda I - S)^{-1}$

is called the resolvent of S. The basic properties of generators and their resolvents are contained in the following:

Proposition 3.1.6. *Let U be a $\sigma(X, F)$-continuous semigroup on the Banach space X, S the generator of U and M, β constants such that*

$$\|U_t\| \leq M \exp\{\beta t\}, \qquad t \in \mathbb{R}_+ .$$

It follows that S is $\sigma(X, F)$-densely defined and $\sigma(X, F)$–$\sigma(X, F)$-closed. If Re $\lambda > \beta$ *then the range $R(\lambda I - S)$ of $\lambda I - S$ satisfies*

$$R(\lambda I - S) = X$$

and for $A \in D(S)$

$$\|(\lambda I - S)(A)\| \geq M^{-1}(\mathrm{Re}\,\lambda - \beta)\|A\|.$$

The resolvent of S is given by the Laplace transform

$$(\lambda I - S)^{-1}A = \int_0^\infty dt\, e^{-\lambda t} U_t A$$

for all $A \in X$ and Re $\lambda > \beta$.

PROOF. If Re $\lambda > \beta$ it follows by Proposition 3.1.4 that we may define a bounded, $\sigma(X, F)$–$\sigma(X, F)$-continuous operator R_λ on X by

$$R_\lambda A = \int_0^\infty ds\, e^{-\lambda s} U_s A.$$

For $t \geq 0$ one has

$$\begin{aligned} \frac{1}{t}(U_t - I)R_\lambda A &= \frac{1}{t}\int_0^\infty ds\, e^{-\lambda s}(U_{s+t} - U_s)A \\ &= \frac{1}{t}\int_0^\infty ds\, (e^{-\lambda(s-t)} - e^{-\lambda s})U_s A - \frac{1}{t}\int_0^t ds\, e^{-\lambda(s-t)} U_s A \\ &\underset{t \to 0}{\longrightarrow} \lambda R_\lambda A - A, \end{aligned}$$

where the first term converges in norm and the second in the $\sigma(X, F)$-topology. It follows that $R_\lambda A \in D(S)$ and

$$(\lambda I - S)R_\lambda A = A$$

for $A \in X$. Since $U_t R_\lambda = R_\lambda U_t$ and R_λ is $\sigma(X, F)$–$\sigma(X, F)$-continuous, it follows that S commutes with R_λ in the sense that

$$R_\lambda(\lambda I - S)A = (\lambda I - S)R_\lambda A = A$$

for $A \in D(S)$. Hence $\lambda \in r(S)$ and $(\lambda I - S)^{-1} = R_\lambda$. The $\sigma(X, F)$–$\sigma(X, F)$-continuity of $(\lambda I - S)^{-1}$ then implies that $\lambda I - S$, and hence S, is $\sigma(X, F)$–$\sigma(X, F)$-closed. Now for any $A \in X$ and $\eta \in F$ one can use the same reasoning as in the proof of Proposition 2.5.22 to deduce that

$$\lim_{n \to \infty} \eta(nR_n A) = \eta(A).$$

Since $D(S) = R(R_n) = R(nR_n)$ it follows that $D(S)$ is $\sigma(X, F)$-dense in X. Finally, note that for $\eta \in F$, $A \in X$, one has

$$\begin{aligned}|\eta(R_\lambda A)| &\leq \int_0^\infty ds\, e^{-s|\mathrm{Re}\,\lambda|}|\eta(U_s A)| \\ &\leq \int_0^\infty ds\, e^{-s|\mathrm{Re}\,\lambda|}Me^{\beta s}\|\eta\|\,\|A\| \\ &\leq M(\mathrm{Re}\,\lambda - \beta)^{-1}\|\eta\|\,\|A\|,\end{aligned}$$

or, alternatively stated, $\|(\lambda I - S)^{-1}\| \leq M(\mathrm{Re}\,\lambda - \beta)^{-1}$. This establishes the bound given in the proposition.

A slight variation of the above argument yields the following often useful fact.

Corollary 3.1.7. *Let U be a $\sigma(X, F)$-continuous semigroup on the Banach space X with generator S. Let D be a subset of the domain $D(S)$, of S, which is $\sigma(X, F)$-dense in X and invariant under U, i.e., $U_t A \in D$ for all $A \in D$ and $t \in \mathbb{R}_+$. It follows that D is a core for S, i.e., the $\sigma(X, F)$–$\sigma(X, F)$-closure of the restriction of S to D is equal to S.*

PROOF. Let $\hat{S}$ denote the closure of S restricted to D. If $R(\lambda I - \hat{S}) = X$ for some λ with $\mathrm{Re}\,\lambda > \beta$ then it follows from Proposition 3.1.6 that $\hat{S} = S$. But for $A \in D$ one can choose Riemann approximants

$$\textstyle\sum_N(A) = \sum_{i=1}^{N} e^{-\lambda t_i}U_{t_i}A(t_{i+1} - t_i),$$

$$\textstyle\sum_N((\lambda I - S)A) = \sum_{i=1}^{N} e^{-\lambda t_i}U_{t_i}(\lambda I - S)A(t_{i+1} - t_i)$$

which converge simultaneously to $(\lambda I - S)^{-1}A$ and A. Now $\sum_N(A) \in D$ because of the invariance of D under U and

$$\textstyle(\lambda I - S)\sum_N(A) = \sum_N((\lambda I - S)A).$$

Thus $\sum_N(A) \to (\lambda I - S)^{-1}A$ and $(\lambda I - S)\sum_N(A) \to A$. Therefore $A \in R(\lambda I - \hat{S})$ for each $A \in D$. But $(\lambda I - S)^{-1}$ is $\sigma(X, F)$–$\sigma(X, F)$-continuous and hence $R(\lambda I - \hat{S})$ is $\sigma(X, F)$-closed. Thus $R(\lambda I - \hat{S}) = X$ by the density of D.

Under some additional assumptions Proposition 3.1.6 implies that a $\sigma(X, F)$-continuous semigroup U is actually continuous in the $\tau(X, F)$-, or Mackey, topology. The additional assumption is an equicontinuity property which states that there exists some β' such that for each $\sigma(F, X)$-compact subset $K \subseteq F$ the subset

$$K' = \{e^{-\beta' t}U_t{}^*\eta;\, \eta \in K, t \geq 0\}$$

has a $\sigma(F, X)$-compact closure. Again $U_t{}^*$ denotes the dual of U_t acting on F. If $F = X^*$ then this equicontinuity is automatically fulfilled. One chooses β' and M such that $\|U_t\| \leq M\exp\{\beta' t\}$ and then the $\sigma(X^*, X)$-compactness property of K' follows from the Alaoglu–Bourbaki theorem. Another instance when the equicontinuity is fulfilled is when β' is large enough that

$$\|e^{-\beta' t}U_t\| \to 0$$

as $t \to \infty$, and $\eta, t \mapsto U_t{}^*\eta$ is jointly continuous in the $\sigma(F, X)$-topology.

In order to state the general result concerning coincidence of continuity properties it is convenient to introduce the concept of $\tau(X, F)$-equicontinuity as follows. A family $\{T_\alpha\}$ of operators, on the Banach space X, is defined to be *$\tau(X, F)$-equicontinuous* if for each seminorm p_K in the definition of the $\tau(X, F)$-topology there exists a seminorm $p_{K'}$ such that

$$p_K(T_\alpha A) \leq p_{K'}(A)$$

for all $A \in X$ and for all α.

The general result is the following:

Corollary 3.1.8. *Let U be a $\sigma(X, F)$-continuous semigroup, on the Banach space X, with generator S. Assume there is a $\beta' \geq 0$ such that $\{U_t e^{-\beta' t}\}_{t \geq 0}$ is a $\tau(X, F)$-equicontinuous family. It follows that $t \mapsto U_t$ is $\tau(X, F)$-continuous and if $A \in D(S)$ then*

$$\lim_{t \to 0} \frac{(U_t - I)A}{t} = SA,$$

where the limit is in the $\tau(X, F)$-topology. In particular, each weakly continuous semigroup is strongly continuous and its weak and strong generators coincide.

PROOF. One has

$$\eta(U_{t_2}A - U_{t_1}A) = \int_{t_1}^{t_2} ds\, \eta(U_s SA)$$

for all $A \in D(S)$, and $\eta \in F$. Hence

$$p_K((U_{t_1} - U_{t_2})A) \leq p_{K'}(SA) \int_{t_1}^{t_2} dt\, e^{\beta' t}.$$

Now for $A \in D(S)$ and $B \in X$ one has

$$\begin{aligned} p_K((U_{t_1} - U_{t_2})B) &\leq p_K((U_{t_1} - U_{t_2})A) + p_K(U_{t_1}(B - A)) + p_K(U_{t_2}(B - A)) \\ &\leq \int_{t_1}^{t_2} dt\, e^{\beta' t} p_{K'}(SA) + (e^{\beta' t_1} + e^{\beta' t_2}) p_{K'}(B - A). \end{aligned}$$

But $D(S)$ is a $\sigma(X, F)$-dense subspace of X and hence is $\tau(X, F)$-dense. Thus this estimate implies $\tau(X, F)$-continuity of U.

Next if $A \in D(S)$ then

$$\frac{(U_t - I)}{t} A = \frac{1}{t} \int_0^t ds\, U_s SA.$$

But the $\tau(X, F)$-limit of the right-hand side exists as $t \to 0$ because of the $\tau(X, F)$-continuity of U and hence the $\tau(X, F)$-limit of the left-hand side also exists.

The final statement follows by choosing $F = X^*$. In this case the Mackey topology $\tau(X, F)$ coincides with the norm topology and the seminorm assumption corresponds to a bound $\|U_t A\| \le M \exp\{\beta' t\} \|A\|$. But this bound was established in Proposition 3.1.3.

Note that the resolvent bound

$$\|(\lambda I - S)^{-1}\| \le M(\text{Re}\,\lambda - \beta)^{-1}, \qquad \text{Re}\,\lambda > \beta,$$

obtained in Proposition 3.1.6 takes the particularly easy form

$$\|(\lambda I - S)^{-1}\| \le 1/\text{Re}\,\lambda$$

whenever U is a C_0-semigroup of contractions and this obviously implies

$$\|(\lambda I - S)^{-n}\| \le 1/(\text{Re}\,\lambda)^n$$

for all $n = 1, 2, \ldots$. The general situation is, however, more complicated. One can derive the Laplace transform relations

$$(\lambda I - S)^{-n}A = \int_0^\infty dt\, e^{-\lambda t} \frac{t^{n-1}}{(n-1)!} U_t A, \qquad \text{Re}\,\lambda > \beta,$$

and then one has

$$\begin{aligned} \|(\lambda I - S)^{-n}\| &\le \int_0^\infty dt\, e^{-t\,\text{Re}\,\lambda} \frac{t^{n-1}}{(n-1)!} M e^{\beta t} \\ &= M(\text{Re}\,\lambda - \beta)^{-n}. \end{aligned}$$

These bounds do not follow by iteration of the bound for $n = 1$, unless $M = 1$, because M only occurs linearly.

As a final remark about Proposition 3.1.6 we note that the resolvent set $r(S)$ of the generator S contains the half-plane $\text{Re}\,\lambda > \beta$. Now if S is the generator of a C_0-group U with $\|U_t\| \le Me^{\beta t}$ then both S and $-S$ generate C_0-semigroups, the semigroups $\{U_t\}_{t \ge 0}$ and $\{U_{-t}\}_{t \ge 0}$, respectively, and hence $r(S)$ contains both the half planes $\text{Re}\,\lambda > \beta$ and $\text{Re}\,\lambda < -\beta$. Thus the spectrum $\sigma(S)$ of S lies in the strip $|\text{Re}\,\lambda| \le \beta$ and if U is a group of isometries the spectrum of the generator must lie on the imaginary axis.

After this preliminary analysis of generators we turn to the more interesting problem of constructing semigroups from generators. We will examine two cases, $F = X^*$ and $F = X_*$, corresponding to weakly (strongly) continuous and weakly* continuous semigroups. Actually the discussion of the two cases is very similar and many of the features of the latter case follow by duality from the former. For this reason the following elementary result is often useful.

Lemma 3.1.9. *Let S be an operator on a Banach space X and F a norm-closed subspace of X^* such that*

$$\|A\| = \sup\{|\eta(A)|;\, \eta \in F,\, \|\eta\| \le 1\}.$$

If S is contained in the dual (adjoint) of a $\sigma(F, X)$-densely defined operator S^*, on F, then S is $\sigma(X, F)$–$\sigma(X, F)$-closable.

Moreover, the following conditions are equivalent:

(1) *S is $\sigma(X, F)$-densely defined, and $\sigma(X, F)$–$\sigma(X, F)$-closed;*
(2) *S is the dual (adjoint) of a $\sigma(F, X)$-densely defined, $\sigma(F, X)$–$\sigma(F, X)$-closed, operator S^* on F.*

If these conditions are fulfilled, and S is bounded, then $\|S\| = \|S^*\|$.

PROOF. (1) $\Rightarrow$ (2) Let $G(S) = \{(A, SA); A \in D(S)\}$ denote the graph of S and consider the orthogonal complement $G(S)^\perp = \{(\omega_1, \omega_2)\}$, of $G(S)$, in $F \times F$. We first claim that $G = \{(-\omega_2, \omega_1); (\omega_1, \omega_2) \in G(S)^\perp\}$ is the graph of an operator S^*, on F. For this it suffices that $\omega_2 = 0$ imply $\omega_1 = 0$. But this follows from the orthogonality relation

$$\omega_1(A) + \omega_2(SA) = 0$$

and the density of $D(S)$. Next remark that $G(S)^\perp$, and hence G, is $\sigma(F, X) \times \sigma(F, X)$-closed by definition. Thus S^* is $\sigma(F, X)$–$\sigma(F, X)$-closed. Finally, if S^* is not $\sigma(F, X)$-densely defined then there must be a nonzero element of $G^\perp$ of the form $(-B, 0)$, i.e., there is a $B \in X$ with $B \neq 0$ such that $(0, B) \in G(S)$. But this contradicts the linearity of S and hence S^* must be densely defined.

(2) $\Rightarrow$ (1) The argument is identical to the preceding.

Now let us return to the first statement of the lemma. Because $D(S^*)$ is $\sigma(F, X)$-dense, the dual S^{**}, of S^*, is well defined as a $\sigma(X, F)$–$\sigma(X, F)$-closed operator on X. But S^{**} is then a closed extension of S, i.e., S is closable. Finally, the equality of the norms for bounded operators follows from

$$\|A\| = \sup\{|\eta(A)|; \eta \in F, \|\eta\| \leq 1\}$$

by a straightforward argument.

The main problem of this section is the construction of C_0- and $C_0{}^*$-semigroups. The problem consists of characterizing those operators which can be exponentiated in a suitable form and we will examine various algorithms for the exponential both here and in Section 3.1.3. We begin with a result which characterizes the generator S of a semigroup of contractions by properties of its resolvent. The algorithm

$$e^{tx} = \lim_{n \to \infty} (1 - tx/n)^{-n}$$

for the numerical exponential can be extended to an operator relation if the "resolvent" $(I - tS/n)^{-n}$ has suitable properties. The definition of the resolvent of a closed operator S requires two pieces of information. Firstly, one must know that the range of $(I - tS/n)$ is equal to the whole space in order that $(I - tS/n)^{-1}$ should be everywhere defined and, secondly, one needs a bound on $\|(I - tS/n)^{-n}\|$. For later purposes it is useful to distinguish these two separate properties.

Theorem 3.1.10 (Hille–Yosida theorem). *Let S be an operator on the Banach space X. If $F = X^*$, or $F = X_*$, then the following conditions are equivalent;*

(1) *S is the infinitesimal generator of a $\sigma(X, F)$-continuous semigroup of contractions U;*

(2) *S is $\sigma(X, F)$-densely defined, and $\sigma(X, F)$–$\sigma(X, F)$-closed. For $\alpha \geq 0$*

$$\|(I - \alpha S)A\| \geq \|A\|, \qquad A \in D(S),$$

and for some $\alpha > 0$

$$R(I - \alpha S) = X.$$

If these conditions are satisfied then the semigroup is defined in terms of S by either of the limits

$$\begin{aligned} U_t A &= \lim_{\varepsilon \to 0} \exp\{tS(I - \varepsilon S)^{-1}\}A \\ &= \lim_{n \to \infty} (I - tS/n)^{-n} A, \end{aligned}$$

where the exponential of the bounded operator $S(I - \varepsilon S)^{-1}$ is defined by power series expansion. The limits exist in the $\sigma(X, F)$-topology, uniformly for t in compacts, and if A is in the norm closure $\overline{D(S)}$ of $D(S)$, the limits exist in norm.

PROOF. Proposition 3.1.6 established that (1) $\Rightarrow$ (2). To prove the converse we construct the associated semigroup with the aid of the first algorithm of the theorem. Condition (2) implies that $(I - \varepsilon S)^{-1}$ is a bounded, $\sigma(X, F)$-continuous operator with $\|(I - \varepsilon S)^{-1}\| \leq 1$ at the point $\varepsilon = \alpha_0$ for which $R(I - \alpha_0 S) = X$. But a perturbation argument using the Neumann series

$$(I - \alpha S)^{-1} = \left(\frac{\alpha_0}{\alpha}\right) \sum_{n \geq 0} \left(\frac{\alpha - \alpha_0}{\alpha}\right)^n (I - \alpha_0 S)^{-n-1}$$

then establishes that $R(I - \varepsilon S) = X$ for all $\varepsilon > 0$.

Now we distinguish between the two cases of a C_0- or a C_0^*-semigroup, i.e., $F = X^*$ or $F = X_*$.

C_0-case: $F = X^*$. First we set $S_\varepsilon = S(I - \varepsilon S)^{-1}$ and use the relation $S_\varepsilon = -\varepsilon^{-1}(I - (I - \varepsilon S)^{-1})$ to deduce that

$$\begin{aligned} \|\exp\{tS_\varepsilon\}\| &\leq \exp\{-t\varepsilon^{-1}\} \sum_{n \geq 0} \frac{(t\varepsilon^{-1})^n}{n!} \|(I - \varepsilon S)^{-n}\| \\ &\leq 1 \end{aligned}$$

for $t \geq 0$. Thus the $U_t^\varepsilon = \exp\{tS_\varepsilon\}$ are uniformly continuous contraction semigroups. Moreover the bounded operators S_ε and S_δ commute and

$$\begin{aligned} \|U_t^\varepsilon A - U_t^\delta A\| &= \left\| \int_0^1 ds \frac{d}{ds} e^{t(sS_\varepsilon + (1-s)S_\delta)} A \right\| \\ &= \left\| t \int_0^1 ds\, e^{tsS_\varepsilon} e^{t(1-s)S_\delta} (S_\varepsilon - S_\delta) A \right\| \leq t\|(S_\varepsilon - S_\delta)A\| \end{aligned} \tag{$*$}$$

for all $t \geq 0$. Next note that if $A \in D(S)$ then

$$\begin{aligned}\|(I - \varepsilon S)^{-1}A - A\| &= \varepsilon\|(I - \varepsilon S)^{-1}SA\| \\ &\leq \varepsilon\|SA\|.\end{aligned}$$

Thus the uniformly bounded family of operators $(I - \varepsilon S)^{-1}$ converges strongly to the identity on the dense set $D(S)$. Hence the operators converge strongly to the identity and one concludes from the relation

$$(S_\varepsilon - S)A = ((I - \varepsilon S)^{-1} - I)SA$$

that $S_\varepsilon A$ converges in norm to SA for all $A \in D(S)$. Appealing to $(*)$ we conclude that $\{U_t^\varepsilon A\}_{\varepsilon \geq 0}$ is norm convergent, uniformly for t in compacts, for $A \in D(S)$. Using uniform boundedness, i.e., $\|U_t^\varepsilon\| \leq 1$, one concludes that $\{U_t^\varepsilon\}_{\varepsilon \geq 0}$ converges strongly on $\overline{D(S)}$, uniformly for t in compacts. If $U = \{U_t\}_{t \geq 0}$ denotes the strong limit it readily follows that U is a C_0-semigroup of contractions.

One establishes straightforwardly that

$$\frac{(U_t^\varepsilon - I)A}{t} = \frac{1}{t}\int_0^t ds\; U_s^\varepsilon S_\varepsilon A$$

for all $A \in X$. But if $A \in D(S)$ one obtains the relation

$$\frac{(U_t - I)A}{t} = \frac{1}{t}\int_0^t ds\; U_s SA$$

by strong limits. Therefore

$$\left\|\frac{(U_t - I)A}{t} - SA\right\| \leq \sup_{0 \leq s \leq t} \|(U_s - I)SA\|$$

and it follows from the strong continuity of U that its generator $\hat{S}$ is an extension of S. But this implies that $(I - \alpha\hat{S})^{-1}$ is an extension of $(I - \alpha S)^{-1}$ for all $\alpha > 0$. As, however, the latter operator is everywhere defined it is impossible that $\hat{S}$ is a strict extension of S. One must have $\hat{S} = S$.

This completes the proof for $F = X^*$ with the exception of the second algorithm, which we will establish later.

$C_0{}^*$-*case*: $F = X_*$. Condition (2) of the theorem and Lemma 3.1.9 imply that S^*, on X_*, is weakly closed and has a weakly dense domain of definition. A second application of Lemma 3.1.9 gives

$$\|(I - \alpha S^*)^{-1}\| = \|(I - \alpha S)^{-1*}\| = \|(I - \alpha S)^{-1}\| \leq 1$$

for $\alpha \geq 0$. The C_0-version of the theorem, i.e., $F = X^*$, then implies that S^* is the generator of a weakly continuous semigroup $U_t{}^*$ of contractions on $F = X_*$. Let U_t be the dual group, on X, of $U_t{}^*$. Clearly, U_t is a $\sigma(X, F)$-continuous semigroup of contractions on X. Let T be the generator of U. Using Proposition 3.1.6 with $\lambda > 0$, $\eta \in X_*$, and $A \in X$, one has

$$\begin{aligned}\eta((\lambda I - T)^{-1}A) &= \int_0^\infty dt\; e^{-\lambda t}\eta(U_t A) \\ &= \int_0^\infty dt\; e^{-\lambda t}(U_t{}^*\eta)(A) \\ &= ((\lambda I - S^*)^{-1}\eta)(A) \\ &= \eta((\lambda I - S)^{-1}A).\end{aligned}$$

It follows that $(\lambda I - T)^{-1} = (\lambda I - S)^{-1}$, i.e., $T = S$, and S is the generator of U_t.

Next let us derive the second exponential algorithm of the theorem. For this the following simple lemma is practical.

Lemma 3.1.11. *Let T be a bounded operator on the Banach space X with $\|T\| \leq 1$. It follows that*

$$\|(e^{n(T-I)} - T^n)A\| \leq \sqrt{n}\|(T - I)A\|$$

for all $A \in X$, and all positive integers n.

PROOF. The difference on the left is estimated by

$$\begin{aligned}\|(e^{n(T-I)} - T^n)A\| &\leq e^{-n} \sum_{m\geq 0} \frac{n^m}{m!} \|(T^m - T^n)A\| \\ &\leq e^{-n} \sum_{m\geq 0} \frac{n^m}{m!} \|(T^{|m-n|} - I)A\| \\ &\leq \|(T - I)A\| e^{-n} \sum_{m\geq 0} \frac{n^m}{m!} |m - n|.\end{aligned}$$

An easy application of the Cauchy–Schwarz inequality then gives

$$e^{-2n}\left(\sum_{m\geq 0} \frac{n^m}{m!}|m - n|\right)^2 \leq e^{-n} \sum_{m\geq 0} \frac{n^m}{m!}(m - n)^2 = n.$$

Combination of these estimates gives the required bound.

The proof of Theorem 3.1.10 is now completed by choosing $A \in D(S)$ and $T = (I - tS/n)^{-1}$ in Lemma 3.1.11. Thus

$$\begin{aligned}\left\|\left(e^{tS(1-tS/n)^{-1}} - \left(I - \frac{tS}{n}\right)^{-n}\right)A\right\| &\leq \left(\frac{t}{\sqrt{n}}\right)\left\|\left(I - \frac{tS}{n}\right)^{-1} SA\right\| \\ &\leq \left(\frac{t}{\sqrt{n}}\right)\|SA\|.\end{aligned}$$

This establishes that

$$\lim_{n\to\infty} \left\| U_t A - \left(I - \frac{tS}{n}\right)^{-n} A \right\| = 0$$

for all $A \in D(S)$, uniformly for t in compacts, and the desired conclusion follows from a uniform boundedness and density argument.

Finally, for the case $F = X_*$ one notes that $\overline{D(S)}$ is invariant under U_t and one demonstrates, as in Corollary 3.1.8, that the restriction of U_t to $\overline{D(S)}$ is strongly continuous. The convergence in norm of the limits for $U_t A$ with $A \in \overline{D(S)}$ then follows from the C_0-version of the theorem. The weak* convergence for general A follows from the strong convergence of the corresponding limits for $U_t{}^*\eta$ for all $\eta \in X_*$.

The Hille–Yosida theorem has a variant which applies to general $\sigma(X, F)$-continuous semigroups. Recall that a family $\{T_\alpha\}$ of operators was defined to be $\tau(X, F)$-equicontinuous if for each seminorm p_K in the definition of the $\tau(X, F)$-topology there exists a seminorm $p_{K'}$ such that

$$p_K(T_\alpha A) \le p_{K'}(A)$$

for all $A \in X$ and all α.

Corollary 3.1.12. *Let S be a $\sigma(X, F)$–$\sigma(X, F)$-closed operator on X. Then the conditions* (1) *and* (2) *are equivalent, and conditions* (1a) *and* (2a) *are equivalent*:

(1) *S is the generator of a $\sigma(X, F)$-continuous semigroup U of contractions on X such that $\{U_t\}_{t\ge 0}$ is $\tau(X, F)$-equicontinuous;*
(1a) *S is the generator of a $\sigma(X, F)$-continuous semigroup U of contractions such that $\{U_t{}^*\}_{t\ge 0}$ is $\tau(F, X)$-equicontinuous.*
(2) $\|(I - \alpha S)^{-1}\| \le 1$ *for* $\alpha \ge 0$ *and* $\{(I - \alpha S)^{-m};\ \alpha \in [0, 1],\ m = 1, 2, \ldots\}$ *is $\tau(X, F)$-equicontinuous;*
(2a) $\|(I - \alpha S)^{-1}\| \le 1$ *for* $\alpha \ge 0$ *and* $\{(I - \alpha S^*)^{-m};\ \alpha \in [0, 1],\ m = 1, 2, \ldots\}$ *is $\tau(F, X)$-equicontinuous.*

In these cases we have

$$U_t A = \lim_{\varepsilon\to 0} \exp\{tS(I - \varepsilon S)^{-1}\}A = \lim_{n\to\infty} \left(I - \frac{tS}{n}\right)^{-n} A,$$

where the limit exists in the $\tau(X, F)$-topology in the case (1)(⇔(2)) *and in the $\sigma(X, F)$-topology in the case* (1*a*) (⇔(2*a*)).

PROOF. The proof of the equivalence of conditions (1) and (2) is essentially identical to the $F = X^*$ case of the Hille–Yosida theorem, and the equivalence of (1a) and (2a) follows as in the $F = X_*$ case. The only minor change is that norm estimates are replaced by estimates with respect to the seminorms and equicontinuity replaces uniform boundedness.

The Hille–Yosida theorem characterizes generators by properties of their resolvents $(I - \alpha S)^{-1}$. We next examine an alternative description for C_0-semigroups based on a notion of dissipativity.

First note that if A is an element of a Banach space X and if $\eta \in X^*$ satisfies $\eta(A) = \|\eta\|\ \|A\|$ then η is called a *tangent functional* at A. The Hahn–Banach theorem ensures that for each $A \in X$ there exists at least one nonzero tangent functional at A.

Definition 3.1.13. An operator S, with domain $D(S)$ on the Banach space X, is called *dissipative* if for each $A \in D(S)$ there exists a nonzero tangent functional η, at A, such that

$$\mathrm{Re}\ \eta(SA) \le 0.$$

If both $\pm S$ are dissipative then S is called *conservative.*

To explain the origin of these notions, suppose that U is a C_0-semigroup of contractions with generator S and that η is a tangent functional at $A \in D(S)$. Therefore $|\eta(U_s A)| \leq \|\eta\| \, \|A\|$ for all $s \geq 0$. But this implies that

$$\operatorname{Re} \eta(U_s A) \leq \operatorname{Re} \eta(U_0 A)$$

and hence

$$\frac{d}{ds} \operatorname{Re} \eta(U_s A)|_{s=0} \leq 0.$$

This last condition is exactly the condition of dissipativity

$$\operatorname{Re} \eta(SA) \leq 0$$

and this demonstrates that the generator S of a C_0-contraction semigroup is dissipative. The same argument also establishes that the generator of a C_0-group of isometries is conservative.

Although dissipativity is useful in the discussion of C_0-semigroups of contractions it is not really suited to general $\sigma(X, F)$-continuous semigroups. If one attempts to modify the definition and only consider tangent functionals $\eta \in F$ then problems of existence occur. If, however, one insists on dissipativity with respect to X^* then it is not evident that the generator S of the $\sigma(X, F)$-semigroups of contractions U is dissipative because the function $s \mapsto \eta(U_s A)$ with $\eta \in X^*$ and $A \in D(S)$ is not necessarily differentiable.

Next we derive alternative characterizations of dissipativity which relate to contractivity of the resolvent.

Proposition 3.1.14. *Let S be an operator on the Banach space X. Then the following conditions are equivalent*:

(1) [(1′)]
$$\|(I - \alpha S)(A)\| \geq \|A\|$$

for all $A \in D(S)$ and all $\alpha > 0$ (for all small $\alpha \geq 0$);

(2)
$$\operatorname{Re} \eta(SA) \leq 0$$

for one nonzero tangent functional η at each $A \in D(S)$.

Moreover, if S is norm-densely defined these conditions are equivalent to the following:

(3)
$$\operatorname{Re} \eta(SA) \leq 0$$

for all tangent functionals η at each $A \in D(S)$.

PROOF. (1′) $\Rightarrow$ (2) Set $B = SA$ and for each small α choose a nonzero tangent functional η_α at $A - \alpha B$. One can assume $\|\eta_\alpha\| = 1$, then Condition (1′) gives

$$\begin{aligned} \|A\| &\leq \|A - \alpha B\| \\ &= \operatorname{Re} \eta_\alpha(A - \alpha B) \\ &= \operatorname{Re} \eta_\alpha(A) - \alpha \operatorname{Re} \eta_\alpha(B) \\ &\leq \operatorname{Re} \eta_\alpha(A) + \alpha\|B\|. \end{aligned}$$

Now the unit ball of X^* is weakly* compact by the Alaoglu–Bourbaki theorem. Thus there exists a weakly* convergent subnet $\eta_{\alpha'}$ whose limit, as α' tends to zero, we denote by η. Hence one deduces from the foregoing inequality that

$$\|A\| \leq \operatorname{Re} \eta(A).$$

But since $\|\eta_\alpha\| = 1$ one has $\|\eta\| \leq 1$ and then

$$\operatorname{Re} \eta(A) \leq \|\eta\| \|A\| \leq \|A\|.$$

Hence $\operatorname{Re} \eta(A) = \|A\|$ and $\|\eta\| = 1$. This proves that η is a normalized tangent functional at A. But one also has

$$\begin{aligned} \|A\| &\leq \operatorname{Re} \eta_\alpha(A) - \alpha \operatorname{Re} \eta_\alpha(B) \\ &\leq \|A\| - \alpha \operatorname{Re} \eta_\alpha(B). \end{aligned}$$

Hence in the limit as α' tends to zero one finds

$$0 \geq \operatorname{Re} \eta(B) = \operatorname{Re} \eta(SA),$$

i.e., Condition (2) is satisfied.

(2) $\Rightarrow$ (1). Let η be a nonzero tangent functional at $A \in D(S)$ satisfying

$$\operatorname{Re} \eta(SA) \leq 0.$$

Then

$$\begin{aligned} \|\eta\| \|A\| &= \operatorname{Re} \eta(A) \\ &\leq \operatorname{Re} \eta((I - \alpha S)A) \\ &\leq \|\eta\| \|(I - \alpha S)(A)\| \end{aligned}$$

for all $\alpha \geq 0$. Dividing by $\|\eta\|$ gives the desired result.

(1) $\Rightarrow$ (1′). This is evident.

Finally, (3) $\Rightarrow$ (2) and it remains to prove (1) $\Rightarrow$ (3) under the assumption that $D(S)$ is norm-dense.

Now if $A, B \in D(S)$ and η is a normalized tangent functional at A one first has

$$\begin{aligned} \|(I + \alpha S)(A)\| &\geq \operatorname{Re} \eta((I + \alpha S)A) \\ &= \|A\| + \alpha \operatorname{Re} \eta(SA). \end{aligned}$$

Therefore

$$\operatorname{Re} \eta(SA) \leq \limsup_{\alpha \to 0^+} (\|(I + \alpha S)(A)\| - \|A\|)/\alpha.$$

But second for all $B \in D(S)$ one has

$$\begin{aligned} \|(I + \alpha S)(A)\| &\leq \|A + \alpha B\| + \alpha \|B - SA\| \\ &\leq \|(I - \alpha S)(A + \alpha B)\| + \alpha \|B - SA\| \\ &\leq \|A\| + 2\alpha \|B - SA\| + \alpha^2 \|SB\| \end{aligned}$$

for all small $\alpha > 0$ by Condition (1). Therefore by combination of these results

$$\operatorname{Re} \eta(SA) \leq 2\|B - SA\|.$$

But since $D(S)$ is norm-dense we may choose B arbitrarily close to SA and deduce that $\operatorname{Re} \eta(SA) \leq 0$, i.e., Condition (3) is satisfied.

The alternative characterization of dissipativity given by Proposition 3.1.14 is particularly useful for the discussion of closure properties.

Proposition 3.1.15. *Let S be a norm-densely defined dissipative operator on the Banach space X. Then*

(1) *S is norm-closable;*
(2) *the closure $\bar{S}$ of S is dissipative;*
(3) *if $\bar{S}$ generates a C_0-contraction semigroup it has no proper dissipative extension.*

PROOF. (1) Assume $A_n \in D(S)$ and $A_n \to 0$, $SA_n \to A$. We must prove that $A = 0$. But

$$\begin{aligned}\|A_n + \alpha B\| &\leq \|(I - \alpha S)(A_n + \alpha B)\| \\ &\leq \|A_n\| + \alpha\|B - SA_n\| + \alpha^2\|SB\|\end{aligned}$$

for $B \in D(S)$ and $\alpha > 0$ by Condition (1) of Proposition 3.1.14. Thus taking the limit $n \to \infty$, then dividing by α, and subsequently taking the limit $\alpha \to 0$, gives

$$\|B\| \leq \|B - A\|.$$

Since this is true for all B in the norm-dense set $D(S)$ one must have $A = 0$.

(2) If $A \in D(\bar{S})$ then there are $A_n \in D(S)$ such that $A_n \to A$ and $SA_n \to \bar{S}A$. Therefore if $\alpha > 0$

$$\begin{aligned}\|A\| &= \lim_{n\to\infty} \|A_n\| \\ &\leq \lim_{n\to\infty} \|(I - \alpha S)(A_n)\| = \|(I - \alpha\bar{S})(A)\|\end{aligned}$$

and $\bar{S}$ is dissipative by Proposition 3.1.14.

(3) Let T be a dissipative extension of $\bar{S}$ then since $\bar{S}$ is a generator

$$X = (I - \alpha\bar{S})(D(\bar{S})) \subseteq (I - \alpha T)(D(T)) \subseteq X$$

for each $\alpha > 0$ and hence $R(I - \alpha T) = X$. Now suppose T is a strict extension of $\bar{S}$ and choose $A \in D(T)$ such that $A \notin D(\bar{S})$. Then there must exist a $B \in D(\bar{S})$ such that

$$(I - \alpha T)(A) = (I - \alpha\bar{S})(B) = (I - \alpha T)(B).$$

Hence $(I - \alpha T)(A - B) = 0$. But since T is dissipative $A = B$, by Proposition 3.1.14, which is a contradiction. Thus no strict dissipative extension exists.

Now one has the following characterization of generators of C_0-contraction semigroups which is in part a reformulation of the Hille–Yosida theorem.

Theorem 3.1.16 (Lumer–Phillips theorem). *Let S be an operator on the Banach space X and consider the following conditions*

(1) *S is the generator of a C_0-semigroup of contractions;*
(2) *S is norm-densely defined, dissipative, and*

$$R(I - \alpha S) = X$$

for some $\alpha > 0$;

(3) *S and its adjoint S^*, on X^*, are norm-densely defined, norm-closed, dissipative, operators.*

Then (1)$\Leftrightarrow$(2)$\Rightarrow$(3) and if X is reflexive, i.e., if $X = X^{**}$, then all three conditions are equivalent.

PROOF. The equivalence (1)$\Leftrightarrow$(2) follows from Theorem 3.1.10, Proposition 3.1.14, and the discussion preceding Proposition 3.1.14.

(3)$\Rightarrow$(2). Suppose there is a nonzero $\eta \in X^*$ such that $\eta((I - \alpha S)(A)) = 0$ for all $A \in D(S)$ and some $\alpha > 0$. Then

$$|\eta(SA)| = \alpha^{-1}|\eta(A)| \leq \alpha^{-1}\|\eta\| \|A\|.$$

Hence $\eta \in D(S^*)$ and since $D(S)$ is norm-dense $(I - \alpha S^*)\eta = 0$. But dissipativity of S^* implies $\|\eta\| \leq \|(I - \alpha S^*)\eta\|$ by Proposition 3.1.14. Thus $\eta = 0$ and $R(I - \alpha S) = X$.

Finally, if S generates the C_0-contraction semigroup U then S^* generates the dual contraction semigroup U^*. But if X is reflexive U^* is weakly, hence strongly, continuous and Condition (3) follows from Condition (1) by the Hille–Yosida theorem and Proposition 3.1.14.

Next we consider the characterization of generators in terms of analytic elements. These elements are introduced by the following definition.

Definition 3.1.17. Let S be an operator on the Banach space X. An element $A \in X$ is defined to be an *analytic element* (*entire analytic element*) for S if $A \in D(S^n)$, for all $n = 1, 2, \ldots$, and if

$$\sum_{n \geq 0} \frac{t^n}{n!} \|S^n A\| < +\infty$$

for some $t > 0$ (all $t > 0$).

We will see below that this notion is particularly useful for the characterization of the generators S of $\sigma(X, F)$-continuous groups of isometries U. But we first remark that in this context the above notion of analyticity is identical to the notion of analyticity for U introduced by Definition 2.5.20. For example, if A is analytic for the generator S of U then the function

$$f(t + z) = \sum_{n \geq 0} \frac{z^n}{n!} U_t S^n A$$

is defined for all $t \in \mathbb{R}$ and all z in the radius of analyticity of $\sum_{n \geq 0}(z^n/n!)\|S^n A\|$. But this function satisfies the criteria of Definition 2.5.20 and hence A is analytic for U. Conversely, if A is strongly (weakly) analytic for U in the strip $\{z; |\text{Im } z| < \lambda\}$ then the usual Cauchy estimates give

$$\left\| \frac{d^n}{dt^n} U_t A \right\| = \|U_t S^n A\| = \|S^n A\| \leq \frac{n!\, M}{\lambda^n}$$

for some M. Thus

$$\sum_{n \geq 0} \frac{|z|^n}{n!} \|S^n A\| \leq M \sum_{n \geq 0} \left(\frac{|z|}{\lambda}\right)^n < +\infty$$

for $|z| < \lambda$.

The equivalence of the two notions of analyticity for a $\sigma(X, F)$-continuous group of isometries U and its generator S allows one to conclude from Corollary 2.5.23 that S has a $\sigma(X, F)$-dense set of entire analytic elements. But the situation for semigroups can be quite different. There are C_0-semigroups for which zero is the only analytic element. For example, let $C_0(\mathbb{R})$ be the Banach space of continuous functions on $\mathbb{R}$ vanishing at infinity, equipped with the supremum norm and let $C_0(\mathbb{R}_+)$ be the Banach subspace consisting of $f \in C_0(\mathbb{R})$ such that $f(x) = 0$ for $x \leq 0$. Define the C_0-group T of translations on $C_0(\mathbb{R})$ by

$$(T_t f)(x) = f(x - t), \qquad x \in \mathbb{R}, \quad t \in \mathbb{R},$$

and the C_0-semigroup U of right translations on $C_0(\mathbb{R}_+)$ by

$$(U_t f)(x) = f(x - t), \qquad x \in \mathbb{R}, \quad t \in \mathbb{R}_+.$$

A function $f \in C_0(\mathbb{R})$ is analytic for T if it is analytic in the usual sense. Now suppose $f \in C_0(\mathbb{R}_+)$ is analytic for U. It follows that f must be analytic for T but this implies that f is analytic in the usual sense. Since $f(x) = 0$ for $x \leq 0$ it then follows that $f = 0$. Thus U has no nonzero analytic elements.

The primary interest of analytic elements is their utility in the construction of semigroups by power series expansion. But it is also essential to have some form of dissipativity to control growth properties.

Theorem 3.1.18. *Let S be a $\sigma(X, F)$-$\sigma(X, F)$-closed operator on X satisfying*

$$\|(I - \alpha S)A\| \geq \|A\|$$

for all $A \in D(S)$ and all small $\alpha > 0$. Further let $\bar{X}_a$ denote the norm-closure of the set X_a of analytic elements of S.

It follows that there exists a C_0-semigroup of contractions U on $\bar{X}_a$ with generator S_a such that $S_a \subseteq S$ and X_a is a core for S_a.

Moreover, if $F = X^*$ or $F = X_*$ and the unit ball of X_a is $\sigma(X, F)$-dense in the unit ball of X then U extends to a $\sigma(X, F)$-continuous semigroup of contractions on X with generator S.

Proof. Let A be an analytic element for S and t_A the radius of the convergence of the series expansion

$$\sum_{m \geq 0} \frac{t^m}{m!} \|S^m A\|.$$

Next note that

$$\left(I + \frac{tS}{n}\right)^n A = \sum_{m=0}^{n} \frac{t^m}{m!} S^m A c_{n,m}$$

where $c_{n,0} = 1$ and

$$c_{n,m} = \prod_{p=0}^{m-1} (1 - p/n) \qquad \text{if} \quad m \geq 1.$$

Clearly $c_{n,m} \leq 1$ but a simple inductive argument establishes that $c_{n,m} \geq 1 - m(m-1)/2n$. Hence

$$\lim_{n\to\infty} \left\| \left(I + \frac{tS}{n}\right)^n A - \sum_{m=0}^{n} \frac{t^m}{m!} S^m A \right\| = 0$$

for $|t| < t_A$. Thus defining $U_t A$ for $|t| < t_A$ by

$$U_t A = \sum_{m\geq 0} \frac{t^m}{m!} S^m A$$

one also has the identification

$$U_t A = \lim_{n\to\infty} \left(I + \frac{tS}{n}\right)^n A.$$

Next remark that dissipativity of S implies

$$\left\| \left(I - \frac{tS}{n}\right)^n B \right\| \geq \|B\|$$

for all $B \in D(S^n)$ and $t > 0$ by Proposition 3.1.14. Therefore setting $B = (I + tS/n)^n A$ one finds

$$\begin{aligned} \left\| \left(I + \frac{tS}{n}\right)^n A \right\| &\leq \left\| \left(I - \frac{t^2 S^2}{n^2}\right)^n A \right\| \\ &\leq \|A\| + \sum_{m=1}^{n} \frac{(2t)^{2m}}{m!} \|S^{2m} A\| d_{n,m} \end{aligned}$$

where

$$d_{n,m} = c_{n,m} \frac{2m!}{2^{2m} n^m m!} \leq \frac{m!}{n^m} \leq \left(\frac{m}{n}\right)^m.$$

But $d_{n,m} \to 0$ as $n \to \infty$ for each fixed m. Hence if $0 < t < t_A/2$ then

$$\|U_t A\| = \lim_{n\to\infty} \left\| \left(I + \frac{tS}{n}\right)^n A \right\| \leq \|A\|.$$

Now note that

$$S \sum_{m=0} \frac{t^m}{m!} S^m A = \sum_{m=0}^{n} \frac{t^m}{m!} S^m S A.$$

Hence, since S is $\sigma(X, F)$–$\sigma(X, F)$-closed, $U_t A \in D(S)$ and $SU_t A = U_t SA$ for $|t| < t_A$. Therefore

$$\|SU_t A\| = \|U_t SA\| \leq \|SA\|$$

for $0 < t < t_A/2$. Iteration of this argument establishes that if $0 < t < t_A/2$ then $U_t A$ is an analytic element for S with associated radius of convergence given by $t_{U_t A} = t_A$. Thus it is possible to iterate the definition of U_t,

$$U_{t+s} A = U_t(U_s A) = \sum_{m\geq 0} \frac{t^m}{m!} S^m(U_s A)$$

for $|t| < t_A$ and $0 < s < t_A/2$ and consequently deduce that $\|U_t A\| \leq \|A\|$ for $0 < t < t_A$.

Repeating this argument one may define $U_t A$ for all $t > 0$ by

$$U_t A = (U_{t/n})^n A$$

where n is chosen so that $n > 2t/t_A$. It is then easy to establish that this definition is independent of the choice of n and that

$$U_s U_t A = U_{s+t} A$$

for all $s, t > 0$. Now since $\|U_t A\| \leq \|A\|$ for all $t > 0$, by construction, each U_t can be extended to $\bar{X}_a$ by continuity and then $U = \{U_t\}_{t \geq 0}$ is a semigroup of contractions. But for $A \in X_a$ one has

$$\lim_{t \to 0} \|U_t A - A\| = 0$$

and hence U is a C_0-semigroup on $\bar{X}_a$. If S_a denotes the generator of U then it is clear that $S_a|_{X_a} = S|_{X_a}$. But X_a is a core for S_a, by Corollary 3.1.7, and hence $S_a \subseteq S$. This proves the first statement of the theorem.

Next since $S_a \subseteq S$ one has

$$X_a \subseteq \bar{X}_a = (I - \alpha S_a)(D(S_a)) \subseteq (I - \alpha S)(D(S)),$$

for all $\alpha > 0$. Hence if $F = X^*$ and $\bar{X}_a = X$ then S and S_a both generate C_0-semigroups of contractions, by the Hille–Yosida theorem, and $S = S_a$ by Proposition 3.1.15(3).

Alternatively, assume $F = X_*$ and the unit ball of X_a is $\sigma(X, F)$-dense in the unit ball of X. Then for each $A \in X$ we may choose $A_\gamma \in X_a$ such that A_γ converges in the weak* topology to A and $\|A_\gamma\| \leq \|A\|$. But the foregoing conclusion asserts the existence of $B_\gamma \in D(S)$ such that $A_\gamma = (I - \alpha S)B_\gamma$. Moreover,

$$\|B_\gamma\| \leq \|(I - \alpha S)B_\gamma\| = \|A_\gamma\| \leq \|A\|.$$

Thus $\{\|B_\gamma\|\}$ is uniformly bounded. But the unit ball of X is weak* compact by the Alaoglu–Bourbaki theorem and hence there exists a weak*-convergent subnet $B_{\gamma'}$ of B_γ. Then $B_{\gamma'} \to B$ and $A_{\gamma'} = (I - \alpha S)(B_{\gamma'}) \to A$ where both limits are in the weak* topology. But S is weakly* closed and so $B \in D(S)$ and $(I - \alpha S)(B) = A$, i.e., $R(I - \alpha S) = X$. Then $S = S_a$ is a generator as before.

We have already remarked that the infinitesimal generator of a $\sigma(X, F)$-continuous group of isometries has a $\sigma(X, F)$-dense set of entire analytic elements by Corollary 2.5.23. Combination of this remark with Theorem 3.1.18 immediately gives another characterization of the generators of $\sigma(X, F)$-groups of isometries, if $F = X^*$ or $F = X_*$.

Corollary 3.1.19. *Let S be an operator on the Banach space X. If $F = X^*$ or $F = X_*$ then the following conditions are equivalent*:

(1) *S is the generator of a $\sigma(X, F)$-continuous group of isometries U;*
(2) *S is $\sigma(X, F)$-densely defined, $\sigma(X, F)$–$\sigma(X, F)$-closed*

$$\|(I - \alpha S)(A)\| \geq \|A\|$$

for all $\alpha \in \mathbb{R}$ and $A \in D(S)$, and either
(A1): $R(I - \alpha S) = X$ for all $\alpha \in \mathbb{R}$ (for one $\alpha > 0$ and one $\alpha < 0$), or
(A2): the unit ball of the set X_a of analytic elements for S is $\sigma(X, F)$-dense in the unit ball of X.

The equivalence of (1) and (2) + (A1) is the Hille–Yosida theorem applied to $\pm S$, and (1) $\Rightarrow$ (A2) by Proposition 2.5.22. But (2) + (A2) $\Rightarrow$ (1) by application of Theorem 3.1.18 to both $\pm S$.

There is one other useful consequence of the construction of U given in the proof of Theorem 3.1.18. This is an alternative version of Corollary 3.1.7.

Corollary 3.1.20. *Let X be a Banach space and U a $\sigma(X, F)$-continuous semigroup of contractions with generator S. Let D be a subspace of $D(S)$ which is $\sigma(X, F)$-dense in X, consists of analytic elements for S, and is such that $SD \subseteq D$. It follows that D is a core for S.*

The foregoing notions of tangent functional, dissipative operator, analytic element, etc. are most easily illustrated in Hilbert space.

EXAMPLE 3.1.21. Let $X = \mathfrak{H}$ be a Hilbert space. Then $\mathfrak{H} = \mathfrak{H}^*$ and the unique normalized tangent functional at ψ is $\eta = \psi/\|\psi\|$. Thus dissipativity of S is equivalent to $\mathrm{Re}(\psi, S\psi) \leq 0$ for all $\psi \in D(S)$, or

$$(\psi, S\psi) + (S\psi, \psi) \leq 0.$$

In particular S is conservative if, and only if,

$$(\psi, S\psi) + (S\psi, \psi) = 0.$$

Setting $S = iH$ and using the standard polarization identities one concludes that $S = iH$ is conservative if, and only if, H is symmetric, i.e.,

$$(H\varphi, \psi) = (\varphi, H\psi), \qquad \varphi, \psi \in D(H) = D(S).$$

Thus, Corollary 3.1.19 states that S is the infinitesimal generator of a strongly continuous one-parameter group of isometries (unitary operators) if, and only if, $S = iH$ where H is a densely defined symmetric operator satisfying

either $R(1 + i\alpha H) = \mathfrak{H}, \qquad \alpha \in \mathbb{R}$,

or $\quad$ H possesses a dense set of analytic elements.

These are the well-known conditions for H to be selfadjoint.

Although we have only given characterizations of generators of isometric groups and contraction semigroups, most of the results have extensions to general groups and semigroups although the growth properties of $\|U_t\|$ cause some complication. For example, Theorem 3.1.19 has the following analogue:

Theorem 3.1.22. *An operator S on the Banach space X is the infinitesimal generator of a $\sigma(X, F)$-continuous group, where $F = X^*$ or $F = X_*$, if, and only if, S is $\sigma(X, F)$-densely defined, $\sigma(X, F)$–(X, F)-closed, and satisfies the following conditions:*

(1) *there is an $M \geq 1$ and $\beta \geq 0$ such that*

$$\|(I - \alpha S)^n A\| \geq M^{-1}(1 - \alpha\beta)^n \|A\|$$

for all $A \in D(S^n)$, all $|\alpha|\beta < 1$, and all $n = 1, 2, \ldots$;

(2) *either*

$$R(I - \alpha S) = X$$

for all $|\alpha|\beta < 1$ (*for one* $0 < \alpha\beta < 1$ *and one* $-1 < \alpha\beta < 0$), *or* S *possesses a* $\sigma(X, F)$*-dense set of analytic elements, and any element in* X *can be approximated by a uniformly bounded net of analytic elements.*

Note that in this case the group U_t has the growth property $\|U_t\| \leq \tilde{M}e^{\beta|t|}$, and any element $A \in X$ is the limit of a sequence A_n of analytic elements with $\|A_n\| \leq M\|A\|$.

We conclude by describing an aspect of C_0*-groups, and semigroups, which is not generally shared by their C_0-counterparts.

Proposition 3.1.23. *Let* U *be a* C_0**-semigroup, on the Banach space* X, *with infinitesimal generator* S. *The following conditions are equivalent*:

(1) $A \in D(S)$;
(2) $\sup_{0<t\leq 1} \|(U_t - I)A\|/t < +\infty$.

PROOF. (1) ⇒ (2) This implication is true for general $\sigma(X, F)$-continuous semigroups. Take $A \in D(S)$. Then

$$(U_t - I)(A) = \int_0^t ds\, U_s SA.$$

Thus

$$\|(U_t - I)(A)\| \leq \int_0^t ds\, Me^{\beta s}\|SA\|$$

$$\leq tMe^{\beta}\|SA\|$$

for $t \in [0, 1]$.

(2) ⇒ (1) The unit ball of X is weakly* compact. Thus if

$$\sup_{0<t\leq 1} \frac{\|(U_t - I)A\|}{t} = M < +\infty$$

there must exist a net $t_\alpha \in \langle 0, 1]$ such that $t_\alpha \to 0$ and $(U_{t_\alpha} - I)A/t_\alpha$ converges in the weak* topology to an element $B \in X$. Thus

$$\lim_\alpha \frac{\eta((U_{t_\alpha} - I)A)}{t_\alpha} = \eta(B)$$

for all $\eta \in X_*$. But if $\eta \in D(S^*)$ then

$$S^*\eta(A) = \lim_\alpha \frac{((U^*_{t_\alpha} - I)\eta)(A)}{t_\alpha} = \eta(B).$$

Hence $A \in D(S^{**}) \cap X = D(S)$ and $B = S^{**}A = SA$.

EXAMPLE 3.1.24. Let H be a selfadjoint operator on the Hilbert space $\mathfrak{H}$ and let $U = \{U_t\}_{t \in \mathbb{R}}$ be the unitary group, $U_t = \exp\{iHt\}$, generated by iH. This group is a C_0-group but, because $\mathfrak{H}$ is self-dual, it is also a $C_0{}^*$-group. Therefore $\psi \in D(H)$ if, and only if,

$$\sup_{t \in \mathbb{R}_+} \frac{\|(e^{iHt} - \mathbb{1})\psi\|}{t} < +\infty.$$

EXAMPLE 3.1.25. Let $X = C_0(\mathbb{R})$ with the supremum norm. Translations act as a C_0-group U on X and if $f \in C_0(\mathbb{R})$ is absolutely continuous with derivative $f' \in L^\infty(\mathbb{R})$, but $f' \notin C_0(\mathbb{R})$ then

$$\begin{aligned} \|(U_t f - f)/t\| &\leq \sup_s \sup_x |(f(x - s) - f(x))/s| \\ &\leq \operatorname{ess.\,sup}_x |f'(x)| = \|f'\|_\infty . \end{aligned}$$

Thus the criterion of Proposition 3.1.23 is satisfied but f is not in the domain of the generator S of U. In fact, $D(S) = \{f\,; f \in C_0(\mathbb{R}),\ f' \in C_0(\mathbb{R})\}$. This demonstrates that the conditions of Proposition 3.1.23 are not equivalent for all C_0-groups.

3.1.3. Convergence Properties

In the preceding subsections we examined the existence and construction of various groups and semigroups, and next we analyze their stability properties. There are several aspects of the notion of stability, convergence, perturbation, approximation, etc. We will approach the problem in three distinct ways in the subsequent three subsections. First we consider convergence properties and use these to extend our previous results on group construction.

In Section 3.1.1 we showed that $U = \{U_t\}_{t \in \mathbb{R}}$ is a uniformly continuous group, on a Banach space X if, and only if, its generator S is bounded. Moreover, we saw that two such groups are close in norm if, and only if, their generators are close in norm. There is an analogue of this result for $\sigma(X, F)$-continuous groups which characterizes convergence of the groups by convergence of the resolvents of their generators.

Theorem 3.1.26. *Let U_n and U be $\sigma(X, F)$-continuous semigroups, on the Banach space X, with generators S_n and S. Assume that there is a $\beta \geq 0$ such that the family*

$$\{e^{-\beta t} U_{n,t};\, t \geq 0,\, n = 1, 2, \ldots\} \cup \{e^{-\beta t} U_t;\, t \geq 0\}$$

is $\tau(X, F)$-equicontinuous. The following four conditions are equivalent:

(1a) [(1b)] $\lim_{n\to\infty} (\lambda I - S_n)^{-1} A = (\lambda I - S)^{-1} A$, $A \in X$, *in the $\tau(X, F)$-topology, for some λ with* Re $\lambda > \beta$ [*uniformly for all λ with* Re $\lambda > \beta + \varepsilon$];

(2a) [(2b)] $\lim_{n\to\infty} U_{n,t} A = U_t A$, $A \in X$, *in the $\tau(X, F)$-topology, for all $t \in \mathbb{R}_+$* [*uniformly for t in any finite interval of $\mathbb{R}_+$*].

PROOF. Clearly, (2b) $\Rightarrow$ (2a) and (1b) $\Rightarrow$ (1a). We will show that (2a) $\Rightarrow$ (1b) and (1a) $\Rightarrow$ (2b).

First note that by replacing $U_{n,t}$, and U_t, by $U_{n,t}\exp\{-\beta t\}$, and $U_t\exp\{-\beta t\}$, one may assume $\beta = 0$. Also note that $t \mapsto U_{n,t}A$ and $t \mapsto U_tA$ are $\tau(X, F)$-continuous for all $A \in X$, $n = 1, 2, \ldots$, by Corollary 3.1.8.

(2a) $\Rightarrow$ (1b) From Proposition 3.1.6 one has

$$(\lambda I - S_n)^{-1}A - (\lambda I - S)^{-1}A = \int_0^\infty dt\, e^{-\lambda t}(U_{n,t} - U_t)A.$$

Thus invoking the continuity property established in Corollary 3.1.8 one finds

$$p_K((\lambda I - S_n)^{-1}A - (\lambda I - S)^{-1}A) \leq \int_0^\infty dt\, e^{-t|\mathrm{Re}\,\lambda|}p_K((U_{n,t} - U_t)A)$$

for any semi-norm p_K. Now condition (1b) follows from the Lebesgue dominated convergence theorem.

(1a) $\Rightarrow$ (2b) First note that if $\mathrm{Re}\,\lambda > 0$ then $R((\lambda I - S)^{-1}) = D(S)$ is $\sigma(X, F)$-dense and the family of differences $U_{n,t} - U_t$ is equicontinuous. Hence it suffices to prove that

$$\lim_{n\to\infty} U_{n,t}(\lambda I - S)^{-1}A = U_t(\lambda I - S)^{-1}A, \qquad A \in X,$$

in the $\tau(X, F)$-topology, uniformly on the finite intervals of $\mathbb{R}_+$. Next remark that

$$\begin{aligned}(U_{n,t} - U_t)(\lambda I - S)^{-1}A &= U_{n,t}((\lambda I - S)^{-1} - (\lambda I - S_n)^{-1})A \\ &\quad + (\lambda I - S_n)^{-1}(U_{n,t} - U_t)A \\ &\quad + ((\lambda I - S_n)^{-1} - (\lambda I - S)^{-1})U_tA.\end{aligned}$$

Hence for any semi-norm p_K one has

$$\begin{aligned}p_K((U_{n,t} - U_t)(\lambda I - S)^{-1}A) &\leq p_{K'}(((\lambda I - S)^{-1} - (\lambda I - S_n)^{-1})A) \\ &\quad + p_K((\lambda I - S_n)^{-1}(U_{n,t} - U_t)A) \\ &\quad + p_K(((\lambda I - S_n)^{-1} - (\lambda I - S)^{-1})U_tA)\end{aligned}$$

by equicontinuity. The first term converges to zero by assumption. We discuss the other terms separately.

The resolvent formula and the equicontinuity of the semigroups $U_{n,t}$ and U_t imply that the resolvents are equicontinuous, e.g., for each p_K there is a $p_{K'}$ such that

$$p_K((\lambda I - S_n)^{-1}B) \leq p_{K'}(B).$$

Thus by Corollary 3.1.8

$$t \in \mathbb{R}_+ \mapsto p_K(((\lambda I - S_n)^{-1} - (\lambda I - S)^{-1})U_tA)$$

is continuous uniformly in n. But this function converges pointwise to zero as $n \to \infty$ and hence it converges to zero uniformly on the finite intervals of $\mathbb{R}$. It remains to examine the second term.

Note first that the product of two equicontinuous families is equicontinuous. Thus for the second term it suffices to prove convergence of the correct type on a $\tau(X, F)$-dense set of A, e.g., $D(S)$. Now

$$\begin{aligned}
&(\lambda I - S_n)^{-1}(U_{n,t} - U_t)(\lambda I - S)^{-1}A \\
&= -\int_0^t ds \frac{d}{ds} U_{n,t-s}(\lambda I - S_n)^{-1}U_s(\lambda I - S)^{-1}A \\
&= -\int_0^t ds\, U_{n,t-s}\{(\lambda I - S)^{-1} - (\lambda I - S_n)^{-1}\}U_s A
\end{aligned}$$

and hence

$$\begin{aligned}
&p_K((\lambda I - S_n)^{-1}(U_{n,t} - U_t)(\lambda I - S)^{-1}A) \\
&\leq \int_0^t ds\, p_{K'}(\{(\lambda I - S)^{-1} - (\lambda I - S_n)^{-1}\}U_s A).
\end{aligned}$$

The desired convergence now follows from the above discussion of the third term and the Lebesgue dominated convergence theorem.

The characterization of convergence of semigroups provided by Theorem 3.1.26 suffers from two major defects. Firstly, it is necessary to assume that the limit is a semigroup. It is possible that $U_{n,t}$ converges to a limit U_t in the $\tau(X, F)$-topology for all $t \in \mathbb{R}_+$ but U is not a continuous semigroup. Continuity would follow if $U_{n,t}A \to A$ as $t \to 0$ uniformly in n, for all $A \in X$. Similarly, if $(I - \varepsilon S_n)^{-1} \to r(\varepsilon)$ in the $\sigma(X, F)$-topology for all $\varepsilon \in \mathbb{R}_+$ then $r(\varepsilon)$ is not necessarily the resolvent of a generator. But it does have this property if $(I - \varepsilon S_n)^{-1}A \to A$ as $\varepsilon \to 0$ uniformly in n, for all $A \in X$. The second defect of Theorem 3.1.26 is that resolvent convergence is an implicit property of the generators, which is often difficult to verify. It is natural to attempt to find alternative characterizations which involve the generators in a more explicit manner. The difficulty here is that the domains $D(S_n)$ of the various generators might very well be mutually disjoint. Thus any direct form of operator comparison is impossible. One way to avoid this problem, at least for C_0-semigroups of contractions, is to consider the notion of graph convergence.

Recall that if S_n is a sequence of operators on the Banach space X then the graphs $G(S_n)$, of S_n, are defined as the subspaces of $X \times X$ formed by the pairs $(A, S_n A)$ with $A \in D(S_n)$. Now consider all sequences $A_n \in D(S_n)$ such that

$$\lim_{n\to\infty} \|A_n - A\| = 0, \qquad \lim_{n\to\infty} \|S_n A_n - B\| = 0$$

for some pair $(A, B) \in X \times X$. The pairs (A, B) obtained in this manner form a graph G, a subspace of $X \times X$, and we introduce the notation $D(G)$ for the set of A such that $(A, B) \in G$ for some B. Similarly, $R(G)$ is the set of B such that $(A, B) \in G$ for some A. Moreover, we write

$$G = \lim_{n\to\infty} G(S_n).$$

In general, the graph G is not the graph of an operator but under special circumstances this can be the case. Thus if there exists an operator S such that $G = G(S)$ we write

$$S = \operatorname{graph} \lim_{n \to \infty} S_n.$$

Clearly, $D(S) = D(G)$ and $R(S) = R(G)$.

The next lemma establishes conditions under which a sequence of operators has an operator as graph limit.

Lemma 3.1.27. *Let S_n be a sequence of operators, on the Banach space X, and assume that*

$$\|(I - \alpha S_n)A\| \geq \|A\|$$

for all $A \in D(S_n)$, all $n \geq 0$, and all $\alpha \in [0, 1]$. Further, let G denote the graph given by

$$G = \lim_{n \to \infty} G(S_n)$$

and assume that $D(G)$ is norm dense in X. It follows that G is a graph of a norm-closable, norm-densely defined operator S and

$$\|(I - \alpha S)A\| \geq \|A\|$$

for all $A \in D(S)$.

PROOF. First suppose that $A_n \in D(S_n)$ and $B \in X$ satisfy

$$\lim_{n \to \infty} \|A_n\| = 0, \qquad \lim_{n \to \infty} \|S_n A_n - B\| = 0.$$

To deduce that G is the graph of an operator S it suffices to prove that $B = 0$. Now for every pair $(A', B') \in G$ there exist $A_n' \in D(S_n)$ such that $A_n' \to A'$ and $S_n A_n' \to B'$. Moreover,

$$\|(I - \alpha S_n)(A_n + \alpha A_n')\| \geq \|A_n + \alpha A_n'\|.$$

Taking the limit of both sides of this inequality and subsequently dividing by α gives

$$\|A' - B - \alpha B'\| \geq \|A'\|$$

for all $\alpha \in [0, 1]$. Taking the limit of α to zero gives

$$\|A' - B\| \geq \|A'\|.$$

But this is true for all A' in the norm dense set $D(G)$ and hence $B = 0$.

Thus G is the graph of a norm-densely defined operator S and the inequality follows by graph convergence of the corresponding inequalities for S_n. The argument of the foregoing paragraph with S_n replaced by S then proves that S is norm closable.

Our next aim is to show that graph convergence of generators can be used to characterize strong convergence of sequences of C_0-semigroups.

Theorem 3.1.28. *Let U_n be a sequence of C_0-semigroups of contractions on the Banach space X, with generators S_n and define the graph G_α by*

$$G_\alpha = \lim_{n\to\infty} G(I - \alpha S_n).$$

The following conditions are equivalent:

(1) *there exists a C_0-semigroup U such that*

$$\lim_{n\to\infty} \|(U_{n,t} - U_t)A\| = 0$$

for all $A \in X$, $t \in \mathbb{R}_+$, uniformly for t in any finite interval of $\mathbb{R}_+$;

(2) *the sets $D(G_\alpha)$ and $R(G_\alpha)$ are norm dense in X for some $\alpha > 0$.*

If these conditions are satisfied then G_α is the graph of $I - \alpha S$, where S is the generator of U.

PROOF. (1) $\Rightarrow$ (2) For $A \in X$ and $\alpha > 0$ define A_n by $A_n = (I - \alpha S_n)^{-1}A$. Then Theorem 3.1.26 implies that $A_n \to (I - \alpha S)^{-1}A$, where S is the generator of U. Moreover, $(I - \alpha S_n)A_n = A$ and hence $G_\alpha \supseteq G(I - \alpha S)$. Thus $D(G_\alpha)$ and $R(G_\alpha)$ are norm dense by Theorem 3.1.10.

(2) $\Rightarrow$ (1) Proposition 3.1.6 implies that $\|(I - \alpha S_n)A\| \geq \|A\|$ for all $A \in D(S_n)$. Thus Lemma 3.1.27 implies the existence of a norm-closable, norm-densely defined operator S such that $G_\alpha = G(I - \alpha S)$ and $\|(I - \alpha S)A\| \geq \|A\|$. The same inequality is then valid for the closure of S and $R(I - \alpha\bar{S})$ is norm closed. But $R(I - \alpha\bar{S}) = \overline{R(G_\alpha)} = X$. Hence $\bar{S}$ is the generator of a C_0-semigroup U of contractions by Theorem 3.1.10. But if $A_n \to A$ and $B_n = (I - \alpha S_n)A_n \to B = (I - \alpha S)A$ then

$$\begin{aligned}\|((I - \alpha S_n)^{-1} - (I - \alpha\bar{S})^{-1})B\| &= \|(I - \alpha S_n)^{-1}(B - B_n) + A_n - A\| \\ &\leq \|B - B_n\| + \|A - A_n\|.\end{aligned}$$

Thus as $R(I - \alpha\bar{S}) = X$ the resolvent of S_n converges strongly to the resolvent of $\bar{S}$ and U_n converges to U by Theorem 3.1.26. But the resolvent convergence also implies that G_α is closed and hence $S = \bar{S}$.

One simple situation in which the implication (2) $\Rightarrow$ (1) of Theorem 3.1.28 can be applied is if S_n, and S, are generators of C_0-contraction semigroups and there exists a core D, of S, such that

$$D \subseteq \bigcup_m \left(\bigcap_{n\geq m} D(S_n)\right)$$

and

$$\lim_{n\to\infty} \|(S_n - S)A\| = 0$$

for all $A \in D$. It then follows that S is the graph limit of the S_n.

EXAMPLE 3.1.29. Let $\mathfrak{H} = L^2(\mathbb{R}^\nu)$, the Hilbert space of square integrable functions over $\mathbb{R}^\nu$, and S the usual selfadjoint Laplacian operator

$$(S\psi)(x) = -\nabla_x{}^2\psi(x).$$

It is well known that the space of infinitely often differentiable functions with compact support form a core D of S. Next for each bounded open set $\Lambda \subseteq \mathbb{R}^{\nu}$ let S_Λ denote any selfadjoint extension of S restricted to the infinitely often differentiable functions with support in Λ. There are many such extensions each of which corresponds to a choice of boundary conditions. If Λ_n is an increasing sequence such that any open bounded set Λ is contained in Λ_n, for n sufficiently large, then

$$D \subseteq \bigcup_m \left(\bigcap_{n \geq m} D(S_{\Lambda_n}) \right)$$

by definition. Hence

$$\lim_{n \to \infty} \|(e^{itS_{\Lambda_n}} - e^{itS})\psi\| = 0$$

for all $\psi \in \mathfrak{H}$, uniformly for t in finite intervals of $\mathbb{R}$. Arguing by contradiction one then deduces that the net of unitary groups e^{itS_Λ} converges strongly to the group e^{itS}.

A number of classical theorems in analysis can also be derived from the theory of semigroup convergence. As an illustration remark that if S generates the C_0-semigroup of contractions U on X then the bounded operators $S_h = (U_h - 1)/h$ satisfy

$$\lim_{h \to 0} \|(S_h - S)A\| = 0$$

for all $A \in D(S)$ by definition. But the uniformly continuous semigroups

$$\begin{aligned} U_t^{(h)} &= \sum_{n \geq 0} \frac{t^n}{n!} S_h^n \\ &= e^{-t/h} \sum_{n \geq 0} \frac{(t/h)^n}{n!} U_{nh}, \qquad h > 0 \end{aligned}$$

are contractive and hence

$$\lim_{h \to 0} \|(U_t^{(h)} - U_t)A\| = 0$$

for all $A \in X$, uniformly for t in any finite interval of $\mathbb{R}_+$, by Theorem 3.1.26 or Theorem 3.1.28. Now suppose $X = C_0(0, \infty)$, the continuous functions on $[0, \infty\rangle$ which vanish at infinity equipped with the supremum norm, and let U be the C_0-semigroup of right translations,

$$(U_t f)(x) = f(x + t).$$

Then

$$\begin{aligned} f(x + t) &= \lim_{h \to 0} (U_t^{(h)} f)(x) \\ &= \lim_{h \to 0} \sum_{n \geqslant 0} \frac{(t/h)^n}{n!} (\Delta_h^n f)(x) \end{aligned}$$

where

$$(\Delta_h^n f)(x) = ((U_h - I)^n f)(x)$$

$$= \sum_{m=0}^{n} (-1)^{n-m} \quad {}^nC_m f(x + mh)$$

and the limit is uniform for $x \in [0, \infty\rangle$ and t in any finite interval of $[0, \infty\rangle$. This is a generalization of Taylor's theorem. But setting $x = 0$ one also deduces that for each $\varepsilon > 0$ and each finite interval $[0, \mu]$ one can choose N and h such that

$$\left| f(t) - \sum_{n=0}^{N} \frac{(t/h)^n}{n!} (\Delta_h^n f)(0) \right| < \varepsilon$$

for all t in $[0, \mu]$. This is an explicit version of the Stone–Weierstrass theorem.

Another version of the Stone–Weierstrass theorem can be obtained by considering the operators

$$F_s(u) = (1 - s)I + sU_{u/s}$$

with $s > 0$, $u \geq 0$, and setting

$$S_n = \frac{(F_s(s/n) - I)}{(s/n)}$$

$$= \frac{(U_{1/n} - I)}{(1/n)}.$$

Again the semigroups $U_t^{(n)} = \exp\{tS_n\}$ are contractive. But since the semigroups U and $U^{(n)}$ commute one estimates by use of the formula

$$(U_t - U_t^{(n)})A = \int_0^t du \frac{d}{du} (U_u U_{t-u}^{(n)})A$$

that

$$\|(U_t - U_t^{(n)})A\| \leq t\|(S_n - S)A\|$$

for all $A \in D(S)$. Furthermore, by Lemma 3.1.11 one has

$$\left\| \left(U_t^{(n)} - F_t\left(\frac{t}{n}\right)^n \right) A \right\| \leq \sqrt{n} \left\| \left(F_t\left(\frac{t}{n}\right) - I \right) A \right\|$$

$$= tn^{-1/2}\|S_n A\|,$$

where one must assume $t \in [0, 1]$ to ensure that $F_t(t/n)$ is a contraction. A combination of these estimates gives

$$\|(U_t - ((1 - t)I + tU_{1/n})^n)A\| \leq t\|(S_n - S)A\| + tn^{-1/2}\|S_n A\|.$$

provided $t \in [0, 1]$. Thus

$$\lim_{n \to \infty} \|U_t A - ((1 - t)I + tU_{1/n})^n A\| = 0$$

for all $A \in D(S)$, and hence for all $A \in X$, uniformly for $t \in [0, 1]$.

Therefore, specializing to the case $X = C_0(0, \infty)$ and U right translations one deduces, for each $f \in C_0(0, \infty)$, that

$$f(x+t) = \lim_{n\to\infty} \sum_{m=0}^{n} {}^nC_m (1-t)^{n-m} t^m f\left(x + \frac{m}{n}\right)$$

uniformly for $x \in [0, \infty\rangle$, and $t \in [0, 1]$. In particular if $f \in C[0, 1]$

$$f(t) = \lim_{n\to\infty} \sum_{m=0}^{n} {}^nC_m (1-t)^{n-m} t^m f\left(\frac{m}{n}\right)$$

uniformly in t. This is Bernstein's version of the Stone–Weierstrass theorem.
The next theorem is a generalization of this last form of approximation.

Theorem 3.1.30. *Let S be the infinitesimal generator of a C_0-semigroup of contractions U on the Banach space X. Let $t \in \mathbb{R}_+ \mapsto F(t) \in \mathscr{L}(X)$ be a function with values in the bounded operators on X such that*

$$F(0) = I, \qquad \|F(t)\| \le 1,$$

$$\lim_{t\to 0} \left\| \frac{(F(t) - I)A}{t} - SA \right\| = 0$$

for all A in a core D of S. It follows that

$$\lim_{n\to\infty} \|U_t A - F(t/n)^n A\| = 0,$$

for all $A \in X$, uniformly for t in any finite interval of $\mathbb{R}_+$.

PROOF. First note that

$$t \in \mathbb{R}_+ \mapsto U_t^{(s)} = \exp\left\{\left(\frac{t}{s}\right)(F(s) - I)\right\}$$

is a contraction semigroup, for each fixed $s > 0$, with infinitesimal generator $S_s = (F(s) - I)/s$. The contraction property follows because

$$\|U_t^{(s)}\| \le e^{t/s} \sum_{m\ge 0} \frac{(t/s)^m}{m!} \|F(s)\|^m \le 1.$$

But S_s converges to S on the core D and we can apply the remark preceding Example 3.1.29 to deduce that

$$\lim_{s\to 0} \|U_t A - U_t^{(s)} A\| = 0$$

for all $A \in X$ uniformly for t in any finite interval of $\mathbb{R}_+$. Hence

$$\lim_{n\to\infty} \|U_t A - U_t^{(t/n)} A\| = 0$$

for all $A \in X$ uniformly for t in any finite interval of $\mathbb{R}_+$. But from Lemma 3.1.11 one has

$$\left\| U_t^{(t/n)} A - F\left(\frac{t}{n}\right)^n A \right\| \leq \frac{tn^{-1/2} \|(F(t/n) - I)A\|}{(t/n)}.$$

Thus

$$\lim_{n \to \infty} \left\| U_t A - F\left(\frac{t}{n}\right)^n A \right\| = 0$$

for all A in the dense set D, and hence for all $A \in X$, uniformly for t in any finite subinterval of $\mathbb{R}_+$.

Note that the choice

$$F(t) = (I - tS)^{-1}$$

satisfies the criteria of Theorem 3.1.30 and reestablishes the result, obtained in Theorem 3.1.10, that

$$\lim_{n \to \infty} \left\| U_t A - \left(I - \frac{tS}{n}\right)^{-n} A \right\| = 0.$$

A more general result along these same lines is the following:

Corollary 3.1.31. *Let U and V be C_0-semigroups of contractions on the Banach space X, with generators S and T. Assume that $S + T$ is norm closable and that its closure is the generator of a C_0-semigroup of contractions W; then*

$$\lim_{n \to \infty} \|W_t A - (U_{t/n} V_{t/n})^n A\| = 0,$$

$$\lim_{n \to \infty} \left\| W_t A - \left(\left(I - \frac{tS}{n}\right)^{-1} \left(I - \frac{tT}{n}\right)^{-1}\right)^n A \right\| = 0.$$

PROOF. Let $F(t) = U_t V_t$; then $F(0) = I$, $\|F(t)\| \leq 1$, and for each $A \in D(S + T)$ one has

$$\frac{(F(t) - I)A}{t} = \frac{((U_t(V_t - 1)A + (U_t - I)A)}{t}.$$

Hence

$$\lim_{t \to 0} \left\| \frac{(F(t) - I)A}{t} - (S + T)A \right\| = 0$$

for all A in the core $D(S + T)$ of the generator of W. A similar conclusion is true if one chooses $F(t) = (I - tS)^{-1}(I - tT)^{-1}$. Hence it follows from Theorem 3.1.30 that

$$W_t A = \lim_{n \to \infty} (U_{t/n} V_{t/n})^n A = \lim_{n \to \infty} \left(\left(I - \frac{tS}{n}\right)^{-1} \left(I - \frac{tT}{n}\right)^{-1}\right)^n A.$$

3.1.4. Perturbation Theory

If S is the generator of a semigroup U of bounded operators on the Banach space X and P is an operator on X then it is natural to examine the properties of P necessary for $S + P$ to be a generator. If U is uniformly continuous then S is bounded, by Proposition 3.1.1, and for $S + P$ to generate a uniformly continuous semigroup it is necessary, and sufficient, that P be bounded. If U is a C_0- or $C_0{}^*$-semigroup then the question is much more complicated and there exist only partial answers in special cases.

The first result involves C_0-semigroups and relatively bounded perturbations.

Theorem 3.1.32. *Let S be the generator of a C_0-semigroup of contractions, on the Banach space X, and P an operator with $D(P) \supseteq D(S)$ satisfying*

$$\|PA\| \leq a\|A\| + b\|SA\|$$

for all $A \in D(S)$, some $a \geq 0$, and some $b < 1$. If P or $S + P$ is dissipative then $S + P$ generates a C_0-semigroup of contractions.

PROOF. First note that the discussion following Definition 3.1.13 established that Re $\eta(SA) \leq 0$ for all tangent functionals η at $A \in D(S)$. But if P is dissipative then Re $\eta(PA) \leq 0$ by Condition 3 of Proposition 3.1.14, because the assumption $D(P) \supseteq D(S)$ automatically implies $D(P)$ is norm-dense. Consequently, if $\lambda \geq 0$ then Re $\eta((S + \lambda P)A) \leq 0$ and $S + \lambda P$ is dissipative. In the case that $S + P$ is dissipative one reaches the same conclusion by noting that

$$\text{Re } \eta((S + \lambda P)A) = \lambda \text{ Re } \eta((S + P)A) + (1 - \lambda) \text{ Re } \eta(SA) \leq 0.$$

Hence in both cases one deduces from Propositions 3.1.14 and 3.1.15 that $S + \lambda P$ is closable and satisfies the estimate

$$\|(I - \alpha(S + \lambda P))A\| \geq \|A\|$$

for all $A \in D(S)$ and $\alpha > 0$.

Next we exploit the relative bound.

First note that $\|(I - \alpha S)^{-1}\| \leq 1$ for all $\alpha \geq 0$ by Proposition 3.1.6. Hence

$$\begin{aligned}\|\alpha P(I - \alpha S)^{-1}A\| &\leq a\|\alpha(I - \alpha S)^{-1}A\| + b\|(I - (I - \alpha S)^{-1})A\| \\ &\leq (a\alpha + 2b)\|A\|.\end{aligned}$$

Thus, if $0 \leq \lambda_1 < (2b)^{-1}$ one may choose α_0 such that $\lambda_1(\alpha a + 2b) < 1$ for $0 \leq \alpha < \alpha_0$, and then the operator $P_\alpha = \lambda_1 \alpha P(I - \alpha S)^{-1}$ is bounded with $\|P_\alpha\| < 1$. Hence $I - P_\alpha$ has a bounded inverse. Next note that the identity

$$I - \alpha(S + \lambda_1 P) = (I - P_\alpha)(I - \alpha S)$$

and the property $R(I - \alpha S) = X$ imply that

$$\begin{aligned}R(I - \alpha(S + \lambda_1 P)) &= R(I - P_\alpha) \\ &= D((I - P_\alpha)^{-1}) = X.\end{aligned}$$

An application of Theorem 3.1.16 establishes that $S + \lambda_1 P$ is the generator of a C_0-semigroup of contractions.

To continue the proof we remark that

$$\|PA\| \leq a\|A\| + b\|(S + \lambda_1 P)A\| + b\lambda_1\|PA\|$$

and, since $\lambda_1 \leq (2b)^{-1}$, one has

$$\|PA\| \leq 2a\|A\| + 2b\|(S + \lambda_1 P)A\|.$$

We may now choose $0 \leq \lambda_2 < (4b)^{-1}$ and repeat the above argument to deduce that $S + (\lambda_1 + \lambda_2)P$ is the generator of a C_0-semigroup of contractions. Iteration of this argument n times proves that $S + \lambda P$ is a generator for all $0 \leq \lambda < (1 - 2^{-n})/b$. Choosing n sufficiently large, but finite, one deduces the desired result for $S + P$.

There does not appear to be a $C_0{}^*$-analogue of Theorem 3.1.32 but in the simplest example, when P is bounded, one may draw a conclusion.

Let P be a bounded operator which is closed in the $\sigma(X, X_*)$-topology on X. Then P is the dual of a bounded norm-closed operator P^*, on X_*, by Lemma 3.1.9. Moreover, $P^* - \|P\|I$ is the bounded generator of a uniformly continuous contraction semigroup on X_*, by Proposition 3.1.1. In particular, $P^* - \|P\|I$ is dissipative on X_*. Next, let U be a $C_0{}^*$-semigroup of contractions on X with generator S. Reapplying Lemma 3.1.9 one deduces that U is the dual of a C_0-semigroup of contractions U^*, on X_*, with dual generator S^*. Therefore Theorem 3.1.32 implies that $S^* + P^* - \|P\|I$ is the generator of a C_0-semigroup of contractions on X_*. By duality $S + P$ is the generator of a $C_0{}^*$-semigroup U^P on X with the growth bound $\|U_t{}^P\| \leq \exp\{\|P\|t\}$.

An alternative way of examining bounded perturbations is provided by the following:

Theorem 3.1.33. *Let U be a $\sigma(X, F)$-continuous semigroup on the Banach space X with generator S and let P be a $\sigma(X, F)$–$\sigma(X, F)$-closed, bounded operator on X. If $F = X^*$, or $F = X_*$, then $S + P$ generates a $\sigma(X, F)$-continuous semigroup U^P such that*

$$U_t{}^P A = U_t A + \sum_{n \geq 1} \int_{0 \leq t_1 < \cdots \leq t_n \leq t} dt_1 \cdots dt_n \, U_{t_1} P U_{t_2 - t_1} P \cdots U_{t_n - t_{n-1}} P U_{t - t_n} A$$

for all $A \in X$. Here the integrals exist in the norm topology if $F = X^*$ and in the $\sigma(X, X_*)$-topology if $F = X_*$, and in either case the integrals define a series of bounded operators in $\mathscr{L}(X)$ which converges in norm. If $\|U_t\| \leq Me^{\beta t}$ then

$$\|U_t{}^P - U_t\| \leq Me^{\beta t}(e^{M\|P\|t} - 1).$$

PROOF. *C_0-case*: $F = X^*$. Let $U_t^{(n)}$ denote the nth term of the perturbation series for U_t^P. Since U is strongly continuous and

$$U_t^{(0)} = U_t, \qquad U_t^{(n)} = \int_0^t dt_1\, U_{t_1} P U_{t-t_1}^{(n-1)} \tag{$*$}$$

it follows by an inductive argument that the $U_t^{(n)}$ are well defined and strongly continuous. One estimates easily that

$$\begin{aligned}\|U_t^{(n)}A\| &\leq \int_{0 \leq t_1 \leq \cdots \leq t_n \leq t} dt_1 \cdots dt_n \|U_{t_1}\| \, \|U_{t_2-t_1}\| \cdots \|U_{t-t_n}\| \, \|P\|^n \|A\| \\ &\leq \frac{t^n}{n!} M^{n+1} e^{\beta t} \|P\|^n \|A\|.\end{aligned}$$

The norm convergence of the perturbation series and the last statement of the theorem follow immediately.

Next note that the recursion relations $(*)$ imply the integral equation

$$U_t^P = U_t + \int_0^t ds\, U_s P U_{t-s}^P \tag{$**$}$$

and hence

$$\begin{aligned}U_{t_1}^P U_{t_2}^P &= U_{t_1} U_{t_2}^P + \int_0^{t_1} ds\, U_s P U_{t_1-s}^P U_{t_2}^P \\ &= U_{t_1+t_2} + \int_0^{t_2} ds\, U_{t_1+s} P U_{t_2-s}^P + \int_0^{t_1} ds\, U_s P U_{t_1-s}^P U_{t_2}^P \\ &= U_{t_1+t_2}^P + \int_0^{t_1} ds\, U_s P \{U_{t_1-s}^P U_{t_2}^P - U_{t_1+t_2-s}^P\}\end{aligned}$$

Thus the family of functions $\lambda \in \mathbb{C} \mapsto F_{t_1}(\lambda) = U_{t_1}^{\lambda P} U_{t_2}^{\lambda P} - U_{t_1+t_2}^{\lambda P}$ is entire analytic and satisfies the homogeneous integral equations

$$F_t(\lambda) = \lambda \int_0^t ds\, U_s P F_{t-s}(\lambda)$$

A Taylor series argument then establishes that $F_t(\lambda) = 0$, i.e., the semi-group property $U_{t_1}^P U_{t_2}^P = U_{t_1+t_2}^P$ is valid. Clearly, $U_0{}^P = I$ and it remains to identify the generator T of U^P.

If $0 < \alpha < (\beta + M\|P\|)^{-1}$ then

$$(I - \alpha T)^{-1} = \int_0^\infty dt\, e^{-t} U_{\alpha t}^P$$

by Proposition 3.1.6. Using the integral equation $(**)$ one finds that

$$\begin{aligned}(I - \alpha T)^{-1} &= \int_0^\infty dt\, e^{-t} U_{\alpha t} + \alpha \int_0^\infty dt \int_0^t ds\, e^{-t} U_{\alpha s} P U_{\alpha(t-s)}^P \\ &= \int_0^\infty dt\, e^{-t} U_{\alpha t} + \alpha \int_0^\infty dt\, e^{-t} U_{\alpha t} P \int_0^\infty ds\, e^{-s} U_{\alpha s}^P \\ &= (I - \alpha S)^{-1} + \alpha (I - \alpha S)^{-1} P (I - \alpha T)^{-1}.\end{aligned}$$

This establishes that

$$(I - \alpha(S + P))(I - \alpha T)^{-1} = I.$$

But $\|(I - \alpha S)^{-1}\| \leq M(1 - \alpha\beta)^{-1}$ by Proposition 3.1.6 and hence our choice of α is such that

$$\alpha\|P\| \leq \|(I - \alpha S)^{-1}\|^{-1}.$$

It follows that $(I - \alpha(S + P))$ is invertible, by a power series expansion, with bounded inverse and hence $(I - \alpha(S + P))^{-1} = (I - \alpha T)^{-1}$. Therefore $T = S + P$.

C_0-case*: $F = X_*$. As P is $\sigma(X, X_*)$–$\sigma(X, X_*)$-closed and bounded it is the dual of a bounded operator P^*, on X_*, by Lemma 3.1.9. Further, U is the dual of a C_0-semigroup U^*, on X_*, with generator S^*, dual to S. The previous argument then establishes that $S^* + P^*$ generates a C_0-semigroup, on X_*, and by duality $S + P$ generates a C_0*-semigroup, on X. The perturbation series follows by transposition of the corresponding series on X_*.

The above theorem establishes that generators of C_0- and C_0*-semigroups are stable under the addition of a bounded perturbation. A similar result for contraction semigroups follows, of course, from Theorem 3.1.32. The "time-dependent" series expansion does, however, have advantages and the conclusions of Theorem 3.1.33 can be extended to certain unbounded perturbations. Essentially, the perturbation must be such that the series for U^P is well defined and convergent. This even includes cases where the relative boundedness of Theorem 3.1.32 does not hold, and this has been useful in constructive quantum field theory. Another example of the power of the expansion is provided by the analysis of dissipative operators on Banach spaces which leave an increasing sequence of closed subspaces of the domain approximately invariant in a sense which is made precise in the following theorem.

Theorem 3.1.34. *Let S be a dissipative operator on a Banach space X, and assume that there exists an increasing sequence X_n of closed subspaces of X such that the closure of $\bigcup_n X_n$ is X and*

$$\bigcup_n X_n \subseteq D(S).$$

Furthermore, assume that there exist linear operators

$$S_{n,m};\ D(S_{n,m}) = X_n \mapsto X_{n+m}$$

and numbers $M, \alpha > 0$ such that

$$\|S|_{X_n} - S_{n,m}\| \leq Mne^{-\alpha m}$$

for $n = 1, 2, \ldots, m = 0, 1, \ldots$. It follows that $\bigcup_n X_n$ is a core for S, and the closure $\bar{S}$ of S is the generator of a semigroup of contractions on X.

If, furthermore, $S_{n,0}$ is dissipative for all n and

$$S_{n,m} = S_{n+m,0}|_{X_n}$$

for all n, m then it follows that

$$e^{t\bar{S}}A = \lim_{n \to \infty} e^{tS_{n,0}} A$$

for all $A \in \bigcup_n X_n$, uniformly for t in compacts.

PROOF. By Proposition 3.1.14 and Theorem 3.1.16 it is enough for the first part of the theorem to show that $(\lambda I - S)(\bigcup_n X_n)$ is dense in X for some $\lambda > 0$. It follows from the same results that we may assume that S is the closure of its restriction to $\bigcup_n X_n$. If the condition of the second part of the theorem is fulfilled, then

$$\lim_{m \to \infty} \|S|_{X_n} - S_{n+m,0}|_{X_n}\| = 0$$

and hence the last statement follows from a small extension of Theorem 3.1.28 (see Notes and Remarks).

We will only prove the theorem in the slightly simpler case that

$$SX_n \subseteq X_{n+1}$$

because the idea of the proof is clearer in this case. The extension to the general case consists mainly of putting more iterations into the perturbation expansion (see Notes and Remarks).

To fix notations, set $S_{n,0} = S_n$ and scale S so that

$$\|S|_{X_n} - S_n\| \leq n/2.$$

Observation (1): *The operator $S_n - (n/2)I$ is bounded and dissipative on X_n.*

PROOF. Since $S|_{X_n}$ is closed, and hence bounded, it follows that S_n is bounded. Let $A \in X_n \backslash \{0\}$. As S is dissipative there exists an $f \in X^*$ with $\|f\| = 1$ and $f(A) = \|A\|$ such that $\text{Re } f(SA) \leq 0$. But then

$$\begin{aligned}\text{Re} f\left(\left(S_n - \frac{n}{2} I\right)A\right) &= \text{Re } f((S_n - S)A) + \text{Re } f(SA) - \frac{n}{2} \|A\| \\ &\leq \frac{n}{2} \|A\| + 0 - \frac{n}{2} \|A\| = 0.\end{aligned}$$

By Observation (1) we can replace S_n by $S_n - (n/2)I$ and assume that all S_n are dissipative and

$$\|S|_{X_n} - S_n\| \leq n.$$

Then e^{tS_n} is a norm-continuous semigroup of contractions on X_n for each n, and we can define a version of the time-dependent perturbation expansion.

If $A \in X_m$, define inductively

$$\sigma_t^{(0)}(A) = A,$$

$$\sigma_t^{(1)}(A) = \int_0^t ds \, e^{(t-s)S_{m+1}} S \sigma_s^{(0)}(A),$$

and

$$\sigma_t^{(n)}(A) = \int_0^t ds\, e^{(t-s)S_{m+n}}(S - S_{m+n-1})\sigma_s^{(n-1)}(A)$$

for $n \geq 2$. Since the restriction of S to each X_n is bounded and $SX_n \subseteq X_{n+1}$, these definitions make sense as Riemann integrals and $t \mapsto \sigma_t^{(n)}(A)$ is a norm-continuous curve in X_{m+n} for $n = 0, 1, \ldots$. If $K = \|S|_{X_m}\|$, then the following estimates follow by induction, using $\|S|_{X_n} - S_n\| \leq n$.

Observation (2).

$$\|\sigma_t^{(0)}(A)\| \leq \|A\|$$

and

$$\|\sigma_t^{(n)}(A)\| \leq \frac{(m + n - 1)!}{n!m!} t^n K\|A\|$$

$$\leq (m + n - 1)^m t^n K\|A\|$$

for $n = 1, 2, \ldots,$ *and* $t \geq 0$.

Define

$$\tau_t^{(k)}(A) = \sum_{n=0}^{k} \sigma_t^{(n)}(A).$$

It follows from Observation (2) that

$$\tau_t(A) = \lim_{k\to\infty} \tau_t^{(k)}(A)$$

exists in norm for $t \in [0, 1\rangle$, uniformly for t in compacts. (At this point $\tau_t(A)$ could depend on m in addition to A and t, and we assume that a fixed $m = m(A)$ is chosen for each A; *a posteriori* $\tau_t(A)$ is nothing but $e^{t\bar{S}}A$.)

Note that

$$\sigma_{t+\delta}^{(n)}(A) - \sigma_t^{(n)}(A)$$

$$= \int_t^{t+\delta} ds\, e^{(t+\delta-s)S_{m+n}}(S - S_{m+n-1})\sigma_s^{(n-1)}(A)$$

$$+ \int_0^t ds\, (e^{(t+\delta-s)S_{m+n}} - e^{(t-s)S_{m+n}})(S - S_{m+n-1})\sigma_s^{(n-1)}(A).$$

and as $(S - S_{m+n-1})|_{X_{m+n-1}}$ and S_{m+n} are bounded, it follows that $t \mapsto \sigma_t^{(n)}(A)$ is differentiable with derivatives given by the following:

$$\frac{d}{dt}\sigma_t^{(0)}(A) = 0;$$

$$\frac{d}{dt}\sigma_t^{(1)}(A) = S\sigma_t^{(0)}(A) - S_{m+1}\sigma_t^{(1)}(A);$$

and

$$\frac{d}{dt}\sigma_t^{(n)}(A) = (S - S_{m+n-1})\sigma_t^{(n-1)}(A) + S_{m+n}\sigma_t^{(n)}(A).$$

It follows by cancellation that

$$\frac{d}{dt}(\tau_t^{(n)}(A)) - S(\tau_t^{(n)}(A)) = (S_{m+n} - S)\sigma_t^{(n)}(A).$$

But by Observation (2) we have the estimate

$$\|(S_{m+n} - S)\sigma_t^{(n)}(A)\| \leq (m+n)^{m+1}t^n K\|A\|$$

since $\sigma_t^{(n)}(A) \in X_{m+n}$, and hence, integrating the relation above,

$$\left\| \tau_t^{(n)}(A) - A - S\left(\int_0^t ds\, \tau_s^{(n)}(A)\right)\right\| \leq \frac{(m+n)^{m+1}}{n+1} t^{n+1}K\|A\|.$$

As S is closed, and

$$\tau_t(A) = \lim_{n\to\infty} \tau_t^{(n)}(A)$$

exists for $t \in [0, 1\rangle$, uniformly for t in compacts, the following observation is valid.

Observation (3).

$$\int_0^t ds\, \tau_s(A) \in D(S)$$

for $t \in [0, 1\rangle$, *and*

$$\tau_t(A) - A = S\left(\int_0^t ds\, \tau_s(A)\right).$$

We next deduce

Observation (4). $t \in [0, 1\rangle \mapsto \|\tau_t(A)\|$ *is nonincreasing in* t.

PROOF. Define regularized elements

$$\tau_t(A_\varepsilon) = \int_0^\varepsilon ds\, \tau_{t+s}(A)$$

for $t \in [0, 1 - \varepsilon\rangle$. (Warning: at this stage the notation $\tau_t(A_\varepsilon)$ should not be taken too literally; $t \mapsto \tau_t(A_\varepsilon)$ should only be viewed as a curve in X.) It follows that $t \mapsto \tau_t(A_\varepsilon)$ is differentiable (with derivative

$$\frac{d}{dt}\tau_t(A_\varepsilon) = \tau_{t+\varepsilon}(A) - \tau_t(A).)$$

Furthermore, by Observation (3)

$$\begin{aligned}\tau_{t+\delta}(A_\varepsilon) - \tau_t(A_\varepsilon) &= \int_0^\varepsilon ds\, (\tau_{t+\delta+s}(A) - \tau_{t+s}(A)) \\ &= \int_0^\varepsilon ds\, S\left(\int_{t+s}^{t+\delta+s} du\, \tau_u(A)\right) \\ &= S\left(\int_0^\varepsilon ds \int_{t+s}^{t+\delta+s} du\, \tau_u(A)\right).\end{aligned}$$

Dividing by δ, and letting $\delta \to 0$, we use the closedness of S to derive

$$\frac{d}{dt}\tau_t(A_\varepsilon) = S(\tau_t(A_\varepsilon)).$$

But as S is dissipative it follows from Lemma 3.1.15 that

$$\|\tau_t(A_\varepsilon) - \delta S(\tau_t(A_\varepsilon))\| \geq \|\tau_t(A_\varepsilon)\|$$

for all $\delta \geq 0$, i.e.,

$$\left\|\tau_t(A_\varepsilon) - \delta\frac{d}{dt}\tau_t(A_\varepsilon)\right\| \geq \|\tau_t(A_\varepsilon)\|$$

and hence

$$\|\tau_{t-\delta}(A_\varepsilon)\| + o(\delta) \geq \|\tau_t(A_\varepsilon)\|,$$

or

$$\frac{1}{\delta}\{\|\tau_{t-\delta}(A_\varepsilon)\| - \|\tau_t(A_\varepsilon)\|\} \geq o(1).$$

It follows that $t \mapsto \|\tau_t(A_\varepsilon)\|$ must be nonincreasing, and as

$$\tau_t(A) = \lim_{\varepsilon\to 0}\frac{1}{\varepsilon}\tau_t(A_\varepsilon)$$

one also concludes that $t \to \|\tau_t(A)\|$ is nonincreasing.

To finish the proof of Theorem 3.1.34, assume *ad absurdum* that $R(\lambda I - S)$ is not dense for some $\lambda > 0$. Then there exists by Hahn–Banach a linear functional $f \in X^*$ such that $\|f\| = 1$ and

$$f((\lambda I - S)B) = 0$$

for all $B \in D(S)$. In particular, Observation (3) implies that

$$\begin{aligned}\lambda f\left(\int_0^t ds\, \tau_s(A)\right) &= f\left(S\left(\int_0^t ds\, \tau_s(A)\right)\right) \\ &= f(\tau_t(A)) - f(A),\end{aligned}$$

or

$$f(\tau_t(A)) = f(A) + \lambda\int_0^t ds\, f(\tau_s(A)).$$

It follows that

$$f(\tau_t(A)) = e^{\lambda t} f(A)$$

for $t \in [0, 1\rangle$. But as $\bigcup_n X_n$ is dense in X, we can find an $A \in \bigcup_n X_n$ such that

$$e^{\lambda/2} |f(A)| > \|A\|$$

and hence

$$\|\tau_{1/2}(A)\| \geq |f(\tau_{1/2}(A))| = e^{\lambda/2} |f(A)| > \|A\|.$$

This contradicts Observation (4), and the contradiction finishes the proof of Theorem 3.1.34.

Remarks. The differential equation

$$\frac{d}{dt} \tau_t(A_\varepsilon) = S(\tau_t(A_\varepsilon))$$

established during the proof, implies that $\tau_t(A_\varepsilon) = e^{tS} A_\varepsilon$, and by continuity it follows that the perturbation expansion

$$e^{tS} A = \sum_{n=0}^{\infty} \sigma_t^{(n)}(A)$$

is valid for $t \in [0, 1\rangle$.

One can show by explicit counterexamples that Theorem 3.1.34 does not extend to the case

$$SX_n \subseteq X_{n+1}, \qquad \|S|_{X_n} - S_{n,0}\| \leq Mn^2$$

even when S is a symmetric operator on a Hilbert space X.

By a much simpler, but less explicit, argument, one can prove the analogue of the first part of Theorem 3.1.34 in the case

$$\|S|_{X_n} - S_{n,0}\| \leq M$$

for $n = 1, 2, \ldots$. First, note that by Observation (1) we can assume that each $S_n = S_{n,0}$ is dissipative, and hence

$$\|(\lambda I - S_n)^{-1}\| \leq \lambda^{-1}$$

for $\lambda > 0$ by Lemma 3.1.15. Now let $\lambda > M$, and let $f \in X^*$ be a functional vanishing on $R(\lambda I - S)$. In particular, for $A \in X_n$ we have

$$\begin{aligned} |f(A)| &= |f((\lambda I - S_n)(\lambda I - S_n)^{-1} A)| \\ &= |f((S - S_n)(\lambda I - S_n)^{-1} A)| \\ &\leq \|f\| M \lambda^{-1} \|A\|. \end{aligned}$$

Since $\bigcup_n X_n$ is dense it follows that

$$\|f\| \leq M\lambda^{-1} \|f\|$$

and hence $f = 0$, and $R(\lambda I - S)$ is dense.

3.1.5. Approximation Theory

In the previous subsection we examined the stability of semigroup generators under perturbations and established stability for bounded perturbations. Now if U is a C_0-semigroup and U^P the perturbed semigroup obtained by adding a bounded perturbation P to the generator of U then Theorem 3.1.33 demonstrates that

$$\|U_t - U_t{}^P\| = O(t)$$

as t tends to zero.[1] The norm distance between two groups, or semigroups, is a convenient measure of their proximity and this example shows that groups which have neighboring generators have a norm distance which is small for small t. We next examine the interrelationships between semigroups which are close for small t. The first result states that the semigroups cannot be too close without being equal.

Proposition 3.1.35. *Let U and V be two $\sigma(X, F)$-continuous semigroups on the Banach space X. If*

$$\|U_t - V_t\| = o(t)$$

as $t \to 0$, then $U = V$.

PROOF. Let S and T denote the generators of U and V, respectively. Choose $A \in D(T)$ and note that

$$\left|\omega\left(\frac{1}{t}(U_t - I)A\right) - \omega\left(\frac{1}{t}(V_t - I)A\right)\right| = o(1)$$

for all $\omega \in F$, as $t \to 0$. Hence $A \in D(S)$, by definition, and $SA = TA$, i.e., $S \supseteq T$. Interchanging the roles of S and T one deduces that $S \subseteq T$ and hence $S = T$. Now, using the expansions $U_t = \lim_{n\to\infty}(I - tS/n)^{-n}A$, $V_t = \lim_{n\to\infty}(I - tT/n)^{-n}A$, which are valid for $A \in D(S) = D(T)$ by Theorem 3.1.10, one concludes that $U_t|_{D(S)} = V_t|_{D(S)}$ and hence $U_t = V_t$ by $\sigma(X, F)$-continuity.

The next result shows, however, that if two C_0- or $C_0{}^*$-groups are moderately close for small t then they are related in a particularly simple manner.

Theorem 3.1.36. *Let U and V be two C_0- or $C_0{}^*$-groups, on the Banach space X, with generators S and T, respectively. The following conditions are equivalent:*

(1) *there exist $\varepsilon_1 > 0$ and $\delta_1 > 0$ such that*

$$\|U_t V_{-t} - I\| \le 1 - \varepsilon_1$$

for $0 \le t \le \delta_1$;

[1]
$$\|U_t - V_t\| = o(t) \text{ means } \lim_{t\to 0}\|U_t - V_t\|/t = 0,$$

and

$$\|U_t - V_t\| = O(t) \text{ means } \limsup_{t\to 0}\|U_t - V_t\|/t < +\infty.$$

(2) *there exist $\varepsilon_2 > 0$, $\delta_2 > 0$, and bounded operators P, W such that W has a bounded inverse,*

$$S = W(T + P)W^{-1},$$

and

$$\|U_t W^{-1} U_{-t} W - I\| \leq 1 - \varepsilon_2$$

for all $0 \leq t \leq \delta_2$.

If these conditions are satisfied then W may be defined by

$$W = \frac{1}{\delta_1} \int_0^{\delta_1} ds\, U_s V_{-s}$$

and hence $\|I - W\| \leq 1 - \varepsilon_1$. Moreover,

$$\|U_t W^{-1} U_{-t} W - I\| = \|U_t V_{-t} - I\| + O(t), \qquad t \to 0,$$

and

$$\|U_t W V_{-t} - W\| = O(t), \qquad t \to 0.$$

PROOF. (1) $\Rightarrow$ (2) Define W by

$$W = \frac{1}{\delta_1} \int_0^{\delta_1} ds\, U_s V_{-s}$$

(where the integral is the strong integral of a strongly continuous function in the C_0-case and is the dual of a strong integral in the $C_0{}^*$-case). It follows that $\|I - W\| \leq 1 - \varepsilon_1$, and hence W has a bounded inverse. Next introduce X_t by

$$X_t = W^{-1} U_t W V_{-t}.$$

One calculates that

$$\frac{1}{h}(X_{t+h} - X_t) = \frac{1}{\delta_1 h} W^{-1} \int_{\delta_1}^{\delta_1 + h} ds\, U_{s+t} V_{-s-t} - \frac{1}{\delta_1 h} W^{-1} \int_0^h ds\, U_{s+t} V_{-s-t}.$$

This implies that X_t is strongly differentiable in the C_0-case, and weakly* differentiable in the $C_0{}^*$-case. The derivative in both cases is given by

$$\frac{dX_t}{dt} = \frac{W^{-1} U_t (U_{\delta_1} V_{-\delta_1} - I) V_{-t}}{\delta_1}.$$

Next remark that

$$\frac{(U_t - I)WA}{t} = \frac{W(V_t - I)A}{t} + \frac{W(X_t - I)V_t A}{t}.$$

If $A \in D(T)$ and $t \to 0$ then the right-hand side converges, either strongly or in the weak* topology. Hence $WA \in D(S)$ and

$$SWA = W(T + P)A,$$

where

$$P = \left.\frac{dX_t}{dt}\right|_{t=0} = \frac{W^{-1}(U_{\delta_1} V_{-\delta_1} - I)}{\delta_1}.$$

Similarly, if $A \in D(S)$ then $W^{-1}A \in D(T)$ and

$$W^{-1}SA = (T + P)W^{-1}A.$$

Thus $D(S) = WD(T)$ and $S = W(T + P)W^{-1}$.

Finally, one computes that

$$\begin{aligned} U_t W^{-1} U_{-t} W - I &= (U_t V_{-t} - I)(V_t W^{-1} U_{-t} W - I) \\ &\quad + \frac{1}{\delta_1} W^{-1} \int_0^t ds\, U_s(I - U_{\delta_1} V_{-\delta_1}) V_{-s} V_t W^{-1} U_{-t} W \\ &\quad + (U_t V_{-t} - I). \end{aligned}$$

Now note that $t \mapsto W^{-1} U_t W$ is the group with generator $T + P$ and hence

$$\|V_t W^{-1} U_{-t} W - I\| = O(t), \qquad t \to 0, \tag{$*$}$$

by Theorem 3.1.33. Thus

$$\|U_t W^{-1} U_{-t} W - I\| = \|U_t V_{-t} - I\| + O(t). \tag{$**$}$$

(2) $\Rightarrow$ (1) Define Q by $Q = -WPW^{-1}$. One has $T = W^{-1}(S + Q)W$ and if $\hat{U}$ is the group generated by $S + Q$ then $\hat{U}_t = WV_tW^{-1}$. Another application of Theorem 3.1.33 implies that $\|U_t \hat{U}_{-t} - I\| = O(t)$ as $t \to 0$. But the identity

$$U_t V_{-t} - I = U_t W^{-1} U_{-t}(U_t \hat{U}_{-t} - I)W + (U_t W^{-1} U_{-t} W - I)$$

then implies

$$\|U_t V_{-t} - I\| = \|U_t W^{-1} U_{-t} W - I\| + O(t). \tag{$***$}$$

The last statements of the theorem follow from ($*$), ($**$), and ($***$).

Although we only stated the above theorem for C_0- and $C_0{}^*$-groups there is a version which is valid for $\sigma(X, F)$-continuous groups if $\{U_t\}_{|t| \le 1}$ and $\{V_t\}_{|t| \le 1}$ are $\tau(X, F)$-equicontinuous. In fact it would suffice for the dual groups $U_t{}^*$ and $V_t{}^*$ to have the analogous $\tau(F, X)$-equicontinuity property. The only point where the continuity plays an important role is in the formation of W. But if, for example, $\{U_t\}_{|t| \le 1}$ and $\{V_t\}_{|t| \le 1}$ are equicontinuous and p_K is a $\tau(X, F)$-seminorm then

$$p_K((U_t V_{-t} - U_s V_{-s})A) \le p_{K'}((V_{-t} - V_{-s})A) + p_{K'}((U_t - U_s)V_{-s}A)$$

for all A and $|t|, |s| \le 1$. Hence $t \in [-1, 1] \mapsto U_t V_{-t} A$ is $\tau(X, F)$-continuous.

The preceding theorem establishes that two groups which are close in norm for all sufficiently small t can only differ by a twist, and a boost. The boost is a bounded perturbation of the generator and the twist is a map $U_t \mapsto WU_tW^{-1}$. Note that if $\|U_t - V_t\| = o(1)$, or $O(t^\alpha)$, as $t \to 0$ then $\|U_t - WU_tW^{-1}\| = o(1)$, or $O(t^\alpha)$, as $t \to 0$. Thus the twist almost leaves U invariant. We will see, however, by examples that the twist cannot be omitted in general. Before examining this problem we draw one simple and amusing corollary of the theorem.

Corollary 3.1.37. *If U is a C_0- or $C_0{}^*$-group, on the Banach space X then the following conditions are equivalent*:

(1) *there exist $\varepsilon > 0$ and $\delta > 0$ such that*
$$\|U_t - I\| \leq 1 - \varepsilon$$
for all $0 \leq t < \delta$;

(2) $$\lim_{t\to 0} \|U_t - I\| = 0.$$

PROOF. This follows from Theorem 3.1.36 by setting $V = I$ and remarking that condition (1) then implies that the generator of U is bounded. The uniform continuity of U then follows from Proposition 3.1.1.

Perturbation theory (Theorem 3.1.33) showed that two semigroups whose generators differ only by a bounded operator are close in norm of order t as t tends to zero. This characterizes bounded perturbations for $C_0{}^*$-semigroups.

Theorem 3.1.38. *Let U and V be two $C_0{}^*$-semigroups on the Banach space X, with generators S and T, respectively. The following conditions are equivalent*:

(1) $\|U_t - V_t\| = O(t)$ *as* $t \to 0$;
(2) $S = T + P$ *where P is a bounded operator.*

PROOF. (1) $\Rightarrow$ (2) First note that it follows from Theorem 3.1.33, and its proof, that $S = T + P$ with P bounded if, and only if
$$U_t - V_t = \int_0^t ds\, U_s P V_{t-s}. \tag{$*$}$$

Next remark that by assumption there are constants M, $\delta > 0$ such that
$$\|U_t - V_t\| \leq Mt$$
for $0 \leq t \leq \delta$. But the unit ball of X is compact in the weak* topology, i.e., the $\sigma(X, X_*)$-topology, by the Alaoglu–Bourbaki theorem. Thus there exists a net $t_\alpha \to 0$ such that $(U_{t_\alpha} - V_{t_\alpha})/t_\alpha$ is weakly* convergent, i.e., $\omega((U_{t_\alpha} - V_{t_\alpha})A)/t_\alpha$ converges as $t_\alpha \to 0$ for all $A \in X$ and $\omega \in X_*$. Consequently, there exists a bounded operator P satisfying $\|P\| \leq M$ such that $\omega((U_{t_\alpha} - V_{t_\alpha})A)/t_\alpha \to \omega(PA)$. (In principle P depends upon the net t_α but the subsequent argument shows that this is not the case.)

Now using the semigroup property and a change of variables one finds, that
$$\int_0^t ds\, \omega(U_s(U_u - V_u)V_{t-s}A) = \int_{t-u}^t ds\, \omega(U_{s+u}V_{t-s}A) - \int_{-u}^0 ds\, \omega(U_{s+u}V_{t-s}A)$$
for each $A \in X$ and $\omega \in X_*$. Therefore, replacing u by t_α and taking the limit one deduces with the aid of $\sigma(X, X_*)$-continuity of V, $\sigma(X_*, X)$-continuity of U^*, and the Lebesgue dominated convergence theorem that
$$\int_0^t ds\, \omega(U_s P V_{t-s}A) = \omega((U_t - V_t)A).$$

Hence one obtains (*) and concludes that $S = T + P$. (Note that this then implies that $(U_t - V_t)A/t$ in fact converges in the weak* topology as $t \to 0$ for all $A \in X$ and hence P does not depend upon the choice of the net t_α.)

(2) ⇒ (1) If $A \in D(S) = D(T)$ then $V_{t-s}A \in D(S)$ and

$$\omega((U_t - V_t)A) = \int_0^t ds\, \omega(U_s(S - T)V_{t-s}A)$$
$$= \int_0^t ds\, \omega(U_s P V_{t-s}A)$$

for each $\omega \in X_*$. Therefore

$$\|U_t - V_t\| \leq t\|P\| \sup_{0 \leq s \leq t} \|U_s\| \|V_{t-s}\|.$$

But there exist M and β such that $\|U_t\| \leq M \exp\{\beta t\}$ and $\|V_t\| \leq M \exp\{\beta t\}$ by Proposition 3.1.3. Hence

$$\|U_t - V_t\| \leq t\|P\|M \exp\{\beta t\} = O(t) \qquad \text{as} \quad t \to 0.$$

This theorem is not necessarily true for C_0-groups, as the following example demonstrates.

EXAMPLE 3.1.39. Let $X = C_0(\mathbb{R})$ be the continuous functions on the real line equipped with the usual supremum norm. Let F be the operator of multiplication by a real function f which is nondifferentiable at a dense set of points, and uniformly Hølder continuous in the sense that

$$|f(s) - f(t)| \leq c|s - t|.$$

Define W by $W = \exp\{iF\}$. Next let U be the C_0-group of translations,

$$(U_t\psi)(x) = \psi(x - t), \qquad t \in \mathbb{R}, \psi \in X,$$

and define a second C_0-group by $V_t = W^{-1}U_tW$. It is easily calculated that

$$\|U_t - V_t\| \leq c|t|.$$

But the generator S of U is a differentiation operator, the generator T of V is $W^{-1}SW$, and $D(S) \cap D(T) = \{0\}$ because f is nondifferentiable at a dense set of points.

This example demonstrates that Theorem 3.1.38 is in general false for C_0-groups. It also demonstrates that the twist of Theorem 3.1.36 cannot always be omitted. Nevertheless, if X is reflexive, i.e., if $X = X^{**}$, then the dual X^* of X is its predual and the weak* and weak topologies coincide. Thus every C_0-group U on X is a C_0^*-group by Definition 3.1.2. Therefore two C_0-semigroups U and V on a reflexive Banach space X satisfy

$$\|U_t - V_t\| = O(t) \qquad \text{as} \quad t \to 0,$$

if, and only if, their generators S and T satisfy $S = T + P$ where P is bounded.

Theorems 3.1.36 and 3.1.38 describe the basic relationships between two groups that are close in norm for small t. It is also natural to examine weaker measures of proximity and in the $C_0{}^*$-case it is possible to characterize relatively bounded perturbations of the type occurring in Theorem 3.1.32.

Theorem 3.1.40. *Let U and V be $C_0{}^*$-semigroups, on the Banach space X, with generators S and T, respectively. The following conditions are equivalent:*

(1)
$$\|(U_t - V_t)A\| = O(t), \qquad t \to 0,$$

for all $A \in D(T)$;

(2)
$$\|(U_t - V_t)(I - \alpha T)^{-1}\| = O(t), \qquad t \to 0,$$

for all α in an interval $0 < \alpha < \delta$;

(3) *$D(S) \supseteq D(T)$ and there exist constants $a, b \geq 0$ such that*

$$\|(S - T)A\| \leq a\|A\| + b\|TA\|$$

for all $A \in D(T)$.

PROOF. (1) $\Rightarrow$ (3) First note that $A \in D(S)$ if, and only if, $\|(U_t - I)A\| = O(t)$ as $t \to 0$, by Proposition 3.1.23. Thus if $A \in D(T)$ then

$$\frac{\|(U_t - I)A\|}{t} \leq \frac{\|(V_t - I)A\|}{t} + \frac{\|(U_t - V_t)A\|}{t} = O(1)$$

and $A \in D(S)$. Therefore $D(S) \supseteq D(T)$. Next note that S and T are both weakly* closed and hence strongly closed. Consider the graph $G(T) = \{(A, TA)\}$ of T equipped with the norm $\|(A, TA)\| = \|A\| + \|TA\|$. The graph $G(T)$ is a closed subspace of $X \times X$ and the mapping

$$(A, TA) \mapsto SA$$

is a linear operator from $G(T)$ into X. But this operator is closed because if (A_n, TA_n) converges in $G(T)$ and SA_n converges in X one has $\|A_n - A\| \to 0$, for some A, and $\|SA_n - B\| \to 0$. Thus $B = SA$ because S is closed. The closed graph theorem now implies that this mapping must be bounded. Hence there is a constant such that

$$\|SA\| \leq c(\|A\| + \|TA\|).$$

Finally,

$$\|(S - T)A\| \leq c\|A\| + (c + 1)\|TA\|.$$

(3) $\Rightarrow$ (2) If $A \in D(T)$ then $V_t A \in D(T) \subseteq D(S)$ and

$$\omega((U_t - V_t)A) = \int_0^t ds\, \omega(U_s(S - T)V_{t-s}A)$$

for all $\omega \in X_*$. One easily estimates that

$$\|(U_t - V_t)A\| \leq t \sup_{0 \leq s \leq t} \|U_s\| \, \|V_{t-s}\|(a\|A\| + b\|TA\|).$$

But there are constants M, β such that $\|U_t\| \leq M \exp\{\beta t\}$ and $\|V_t\| \leq M \exp\{\beta t\}$, and so one finds

$$\|(U_t - V_t)A\| \leq tM^2 e^{\beta t}(a\|A\| + b\|TA\|).$$

Finally, if $A = (I - \alpha T)^{-1}B$ then

$$\|(U_t - V_t)(I - \alpha T)^{-1}B\| \leq tM^2 e^{\beta t}(a\|(I - \alpha T)^{-1}B\| + b\alpha^{-1}\|(I - (I - \alpha T)^{-1})B\|)$$

Thus estimating the resolvent with Proposition 3.1.6, one finds for $0 < \alpha < \beta^{-1}$

$$\begin{aligned}\|(U_t - V_t)(I - \alpha T)^{-1}\| &\leq tM^2 e^{\beta t}(aM(1 - \alpha\beta)^{-1} + b\alpha^{-1}(1 + M(1 - \alpha\beta)^{-1})) \\ &= O(t).\end{aligned}$$

(2) $\Rightarrow$ (1) This follows immediately because $D(T) = R((i - \alpha T)^{-1})$.

3.2. Algebraic Theory

In this section we principally analyze one-parameter groups of *-automorphisms of C^*-algebras and von Neumann algebras, and use the algebraic structure to extend the results of Section 3.1. In applications to mathematical physics this analysis is useful for the discussion of dynamical development or for the examination of one-parameter symmetry groups. In such contexts, however, it is natural to first study more general forms of transformations than *-automorphisms.

Symmetries of a physical system can be described in two complementary but distinct manners. Either one may view a symmetry as an invariance of the states of the system under a transformation of the observing apparatus or as an invariance of observations under a transformation of the states. If one adopts the position that the observables of a physical system are described by elements of an algebra $\mathfrak{A}$ and physically realizable states are represented by mathematical states ω, over $\mathfrak{A}$, then these two descriptions of a symmetry are related by a duality. A symmetry can be described as a transformation of the algebra $\mathfrak{A}$ with the states $E_{\mathfrak{A}}$ fixed or, by duality, as a transformation of $E_{\mathfrak{A}}$ with $\mathfrak{A}$ fixed. In either case a condition of positivity plays an important role. The positivity of the states over the algebra of observables is related to a probabilistic interpretation of the results of observations, and the conservation of positivity under a symmetry transformation corresponds to a conservation of probability. Similarly, the spectral values of selfadjoint observables correspond to the physically attainable values of these observables, and transformations which send positive elements into positive elements have a certain physical coherence. Thus we begin by examining positive maps of algebras and characterize the special subclasses of *-automorphisms. Subsequently we return to the examination of automorphism groups.

3.2.1. Positive Linear Maps and Jordan Morphisms

In this subsection we examine the various types of positive linear maps φ of C^*-algebras and von Neumann algebras $\mathfrak{A}$ alluded to in the introduction. Particular examples of such maps are *-automorphisms and *-antiautomorphisms. Other examples occur as duals of maps $\varphi^*; \mathfrak{A}^* \mapsto \mathfrak{A}^*$ which have the property that φ^* maps states into states. We will describe in more details these and other characteristic requirements on φ which are natural on physical grounds, and show that they are equivalent. We also show that these requirements imply that φ splits into a morphism and an antimorphism.

This type of general study originated in mathematical physics with Wigner's formulation of symmetries in quantum mechanics. The basic states of a quantum mechanical system are represented by rays of unit vectors in a Hilbert space $\mathfrak{H}$. If $\psi \in \mathfrak{H}$ and $\|\psi\| = 1$ then the corresponding ray $\hat{\psi}$ is defined as the set of vectors of the form $e^{i\alpha}\psi$ with $\alpha \in \mathbb{R}$. The numbers

$$p(\hat{\varphi}, \hat{\psi}) = |(\varphi, \psi)|^2$$

are obviously independent of the choice of representatives φ, ψ of the rays $\hat{\varphi}$, $\hat{\psi}$, and they give the transition probability for the system to pass from the state φ to the state ψ. Wigner formulated a symmetry of the system as a one-to-one map of the rays onto the rays which preserves the transition probabilities $p(\hat{\varphi}, \hat{\psi})$. This notion can also be reexpressed in terms of projectors. If $\mathscr{E}(\mathfrak{H})$ is the set of all rank one projectors on $\mathfrak{H}$ then a Wigner symmetry is a one-to-one map $E \in \mathscr{E}(\mathfrak{H}) \mapsto \alpha(E) \in \mathscr{E}(\mathfrak{H})$ such that

$$\operatorname{Tr}(\alpha(E_\varphi)\alpha(E_\psi)) = \operatorname{Tr}(E_\varphi E_\psi), \tag{$*$}$$

where E_φ and E_ψ denote the projectors with φ and ψ in their respective ranges. The map α may be extended by linearity to all finite-rank operators and then by continuity to the C^*-algebra, $\mathscr{LC}(\mathfrak{H})$, of all compact operators on $\mathfrak{H}$. Note that as α satisfies the invariance property $(*)$ this extension is coherent, i.e., if $A = 0$ then $\alpha(A) = 0$. Moreover, it follows that it maps the density matrices ρ, described in Section 2.6.2, into density matrices. Moreover, if $\rho = \lambda\rho_1 + (1 - \lambda)\rho_2$ with $0 \leq \lambda \leq 1$ then $\alpha(\rho) = \lambda\alpha(\rho_1) + (1 - \lambda)\alpha(\rho_2)$. Thus α defines a map of the states of the C^*-algebra $\mathscr{LC}(\mathfrak{H})$, or the normal states of the von Neumann algebra $\mathscr{L}(\mathfrak{H})$ of all bounded operators, by

$$\omega(A) = \operatorname{Tr}(\rho A) \mapsto (\alpha^*\omega)(A) = \operatorname{Tr}(\alpha(\rho)A),$$

and this map is affine, i.e.,

$$\alpha^*(\lambda\omega_1 + (1 - \lambda)\omega_2) = \lambda\alpha^*\omega_1 + (1 - \lambda)\alpha^*\omega_2.$$

Thus Wigner symmetries fall into the class of maps that we consider. In fact, these symmetries $\hat{\psi} \mapsto \alpha\hat{\psi}$ have a very simple structure because it turns out that

$$\alpha\hat{\psi} = U\hat{\psi},$$

where U is a unitary, or antiunitary, operator which is uniquely determined up to a phase. This characterization results from the general theory, e.g., Theorem 3.2.8, and we discuss it as a particular example later in the subsection.

We begin with the formal definition of the various new forms of maps which will be of interest.

Definition 3.2.1. Let $\mathfrak{A}$, $\mathfrak{B}$ be C^*-algebras and $\varphi; \mathfrak{A} \mapsto \mathfrak{B}$ a linear map. We introduce the following definitions:

(1) φ is a *positive map* if $\varphi(\mathfrak{A}_+) \subseteq \mathfrak{B}_+$;
(2) φ is a *Jordan morphism* if

$$\varphi(A^*) = \varphi(A)^*,$$
$$\varphi(\{A, B\}) = \{\varphi(A), \varphi(B)\}$$

for all $A, B \in \mathfrak{A}$, where $\{A, B\} = AB + BA$;
(3) φ is an *antimorphism* if

$$\varphi(A^*) = \varphi(A)^*$$
$$\varphi(AB) = \varphi(B)\varphi(A)$$

for all $A, B \in \mathfrak{A}$;
(4) φ is an *isometry* if $\|\varphi(A)\| = \|A\|$ for all $A \in \mathfrak{A}$;
(5) φ is an *order isomorphism* if φ^{-1} exists and both φ and φ^{-1} are positive maps;
(6) φ is a *Jordan isomorphism* if φ^{-1} exists and φ is a Jordan morphism (and thus φ^{-1} is a Jordan morphism);
(7) φ is an *antiisomorphism* if φ^{-1} exists and φ is an antimorphism.

We also use some related concepts such as order automorphism, Jordan automorphism, and antiautomorphism, whose definitions should be self-explanatory. Some relations between these concepts are immediate, for example, both morphisms and antimorphisms are Jordan morphisms. If φ is a Jordan morphism and $A \in \mathfrak{A}$ is selfadjoint, then $\varphi(A^2) = \varphi(A)^2$, hence φ is a positive map. Note that there is a close connection between antimorphisms and antilinear morphisms; the antilinear antiautomorphism $A \mapsto A^*$ transforms the former into the latter and conversely.

We next turn to a characterization of Jordan morphisms, showing that they are obtained by "adding" a morphism and an antimorphism.

Proposition 3.2.2. *Let $\mathfrak{A}$ be a C^*-algebra, $\varphi; \mathfrak{A} \mapsto \mathscr{L}(\mathfrak{H})$ a Jordan morphism, and let $\mathfrak{B}$ be the C^*-algebra generated by $\varphi(\mathfrak{A})$. It follows that there exists a projection $E \in \mathfrak{B}' \cap \mathfrak{B}''$ such that*

$$A \mapsto \varphi(A)E$$

is a morphism, and

$$A \mapsto \varphi(A)(\mathbb{1} - E)$$

is an antimorphism.

In particular, if π is an irreducible representation of $\mathfrak{B}$, $\pi \circ \varphi$ is either a morphism or an antimorphism.

As the complete proof of this proposition is somewhat lengthy we only indicate the main lines of the argument.

Observation (1). *If $A, B \in \mathfrak{A}$ then*

$$\varphi(ABA) = \varphi(A)\varphi(B)\varphi(A).$$

This is established by remarking that $\varphi((A + B)^3) = \varphi(A + B)^3$, $\varphi(A^3) = \varphi(A)^3$, etc., and hence one computes that

$$\varphi(ABA + BAB) = \varphi(A)\varphi(B)\varphi(A) + \varphi(B)\varphi(A)\varphi(B).$$

A similar computation using $\varphi((A - B)^3) = \varphi(A - B)^3$ gives

$$\varphi(ABA - BAB) = \varphi(A)\varphi(B)\varphi(A) - \varphi(B)\varphi(A)\varphi(B)$$

and the desired result follows by addition.

Observation (2). *If $A, B, C \in \mathfrak{A}$ then*

$$\varphi(ABC + CBA) = \varphi(A)\varphi(B)\varphi(C) + \varphi(C)\varphi(B)\varphi(A).$$

This follows from the identity

$$ABC + CBA = (A + C)B(A + C) - ABA - CBC$$

and Observation (1).

Observation (3). *If $A, B \in \mathfrak{A}$ then*

$$[\varphi(AB) - \varphi(A)\varphi(B)][\varphi(AB) - \varphi(B)\varphi(A)] = 0.$$

This follows from explicit multiplication and successive applications of Observations (1) and (2):

$$\begin{aligned} &[\varphi(AB) - \varphi(A)\varphi(B)][\varphi(AB) - \varphi(B)\varphi(A)] \\ &= \varphi((AB)^2) + \varphi(ABBA) - \varphi(A)\varphi(B)\varphi(AB) - \varphi(AB)\varphi(B)\varphi(A) \\ &= \varphi((AB)^2 + (AB)(BA)) - \varphi((AB)(AB) + (AB)(BA)) \\ &= 0. \end{aligned}$$

Now using Observation (3) and manipulating with matrix units one can prove Proposition 3.2.2 if $\mathfrak{A}$ is a von Neumann algebra of the form $\mathfrak{A} \otimes M_n$, where M_n are the $n \times n$ matrices and $n \geq 2$. In fact, this argument is of an algebraic–combinatoric nature and one can prove the result for matrices over more general rings than von Neumann algebras. Next if $\mathfrak{A}$ is a general von Neumann algebra, then there exists a sequence $\{E_n\}_{n \geq 1}$ of mutually orthogonal projections in the center $\mathfrak{A} \cap \mathfrak{A}'$ such that $\sum_n E_n = \mathbb{1}$, $\mathfrak{A}E_1$ is abelian, and $\mathfrak{A}E_n = \mathfrak{M}_n \otimes M_n$, where $\mathfrak{M}_n$ is a von Neumann algebra. This establishes the proposition if $\mathfrak{A}$ is a von Neumann algebra. Finally, if $\mathfrak{A}$ is a C^*-algebra one first extends φ to $\mathfrak{A}^{**}$, where $\mathfrak{A}^{**}$ is the von Neumann algebra generated by $(\bigoplus_{\omega \in E_{\mathfrak{A}}} \pi_\omega)(\mathfrak{A})$, and then applies the von Neumann algebra result.

We now come to the first major result of this subsection, a characterization of Jordan automorphisms and their duals.

Theorem 3.2.3. *Let $\mathfrak{A}$, $\mathfrak{B}$ be C^*-algebras with identities and assume that $\mathfrak{B}$ acts nondegenerately on a Hilbert space. Let φ; $\mathfrak{A} \mapsto \mathfrak{B}$ be a linear map such that $\varphi(\mathbb{1}) = \mathbb{1}$ and φ^{-1} exists. The following conditions are equivalent:*

(1) *φ is a Jordan isomorphism;*
(2) *there exists a projection $E \in \mathfrak{B}' \cap \mathfrak{B}''$ such that $A \mapsto \varphi(A)E$ is a morphism and $A \mapsto \varphi(A)(\mathbb{1} - E)$ is an antimorphism;*
(3) (i) *if $U \in \mathfrak{A}$ is unitary, $\varphi(U) \in \mathfrak{B}$ is unitary,*
(ii) *if $A = A^*$ then $\varphi(|A|) = |\varphi(A)|$,*
(iii) *if A is invertible then $\varphi(A)$ is invertible and $\varphi(A)^{-1} = \varphi(A^{-1})$;*
(4) *φ is an isometry;*
(5) *φ is an order isomorphism;*
(6) *the dual φ^*; $\mathfrak{B}^* \mapsto \mathfrak{A}^*$ of φ has the property that*

$$\varphi^*(E_{\mathfrak{B}}) = E_{\mathfrak{A}},$$

where $E_{\mathfrak{A}}$ and $E_{\mathfrak{B}}$ are the sets of states over $\mathfrak{A}$ and $\mathfrak{B}$, respectively.

We will prove this theorem with the aid of a series of propositions, some of which are of interest in their own right. The first proposition is sometimes referred to as the generalized Schwarz inequality because it reduces to the latter in the case of states.

Proposition 3.2.4. *Let $\mathfrak{A}$, $\mathfrak{B}$ be C^*-algebras with identity, φ; $\mathfrak{A} \mapsto \mathfrak{B}$ a positive map such that $\varphi(\mathbb{1}) = \mathbb{1}$, and $A \in \mathfrak{A}$ a normal element, i.e., $AA^* = A^*A$. It follows that*

$$\varphi(A^*A) \geq \varphi(A)^*\varphi(A).$$

Proof. Because A is normal, we may assume that $\mathfrak{A}$ is abelian. Then $\mathfrak{A} = C(X)$, where X is a compact Hausdorff space (Theorem 2.1.11). One may assume that the C^*-algebra $\mathfrak{B}$ acts on a Hilbert space $\mathfrak{H}$.

We first prove that

$$\sum_{ij} (\xi_i, \varphi(A_i^*A_j)\xi_j) \geq 0$$

for any pair of finite sequences $\xi_1, \xi_2, \ldots, \xi_n \in \mathfrak{H}$; $A_1, A_2, \ldots, A_n \in \mathfrak{A}$. By the Riesz representation theorem, there exist Baire measures $d\mu_{\xi_i, \xi_j}$ on X of finite total variation such that

$$(\xi_i, \varphi(A)\xi_j) = \int_X d\mu_{\xi_i, \xi_j}(x)\, A(x)$$

for all $A \in \mathfrak{A} = C(X)$. Let $d\mu = \sum_{ij} d|\mu_{\xi_i, \xi_j}|$. Then μ is a positive, finite Baire measure, and by the Radon–Nikodym theorem there exists μ-measurable functions h_{ξ_i, ξ_j} on X such that

$$d\mu_{\xi_i, \xi_j}(x) = h_{\xi_i, \xi_j}(x)\, d\mu(x).$$

If $\lambda_1, \ldots, \lambda_n \in \mathbb{C}$, we have that

$$\sum_{ij} \bar{\lambda}_i \lambda_j\, d\mu_{\xi_i, \xi_j} = d\mu_{\xi, \xi},$$

where $\xi = \sum_i \lambda_i \xi_i$. The measure $d\mu_{\xi,\xi}$ is positive, and

$$d\mu_{\xi,\xi}(x) = \left(\sum_{ij} \bar{\lambda}_i \lambda_j h_{\xi_i,\xi_j}(x)\right) d\mu(x).$$

It follows that

$$\sum_{ij} \bar{\lambda}_i \lambda_j h_{\xi_i,\xi_j}(x) \geq 0$$

for all $\lambda_1, \ldots, \lambda_n \in \mathbb{C}$, and μ-almost all $x \in X$. But this implies that

$$\sum_{ij} (\xi_i, \varphi(A_i^* A_j)\xi_j) = \int d\mu(x) \left\{\sum_{ij} \overline{A_i(x)} A_j(x) h_{\xi_i,\xi_j}(x)\right\} \geq 0$$

for all $A_1, \ldots, A_n \in \mathfrak{A}$.

Now, equip the vector space tensor product $\mathfrak{A} \otimes \mathfrak{H}$ of $\mathfrak{A}$ and $\mathfrak{H}$ with the sesquilinear form

$$\left(\sum_i A_i \otimes \xi_i, \sum_j B_j \otimes \eta_j\right) = \sum_{ij} (\xi_i, \varphi(A_i^* B_j)\eta_j).$$

This form is, in fact, well defined, and is a positive semidefinite inner product by the inequality we just have proved. Let $\mathcal{N}$ denote the set of null vectors with respect to the form, and let $\mathfrak{K}$ denote the closure of $\mathfrak{A} \otimes \mathfrak{H}/\mathcal{N}$. Then $\mathfrak{K}$ is a Hilbert space, and V defined by $V\psi = \mathbb{1} \otimes \psi + \mathcal{N}$ is a linear isometry of $\mathfrak{H}$ into $\mathfrak{K}$. Let π be the representation induced on $\mathfrak{K}$ by the representation π' of $\mathfrak{A}$ on $\mathfrak{A} \otimes \mathfrak{H}$ defined by

$$\pi'(A)\left(\sum_i B_i \otimes \psi_i\right) = \sum_i AB_i \otimes \psi_i.$$

An easy calculation shows that π is a representation of $\mathfrak{A}$ and that

$$\varphi(A) = V^* \pi(A) V$$

for all $A \in \mathfrak{A}$. (Note that this is a generalization of the GNS construction.) Now, for $A \in \mathfrak{A}$ and $\psi \in \mathfrak{H}$ we have

$$\begin{aligned}
(\psi, \varphi(A^*A)\psi) &= (\psi, V^*\pi(A)^*\pi(A)V\psi) \\
&= \|\pi(A)V\psi\|^2 \\
&\geq \|V^*\pi(A)V\psi\|^2 \\
&= (\psi, \varphi(A)^*\varphi(A)\psi).
\end{aligned}$$

Hence

$$\varphi(A^*A) \geq \varphi(A)^*\varphi(A).$$

Lemma 2.2.14 and its proof imply that if $\mathfrak{A}$ is a C^*-algebra with identity then any element $A \in \mathfrak{A}$ with $\|A\| \leq 1/2$ is a convex combination of unitary elements in $\mathfrak{A}$. More generally, it can be shown that any $A \in \mathfrak{A}$ with $\|A\| < 1$ is a convex combination of unitary elements. However, we need only the following weaker result.

Proposition 3.2.5. *If $\mathfrak{A}$ is a C^*-algebra with identity then the unit ball in $\mathfrak{A}$ is the closed convex hull of the unitary elements in $\mathfrak{A}$.*

PROOF. If $A \in \mathfrak{A}$ and $\|A\| < 1$ then $\mathbb{1} - AA^*$ is positive and invertible, so the element

$$f(A, \lambda) = (\mathbb{1} - AA^*)^{-1/2}(\mathbb{1} + \lambda A)$$

exists in $\mathfrak{A}$ and is invertible for all $\lambda \in \mathbb{C}$ with $|\lambda| = 1$. We have

$$\begin{aligned} A^*(\mathbb{1} - AA^*)^{-1} &= A^*\left(\sum_{n\geq 0}(AA^*)^n\right) \\ &= \left(\sum_{n\geq 0}(A^*A)^n\right)A^* = (\mathbb{1} - A^*A)^{-1}A^*, \end{aligned}$$

whence

$$\begin{aligned} f(A, \lambda)^*f(A, \lambda) + \mathbb{1} &= (\mathbb{1} + \bar{\lambda}A^*)(\mathbb{1} - AA^*)^{-1}(\mathbb{1} + \lambda A) + \mathbb{1} \\ &= (\mathbb{1} - AA^*)^{-1} + (\mathbb{1} - A^*A)^{-1}\bar{\lambda}A^* + (\mathbb{1} - AA^*)^{-1}\lambda A \\ &\quad + (\mathbb{1} - A^*A)^{-1}. \end{aligned}$$

This expression is unchanged under the transformation $A \mapsto A^*$, $\lambda \mapsto \bar{\lambda}$, and we conclude that

$$f(A, \lambda)^*f(A, \lambda) = f(A^*, \bar{\lambda})^*f(A^*, \bar{\lambda}).$$

It follows that the element

$$U_\lambda = f(A, \lambda)f(A^*, \bar{\lambda})^{-1}$$

is unitary when $|\lambda| = 1$.

The function

$$U(\lambda) = (\mathbb{1} - AA^*)^{-1/2}(\lambda\mathbb{1} + A)(\mathbb{1} + \lambda A^*)^{-1}(\mathbb{1} - A^*A)^{1/2}$$

is analytic in a neighborhood of the closed unit disc and $U(\lambda) = \lambda U_{\bar{\lambda}}$ when $|\lambda| = 1$. Moreover,

$$\begin{aligned} U(0) &= (\mathbb{1} - AA^*)^{-1/2}A(\mathbb{1} - A^*A)^{1/2} \\ &= (\mathbb{1} - AA^*)^{-1/2}(\mathbb{1} - AA^*)^{1/2}A = A. \end{aligned}$$

By Cauchy's integral formula,

$$A = (2\pi)^{-1}\int_0^{2\pi} U(e^{it})\,dt,$$

where the integral exists as a Riemann integral. It follows that the open unit ball in $\mathfrak{A}$ is contained in the closed convex hull of the unitary elements in $\mathfrak{A}$, which establishes the proposition.

Proposition 3.2.5 has the following corollary, which we have already proved in the case of states (Proposition 2.3.11).

Corollary 3.2.6. *Let φ be a linear map between two C^*-algebras $\mathfrak{A}$ and $\mathfrak{B}$ with identity and assume $\varphi(\mathbb{1}) = \mathbb{1}$. It follows that φ is positive if, and only if, $\|\varphi\| = 1$.*

PROOF. If $\|\varphi\| = 1$ and ω is a state of $\mathfrak{B}$, then $\omega \circ \varphi(\mathbb{1}) = 1$ and $\|\omega \circ \varphi\| \leq \|\omega\| \, \|\varphi\| = 1$, hence $\omega \circ \varphi$ is a state on $\mathfrak{A}$ by Proposition 2.3.11. It follows that φ is positive.

Conversely, if φ is positive and if $A = A^* \in \mathfrak{A}$ we have $-\|A\|\mathbb{1} \leq A \leq \|A\|\mathbb{1}$, and $-\|A\|\mathbb{1} \leq \varphi(A) \leq \|A\|\mathbb{1}$. Hence $\|\varphi(A)\| \leq \|A\|$ if $A = A^*$ and $\|\varphi(A)\| \leq 2\|A\|$ for general A by decomposition into selfadjoint elements. This proves that φ is continuous. Now if $U \in \mathfrak{A}$ is unitary it follows from Proposition 3.2.4 that

$$\begin{aligned} \|\varphi(U)\|^2 &= \|\varphi(U)^*\varphi(U)\| \\ &\leq \|\varphi(U^*U)\| = \|\varphi(\mathbb{1})\| = 1. \end{aligned}$$

The continuity of φ and Proposition 3.2.5 then imply that $\|\varphi\| = 1$.

PROOF OF THEOREM 3.2.3. (1) $\Rightarrow$ (2) follows from Proposition 3.2.2. and (2) $\Rightarrow$ (3) is trivial. Next note that (3ii) implies φ and φ^{-1} are positive. Hence it follows from Corollary 3.2.6 that $\|\varphi\| \leq 1$, $\|\varphi^{-1}\| \leq 1$, and φ is an isometry. This establishes that (3) $\Rightarrow$ (4). The equivalence (4) $\Leftrightarrow$ (5) follows from Corollary 3.2.6, while (5) $\Leftrightarrow$ (6) is trivial. Finally, we prove (5) $\Rightarrow$ (1). If $A = A^* \in \mathfrak{A}$, Proposition 3.2.4 applied to φ and φ^{-1} gives

$$\begin{aligned} \varphi(A^2) &\geq \varphi(A)^2 \\ &= \varphi(\varphi^{-1}(\varphi(A)^2)) \\ &\geq \varphi(\varphi^{-1}\varphi(A)^2) = \varphi(A^2). \end{aligned}$$

Hence $\varphi(A^2) = \varphi(A)^2$ if $A = A^*$. For $A = A^* \in \mathfrak{A}$, $B = B^* \in \mathfrak{A}$ we use

$$(A + B)^2 - A^2 - B^2 = AB + BA = \{A, B\}$$

to conclude that

$$\varphi(\{A, B\}) = \{\varphi(A), \varphi(B)\}.$$

This relation for general A and B then follows by linearity.

Theorem 3.2.3 has various interesting consequences for affine maps of states and for one-parameter groups of automorphisms. The rest of this subsection is devoted to their deduction and illustration. Although the deduction of these implications is basically straightforward it does need a certain amount of extra technical machinery. First it is convenient for the analysis of state maps to extend these maps to the whole dual, or predual. For this purpose we need some information concerning a decomposition of linear functionals on C^*-algebras and von Neumann algebras.

We begin by defining a functional η on a C^*-algebra $\mathfrak{A}$ to be hermitian if $\eta(A^*) = \overline{\eta(A)}$. It is clear that states are hermitian and that hermitian functionals are uniquely determined by their restriction to the real space $\mathfrak{A}_{sa}$ of selfadjoint elements of $\mathfrak{A}$.

Proposition 3.2.7 (Jordan decomposition). *If $\mathfrak{A}$ is a C^*-algebra and $\eta \in \mathfrak{A}^*$ then η has a unique decomposition $\eta = \eta_1 + i\eta_2$ into two hermitian functionals, where*

$$\eta_1(A) = (\eta(A) + \overline{\eta(A^*)})/2 \quad \text{and} \quad \eta_2(A) = (\eta(A) - \overline{\eta(A^*)})/2i.$$

If $\mathfrak{A} = \mathfrak{M}$ is a von Neumann algebra, and $\eta \in \mathfrak{M}_*$ then $\eta_1, \eta_2 \in \mathfrak{M}_*$.

If $\mathfrak{M}$ is a von Neumann algebra and $\eta \in \mathfrak{M}_*$ is hermitian then there exists a unique pair of elements $\eta_1, \eta_2 \in \mathfrak{M}_{*+}$ such that

$$\eta = \eta_1 - \eta_2,$$
$$\|\eta\| = \|\eta_1\| + \|\eta_2\|.$$

If $\mathfrak{A}$ is a C^*-algebra, set $\mathfrak{A}'' = (\bigoplus_{\omega \in E_{\mathfrak{A}}} \pi_\omega)(\mathfrak{A})''$ then any element $\eta \in \mathfrak{A}^*$ extends uniquely to a σ-weakly continuous linear functional on the von Neumann algebra $\mathfrak{A}''$, and $\mathfrak{A}''$ identifies with the bidual $\mathfrak{A}^{**}$ of $\mathfrak{A}$. If $\eta \in \mathfrak{A}^*$ is hermitian then there exists a unique pair of elements $\eta_1, \eta_2 \in \mathfrak{A}_+^*$ such that

$$\eta = \eta_1 - \eta_2,$$
$$\|\eta\| = \|\eta_1\| + \|\eta_2\|.$$

PROOF. The existence and uniqueness of the decomposition into hermitian functionals is straightforward. In the case that $\eta \in \mathfrak{M}_*$, we have that $\eta(A) = \sum_i (\xi_i, A\psi_i)$. Thus $\overline{\eta(A^*)} = \sum_i (\psi_i, A\xi_i)$. Consequently, $\eta_1, \eta_2 \in \mathfrak{M}_*$.

Now, assume that $\eta \in \mathfrak{M}_*$ is hermitian, and $\|\eta\| = 1$. Since $\mathfrak{M}_1$ is σ-weakly compact there exists an $A \in \mathfrak{M}_1$ such that $\eta(A) = 1 = \eta(A^*)$. Thus, replacing A by $(A + A^*)/2$ we may assume $A = A^*$.

Let $W^*(A)$ be the abelian von Neumann algebra generated by A and $\mathbb{1}$. Then $W^*(A)$ is, by the Gelfand representation, an algebra of the form $C(X)$, where X is a compact Hausdorff space. The set of $B \in C(X)_{\text{sa}\,1}$ such that $\eta(B) = 1$ is nonempty and σ-weakly closed, thus σ-weakly compact. By the Krein–Milman theorem this set has an extremal point B. Using $\|\eta\| = 1$ it is easy to see that B is also extremal in $C(X)_{\text{sa}\,1}$. Therefore the representative of B, in $C(X)$, can only assume values in the extremal points of the interval $[-1, 1]$. Thus $B = P_1 - P_2$, where P_1, P_2 are projections, $P_1P_2 = 0$ and $P_1 + P_2 = \mathbb{1}$.

Next, define

$$\eta_1(A) = \eta(P_1A) \qquad \eta_2(A) = -\eta(P_2A)$$

for all $A \in \mathfrak{M}$. Then η_1, η_2 are σ-weakly continuous and $\eta = \eta_1 - \eta_2$. Furthermore,

$$(\eta_1 + \eta_2)(\mathbb{1}) = \eta(P_1 - P_2) = 1$$

and

$$|(\eta_1 + \eta_2)(A)| = |\eta((P_1 - P_2)A)| \le \|\eta\|\,\|P_1 - P_2\|\,\|A\|.$$

Hence $\|\eta_1 + \eta_2\| = 1$, and $\eta_1 + \eta_2$ is a state by Proposition 2.3.11.

Now the norm of η_1 on $P_1\mathfrak{M}P_1$ is $\eta(P_1)$ because if $\eta(P_1) < \eta(A)$, where $A \in (P_1\mathfrak{M}P_1)_1$, then $\|A - P_2\| \le 1$ and $\eta(A - P_2) = \eta(A) - \eta(P_2) > \eta(P_1 - P_2) = 1$, which is a contradiction. Hence by Proposition 2.3.11, $\eta_1 \ge 0$ on $P_1\mathfrak{M}P_1$. Since $\eta_1 = (\eta + (\eta_1 + \eta_2))/2$, η_1 is hermitian. Hence

$$\eta(AP_1) = \overline{\eta(P_1A^*)} = \overline{\eta_1(A^*)} = \eta_1(A) = \eta(P_1A)$$

and

$$\eta_1(A) = \eta(P_1A) = \eta(P_1{}^2A) = \eta(P_1AP_1) = \eta_1(P_1AP_1).$$

It follows that η_1 is positive on all of $\mathfrak{M}$. Similarly, η_2 is positive and $\|\eta_2\| = -\eta(P_2) = \eta_2(P_2)$. Hence $\|\eta_1\| + \|\eta_2\| = \eta(P_1 - P_2) = 1$.

For any $\omega \in \mathfrak{M}_{*+}$ let $S(\omega)$ be the smallest projection in $\mathfrak{M}$ such that $\omega(S(\omega)) = \|\omega\|$ (equivalently, $S(\omega)$ is the largest projection in $\mathfrak{M}$ such that $\omega(\mathbb{1} - S(\omega)) = 0$; $S(\omega)$ is called the *support of* ω). To show uniqueness of the decomposition of η, set

$$\eta = \eta_1' - \eta_2' = \eta_1 - \eta_2, \qquad \|\eta_1'\| + \|\eta_2'\| = 1, \qquad \eta_1', \eta_2' \in \mathfrak{M}_{*+}.$$

Then

$$\eta_1(\mathbb{1}) + \eta_2(\mathbb{1}) = 1 = \eta_1'(\mathbb{1}) + \eta_2'(\mathbb{1}),$$
$$\eta_1(\mathbb{1}) - \eta_2(\mathbb{1}) = \eta(\mathbb{1}) = \eta_1'(\mathbb{1}) - \eta_2'(\mathbb{1});$$

hence

$$\eta_i(\mathbb{1}) = \eta_i'(\mathbb{1}), \qquad i = 1, 2.$$

Since $S(\eta_1) \leq P_1$ one has

$$\begin{aligned}\|\eta_1'\| = \eta_1'(\mathbb{1}) = \eta_1(\mathbb{1}) = \|\eta_1\| &= \eta_1(S(\eta_1)) \\ &= \eta(S(\eta_1)) = \eta_1'(S(\eta_1)) - \eta_2'(S(\eta_1)).\end{aligned}$$

Thus $\eta_2'(S(\eta_1)) = 0$ and $S(\eta_1') \leq S(\eta_1)$. Hence

$$\begin{aligned}\eta_1(A) &= \eta(S(\eta_1)A) = \eta_1'(S(\eta_1)A) - \eta_2'(S(\eta_1)A) \\ &= \eta_1'(S(\eta_1)A) = \eta_1'((S(\eta_1) - S(\eta_1'))A) + \eta_1'(S(\eta_1')A). \\ &= \eta_1'(A).\end{aligned}$$

Thus $\eta_1 = \eta_1'$, and so $\eta_2 = \eta_2'$.

We now turn to the C^*-part of the proposition. By Theorem 2.3.15, the set $B_{\mathfrak{A}}$ of positive linear functionals over $\mathfrak{A}$ with norm less than or equal to one is weakly* compact. Hence the set $\mathfrak{K}$ of convex combinations of elements in $B_{\mathfrak{A}}$ and $-B_{\mathfrak{A}}$ is weakly* compact. If $A = A^* \in \mathfrak{A}$ it follows by a straightforward extension of Lemma 2.3.23 that

$$\begin{aligned}\|A\| &= \sup\{|\omega(A)|; \omega \in E_{\mathfrak{A}}\}, \\ &= \sup\{\eta(A); \eta \in \mathfrak{K}\}.\end{aligned}$$

We will show that any hermitian functional η such that $\|\eta\| \leq 1$ lies in $\mathfrak{K}$. Assume the contrary and then, by the Hahn–Banach theorem applied to the real space $\mathfrak{A}_{sa}$, there exists an $A \in \mathfrak{A}_{sa}$ and an $\alpha \in \mathbb{R}$ such that $\eta(A) > \alpha$ and $\varphi(A) \leq \alpha$ for all $\varphi \in \mathfrak{K}$. Since $-\mathfrak{K} = \mathfrak{K}$ this implies that $|\varphi(A)| \leq \alpha$ for all $\varphi \in \mathfrak{K}$. Hence $\|A\| \leq \alpha$. This contradicts $\eta(A) > \alpha$, and thus $\eta \in \mathfrak{K}$. We have thus shown that any hermitian functional η on $\mathfrak{A}$ is of the form $\eta = \eta_1 - \eta_2$, where $\eta_i \in \mathfrak{A}^*_+$. Now, clearly, η_1 and η_2 have σ-weakly continuous extensions $\hat{\eta}_1$ and $\hat{\eta}_2$ to $\mathfrak{A}'' = (\bigoplus_{\omega \in E_{\mathfrak{A}}} \pi_\omega)(\mathfrak{A})''$. Thus η has a σ-weakly continuous extension $\hat{\eta}$. But this extension is unique because $\mathfrak{A}$ is σ-weakly dense in $\mathfrak{A}''$, and $\|\hat{\eta}\| = \|\eta\|$ by the Kaplansky density theorem (Theorem 2.4.16). The C^*-version of the proposition now follows from the von Neumann version.

Now we return to the examination of the implications of Theorem 3.2.3. We first examine affine maps of states and identify these as duals of Jordan automorphisms. This duality reflects the two possible methods of describing symmetry transformations in physical applications and the following result proves the equivalence of these methods.

Theorem 3.2.8. *Let $\mathfrak{M}$ be a von Neumann algebra and $N_{\mathfrak{M}} = E_{\mathfrak{M}} \cap \mathfrak{M}_*$ the normal states on $\mathfrak{M}$. Further, let φ_* be an affine, invertible map from $N_{\mathfrak{M}}$ onto $N_{\mathfrak{M}}$, i.e.,*

$$\varphi_*(\lambda\omega_1 + (1 - \lambda)\omega_2) = \lambda\varphi_*(\omega_1) + (1 - \lambda)\varphi_*(\omega_2)$$

for all $\lambda \in [0, 1]$, $\omega_1, \omega_2 \in N_{\mathfrak{M}}$. It follows that there exists a unique Jordan automorphism φ of $\mathfrak{M}$ such that

$$(\varphi_*\omega)(A) = \omega(\varphi(A))$$

for all $\omega \in N_{\mathfrak{M}}$, $A \in \mathfrak{M}$.

PROOF. φ_* extends uniquely to an invertible map of $\mathfrak{M}_{*+}$ onto $\mathfrak{M}_{*+}$ by $\varphi_*(\lambda\omega) = \lambda\varphi_*(\omega)$, $\lambda \in \mathbb{R}_+$. The extended map satisfies $\varphi_*(\lambda_1\omega_1 + \lambda_2\omega_2) = \lambda_1\varphi_*(\omega_1) + \lambda_2\varphi_*(\omega_2)$. Using the Jordan decomposition of Proposition 3.2.7, φ_* and φ_*^{-1} extend uniquely by linearity to invertible linear maps of $\mathfrak{M}_*$ onto $\mathfrak{M}_*$, and the extended maps are positive and of norm less than two. Hence $\varphi = \varphi_*{}^*$ exists as an invertible map on $\mathfrak{M}$, and φ and φ^{-1} are positive.

If $\omega \in N_{\mathfrak{M}}$ we have $\omega(\varphi(\mathbb{1})) = (\varphi_*\omega)(\mathbb{1}) = 1 = \omega(\mathbb{1})$. Thus $\varphi(\mathbb{1}) = \mathbb{1}$ and φ is a Jordan automorphism by Theorem 3.2.3.

Note that the result of Theorem 3.2.8 can be restated in a C^*-algebra version. Let $\mathfrak{A}$ be a C^*-algebra, π a representation of $\mathfrak{A}$ and N_π the π-normal states (see Definition 2.4.25). It follows that every affine, invertible map φ_* from N_π onto N_π defines by duality a Jordan automorphism φ of the von Neumann algebra generated by π. In particular, if $\pi = \bigoplus_{\omega \in E_{\mathfrak{A}}} \pi_\omega$ we have $N_\pi = E_{\mathfrak{A}}$ so every affine, invertible map φ_* of $E_{\mathfrak{A}}$ defines a Jordan automorphism of $\mathfrak{A}'' = (\bigoplus_{\omega \in E_{\mathfrak{A}}} \pi_\omega)(\mathfrak{A})''$. Our next aim is to demonstrate that if one places extra continuity restrictions on φ_* then the dual Jordan automorphism φ actually gives a Jordan automorphism of $\pi(\mathfrak{A})$. Thus if π is faithful φ corresponds to a Jordan automorphism of the abstract C^*-algebra $\mathfrak{A}$. The additional continuity conditions require that φ_* map pairs of neighboring states into pairs of neighboring states. The notion of neighborhood, or closeness, is measured by the weakly* uniform structure. In a physical interpretation, where the values taken by the states represent the results of physical measurements, the continuity requirements demand that the symmetry transformation does not introduce radical differences between similar systems.

In the C^*-version of Theorem 3.2.8 that we just mentioned the initial state map is only defined on the subset formed by the π-normal states. In many applications it is also natural to consider special subsets of states, e.g., locally normal states over quasi-local algebras, and this motivated our restriction to $\sigma(X, F)$-continuous semigroups, etc., in Section 3.1. If the algebraic description is not to be partially redundant it is, however, necessary that the subsets of states should be determining in a certain sense, for the algebra. The following notion of a full set of states is natural in this context.

Definition 3.2.9. A *full family of states S, over a C^*-algebra $\mathfrak{A}$*, is defined to be a convex subset of the states $E_{\mathfrak{A}}$, over $\mathfrak{A}$, which has the property that if $\omega(A) \geq 0$ for all $\omega \in S$ then $A \geq 0$.

Note, in particular, that the normal states over a von Neumann algebra is a full set of states. Thus if π is a faithful representation of a C^*-algebra $\mathfrak{A}$ then the π-normal states also form a full set.

The C^*-generalization of Theorem 3.2.8 will be expressed in terms of mappings of full sets of states. These sets have a simple characterization, which is slightly complicated to deduce, as weakly* dense subsets of the set of all states. As a preliminary we first derive this useful fact:

Proposition 3.2.10. *Let S be a convex subset of the states over a C^*-algebra $\mathfrak{A}$. The following conditions are equivalent:*

(1) *S is full;*
(2) *S is weakly* dense.*

PROOF. (1) ⇒ (2) First we argue that if $A \geq 0$ then

$$\sup_{\omega \in S} \omega(A) = \|A\|.$$

Assume the converse and consider the states $\omega \in S$ as probability measures $d\mu_\omega$ on the spectrum $\sigma(A)$ of the abelian algebra $C_0(\sigma(A))$ generated by A (see Theorem 2.1.11B). If $\omega(A) \leq \lambda < \|A\|$ for all $\omega \in S$ then the representing measures must have a weight inferior to $(\|A\| - \lambda)/(\|A\| - \lambda/2) < 1$ on $[\|A\| - \lambda/2, \|A\|] \cap \sigma(A)$. Thus there exists a real function $f \in C_0(\sigma(A))$ such that $f(t) < 0$ for $t \geq \|A\| - \lambda/2$ but

$$\omega(f(A)) = \int d\mu_\omega(t)\, f(t) \geq 0$$

for all $\omega \in S$. But $f(A)$ is not positive and this contradicts the fullness of S.

Secondly, we claim that if $A = A^*$, $\lambda \in \sigma(A)$, and $\varepsilon > 0$ then there exists an $\omega \in S$ such that $|\omega(A) - \lambda| < \varepsilon$. If $\lambda \neq 0$ we choose a positive function $f \in C_0(\sigma(A))$ with $\|f\|_\infty = 1$ and $f(t) = 0$ for $|t - \lambda| > \varepsilon'$, where $\varepsilon'(1 + 2\|A\|) = \varepsilon$. By the previous argument one may find an $\omega \in S$ such that $\omega(f(A)) \geq 1 - \varepsilon'$. Thus the representing measure must have weight superior to $1 - \varepsilon$ on $[\lambda - \varepsilon', \lambda + \varepsilon']$. Therefore

$$\begin{aligned} |\omega(A) - \lambda| &= \left| \int d\mu_\omega(t)\, (t - \lambda) \right| \\ &\leq \left| \int_{|t - \lambda| \leq \varepsilon'} d\mu_\omega(t)\, (t - \lambda) \right| + \left| \int_{|t - \lambda| > \varepsilon'} d\mu_\omega(t)\, (t - \lambda) \right| \\ &= \varepsilon'(1 + 2\|A\|) = \varepsilon. \end{aligned}$$

Now if $\mathfrak{A}$ contains an identity then one may replace $C_0(\sigma(A))$ by $C(\sigma(A))$ and the argument also works for $\lambda = 0$. Furthermore, if $\mathbb{1} \notin \mathfrak{A}$ but zero is not an isolated point in $\sigma(A)$ then the claim follows by approximating zero with $\lambda \in \sigma(A)\backslash\{0\}$. Finally, if $\mathbb{1} \notin \mathfrak{A}$ and zero is an isolated point in $\sigma(A)$ we choose $f \in C_0(\sigma(A))$ such that $f(t) = 1$

for $t \in \sigma(A)\backslash\{0\}$ and set $P = f(A)$. Thus $P \in \mathfrak{A}$ is a projector and $P \neq \mathbb{1}$. Consequently there is a $B \in \mathfrak{A}$ such that $BP - B \neq 0$ and hence $C = (BP - B)^*(BP - B)/\|BP - B\|^2$ is positive, $\|C\| = 1$, and $CP = 0 = PC$. But for $\varepsilon > 0$ the first part of the proof established the existence of an $\omega \in S$ such that $\omega(C) \geq 1 - \varepsilon^2$ and if $\tilde{\omega}$ is the extension of ω to $\tilde{\mathfrak{A}} = \mathbb{C}\mathbb{1} + \mathfrak{A}$ one then has

$$\begin{aligned} |\omega(P)|^2 &= |\tilde{\omega}(P(\mathbb{1} - C))|^2 \\ &\leq \omega(P)\tilde{\omega}((\mathbb{1} - C)^2) \\ &\leq \|\mathbb{1} - C\|\tilde{\omega}(\mathbb{1} - C) < \varepsilon^2. \end{aligned}$$

Finally, P was chosen such that

$$-\|A\|P \leq A \leq \|A\|P$$

and hence

$$|\omega(A)| \leq \varepsilon\|A\|.$$

We conclude the proof by assuming that the full set of states S is not weakly* dense and then deriving contradiction.

It $\omega \in E_{\mathfrak{A}}$ is not in the weak* closure of the convex set S then the Hahn–Banach theorem, applied to the pair $\mathfrak{A}^*_{\text{sa}}$ and $\mathfrak{A}_{\text{sa}}$ implies the existence of an $A = A^* \in \mathfrak{A}$ such that $\omega(A) > 1$ and $\omega'(A) \leq 1$ for all $\omega' \in S$. Set $\varepsilon = \omega(A) - 1 > 0$ and let $A = A_+ - A_-$ be the decomposition of A into positive, and negative, parts A_+ and A_-. Let E_+ be the range projection of A_+ in

$$\mathfrak{A}'' = \left(\bigoplus_{\omega \in E_{\mathfrak{A}}} \pi_\omega\right)(\mathfrak{A})''$$

and let $E_- = \mathbb{1} - E_+$. It follows that $\omega(E_+) + \omega(E_-) = 1$ and hence

$$0 \leq \omega(A_+) = \omega(E_+ A_+ E_+) \leq \|A_+\|\omega(E_+).$$

Similarly,

$$0 \leq \omega(A_-) \leq \|A_-\|\omega(E_-).$$

But the sets $\{\omega'(A_\pm);\, \omega' \in S\}$ are dense in the convex closures of $\sigma(A_\pm)$, respectively, by the preceding argument. Moreover, $\omega(A_\pm)/\|A_\pm\|$ lie in the respective convex closures because of the above estimates. Hence there exist states $\omega_\pm \in S$ such that

$$|\omega(E_\pm)\omega_\pm(A_\pm) - \omega(A_\pm)| < \frac{\varepsilon}{2}.$$

Now set $\omega' = \omega(E_+)\omega_+ + \omega(E_-)\omega_-$. It follows that $\omega' \in S$ and

$$|\omega'(A) - \omega(A)| < \varepsilon.$$

But since $\omega(A) = 1 + \varepsilon$ and $\omega'(A) \leq 1$ this is a contradiction and hence S must be weakly* dense.

(2) $\Rightarrow$ (1) If S is weakly* dense and $A = A^*$ is not positive then there is certainly an $\omega \in S$ such that $\omega(A) < 0$. Thus S must be full.

After this rather lengthy preliminary examination of subsets of states we return to the examination of affine state maps. The principal C^*-algebraic result is the following extension of Theorem 3.2.8.

Theorem 3.2.11. *Let S be a full set of states over a C^*-algebra $\mathfrak{A}$ and φ_* an affine invertible map from S onto S. Further assume that for each $A \in \mathfrak{A}$ there exists a $\sigma(\mathfrak{A}, \mathfrak{A}^*)$-compact subset K_A of $\mathfrak{A}$ such that*

$$\begin{aligned} |(\varphi_*(\omega_1) - \varphi_*(\omega_2))(A)| &\leq \sup_{B \in K_A} |(\omega_1 - \omega_2)(B)| \\ |(\varphi_*^{-1}(\omega_1) - \varphi_*^{-1}(\omega_2))(A)| &\leq \sup_{B \in K_A} |(\omega_1 - \omega_2)(B)| \end{aligned} \qquad (*)$$

for all pairs $\omega_1, \omega_2 \in S$. It follows that there exists a unique Jordan automorphism φ of $\mathfrak{A}$ such that

$$(\varphi_* \omega)(A) = \omega(\varphi(A))$$

for all $\omega \in S$ and $A \in \mathfrak{A}$.

Proof. First note that the convex set S is weakly* dense in $E_{\mathfrak{A}}$ and hence it is dense in the Mackey topology $\tau(\mathfrak{A}^*, \mathfrak{A})$ restricted to $E_{\mathfrak{A}}$. Thus φ_* and φ_*^{-1}, extend uniquely by continuity to affine maps of $E_{\mathfrak{A}}$ such that the continuity estimates $(*)$ extend to all pairs $\omega_1, \omega_2 \in E_{\mathfrak{A}}$. Furthermore, one may use the Jordan decomposition of Proposition 3.2.7 to extend φ_* to a linear bounded map of $\mathfrak{A}^*$.

Next assume that $A = A^* \in \mathfrak{A}$. If $\omega_1, \omega_2 \in E_{\mathfrak{A}}$ then the difference $\omega_1 - \omega_2$ is hermitian and one has the estimate

$$\begin{aligned} |(\omega_1 - \omega_2)(B)| &\leq \left|(\omega_1 - \omega_2)\left(\frac{(B + B^*)}{2}\right)\right| \\ &\quad + \left|(\omega_1 - \omega_2)\left(\frac{(B - B^*)}{2i}\right)\right| \\ &\leq 2 \sup\left\{(\omega_1 - \omega_2)\left(\pm \frac{(B + B^*)}{2}\right), (\omega_1 - \omega_2)\left(\pm \frac{(B - B^*)}{2i}\right)\right\} \end{aligned}$$

for any B in the compact K_A. By the successive replacements of K_A by $K_A + K_A{}^*$, then by the selfadjoint part of K_A, and finally by $K_A \cup (-K_A)$, we may assume that K_A is a compact, balanced, subset of $\mathfrak{A}_{sa}$ and

$$\pm(\varphi_*(\omega_1) - \varphi_*(\omega_2))(A) \leq \sup_{B \in K_A} (\omega_1 - \omega_2)(B).$$

Now let $\eta \in \mathfrak{A}^*$ be any hermitian functional with $\eta(\mathbb{1}) = 0$. If $\eta = \eta_1 - \eta_2$ is the Jordan decomposition of η then

$$\|\eta_1\| = \eta_1(\mathbb{1}) = \eta_2(\mathbb{1}) = \|\eta_2\|.$$

Applying the above inequality to $(\eta_1 - \eta_2)/\|\eta_1\|$ one obtains

$$\pm(\varphi_* \eta)(A) \leq \sup_{B \in K_A} \eta(B).$$

Now if ω_0 is a fixed state on $\mathfrak{A}$ then any hermitian functional η, on $\mathfrak{A}$, has a unique decomposition

$$\eta = \lambda\omega_0 + \eta',$$

where $\lambda = \eta(\mathbb{1}) \in \mathbb{R}$ and η' is a hermitian functional such that $\eta'(\mathbb{1}) = 0$. We take ω_0 to be a weak* limit of a subnet of the sequence

$$\omega_n = \frac{1}{2n+1} \sum_{k=-n}^{n} \varphi_*{}^k \omega,$$

where $\omega \in E_{\mathfrak{A}}$. One has

$$\varphi_* \omega_0 = \omega_0 .$$

Now, for $A = A^*$ given, we may choose K_A such that $A \in K_A$. Then

$$\begin{aligned}(\varphi_* \eta)(A) &= \lambda(\varphi_* \omega_0)(A) + (\varphi_* \eta')(A) \\ &\leq \lambda \omega_0(A) + \sup_{B \in K_A} \eta'(B) \\ &\leq \sup_{B \in K_A} \lambda \omega_0(B) + \sup_{B \in K_A} \eta'(B).\end{aligned}$$

Since K_A is compact there exist operators $B', B'' \in K_A$ such that

$$\sup_{B \in K_A} \lambda \omega_0(B) = \lambda \omega_0(B'),$$

and

$$\sup_{B \in K_A} \eta'(B) = \eta'(B'').$$

Now put

$$C' = \omega_0(B')\mathbb{1}, \qquad C'' = B'' - \omega_0(B'')\mathbb{1}.$$

If $M = \sup_{B \in K_A} \|B\|$, then

$$C' + C'' \in K'_A \equiv K_A + [-2M, 2M]\mathbb{1},$$

and K'_A is still compact, and depends only on K_A, not on η. Furthermore

$$\begin{aligned}\lambda \omega_0(C') &= \lambda \omega_0(B'), \\ \eta'(C') &= 0, \\ \lambda \omega_0(C'') &= 0, \\ \eta'(C'') &= \eta'(B''),\end{aligned}$$

and the above estimate implies

$$\begin{aligned}(\varphi_* \eta)(A) &\leq \lambda \omega_0(B') + \eta'(B'') \\ &= (\lambda \omega_0 + \eta')(C' + C'') \\ &= \eta(C' + C'') \\ &\leq \sup_{C \in K'_A} \eta(C).\end{aligned}$$

Hence $\eta \mapsto (\varphi_* \eta)(A)$ is a $\tau(\mathfrak{A}^*_{sa}, \mathfrak{A}_{sa})$-continuous functional in $\mathfrak{A}^*_{sa}$, the hermitian functionals on $\mathfrak{A}$, i.e., the dual of the real Banach space $\mathfrak{A}_{sa}$. By the Mackey–Arens theorem there then exists a $B \in \mathfrak{A}_{sa}$ such that

$$(\varphi_* \eta)(A) = \eta(B), \qquad \eta \in \mathfrak{A}^*_{sa}.$$

We now define φ by $B = \varphi(A)$.

A similar discussion of φ_*^{-1} shows that φ is invertible and hence $\varphi(\mathfrak{A}) = \mathfrak{A}$. φ and φ^{-1} are clearly positive maps. Now noting that

$$\omega(\varphi(\mathbb{1})) = (\varphi_*\omega)(\mathbb{1}) = 1 = \omega(\mathbb{1})$$

one has $\varphi(\mathbb{1}) = \mathbb{1}$ and φ is a Jordan automorphism of $\mathfrak{A}$ by Theorem 3.2.3.

Up to this stage in our analysis of positive maps and Jordan automorphisms we have concentrated on individual maps. We next consider one-parameter groups of such maps. All the foregoing results have natural extensions, but the next point we want to make is that continuity properties of the group can strengthen the previous conclusions. In particular, a one-parameter group of affine state maps with suitably strong continuity properties gives rise to a group of *-automorphisms and not merely Jordan automorphisms. This is part of the content of the following corollary.

Corollary 3.2.12. *Let $t \mapsto \alpha_t$ be a strongly continuous one-parameter group of maps of a C^*-algebra $\mathfrak{A}$ with identity $\mathbb{1}$, and assume that $\alpha_t(\mathbb{1}) = \mathbb{1}$ for all $t \in \mathbb{R}$. The following conditions are equivalent:*

(1) *each α_t is a *-automorphism of $\mathfrak{A}$;*
(2) *$\|\alpha_t\| \leq 1$ for all $t \in \mathbb{R}$;*
(3) *$\alpha_t(\mathfrak{A}_+) \subseteq \mathfrak{A}_+$ for all $t \in \mathbb{R}$;*
(4) *$\alpha_t{}^*(E_{\mathfrak{A}}) \subseteq E_{\mathfrak{A}}$ for all $t \in \mathbb{R}$.*

PROOF. By Theorem 3.2.3 it suffices to show that a strongly continuous group α_t of Jordan automorphisms, of $\mathfrak{A}$, is a group of *-automorphisms. But if π is an irreducible representation of $\mathfrak{A}$ then $\pi \circ \alpha_t$ is a morphism, or an antimorphism, of $\mathfrak{A}$, for each t, by Proposition 3.2.2. Since $t \mapsto \pi(\alpha_t(A))$ is continuous in t for each $A \in \mathfrak{A}$, it follows easily that the set $\mathscr{U}$ (respectively $\mathscr{V}$) such that $\pi \circ \alpha_t$ is a morphism (respectively antimorphism) is closed (is closed). Hence $\mathscr{U}$ and $\mathscr{V}$ are both open and closed.

Because $\mathbb{R}$ is connected and $0 \in \mathscr{U}$ it follows that $\mathscr{U} = \mathbb{R}$ and each $\pi \circ \alpha_t$ is a morphism. Since the direct sum of the irreducible representations of $\mathfrak{A}$ is a faithful representation (Lemma 2.3.23), it follows that each α_t is a *-automorphism.

The situation is similar for some special von Neumann algebras and σ-weakly continuous groups.

Corollary 3.2.13. *Let $t \mapsto \alpha_t$ be a σ-weakly continuous one-parameter group of maps of a von Neumann algebra $\mathfrak{M}$ such that $\alpha_t(\mathbb{1}) = \mathbb{1}$ for each $t \in \mathbb{R}$. Assume that $\mathfrak{M}$ is a factor or that $\mathfrak{M}$ is abelian. Let $N_{\mathfrak{M}} = \mathfrak{M}_* \cap E_{\mathfrak{M}}$ be the normal states on $\mathfrak{M}$. The following conditions are equivalent:*

(1) *α_t is a *-automorphism of $\mathfrak{M}$ for each $t \in \mathbb{R}$;*
(2) *$\|\alpha_t\| \leq 1$ for all $t \in \mathbb{R}$;*
(3) *$\alpha_t(\mathfrak{M}_+) \subseteq \mathfrak{M}_+$ for all $t \in \mathbb{R}$;*
(4) *$\alpha_t{}^*(N_{\mathfrak{M}}) \subseteq N_{\mathfrak{M}}$ for all $t \in \mathbb{R}$.*

PROOF. (1) $\Rightarrow$ (2) $\Leftrightarrow$ (3) is clear from Theorem 3.2.3. But condition (3) implies that $\alpha_t(\mathfrak{M}_+) = \mathfrak{M}_+$, $\alpha_t^{-1}(\mathfrak{M}_+) = \mathfrak{M}_+$. Hence if A_α is an increasing net in $\mathfrak{M}_+$ converging to A then $\alpha_t(A_\alpha)$ converges to $\alpha_t(A)$. Thus (3) $\Rightarrow$ (4) while (4) $\Rightarrow$ (3) is trivial. To establish (3) $\Rightarrow$ (1) it suffices to show that a σ-weakly continuous group of Jordan automorphisms of $\mathfrak{M}$ is a group of automorphisms. If $\mathfrak{M}$ is abelian this is evident. If $\mathfrak{M}$ is a factor Proposition 3.2.2 implies that each α_t is either an automorphism or an antiautomorphism. By the connectedness of $\mathbb{R}$ and the argument in the proof of the previous corollary it follows that each α_t is an automorphism.

Note that Corollary 3.2.13 is not true for general $\mathfrak{M}$. It is possible to construct counterexamples (see Notes and Remarks).

As an illustration of these results we classify all the Jordan automorphisms of $\mathscr{L}(\mathfrak{H})$ and thus all order automorphisms, or isometries, φ with $\varphi(\mathbb{1}) = \mathbb{1}$. This classification then solves the problem of characterizing the Wigner symmetries discussed in the introductory paragraphs of this subsection. In the introduction we established that each Wigner symmetry defines an affine invertible map φ_* of the normal states of the von Neumann algebra $\mathscr{L}(\mathfrak{H})$. But Theorem 3.2.8 then implies that φ_* is the dual of a Jordan automorphism φ of $\mathscr{L}(\mathfrak{H})$. The following example elucidates the action of φ on $\mathscr{L}(\mathfrak{H})$, and the resulting action of the Wigner symmetry on $\mathfrak{H}$.

EXAMPLE 3.2.14. The algebra $\mathscr{L}(\mathfrak{H})$ of all bounded operators on the Hilbert space $\mathfrak{H}$ is a factor and any Jordan automorphism α is either an automorphism or an antiautomorphism by Theorem 3.2.3. Assume first that α is an automorphism. Let Ω be a fixed unit vector in $\mathfrak{H}$, and $E \in \mathscr{L}(\mathfrak{H})$ the orthogonal projection with range $\mathbb{C}\Omega$. Since E is a minimal nonzero projection in $\mathscr{L}(\mathfrak{H})$, $F = \alpha^{-1}(E)$ also has this property. Hence $F\mathfrak{H} = \mathbb{C}\xi$, where ξ is a unit vector. Next define an operator U on $\mathfrak{H}$ by

$$UA\xi = \alpha(A)\Omega, \qquad A \in \mathscr{L}(\mathfrak{H}).$$

Then

$$\begin{aligned} \|A\xi\| &= \|AF\xi\| = \|AF\| = \|\alpha(AF)\| \\ &= \|\alpha(A)E\| = \|\alpha(A)E\Omega\| = \|\alpha(A)\Omega\|. \end{aligned}$$

This shows that the definition of U is consistent and that U is an isometry. Since the range of U is $\mathscr{L}(\mathfrak{H})\Omega = \mathfrak{H}$ it follows that U is unitary, and $U^* = U^{-1}$ is given by

$$U^*A\Omega = \alpha^{-1}(A)\xi, \qquad A \in \mathscr{L}(\mathfrak{H}).$$

Hence for $A, B \in \mathscr{L}(\mathfrak{H})$,

$$UAU^*B\Omega = UA\alpha^{-1}(B)\xi = \alpha(A)B\Omega,$$

i.e. $\alpha(A) = UAU^*$. It follows that the automorphisms of $\mathscr{L}(\mathfrak{H})$ are just the inner automorphisms, and the group $\mathrm{Aut}(\mathscr{L}(\mathfrak{H}))$ of automorphisms is isomorphic to the group of unitary operators in $\mathscr{L}(\mathfrak{H})$ modulo the circle group $\{\lambda\mathbb{1}; |\lambda| = 1\}$.

Let $\{E_{ij}\}_{i,j}$ be a complete set of matrix units in $\mathscr{L}(\mathfrak{H})$ and define

$$\sigma_0(E_{ij}) = E_{ji}.$$

Then by straightforward computations σ_0 extends to an antiautomorphism of $\mathscr{L}(\mathfrak{H})$ such that $\sigma_0{}^2 = \iota$. Now if α is an arbitrary antiautomorphism of $\mathscr{L}(\mathfrak{H})$ then $\alpha \circ \sigma_0$ is an automorphism. Thus there exists a unitary $U \in \mathscr{L}(\mathfrak{H})$ such that

$$\alpha(A) = U\sigma_0(A)U^*.$$

This classifies the antiautomorphisms of $\mathscr{L}(\mathfrak{H})$.

One can also exploit this construction to classify the continuous groups α_t of *-automorphisms or *-antiautomorphisms. For each t one chooses a vector ξ_t in the range of $\alpha_{-t}(E)$ and then defines U_t by

$$U_t A\xi_t = \alpha_t(A)\Omega.$$

But the choice of ξ_t is arbitrary up to a phase and consequently the U_t only form a group up to a phase, i.e., $U_s U_t U_{-s-t} = \exp\{i\gamma(s, t)\}$ for some function $\gamma(s, t) \in \mathbb{R}$. Nevertheless, one may argue that if α_t is weakly*-continuous then the phases of the ξ_t can be chosen in a coherent way which ensures that the corresponding U_t form a one-parameter group which is weakly, and hence strongly, continuous in t. This result will actually be derived by a slightly different method in the following subsection (Example 3.2.35).

As a final description of Wigner symmetries of the Hilbert space $\mathfrak{H}$ we remark that each such symmetry α extends to either an automorphism or an antiautomorphism φ of $\mathscr{L}(\mathfrak{H})$, and the action of α on $\mathfrak{H}$ is then determined by the action of φ on the rank one projectors in $\mathscr{L}(\mathfrak{H})$. One always has

$$\alpha(\psi) = U\psi,$$

where U is either unitary or antiunitary, and is uniquely determined up to a phase. The discussion of the action of one-parameter groups of Wigner symmetries will also be given in the next subsection (after Example 3.2.35).

We now leave these illustrative examples and return to a further discussion of Jordan automorphisms of a von Neumann algebra $\mathfrak{M}$. For this purpose it is useful to consider $\mathfrak{M}$ in standard form. Thus in the rest of this subsection we assume that $\mathfrak{M}$ has a cyclic and separating vector Ω and let Δ, J, $\mathscr{P}$ denote the corresponding modular operator, modular conjugation, and natural positive cone. Recall that $\mathscr{P}$ is defined as the closure of the set $\{Aj(A)\Omega;\ A \in \mathfrak{M}\}$, where $j(A) = JAJ$; it can alternatively be characterized as the closure of the set $\{\Delta^{1/4}A\Omega;\ A \in \mathfrak{M}_+\}$ (Proposition 2.5.26). It is also of importance that an element $\xi \in \mathscr{P}$ is cyclic, and separating, for $\mathfrak{M}$ if it is either cyclic or separating. Moreover, if $\xi \in \mathscr{P}$ is cyclic and separating then the conjugation J_ξ, corresponding to the pair $\{\mathfrak{M}, \xi\}$, is equal to J and the corresponding positive cone $\mathscr{P}_\xi$ is equal to $\mathscr{P}$ (Proposition 2.5.30).

The cone $\mathscr{P}$ was used in Section 2.5 to establish that each *-automorphism α of $\mathfrak{M}$ could be unitarily implemented, i.e., there exists a unitary $U(\alpha)$ such that $\alpha(A) = U(\alpha)AU(\alpha)^*$ for all $A \in \mathfrak{M}$. We now use this to derive a characterization of the Jordan automorphisms of $\mathfrak{M}$.

Theorem 3.2.15. *Let $\mathfrak{M}$ be a von Neumann algebra with a cyclic and separating vector Ω and let Δ, J, $\mathscr{P}$ denote the associated modular operator, modular conjugation, and natural positive cone. If U is any unitary operator* *such that* $U\mathscr{P} = \mathscr{P}$ *then there exists a unique Jordan automorphism* α *of* $\mathfrak{M}$ *such that*

$$(\xi, \alpha(A)\xi) = (\xi, UAU^*\xi) \qquad (*)$$

for all $A \in \mathfrak{M}$ *and* $\xi \in \mathscr{P}$.

Conversely, if α *is a Jordan automorphism of* $\mathfrak{M}$ *then there exists a unique unitary operator* U *such that* $U\mathscr{P} = \mathscr{P}$ *and relation* $(*)$ *is again valid.*

If, in either of these cases, $E \in \mathfrak{M} \cap \mathfrak{M}'$ *is a projector such that* $A \in \mathfrak{M} \mapsto \alpha(A)E$ *is a morphism, and* $A \in \mathfrak{M} \mapsto \alpha(A)(\mathbb{1} - E)$ *is an antimorphism (see Proposition* 3.2.2) *then*

$$UAU^* = \alpha(A)E + J\alpha(A^*)J(\mathbb{1} - E).$$

The proof essentially relies upon the earlier results of Section 2.5 and some further analysis of the geometry of the cone $\mathscr{P}$. In particular, one needs the following invariance properties of $\mathscr{P}$.

Lemma 3.2.16. *Let* $\mathfrak{M}$ *be a von Neumann algebra with a cyclic and separating vector* Ω, *and let* Δ, J, $\mathscr{P}$ *denote the associated modular operator, modular conjugation, and natural positive cone. If* $A \in \mathfrak{M} \cap \mathfrak{M}'$ *one has*

$$\Delta^{it} A \Delta^{-it} = A, \qquad JAJ = A^*.$$

If $E \in \mathfrak{M} \cap \mathfrak{M}'$ *is a projector and we set* $\mathfrak{N} = \mathfrak{M}E + \mathfrak{M}'(\mathbb{1} - E)$ *then* $\mathfrak{N}$ *is a von Neumann algebra and* Ω *is cyclic and separating for* $\mathfrak{N}$. *The associated* Δ_0, J_0, *and* $\mathscr{P}_0$ *satisfy*

$$\Delta_0 = \Delta E + \Delta^{-1}(\mathbb{1} - E), \qquad J_0 = J, \qquad \mathscr{P}_0 = \mathscr{P}.$$

Proof. Assume that $E \in \mathfrak{M} \cap \mathfrak{M}'$ is a projector. Then with the notation of Section 2.5 we have for all $A \in \mathfrak{M}$, $A' \in \mathfrak{M}'$,

$$SEA\Omega = A^*E\Omega = EA^*\Omega = ESA\Omega,$$
$$FEA'\Omega = EFA'\Omega.$$

Hence by closure, using Proposition 2.5.11, one has

$$J\Delta^{1/2}E = EJ\Delta^{1/2}, \qquad \Delta^{1/2}JE = E\Delta^{1/2}J.$$

Next note that

$$\Delta E = \Delta^{1/2}JJ\Delta^{1/2}E = E\Delta^{1/2}JJ\Delta^{1/2} = E\Delta.$$

Since E is bounded, it follows that E commutes with Δ^{it}, and then with $\Delta^{1/2}$. Thus

$$EJ\Delta^{1/2} = J\Delta^{1/2}E = JE\Delta^{1/2}.$$

It follows that $JE = EJ$. The first statement of the lemma now follows by approximating A with linear combinations of projections.

The set $\mathfrak{N} = \mathfrak{M}E + \mathfrak{M}'(\mathbb{1} - E)$ is clearly a von Neumann algebra with $\mathfrak{N}' = \mathfrak{M}'E + \mathfrak{M}(\mathbb{1} - E)$, and an easy argument establishes that Ω is separating and cyclic for $\mathfrak{N}$. Now note that $E\Omega$ is cyclic and separating for $\mathfrak{M}E$ on $E\mathfrak{H}$. It follows readily from Propositions 2.5.11 and 2.5.26 that

$$\begin{aligned} J_0 &= J_0 E + J_0(\mathbb{1} - E) = JE + J(\mathbb{1} - E) = J, \\ \Delta_0 &= \Delta E + \Delta^{-1}(\mathbb{1} - E), \\ \mathscr{P}_0 &= E\mathscr{P}_0 + (\mathbb{1} - E)\mathscr{P}_0 = E\mathscr{P} + (\mathbb{1} - E)\mathscr{P} = \mathscr{P}. \end{aligned}$$

PROOF OF THEOREM 3.2.15. Let us now return to the proof of Theorem 3.2.15. We begin with the second statement of the theorem. Remark that Proposition 3.2.2 guarantees the existence of a projector $E \in \mathfrak{M} \cap \mathfrak{M}'$ such that $A \mapsto \alpha(A)E$ is a morphism and $A \mapsto \alpha(A)(\mathbb{1} - E)$ is an antimorphism. Now define a map $\beta; \mathfrak{M} \mapsto \mathfrak{N} \equiv \mathfrak{M}E + \mathfrak{M}'(\mathbb{1} - E)$ by

$$\beta(A) = \alpha(A)E + J\alpha(A^*)J(\mathbb{1} - E).$$

If $F \in \mathfrak{M} \cap \mathfrak{M}'$ is another projector with the same properties as E it follows readily that $P = E + F - EF$ has the property that $P\mathfrak{M} = P\mathfrak{M}'$ is abelian; thus $\mathfrak{N}$ is uniquely determined by α and so, by Lemma 3.2.16, β is uniquely determined. β is a morphism, e.g.,

$$\begin{aligned} \beta(AB) &= (\alpha(A)E)(\alpha(B)E) \\ &\quad + J\alpha(A^*)\alpha(B^*)(\mathbb{1} - E)J \\ &= \beta(A)\beta(B), \end{aligned}$$

where we used $JE = EJ$. Since β maps $\mathfrak{M}\alpha^{-1}(E)$ (resp. $\mathfrak{M}\alpha^{-1}(\mathbb{1} - E)$) isomorphically into $\mathfrak{M}E$ (resp. $\mathfrak{M}'E$) β is an isomorphism. The natural cones associated with the pairs $\{\mathfrak{M}, \Omega\}$ and $\{\mathfrak{N}, \Omega\}$ are identical by Lemma 3.2.16. Thus by Corollary 2.5.32 there exists a unique unitary $U \equiv U(\alpha)$ such that $U\mathscr{P} = \mathscr{P}$ and $\beta(A) = UAU^*$. Finally, the connection between $U = U(\alpha)$ and α is established by remarking that for $\xi \in \mathscr{P}$ one has $J\xi = \xi$ (Proposition 2.5.26(4)), and

$$\begin{aligned} (\xi, UAU^*\xi) &= (\xi, \beta(A)\xi) \\ &= (\xi, \alpha(A)E\xi) + (\alpha(A^*)(\mathbb{1} - E)\xi, \xi) \\ &= (\xi, \alpha(A)E\xi) + (\xi, \alpha(A)(\mathbb{1} - E)\xi) \\ &= (\xi, \alpha(A)\xi). \end{aligned}$$

This simultaneously establishes the last statement of the theorem. To prove the first statement, we need the following characterization of the faces of $\mathscr{P}$.

Lemma 3.2.17. *Let $\mathfrak{M}$ be in standard form and take $\xi \in \mathscr{P}$. The following conditions are equivalent:*

(1) *ξ is cyclic and separating;*
(2) *the set $Q_\xi = \{\eta \in \mathscr{P}; \lambda\eta \le \xi$ for some $\lambda > 0\}$ is dense in $\mathscr{P}$.*

If these conditions are satisfied then $\eta \le \xi$ if, and only if $\eta = \Delta_\xi^{1/4}A\xi$ for some $A \in \mathfrak{M}_+$ with $\|A\| \le 1$.

PROOF. (1) ⇒ (2) If ξ is separating the final statement of the lemma has been established in Lemma 2.5.40. Moreover, this lemma yields $Q_\xi = \Delta_\xi^{1/4}\mathfrak{M}_+\xi$, which is dense in $\mathscr{P}_\xi = \mathscr{P}$, by Propositions 2.5.26(1) and 2.5.30(2).

(2) ⇒ (1) Assume Q_ξ is dense but ξ is not separating. Thus there must exist a nonzero projector $E \in \mathfrak{M}$ such that $E\xi = 0$. But then $Ej(E)\xi = 0$, and because $Ej(E)\mathscr{P} \subseteq \mathscr{P}$ (Proposition 2.5.26(5)) one also concludes that $Ej(E)\eta = 0$ for all $\eta \leq \xi$. It follows from the density of Q_ξ in $\mathscr{P}$ and Proposition 2.5.28(4) that $Ej(E) = 0 = j(E)E$. Next let F denote the projector with range $[\mathfrak{M}\mathfrak{M}'E\mathfrak{H}]$ or, equivalently, with range $[\mathfrak{M}'\mathfrak{M}j(E)\mathfrak{H}]$. One has $F \in \mathfrak{M} \cap \mathfrak{M}'$ and $F \geq E$. But $0 = \mathfrak{M}'Ej(E)\mathfrak{M} = E\mathfrak{M}'\mathfrak{M}j(E)$ and hence $EF = 0$. Thus $E = EF = 0$ which is a contradiction. Therefore ξ must be separating and then it is cyclic by Proposition 2.5.30.

PROOF OF THEOREM 3.2.15. Let us now return to the proof of the first statement of Theorem 3.2.15. Let $\xi \in \mathscr{P}$ be any separating and cyclic vector and set $\eta = U^*\xi$. By Lemma 3.2.17 one has $\xi \geq \Delta_\xi^{1/4}A\xi/\|A\|$ for all $A \in \mathfrak{M}_+$. Therefore $\eta \geq U^*\Delta_\xi^{1/4}A\xi/\|A\|$ for all $A \in \mathfrak{M}_+$ and η is cyclic and separating by another application of the lemma. Moreover, for all $A \in \mathfrak{M}_+$ there must exist an $\alpha_\xi(A) = \alpha(A) \in \mathfrak{M}_+$ with $\|\alpha(A)\| \leq \|A\|$ such that

$$U\Delta_\eta^{1/4}A\eta = \Delta_\xi^{1/4}\alpha(A)\xi.$$

The map α can now be extended by linearity to the whole of $\mathfrak{M}$, and the uniqueness of the hermitian and orthogonal decompositions ensures that this extension is well defined. Thus one arrives at a positive map α of $\mathfrak{M}$ such that $\alpha(\mathbb{1}) = \mathbb{1}$. Repeating this construction with U replaced by U^*, one establishes that α^{-1} exists and is a positive map. Thus α is an order isomorphism of $\mathfrak{M}$, and since (5) ⇒ (1) in Theorem 3.2.3, α is a Jordan automorphism. Let $U(\alpha)$ be the unitary element associated to α by the second part of Theorem 3.2.15. We have

$$\begin{aligned}(\xi, \alpha(A)\xi) &= (\xi, \Delta_\xi^{1/4}\alpha(A)\xi)\\ &= (\xi, U\Delta_\eta^{1/4}A\eta)\\ &= (\eta, A\eta).\end{aligned}$$

It follows that $U(\alpha)^*\xi = \eta = U^*\xi$. Hence, by Theorem 2.5.31 and Proposition 2.5.30,

$$U(\alpha)\Delta^{1/4}A\eta = \Delta_\xi^{1/4}\alpha(A)\xi = U\Delta_\eta^{1/4}A\eta$$

and so, finally, $U(\alpha) = U$. This shows that α is independent of ξ, and the relation between U and α follows from the second statement of the theorem, which was established above.

The characterization of Jordan automorphisms provided by Theorem 3.2.15 has certain drawbacks insofar as it involves the natural cone $\mathscr{P}$. This cone consists of the closure of the set $\Delta^{1/4}\mathfrak{M}_+\Omega$ and although $\mathfrak{M}$ and Ω are primary objects in the theory the modular operator Δ is a derived object which is not always easy to construct. Thus the utility of the criterion $U\mathscr{P} = \mathscr{P}$ for Jordan automorphisms is limited by the lack of access to the modular operator Δ, and hence to the cone $\mathscr{P}$. It is natural to ask whether the cone $\mathfrak{M}_+\Omega$, or its closure, could be substituted in the foregoing criterion, and this is the next object of our investigation. The next theorem establishes that the primary cone $\mathfrak{M}_+\Omega$ can indeed be used to distinguish Jordan automorphisms if these automorphisms satisfy subsidiary invariance conditions.

Theorem 3.2.18. *Let $\mathfrak{M}$ be a von Neumann algebra with a cyclic and separating vector Ω and let σ_t denote the associated modular automorphism group. If U is a unitary operator such that $U\Omega = \Omega$ and*

$$U\mathfrak{M}_+\Omega \subseteq \overline{\mathfrak{M}_+\Omega}, \qquad U^*\mathfrak{M}_+\Omega \subseteq \overline{\mathfrak{M}_+\Omega}$$

then there exists a unique Jordan automorphism α of $\mathfrak{M}$ such that

$$UA\Omega = \alpha(A)\Omega$$

for all $A \in \mathfrak{M}$, and α has the invariance properties

$$(\Omega, A\Omega) = (\Omega, \alpha(A)\Omega), \qquad \sigma_t(\alpha(A)) = \alpha(\sigma_t(A)) \tag{$*$}$$

for all $A \in \mathfrak{M}$ and $t \in \mathbb{R}$.

Conversely, if α is a Jordan automorphism satisfying the conditions $(*)$ then there exists a unique unitary U such that $U\Omega = \Omega$,

$$U\mathfrak{M}_+\Omega \subseteq \mathfrak{M}_+\Omega, \qquad U^*\mathfrak{M}_+\Omega \subseteq \mathfrak{M}_+\Omega,$$

and

$$UA\Omega = \alpha(A)\Omega$$

for all $A \in \mathfrak{M}$. Furthermore, the U obtained in this manner is identical with the U of Theorem 3.2.15.

The first essential ingredient in the proof is the relationship between U and α and for this it is necessary to prove that U maps the cone $\mathfrak{M}_+\Omega$ into itself. This is established by the following lemma.

Lemma 3.2.19. *Let $\mathfrak{M}$ be a von Neumann algebra with cyclic and separating vector Ω and T a bounded operator such that $T\Omega = \Omega = T^*\Omega$ and*

$$T\mathfrak{M}_+\Omega \subseteq \overline{\mathfrak{M}_+\Omega}.$$

It follows that

$$T\mathfrak{M}_+\Omega \subseteq \mathfrak{M}_+\Omega.$$

Proof. We first establish that

$$T^*\mathfrak{M}'_+\Omega \subseteq \mathfrak{M}'_+\Omega.$$

Assume $A' \in \mathfrak{M}'_+$ and $A \in \mathfrak{M}_+$. One then has

$$(T^*A'\Omega, A\Omega) = (A'\Omega, TA\Omega) \geq 0$$

because $TA\Omega \in \overline{\mathfrak{M}_+\Omega}$. Choose $A_n \in \mathfrak{M}_+$ such that

$$A_n\Omega \to TA\Omega$$

and then one finds

$$\begin{aligned}(T^*A'\Omega, A\Omega) &= \lim_n (A_n^{1/2}\Omega, A'A_n^{1/2}\Omega)\\ &\leq \|A'\| \lim_n (\Omega, A_n\Omega)\\ &= \|A'\|(\Omega, TA\Omega)\\ &= \|A'\|(\Omega, A\Omega).\end{aligned}$$

Hence by Theorem 2.3.19 there exists a $B' \in \mathfrak{M}'_+$ such that $\|B'\| \leq \|A'\|$ and

$$(T^*A'\Omega, A\Omega) = (B'\Omega, A\Omega)$$

for all $A \in \mathfrak{M}_+$. Therefore $T^*A'\Omega = B'\Omega$ and $T^*\mathfrak{M}'_+\Omega \subseteq \mathfrak{M}'_+\Omega$. Applying the same argument but with $\mathfrak{M}$ and $\mathfrak{M}'$ interchanged one concludes that $T\mathfrak{M}_+\Omega \subseteq \mathfrak{M}_+\Omega$.

PROOF OF THEOREM 3.2.18. Now let us return to the proof of the theorem. If $A \in \mathfrak{M}_+$ then Lemma 3.2.19 establishes the existence of a unique $\alpha(A) \in \mathfrak{M}_+$ such that

$$UA\Omega = \alpha(A)\Omega$$

and $\alpha(\mathbb{1}) = \mathbb{1}$. Since U^* also maps $\mathfrak{M}_+\Omega$ into $\mathfrak{M}_+\Omega$ the linear extension of α to $\mathfrak{M}$ is an order automorphism and hence a Jordan automorphism by Theorem 3.2.3. Now clearly

$$(\Omega, A\Omega) = (\Omega, UA\Omega) = (\Omega, \alpha(A)\Omega).$$

Next, if $S = J\Delta^{1/2}$ then

$$\begin{aligned}USA\Omega &= UA^*\Omega\\ &= \alpha(A)^*\Omega = SUA\Omega.\end{aligned}$$

It follows by closure that

$$UJ\Delta^{1/2} = J\Delta^{1/2}U$$

and by the uniqueness of the polar decomposition

$$UJ = JU, \qquad U\Delta^{1/2} = \Delta^{1/2}U.$$

Since U is bounded, it follows that U commutes with all bounded functions of $\Delta^{1/2}$ and hence

$$\begin{aligned}\alpha(\sigma_t(A))\Omega &= U\Delta^{it}A\Omega\\ &= \Delta^{it}UA\Omega = \sigma_t(\alpha(A))\Omega.\end{aligned}$$

As Ω is separating, α and σ_t must commute.

To prove the second statement we first invoke Theorem 3.2.15 to establish the existence of a unitary U such that $U\mathscr{P} = \mathscr{P}$, etc. But the invariance condition

$$(\Omega, A\Omega) = (\Omega, \alpha(A)\Omega) = (U^*\Omega, AU^*\Omega)$$

and Theorem 2.5.31 imply that $U\Omega = \Omega$. Finally, the action of U derived in Theorem 3.2.15, $U\Delta^{1/4}A\Omega = \Delta^{1/4}\alpha(A)\Omega$, and the commutation of α and σ_t give

$$U\Delta^{it}\Delta^{1/4}A\Omega = \Delta^{1/4}\alpha(\sigma_t(A))\Omega = \Delta^{1/4}\sigma_t(\alpha(A))\Omega = \Delta^{it}U\Delta^{1/4}A\Omega.$$

But $\Delta^{1/4}\mathfrak{M}\Omega$ is dense by Propositions 2.5.26 and 2.5.28. Therefore U commutes with Δ^{it}. Finally, choosing A to be an entire analytic element with respect to the modular group σ_t one easily deduces that $U\Delta^{1/4}A\Omega = \Delta^{1/4}UA\Omega$. Therefore $\Delta^{1/4}UA\Omega = \Delta^{1/4}\alpha(A)\Omega$ and hence

$$UA\Omega = \alpha(A)\Omega.$$

By density of the entire elements and Kaplansky's density theorem (Theorem 2.4.16) one finally concludes that

$$U\mathfrak{M}_+\Omega \subseteq \mathfrak{M}_+\Omega$$

and similarly for U^*.

The criterion provided by Theorem 3.2.18 will be useful in Section 3.2.5 for the description of automorphism groups in the presence of invariant states.

Note that the second invariance condition in $(*)$, $\sigma_t \circ \alpha = \alpha \circ \sigma_t$, is necessary for the theorem to be true because there exist unitary elements U such that $U\Omega = \Omega$, $U\mathscr{P} = \mathscr{P}$, but $U\mathfrak{M}_+\Omega \nsubseteq \mathfrak{M}_+\Omega$. A simple example is given by defining $\mathfrak{N} = \mathfrak{M} \oplus \mathfrak{M}'$, $\Omega_\mathfrak{N} = \Omega \oplus \Omega$, and

$$U = \begin{pmatrix} 0 & 1 \\ 1 & 0 \end{pmatrix}.$$

One then has $U\mathfrak{N}U^* = \mathfrak{N}'$, $U\Omega_\mathfrak{N} = \Omega_\mathfrak{N}$ and hence $U\mathscr{P} = \mathscr{P}$ for the natural cone corresponding to $\{\mathfrak{N}, \Omega_\mathfrak{N}\}$. But $\overline{U\mathfrak{N}_+\Omega} = \overline{\mathfrak{N}'_+\Omega} = \overline{\Delta^{1/2}\mathfrak{N}_+\Omega}$ and this latter cone is usually distinct from $\mathfrak{N}_+\Omega$.

To conclude this subsection we remark that there are other interesting positive maps which we have not examined. Corollary 3.2.12 established that each strongly continuous one-parameter group of positive, identity-preserving maps of a C^*-algebra is automatically a group of $*$-automorphisms. This is not necessarily the case if one considers semigroups instead of groups. There exist strongly continuous semigroups $\{\alpha_t\}_{t \in \mathbb{R}_+}$ of the above type, which satisfy the generalized Schwarz inequality

$$\alpha_t(A^*A) \geq \alpha_t(A)^*\alpha_t(A)$$

and which do not extend to groups of $*$-automorphisms. The following example, which occurs in the theory of diffusion, illustrates this structure.

EXAMPLE 3.2.20. Let $\mathfrak{A}$ be the C^*-algebra $C_0(\mathbb{R}) + \mathbb{C}1$ of bounded, continuous, complex-valued, functions over the real line, equipped with the supremum norm. Define $t \in \mathbb{R} \mapsto \alpha_t$ by

$$\begin{aligned} (\alpha_t f)(x) &= (2\pi t)^{-1/2} \int dy\, e^{-(x-y)^2/2t} f(y), \qquad && t > 0, \\ &= f(x), && t = 0. \end{aligned}$$

By standard reasoning this is a C_0-semigroup with infinitesimal generator $-d^2/dt^2$. Because the kernel of α is positive the semigroup is positivity preserving and because $\mathfrak{A}$ is abelian the generalized Schwarz inequality is valid by Proposition 3.2.4.

3.2.2. General Properties of Derivations

In the previous subsection we characterized one-parameter groups of $*$-automorphisms of C^*- and von Neumann algebras by various properties of positivity preservation. We next discuss the characterization of these groups in terms of their infinitesimal generators. In the Banach space discussion of Section 3.1 we saw that there are several types of groups which naturally occur. The various groups differ in their continuity properties. Now, however, there is also a possible variation of algebraic structure. Our discussion is aimed to cover uniformly, strongly, and weakly* continuous groups of both C^*- and von Neumann algebras $\mathfrak{A}$. Note, however, that the concept of a $C_0{}^*$-group, i.e., a weakly* continuous group, of an algebra $\mathfrak{A}$ is only defined when $\mathfrak{A}$ has a predual. But in this case $\mathfrak{A}$ is automatically a von Neumann algebra by Sakai's theorem (Section 2.4.3). Moreover, one may demonstrate (see Example 3.2.36 for the $\mathscr{L}(\mathfrak{H})$ case) that a C_0-group, i.e., a strongly continuous group, of $*$-automorphisms of a von Neumann algebra $\mathfrak{M}$ is automatically uniformly continuous. Thus C_0-groups are appropriate to C^*-algebras, $C_0{}^*$-groups to von Neumann algebras, and uniformly continuous groups to both structures.

In this subsection we principally derive algebraic properties of the infinitesimal generators of automorphism groups. These properties are of two types. One may derive conditions for closability, etc., and one may also develop a functional analysis of the domains of generators. Both these facets are of interest and use in the subsequent characterizations of groups and the analysis of their stability. The main tool in this investigation is the derivation property of the generators which reflects, infinitesimally, the property

$$\alpha_t(AB) = \alpha_t(A)\alpha_t(B)$$

of the automorphism group $t \in \mathbb{R} \mapsto \alpha_t$. The continuity of the group affects the topological properties of the generator but most of the results of this subsection are based on the analysis of generators of C_0-groups, i.e., norm-closed, norm-densely defined operators.

The definition which is basic to the analysis of generators is the following:

Definition 3.2.21. A (*symmetric*) *derivation* δ of a C^*-algebra $\mathfrak{A}$ is a linear operator from a $*$-subalgebra $D(\delta)$, the domain of δ, into $\mathfrak{A}$ with the properties that

(1) $\delta(A)^* = \delta(A^*)$, $A \in D(\delta)$,
(2) $\delta(AB) = \delta(A)B + A\delta(B)$, $A, B \in D(\delta)$.

Often the term derivation is used for operators with property (2) but without the symmetry property, $\delta(A)^* = \delta(A^*)$. We will only be interested in symmetric derivations and we sometimes drop the qualification symmetric. Obviously a nonsymmetric derivation with a selfadjoint domain can always be decomposed as a sum $\delta = \delta_1 + i\delta_2$ of symmetric derivations, e.g., $\delta_1(A) = (\delta(A) + \delta(A^*)^*)/2$, $\delta_2(A) = (\delta(A) - \delta(A^*)^*)/2i$. Hence from the structural point of view it suffices to a certain extent to examine the symmetric derivations. Note that if the identity $\mathbb{1} \in D(\delta)$ then $\delta(\mathbb{1}) = 0$ because of the relation $\delta(\mathbb{1}) = \delta(\mathbb{1}^2) = 2\delta(\mathbb{1})$.

Derivations arise as infinitesimal generators of continuous groups of *-automorphisms, $t \in \mathbb{R}$, $A \in \mathfrak{A} \mapsto \tau_t(A) \in \mathfrak{A}$. The two defining properties originate by differentiation, in the topology dictated by the continuity of τ, of the relations

$$\tau_t(A)^* = \tau_t(A^*), \qquad \tau_t(AB) = \tau_t(A)\tau_t(B).$$

Of course, generators have many auxiliary properties which stem from the Banach space structure of $\mathfrak{A}$. These properties, closedness, dissipativeness, etc., have already been extensively discussed in Section 3.1. Our first aim is to establish a property which ensures the dissipativeness of a derivation. In fact, we will discuss a wider class of operators which can be regarded as prototypes of generators of positivity preserving semigroups (see Example 3.2.20 and the remarks preceding this example).

Proposition 3.2.22. *Let $\mathfrak{A}$ be a C^*-algebra with identity $\mathbb{1}$ and δ an operator from a *-subalgebra $D(\delta)$, of $\mathfrak{A}$, into $\mathfrak{A}$ such that*

(1) $\mathbb{1} \in D(\delta)$,
(2) *if $A \in D(\delta)$ and $A \geq 0$ then $A^{1/2} \in D(\delta)$,*
(3) *if $A \in D(\delta)$ then $\delta(A)^* = \delta(A^*)$ and $\delta(A^*A) \geq \delta(A^*)A + A^*\delta(A)$.*

It follows that δ is dissipative.

PROOF. If $A \in D(\delta)$ then $A^*A \in D(\delta)$. Let η be a tangent functional at A^*A and for convenience normalize η so that $\|\eta\| = 1$. We first claim that η is positive and, by Proposition 2.3.11, it suffices to show that $\eta(\mathbb{1}) = 1$. But if $\eta(\mathbb{1}) = \alpha + i\beta$ we use the fact that $\|\mathbb{1} - 2A^*A/\|A\|^2\| \leq 1$ to deduce that

$$\alpha^2 + \beta^2 = |\eta(\mathbb{1})|^2 \leq 1,$$
$$(\alpha - 2)^2 + \beta^2 = |\eta(\mathbb{1} - 2A^*A/\|A\|^2)|^2 \leq 1.$$

Therefore $\alpha = 1$ and $\beta = 0$.

Next define $\eta_A \in \mathfrak{A}^*$ by $\eta_A(B) = \eta(A^*B)$, for $B \in \mathfrak{A}$, and note that $\|\eta_A\| \leq \|A\|$ by the Cauchy–Schwarz inequality. But one also has

$$\|A\|^2 = \eta(A^*A) = \eta_A(A) \leq \|\eta_A\| \, \|A\|$$

and hence η_A is a tangent functional at A. Now as η is positive one has

$$\overline{\eta_A(B)} = \overline{\eta(A^*B)} = \eta(B^*A).$$

Therefore

$$\begin{aligned} 2 \operatorname{Re} \eta_A(\delta(A)) &= \eta(A^*\delta(A)) + \eta(\delta(A^*)A) \\ &\leq \eta(\delta(A^*A)) \\ &= -\eta(\delta(B^2)) + \|A\|^2\eta(\delta(\mathbb{1})), \end{aligned}$$

where $B = (\|A\|^2\mathbb{1} - A^*A)^{1/2}$. But

$$\delta(\mathbb{1}) = \delta(\mathbb{1}^2) \geq \delta(\mathbb{1})\mathbb{1} + \mathbb{1}\delta(\mathbb{1}) = 2\delta(\mathbb{1})$$

and hence $\delta(\mathbb{1}) \leq 0$. Moreover,

$$-\eta(\delta(B^2)) \leq -\eta(\delta(B)B) - \eta(B\delta(B)) = 0,$$

where the last equality follows from another application of the Cauchy–Schwarz inequality, e.g.,

$$\begin{aligned} |\eta(\delta(B)B)|^2 &\leq \eta(\delta(B)^2)\eta(B^2) \\ &= \eta(\delta(B)^2)(\|A\|^2\eta(\mathbb{1}) - \eta(A^*A)) \\ &= 0. \end{aligned}$$

Combining these estimates gives $\operatorname{Re} \eta_A(\delta(A)) \leq 0$ and δ is dissipative.

The property of dissipativeness was discussed in Section 3.1.2 in connection with C_0-semigroups. It is of interest because it has several immediate implications. For example, a norm-densely defined dissipative operator δ is norm closable and satisfies

$$\|(I - \alpha\delta)(A)\| \geq \|A\|$$

for all $A \in D(\delta)$ and $\alpha \geq 0$, by Propositions 3.1.14 and 3.1.15. The only problem with Proposition 3.2.22 is that the domain of a derivation is not generally invariant under the square root operation. If δ is norm closed then the functional analysis that we derive later in this subsection can be used to establish that $D(\delta)$ is invariant under the formation of square roots of positive invertible elements. Nevertheless, one can show that if δ is norm closed and $D(\delta)$ is invariant under the square root operation of positive elements then δ is automatically bounded. We will not prove this last statement but instead consider the related case that $D(\delta) = \mathfrak{A}$. In this situation the domain does have the required invariance and one immediately has the following:

Corollary 3.2.23. *Let δ be an everywhere defined derivation from a C^*-algebra $\mathfrak{A}$ into a larger C^*-algebra $\mathfrak{B}$. It follows that δ is bounded.*

PROOF. If $\mathfrak{A}$ does not have an identity then extend δ to $\tilde{\mathfrak{A}} = \mathbb{C}\mathbb{1} + \mathfrak{A}$ by setting $\delta(\alpha\mathbb{1} + A) = \delta(A)$. In both cases δ satisfies the conditions of the Proposition 3.2.22 and Proposition 3.1.14 implies that δ is norm closable, hence norm closed. But an everywhere defined norm-closed operator is automatically bounded by the closed graph theorem.

This last conclusion can be strengthened in the case of a von Neumann algebra. If δ is a derivation of a von Neumann algebra and there exists a C^*-subalgebra $\mathfrak{A}$ of $\mathfrak{M}$ such that $\mathfrak{A} = D(\delta)$, and $\mathfrak{A}$ is weakly dense in $\mathfrak{M}$, then δ has a bounded extension to $\mathfrak{M}$. This follows directly from Corollary 3.2.23 and the following extension results.

Proposition 3.2.24. *Let $\mathfrak{A}$ be a C^*-algebra of bounded operators on the Hilbert space $\mathfrak{H}$. Further, let δ be a derivation from $\mathfrak{A}$ into the weak closure $\mathfrak{M}$ of $\mathfrak{A}$, i.e., $\mathfrak{A} = D(\delta)$ and $\delta(\mathfrak{A}) \subseteq \mathfrak{M}$. It follows that δ has a unique σ-weakly closed bounded extension $\tilde{\delta}$ to $\mathfrak{M}$ and $\tilde{\delta}$ is a derivation of $\mathfrak{M}$ with $\|\tilde{\delta}\| = \|\delta\|$.*

PROOF. Firstly, note that δ is bounded by Corollary 3.2.23. Secondly, remark that if A is a positive element of $\mathfrak{A}$ then

$$\begin{aligned}|\omega(\delta(A))| &= |\omega(\delta(A^{1/2})A^{1/2} + A^{1/2}\delta(A^{1/2}))| \\ &\leq 2\|\delta\|(\|A\|\omega(A))^{1/2}\end{aligned}$$

for each normal state over $\mathfrak{M}$. Moreover, if A is a positive element of $\mathfrak{M}$ then the Kaplansky density theorem implies the existence of $A_\alpha \in \mathfrak{A}$ such that $A_\alpha \geq 0$, $\|A_\alpha\| \leq \|A\|$, and $A_\alpha \to A$ σ-weakly, i.e., in the $\sigma(\mathfrak{M}, \mathfrak{M}_*)$-topology. Therefore the above inequality and the Jordan decomposition (Proposition 3.2.7) establish that $\delta(A_\alpha)$ converges σ-weakly. We define $\tilde{\delta}(A)$ by $\tilde{\delta}(A) = \lim \delta(A_\alpha)$. Next note that each $A \in \mathfrak{M}$ has a decomposition $A = A_1 - A_2 + i(A_3 - A_4)$ in terms of four positive elements A_i with $\|A_i\| \leq \|A\|$ and hence δ extends by linearity to an operator $\tilde{\delta}$ on $\mathfrak{M}$. It follows immediately that $\tilde{\delta}$ is a derivation of $\mathfrak{M}$ but $\|\tilde{\delta}\| = \|\delta\|$ by another application of the Kaplansky density theorem.

There are other applications of Proposition 3.2.22 which go beyond the realm of bounded derivations. Many derivations of UHF algebras also have domains which are invariant under the square root operation. We illustrate this by the following example.

EXAMPLE 3.2.25. Let $\mathfrak{A}$ denote a UHF algebra as introduced in Example 2.6.12. Thus $\mathfrak{A}$ is the norm closure of a family $\{\mathfrak{A}_\Lambda\}_{\Lambda \in I_f}$, of full-matrix subalgebras $\mathfrak{A}_\Lambda$, where I_f denotes the finite subsets of an index set I. If $\Lambda_1 \cap \Lambda_2 = \emptyset$ then $\mathfrak{A}_{\Lambda_1}$ and $\mathfrak{A}_{\Lambda_2}$ commute. Now let $\{\Lambda_n\}_{n \geq 1}$ be any increasing family of subsets of I_f such that $\bigcup_n \Lambda_n = I$ and choose elements $H_n = H_n^* \in \mathfrak{A}_{\Lambda_n}$ such that $H_n - H_{n-1}$ commutes with $\mathfrak{A}_{\Lambda_{n-1}}$. One can define a symmetric derivation δ of $\mathfrak{A}$ by

$$D(\delta) = \bigcup_{\Lambda \in I_f} \mathfrak{A}_\Lambda$$

and

$$\delta(A) = i \lim_{n \to \infty} [H_n, A], \qquad A \in D(\delta),$$

because the commutativity condition for $H_n - H_{n-1}$ ensures that the limit exists. But $D(\delta)$ is invariant under the square root operation because each $\mathfrak{A}_\Lambda$ has this invariance.

It is interesting to note that this construction has a converse. If δ is a derivation with

$$D(\delta) = \bigcup_{\Lambda \in I_f} \mathfrak{A}_\Lambda$$

and e_{ij} is a set of matrix units for $\mathfrak{A}_\Lambda$ then

$$\delta(A) = i[H_\Lambda, A]$$

for all $A \in \mathfrak{A}_\Lambda$, where

$$H_\Lambda = \frac{1}{i} \sum_j \delta(e_{j1}) e_{1j}.$$

Next we consider other criteria for closability of derivations which have wider applicability than Proposition 3.2.22 or Lemma 3.1.27.

Proposition 3.2.26. *Let δ be a norm-densely defined derivation of a C^*-algebra $\mathfrak{A}$. For δ to be norm closable it suffices that there exists a state ω, over $\mathfrak{A}$, such that*

(1) $|\omega(\delta(A))| \leq L\|A\|$ *for all* $A \in D(\delta)$ *and some* $L \geq 0$;
(2) *the representation* $(\mathfrak{H}_\omega, \pi_\omega)$, *of* $\mathfrak{A}$, *associated with* ω *is faithful.*

PROOF. Lemma 3.1.9 establishes that the norm-densely defined operator δ, on $\mathfrak{A}$, is norm closable (i.e., $\sigma(\mathfrak{A}, \mathfrak{A}^*)$–$\sigma(\mathfrak{A}, \mathfrak{A}^*)$-closable) if, and only if, the dual δ^* on $\mathfrak{A}^*$ is weak*-densely defined. Thus we concentrate on proving this latter property of δ^*.

First note that by definition $\omega \in D(\delta^*)$ and

$$(\delta^*\omega)(A) = \omega(\delta(A)).$$

Moreover, the derivation property gives the identity

$$\omega(A\delta(B)C) = \omega(\delta(ABC)) - \omega(\delta(A)BC) - \omega(AB\delta(C))$$

for all $A, B, C \in D(\delta)$, and hence

$$|\omega(A\delta(B)C)| \leq \|B\|(L\|A\|\|C\| + \|\omega\|(\|\delta(A)\|\|C\| + \|A\|\|\delta(C)\|))$$

Therefore the functional $\omega_{A,C}$ defined by

$$B \mapsto \omega_{A,C}(B) = \omega(ABC)$$

is also in the domain of δ^*. Thus $D(\delta^*)$ contains the subspace of $\mathfrak{A}^*$ spanned by the set

$$\{\omega_{A,C};\ A, C \in D(\delta)\}.$$

But this set is $\sigma(\mathfrak{A}^*, \mathfrak{A})$-dense because if this were not the case the Hahn–Banach theorem would imply the existence of a nonzero $B \in \mathfrak{A}$ such that

$$\omega_{A,C}(B) = 0$$

for all $A, C \in D(\delta)$. The density of $D(\delta)$ and representation theory would then imply that $\pi_\omega(B) = 0$, which contradicts the faithfulness of $(\mathfrak{H}_\omega, \pi_\omega)$. Thus $D(\delta^*)$ is $\sigma(\mathfrak{A}^*, \mathfrak{A})$-dense and δ is norm closable.

Note that the proof of Proposition 3.2.26 depends upon a duality argument which is not restricted to the norm topology. If $\mathfrak{A}$ is a von Neumann algebra one can apply exactly the same argument to the weak* topology to obtain the following result.

Corollary 3.2.27. *Let δ be a $\sigma(\mathfrak{M}, \mathfrak{M}_*)$-densely defined derivation of the von Neumann algebra $\mathfrak{M}$. For δ to be $\sigma(\mathfrak{M}, \mathfrak{M}_*)$–$\sigma(\mathfrak{M}, \mathfrak{M}_*)$-closable, it suffices that there exists a normal state ω over $\mathfrak{M}$ such that*

(1) *$\omega \circ \delta$ extends to a bounded σ-weakly continuous functional on $\mathfrak{M}$;*
(2) *the representation $(\mathfrak{H}_\omega, \pi_\omega)$, of $\mathfrak{M}$, associated with ω is faithful.*

If one replaces the norm continuity property

$$|\omega(\delta(A))| \leq L\|A\|,$$

which ensures closability of a derivation, by a stronger form of continuity then one can deduce a more detailed description of the action of δ on the representation $(\mathfrak{H}_\omega, \pi_\omega)$.

Proposition 3.2.28. *Let δ be a symmetric derivation defined on a *-subalgebra $\mathfrak{D}$ of the bounded operators on a Hilbert space $\mathfrak{H}$. Let $\Omega \in \mathfrak{H}$ be a unit vector cyclic for $\mathfrak{D}$ in $\mathfrak{H}$ and denote the corresponding state by ω, i.e.,*

$$\omega(A) = (\Omega, A\Omega), \qquad A \in \mathfrak{D}.$$

Consider the following conditions:

(1) $$|\omega(\delta(A))| \leq L(\omega(A^*A) + \omega(AA^*))^{1/2}$$

for all $A \in \mathfrak{D}$ and some $L \geq 0$;

(2) *there exists a symmetric operator H, on $\mathfrak{H}$, with the properties*

$$D(H) = \mathfrak{D}\Omega,$$
$$\delta(A)\psi = i[H, A]\psi$$

for all $A \in \mathfrak{D}$ and $\psi \in D(H)$.

It follows that (1) $\Rightarrow$ (2) and if $\mathfrak{D}$ contains the identity $\mathbb{1}$, then (2) $\Rightarrow$ (1) and H may be chosen such that

$$\|H\Omega\|^2 \leq \frac{L^2}{2}.$$

PROOF. (2) $\Rightarrow$ (1) Assume that $\mathbb{1} \in \mathfrak{D}$ and thus $\Omega \in D(H)$. For $A \in \mathfrak{D}$ one has

$$\begin{aligned} |\omega(\delta(A))| &= |(H\Omega, A\Omega) - (A^*\Omega, H\Omega)| \\ &\leq \|H\Omega\|(\|A\Omega\| + \|A^*\Omega\|) \end{aligned}$$

by the Cauchy–Schwarz inequality. But then

$$\begin{aligned} |\omega(\delta(A))|^2 &\leq \|H\Omega\|^2(\sqrt{\omega(A^*A)} + \sqrt{\omega(AA^*)})^2 \\ &\leq 2\|H\Omega\|^2(\omega(A^*A) + \omega(AA^*)). \end{aligned}$$

(1) $\Rightarrow$ (2) Consider the Hilbert space

$$\mathfrak{H}_+ = \mathfrak{H} \oplus \overline{\mathfrak{H}},$$

where $\overline{\mathfrak{H}}$ is the conjugate space of $\mathfrak{H}$ (see footnote on page 70). Let $\hat{\mathfrak{H}}_+$ denote the subspace of $\mathfrak{H}_+$ spanned by vectors of the form $\{A\Omega, A^*\Omega\}$, with $A \in \mathfrak{D}$. Define a linear functional η on $\hat{\mathfrak{H}}_+$ by

$$\eta(\{A\Omega, A^*\Omega\}) = i\omega(\delta(A)).$$

One has

$$|\eta(\{A\Omega, A^*\Omega\})| \leq L\|\{A\Omega, A^*\Omega\}\|$$

by assumption. Hence η is well defined and $\|\eta\| \leq L$. The Riesz representation theorem then implies the existence of a vector $\{\varphi, \psi\}$ in the closure of $\hat{\mathfrak{H}}_+$ such that

$$\begin{aligned} i\omega(\delta(A)) &= \eta(\{A\Omega, A^*\Omega\}) \\ &= (\{\varphi, \psi\}, \{A\Omega, A^*\Omega\}) \\ &= (\varphi, A\Omega) + (A^*\Omega, \psi). \end{aligned}$$

Next, using the symmetry of δ one finds

$$\begin{aligned} i\omega(\delta(A)) &= -\overline{i\omega(\delta(A^*))} \\ &= -[\overline{(\varphi, A^*\Omega)} + \overline{(A\Omega, \psi)}] \\ &= -[(\psi, A\Omega) + (A^*\Omega, \varphi)]. \end{aligned}$$

Taking the average of these two expressions one finds

$$i^{-1}\omega(\delta(A)) = (\Omega_\delta, A\Omega) - (A^*\Omega, \Omega_\delta),$$

where

$$\Omega_\delta = \frac{\psi - \varphi}{2}.$$

Next define the operator H by $D(H) = \mathfrak{D}\Omega$ and

$$HA\Omega = i^{-1}\delta(A)\Omega + A\Omega_\delta, \qquad A \in \mathfrak{A}.$$

One calculates that

$$\begin{aligned} &(HA\Omega, B\Omega) - (A\Omega, HB\Omega) \\ &= -i^{-1}\omega(\delta(A^*)B) - i^{-1}\omega(A^*\delta(B)) + (\Omega_\delta, A^*B\Omega) - (\Omega, A^*B\Omega_\delta) \\ &= -i^{-1}\omega(\delta(A^*B)) + i^{-1}\omega(\delta(A^*B)) = 0. \end{aligned}$$

Because $\mathfrak{D}\Omega$ is dense in $\mathfrak{H}$ this shows that H is well defined, i.e., $A\Omega = 0$ implies $HA\Omega = 0$, and that H is symmetric. But for $A, B \in \mathfrak{D}$ one has

$$\begin{aligned} \delta(A)B\Omega &= \delta(AB)\Omega - A\delta(B)\Omega \\ &= iHAB\Omega - AB\Omega_\delta - AiHB\Omega + AB\Omega_\delta \\ &= i[H, A]B\Omega. \end{aligned}$$

Finally, the bound on $\|H\Omega\|$ follows from $\delta(\mathbb{1}) = 0$ and the calculation

$$\begin{aligned} \|H\Omega\|^2 &= \|\Omega_\delta\|^2 \\ &\leq \frac{\|\varphi - \psi\|^2 + \|\varphi + \psi\|^2}{4} \\ &= \frac{\|\varphi\|^2 + \|\psi\|^2}{2} \\ &= \frac{\|\eta\|^2}{2} \leq \frac{L^2}{2}. \end{aligned}$$

Proposition 3.2.28 is particularly useful for the discussion of invariant states. If $t \in \mathbb{R} \mapsto \alpha_t$ is a one-parameter group of *-automorphisms of the C^*-algebra $\mathfrak{A}$ and ω is a state then ω is invariant under α whenever

$$\omega(\alpha_t(A)) = \omega(A)$$

for all $t \in \mathbb{R}$ and $A \in \mathfrak{A}$. If the group α is strongly continuous with generator δ then this invariance condition is equivalent to the infinitesimal condition

$$\omega(\delta(A)) = 0$$

for all $A \in D(\delta)$. Thus the proposition applies to δ acting on the representation $(\mathfrak{H}_\omega, \pi_\omega, \Omega_\omega)$ associated with ω. Note that in this case the automorphism group α is implemented by a one-parameter group of unitary operators U_t, on $\mathfrak{H}_\omega$, in the form

$$\pi_\omega(\alpha_t(A)) = U_t \pi_\omega(A) U_t^*.$$

This follows from Corollary 2.3.17. The same corollary establishes that the U may be chosen such that $U_t\Omega_\omega = \Omega_\omega$ for all $t \in \mathbb{R}$, and then the simple estimate

$$\|U_t \pi_\omega(A)\Omega_\omega - \pi_\omega(A)\Omega_\omega\| \leq \|\alpha_t(A) - A\|$$

shows that $\{U_t\}_{t\in\mathbb{R}}$ is in fact strongly continuous. If H is the selfadjoint generator of U one then has $\pi_\omega(D(\delta))\Omega_\omega \subseteq D(H)$ and

$$\pi_\omega(\delta(A))\psi = i[H, \pi_\omega(A)]\psi$$

for all $\psi \in \pi_\omega(D(\delta))\Omega_\omega$. Thus the symmetric operator H occurring in condition (2) of Proposition 3.2.28 can be chosen selfadjoint. This is not the case, however, for a general δ.

In analogy with the foregoing example of a generator δ we will say that a state ω, over a C^*-algebra $\mathfrak{A}$, is invariant under a derivation δ, of $\mathfrak{A}$, whenever the infinitesimal condition $\omega(\delta(A)) = 0$, for $A \in D(\delta)$, is satisfied.

After this discussion of closability criteria we now examine various domain properties of closed derivations which are of a purely algebraic nature. These properties have many subsequent applications in the general analysis of derivations. For orientation let us begin by examining the case of an abelian algebra.

If δ is a norm-closed derivation of the abelian C^*-algebra $\mathfrak{A}$ and $A = A^* \in D(\delta)$, the domain of δ, then for any polynomial P one has by simple computation $P(A) \in D(\delta)$ and

$$\delta(P(A)) = \delta(A)P'(A).$$

The prime denotes a derivative. Now let f be a once continuously differentiable function. One may choose polynomials P_n such that $P_n \to f$ and $P_n' \mapsto f'$ uniformly on the spectrum of A. Thus by the foregoing $\delta(P_n(A)) = \delta(A)P_n'(A)$ converges to $\delta(A)f'(A)$. Hence, because δ is closed, $f(A) \in D(\delta)$, and

$$\delta(f(A)) = \delta(A)f'(A).$$

Now we wish to examine domain properties of this type for a general C^*-algebra and establish sufficient conditions on functions f such that $f(A) \in D(\delta)$ whenever $A = A^* \in D(\delta)$. There are two approaches to this type of functional analysis, Fourier analysis, or complex analysis, but the following result is useful in both cases.

Proposition 3.2.29. *Let δ be a norm-closed derivation of a C^*-algebra $\mathfrak{A}$ with identity $\mathbb{1}$. If $A = A^* \in D(\delta)$ and $\lambda \notin \sigma(A)$, the spectrum of A, then $A(\lambda\mathbb{1} - A)^{-1} \in D(\delta)$ and*

$$\delta(A(\lambda\mathbb{1} - A)^{-1}) = \lambda(\lambda\mathbb{1} - A)^{-1}\delta(A)(\lambda\mathbb{1} - A)^{-1}.$$

If, moreover, $\mathbb{1} \in D(\delta)$ then $(\lambda\mathbb{1} - A)^{-1} \in D(\delta)$ and

$$\delta((\lambda\mathbb{1} - A)^{-1}) = (\lambda\mathbb{1} - A)^{-1}\delta(A)(\lambda\mathbb{1} - A)^{-1}.$$

PROOF. Set $A_\lambda = A(\lambda\mathbb{1} - A)^{-1}$. If $|\lambda|$ is larger than the spectral radius of A then the Neumann series

$$A_\lambda = \sum_{n \ge 0} (A/\lambda)^{n+1}$$

converges in norm and $\|A_\lambda\| \le \|A\|(|\lambda| - \|A\|)^{-1}$. But $A^{n+1} \in D(\delta)$,

$$\delta(A^{n+1}) = \sum_{p=0}^{n} A^p\delta(A)A^{n-p}$$

and the double series

$$\lambda^{-1} \sum_{n \ge 0} \sum_{p=0}^{n} \left(\frac{A}{\lambda}\right)^p \delta(A) \left(\frac{A}{\lambda}\right)^{n-p}$$

also converges in norm. Thus $A_\lambda \in D(\delta)$ because δ is norm closed. Moreover, one has

$$\begin{aligned}\delta(A_\lambda) &= \lambda^{-1} \sum_{p \ge 0} \left(\frac{A}{\lambda}\right)^p \delta(A) \sum_{n \ge 0} \left(\frac{A}{\lambda}\right)^n \\ &= \lambda(\lambda\mathbb{1} - A)^{-1}\delta(A)(\lambda\mathbb{1} - A)^{-1}.\end{aligned}$$

Next assume $|\lambda_0| > \|A\|$, $A_{\lambda_0} \in D(\delta)$, and

$$\left|\frac{\lambda - \lambda_0}{\lambda}\right| < \|A_{\lambda_0}\|^{-1} = \inf_{\gamma \in \sigma(A)} |\gamma(\lambda_0 - \gamma)^{-1}|^{-1}.$$

One then has

$$A_\lambda = \left(\frac{\lambda_0}{\lambda_0 - \lambda}\right) \sum_{n \ge 0} \left(\left(\frac{\lambda_0 - \lambda}{\lambda}\right) A_{\lambda_0}\right)^{n+1}.$$

By the same argument $A_\lambda \in D(\delta)$ and the action of δ is calculated in an identical manner:

$$\begin{aligned}\delta(A_\lambda) &= \left(\frac{\lambda_0}{\lambda}\right) \sum_{p \ge 0} \left(\left(\frac{\lambda_0 - \lambda}{\lambda}\right) A_{\lambda_0}\right)^p \delta(A_{\lambda_0}) \sum_{n \ge 0} \left(\left(\frac{\lambda_0 - \lambda}{\lambda}\right) A_{\lambda_0}\right)^n \\ &= \left(\frac{\lambda_0}{\lambda}\right)\left(\frac{\lambda}{\lambda_0}\right)(\lambda_0\mathbb{1} - A)(\lambda\mathbb{1} - A)^{-1}\delta(A_{\lambda_0})\left(\frac{\lambda}{\lambda_0}\right)(\lambda_0\mathbb{1} - A)(\lambda\mathbb{1} - A)^{-1} \\ &= \lambda(\lambda\mathbb{1} - A)^{-1}\delta(A)(\lambda\mathbb{1} - A)^{-1}.\end{aligned}$$

An analytic continuation argument then allows one to conclude that $A(\lambda\mathbb{1} - A)^{-1} \in D(\delta)$ for all $\lambda \notin \sigma(A)$.

The second statement of the proposition follows by a similar but simpler argument.

This proposition has an immediate corollary which concerns the identity element.

Corollary 3.2.30. *Let δ be a norm-closed derivation of a C^*-algebra $\mathfrak{A}$ with identity $\mathbb{1}$. The following conditions are equivalent:*

(1) *there is a positive invertible $A \in D(\delta)$;*
(2) $\mathbb{1} \in D(\delta)$.

In particular, if δ is norm-densely defined then $\mathbb{1} \in D(\delta)$

PROOF. (2) $\Rightarrow$ (1) This follows by taking $A = \mathbb{1}$.

(1) $\Rightarrow$ (2) It follows from Proposition 3.2.29 that $A(\varepsilon\mathbb{1} + A)^{-1} \in D(\delta)$ for all $\varepsilon > 0$ and

$$\delta(A(\varepsilon\mathbb{1} + A)^{-1}) = -\varepsilon(\varepsilon\mathbb{1} + A)^{-1}\delta(A)(\varepsilon\mathbb{1} + A)^{-1}.$$

Now $A(\varepsilon\mathbb{1} + A)^{-1}$ converges in norm to $\mathbb{1}$ but

$$\overline{\lim_{\varepsilon\to 0}}\|\delta(A(\varepsilon\mathbb{1} + A)^{-1})\| \le \overline{\lim_{\varepsilon\to 0}}\, \varepsilon\|A^{-1}\|^2\|\delta(A)\| = 0.$$

Thus $\mathbb{1} \in D(\delta)$, because δ is norm closed, and $\delta(\mathbb{1}) = 0$. If δ is norm-densely defined, choose $B \in D(\delta)$ such that $B = B^*$ and $\|\mathbb{1} - B\| < 1$. This is possible because $D(\delta)$ is norm dense. But it follows easily that B is positive and invertible.

Corollary 3.2.30 simplifies the analysis of norm-closed, norm-densely defined derivations δ because it implies that one may always assume the existence of an identity $\mathbb{1}$ such that $\mathbb{1} \in D(\delta)$. If the C^*-algebra $\mathfrak{A}$ does not contain an identity then the derivation δ may be extended to a norm-closed derivation $\tilde{\delta}$ of the algebra $\tilde{\mathfrak{A}} = \mathbb{C}\mathbb{1} + \mathfrak{A}$ by setting $D(\tilde{\delta}) = \mathbb{C}\mathbb{1} + D(\delta)$ and

$$\tilde{\delta}(\lambda\mathbb{1} + A) = \delta(A), \qquad A \in D(\delta), \lambda \in \mathbb{C}.$$

Thus in both cases, $\mathbb{1} \in \mathfrak{A}$ or $\mathbb{1} \notin \mathfrak{A}$, one can reduce the situation to that of $\mathbb{1} \in D(\delta) \subseteq \mathfrak{A}$.

Now we examine more detailed domain properties. We always examine norm-closed derivations δ of a C^*-algebra but of course this automatically includes weakly *closed derivations of von Neumann algebras. There are two different approaches. If $A = A^* \in D(\delta)$ and $z \in \mathbb{C} \mapsto f(z) \in \mathbb{C}$ is a function analytic in an open simply connected set Σ_f containing $\sigma(A)$ then one may form $f(A)$ by use of the Cauchy representation

$$f(A) = (2\pi i)^{-1} \int_C d\lambda\, f(\lambda)(\lambda\mathbb{1} - A)^{-1},$$

where C is a simple rectifiable curve contained in Σ_f with $\sigma(A)$ contained in its interior. The integral is defined as a norm limit of Riemann sums

$$\Sigma_N(f) = (2\pi i)^{-1} \sum_{i=1}^{N} (\lambda_i - \lambda_{i-1}) f(\lambda_i)(\lambda_i \mathbb{1} - A)^{-1}.$$

Now by Proposition 3.2.29 one has $A\Sigma_N(f) \in D(\delta)$ and

$$\delta(A\Sigma_N(f)) = (2\pi i)^{-1} \sum_{i=1}^{N} (\lambda_i - \lambda_{i-1}) f(\lambda_i)\lambda_i(\lambda_i \mathbb{1} - A)^{-1}\delta(A)(\lambda_i \mathbb{1} - A)^{-1}.$$

But this latter sum converges in norm. Hence $Af(A) \in D(\delta)$ and

$$\delta(Af(A)) = (2\pi i)^{-1} \int_C d\lambda\, f(\lambda)\lambda(\lambda\mathbb{1} - A)^{-1}\delta(A)(\lambda\mathbb{1} - A)^{-1}.$$

Similarly, if $\mathbb{1} \in D(\delta)$ one concludes that $f(A) \in D(\delta)$ and

$$\delta(f(A)) = (2\pi i)^{-1} \int_C d\lambda\, f(\lambda)(\lambda\mathbb{1} - A)^{-1}\delta(A)(\lambda\mathbb{1} - A)^{-1}.$$

In particular, this type of argument shows that if $A \in D(\delta)$ is positive and invertible then $A^{1/2} \in D(\delta)$.

The second approach, which is more efficient, proceeds by Fourier analysis and the following lemma is basic.

Lemma 3.2.31. *Let δ be a norm-closed derivation of C^*-algebra $\mathfrak{A}$ with identity $\mathbb{1}$ and assume $\mathbb{1} \in D(\delta)$. If $A = A^* \in D(\delta)$ and $U_z = \exp\{zA\}$ for $z \in \mathbb{C}$ then $U_z \in D(\delta)$ and*

$$\delta(U_z) = z \int_0^1 dt\, U_{tz}\delta(A)U_{(1-t)z}$$

Proof. First remark that the exponential function may be defined by power series expansion and thus the integrand $U_{tz}\delta(A)U_{(1-t)z}$ of the above integral is norm continuous. Thus the integral may be understood in the Riemann sense. Next it is useful to define U_z by the alternative algorithm

$$\lim_{n\to\infty} \left\| U_z - \left(\mathbb{1} - \frac{zA}{n}\right)^{-n} \right\| = 0,$$

which may be established by manipulation with norm-convergent power series.

Next choose $n > |z|\,\|A\|$ and note that $(\mathbb{1} - zA/n)^{-1} \in D(\delta)$ by Proposition 3.2.29; further,

$$\begin{aligned}\delta\left(\left(\mathbb{1} - \frac{zA}{n}\right)^{-n}\right) &= \sum_{m=0}^{n-1} \left(\mathbb{1} - \frac{zA}{n}\right)^{-m} \delta\left(\left(\mathbb{1} - \frac{zA}{n}\right)^{-1}\right)\left(\mathbb{1} - \frac{zA}{n}\right)^{-n+m+1} \\ &= \left(\frac{z}{n}\right) \sum_{m=0}^{n-1} \left(\mathbb{1} - \frac{zA}{n}\right)^{-m-1} \delta(A)\left(\mathbb{1} - \frac{zA}{n}\right)^{-n+m}.\end{aligned}$$

The second equality follows from a second application of Proposition 3.2.29. Now the usual estimates associated with the Riemann integral show that the right-hand side of this last equation converges in norm to the integral

$$z \int_0^1 dt\, U_{tz}\delta(A)U_{(1-t)z}$$

and the statement of the lemma follows because δ is norm closed.

Note that we could have dropped the assumption $\mathbb{1} \in D(\delta)$ in the lemma and nevertheless arrived at the weaker conclusion $AU_z \in D(\delta)$. This remark is also valid for the following theorem which develops the Fourier analysis of $D(\delta)$.

Theorem 3.2.32. *Let δ be a norm-closed derivation of a C^*-algebra $\mathfrak{A}$ with identity $\mathbb{1}$ and assume $\mathbb{1} \in D(\delta)$. Further, let f be a function of one real variable satisfying*

$$|\hat{f}| = (2\pi)^{-1/2} \int dp |\hat{f}(p)||p| < \infty,$$

where $\hat{f}$ is the Fourier transform of f. It follows that if $A = A^* \in D(\delta)$ then $f(A) \in D(\delta)$ and

$$f(A) = (2\pi)^{-1/2} \int dp\, \hat{f}(p)e^{ipA},$$

$$\delta(f(A)) = i(2\pi)^{-1/2} \int dp\, \hat{f}(p)p \int_0^1 dt\, e^{itpA}\delta(A)e^{i(1-t)pA}.$$

Hence

$$\|\delta(f(A))\| \leq |\hat{f}|\, \|\delta(A)\|.$$

PROOF. First assume $\hat{f}$ to be continuous. One may approximate $f(A)$ by Riemann sums

$$\Sigma_N(f) = (2\pi)^{-1/2} \sum_{i=1}^{N} (p_i - p_{i-1})\hat{f}(p_i)e^{ip_iA}$$

and then $\Sigma_N(f) \in D(\delta)$ and

$$\delta(\Sigma_N(f)) = (2\pi)^{-1/2} \sum_{i=1}^{N} (p_i - p_{i-1})\hat{f}(p_i)ip_i \int_0^1 dt\, e^{itp_iA}\delta(A)e^{i(1-t)p_iA}$$

by Lemma 3.2.31. Both these sums converge in norm and as δ is normclosed the desired domain property, and action of δ, follow directly. The norm estimate $\|\delta(f(A))\| \leq |\hat{f}|\, \|\delta(A)\|$ is evident and the result for general f then follows by a continuity argument using this estimate.

This result almost reproduces the result obtained for abelian C^*-algebras at the beginning of this subsection because the class of functions over the spectrum $\sigma(A)$, of A, which have extensions f to $\mathbb{R}$ with $|\hat{f}| < \infty$ is almost the set of once-continuously differentiable functions. In fact this class contains the twice-continuously differentiable functions.

Corollary 3.2.33. *Let δ be a norm-closed derivation of a C^*-algebra $\mathfrak{A}$ with identity $\mathbb{1}$ and assume $\mathbb{1} \in D(\delta)$. Take $A = A^* \in D(\delta)$ and let f be a function of one real variable which is twice-continuously differentiable with compact support. It follows that $f(A) \in D(\delta)$ and*

$$\|\delta(f(A))\| \leq \|\delta(A)\| \left(\frac{\pi}{2} \int dx \left| \frac{d^2f}{dx^2}(x) - \frac{d}{dx} f(x) \right|^2 \right)^{1/2}.$$

PROOF. By Theorem 3.2.32, $f(A) \in D(\delta)$ and

$$\begin{aligned}
\|\delta(f(A))\| &\leq \|\delta(A)\|(2\pi)^{-1/2} \int dp\, |p|\, |\hat{f}(p)| \\
&\leq \|\delta(A)\|(2\pi)^{-1/2} \int dp\, |p + i|^{-1} |p^2 + ip|\, |\hat{f}(p)| \\
&\leq \|\delta(A)\| \left(\frac{\pi}{2} \int dp\, |(p^2 + ip)\hat{f}(p)|^2 \right)^{1/2},
\end{aligned}$$

where the last step uses the Cauchy–Schwarz inequality. Application of Parseval's equality for square-integrable functions gives

$$\|\delta(f(A))\| \leq \|\delta(A)\| \left(\frac{\pi}{2} \int dx \left| \frac{d^2f}{dx^2}(x) - \frac{d}{dx} f(x) \right|^2 \right)^{1/2}.$$

Although the functions covered by Theorem 3.2.32 almost include the functions which are once-continuously differentiable on $\sigma(A)$ this is not quite the case. What is perhaps more surprising is that the result of the theorem cannot be extended to all such functions. It is possible to construct a norm-closed, norm-densely defined derivation δ of a C^*-algebra $\mathfrak{A}$ for which there exists an $A = A^* \in D(\delta)$ and a function f which is once-continuously differentiable on an interval containing the spectrum of A but for which $f(A) \notin D(\delta)$. The construction of this example is, however, quite complicated. It illustrates that the properties of derivations of general C^*-algebras differ radically in structure from those of an abelian C^*-algebra.

To conclude this subsection we will use the above techniques to discuss derivations and automorphism groups of the algebra $\mathcal{L}(\mathfrak{H})$ of bounded operators on the Hilbert space $\mathfrak{H}$. There are two perturbation techniques which are useful in this respect and are of a general nature.

First let α_t be a C_0-group, or $C_0{}^*$-group of $*$-automorphisms of the C^*-algebra $\mathfrak{A}$ with generator δ. If δ_P is a bounded derivation of $\mathfrak{A}$ then $\delta + \delta_P$ is the generator of a C_0- or $C_0{}^*$-group $\alpha_t{}^P$ of $*$-mappings of the Banach space $\mathfrak{A}$

by Theorem 3.1.33. But α_t^P is actually a group of *-automorphisms because for $A, B \in D(\delta)$ one calculates, with the aid of the derivation property, that

$$\frac{d}{dt}\alpha_{-t}^P(\alpha_t^P(A)\alpha_t^P(B)) = 0.$$

Hence $\alpha_t^P(AB) = \alpha_t^P(A)\alpha_t^P(B)$.

Secondly, let δ be a derivation of a C^*-algebra and assume there is a projector $E \in D(\delta)$. One can then define a bounded derivation δ_E by

$$\delta_E(A) = i[H_E, A], \qquad H_E = i\delta(E)E - iE\delta(E),$$

for all $A \in \mathfrak{A}$. Now consider the sum $\delta^E = \delta + \delta_E$ on $D(\delta)$. One has

$$\begin{aligned}\delta^E(E) &= \delta(E) + 2E\delta(E)E - \delta(E)E - E\delta(E)\\ &= \delta(E - E^2) + 2E\delta(E)E\\ &= 0,\end{aligned}$$

where we have used $E^2 - E = 0$ and hence $E\delta(E^2 - E)E = E\delta(E)E = 0$. Thus if δ is the generator of a group of *-automorphisms α_t then δ^E is the generator of a perturbed group α_t^E which satisfies $\alpha_t^E(E) = E$.

Now we turn to the examination of $\mathscr{L}(\mathfrak{H})$.

EXAMPLE 3.2.34. Let δ be a norm-closed symmetric derivation of the algebra $\mathscr{L}(\mathfrak{H})$ of all bounded operators on the Hilbert space $\mathfrak{H}$. Assume that $D(\delta)$ is dense in the weak (strong) operator topology on $\mathscr{L}(\mathfrak{H})$ and $D(\delta) \cap \mathscr{LC}(\mathfrak{H}) \neq \{0\}$. We first claim that $D(\delta)$ contains a rank one projector. To establish this we choose a nonzero $B \in D(\delta) \cap \mathscr{LC}(\mathfrak{H})$ and form $C = B^*B$. Now if C has an eigenvalue c and if E_c is the associated finite-rank projector it follows that

$$E_c = (2\pi c i)^{-1}\int d\lambda\, C(\lambda\mathbb{1} - C)^{-1},$$

where the integral is around a closed curve which contains the isolated point c of the spectrum of C. Now by Proposition 3.2.29 and the subsequent discussion $E_c \in D(\delta)$. Next let P be a rank one projector such that $E_cPE_c = P$ and choose $A_n \in D(\delta)$ such that $A_n = A_n^*$ and $A_n \to P$ strongly. It follows that $\|E_cA_nE_c - E_cPE_c\| \to 0$ and hence $\|E_cA_nE_c - P\| \to 0$. Thus for n sufficiently large $E_cA_nE_c \in D(\delta)$ has a simple eigenvalue in the neighborhood of one. If E is the associated spectral projector, then $E \in D(\delta)$ by a second contour integral argument. Let Ω be a unit vector in the range of E and consider the vector state $\omega = \omega_\Omega$. One has

$$\begin{aligned}|\omega(\delta(A))| &= |\omega(E\delta(A)E)|\\ &= |\omega(\delta(EAE)) - \omega(\delta(E)AE) - \omega(EA\delta(E))|\\ &\leq |\omega(\delta(\omega(A)E))| + |\omega(\delta(E)A)| + |\omega(A\delta(E))|\\ &\leq 3\|\delta(E)\|\{\omega(A^*A) + \omega(AA^*)\}^{1/2}.\end{aligned}$$

Therefore Proposition 3.2.28 establishes the existence of a symmetric operator H such that $D(H) = D(\delta)\Omega$ and

$$\delta(A) = i[H, A]$$

for all $A \in D(\delta)$. In particular, every weakly* (norm-) closed, weakly* densely (norm-densely) defined derivation of $\mathscr{L}(\mathfrak{H})$ with $D(\delta) \cap \mathscr{L}\mathscr{C}(\mathfrak{H}) \neq \{0\}$ has this form. In the next example we will see that this class contains all the generators of weakly* continuous one-parameter automorphism groups.

A variation of the above argument shows that if δ is a derivation of $\mathscr{L}\mathscr{C}(\mathfrak{H})$ such that $D(\delta)$ contains a finite-rank operator, then δ is norm closable and extends to a σ-weakly closed derivation of $\mathscr{L}(\mathfrak{H})$. To prove this, one first exploits that $D(\delta)$ is a dense *-algebra in $\mathscr{L}\mathscr{C}(\mathfrak{H})$ to produce a one-dimensional projection E in $D(\delta)$, and then uses the estimate on $\omega(\delta(A))$ derived above together with Corollary 3.2.27. The same type of argument shows that any derivation of $\mathscr{L}(\mathfrak{H})$ with a finite-rank operator in its domain is σ-weakly closable.

This discussion of the derivations of $\mathscr{L}(\mathfrak{H})$ allows us to complete the characterizations of one-parameter groups of *-automorphisms begun in Example 3.2.14. In this example we showed that every group of this type is represented by a family of unitary operators U_t on $\mathfrak{H}$ in the form

$$\alpha_t(A) = U_t A U_t^*$$

but it remains to show that continuity of α_t implies that the U_t may be chosen to be a continuous one-parameter group.

EXAMPLE 3.2.35. Let α_t be a weakly* continuous one-parameter group of *-automorphisms of $\mathscr{L}(\mathfrak{H})$ and let δ denote the weak* generator of α_t. It follows that δ is weakly* closed, hence norm-closed, and $D(\delta)$ is dense in the weak operator topology on $\mathscr{L}(\mathfrak{H})$. Now $\alpha_t(\mathscr{L}\mathscr{C}(\mathfrak{H})) = \mathscr{L}\mathscr{C}(\mathfrak{H})$ for all t by Example 3.2.14, and the restriction of $t \mapsto \alpha_t$ to $\mathscr{L}\mathscr{C}(\mathfrak{H})$ is weakly continuous by Proposition 2.6.13. Hence $D(\delta) \cap \mathscr{L}\mathscr{C}(\mathfrak{H})$ is norm dense in $\mathscr{L}\mathscr{C}(\mathfrak{H})$. Thus by the arguments of Example 3.2.34 there is a rank one projector $E \in D(\delta)$. Again form $\delta^E = \delta + \delta_E$ in the manner described prior to Example 3.2.34 and remark that δ^E is the generator of a perturbed group, α_t^E, of *-automorphisms by the same discussion. Moreover, $\alpha_t^E(E) = E$. Let Ω be a unit vector in the range of E and define operators U_t^E on $\mathfrak{H}$ by

$$U_t^E A\Omega = \alpha_t^E(A)\Omega.$$

One easily checks that this is a consistent definition of a unitary group, e.g.,

$$\begin{aligned}(U_t^E A\Omega, U_s^E B\Omega) &= \omega_\Omega(\alpha_t^E(A^*\alpha_{s-t}^E(B))) \\ &= (A\Omega, \alpha_{s-t}^E(B)\Omega) = (A\Omega, U_{s-t}^E B\Omega),\end{aligned}$$

where we have used the invariance of ω_Ω, under α_t^E, which follows from $\alpha_t^E(E) = E$. But U_t^E is strongly continuous because

$$\|(U_t^E - I)A\Omega\|^2 = 2\omega_\Omega(A^*A) - \omega_\Omega(A^*\alpha_t^E(A)) - \omega_\Omega(\alpha_t^E(A^*)A).$$

Let H^E denote the self adjoint generator of U_t^E and define $H = H^E - H_E$, where H_E was given prior to Example 3.2.34. The operator H is selfadjoint and if $U_t = e^{itH}$ one checks that $\alpha_t(A) = U_t A U_t^*$, e.g., the generator of $t \mapsto U_t A U_t^*$ is $\delta^E - \delta_E = \delta$.

Thus we have established that each C_0*-group of *-automorphisms α_t, of $\mathscr{L}(\mathfrak{H})$, has the form $\alpha_t(A) = U_t A U_t^*$, where U_t is a strongly continuous group of unitary operators on $\mathfrak{H}$.

Let us now return to the discussion of Wigner symmetries of the Hilbert space $\mathfrak{H}$. Let $\hat{\psi}$ denote a ray in $\mathfrak{H}$ and $\hat{\psi} \mapsto \hat{\alpha}_t(\hat{\psi})$ a one-parameter group of Wigner symmetries. It follows from Example 3.2.14 and the associated discussion that $\hat{\alpha}_t$ induces a one-parameter group of *-automorphisms, or antilinear *-automorphisms α_t of $\mathscr{L}(\mathfrak{H})$ and that these maps are implemented as

$$\alpha_t(A) = U_t A U_t^*,$$

where the U_t are unitary, or antiunitary, operators which are determined uniquely up to a phase. Now if the symmetry group $\hat{\alpha}_t$ is continuous in the sense that

$$p(\hat{\psi} - \hat{\alpha}_t(\hat{\psi}), \hat{\psi} - \hat{\alpha}_t(\hat{\psi})) = \mathrm{Tr}((E_\psi - \alpha_t(E_\psi))^2) \underset{t=0}{\rightarrow} 0,$$

where E_ψ is the projector with ψ in its range, then it is easily checked that α_t is weakly* continuous. Thus each α_t is a *-automorphism by Corollary 3.2.13, and Example 3.2.35 implies that the phases of U_t may be chosen so that the U_t are strongly continuous and satisfy the group law

$$U_s U_t = U_{s+t}.$$

Thus all continuous one-parameter groups of Wigner symmetries are implemented by strongly continuous one-parameter groups of unitary operators.

We conclude with a discussion of C_0-groups α_t of *-automorphisms of $\mathscr{L}(\mathfrak{H})$.

EXAMPLE 3.2.36. Let $\alpha_t(A) = U_t A U_t^*$ be a one-parameter group of *-automorphisms of $\mathscr{L}(\mathfrak{H})$ and assume that α_t is strongly continuous, i.e.,

$$\lim_{t \to 0} \|\alpha_t(A) - A\| = 0, \qquad A \in \mathscr{L}(\mathfrak{H}).$$

Next let H denote the infinitesimal generator of the group U_t, and E_H the spectral family of H. Assume that the spectrum of H is unbounded. If $\varepsilon > 0$, $\delta > 0$ it is easily argued that one may choose $\{a_n\}_{n \geq 0}$ and a such that the intervals $I_n = [a_n, a_n + a]$ are disjoint, $E_H(I_n)\mathfrak{H}$ is nonempty, and

$$\sup_n |e^{i(a_n - a_{n+1})t} - 1| \geq 2 - \delta, \qquad |e^{ita} - 1| < \tfrac{1}{2}$$

for all $|t| < \varepsilon$. Now choose unit vectors $\psi_n \in E_H(I_n)\mathfrak{H}$ and define V by

$$V\psi = \sum_{n \geq 0} \psi_n(\psi_{n+1}, \psi).$$

One has $\|V\| = 1$ and

$$(e^{i(a_n - a_{n+1})t} - 1)\psi_n = (\alpha_t(V) - V)\psi_{n+1} - (U_t - e^{ia_n t})\psi_n(\psi_{n+1}, U_{-t}\psi_{n+1})$$
$$- e^{ia_n t}\psi_n(\psi_{n+1}, (U_{-t} - e^{-ia_{n+1}t})\psi_{n+1})$$

and hence

$$2 \leq \|\alpha_t(V) - V\| + 1 + \delta$$

for all $|t| < \varepsilon$, which is a contradiction. Hence H must have bounded spectrum and α_t must be uniformly continuous because $\|\alpha_t(A) - A\| \leq 2\|A\|(\exp\{|t|\,\|H\|\} - 1)$.

3.2.3. Spectral Theory and Bounded Derivations

In the next three subsections we study the characterization of those derivations which are generators of automorphism groups of C^*-algebras and von Neumann algebras. In the corresponding Banach space study of Section 3.1 we began with the simplest case of uniformly continuous groups and established that this continuity property was equivalent to the boundedness of the generator. Thus it is natural to begin with an examination of bounded derivations. We have already seen in Corollary 3.2.23 that an everywhere defined derivation of a C^*-algebra $\mathfrak{A}$ is bounded and one of the principal results of this subsection is to establish that the most general bounded derivation δ of a von Neumann algebra $\mathfrak{M}$ is of the form $\delta(A) = i[H, A]$ for some $H \in \mathfrak{M}$. Our proof of this fact will use the so-called spectral theory of one-parameter groups of automorphisms of $\mathfrak{A}$, or $\mathfrak{M}$. This use of spectral theory has the disadvantage that the proof of the derivation theorem is rather protracted but it has the great advantage that it allows the deduction of more general results for groups whose generators are semibounded in a suitable sense.

An automorphism α of $\mathfrak{A}$, or $\mathfrak{M}$, is called *inner* if there exists a unitary element $U \in \mathfrak{A}$, or $U \in \mathfrak{M}$, such that $\alpha(A) = UAU^*$. If $t \mapsto \alpha_t$ is a one-parameter group of automorphisms then the set of t such that α_t is inner clearly forms a subgroup of $\mathbb{R}$. Spectral theory is of importance for the analysis of this subgroup and is generally of use for the analysis of inner automorphisms. The derivation theorem mentioned in the previous paragraph illustrates one aspect of this analysis and the examination of automorphism groups

$$\alpha_t(A) = U_t A U_t^*$$

for which the unitary group has a positive generator illustrates another aspect. (For further results, see Notes and Remarks.) For later applications to gauge groups we develop spectral theory for general locally compact abelian groups.

Throughout the rest of this section we will use the following notation.

G is locally compact abelian group with Haar measure dt;
$\hat{G}$ is the dual group of G, i.e., the character group of G with the topology introduced in Section 2.7.1.

$L^1(G)$ is the group algebra of G, i.e., the set of L^1-functions on G with algebraic operations defined by

$$f * g(t) = \int ds\, f(t - s)g(s),$$

$$f^*(t) = \overline{f(-t)}.$$

$f * g$ is called the *convolution product* of f and g.

It is well known that there is a one-to-one correspondence between the characters of $L^1(G)$ and characters $t \mapsto (\gamma, t)$ of G given by the Fourier transform

$$f \in L^1(G) \mapsto \hat{f}(\gamma) = \int dt\, f(t)\overline{(\gamma, t)}.$$

Thus the Gelfand transformation coincides with the Fourier transformation in this case, and realizes $L^1(G)$ as a dense *-subalgebra of the C^*-algebra $C_0(\hat{G})$.

There is a bijective correspondence between closed ideals $\mathfrak{J}$ in $C_0(\hat{G})$ and closed subsets $K \subseteq \hat{G}$ given by

$$K = \{\gamma; \hat{f}(\gamma) = 0 \text{ for all } \hat{f} \in \mathfrak{J}\}.$$

Thus there is a mapping from the closed *-ideals $\mathfrak{J} \in L^1(G)$ onto the closed subsets $K \subseteq \hat{G}$ given by

$$K = \{\gamma; \hat{f}(\gamma) = 0 \text{ for all } f \in \mathfrak{J}\}.$$

This correspondence is not one-to-one in general, but by the Tauberian theorem there is only one ideal corresponding to $K = \emptyset$ or to one-point sets K (and for other special choices of K). A result we will often use is that given a compact set $K \subseteq \hat{G}$ and an open set $W \supseteq K$, there exists a function $f \in L^1(G)$ such that $\hat{f}(\gamma) = 1$ for $\gamma \in K$ and $\hat{f}(\gamma) = 0$ for $\gamma \in \hat{G}\backslash W$.

It is known by the SNAG (Stone–Naimark–Ambrose–Godement) theorem that there is a one-to-one correspondence between continuous unitary representations $\{\mathfrak{H}, U\}$ of G and projection-valued measures dP on $\hat{G}$ with values in $\mathfrak{H}$, given by

$$U_t = \int_{\hat{G}} \overline{(\gamma, t)}\, dP(\gamma).$$

When $G = \mathbb{R}$, this is simply Stone's theorem. In this section we consider partial extensions of the SNAG theorem to $\sigma(X, F)$-continuous representations U of G on the Banach space X, where we assume $\|U_t\| \leq M$ for all $t \in G$ (for a suitable constant M), and the pair (X, F) satisfy the requirements (a), (b), (c) stated at the beginning of Section 3.1.2. By Proposition 3.1.4, such a representation U of G defines a representation U of $L^1(G)$ as $\sigma(X, F)$–$\sigma(X, F)$-closed, norm-bounded operators on X, by

$$U(f) = \int dt\, f(t)U_t, \qquad f \in L^1(G),$$

i.e., $U(f * g) = U(f)U(g)$. We also occasionally use the notation $U_f = U(f)$. Now the basic concepts for the partial development of a spectral theory for U are introduced as follows.

Definition 3.2.37. Let $t \mapsto U_t$ be a $\sigma(X, F)$-continuous representation of G such that $\|U_t\| \leq M$ for all $t \in G$. Let Y be a subset of X. Define

$$\mathfrak{J}_Y{}^U = \{f \in L^1(G); U(f)A = 0 \text{ for all } A \in Y\}.$$

Then $\mathfrak{I}_Y^U$ is a closed *ideal in $L^1(G)$. Next define the *spectrum of* Y as the following closed subset of $\hat{G}$:

$$\sigma_U(Y) = \{\gamma \in \hat{G}; \hat{f}(\gamma) = 0 \text{ for all } f \in \mathfrak{I}_Y^U\}.$$

The *spectrum of* U is defined by

$$\sigma(U) = \sigma_U(X).$$

The *spectral subspace* $X^U(E)$ corresponding to a subset $E \subseteq \hat{G}$ is defined by

$$X^U(E) = \overline{\{A \in X; \sigma_U(A) \subseteq E\}},$$

where the bar denotes closure in the $\sigma(X, F)$-topology. Finally, we define the associated subspace $X_0{}^U(E)$ as the $\sigma(X, F)$-closed linear span of elements of the form $U(f)A$, where $\operatorname{supp} \hat{f} \subseteq E$ and $A \in X$.

Let us now illustrate these various concepts with two examples. First consider a continuous representation of a one-parameter group by unitary operators U_t on a Hilbert space $\mathfrak{H}$. The unitary group has the spectral decomposition

$$U_t = \int dP(p)\, e^{-itp}$$

and for any closed, or open, set $E \subseteq \hat{\mathbb{R}} = \mathbb{R}$

$$\mathfrak{H}^U(E) = P(E)\mathfrak{H}.$$

Thus one has

$$\sigma(U) = \text{support of } P = \sigma(H),$$

where H is the selfadjoint generator of U. Furthermore if, $\psi \in \mathfrak{H}$ then $\sigma(\psi)$ is the smallest closed set $E \subseteq \hat{\mathbb{R}}$ such that $P(E)\psi = \psi$. We have used the identification of $\mathbb{R}$ with $\hat{\mathbb{R}}$ in which $p \in \mathbb{R}$ corresponds to the character $t \mapsto (p, t) = e^{ipt}$. The results we have stated are easy to derive and will be consequences of Lemma 3.2.39 and Proposition 3.2.40.

Although the foregoing example amply illustrates the nature of the spectrum it does not fully illustrate the important algebraic aspect of the spectral subspaces. For this let us specialize to the case that U_t is the group of translations on $L^2(\mathbb{R})$,

$$(U_t\psi)(x) = \psi(x - t), \qquad \psi \in L^2(\mathbb{R}),$$

and consider the algebra $L^\infty(\mathbb{R})$ acting by multiplication on $L^2(\mathbb{R})$. The group of translations also act on $L^\infty(\mathbb{R})$ and one can consider the related spectral subspaces. If E is closed, then $X_0{}^U(E)$ is the closed subspace of $L^\infty(\mathbb{R})$ formed by the $f \in L^\infty(\mathbb{R})$ with $\operatorname{supp} \hat{f} \subseteq E$. Now for $f \in X_0{}^U(E)$ and $\psi \in \mathfrak{H}^U(F)$ consider the product $f\psi$. One has

$$(f\psi)(x) = f(x)\psi(x) = \int dp(\widehat{f\psi})(p)e^{ipx}$$

where

$$(\widehat{f\psi})(p) = \int dq\, \hat{f}(p - q)\hat{\psi}(q).$$

In particular, $\operatorname{supp}\widehat{f\psi} \subseteq \overline{E + F}$. Thus the spectrum of ψ is augmented by multiplication by f and the augmentation corresponds to the addition of values in E. Operators that have this property of increasing spectral values are familiar in various domains of mathematical physics and are called creation operators in field theory, shift operators in various settings of group theory and functional analysis, and occasionally raising and lowering operators. This type of property is the fundamental motivation for the introduction of spectral subspaces and these characteristics of raising spectral values will be developed further in Proposition 3.2.43.

The foregoing example not only illustrates the nature of the spectral subspaces but also introduces one of the basic techniques for their analysis, the convolution product.

We next state some elementary properties on the spectrum of an element.

Lemma 3.2.38. *Let U be a $\sigma(X, F)$-continuous uniformly bounded representation of G, and let $A, B \in X, f \in L^1(G)$. It follows that:*

(1) $\sigma_U(U_t A) = \sigma_U(A)$, $t \in G$;
(2) $\sigma_U(\alpha A + B) \subseteq \sigma_U(A) \cup \sigma_U(B)$;
(3) $\sigma_U(U(f)A) \subseteq \operatorname{supp} \hat{f} \cap \sigma_U(A)$;
(4) *if $f_1, f_2 \in L^1(G)$ and $\hat{f}_1 = \hat{f}_2$ in a neighborhood of $\sigma_U(A)$ then*

$$U(f_1)A = U(f_2)A.$$

PROOF. (1) $U(f)U_t A = U(f_t)A$, where $f_t(s) = f(s - t)$. Since $\hat{f}_t(\gamma) = \overline{(\gamma, t)}\hat{f}(\gamma)$ it follows that $\hat{f}_t(\gamma) = 0$ if and only if $\hat{f}(\gamma) = 0$.

(2) Clearly, $\sigma_U(\alpha A) = \sigma_U(A)$, so we may assume $\alpha = 1$. If $\gamma \notin \sigma_U(A) \cup \sigma_U(B)$, then we can find $f \in \mathfrak{I}_A{}^U$ with $\hat{f}(\gamma) = 1$ and $g \in \mathfrak{I}_B{}^U$ with $\hat{g}(\gamma) = 1$. Now $f * g \in \mathfrak{I}_{A+B}^U$ since

$$\begin{aligned} U(f * g)(A + B) &= U(f)U(g)(A + B) \\ &= U(g)U(f)A + U(f)U(g)B \\ &= 0. \end{aligned}$$

But $(\widehat{f * g})(\gamma) = \hat{f}(\gamma)\hat{g}(\gamma) = 1$; thus $\gamma \notin \sigma_U(A + B)$.

(3) If $U(g)A = 0$ then $U(g)U(f)A = U(f)U(g)A = 0$, hence $\sigma_U(U(f)A) \subseteq \sigma_U(A)$. On the other hand, if $\hat{g}$ vanishes on $\operatorname{supp} \hat{f}$ then $f * g = 0$; thus $U(g)U(f)A = 0$. Thus $\sigma_U(U(f)A) \subseteq \operatorname{supp} \hat{f}$.

(4) Set $g = f_1 - f_2$. We must show that $U(g)A = 0$. But $\hat{g}$ vanishes on a neighborhood of $\sigma_U(A)$; hence, by (3),

$$\sigma_U(U(g)A) \subseteq \operatorname{supp} \hat{g} \cap \sigma_U(A) = \emptyset.$$

It follows that $U(g)A = 0$.

We next turn to properties of the spectral subspaces $X^U(E)$ and $X_0{}^U(E)$.

Lemma 3.2.39. *Let U be a $\sigma(X, F)$-continuous uniformly bounded representation of G, and let E be a subset of $\hat{G}$. It follows that:*

(1) $X_0{}^U(E) \subseteq X^U(E)$;

(2) $U_t X_0{}^U(E) = X_0{}^U(E)$, $\quad U_t X^U(E) = X^U(E)$, $\quad t \in G$;

(3) *if E is closed, then*

$$X^U(E) = \{A \in X;\, \sigma_U(A) \subseteq E\}$$

(without closure);

(4) *if E is open, then*

$$X^U(E) = X_0{}^U(E) = \bigvee \{X^U(K);\, K \subseteq E,\, K \text{ is compact}\},$$

where $\bigvee$ denotes $\sigma(X, F)$-closed linear span;

(5) *if E is closed and N ranges over the open neighborhoods of 0 in $\hat{G}$, then*

$$X^U(E) = \bigcap_N X_0{}^U(E + N).$$

PROOF. (1) follows from Lemma 3.2.38(3).

(2) follows from Lemma 3.2.38(1) and the fact that supp $\hat{f}_t$ = supp $\hat{f}$.

(3) We have to show that the right-hand set is $\sigma(X, F)$-closed. Assume that A is in the $\sigma(X, F)$-closure of this set. If $\gamma \notin E$, pick $f \in L^1(G)$ such that $\hat{f}(\gamma) = 1$ and $\hat{f}$ vanishes in a neighborhood of E. By Lemma 3.2.38(4) we have then that $U(f)B = 0$ for all B in this set. Since $U(f)$ is $\sigma(X, F)$–$\sigma(X, F)$-continuous, $U(f)A = 0$. Hence $\sigma_U(A) \subseteq E$.

(4) Clearly, $X^U(E) \supset \bigvee_{K \subseteq E} X^U(K)$. Conversely, assume $\sigma_U(A) \subseteq E$. It is known that $L^1(G)$ has an approximate identity consisting of functions whose Fourier transforms have compact support, thus $X = \bigvee_{K \subseteq \hat{G}} X^U(K)$. Now A can be approximated by elements of the form $U(f)A$, where $\hat{f}$ has compact support K. But by Lemma 3.2.38(3), $\sigma_U(U(f)A) \subseteq \text{supp}\, \hat{f} \cap \sigma_U(A)$, which is a compact subset of E. Thus $X^U(E) \subseteq \bigvee_{K \subseteq E} X^U(K)$.

Finally, to see that $\bigvee_{K \subseteq E} X^U(K) \subseteq X_0{}^U(E)$ we remark that if $\sigma_U(A) \subseteq K \subseteq E$ it is possible to find an $f \in L^1(G)$ such that $\hat{f} = 1$ on a neighborhood of K and $\hat{f} = 0$ on $\hat{G} \setminus E$. Then $U(f)A = A$ by Lemma 3.2.38(4).

(5) It follows immediately from the definition of $X^U(E)$ that

$$X^U(E) = \bigcap_N X^U(\overline{E + N})$$

if E is closed, and since, by (4),

$$X^U(E) \subseteq X^U(E + N) = X_0{}^U(E + N) \subseteq X^U(\overline{E + N})$$

property (5) follows.

The Tauberian theorem implies that $\sigma_U(A) = \{\gamma\}$ if and only if $U_t A = \overline{(\gamma, t)}A$. We next prove a more general result on spectral concentration. For unitary groups this result is sometimes referred to as the Weyl criterion, and it gives an intuitive explanation of the spectrum.

Proposition 3.2.40. *Let U be a $\sigma(X, F)$-continuous uniformly bounded representation of an abelian locally compact group G. The following conditions are equivalent:*

(1) $\gamma_0 \in \sigma(U)$;
(2) *for all neighborhoods E around γ_0*

$$X_0{}^U(E) \neq \{0\};$$

(3) *for all $\varepsilon > 0$ and all compacts $K \subseteq G$ there exists a compact neighborhood E around γ_0 such that $X^U(E) \neq \{0\}$ and*

$$\|U_t A - \overline{(\gamma_0, t)} A\| \leq \varepsilon \|A\|$$

for all $A \in X^U(E)$, $t \in K$;
(4) *there exists a net (sequence if G is separable) of elements $A_\alpha \in X$ such that $\|A_\alpha\| = 1$ for all α and*

$$\lim_\alpha \|U_t A_\alpha - \overline{(\gamma_0, t)} A_\alpha\| = 0$$

uniformly for t in compacts;
(5) *for all $f \in L^1(G)$ we have*

$$|\hat{f}(\gamma_0)| \leq \|U(f)\|_{\mathscr{L}(X)}.$$

If $G = \mathbb{R}$ and $U_t = \exp(tS)$, then these conditions are equivalent to:

(6) $-i\gamma_0 \in \sigma(S)$,

i.e.,
$$\sigma(S) = -i\sigma(U).$$

PROOF. (1) $\Rightarrow$ (2) If $X^U(E) = 0$ for some open neighborhood E of γ_0, pick an $f \in L^1(G)$ such that $\hat{f}(\gamma_0) = 1$ and $\operatorname{supp} \hat{f} \subseteq E$. Then $U(f) = 0$ by Lemma 3.2.39(4); thus $f \in \mathfrak{I}_X{}^U$ and $\gamma_0 \notin \sigma(U)$.

(2) $\Rightarrow$ (3) Let E_1 be a compact neighborhood of γ_0, and choose an $f \in L^1(G)$ such that $\hat{f}(\gamma) = 1$ for $\gamma \in E_1$. For $s \in K$ define $F(s) \in L^1(G)$ by

$$F(s)(t) = f(t - s) - \overline{(\gamma_0, s)} f(t).$$

Then

$$\widehat{F(s)}(\gamma) = (\overline{(\gamma, s)} - \overline{(\gamma_0, s)}) \hat{f}(\gamma).$$

Now $\widehat{F(s)}(\gamma_0) = 0$, $\|F(s)\|_1$ is bounded uniformly for $s \in K$, and for any $\delta > 0$ there is a compact subset $E \subseteq G$ such that

$$\int_{G \backslash E} dt \, |F(s)(t)| < \delta$$

for all $s \in K$. Therefore, by a standard argument of Fourier analysis (see Notes and Remarks), one may, for each $\varepsilon > 0$, find a $g \in L^1(G)$ such that $\|g\|_1 \leq 1 + \varepsilon$, $\hat{g} = 1$ in a neighborhood of γ_0, $\operatorname{supp} \hat{g} \subset E_1$, and $\|F(s) * g\|_1 < \varepsilon$ for $s \in K$.

Let E be a compact neighborhood of γ_0 contained in the interior of $E_1 \cap E_2$ and let h be a function such that $\hat{h}(\gamma) = 1$ in a neighborhood of E and $\operatorname{supp} \hat{h} \subseteq E_1 \cap E_2$. Then $U(h)A = A$ for $A \in X^U(E)$ (Lemma 3.2.38(4)) and we have for $\gamma \in E_1 \cap E_2$ and $s \in K$ that

$$\begin{aligned} \widehat{F(s) * g * h(\gamma)} &= (\overline{(\gamma, s)} - \overline{(\gamma_0, s)})\hat{f}(\gamma)\hat{g}(\gamma)\hat{h}(\gamma) \\ &= (\overline{(\gamma, s)} - \overline{(\gamma_0, s)})\hat{h}(\gamma), \end{aligned}$$

which implies, for $s \in K$,

$$U(F(s) * g * h) = (U_s - \overline{(\gamma_0, s)})U(h).$$

Hence, for $A \in X^U(E)$ and $s \in K$,

$$\begin{aligned} \|U_s A - \overline{(\gamma_0, s)}A\| &= \|(U_s - \overline{(\gamma_0, s)})U(h)A\| \\ &= \|U(F(s) * g)U(h)A\| \\ &\leq \|F(s) * g\|_1 \|U(h)A\| \\ &\leq \varepsilon\|A\|. \end{aligned}$$

(3) $\Rightarrow$ (4) This is evident.
(4) $\Rightarrow$ (5) For each compact $K \subseteq G$ we have

$$\begin{aligned} \|U(f)A_\alpha - \hat{f}(\gamma_0)A_\alpha\| &= \left\| \int_G dt\, f(t)(U_t A_\alpha - \overline{(\gamma_0, t)}A_\alpha) \right\| \\ &\leq \sup_K \|U_t A_\alpha - \overline{(\gamma_0, t)}A_\alpha\| \, \|f\|_1 + (M + 1) \int_{G \setminus K} dt\, |f(t)|. \end{aligned}$$

Thus, for all $\varepsilon > 0$ there exists an A_α such that $\|U(f)A_\alpha\| \geq |\hat{f}(\gamma_0)| - \varepsilon$.

(5) $\Rightarrow$ (1) If $f \in \mathfrak{I}_X{}^U$ then $U(f) = 0$ by definition, hence $\hat{f}(\gamma_0) = 0$ and $\gamma_0 \in \sigma(U)$.

We conclude by showing the conditions (1)–(5) are equivalent to (6) in the special case.

(6) $\Rightarrow$ (4) If $-i\gamma_0 \in \sigma(S)$ then $i\gamma_0 + S$ is not invertible. Now, for any $\varepsilon > 0$, $-i\gamma_0 + \varepsilon - S$ is invertible and

$$(-i\gamma_0 + \varepsilon - S)^{-1} = \int_0^\infty e^{i\gamma_0 t - \varepsilon t} U_t\, dt$$

by Proposition 3.1.6.

It follows that $\lim_{\varepsilon \to 0} \|(-i\gamma_0 + \varepsilon - S)^{-1}\| = \infty$. Thus there exist A_ε such that $A_\varepsilon \in D(S)$, $\|A_\varepsilon\| = 1$ and $\lim_{\varepsilon \to 0}(-i\gamma_0 + \varepsilon - S)A_\varepsilon = 0$. Thus $\lim_{\varepsilon \to 0}(-i\gamma_0 - S)A_\varepsilon = 0$. We then have

$$\begin{aligned} \|U_t A_\varepsilon - e^{-i\gamma_0 t}A_\varepsilon\| &= \left\| \int_0^t ds \frac{d}{ds}\, U_s e^{-i\gamma_0(t-s)} A_\varepsilon \right\| \\ &= \left\| \int_0^t ds\, U_s(S + i\gamma_0)e^{-i\gamma_0(t-s)} A_\varepsilon \right\| \\ &\leq M|t| \, \|(S + i\gamma_0)A_\varepsilon\|. \end{aligned}$$

(4) ⇒ (6) If A_α is as in (4) we have, for $\varepsilon > 0$,

$$\left\|(-i\gamma_0 + \varepsilon - S)^{-1}A_\alpha - \frac{1}{\varepsilon}A_\alpha\right\|$$

$$= \left\|\int_0^\infty dt\, e^{i\gamma_0 t - \varepsilon t}U_t A_\alpha - \int_0^\infty dt\, e^{-\varepsilon t}A_\alpha\right\| \to 0$$

as $\alpha \to \infty$. Thus $\|(-i\gamma_0 + \varepsilon - S)^{-1}\| \geq 1/\varepsilon$. Hence $(-i\gamma_0 - S)^{-1}$ cannot exist as a bounded operator.

In Proposition 3.1.1 we proved that a one-parameter group is norm continuous if, and only if, its generator is bounded. A similar result involving the spectrum is given as follows:

Proposition 3.2.41. *Let U be a $\sigma(X, F)$-continuous uniformly bounded representation of G. The following conditions are equivalent:*

(1) *$\sigma(U)$ is compact;*
(2) *U is norm continuous.*

PROOF. (1) ⇒ (2) Assume $\sigma(U)$ compact, and choose an $f \in L^1(G)$ such that $\hat{f} = 1$ on a neighborhood of $\sigma(U)$. Then by Lemma 3.2.38(4), $U(f)A = A$ for all $A \in X$. Hence

$$\begin{aligned}\|(U_t - I)A\| &= \|(U_t - I)U(f)A\| \\ &= \|(U(f_t) - U(f))A\| \\ &\leq M\|f_t - f\|_1\|A\|.\end{aligned}$$

Since $\|f_t - f\|_1 \to 0$ as $t \to 0$, the norm continuity follows.

(2) ⇒ (1) When $G = \mathbb{R}$, i.e., $U_t = \exp(tS)$ then it follows from Proposition 3.1.1 that $\|S\| \leq \infty$, and since $\sigma(S) \subseteq [-\|S\|, \|S\|]$ condition (1) follows from Proposition 3.2.40(6). For general G, one notes that if $\{f_\alpha\}$ is any approximate unit for $L^1(G)$ then $\|U(f_\alpha) - I\| \to 0$ in this case. Thus if $\mathfrak{A}$ is the abelian Banach subalgebra of $\mathcal{L}(X)$ generated by $U(L^1(G))$, one has $I \in \mathfrak{A}$ and so $\mathfrak{A}$ has compact spectrum. Now if $\gamma \in \sigma(U)$ it follows from Proposition 3.2.40(5) that

$$\chi_\gamma(U(f)) = \hat{f}(\gamma)$$

defines a character on $\mathfrak{A}$. Conversely, each character χ on $\mathfrak{A}$ defines a character on $L^1(G)$ by composition with U, i.e., there is a $\gamma \in \hat{G}$ such that

$$\chi(U(f)) = \hat{f}(\gamma),$$

and since

$$|\hat{f}(\gamma)| = |\chi(U(f))| \leq \|U(f)\|$$

it follows from Proposition 3.2.40(5) that $\gamma \in \sigma(U)$. Hence $\sigma(U) = \sigma(\mathfrak{A})$, so $\sigma(U)$ is compact.

We next apply the general spectral theory developed in the preceding pages to groups of automorphisms of operator algebras. Lemma 3.2.38 then has an extension due to the richer structure of these algebras.

Lemma 3.2.42. *Let α be a representation of a locally compact abelian group G in the automorphism group of a C^*-algebra $\mathfrak{A}$ or a von Neumann algebra $\mathfrak{M}$. Assume that α is weakly (strongly) continuous in the C^*-case and σ-weakly continuous in the von Neumann case. The following statements are valid (with $\mathfrak{R} = \mathfrak{A}$ or $\mathfrak{R} = \mathfrak{M}$):*

(1) $\sigma_\alpha(A^*) = -\sigma_\alpha(A)$;
(2) $\mathfrak{R}^\alpha(E)^* = \mathfrak{R}^\alpha(-E)$, $E \subseteq \hat{G}$;
(3) $\mathfrak{R}_0{}^\alpha(E_1)\mathfrak{R}_0{}^\alpha(E_2) \subseteq \mathfrak{R}_0{}^\alpha(E_1 + E_2)$ *if E_1, E_2 are open subsets of $\hat{G}$*;
(4) $\mathfrak{R}^\alpha(E_1)\mathfrak{R}^\alpha(E_2) \subseteq \mathfrak{R}^\alpha(\overline{E_1 + E_2})$ *if E_1, E_2 are closed subsets of $\hat{G}$*;
(5) $\sigma_\alpha(AB) \subseteq \overline{\sigma_\alpha(A) + \sigma_\alpha(B)}$.

PROOF. (1) follows from $\operatorname{supp} \hat{\bar{f}} = -\operatorname{supp} \hat{f}$ and $\alpha(f)(A)^* = \alpha(\bar{f})(A^*)$. Property (2) follows from (1). To prove (3), note that by Lemma 3.2.39(4), it is enough to prove that if $f, g \in L^1$, with $\operatorname{supp} \hat{f}$ and $\operatorname{supp} \hat{g}$ compact in E_1, E_2, respectively, and $A, B \in \mathfrak{R}$, then $\alpha(f)(A)\alpha(g)(B) \subseteq \mathfrak{R}^\alpha(E_1 + E_2)$. Thus it suffices to show that if $h \in L^1(G)$ with $\operatorname{supp} \hat{h} \cap (E_1 + E_2) = \emptyset$ then $\alpha(h)(\alpha(f)(A)\alpha(g)(B)) = 0$. But

$$\begin{aligned}\alpha(h)(\alpha(f)(A)\alpha(g)(B)) &= \iiint dr\, ds\, dt\; h(r)f(t)g(s)\alpha_{r+t}(A)\alpha_{r+s}(B)\\ &= \iiint du\, dv\, dw\; h(u)f(v-u)g(w+v-u)\alpha_v(A)\alpha_{w+v}(B).\end{aligned}$$

Now,

$$\int du\; h(u)f(v-u)g(w+v-u) = h * (f \cdot g_w)(v),$$

where $g_w(v) = g(v + w)$. The Fourier transform of the latter function of U is $\hat{h} \cdot (\hat{f} * \hat{g}_w)$. But $\operatorname{supp}(\hat{f} * \hat{g}_w) \subseteq \operatorname{supp} \hat{f} + \operatorname{supp} \hat{g} \subseteq E_1 + E_2$. It follows that $\hat{h} \cdot (\hat{f} * \hat{g}_w) = 0$; thus

$$\alpha(h)(\alpha(f)(A)\alpha(g)(B)) = 0$$

by an application of Fubini's theorem. Finally, (4) follows from (3) and Lemma 3.2.39(5), while (5) follows from (4) and Lemma 3.2.39(3).

Next, we relate the spectral subspaces of a unitarily implemented automorphism group with the spectral subspaces of the unitary group. In particular the next proposition identifies elements of certain spectral subspaces as "creation" operators or "shift" operators on the spectrum of the unitary group.

Proposition 3.2.43. *Let U be a strongly continuous one-parameter group of unitary operators on a Hilbert space $\mathfrak{H}$, with the spectral decomposition*

*$U_t = \int e^{-it\gamma}\, dP(\gamma)$. Let $\beta_t(A) = U_t A U_t^*$ be the σ-weakly continuous one-parameter group of *-automorphisms of $\mathscr{L}(\mathfrak{H})$ implemented by U. The following conditions are equivalent, for all $A \in \mathscr{L}(\mathfrak{H})$ and $\gamma \in \mathbb{R}$:*

(1) $\sigma_\beta(A) \subseteq [\gamma, \infty\rangle$;
(2) $AP([\lambda, \infty\rangle)\mathfrak{H} \subseteq P([\lambda + \gamma, \infty\rangle)\mathfrak{H}$ *for each* $\lambda \in \mathbb{R}$.

PROOF. We first remark that it is easily verified that $P(E)\mathfrak{H} = \mathfrak{H}^U(E)$ for all closed sets $E \subseteq \hat{\mathbb{R}} = \mathbb{R}$.

(1) ⇒ (2) By Lemma 3.2.39 it is enough to show that

$$\beta_f(A)U(g)\psi \in P([\lambda + \gamma - 2\varepsilon, \infty\rangle)\mathfrak{H},$$

where $\psi \in \mathfrak{H}$, $f, g \in L^1(\mathbb{R})$, and $\operatorname{supp} \hat{f} \subseteq \langle \gamma - \varepsilon, \infty\rangle$, $\operatorname{supp} \hat{g} \subseteq \langle \lambda - \varepsilon, \infty\rangle$, i.e., we have to show

$$U(h)\beta_f(A)U(g) = 0,$$

where f and g are as above and $\operatorname{supp} \hat{h} \subseteq \langle -\infty, \lambda + \gamma - 2\varepsilon\rangle$. But

$$U(h)\beta_f(A)U(g) = \iiint dt\, ds\, dr\, h(t)f(s)g(r)U_{t+s}AU_{-s+r}.$$

After the change of variables $u = t$, $v = s + t$, $w = r - s$, and an application of Fubini's theorem, one finds

$$U(h)\beta_f(A)U(g) = \iint k_w(v)U_v AU_w\, dv\, dw,$$

where $k_w(v) = h * (f \cdot g_w)$, $g_w(s) = g(s + w)$. Then $\widehat{k_w} = \hat{h}(\hat{f} * \widehat{g_w})$. But

$$\operatorname{supp} \widehat{g_w} = \operatorname{supp} \hat{g} \subseteq \langle \lambda - \varepsilon, \infty\rangle;$$

thus

$$\begin{aligned} \operatorname{supp}(\hat{f} * \widehat{g_w}) &\subseteq \overline{\operatorname{supp} \hat{f} + \operatorname{supp} \hat{g}} \\ &\subseteq \langle \gamma + \lambda - 2\varepsilon, \infty\rangle. \end{aligned}$$

But the last set is disjoint from $\operatorname{supp} \hat{h}$, and so $\widehat{k_w} = 0$, i.e., $U(h)\beta_f(A)U(g) = 0$.

(2) ⇒ (1) Assume (2) and take $\lambda_0 < \gamma$. Set $\varepsilon = (\gamma - \lambda_0)/5$, and let $f \in L^1(\mathbb{R})$ be such that $\hat{f}(\lambda_0) = 1$, $\operatorname{supp} \hat{f} \in \langle \lambda_0 - \varepsilon, \lambda_0 + \varepsilon\rangle$. We have to show $\beta_f(A) = 0$. But by Lemma 3.2.39 it is enough to show that $\beta_f(A)U(g)\psi = 0$, where $g \in L^1(\mathbb{R})$ is any function such that $\operatorname{supp} \hat{g}$ lies in an interval of the form $\langle \lambda_1 - \varepsilon, \lambda_1 + \varepsilon\rangle$. But since $\sigma_u U(g)\psi \subseteq [\lambda_1 - \varepsilon, \lambda_1 + \varepsilon] \subseteq [\lambda_1 - \varepsilon, \infty\rangle$, it follows from assumption (2) that $\sigma_u \beta_f(A)U(g)\psi \subseteq [\lambda_1 - \varepsilon + \gamma, \infty\rangle$. Thus it is enough to show that $U(h)\beta_f(A)U(g) = 0$ for any h such that $\operatorname{supp} \hat{h} \subseteq [\lambda_1 - 2\varepsilon + \gamma, \infty\rangle$. Using the transformation of variables used in the proof of (1) ⇒ (2) and keeping the same notation, one finds

$$\begin{aligned} \operatorname{supp}(\hat{f} * \widehat{g_w}) &\subseteq \langle \lambda_0 - \varepsilon, \lambda_0 + \varepsilon\rangle + \langle \lambda_1 - \varepsilon, \lambda_1 + \varepsilon\rangle \\ &= \langle \lambda_0 + \lambda_1 - 2\varepsilon, \lambda_0 + \lambda_1 + 2\varepsilon\rangle, \end{aligned}$$

which is disjoint from $[\gamma + \lambda_1 - 2\varepsilon, \infty\rangle$. Therefore $k_w = 0$ and $\beta_f(A) = 0$.

We next derive a result for comparison of two groups of automorphisms. This comparison states that the inclusion of the spectral subspaces of one group within the corresponding spectral subspaces of the second group ensures the equality of the groups. Although we state and derive this result in the von Neumann algebra case, a similar result is also true for strongly continuous one-parameter groups of *-automorphisms of C^*-algebras, by essentially the same proof.

Proposition 3.2.44. *Let α, β be two σ-weakly continuous one-parameter groups of *-automorphisms of a von Neumann algebra $\mathfrak{M}$ and assume that $\mathfrak{M}^\alpha[\lambda, \infty\rangle \subseteq \mathfrak{M}^\beta[\lambda, \infty\rangle$ for all $\lambda \in \mathbb{R}$. It follows that $\alpha_t = \beta_t$ for all $t \in \mathbb{R}$.*

PROOF. For $A \in \mathfrak{M}$, $\eta \in \mathfrak{M}_{*+}$ consider the function

$$f(t, s) = \eta(\beta_t \alpha_s(A)), \qquad t, s \in \mathbb{R}.$$

We will show that there exists a continuous function g on $\mathbb{R}$ such that $f(t, s) = g(t + s)$. It then follows that $f(-t, t) = f(0, 0)$, i.e., $\beta_{-t}\alpha_t(A) = A$, or, equivalently, $\alpha_t(A) = \beta_t(A)$ for all t.

First introduce the notation $f(h)$ by

$$f(h) = \iint dt\, ds\, h(t, s) f(t, s)$$

for all $h \in L^1(\mathbb{R}^2)$. Now if $h_1, h_2 \in L^1(\mathbb{R})$ with $\operatorname{supp} \hat{h}_1 \subseteq \langle\lambda, \infty\rangle$ and $\operatorname{supp} \hat{h}_2 \subseteq \langle-\infty, \lambda\rangle$ then

$$f(h_1 h_2) = \eta(\beta_{h_1}\alpha_{h_2}(A)) = 0$$

because $\alpha_{h_2}(A) \in \mathfrak{M}^\alpha[\lambda, \infty\rangle \subseteq \mathfrak{M}^\beta[\lambda, \infty\rangle$. Also

$$\begin{aligned} \iint dt\, ds\, \bar{h}_2(t)\bar{h}_1(s) f(t, s) &= \overline{\iint dt\, ds\, h_2(t)h_1(s)\overline{f(t, s)}} \\ &= \overline{\iint dt\, ds\, h_2(t)h_1(s)\eta(\beta_t\alpha_s(A^*))} \\ &= 0 \end{aligned}$$

by the same reasoning. But $\operatorname{supp} \hat{\bar{h}}_1 \subseteq \langle-\infty, -\lambda\rangle$, $\operatorname{supp} \hat{\bar{h}}_2 \subseteq \langle-\lambda, \infty\rangle$ and hence one deduces that

$$f\left(\sum_{i=1}^n h_{1i}h_{2i}\right) = 0$$

for all pairs h_{1i}, h_{2i} such that $\operatorname{supp} \hat{h}_{1i} \cap \operatorname{supp} \hat{h}_{2i} = \emptyset$. But because

$$|f(h)| \le \|f\|_\infty \|h\|_1$$

it easily follows that

$$f(h) = 0$$

for all $h \in L^1(\mathbb{R}^2)$ such that the Fourier transform $\hat{h}(p, q)$ is infinitely often differentiable with compact support in the region $p \neq q$.

The proof of the proposition is completed by a change of variables, to $s + t$ and $s - t$, and an application of the following lemma.

Lemma 3.2.45. *Let f be a bounded continuous function of two real variables and for $h \in L^1(\mathbb{R}^2)$ define $f(h)$ by*

$$f(h) = \iint ds\, dt\, f(s, t)h(s, t).$$

Assume that $f(h) = 0$ for all h such that the Fourier transform $\hat{h}(p, q)$ is infinitely often differentiable with compact support in the region $q \neq 0$. It follows that

$$f(s, t) = g(s)$$

for some bounded continuous function g.

PROOF. Firstly, note that

$$f(h) = f(h_\varepsilon),$$

where

$$\hat{h}_\varepsilon(p, q) = \hat{h}(p, q)\hat{\chi}_\varepsilon(q)$$

and $\hat{\chi}_\varepsilon$ is an infinitely often differentiable function of compact support which takes the value one for $|q| < \varepsilon$. Secondly, remark that

$$|f(h)| \leq \|f\|_\infty \|h\|_1.$$

Next let $h(s, t)$ be differentiable in t with differential $h'(s, t) = \partial h(s, t)/\partial t \in L^1(\mathbb{R}^2)$. One then has

$$|f(h')| \leq \|f\|_\infty \|h_\varepsilon'\|_1 \leq \|f\|_\infty \|h\|_1 \|\chi_\varepsilon'\|_1$$

where we have used

$$\begin{aligned} \|h_\varepsilon'\| &= \iint ds\, dt\, |h_\varepsilon'(s, t)| \\ &= \iint ds\, dt \left| \int dt'\, h(s, t')\chi_\varepsilon'(t - t') \right| \leq \|h\|_1 \|\chi_\varepsilon'\|_1. \end{aligned}$$

Next choose χ such that $\hat{\chi}$ is infinitely often differentiable with compact support and $\hat{\chi}(q) = 1$ for $|q| < 1$. If one defines $\hat{\chi}_\varepsilon(q) = \hat{\chi}(q/\varepsilon)$ then $\hat{\chi}_\varepsilon$ satisfies the above requirements and one easily checks that

$$\|\chi_\varepsilon'\|_1 = \varepsilon \|\chi'\|_1.$$

Taking the limit of ε to zero one deduces that $f(h') = 0$.

Finally, if h is a function whose Fourier transform is infinitely often differentiable with compact support and if h_a is defined by $h_a(s, t) = h(s, t + a)$ one has

$$f(h_a) - f(h) = \int_0^a db\, f(h_b') = 0.$$

This last relation can be alternatively expressed as

$$\iint ds\, dt\, (f(s, t) - f(s, t + a))h(s, t) = 0$$

and as $f \in L^\infty(\mathbb{R}^2)$ and h ranges over a dense subset of $L^1(\mathbb{R}^2)$ one has

$$f(s, t) = f(s, t - a) = f(s, 0).$$

After these extensive preliminaries on spectral analysis we return to the topic discussed at the beginning of the subsection, the characterization of uniformly continuous one-parameter groups of automorphisms of von Neumann algebras. Proposition 3.2.41 establishes that these groups are characterized by the condition of compact spectrum and we now approach the problem from this point of view. The next theorem, which is the principal result of this section, gives a description of groups whose spectrum has a semiboundedness property and the norm-continuous groups are contained as a special case. The theorem gives information of several different types. Firstly, it implies that each group whose spectrum has the semiboundedness property is automatically implemented by a group of unitary operators. Secondly, it establishes that the unitary operators may be chosen in the algebra, and thirdly, it demonstrates that the unitary group has a semibounded spectrum and that the spectral subspace of the unitary group is intimately related to the spectral subspaces of the automorphism group.

Theorem 3.2.46 (Borchers–Arveson theorem). *Let $t \mapsto \alpha_t$ be a σ-weakly continuous one-parameter group of *-automorphisms of a von Neumann algebra $\mathfrak{M}$. The following conditions are equivalent*:

(1) *there is a strongly continuous one-parameter unitary group $t \mapsto U_t \in \mathscr{L}(\mathfrak{H})$ with nonnegative spectrum such that*

$$\alpha_t(A) = U_t A U_t^*$$

for all $A \in \mathfrak{M}$, $t \in \mathbb{R}$;

(2) *there is a strongly continuous one-parameter unitary group $t \mapsto U_t$ in $\mathfrak{M}$ with nonnegative spectrum such that $\alpha_t(A) = U_t A U_t^*$ for all $A \in \mathfrak{M}$, $t \in \mathbb{R}$;*

(3)
$$\bigcap_{t \in \mathbb{R}} [\mathfrak{M}^\alpha[t, \infty\rangle \mathfrak{H}] = \{0\}.$$

Moreover, if these conditions are satisfied, one may take for U the group

$$U_t = \int_{-\infty}^{\infty} e^{-itp} dP(p),$$

where $P(\cdot)$ is the unique projection valued measure on $\mathbb{R}$ satisfying

$$P[t, \infty\rangle \mathfrak{H} = \bigcap_{s<t} [\mathfrak{M}^\alpha[s, \infty\rangle \mathfrak{H}].$$

PROOF. (2) $\Rightarrow$ (1) is trivial.

(1) $\Rightarrow$ (3) Let P be the projection-valued measure associated with U. Then $P([t, \infty\rangle) = \mathbb{1}$ when $t \leq 0$ and

$$\bigcap_{t \in \mathbb{R}} P([t, \infty\rangle)\mathfrak{H} = \{0\}.$$

But by Proposition 3.2.43, one has

$$\begin{aligned}\mathfrak{M}^{\alpha}[t, \infty\rangle\mathfrak{H} &= \mathfrak{M}^{\alpha}[t, \infty\rangle P[0, \infty\rangle\mathfrak{H} \\ &\subseteq P[t, \infty\rangle\mathfrak{H}.\end{aligned}$$

Condition (3) follows immediately.

(3) $\Rightarrow$ (2) For each $t \in \mathbb{R}$, define

$$Q_t = \bigcap_{s<t} [\mathfrak{M}^{\alpha}[s, \infty\rangle\mathfrak{H}].$$

Then Q_t is a decreasing family of projections which is left continuous in t and such that $Q_t \to 0$ strongly as $t \to \infty$. Moreover, $Q_t = \mathbb{1}$ for $t \leq 0$. Thus there is a unique projection-valued measure P on $\mathbb{R}$ such that $P[t, \infty\rangle = Q_t$ for all t.

Now since $[\mathfrak{M}^{\alpha}[t, \infty\rangle\mathfrak{H}] \in \mathfrak{M}'' = \mathfrak{M}$ it follows that $P[t, \infty\rangle \in \mathfrak{M}$ for all t. Therefore

$$U_t = \int e^{-itp}\, dP(p) \in \mathfrak{M}.$$

But U has nonnegative spectrum because $P[0, \infty\rangle = \mathbb{1}$. Next define an automorphism group β_t of $\mathfrak{M}$ by $\beta_t(A) = U_t A U_t{}^*$. By Lemma 3.2.42 one has

$$\mathfrak{M}^{\alpha}[s, \infty\rangle\mathfrak{M}^{\alpha}[t, \infty\rangle \subseteq \mathfrak{M}^{\alpha}[t + s, \infty\rangle$$

for all $t \in \mathbb{R}$. Therefore

$$\mathfrak{M}^{\alpha}[s, \infty\rangle P[t, \infty\rangle\mathfrak{H} \subseteq P[s + t, \infty\rangle\mathfrak{H}$$

for all $t \in \mathbb{R}$. It then follows from Proposition 3.2.43 that

$$\mathfrak{M}^{\alpha}[s, \infty\rangle \subseteq \mathfrak{M}^{\beta}[s, \infty\rangle$$

for all $s \in \mathbb{R}$. But then Proposition 3.2.44 implies $\alpha_t = \beta_t$ for all $t \in \mathbb{R}$.

As an immediate corollary of Theorem 3.2.46 we get the so-called derivation theorem, which we state both in a von Neumann and in a C^*-version. In the C^*-case the result can be strengthened under suitable auxiliary conditions (see Notes and Remarks).

Corollary 3.2.47. *Let δ be an everywhere defined, hence bounded, symmetric derivation of a von Neumann algebra $\mathfrak{M}$. It follows that there exists an $H = H^* \in \mathfrak{M}$ with $\|H\| \leq \|\delta\|/2$ such that*

$$\delta(A) = i[H, A]$$

for all $A \in \mathfrak{M}$.

PROOF. The derivation δ is bounded by Corollary 3.2.23, and one may introduce the norm-continuous group α_t of *-automorphisms of $\mathfrak{M}$ by

$$\alpha_t(A) = e^{t\delta}(A) = \sum_{n \geq 0} \frac{t^n}{n!} \delta^n(A).$$

It follows from Proposition 3.2.40 that

$$\sigma(\alpha) = i\sigma(\delta)$$

and then it follows from an application of condition (3) of the same proposition that $\mathfrak{M}^\alpha[t, \infty\rangle = \{0\}$ whenever $t > \|\delta\|$. Thus the Borchers–Arveson theorem implies the existence of a strongly continuous one-parameter group of unitary operators U_t with a positive self adjoint generator $H_0 \in \mathfrak{M}$ such that

$$\alpha_t(A) = U_t A U_t^*$$

for all $A \in \mathfrak{M}$. Hence by differentiation

$$\delta(A) = i[H_0, A].$$

But the explicit construction of U_t in the Borchers–Arveson theorem establishes that $\sigma(H_0) \subseteq [0, \|\delta\|]$. Introducing $H = H_0 - \|\delta\|\mathbb{1}/2$ one has $\|H\| \leq \|\delta\|/2$ and

$$\delta(A) = i[H, A].$$

One C^*-algebraic version of this result is the following:

Corollary 3.2.48. *Let δ be an everywhere defined, hence bounded, symmetric derivation of a C^*-algebra $\mathfrak{A}$. It follows that for any representation π of $\mathfrak{A}$ there exists an $H = H^* \in \pi(\mathfrak{A})''$ such that $\|H\| \leq \|\delta\|/2$ and*

$$\pi(\delta(A)) = i[H, \pi(A)]$$

for all $A \in \mathfrak{A}$.

PROOF. First we show that if $\mathfrak{J} = \ker \pi$ then $\delta(\mathfrak{J}) \subseteq \mathfrak{J}$. It is enough to show that $\delta(A) \subseteq \mathfrak{J}$ for $A \in \mathfrak{J}_+$. But then $A = B^2$ with $B \in \mathfrak{J}_+$. Hence $\delta(A) = \delta(B)B + B\delta(B) \subseteq \mathfrak{J}$ since $\mathfrak{J}$ is an ideal. Hence $\delta(\mathfrak{J}) \subseteq \mathfrak{J}$ and one may consistently define a map $\hat{\delta}$ of $\pi(\mathfrak{A})$ by $\hat{\delta}(\pi(A)) = \pi(\delta(A))$. $\hat{\delta}$ is a derivation of $\pi(\mathfrak{A})$ and, by Proposition 3.2.24, $\hat{\delta}$ has a unique σ-weakly closed extension $\tilde{\delta}$ to $\pi(\mathfrak{A})''$ such that $\|\hat{\delta}\| = \|\tilde{\delta}\| \leq \|\delta\|$. The corollary now follows from Corollary 3.2.47.

Although we have only given the infinitesimal generator description of groups with compact spectrum one can also draw similar conclusions from Theorem 3.2.46 for groups with the semiboundedness property of the spectrum. These groups have generators of the form

$$\delta(A) = i[H, A],$$

where the selfadjoint operator H may be chosen to be positive and also affiliated to the von Neumann algebra $\mathfrak{M}$. We conclude by remarking that this choice of H which ensures that the implementing group $U_t = e^{itH}$ is contained in $\mathfrak{M}$ is not always the most natural choice. If $\mathfrak{M}$ is in standard form it follows from Corollary 2.5.32 that there exists a second unitary group V_t such that

$$\alpha_t(A) = V_t A V_t^* = U_t A U_t^*.$$

The group V_t is introduced by the aid of the natural cone $\mathscr{P}$ and is uniquely defined by the requirement $V_t\mathscr{P} \subseteq \mathscr{P}$. The "inner" group U_t and the "natural cone" group V_t are, however, distinct. One has in fact that $t \mapsto U_t J U_t J$ is a unitary group such that $U_t J U_t J \mathscr{P} = \mathscr{P}$ and $U_t J U_t J A J U_{-t} J U_{-t} = \alpha_t(A)$, hence

$$V_t = U_t J U_t J,$$

which means that the generator K of V_t is

$$K = H - JHJ.$$

3.2.4. Derivations and Automorphism Groups

The principal interest of symmetric derivations, and the main motivation for their analysis, arises from the fact that they occur as the generators of one-parameter groups of automorphisms. In this subsection we examine the characterizations of those derivations which are generators of automorphism groups. The simplest criterion of this sort comes from Proposition 3.1.1 and Corollary 3.2.23 and identifies the generators of norm-continuous groups.

Corollary 3.2.49. *Let $\mathfrak{A}$ be a C^*-algebra, and let δ be a linear operator on $\mathfrak{A}$. Then the following conditions are equivalent:*

(1) *δ is an everywhere defined symmetric derivation of $\mathfrak{A}$;*
(2) *δ is the generator of a norm-continuous one-parameter group $t \mapsto \tau_t$ of *-automorphisms of $\mathfrak{A}$.*

In this case, if π is any representation of $\mathfrak{A}$ there exists an $H = H^* \in \pi(\mathfrak{A})''$ such that

$$\pi(\tau_t(A)) = e^{itH}\pi(A)e^{-itH}$$

for all $A \in \mathfrak{A}$ and $t \in \mathbb{R}$.

The last statement of the corollary follows from Corollary 3.2.48.

Thus in the special case of norm continuous groups, the algebraic structure ensures that the automorphisms are unitarily implemented in all representations π. Moreover, the implementing unitary group U_t can be chosen as elements of $\pi(\mathfrak{A})''$. This last property is, however, very particular and cannot be expected to be generally valid. For example, if $\mathfrak{A}$ is abelian the corollary implies that there are no nontrivial norm-continuous groups of automorphisms and no nontrivial bounded derivations. It is nevertheless easy to construct examples of strongly continuous groups with unbounded derivations as generators, e.g., translations on $C_0(\mathbb{R})$ are generated by the operation of differentiation. For such general derivations we can obtain criteria which characterize generators by combining the Banach space theory of Section 3.1.2 with the theory for positive maps of Section 3.2.1. We first state the resulting twenty three criteria for C^*-algebras with identities.

Theorem 3.2.50. *Let $\mathfrak{A}$ be a C^*-algebra with identity $\mathbb{1}$, and let δ be a norm-densely defined norm-closed operator on $\mathfrak{A}$ with domain $D(\delta)$. It follows that δ is the generator of a strongly continuous one-parameter group of *-automorphisms of $\mathfrak{A}$ if, and only if, it satisfies at least one of the criteria in each of the families* (A), (B), *and* (C) *with the exception of the four combinations* ((A2), (B2), (C2)) *and* ((A3), (Bj), (C2)) *for $j = 1, 2, 3$:*

(A1) *$D(\delta)$ is a *-algebra and δ is a symmetric derivation;*
(A2) *$\mathbb{1} \in D(\delta)$ and $\delta(\mathbb{1}) = 0$;*
(A3) *$\mathbb{1} \in D(\delta)$ and $\delta(\mathbb{1})$ is selfadjoint;*
(B1) *$(I + \alpha\delta)(D(\delta)) = \mathfrak{A}$, $\alpha \in \mathbb{R}\backslash\{0\}$;*
(B2) *δ has a dense set of analytic elements;*
(B3) *The selfadjoint analytic elements for δ are dense in $\mathfrak{A}_{sa}$;*
(C1) *$\|(I + \alpha\delta)(A)\| \geq \|A\|$, $\alpha \in \mathbb{R}$, $A \in D(\delta)$;*
(C2) *$(I + \alpha\delta)(A) \geq 0$ implies that $A \geq 0$ for all $\alpha \in \mathbb{R}$, $A \in D(\delta)$;*
(C3) *δ is conservative.*

After removal of the conditions (A2), (A3), *and* (C2), *the statement of the theorem remains valid when $\mathfrak{A}$ does not have an identity.*

PROOF. Assume first that δ is the generator of a strongly continuous one-parameter group $t \mapsto \tau_t$ of *-automorphisms of $\mathfrak{A}$. Then (A1), (A2), and thus (A3), follow from the discussion in the introduction to Section 3.2.2. Each τ_t is an isometry by Corollary 2.3.4 and hence the conditions (B1), (B2), (C1), (C3) follow from Theorems 3.1.16 and 3.1.17. The condition (B3) follows from (B2) since δ is a symmetric operator. Since each τ_t is positivity preserving, (C2) is a consequence of the Laplace transform formula of Proposition 3.1.6:

$$(I + \alpha\delta)^{-1}(A) = \int_0^\infty dt\, e^{-t}\tau_{-\alpha t}(A).$$

We next prove the sufficiency of any triple ((Ai), (Bj), (Ck)) of conditions with the four exceptions mentioned in the theorem, when $\mathfrak{A}$ has an identity $\mathbb{1}$. First note that (A1) $\Rightarrow$ (A2) by Corollary 3.2.30 and (A2) $\Rightarrow$ (A3) trivially, so we may assume $i = 3$ except for ((A1), (B2), (C2)), ((A2), (B1), (C2)) and ((A2), (B3), (C2)). Since (B3) $\Rightarrow$ (B2) trivially, and (C1) $\Leftrightarrow$ (C3) by Proposition 3.1.14, any of the six pairs (Bj, Ck), where $j = 1, 2, 3$, $k = 1, 3$ suffices for δ to be a generator of a strongly continuous one-parameter group τ_t of isometries, by Theorem 3.1.19. But (A3) implies that $\tau_t(\mathbb{1}) = \mathbb{1} + t\delta(\mathbb{1}) + o(t)$, and as $\|\tau_t(\mathbb{1})\| = 1$ and $\delta(\mathbb{1})$ is selfadjoint it follows by applying spectral theory for small positive t that $\sigma(\delta(\mathbb{1})) \in \langle -\infty, 0]$, and correspondingly small negative t gives $\sigma(\delta(\mathbb{1})) \in [0, +\infty\rangle$. Thus $\sigma(\delta(\mathbb{1})) = 0$, and as $\delta(\mathbb{1})$ is selfadjoint it follows that $\delta(\mathbb{1}) = 0$, i.e., (A2) is valid. But this implies that $\tau_t(\mathbb{1}) = \mathbb{1}$ for all t, thus each τ_t is a *-automorphism by Corollary 3.2.12.

We now consider the triple ((A2), (B1), (C2)). If (C2) holds and $(I + \alpha\delta)(A) = 0$ then $\pm A \geq 0$. Consequently, $A = 0$. Thus (B1) implies the existence of $(I + \alpha\delta)^{-1}$. The resolvent $(I + \alpha\delta)^{-1}$ is positivity preserving by (C2) and $(I + \alpha\delta)^{-1}(\mathbb{1}) = \mathbb{1}$ by (A2). Hence $\|(I + \alpha\delta)^{-1}\| \leq 1$ by Corollary 3.2.6, and (C1) is true. But we have seen above that ((A2), (B1), (C1)) implies that δ is a generator.

We next consider the triple ((A1), (B2), (C2)). Let $D(\delta)_a$ be the analytic elements in $D(\delta)$. For $A \in D(\delta)_a$, define

$$\tau_t(A) = \sum_{n \geq 0} \frac{t^n}{n!} \delta^n(A) = \lim_{n \to \infty} \left(I + \frac{t}{n} \delta\right)^n (A)$$

for $|t| < t_A$, where t_A is the radius of convergence of the series

$$\sum_{n \geq 0} \frac{t^n}{n!} \|\delta^n(A)\|.$$

If $\tau_t(A) \geq 0$ for some t then for any $\varepsilon > 0$ there is an n such that $(I + (t/n)\delta)^n(A + \varepsilon\mathbb{1}) \geq 0$. Here we have used (A2). But then (C2) implies $A + \varepsilon\mathbb{1} \geq 0$ and we conclude that $A \geq 0$. Next note that since δ is a symmetric derivation, $D(\delta)_a$ is a *-algebra (and one has $t_{AB} = \inf\{t_A, t_B\}$, $t_{A^*} = t_A$). Further, one finds

$$A^*A = \tau_t(\tau_{-t}(A^*A)) \geq 0$$

for $|t| < t_{A^*A}/2 = t_A/2$. Therefore one may conclude that $\tau_{-t}(A^*A) \geq 0$ for $|t| < t_A/2$. Applying this argument to the positive analytic element $\mathbb{1} - A^*A/\|A\|^2$ and using $\tau_t(\mathbb{1}) = \mathbb{1}$ one deduces that

$$0 \leq \tau_t(A^*A) \leq \|A\|^2\mathbb{1}, \qquad |t| < \frac{t_A}{2}.$$

Since δ is a symmetric derivation one has that $\tau_t(A^*A) = \tau_t(A)^*\tau_t(A)$ within the range $|t| < t_A/2$ and one concludes

$$\|\tau_t(A)\| \leq \|A\|, \qquad |t| < \frac{t_A}{2}.$$

Finally, this contractive property and the group relation imply the isometry condition

$$\|\tau_t(A)\| = \|A\|, \qquad |t| < \frac{t_A}{2}.$$

Reasoning as in the proof of Theorem 3.1.19, one deduces that τ_t extends to a strongly continuous group of isometries of $\mathfrak{A}$ with generator δ. Calculating with analytic elements, one deduces that τ_t is a group of *-automorphisms.

The only remaining combination is ((A2), (B3), (C2)). This can be dealt with by a slight modification of the discussion of the previous combination, i.e., if $A = A^*$ is analytic and $\alpha\mathbb{1} \leq A \leq \beta\mathbb{1}$, then $\alpha\mathbb{1} \leq \alpha_t(A) \leq \beta\mathbb{1}$ for $|t| < t_A/2$, hence τ_t is isometric on the selfadjoint elements in $D(\delta)_a$, etc.

Using Theorem 3.1.19 and calculating with analytic elements, one establishes the sufficiency of all the conditions ((A1), (Bj), (Ck)), $i = 1, 2, 3$, $k = 1, 3$, when $\mathfrak{A}$ does not have an identity.

Theorem 3.2.50 of course has an analogue for von Neumann algebras and σ-weakly continuous groups.

Theorem 3.2.51. *Let $\mathfrak{M}$ be a von Neumann algebra, and assume that $\mathfrak{M}$ is abelian, or that $\mathfrak{M}$ is a factor. Let δ be a $\sigma(\mathfrak{M}, \mathfrak{M}_*)$-densely defined, $\sigma(\mathfrak{M}, \mathfrak{M}_*)$–$\sigma(\mathfrak{M}, \mathfrak{M}_*)$-closed, operator on $\mathfrak{M}$ with domain $D(\delta)$ containing $\mathbb{1}$. It follows that δ is the generator of a σ-weakly continuous one-parameter group of *-automorphisms of $\mathfrak{M}$ if, and only if, it satisfies at least one of the criteria in each of the families* (A), (B), *and* (C) *with the exception of the combination* ((A2), (B2), (C2)):

(A1) *$D(\delta)$ is a *-algebra and δ is a symmetric derivation;*
(A2) $\delta(\mathbb{1}) = 0$;
(B1) $(I + \alpha\delta)(D(\delta)) = \mathfrak{M}$, $\alpha \in \mathbb{R}\backslash\{0\}$;
(B2) *the unit sphere in the space of analytic elements for δ is σ-weakly dense in the unit sphere of $\mathfrak{M}$;*
(C1) $\|(I + \alpha\delta)(A)\| \geq \|A\|$, $\alpha \in \mathbb{R}$, $A \in D(\delta)$;
(C2) $(I + \alpha\delta)(A) \geq 0$ *implies that* $A \geq 0$ *for all* $\alpha \in \mathbb{R}$, $A \in D(\delta)$.

For general von Neumann algebras $\mathfrak{M}$, the theorem remains true if (A2) *is removed and* (B2) *replaced by the weaker condition*

(B2′) *the set of analytic elements for δ is σ-weakly dense in $\mathfrak{M}$.*

PROOF. First note that an automorphism τ of a von Neumann algebra is automatically a σ-weakly continuous isometry, by Corollary 2.3.4 and Theorem 2.4.23. Therefore a σ-weakly continuous group of automorphisms of a von Neumann algebra is automatically a C_0*-group of isometries, in the sense of Definition 3.1.2., because $\mathfrak{M}$ is the dual of $\mathfrak{M}_*$ by Proposition 2.4.18. Thus the general theory of C^*-groups of isometries, and in particular Theorem 3.1.19, is applicable. Using Corollary 3.2.13 the proof of the von Neumann theorem now goes approximately as in the case of a C^*-algebra and we omit the details. The only additional remark of relevance is that if δ is a derivation, then the analytic elements for δ form a *-algebra. Hence (B2′) implies (B2) in this case because of Kaplansky's density theorem (Theorem 2.4.16).

After this description of generators and groups for general algebras, we illustrate the structure by examination of a particular class of algebras.

In the remainder of this subsection we continue the study of derivations on UHF algebras initiated in Example 3.2.25. It can be shown, using functional analysis on the domain as in Example 3.2.34, that if δ is a closed derivation of an UHF algebra $\mathfrak{A}$, then there exists an increasing sequence $\{\mathfrak{A}_n\}$ of full matrix subalgebras of $\mathfrak{A}$ with the same identity as $\mathfrak{A}$ such that $\bigcup_n \mathfrak{A}_n \subseteq D(\delta)$ and $\bigcup_n \mathfrak{A}_n$ is dense in $\mathfrak{A}$. If δ is a generator $\bigcup_n \mathfrak{A}_n$ may even be chosen such that $\bigcup_n \mathfrak{A}_n$ consists of analytic elements for δ, but it is not known if $\bigcup_n \mathfrak{A}_n$ can be chosen to be a core for δ in this case. This motivates the following proposition.

Proposition 3.2.52. *If* $\mathfrak{A} = \overline{\bigcup_{n \geq 0} \mathfrak{A}_n}$ *is a UHF algebra and* δ *is a derivation of* $\mathfrak{A}$ *with domain* $D(\delta) = \bigcup_n \mathfrak{A}_n$ *then* δ *is closable and there exist elements* $H_n = H_n^* \in \mathfrak{A}$ *such that*

$$\delta(A) = \delta_n(A) \equiv [iH_n, A]$$

for all $A \in \mathfrak{A}_n$, $n = 1, 2, \ldots$.

If $(I \pm \delta)(\bigcup_n \mathfrak{A}_n)$ *are dense in* $\mathfrak{A}$ *then the closure* $\bar{\delta}$ *of* δ *is the generator of a strongly continuous one-parameter group* τ_t *of* $*$*-automorphisms of* $\mathfrak{A}$. *Moreover,*

$$\tau_t(A) = \lim_{n \to \infty} e^{t\delta_n}(A)$$

for all $A \in \mathfrak{A}$, *where the limit exists in norm, uniformly for t in finite intervals.*

PROOF. The existence of H_n was shown in Example 3.2.25, while $\pm\delta$, and thus $\pm\bar{\delta}$, are dissipative by Proposition 3.2.22. Lemma 3.1.15 then implies $\|(I + \alpha\bar{\delta})(A)\| \geq \|A\|$ for all $A \in D(\bar{\delta})$, $\alpha \in \mathbb{R}$. Using the closedness of $\bar{\delta}$ one deduces $(I \pm \bar{\delta})(D(\bar{\delta})) = \mathfrak{A}$; thus $(I + \alpha\bar{\delta})^{-1}$ exists and $\|(I + \alpha\bar{\delta})^{-1}\| \leq 1$ by the usual Neumann series argument. By Theorem 3.2.50 ((A1), (B1), (C1)), $\bar{\delta}$ is a generator of a group τ_t of $*$-automorphisms. Since $\bigcup_n \mathfrak{A}_n$ is a core for $\bar{\delta}$ and $\delta_m(A) \to \delta(A)$ for $A \in \bigcup \mathfrak{A}_n$, it follows from Theorem 3.1.28 that

$$\tau_t(A) = \lim_{n \to \infty} e^{t\delta_n} A$$

for all $A \in \mathfrak{A}$, uniformly for t in compacts.

We next study a condition on the sequence H_n in $\mathfrak{A}$ which ensures that $(I + \delta)(\bigcup_n \mathfrak{A}_n)$ is dense. The physical interpretation of this condition is that the surface energy of the finite subsystems grows linearly with the subsystem (see Chapter 6).

Theorem 3.2.53. *Adopt the same notation as in Proposition* 3.2.52, *and assume that there exists a double sequence*

$$K_{n,m} \in \mathfrak{A}_{n+m}$$

$n = 1, 2, \ldots, m = 0, 1, \ldots,$ *and constants* $M, \alpha > 0$ *such that*

$$\|H_n - K_{n,m}\| \leq Mne^{-\alpha m}.$$

It follows that $(I \pm \delta)(\bigcup_n \mathfrak{A}_n)$ *is dense in* $\mathfrak{A}$, *and hence the closure* $\bar{\delta}$ *of* δ *is the generator of a strongly continuous one-parameter group* τ *of* $*$*-automorphisms of* $\mathfrak{A}$. *Moreover,*

$$\tau_t(A) = \lim_{n \to \infty} e^{itH_n} A e^{-itH_n}$$

for all $A \in \mathfrak{A}$, *where the limit exists in norm, uniformly for t in finite intervals.*

This theorem is an immediate consequence of Proposition 3.2.52, Theorem 3.1.34 and the estimate

$$\|\delta|_{\mathfrak{A}_n} - \delta_{iK_{n,m}}|_{\mathfrak{A}_n}\| \leq 2\|H_n - K_{n,m}\|.$$

3.2.5. Spatial Derivations and Invariant States

The discussion of bounded derivations of a von Neumann algebra $\mathfrak{M}$ in Section 3.2.3 established that each such derivation has the form

$$\delta(A) = i[H, A] \tag{$*$}$$

for some bounded operator H and H could even be chosen to be an element of $\mathfrak{M}$. More generally if $t \in \mathbb{R} \mapsto U_t$ is a strongly continuous one-parameter group of unitary operators and $U_t \mathfrak{M} U_{-t} \subseteq \mathfrak{M}$ for all $t \in \mathbb{R}$ then the family of mappings

$$A \in \mathfrak{M} \mapsto \alpha_t(A) = U_t A U_t{}^*$$

is a weak *-continuous group of *-automorphisms of $\mathfrak{M}$ whose generator δ is of the form

$$\delta(A) = i[H, A].$$

In this latter case H is the selfadjoint generator of the unitary group U_t. The Borchers–Arveson theorem (Theorem 3.2.46) demonstrated that if H is semibounded then it can be arranged that H is affiliated with $\mathfrak{M}$.

These examples motivate the study of general derivations of the form $(*)$. We call such derivations spatial.

Definition 3.2.54. A symmetric derivation δ of a C^*-algebra of bounded operators on a Hilbert space $\mathfrak{H}$ is said to be *spatial* if there exists a symmetric operator H, on $\mathfrak{H}$, with domain $D(H)$ such that $D(\delta)D(H) \subseteq D(H)$ and

$$\delta(A) = i[H, A], \qquad A \in D(\delta),$$

on $D(H)$. We say that H *implements* δ.

Note that Proposition 3.2.28 already established a criterion for a derivation to be spatial. It suffices that there exist a state $\omega(A) = (\Omega, A\Omega)$, over $\mathfrak{A}$, such that Ω is cyclic for $\mathfrak{A}$ and

$$|\omega(\delta(A))|^2 \leq L\{\omega(A^*A) + \omega(AA^*)\}$$

for all $A \in D(\delta)$ and some $L \geq 0$. In particular, this criterion applies if ω is invariant under δ, i.e.,

$$\omega(\delta(A)) = 0$$

for all $A \in D(\delta)$. If, however, ω is invariant then δ is not only spatial but H may be chosen so that $\Omega \in D(H)$ and

$$H\Omega = 0.$$

Invariant states are of particular importance in physical applications and they provide the second topic of this subsection.

In general, there are several inequivalent ways of defining a spatial derivation δ. The different possibilities occur because the commutators $[H, A]$, with H unbounded, are not unambiguously defined and various conventions can be adopted to give meaningful definitions. Each of these conventions leads to a distinct definition of a spatial derivation. If, however, δ is implemented by a selfadjoint H then there is no ambiguity as the following result demonstrates.

Proposition 3.2.55. *Let H be a selfadjoint operator on a Hilbert space $\mathfrak{H}$ and let*

$$\alpha_t(A) = e^{itH}Ae^{-itH}, \quad A \in \mathscr{L}(\mathfrak{H}),$$

be the corresponding one-parameter group of automorphisms of $\mathscr{L}(\mathfrak{H})$. Denote by δ the infinitesimal generator of α. For $A \in \mathscr{L}(\mathfrak{H})$ given, the following conditions are equivalent:

(1) *$A \in D(\delta)$;*
(2) *there exists a core D for H such that the sesquilinear form*

$$\psi, \varphi \in D \times D \mapsto i(H\psi, A\varphi) - i(\psi, AH\varphi)$$

is bounded;
(3) *there exists a core D for H such that $AD \subseteq D$ and the mapping*

$$\psi \in D \mapsto i[H, A]\psi$$

is bounded.

If condition (2) is valid the bounded operator associated with the sesquilinear form is $\delta(A)$. Similarly, the bounded mapping of condition (3) defines $\delta(A)$.

PROOF. A closure argument shows that conditions (2) and (3) imply the corresponding conditions with $D = D(H)$, while the converse is trivial. Thus we may assume that $D = D(H)$ in the following.

(1) $\Rightarrow$ (2) Assume $A \in D(\delta)$ and $\psi, \varphi \in D(H)$; then

$$\begin{aligned}(\psi, \delta(A)\varphi) &= \lim_{t \to 0} \frac{1}{t} \{(\psi, e^{itH}Ae^{-itH}\varphi) - (\psi, A\varphi)\} \\ &= \lim_{t \to 0} \left(\frac{1}{t}(e^{-itH} - I)\psi, Ae^{-itH}\varphi\right) \\ &\quad + \lim_{t \to 0} \left(\psi, A\frac{1}{t}(e^{-itH} - I)\varphi\right) \\ &= (-iH\psi, A\varphi) + (\psi, A(-iH)\varphi).\end{aligned}$$

(2) $\Rightarrow$ (3) Assume that there exists a bounded operator B such that

$$(H\psi, A\varphi) = (\psi, AH\varphi) - i(\psi, B\varphi)$$

for $\psi, \varphi \in D(H)$. This relation demonstrates that

$$\psi \mapsto (H\psi, A\varphi)$$

is continuous for fixed $\varphi \in D(H)$. Hence $A\varphi \in D(H^*) = D(H)$ and, further, $(H\psi, A\varphi) = (\psi, HA\varphi)$. Therefore one has

$$(\psi, B\varphi) = i(\psi, [H, A]\varphi).$$

The implication (3) ⇒ (2) is trivial and the proof is completed by the following:

(2) ⇒ (1) Assume that A satisfies condition (2) and that B is the corresponding bounded operator. Using the technique of the first part of the proof one verifies, for $\psi, \varphi \in D(H)$, that

$$\lim_{t \to 0} \frac{(\psi, (\alpha_{s+t}(A) - \alpha_s(A))\varphi)}{t} = (\psi, \alpha_s(B)\varphi).$$

Hence

$$(\psi, (\alpha_s(A) - A)\varphi) = \int_0^s dt\, (\psi, \alpha_t(B)\varphi).$$

Therefore one has

$$\alpha_s(A) = A + \int_0^s dt\, \alpha_t(B)$$

and from this relation it follows that $A \in D(\delta)$ and $\delta(A) = B$.

Next remark that a spatial derivation of a von Neumann algebra which is implemented by a selfadjoint operator H has an extension which generates a weak *-continuous group α_t of *-automorphisms of $\mathscr{L}(\mathfrak{H})$. Explicitly,

$$\alpha_t(B) = e^{itH} B e^{-itH}$$

for all $B \in \mathscr{L}(\mathfrak{H})$. Using the characterizations of generators given in the previous subsection, e.g., Theorem 3.2.51, one can then deduce some of the typical properties of generators.

Corollary 3.2.56. *Let δ be a symmetric, spatial, derivation of a von Neumann algebra $\mathfrak{M}$ on a Hilbert space $\mathfrak{H}$, i.e.,*

$$\delta(A) = i[H, A]$$

for all $A \in D(\delta)$. Assume that H is selfadjoint. It follows that δ is $\sigma(\mathfrak{M}, \mathfrak{M}_*)$-closable and

$$\|(I - \alpha\delta)(A)\| \geq \|A\|$$

for all $A \in D(\delta)$ and $\alpha \in \mathbb{R}$.

After these preliminary remarks we devote the rest of this subsection to the analysis of invariant states. Thus we examine spatial derivations $\delta(A) = i[H, A]$ of an operator algebra $\mathfrak{M}$ with cyclic vector Ω such that $H\Omega = 0$.

The characterizations of generators obtained in Section 3.2.4 are considerably simplified by the existence of an invariant state. A typical C^*-algebraic result is the following:

Corollary 3.2.57. *Let δ be a symmetric derivation of a C^*-algebra $\mathfrak{A}$ and assume there exists a state ω such that $\omega(\delta(A)) = 0$, for all $A \in D(\delta)$, and, furthermore, the associated representation $(\mathfrak{H}, \pi, \Omega)$ is faithful. Also assume*

either $\quad \overline{R(I \pm \delta)} = \mathfrak{A}$,

where the bar denotes norm closure,

or $\quad$ *δ possesses a dense set of analytic elements.*

*It follows that δ is norm closable and its closure $\bar{\delta}$ generates a strongly continuous one-parameter group of *-automorphisms of $\mathfrak{A}$.*

Proof. The derivation δ is norm closable by Proposition 3.2.26 and its closure is spatial by Proposition 3.2.28. If H denotes the symmetric operator which implements $\bar{\delta}$ then $H\Omega = 0$ and hence $\pi(\bar{\delta}(A))\Omega = iH\pi(A)\Omega$ for all $A \in D(\bar{\delta})$. Either of the two assumptions then imply that H is essentially selfadjoint (see Example 3.1.21). Thus applying Corollary 3.2.56 one concludes that $\|(I \pm \bar{\delta})(A)\| \geq \|A\|$ for all $A \in D(\bar{\delta})$ and then Theorem 3.2.50 establishes that $\bar{\delta}$ is a generator.

The last corollary has a quasi-converse. If $\mathfrak{A}$ is a simple C^*-algebra with identity and α_t is a strongly continuous one-parameter group of *-automorphisms with generator δ then there exists a α_t-invariant state, $\omega(\alpha_t(A)) = \omega(A)$, $A \in \mathfrak{A}$. The representation π_ω is automatically faithful, because $\mathfrak{A}$ is simple, and $\omega(\delta(A)) = 0$, $A \in D(\delta)$. The existence of ω follows by taking a mean value of a family of states $\omega_t(A) = \omega_0(\alpha_t(A))$ and this exploits the weak *-compactness of the states of $\mathfrak{A}$.

This corollary has an obvious analogue for von Neumann algebras but in this latter case one can obtain even stronger results.

Proposition 3.2.58. *Let δ be a spatial derivation of a von Neumann algebra $\mathfrak{M}$ implemented by a symmetric operator H and assume that $\mathfrak{M}$ has a cyclic vector Ω such that $H\Omega = 0$. Further assume there exists a *-subalgebra $\mathfrak{D} \subseteq D(\delta)$ such that*

(1) *$\mathfrak{D}$ is strongly dense in $\mathfrak{M}$,*
(2) $\delta(\mathfrak{D}) \subseteq \mathfrak{D}$,
(3) *$\mathfrak{D}\Omega$ consists of analytic elements for H.*

It follows that H is essentially selfadjoint and its closure $\bar{H}$ satisfies

$$e^{it\bar{H}}\mathfrak{M}e^{-it\bar{H}} = \mathfrak{M}, \qquad t \in \mathbb{R}.$$

Proof. The set $\mathfrak{D}\Omega$ is a dense set of analytic elements for H and $H\mathfrak{D}\Omega \subseteq \mathfrak{D}\Omega$. Therefore $\bar{H}$ is selfadjoint (see Example 3.1.21). Next note that the automorphism property follows from

$$e^{it\bar{H}}\mathfrak{M}'e^{-it\bar{H}} = \mathfrak{M}'$$

because one then has

$$[e^{it\bar{H}}Ae^{-it\bar{H}}, A'] = e^{it\bar{H}}[A, e^{-it\bar{H}}A'e^{it\bar{H}}]e^{-it\bar{H}} = 0$$

for all $A \in \mathfrak{M}$ and $A' \in \mathfrak{M}'$. Now for $A, B, C \in \mathfrak{D}$ and $A' \in \mathfrak{M}'$ define g by

$$g(t) = (B\Omega, [A, e^{-it\bar{H}}A'e^{it\bar{H}}]C\Omega).$$

By cyclicity of Ω and the density of $\mathfrak{D}$ it suffices to prove that $g(t) = 0$ for all the possible choices of A, B, C, and A'. But

$$g(t) = (e^{it\bar{H}}A^*B\Omega, A'e^{it\bar{H}}C\Omega) - (e^{it\bar{H}}B\Omega, A'e^{it\bar{H}}AC\Omega)$$

and hence g is analytic by the special choice of A, B, C. One now calculates

$$\begin{aligned}
\frac{d^n g}{dt^n}(0) &= \sum_{k=0}^{n} \binom{n}{k} ((iH)^k A^*B\Omega, A'(iH)^{n-k}C\Omega) \\
&\quad - \sum_{k=0}^{n} \binom{n}{k} ((iH)^{n-k}B\Omega, A'(iH)^k AC\Omega) \\
&= \sum_{k=0}^{n} \binom{n}{k} (A'^*\Omega, \delta^k(B^*A)\delta^{n-k}(C)\Omega) \\
&\quad - \sum_{k=0}^{n} \binom{n}{k} (A'^*\Omega, \delta^{n-k}(B^*)\delta^k(AC)\Omega) \\
&= (A'^*\Omega, \delta^n(B^*AC)\Omega) - (A'^*\Omega, \delta^n(B^*AC)\Omega) \\
&= 0.
\end{aligned}$$

Thus $g(t) = 0$ by analyticity and the automorphism property is established.

The above proposition emphasizes properties of the analytic elements of the implementing operator H but in the remaining analysis of spatial derivations and invariant states the properties of the ranges $R(I \pm iH)$ will be of greater importance.

There are two cases of particular significance in physical applications to the description of dynamics or the description of symmetries.

The first case typically occurs in connection with ground state phenomena. The operator H is interpreted as a Hamiltonian, or energy, operator and is positive either by assumption or construction. The eigenvector Ω then corresponds to the lowest energy state of the system, i.e., the ground state.

The second case does not necessarily involve the interpretation of H as a Hamiltonian and it is no longer supposed to be positive. One assumes, however, that the eigenvector Ω is both cyclic and separating for the von Neumann algebra. This situation occurs in the usual description of finite temperature equilibrium states in statistical mechanics.

We next examine these two cases and derive various criteria for the derivations to be generators of automorphism groups. Although both cases have their own special characteristics the following result emphasizes a unifying feature.

Theorem 3.2.59. *Let $\mathfrak{M}$ be a von Neumann algebra, on the Hilbert space $\mathfrak{H}$, with cyclic vector Ω and δ a spatial derivation of $\mathfrak{M}$ implemented by a selfadjoint operator H such that $\Omega \in D(H)$ and $H\Omega = 0$. Let $D(\bar{\delta})$ denote the set*

$$D(\bar{\delta}) = \{A; A \in \mathfrak{M}, i[H, A] = \bar{\delta}(A) \in \mathfrak{M}\}$$

and assume that $D(\bar{\delta})\Omega$ is a core for H. Further assume that

either $H \geq 0$

or Ω *is separating for* $\mathfrak{M}$.

The following conditions are equivalent:

(1) $e^{itH}\mathfrak{M}e^{-itH} = \mathfrak{M}, t \in \mathbb{R}$;
(2) $e^{itH}\mathfrak{M}_+\Omega \subseteq \mathfrak{M}_+\Omega, t \in \mathbb{R}$;
(3) $e^{itH}\mathfrak{M}_+\Omega \subseteq \overline{\mathfrak{M}_+\Omega}, t \in \mathbb{R}$;
(4) $(I \pm iH)^{-1}\mathfrak{M}_+\Omega \subseteq D(\bar{\delta})_+\Omega$;
(5) $(I \pm iH)^{-1}\mathfrak{M}_+\Omega \subseteq \mathfrak{M}_+\Omega$;
(6) $(I \pm iH)^{-1}\mathfrak{M}_+\Omega \subseteq \overline{\mathfrak{M}_+\Omega}$.

(The bar denotes weak (strong) closure.)

PROOF. Several of the implications are obvious and independent of the particular assumption. For example, (2) $\Rightarrow$ (3) and (4) $\Rightarrow$ (5) $\Rightarrow$ (6) trivially. But $H\Omega = 0$ implies $e^{itH}\Omega = \Omega$ and condition (1) ensures that

$$e^{itH}\mathfrak{M}_+ e^{-itH} = \mathfrak{M}_+.$$

Therefore (1) $\Rightarrow$ (2). Condition (1) also gives

$$\int_0^\infty dt\, e^{-t}e^{\mp itH}\mathfrak{M}_+ e^{\pm itH} \subseteq D(\bar{\delta})_+$$

and application of this relation to Ω gives

$$(I \pm iH)^{-1}\mathfrak{M}_+\Omega \subseteq D(\bar{\delta})_+\Omega.$$

Thus (1) $\Rightarrow$ (4). Next note that condition (6) and a simple application of the Neumann series for the resolvent $(I + i\alpha H)^{-1}$ ensures that

$$(I + i\alpha H)^{-1}\mathfrak{M}_+\Omega \subseteq \overline{\mathfrak{M}_+\Omega}$$

for all $\alpha \in \mathbb{R}$. Therefore one has

$$e^{itH}\mathfrak{M}_+\Omega = \lim_{n\to\infty}(I - itH/n)^{-n}\mathfrak{M}_+\Omega \subseteq \overline{\mathfrak{M}_+\Omega},$$

i.e., (6) $\Rightarrow$ (3).

The proof is now completed in both cases by establishing that (3) $\Rightarrow$ (1). This is the only point at which the core assumption is important. The two cases are handled by distinct arguments and we first consider the case $H \geq 0$.

Case $H \geq 0$, (3) $\Rightarrow$ (1). Let $A \in \mathfrak{M}_{sa}$, $A' \in \mathfrak{M}'_{sa}$ and define a function g by

$$g(t) = (A'\Omega, U_t A\Omega),$$

where $U_t = e^{itH}$. The positivity of H ensures that g has an analytic extension to the upper half-plane.

$$g(t_1 + it_2) = (A'\Omega, U_{t_1} e^{-t_2 H} A\Omega).$$

Further, one has for $t \in \mathbb{R}$

$$\begin{aligned} g(t) &= (A'\Omega, U_t A\Omega) \\ &= \lim_{n \to \infty} (A'\Omega, B_n \Omega) \\ &= \lim_{n \to \infty} (B_n \Omega, A'\Omega) \\ &= (U_t A\Omega, A'\Omega) = \overline{g(t)}, \end{aligned}$$

where the $B_n \in \mathfrak{M}_{sa}$ are chosen such that $B_n \Omega$ converges to $U_t A\Omega$. Such a choice is possible by condition (3). Thus g is real on the real axis and the Schwarz reflection principle assures the existence of an entire function $\mathfrak{G}$ such that $\mathfrak{G}(z) = g(z) = \overline{\mathfrak{G}(\bar{z})}$ for $\operatorname{Im} z \geq 0$. One now has

$$|\mathfrak{G}(z)| = |\mathfrak{G}(\bar{z})| \leq \|A\Omega\| \, \|A'\Omega\|$$

for $\operatorname{Im} z \geq 0$ and hence $\mathfrak{G}$ is constant by Liouville's theorem. In particular, $g(t) = g(0)$ gives

$$(A\Omega, U_{-t} A'\Omega) = (A\Omega, A'\Omega)$$

for all $A \in \mathfrak{M}_{sa}$ and $A' \in \mathfrak{M}'_{sa}$. But each element $A \in \mathfrak{M}$, or $\mathfrak{M}'$, has a decomposition of the form $A = B + iC$ with $B, C \in \mathfrak{M}_{sa}$, or $\mathfrak{M}'_{sa}$. Thus this last relation is valid for all $A \in \mathfrak{M}$ and $A' \in \mathfrak{M}'$, and the cyclicity of Ω for $\mathfrak{M}$ then yields

$$U_{-t} A'\Omega = A'\Omega$$

for all $A' \in \mathfrak{M}'$. In particular, $A'\Omega \in D(H)$ and $HA'\Omega = 0$. Now if $A \in D(\bar{\delta})$ and $\psi \in D(H)$ then $A\psi \in D(H)$. Thus $AA'\Omega \in D(H)$ or, equivalently, $A'A\Omega \in D(H)$. It follows that

$$\begin{aligned} HA'A\Omega &= HAA'\Omega \\ &= [H, A]A'\Omega \\ &= A'[H, A]\Omega \\ &= A'HA\Omega. \end{aligned}$$

Now A' is bounded and $D(\bar{\delta})\Omega$ is a core for H. Therefore one has

$$HA'\psi = A'H\psi$$

for all $\psi \in D(H)$ and hence

$$(e^{itH}A' - A'e^{itH})\psi = i \int_0^t ds \, e^{isH}[H, A']e^{i(t-s)H}\psi = 0.$$

This implies $U_t \in \mathfrak{M}$ and consequently $U_t \mathfrak{M} U_t^* = \mathfrak{M}$.

Case Ω separating. (3) $\Rightarrow$ (1). We will establish the desired implication indirectly by arguing that (3) $\Rightarrow$ (6) $\Rightarrow$ (5) $\Rightarrow$ (4) $\Rightarrow$ (1). The first step again uses the Laplace transform

$$(I \pm iH)^{-1}\mathfrak{M}_+\Omega = \int_0^\infty dt\, e^{-t}e^{\mp itH}\mathfrak{M}_+\Omega \subseteq \overline{\mathfrak{M}_+\Omega}.$$

The second step follows immediately from application of Lemma 3.2.19 to the operators $T_\pm = (I \pm iH)^{-1}$. The crucial third step in this proof is to deduce that (5) $\Rightarrow$ (4). For this we first remark that by decomposing a general $A \in \mathfrak{M}$ as a superposition of positive elements one deduces that

$$(I + iH)^{-1}A\Omega = B\Omega$$

implies that

$$(I + iH)^{-1}A^*\Omega = B^*\Omega.$$

Next choose $A \in \mathfrak{M}_+$ and $C = C^* \in D(\delta)$. Introduce B by

$$(I + iH)^{-1}A\Omega = B\Omega$$

and remark that

$$\delta(C)B\Omega + C(I + iH)B\Omega = (I + iH)CB\Omega.$$

Thus $CB\Omega \in D(H)$ and

$$(I + iH)^{-1}F\Omega = CB\Omega,$$

where

$$F = \delta(C)B + CA.$$

Hence it follows that

$$(I + iH)^{-1}F^*\Omega = BC\Omega$$

i.e., $BC\Omega \in D(H)$ and

$$(I + iH)BC\Omega = AC\Omega + B\delta(C)\Omega.$$

This last relation may be rewritten as

$$i(HB - BH)C\Omega = (A - B)C\Omega.$$

This last equation extends linearly to nonselfadjoint $C \in D(\delta)$ and then it follows from Proposition 3.2.55 that $B \in D(\bar{\delta})_+$, i.e., $(I + iH)^{-1}\mathfrak{M}_+\Omega \subseteq D(\bar{\delta})_+\Omega$.

Applying an identical argument to $(I - iH)^{-1}$ one concludes the validity of condition (4). Now to deduce condition (1) we again remark that each element of $\mathfrak{M}$ is a linear superposition of four positive elements and hence (4) implies

$$(I \pm iH)^{-1}\mathfrak{M}\Omega \subseteq D(\bar{\delta})\Omega.$$

Therefore

$$\mathfrak{M}\Omega \subseteq (I \pm iH)D(\bar{\delta})\Omega = (I \pm \bar{\delta})(D(\bar{\delta}))\Omega \subseteq \mathfrak{M}\Omega.$$

As Ω is separating this last condition is equivalent to

$$(I \pm \bar{\delta})(D(\bar{\delta})) = \mathfrak{M}.$$

But by Corollary 3.2.56 the spatial derivation $\bar{\delta}$ is $\sigma(\mathfrak{M}, \mathfrak{M}_*)$-closable and

$$\|(I + \alpha\bar{\delta})(A)\| \geq \|A\|$$

for all $A \in D(\bar{\delta})$ and $\alpha \in \mathbb{R}$. Combining these conditions one first concludes that the resolvent $(I + \alpha\bar{\delta})^{-1}$ is $\sigma(\mathfrak{M}, \mathfrak{M}_*)$-continuous, therefore $\bar{\delta}$ is $\sigma(\mathfrak{M}, \mathfrak{M}_*)$-closed, and then condition (1) follows from Theorem 3.2.51.

In the course of the above proof we actually established somewhat more for the case $H \geq 0$.

Corollary 3.2.60. *Adopt the general assumptions of Theorem* 3.2.59 *and further assume* $H \geq 0$. *The following conditions are equivalent*:

(1) $e^{itH}\mathfrak{M}e^{-itH} = \mathfrak{M}$, $t \in \mathbb{R}$;
(2) $e^{itH} \in \mathfrak{M}$, $t \in \mathbb{R}$;
(3) $e^{itH}A'\Omega = A'\Omega$, $A' \in \mathfrak{M}'$, $t \in \mathbb{R}$;
(4), (4′) $e^{itH}\mathfrak{M}_{sa}\Omega \subseteq \mathfrak{M}_{sa}\Omega$, $(e^{itH}\mathfrak{M}_{sa}\Omega \subseteq \overline{\mathfrak{M}_{sa}\Omega})$, $t \in \mathbb{R}$.

PROOF. Clearly $(1) \Rightarrow (4) \Rightarrow (4')$ but in the proof of the implication $(3) \Rightarrow (1)$ in Theorem 3.2.59 we actually established that $(4') \Rightarrow (3) \Rightarrow (2) \Rightarrow (1)$.

The conditions given in this corollary are typical of the ground state situation $H \geqslant 0$ and conditions (2) and (3) are certainly not true in the case that Ω is separating.

Next remark that if we assume that Ω is separating in Theorem 3.2.59 then the equivalences $(1) \Leftrightarrow (4)$ and $(2) \Leftrightarrow (3) \Leftrightarrow (5) \Leftrightarrow (6)$ do not use the assumption that $D(\delta)\Omega$ is a core for H. Moreover, without this assumption any of the equivalent conditions (2), (3), (5), (6) implies, by Theorem 3.2.18, that the mapping $A \in \mathfrak{M}_+ \Rightarrow \alpha_t(A) \in \mathfrak{M}_+$ defined by

$$e^{itH}A\Omega = \alpha_t(A)\Omega$$

extends to a one-parameter group of Jordan automorphisms of $\mathfrak{M}$. If $\mathfrak{M}$ is a factor, or if $\mathfrak{M}$ is abelian, then these automorphisms are automatically *-automorphisms by Corollary 3.2.13 and one has

$$\alpha_t(A) = e^{itH}Ae^{-itH}.$$

For general $\mathfrak{M}$ the core condition is, however, crucial in the deduction that α_t is a group of *-automorphisms. One can indeed establish, by construction of counterexamples, that the core condition cannot be omitted. On the other hand, the following theorem (Theorem 3.2.61) demonstrates that the core condition is almost sufficient to establish the automorphism property without the assumed invariance of $\mathfrak{M}_+\Omega$ under e^{itH}. The theorem extends Theorem 3.2.59 in the case that Ω is separating. Note that it is not explicitly assumed that δ is σ-weakly densely defined.

Theorem 3.2.61. *Let $\mathfrak{M}$ be a von Neumann algebra, on a Hilbert space $\mathfrak{H}$, with a cyclic and separating vector Ω. Let H be a selfadjoint operator on $\mathfrak{H}$ such that $H\Omega = 0$ and define*

$$D(\delta) = \{A \in \mathfrak{M};\, i[H, A] \in \mathfrak{M}\}.$$

Let Δ be the modular operator associated with the pair $(\mathfrak{M}, \Omega)$, and let $\mathfrak{H}_\#$ be the graph Hilbert space associated to $\Delta^{1/2}$. The following conditions are equivalent:

(1) $e^{itH}\mathfrak{M}e^{-itH} = \mathfrak{M}$, $t \in \mathbb{R}$;
(2) (a) *$D(\delta)\Omega$ is a core for H,*
  (b) *H and Δ commute strongly, i.e.,*

$$\Delta^{is}H\Delta^{-is} = H, \qquad s \in \mathbb{R};$$

(3) *the restriction of H to $D(\delta)\Omega$ is essentially selfadjoint as an operator on $\mathfrak{H}_\#$.*

Remark. The graph Hilbert space $\mathfrak{H}_\#$ is defined as the linear space $D(\Delta^{1/2})$ equipped with the inner product

$$(\psi, \varphi)_\# = (\Delta^{1/2}\psi, \Delta^{1/2}\varphi) + (\psi, \varphi).$$

Since $\mathfrak{M}\Omega \subseteq D(\Delta^{1/2})$, it follows that the restriction of H to $D(\delta)\Omega$ is a bona fide operator on $\mathfrak{H}_\#$ because

$$D(\delta)\Omega \subseteq \mathfrak{H}_\#,$$
$$iHD(\delta)\Omega = \delta(D(\delta))\Omega \subseteq \mathfrak{M}\Omega \subseteq \mathfrak{H}_\#,$$

where $\delta(B) = i[H, B]$ for $B \in D(\delta)$. This restriction is a symmetric operator on $\mathfrak{H}_\#$, because if J is the modular involution associated with Ω and $S = J\Delta^{1/2}$ then

$$\begin{aligned}(A\Omega, HB\Omega)_\# &= (A\Omega, HB\Omega) + (\Delta^{1/2}A\Omega, \Delta^{1/2}(-i\delta(B))\Omega)\\ &= (HA\Omega, B\Omega) - i(S\delta(B)\Omega, SA\Omega)\\ &= (HA\Omega, B\Omega) - i(\delta(B^*)\Omega, A^*\Omega)\\ &= (HA\Omega, B\Omega) - (HB^*\Omega, A^*\Omega)\\ &= (HA\Omega, B\Omega) - (B^*\Omega, HA^*\Omega)\\ &= (HA\Omega, B\Omega) + (\Delta^{1/2}HA\Omega, \Delta^{1/2}B\Omega)\\ &= (HA\Omega, B\Omega)_\#\end{aligned}$$

for all $A, B \in D(\delta)$. Hence statement (3) makes sense.

PROOF. We will prove (1) $\Rightarrow$ (3) $\Rightarrow$ (2) $\Rightarrow$ (1). The hard part of the proof, (2) $\Rightarrow$ (1), will require some results on modular automorphism groups which will be derived subsequently in Section 5.3.

If (1) holds then it follows from Proposition 3.2.55 and Theorem 3.2.51 that

$$(I + i\alpha H)D(\delta)\Omega = R(I + \alpha\delta)\Omega = \mathfrak{M}\Omega$$

for $\alpha \in \mathbb{R}\backslash\{0\}$. But $\mathfrak{M}\Omega$ is a core for $S = J\Delta^{1/2}$ and hence $\mathfrak{M}\Omega$ is dense in $\mathfrak{H}_\#$. As the restriction of H to $D(\delta)\Omega$ is symmetric on $\mathfrak{H}_\#$, by the preceding remark, it follows from Example 3.1.21 that H is essentially selfadjoint on $D(\delta)\Omega$ as an operator on $\mathfrak{H}_\#$.

(3) $\Rightarrow$ (2) Condition (3) implies that $(I + i\alpha H)D(\delta)\Omega$ is dense in $\mathfrak{H}_{\#}$ for $\alpha \in \mathbb{R}\backslash\{0\}$, and then $(I + i\alpha H)D(\delta)\Omega$ is dense in $\mathfrak{H}$ with respect to the usual norm, i.e., $D(\delta)\Omega$ is a core for H. Since $\delta(A^*) = \delta(A)^*$ for $A \in D(\delta)$, we then obtain

$$(I + i\alpha H)S\xi = S(I + i\alpha H)\xi$$

for $\xi \in D(\delta)\Omega$ and $\alpha \in \mathbb{R}\backslash\{0\}$, where $S = J\Delta^{1/2}$. Thus

$$S(I + i\alpha H)^{-1}\eta = (I + i\alpha H)^{-1}S\eta$$

for all $\eta \in (I + i\alpha H)D(\delta)\Omega$. But condition (3) implies that $(I + i\alpha H)D(\delta)\Omega$ is dense in $D(\Delta^{1/2}) = D(S)$ with respect to the graph norm of $\Delta^{1/2}$. Hence

$$S(I + i\alpha H)^{-1}\eta = (I + i\alpha H)^{-1}S\eta$$

for all $\eta \in D(\Delta^{1/2})$, i.e.,

$$S(I + i\alpha H)^{-1} \supseteq (I + i\alpha H)^{-1}S$$

for any $\alpha \in \mathbb{R}\backslash\{0\}$. Thus S commutes both with $(I + i\alpha H)^{-1}$ and its adjoint $(I - i\alpha H)^{-1}$, and hence the components J and $\Delta^{1/2}$ of the polar decomposition of S commute with $(I + i\alpha H)^{-1}$. In particular, Δ commutes strongly with H.

(2) $\Rightarrow$ (1) If $\tilde{\delta}$ is the generator of the one-parameter group $t \mapsto \tau_t$ on $\mathscr{L}(\mathfrak{H})$ defined by

$$\tau_t(A) = e^{itH}Ae^{-itH}, \qquad A \in \mathscr{L}(\mathfrak{H}),\ t \in \mathbb{R},$$

then it is enough to show that

$$(I + \alpha\tilde{\delta})^{-1}(\mathfrak{M}) \subseteq \mathfrak{M}$$

for all $\alpha \in \mathbb{R}\backslash\{0\}$, because

$$\tau_t(A) = \lim_{n\to\infty}\left(I - \frac{t}{n}\tilde{\delta}\right)^{-n}(A),$$

where the limit exists in the σ-weak topology by Theorem 3.1.10, and this implies that $\tau_t(\mathfrak{M}) \subseteq \mathfrak{M}$ for all $t \in \mathbb{R}$. As δ is σ-weakly closed, it is therefore enough to show that

$$R(I + \alpha\delta) \equiv (I + \alpha\delta)(D(\delta)) \tag{$*$}$$

is σ-weakly dense in $\mathfrak{M}$ for all $\alpha \in \mathbb{R}\backslash\{0\}$. If $t \mapsto \sigma_t^{\omega}$ is the modular group associated with $\mathfrak{M}$ we define the centralizer $\mathfrak{M}_{\omega}$ of the state $\omega(A) = (\Omega, A\Omega)$ as follows:

$$\begin{aligned}\mathfrak{M}_{\omega} &= \{A \in \mathfrak{M};\, \sigma_t^{\omega}(A) = A,\ t \in \mathbb{R}\}\\ &= \{A \in \mathfrak{M};\, \omega(AB) = \omega(BA),\ B \in \mathfrak{M}\}.\end{aligned}$$

(The equivalence of the two definitions of $\mathfrak{M}_{\omega}$ is easy to establish but for systematic reasons we defer the proof to Proposition 5.3.28.)

The theorem is now a consequence of the following lemma in the case that ω is a trace state, i.e., $\omega(AB) = \omega(BA)$ for all $A, B \in \mathfrak{M}$.

Lemma 3.2.62. *Adopt the hypotheses of Theorem* 3.2.61, *together with Condition* (2), *parts* (a) *and* (b). *It follows that*

$$\mathfrak{M}_{\omega} \subseteq R(I + \alpha\delta), \qquad \alpha \in \mathbb{R}\backslash\{0\}.$$

PROOF. We start by proving that $(\mathfrak{M}_\omega \cap R(I + \alpha\delta))\Omega$ is dense in $\mathfrak{M}_\omega\Omega$. Since $D(\delta)\Omega$ is a core for H, the space

$$R(I + \alpha\delta)\Omega = (I + i\alpha H)D(\delta)\Omega$$

is dense in $\mathfrak{H}$. Let $A \in \mathfrak{M}_\omega$ and let $\varepsilon > 0$. Choose $B \in R(I + \alpha\delta)$ such that

$$\|B\Omega - A\Omega\| < \varepsilon.$$

Let M be an invariant mean on the continuous, bounded function $C_b(\mathbb{R})$ from $\mathbb{R}$ to $\mathbb{C}$, i.e., M is a state on $C_b(\mathbb{R})$ which is invariant under translations. Then there exists (see remark to Proposition 4.3.42) a net $\{\lambda_i^\alpha, t_i^\alpha; i = 1, \ldots, n_\alpha\}_\alpha$ such that $\lambda_i^\alpha \geq 0$, $\sum_{i=1}^{n_\alpha} \lambda_i^\alpha = 1$, $t_i^\alpha \in \mathbb{R}$, and

$$M(f) = \lim_\alpha \sum_{i=1}^{n_\alpha} \lambda_i^\alpha f(t_i^\alpha)$$

for all $f \in C_b(\mathbb{R})$. Now, M defines a projection Φ from $\mathfrak{M}$ onto the centralizer $\mathfrak{M}_\omega$ by

$$\varphi(\Phi(C)) = M(\varphi(\sigma^\omega(C)))$$

for all $C \in \mathfrak{M}$, $\varphi \in \mathfrak{M}_*$. The existence of $\Phi(C)$ follows because $\varphi \mapsto M(\varphi(\sigma^\omega(C)))$ is a continuous linear functional on the predual $\mathfrak{M}_*$ of $\mathfrak{M}$, and $\mathfrak{M} = \mathfrak{M}_*{}^*$.

Now as $\Delta^{it}H\Delta^{-it} = H$ for all $t \in \mathbb{R}$, it follows that $R(I + \alpha\delta)$ is invariant under σ_t^ω, and as $R(I + \alpha\delta)$ is σ-weakly closed it follows that

$$\Phi(B) = \lim_\alpha \sum_{i=1}^{n_\alpha} \lambda_i^\alpha \sigma_{t_i^\alpha}^\omega(B) \in R(I + \alpha\delta).$$

Moreover,

$$\begin{aligned}\left\| \sum_{i=1}^{n_\alpha} \lambda_i^\alpha \sigma_{t_i^\alpha}^\omega(B)\Omega - A\Omega \right\| &= \left\| \sum_{i=1}^{n_\alpha} \lambda_i^\alpha \sigma_{t_i^\alpha}^\omega(B - A)\Omega \right\| \\ &= \left\| \sum_i \lambda_i^\alpha \Delta^{it_i^\alpha}(B - A)\Omega \right\| \\ &\leq \|(B - A)\Omega\| \leq \varepsilon\end{aligned}$$

for all α, and hence

$$\|\Phi(B)\Omega - A\Omega\| \leq \varepsilon.$$

Therefore $(\mathfrak{M}_\omega \cap R(I + \alpha\delta))\Omega$ is dense in $\mathfrak{M}_\omega\Omega$.

Next let $A = A^* \in \mathfrak{M}_\omega$. We can find a sequence $A_n \in R(I + \alpha\delta) \cap \mathfrak{M}_\omega$ such that $\|A_n\Omega - A\Omega\| \to 0$. But as Ω is a trace vector for $\mathfrak{M}_\omega$, we also have

$$\|A_n^*\Omega - A\Omega\| = \|A_n^*\Omega - A^*\Omega\| = \|A_n\Omega - A\Omega\| \to 0.$$

Hence, replacing A_n by $(A_n + A_n^*)/2$, we may assume $A_n = A_n^* \in R(I + \alpha\delta)$. Let $B_n = B_n^* \in D(\delta)$ be such that

$$A_n = (I + \alpha\delta)(B_n).$$

As Δ and H commute strongly, we have $\sigma_t^\omega \circ \delta = \delta \circ \sigma_t^\omega$ and therefore $B_n \in \mathfrak{M}_\omega$. Also,

$$B_n\Omega = (I + i\alpha H)^{-1}A_n\Omega \to \psi \equiv (I + i\alpha H)^{-1}A\Omega.$$

Since Ω is separating for $\mathfrak{M}_\omega$ it is cyclic for $\mathfrak{M}_\omega'$, and for each $A' \in \mathfrak{M}_\omega'$ one has $B_n A'\Omega = A'B_n\Omega \to A'\psi$. Thus the graph limit of the sequence $\{B_n\}$ has dense domain. But because B_n is selfadjoint $\|(I \pm iB_n)^{-1}\| \leq 1$, and Lemma 3.1.27 establishes that the graph limit is an operator B.

Now, define $\Omega_n = (I \pm iB_n)^{-1}\Omega$ and note that

$$\begin{aligned} \|\Omega_n - \Omega_m\| &\leq \|(I \pm iB_n)^{-1}(B_n - B_m)(I \pm iB_m)^{-1}\Omega\| \\ &\leq \|(B_n - B_m)(I \pm iB_m)^{-1}\Omega\| \\ &= \|(I \mp iB_m)^{-1}(B_n - B_m)\Omega\| \\ &\leq \|(B_n - B_m)\Omega\|, \end{aligned}$$

where the third step uses the trace property of Ω on $\mathfrak{M}_\omega$. This implies that Ω_n is a Cauchy sequence and hence the identity

$$(I \pm iB_n)A'\Omega_n = A'\Omega,$$

which is valid for $A' \in \mathfrak{M}_\omega'$, implies that $\mathfrak{M}_\omega'\Omega \subseteq R(I \pm iB)$. But it then follows from Theorem 3.1.28 that B is selfadjoint, and $e^{itB_n} \to e^{itB}$ strongly, uniformly for t in compacts. Hence if χ is an infinitely differentiable function with compact support, it follows from the expansion

$$(\chi(B_n) - \chi(B))\xi = \int dt\, \hat{\chi}(t)(e^{itB_n} - e^{itB})\xi$$

that $\chi(B_n) \to \chi(B)$ strongly. Let f be a function such that $f' = \chi$. As $B_n \in D(\delta)$ it follows from Theorem 3.2.32 that $f(B_n) \in D(\delta)$ and

$$\delta(f(B_n)) = \int_{-\infty}^{\infty} dp\, \hat{\chi}(p) \int_0^1 dr\, e^{irpB_n}\delta(B_n)e^{i(1-r)pB_n}.$$

The trace property of Ω on $\mathfrak{M}_\omega$ then implies

$$\begin{aligned} 0 &= (\Omega, \delta(f(B_n))\Omega) \\ &= \left(\Omega, \int_{-\infty}^{\infty} dp\, \hat{\chi}(p)e^{ipB_n}\delta(B_n)\Omega\right) \\ &= (\Omega, \chi(B_n)\delta(B_n)\Omega). \end{aligned}$$

But because $A_n = (I + \alpha\delta)(B_n)$, this implies

$$(\Omega, \chi(B_n)A_n\Omega) = (\Omega, \chi(B_n)B_n\Omega).$$

Hence, taking strong limits,

$$(\Omega, \chi(B)A\Omega) = (\Omega, \chi(B)B\Omega).$$

If χ is positive, we get the estimate

$$\begin{aligned} |(\Omega, \chi(B)B\Omega)| &= |(\chi(B)^{1/2}\Omega, A\chi(B)^{1/2}\Omega)| \\ &\leq \|A\|(\Omega, \chi(B)\Omega). \end{aligned}$$

But as Ω is separating, one deduces from spectral theory that

$$\|B\| \leq \|A\|.$$

Hence we have proved that if $A = A^* \in \mathfrak{M}_\omega$ there exists a $B = B^* \in \mathfrak{M}_\omega$ such that

$$B\Omega = (I + i\alpha H)^{-1}A\Omega.$$

But as $(I + i\alpha H)^{-1}$ commutes with $\Delta^{1/2}$ it follows easily from the relation

$$iHJ\Delta^{1/2}C\Omega = J\Delta^{1/2}iHC\Omega,$$

which is valid for C in the core $D(\delta)\Omega$ of H, that $(I + i\alpha H)^{-1}$ commutes with $S = J\Delta^{1/2}$. But then it follows that $B \in D(\delta)$ and $A = (I + \alpha\delta)(B)$ by the reasoning used to prove (3) $\Rightarrow$ (1) in the case that Ω is separating in Theorem 3.2.59.

We now extend the previous lemma to general eigenelements for the modular automorphism group.

Lemma 3.2.63. *Adopt the hypotheses of Theorem* 3.2.61, *together with Condition* (2), *parts* (a) *and* (b). *If* $A \in \mathfrak{M}$ *and there exists a* $\lambda > 0$ *such that*

$$\sigma_t^\omega(A) = \lambda^{it}A$$

for all $t \in \mathbb{R}$ *then it follows that*

$$A \in R(I + \alpha\delta)$$

for all $\alpha \in \mathbb{R}\backslash\{0\}$.

Proof. Let M_2 be the von Neumann algebra of 2×2 matrices acting on the Hilbert space $\mathfrak{H}_2$ of 2×2 matrices with Hilbert–Schmidt norm $\|A\|_2{}^2 = \sum_{ij}|A_{ij}|^2$. The von Neumann algebra $\mathfrak{M} \otimes M_2$ acts on the Hilbert space $\mathfrak{H} \otimes \mathfrak{H}_2$. Let $\{E_{ij}\}_{i,j=1,2}$ be matrix units for M_2 and define

$$\tilde{\Omega} = \Omega \otimes (E_{11} + \lambda^{1/2}E_{22}) \in \mathfrak{H} \otimes \mathfrak{H}_2.$$

Then $\tilde{\Omega}$ is cyclic and separating for $\mathfrak{M} \otimes M_2$, and the vector functional $\tilde{\omega}$ associated with $\tilde{\Omega}$ is given by

$$\tilde{\omega}\left(\sum_{ij} B_{ij} \otimes E_{ij}\right) = \omega(B_{11}) + \lambda\omega(B_{22})$$

for $B = \sum_{ij} B_{ij} \otimes E_{ij} \in \mathfrak{M} \otimes M_2$.

If $\sigma^{\tilde{\omega}}$ is the modular group associated to $\tilde{\omega}$, it is easy to compute

$$\sigma_t^{\tilde{\omega}}\begin{pmatrix} B_{11} & B_{12} \\ B_{21} & B_{22} \end{pmatrix} = \begin{pmatrix} \sigma_t^\omega(B_{11}) & \lambda^{-it}\sigma_t^\omega(B_{12}) \\ \lambda^{it}\sigma_t^\omega(B_{21}) & \sigma_t^\omega(B_{22}) \end{pmatrix}$$

for $B_{ij} \in \mathfrak{M}$. This computation will be carried out explicitly under more general circumstances in Theorem 5.3.34. In particular, when $\sigma_t^\omega(A) = \lambda^{it}A$ we get

$$\sigma_t^{\tilde{\omega}}(A \otimes E_{12}) = A \otimes E_{12},$$

i.e., $A \otimes E_{12}$ is in the centralizer for $\tilde{\omega}$. Let now $\tilde{\delta}$ be the derivation of $\mathfrak{M} \otimes M_2$ implemented by $\tilde{H} = H \otimes I$. Clearly,

$$D(\tilde{H}) = \left\{\xi = \sum_{ij} \xi_{ij} \otimes E_{ij};\ \xi_{ij} \in D(H)\right\}$$

and

$$\tilde{H}\left(\sum_{ij} \xi_{ij} \otimes E_{ij}\right) = \sum_{ij} H\xi_{ij} \otimes E_{ij}$$

for $\xi_{ij} \in D(H)$. From this one easily verifies that

$$D(\tilde{\delta}) = \left\{B = \sum_{ij} B_{ij} \otimes E_{ij};\ B_{ij} \in D(\delta)\right\}$$

and

$$\tilde{\delta}\left(\sum_{ij} B_{ij} \otimes E_{ij}\right) = \sum_{ij} \delta(B_{ij}) \otimes E_{ij}.$$

Clearly, $\tilde{\Omega} \in D(\tilde{H})$ and $\tilde{H}\tilde{\Omega} = 0$. Moreover,

$$D(\tilde{\delta})\tilde{\Omega} = D(\delta)\Omega \otimes \mathfrak{H}_2,$$

which proves that $D(\tilde{\delta})\tilde{\Omega}$ is a core for $\tilde{H}$. From the form of $\sigma^{\tilde{\omega}}$ given above one computes that $\tilde{\delta} \circ \sigma_t^{\tilde{\omega}} = \sigma_t^{\tilde{\omega}} \circ \tilde{\delta}$ and hence $iH\tilde{\Delta}^{it} = \tilde{\Delta}^{it}iH$, where $\tilde{\Delta}$ is the modular operator associated to the couple $(\mathfrak{M} \otimes M_2, \tilde{\Omega})$. Application of Lemma 3.2.62 to $\mathfrak{M} \otimes M_2$, $\tilde{\Omega}$, $\tilde{H}$ now gives that $A \otimes E_{12} \in R(I + \alpha\tilde{\delta})$ or, equivalently,

$$A \in R(I + \alpha\delta)$$

for $\alpha \in \mathbb{R}\backslash\{0\}$.

We now prove (2) $\Leftrightarrow$ (1) in the special case that $\sigma_T{}^{\omega}$ is inner for some $T \neq 0$.

Lemma 3.2.64. *Adopt the hypotheses of Theorem* 3.2.61, *together with condition* (2), *parts* (a) *and* (b). *Assume in addition that there exists a* $T > 0$ *and a unitary element* $U \in \mathfrak{M}$ *such that*

$$\sigma_T{}^{\omega}(A) = UAU^*, \qquad A \in \mathfrak{M},$$

and assume that $U \in D(\delta)$ *with*

$$\delta(U) = 0.$$

It follows that

$$e^{itH}\mathfrak{M}e^{-itH} = \mathfrak{M}$$

for all $t \in \mathbb{R}$.

Proof. We first assume $U = \mathbb{1}$, i.e., $\sigma_T{}^{\omega}(A) = A$ for all $A \in \mathfrak{M}$. Since σ^{ω} is periodic it follows from Lemma 3.2.39(4) that the eigenspaces

$$\mathfrak{M}_n = \{A \in \mathfrak{M};\ \sigma_t{}^{\omega}(A) = e^{-i(2\pi n/T)t}A\}, \qquad n \in \mathbb{Z},$$

span a σ-weakly dense subalgebra of $\mathfrak{M}$. By Lemma 3.2.63,

$$\bigcup_{n \in \mathbb{Z}} \mathfrak{M}_n \subseteq R(I + \alpha\delta).$$

Hence $R(I + \alpha\delta)$ is dense, and it follows from $(*)$, page 279, that

$$e^{itH}\mathfrak{M}e^{-itH} = \mathfrak{M}, \qquad t \in \mathbb{R}.$$

Assume next that $\sigma_T{}^\omega(A) = UAU^*$ with $\delta(U) = 0$. Since

$$\omega(UAU^*) = \omega(\sigma_T{}^\omega(A)) = \omega(A)$$

for $A \in \mathfrak{M}$, it follows that U belongs to the centralizer $\mathfrak{M}_\omega$ for ω. As $\mathfrak{M}_\omega$ is a von Neumann algebra, there exists an $A = A^* \in \mathfrak{M}_\omega$ with $\|A\| \leq \pi$ such that $U = e^{iA}$. Put

$$B = e^{A/T}.$$

Then B is positive with bounded inverse, $B^{iT} = U$, and $B \in \mathfrak{M}_\omega$. Moreover, as $[H, U] = 0$, we have $[H, B^\beta] = 0$ for all $\beta \in \mathbb{C}$. In particular, $B^\beta \in D(\delta)$. Put

$$\Omega' = B^{-1/2}\Omega.$$

Since $B^{1/2}$ and $B^{-1/2}$ are bounded, it follows that Ω' is cyclic and separating for $\mathfrak{M}$. Put $\omega'(A) = (\Omega', A\Omega')$. As $\sigma_t{}^\omega(B^\beta) = B^\beta$ for all $t \in \mathbb{R}$ and $\beta \in \mathbb{C}$, it is easily verified that

$$\sigma_t(A) = B^{-it}\sigma_t{}^\omega(A)B^{it}$$

defines a one-parameter group of automorphisms of $\mathfrak{M}$, and that ω satisfies the KMS condition with respect to this group in the sense of Definition 5.3.1. It follows from Theorem 5.3.10 that $\sigma = \sigma^{\omega'}$ is the modular group associated to ω'. But then

$$\sigma_T{}^{\omega'}(A) = B^{-iT}\sigma_T{}^\omega(A)B^{iT} = U^*(UAU^*)U = A$$

for all $A \in \mathfrak{M}$, i.e., $\sigma^{\omega'}$ is periodic. Since $B^{1/2}, B^{-1/2} \in D(\delta)$ we have $D(\delta)\Omega' = D(\delta)\Omega$, and hence $D(\delta)\Omega'$ is a core for H. As $\delta(B^{-it}) = 0$ for all t, and $\delta \circ \sigma_t{}^\omega = \sigma_t{}^\omega \circ \delta$ it follows that $\delta \circ \sigma_t^{\omega'} = \sigma_t^{\omega'} \circ \delta$, and hence $H\Delta'^{it} = \Delta'^{it}H$, where Δ' is the modular operator associated with the pair $(\mathfrak{M}, \Omega')$. But now it follows from the first part of the proof of this lemma that

$$e^{itH}\mathfrak{M}e^{-itH} = \mathfrak{M}.$$

We need one more lemma to prove (2) $\Leftrightarrow$ (1) in Theorem 3.2.61. Adopting parts (a) and (b) of condition (2) of that theorem, let $T > 0$ be a fixed real number, and consider the discrete crossed product $\mathfrak{N} = W^*(\mathfrak{M}, \sigma_T{}^\omega)$, where $\omega(A) = (\Omega, A\Omega)$, $A \in \mathfrak{M}$. Let $\tilde{\mathfrak{H}}$ be the Hilbert space

$$\tilde{\mathfrak{H}} = \bigoplus_{n=-\infty}^{\infty} \mathfrak{H}_n,$$

where $\mathfrak{H}_n = \mathfrak{H}$, $n \in \mathbb{Z}$. Then it follows from Definition 2.7.3 that $\mathfrak{N}$ is the von Neumann algebra generated by the operators $\pi(A)$, $A \in \mathfrak{M}$, and U defined on $\tilde{\mathfrak{H}}$ by

$$(\pi(A)\xi)_n = \sigma^\omega_{-nT}(A)\xi_n,$$
$$(U\xi)_n = \xi_{n-1},$$

where $A \in \mathfrak{M}$ and $\xi = (\xi_n)_n \in \tilde{\mathfrak{H}}$. Note that $U\pi(A)U^* = \pi(\sigma_T{}^\omega(A))$ for $A \in \mathfrak{M}$. Consider the vector $\tilde{\Omega} \in \tilde{\mathfrak{H}}$ given by

$$\tilde{\Omega}_n = \begin{cases} \Omega, & n = 0, \\ 0, & n \neq 0, \end{cases}$$

and let $\tilde{\omega}$ be the positive vector functional on $\mathfrak{N}$ defined by $\tilde{\Omega}$.

Lemma 3.2.65. *Let $\mathfrak{M}$ be a von Neumann algebra with a separating and cyclic vector Ω, and define $\mathfrak{N}$, $\tilde{\mathfrak{H}}$, π, U, $\tilde{\Omega}$, and $\tilde{\omega}$ as above. It follows that*

(1) *$\tilde{\Omega}$ is cyclic and separating for $\mathfrak{N}$,*
(2) *The modular automorphism group $\sigma^{\tilde{\omega}}$ associated to $\tilde{\omega}$ is given by*

$$\sigma_t^{\tilde{\omega}}(\pi(A)) = \pi(\sigma_t^{\omega}(A)), \qquad A \in \mathfrak{M},$$
$$\sigma_t^{\tilde{\omega}}(U) = U,$$

(3) *$\sigma_T^{\tilde{\omega}}(B) = UBU^*$, $B \in \mathfrak{N}$,*
(4) *The modular operator $\tilde{\Delta}$ associated with $\tilde{\Omega}$ is given by*

$$(\tilde{\Delta}^{it}\xi)_n = \Delta^{it}\xi_n, \qquad (\xi_n)_n \in \tilde{\mathfrak{H}}.$$

PROOF. (1) Any element A in the algebra $\mathfrak{N}_0$ generated algebraically by $\pi(\mathfrak{M})$ and U has the form

$$A = \sum_{n=-p}^{p} U^n\pi(A_n),$$

where $A_n \in \mathfrak{M}$. Since the maps $A \in \mathfrak{N}_0 \to A_n$ are σ-strongly continuous, and $\mathfrak{N}_0$ is σ-strongly dense in $\mathfrak{N}$, it follows that any $A \in \mathfrak{N}$ has an expansion

$$A \sim \sum_{n=-\infty}^{\infty} U^n\pi(A_n), \qquad A_n \in \mathfrak{M},$$

which converges in the sense that

$$A\xi = \sum_{n=-\infty}^{\infty} U^n\pi(A_n)\xi$$

for all $\xi \in \tilde{\mathfrak{H}}$ with finite support. In particular,

$$A\tilde{\Omega} = \sum_{n=-\infty}^{\infty} U^n\pi(A_n)\tilde{\Omega}.$$

Hence $(A\tilde{\Omega})_n = A_n\Omega$, and therefore $\|A\tilde{\Omega}\|^2 = \sum_{n=-\infty}^{\infty} \|A_n\Omega\|^2$. This proves that $\tilde{\Omega}$ is separating for $\mathfrak{N}$. As $\mathfrak{N}\tilde{\Omega}$ contains all vectors $(\xi_n)_n \in \tilde{\mathfrak{H}}$ with finite support for which $\xi_n \in \mathfrak{M}\Omega$ it follows that $\tilde{\Omega}$ is cyclic for $\mathfrak{N}$.

(2) Let $t \mapsto V_t$ be the strongly continuous one-parameter unitary group on $\tilde{\mathfrak{H}}$ given by

$$(V_t\,\xi)_n = \Delta^{it}\xi_n.$$

It is trivial to check that

$$V_t\pi(A)V_t^* = \pi(\sigma_t^{\omega}(A)),$$
$$V_tUV_t^* = U.$$

Hence V implements a σ-weakly continuous one-parameter group σ of *-automorphisms of $\mathfrak{N}$. Now one easily checks that $\tilde{\omega}$ is KMS with respect to σ on $\mathfrak{N}_0$ (see Definition 5.3.1 and Proposition 5.3.7). But then it follows from Theorem 5.3.10 that σ is the modular automorphism group associated with $\tilde{\omega}$.

(3) For $A \in \mathfrak{M}$ we have by (2) that

$$\sigma_T^{\tilde{\omega}}(\pi(A)) = \pi(\sigma_T^{\omega}(A)) = U\pi(A)U^*.$$

Moreover,

$$\sigma_T^{\tilde{\omega}}(U) = U = UUU^*.$$

Hence

$$\sigma_T^{\tilde{\omega}}(B) = UBU^*$$

for all $B \in \mathfrak{N}$.

(4) Let $A \in \mathfrak{N}_0$, $A = \sum_{n=-p}^{p} U^n\pi(A_n)$. Then, as $(A\tilde{\Omega})_n = A_n\Omega_n$, we have by (2)

$$(\tilde{\Delta}^{it}A\tilde{\Omega})_n = (\sigma_t^{\tilde{\omega}}(A)\tilde{\Omega})_n = \sigma_t^{\omega}(A_n)\Omega = \Delta^{it}A_n\Omega.$$

Hence

$$\tilde{\Delta}^{it} = \bigoplus_{n=-\infty}^{\infty} \Delta^{it}.$$

END OF PROOF OF THEOREM 3.2.61. (2) ⇒ (1) Define a selfadjoint operator $\tilde{H}$ on $\tilde{\mathfrak{H}}$ by

$$D(\tilde{H}) = \left\{(\xi_n)_n \in \tilde{H};\, \xi_n \in D(H),\, \sum_n \|H\xi_n\|^2 < +\infty\right\},$$

$$(\tilde{H}\xi)_n = H\xi_n.$$

Let $\tilde{\delta} = i[\tilde{H}, \cdot]$ be the derivation on $\mathfrak{N}$ implemented by $\tilde{H}$. We shall prove that $\mathfrak{N}$, $\tilde{\Omega}$, $\tilde{H}$, $\tilde{\delta}$ and U satisfy the hypotheses of Lemma 3.2.64. Clearly, $\tilde{H}\tilde{\Omega} = 0$. Since $\delta \circ \sigma_t^{\omega} = \sigma_t^{\omega} \circ \delta$ by (2), condition (b), one has $\pi(A) \in D(\tilde{\delta})$ and

$$\tilde{\delta}(\pi(A)) = \pi(\delta(A))$$

whenever $A \in D(\delta)$. Moreover $U\tilde{H} = \tilde{H}U$, and hence $U \in D(\tilde{\delta})$ with

$$\tilde{\delta}(U) = 0.$$

Therefore $D(\tilde{\delta})$ contains all operators of the form

$$\sum_{n=-p}^{p} U^n\pi(A_n), \qquad A_n \in D(\delta).$$

In particular, $D(\tilde{\delta})\tilde{\Omega}$ contains all vectors $(\xi_n) \in \tilde{\mathfrak{H}}$ of finite support for which $\xi_n \in D(\delta)\Omega$. Hence $D(\tilde{\delta})\tilde{\Omega}$ is a core for $\tilde{H}$. Since H commutes with Δ^{it}, it follows from Lemma 3.2.65(4) that $\tilde{H}$ commutes with $\tilde{\Delta}^{it}$. The same lemma also implies that $\sigma_T^{\tilde{\omega}}$ is implemented by U, and as $[\tilde{H}, U] = 0$ it follows from Lemma 3.2.64 that

$$e^{it\tilde{H}}\mathfrak{N}e^{-it\tilde{H}} = \mathfrak{N}, \qquad t \in \mathbb{R}.$$

Now, let P be the projection from $\tilde{\mathfrak{H}} = \bigoplus_{n=-\infty}^{\infty} \mathfrak{H}_n$ onto the zeroth component $\mathfrak{H}_0 = \mathfrak{H}$. Then for $A \in \mathfrak{N}$, $A \sim \sum_{n=-\infty}^{\infty} U^n\pi(A_n)$, we have

$$PAP^* = A_0.$$

Moreover,

$$e^{itH}P = Pe^{it\tilde{H}}.$$

Now choose $A \in \mathfrak{M}$. Put $B = e^{it\tilde{H}}\pi(A)e^{-it\tilde{H}} \in \mathfrak{N}$ and let

$$B \sim \sum_{n=-\infty}^{\infty} U^n\pi(B_n), \qquad B_n \in \mathfrak{M},$$

be the expansion of B. Then

$$\begin{aligned} e^{itH}Ae^{-itH} &= e^{itH}P\pi(A)P^*e^{-itH} \\ &= Pe^{it\tilde{H}}\pi(A)e^{-it\tilde{H}}P^* \\ &= PBP^* = B_0 \in \mathfrak{M}. \end{aligned}$$

Hence

$$e^{itH}\mathfrak{M}e^{-itH} \subseteq \mathfrak{M}$$

for any $t \in \mathbb{R}$, ending the proof of Theorem 3.2.61.

Remark 3.2.66. One corollary of Theorem 3.2.61, indeed of Lemma 3.2.62, is that if,

(1) $\mathfrak{M}$ is a von Neumann algebra on a Hilbert space $\mathfrak{H}$,
(2) Ω is a cyclic vector defining a trace on $\mathfrak{M}$, i.e.,

$$(\Omega, AB\Omega) = (\Omega, BA\Omega), \qquad A, B \in \mathfrak{M},$$

and,

(3) δ is a spatial derivation of $\mathfrak{M}$ implemented by a selfadjoint operator H such that $H\Omega = 0$ and $D(\delta)\Omega$ is a core for H, then it follows that

$$e^{itH}\mathfrak{M}e^{-itH} = \mathfrak{M}$$

for all $t \in \mathbb{R}$. Remark that Ω is automatically separating for $\mathfrak{M}$ in this case, because if $A\Omega = 0$ and $B \in \mathfrak{M}$ one has

$$\begin{aligned} \|AB\Omega\|^2 &= (\Omega, B^*A^*AB\Omega) \\ &= (\Omega, BB^*A^*A\Omega) = 0 \end{aligned}$$

and the cyclicity of Ω implies $A = 0$. As $\Delta = I$ in this case, this corollary follows already from Lemma 3.2.62. The definition of a derivation used in this corollary can be weakened to allow a mapping $A \in D(\delta) \subseteq \mathfrak{M} \mapsto \delta(A) \in \mathfrak{N}$, where $\mathfrak{N}$ is the set of operators affiliated with $\mathfrak{M}$ and having Ω in their domain. For this generalization it is essential to remark that $\mathfrak{N}$ is a selfadjoint space such that $\mathfrak{N}\mathfrak{M} \subseteq \mathfrak{N}$. The selfadjointness follows by exploiting the trace condition. If $X \in \mathfrak{N}$ and $X = U|X|$ is its polar decomposition then $U \in \mathfrak{M}$ and $|X|$ is affiliated with $\mathfrak{M}$ by Lemma 2.5.8. But if E_n is the spectral projector of $|X|$ corresponding to the interval $[0, n]$ then

$$\|XE_n\Omega\| = \||X|E_n\Omega\| = \|E_n|X|U^*\Omega\| = \|E_nX^*\Omega\|$$

by the trace property of Ω. Hence $\Omega \in D(X^*)$ and $\|X^*\Omega\| = \|X\Omega\|$. Thus $\mathfrak{N}$ is selfadjoint. But it is trivial that $\mathfrak{M}\mathfrak{N} \subseteq \mathfrak{N}$ and hence $\mathfrak{N}\mathfrak{M} \subseteq \mathfrak{N}$ by selfadjointness. Therefore the derivation properties

$$\delta(AB) = \delta(A)B + A\delta(B), \qquad \delta(A^*) = \delta(A)^*$$

are well defined. The proof of this consequence of Theorem 3.2.61 needs a slight extension of Theorem 3.2.59 to establish that the automorphism property is equivalent to the positivity preserving property

$$(I \pm iH)^{-1}\mathfrak{M}_+\Omega \subseteq \overline{\mathfrak{M}_+\Omega}$$

and in the proof of Lemma 3.2.62 one must calculate $\delta(f(B))$ directly in the vector state given by Ω.

If in Theorem 3.2.61 the algebra $\mathfrak{M}$ is abelian then the conclusion is really a statement of global existence for a class of first-order differential equations of the type that occur in the Hamilton–Liouville description of classical mechanics. We illustrate this with the following example.

EXAMPLE 3.2.67. Let X be a locally compact Hausdorff space and μ a probability measure on X. For example, X could represent the phase space of particles in classical mechanics and μ the probability distribution describing their positions and momenta. If $t \mapsto T_t$ is a continuous measure-preserving group of homeomorphisms of X then T_t determines a strongly continuous group of *-automorphisms of $C_0(X)$ by

$$(\alpha_t f)(x) = f(T_t x)$$

for $f \in C_0(X)$. Now realizing $C_0(X)$ as a C^*-algebra of multiplication operators on the Hilbert space $\mathfrak{H} = L^2(X; d\mu)$ one has

$$\alpha_t(f) = U_t f U_t^*,$$

where $t \mapsto U_t$ is the strongly continuous unitary group given by

$$(U_t\psi)(x) = \psi(T_t x)$$

for $\psi \in L^2(X; d\mu)$. Next choosing Ω to be the unit function one has

$$\omega(f) = (\Omega, f\Omega) = \int_X d\mu(x) f(x)$$

and the invariance condition

$$U_t\Omega = \Omega.$$

If δ is the generator of the group α_t then $D(\delta) \subseteq C_0(X) \subseteq L^2(X; d\mu)$ and $D(\delta)$ is invariant under U_t. Hence identifying $D(\delta)$ and $D(\delta)\Omega$ one deduces that $D(\delta)\Omega$ is a core for the self-adjoint generator H of U_t.

More generally, if $t \mapsto T_t$ is a group of measure-preserving Borel automorphisms such that

$$t \mapsto \int_X d\mu(x) f(T_t x) g(x)$$

is continuous for all $f \in L^\infty(X; d\mu)$ and $g \in L^1(X; d\mu)$ then

$$(\alpha_t f)(x) = f(T_t x)$$

is a σ-weakly continuous group of *-automorphisms of the von Neumann algebra $L^\infty(X; d\mu)$. As before one has

$$\alpha_t(f) = U_t f U_t^*$$

and the generators δ of α_t and H of U_t are such that $D(\delta)(=D(\delta)\Omega)$ is a core for H.

Theorem 3.2.61 and Remark 3.2.66 now give a converse statement. If $-i\delta$ is an essentially selfadjoint operator from a domain $D(\delta) \subseteq L^\infty(X; d\mu)$ to $L^2(X; d\mu)$ with the properties

(1) $\delta(fg) = \delta(f)g + f\delta(g), \qquad f, g \in D(\delta),$
(2) $\delta(\bar{f}) = \overline{\delta(f)}, \qquad f \in D(\delta),$
(3) $\Omega \in D(\delta)$

then the theorem implies the existence of a σ-weakly continuous group α_t of *-automorphisms of $L^\infty(X; d\mu)$. Now considering the Borel sets in X as projections in $L^\infty(X; d\mu)$ the group α_t determines a measure-preserving automorphism group T_t of X such that

$$(\alpha_t f)(x) = f(T_t x)$$

for all $f \in L^\infty(X; d\mu)$. Thus Theorem 3.2.61 potentially plays a role in the integration of the equations of motion of classical mechanics.

Finally, we remark that both (a) and (b) of condition (2) in Theorem 3.2.61 are necessary for the automorphism property. Although the formal calculation

$$iHJ\Delta^{1/2}A\Omega = iHA^*\Omega = \delta(A)^*\Omega = J\Delta^{1/2}iHA\Omega$$

indicates that H and Δ commute the domain problems which arise when both H and Δ are unbounded invalidate this conclusion. In fact, if one drops (b) of condition (2) in the statement of the theorem then the conclusion is incorrect as the following example shows.

EXAMPLE 3.2.68. Let Ω_0 be a unit vector in a Hilbert space $\mathfrak{H}_0$ and assume the existence of two von Neumann algebras $\mathfrak{N}_1$ and $\mathfrak{N}_2$ on $\mathfrak{H}_0$ such that $\mathfrak{N}_1 \subseteq \mathfrak{N}_2$, $\mathfrak{N}_1 \neq \mathfrak{N}_2$, and Ω_0 is cyclic and separating for both algebras. (Examples of this type abound in quantum field theory where one encounters quasi-local algebras with a ground state vector which is cyclic and separating for each of the local algebras.) Let $\mathbb{T} = \mathbb{R}/\mathbb{Z}$ be the circle group equipped with Haar measure and define an action of $\mathbb{T}$ on the von Neumann algebra $\mathcal{L}(\mathfrak{H}_0) \bigoplus \mathcal{L}^\infty(\mathbb{T}) = L^\infty(\mathcal{L}(\mathfrak{H}_0); \mathbb{T})$ by

$$(\tau_t f)(s) = f(s - t)$$

for all $f \in L^\infty(\mathcal{L}(\mathfrak{H}_0); \mathbb{T})$. Let $\mathfrak{M}$ be the sub-von Neumann algebra consisting of $f \in L^\infty(\mathcal{L}(\mathfrak{H}_0); \mathbb{T})$ such that

$$f(s) \in \begin{cases} \mathfrak{N}_1 & \text{if } 0 \leq s < \frac{1}{2} \\ \mathfrak{N}_2 & \text{if } \frac{1}{2} \leq s < 1 \end{cases}$$

and realize $\mathfrak{M}$ as an algebra of multiplication operators on $\mathfrak{H} = L^2(\mathfrak{H}_0; \mathbb{T})$ by

$$(f\xi)(s) = f(s)\xi(s)$$

for $f \in L^\infty(\mathscr{L}(\mathfrak{H}_0); \mathbb{T})$, $\xi \in \mathfrak{H}$. Let $\Omega \in \mathfrak{H}$ be the element given by

$$\Omega(s) = \Omega_0$$

for all $s \in \mathbb{T}$. Then Ω is separating and cyclic for $\mathfrak{M}$. Moreover, $U(t)\xi(s) = \xi(s - t)$ defines a unitary representation of $\mathbb{T}$ on $\mathfrak{H}$ such that

$$\tau_t(f) = U_t f U_t^*$$

for $f \in L^\infty(\mathscr{L}(\mathfrak{H}_0); \mathbb{T})$ and $U_t\Omega = \Omega$. Let $\hat{\delta}$ be the infinitesimal generator of τ, and $iH = \mathbb{1} \otimes (-d/dt)$ the generator of U_t. Then $D(\hat{\delta})$ consists of the Hölder-continuous functions from $\mathbb{T}$ into $\mathscr{L}(\mathfrak{H})$, where $\mathscr{L}(\mathfrak{H})$ is equipped with the norm topology. Define δ on $\mathfrak{M}$ as $\hat{\delta}|_{D(\delta)}$, where

$$D(\delta) = \{f; f \in D(\hat{\delta}) \cap \mathfrak{M}, \quad \hat{\delta}(f) \in \mathfrak{M}\}.$$

Then δ is a σ-weakly densely defined derivation on $\mathfrak{M}$ such that

$$\delta(f) = [iH, f]$$

for $f \in \mathfrak{M}$.

Now $D(\delta)\Omega$ contains the set D of all ξ of the form $s \mapsto f(s)\Omega_0$, where $s \mapsto f(s)$ is Hölder continuous and $f(s) \in \mathfrak{N}_1$. As $U(t)D = D$, D is a core for H by Corollary 3.1.7, and hence $D(\delta)\Omega$ is also a core. But clearly, $U_t \mathfrak{M} U_t^* \neq \mathfrak{M}$ unless $t \in \mathbb{Z}$.

3.2.6. Approximation Theory for Automorphism Groups

In Section 3.1 we distinguished between three different aspects of stability of one-parameter groups of automorphisms. We discussed convergence theory, perturbation theory, and approximation theory. All the results of Section 3.1 apply to one-parameter groups of *-automorphisms of topological algebras and the algebraic structure contributes no essentially new feature to the first aspect of stability, i.e., convergence. In Section 5.4 we discuss various algebraic refinements of the time-dependent perturbation series and in this section we show how the results of Section 3.1.5 can be considerably sharpened for σ-weakly continuous one-parameter groups of *-automorphisms of von Neumann algebras. Recall that these groups are $C_0{}^*$-groups of isometries, by Corollary 2.3.4 and Theorem 2.4.23. In particular, Theorem 3.1.38 can be applied for the comparison of two such groups that differ in norm by $O(t)$. The next result strengthens the conclusion of the latter theorem.

Proposition 3.2.69. *Let α_t, β_t be σ-weakly continuous one-parameter groups of *-automorphisms of a von Neumann algebra $\mathfrak{M}$, with generators δ_α and δ_β, respectively. The following conditions are equivalent:*

(1) *$\|\alpha_t - \beta_t\| = O(t)$ as $t \to 0$;*
(2) *$D(\delta_\alpha) = D(\delta_\beta)$ and there exists a bounded derivation δ of $\mathfrak{M}$ such that*

$$\delta_\alpha(A) - \delta_\beta(A) = \delta(A)$$

for all $A \in D(\delta_\alpha) = D(\delta_\beta)$.

Remark. From the proof of this proposition and Corollary 3.2.47, it follows that $\|\delta\| = \overline{\lim}_{t\to 0} \|\alpha_t - \beta_t\|/|t|$. Moreover, there exists an $H = H^* \in \mathfrak{M}$ such that $\|H\| \leq \|\delta\|/2$ and $\delta(A) = i[H, A]$. Note that an elaboration of Example 3.1.39 shows that the theorem is false for C^*-algebras, even for simple C^*-algebras with identity.

PROOF. It follows from Theorem 3.1.38 that (2) $\Rightarrow$ (1) and that (1) implies $D(\delta_\alpha) = D(\delta_\beta)$ and $\delta_\alpha - \delta_\beta$ is norm bounded on $D(\delta_\alpha)$. But $\delta_\alpha - \delta_\beta$ then extends by continuity to a bounded derivation δ_0 from the C^*-algebra $\overline{D(\delta_\alpha)}^{\|\cdot\|}$ into $\mathfrak{M}$. Proposition 3.2.24 then states that δ_0 has a σ-weakly closed extension δ to $\mathfrak{M}$ with $\|\delta\| = \|\delta_0\|$.

We will now consider more general approximations than the $O(t)$ case, and we first study the case of unitary groups on Hilbert spaces. In this case Theorem 3.1.36 can be improved.

Proposition 3.2.70. *Let $U_t = \exp\{itH\}$ and $V_t = \exp\{itK\}$ be strongly continuous unitary groups on a Hilbert space $\mathfrak{H}$. The following conditions are equivalent:*

(1) *there exists $\varepsilon_1 > 0$ and $\delta_1 > 0$, such that*

$$\|U_t - V_t\| \leq \sqrt{2} - \varepsilon_1$$

for $0 \leq t \leq \delta_1$;

(2) *there exists $\varepsilon_2 > 0$, $\delta_2 > 0$, a bounded selfadjoint operator P, and a unitary operator W such that*

$$H = W(K + P)W^*$$

and

$$\|W^*U_tW - U_t\| \leq \sqrt{2} - \varepsilon_2$$

for all $0 \leq t \leq \delta_2$.

If these conditions are satisfied, then W may be chosen as the unitary part of the polar decomposition of

$$\Omega = \frac{1}{\delta_1}\int_0^{\delta_1} ds\, U_s V_{-s}$$

and W satisfies the estimate $\|W - I\| \leq \sqrt{2} - \varepsilon_1$. Moreover,

$$\|W^*U_tW - U_t\| = \|V_t - U_t\| + O(t)$$

and

$$\|W^*U_tW - V_t\| = O(t)$$

as $t \to 0$.

PROOF. First note that (2) $\Rightarrow$ (1) follows essentially from Theorem 3.1.36.

(1) $\Rightarrow$ (2) Some results on the numerical range of an operator are needed. We recall that if A is a bounded operator on a Hilbert space $\mathfrak{H}$ then its numerical range $W(A)$ is the subset of $\mathbb{C}$ given by

$$W(A) = \{(\psi, A\psi); \psi \in \mathfrak{H}, \|\psi\| = 1\}.$$

It can be shown that $W(A)$ is a convex subset of $\mathbb{C}$ whose closure contains the spectrum $\sigma(A)$ of A, and if A is normal, i.e., if $A^*A = AA^*$, then $\overline{W(A)}$ is just the convex closure of $\sigma(A)$. The sector $S(A)$ of A is defined as the closed cone generated by $W(A)$, i.e.,

$$S(A) = \overline{\{(\psi, A\psi); \psi \in \mathfrak{H}\}}.$$

For bounded A it is also known that if the partial isometry U in the polar decomposition $A = U|A|$ is unitary then $\sigma(U) \subseteq S(A)$.

Now, assumption (1) implies by spectral theory that

$$W(U_t V_{-t}) \subseteq \{z; z \in \mathbb{C}, |z| < 1, \operatorname{Re} z \geq \sqrt{2}\varepsilon_1 - \varepsilon_1{}^2/2\}$$

when $0 \leq t \leq \delta_1$. Define

$$\Omega = \frac{1}{\delta_1} \int_0^{\delta_1} ds\, U_s V_{-s}.$$

As the last set above is convex, it follows that

$$W(\Omega) \subseteq \{z; z \in \mathbb{C}, |z| \leq 1, \operatorname{Re} z \geq \sqrt{2}\varepsilon_1 - \varepsilon_1{}^2/2\}.$$

In particular, $0 \notin \overline{W(\Omega)} \supseteq \sigma(\Omega)$, i.e., Ω is invertible with bounded inverse Ω^{-1}. By the proof of Theorem 3.1.36 one has

$$\|U_t \Omega V_{-t} - \Omega\| = O(t) = \|V_t \Omega^* U_{-t} - \Omega^*\|.$$

Hence

$$\begin{aligned} \|V_t \Omega^* \Omega V_{-t} - \Omega^* \Omega\| &= \|(V_t \Omega^* U_{-t})(U_t \Omega V_{-t}) - \Omega^* \Omega\| \\ &\leq \|V_t \Omega^* U_{-t} - \Omega^*\| \, \|U_t \Omega V_{-t}\| + \|U_t \Omega V_{-t} - \Omega\| = O(t). \end{aligned}$$

By Proposition 3.1.23, it follows that $|\Omega|^2$ lies in the domain of the derivation which generates the automorphism group, of $\mathscr{L}(\mathfrak{H})$, implemented by V_t. It then follows from Corollary 3.2.33 that $|\Omega|^{-1}$ lies in the domain of this derivation. Hence

$$\|V_t |\Omega|^{-1} V_{-t} - |\Omega|^{-1}\| = O(t).$$

Now, defining W as the unitary part of the polar decomposition of Ω, one has $W = \Omega|\Omega|^{-1}$ and

$$\begin{aligned} \|U_t W V_{-t} - W\| &= \|(U_t \Omega V_{-t})(V_t |\Omega|^{-1} V_{-t}) - \Omega|\Omega|^{-1}\| \\ &= \|U_t \Omega V_{-t} - \Omega\| \, \||\Omega|^{-1}\| + \|\Omega\| \, \|V_t |\Omega|^{-1} V_{-t} - |\Omega|^{-1}\| \\ &= O(t). \end{aligned}$$

Next introduce $\hat{V}_t = W V_t W^*$ and note that

$$\|U_t - \hat{V}_t\| = O(t).$$

It follows that the generators iH and $WiKW^*$ differ by a bounded skew-adjoint operator (Example 3.1.40). This establishes the relation between H and K.

Next note that

$$\|W^*U_tW - V_t\| = O(t)$$

since $W^*iHW - iK = iP$ is bounded. Moreover, the relation

$$U_t - V_t = U_t - W^*U_tW + W^*U_tW - V_t$$

allows one to conclude that

$$\|U_t - V_t\| = \|U_t - W^*U_tW\| + O(t).$$

Finally, since

$$S(\Omega) \subseteq \{z; z \in \mathbb{C}, \operatorname{Re} z \geq (\sqrt{2}\varepsilon_1 - \varepsilon_1{}^2/2)|\operatorname{Im} z|\}$$

it follows that $S(W)$ is contained in the same cone. Hence $\sigma(W)$ is contained in this cone, and

$$\|W - I\| \leq \sqrt{2} - \varepsilon_1.$$

Our next aim is to derive an analogue of Proposition 3.2.70 for σ-weakly continuous automorphism groups of a von Neumann algebra. The above result for unitary groups will be necessary for this derivation but we will also need three other types of preliminary information of which at least two have interest in their own right. Firstly, we consider the relation between two automorphisms of a von Neumann algebra which are close in norm. Secondly, we derive a measure theoretic result, and thirdly, a cohomological result, for joint unitary implementability of two groups of automorphisms.

Since an automorphism of a von Neumann algebra $\mathfrak{M}$ is an isometry we have $\|\alpha - \beta\| = \|\alpha\beta^{-1} - \iota\|$ for any pair of automorphisms α, β. Hence it suffices to consider the single automorphism $\alpha\beta^{-1}$ when studying implications of closeness in norms. Clearly $\|\alpha - \iota\| \leq \|\alpha\| + \|\iota\| = 2$ and it is remarkable that if $\|\alpha - \iota\| < 2$ then α is inner. As a consequence the set $\operatorname{Inn}(\mathfrak{M})$ of inner automorphisms of $\mathfrak{M}$ is just the connected component of ι in the set $\operatorname{Aut}(\mathfrak{M})$ of automorphisms of $\mathfrak{M}$, when $\operatorname{Aut}(\mathfrak{M})$ is equipped with the norm topology.

Theorem 3.2.71. *Let α be an automorphism of a von Neumann algebra $\mathfrak{M}$, and assume that*

$$\|\alpha - \iota\| < 2.$$

It follows that α is inner.

Moreover, there exists a unitary operator $U \in \mathfrak{M}$ implementing α such that the spectrum of U is contained in the half-plane

$$\{z; \operatorname{Re} z \geq \tfrac{1}{2}(4 - \|\alpha - \iota\|^2)^{1/2}\},$$

i.e., one has the bound

$$\|U - \mathbb{1}\| \leq \{2(1 - \sqrt{1 - \|\alpha - \iota\|^2/4})\}^{1/2}.$$

Remark. The condition on the spectrum, or on the norm, of U can be reinterpreted more geometrically as saying that $\sigma(U)$ lies on the arc of the unit circle symmetric about 1 with endpoints midway between 1 and the points at distance $\|\alpha - \iota\|$ from 1. In fact the estimate on $\|U - \mathbb{1}\|$ is the best possible.

We will not prove Theorem 3.2.71 here, but only remark that if $\|\alpha - \iota\| < \sqrt{3}$, one can show, using Lie algebra techniques, that $\delta = \log \alpha$, defined as a contour integral, is a derivation of $\mathfrak{M}$. It follows by Corollary 3.2.47 that $\delta(A) = [H, A]$ for some $H \in \mathfrak{M}$ and hence $\alpha(A) = \exp(\delta)(A) = e^H A e^{-H}$. If U is the unitary part of the polar decomposition of e^H one then shows easily that $\alpha(A) = UAU^*$.

To prove the theorem for $\|\alpha - \iota\| < 2$ and to obtain the estimate on $\|U - \mathbb{1}\|$ a less straightforward approach is needed (see Notes and Remarks).

If α, β are σ-weakly continuous one-parameter groups of automorphisms of a von Neumann algebra $\mathfrak{M}$, and $\|\alpha_t - \beta_t\| < 2$ for small t, then it follows from Theorem 3.2.71 that $\gamma_t = \beta_t \alpha_{-t}$ is inner for small t, and hence by iterating the cocycle identity,

$$\gamma_{t+s} = \gamma_t(\alpha_t \gamma_s \alpha_{-t}),$$

γ_t is inner for all t. Thus for each t there exists a $W_t \in \mathfrak{M}$ such that $\beta_t(A) = W_t \alpha_t(A) W_t^*$, $A \in \mathfrak{M}$. We would like to choose $t \mapsto W_t$ in such a way that if $t \mapsto U_t$ is a strongly continuous unitary representation implementing α then $t \mapsto W_t U_t$ is a strongly continuous unitary representation, which automatically implements β. In particular, we want

$$W_{t+s} U_{t+s} = W_t U_t W_s U_s$$

or, equivalently,

$$W_{t+s} = W_t \alpha_t(W_s).$$

As a first step, we prove that $t \mapsto W_t$ can be taken to be a Borel map.

Proposition 3.2.72. *Let α_t, β_t be σ-weakly continuous one-parameter groups of *-automorphisms of a von Neumann algebra $\mathfrak{M}$ with separable predual $\mathfrak{M}_*$. Suppose that there exists $0 \leq \varepsilon < 2$ and $\delta > 0$ such that*

$$\|\beta_t - \alpha_t\| < \varepsilon$$

whenever $|t| < \delta$. It follows that there exists a Borel mapping $t \mapsto W_t$ from $\mathbb{R}$ into the unitary group $\mathcal{U}(\mathfrak{M})$, of $\mathfrak{M}$, such that

$$\beta_t(A) = W_t \alpha_t(A) W_t^* \qquad A \in \mathfrak{M},\ t \in \mathbb{R},$$

and

$$\|W_t - \mathbb{1}\| \leq \{2(1 - \sqrt{1 - \varepsilon^2/4})\}^{1/2}$$

for $|t| \leq \delta$.

PROOF. The existence for each t of a W_t with the given properties follows by the remarks before the proposition, and we must explore the Borel property. Recall that a topological space is called Polish if it is homeomorphic to a complete separable metric space. A subset of a Polish space is analytic if it is the continuous image of a Polish space. We will need the following two results:

(1) if X_1 and Y_1 are analytic Borel spaces, and f_1 is a one-to-one Borel map of X_1 onto Y_1, then f_1 is a Borel isomorphism;

(2) if X_2 is a Polish space, Y_2 a Borel space, and $f_2; X_2 \mapsto Y_2$ is a function onto Y_2 such that

 (i) the inverse image by f_2 of each point in Y_2 is a closed set in X_2

 (ii) f_2 maps open sets into Borel sets,

 then it follows that there exists a Borel function $g; Y_2 \mapsto X_2$ such that $f(g(y)) = y$ for all $y \in Y$

(see Notes and Remarks). The function g defined by the second statement is called a Borel cross section.

We will apply these theorems to $X_2 = \mathscr{U}(\mathfrak{M})$, equipped with the strong* topology (which is equivalent with the weak, and the strong, topology by a simple argument). As $\mathfrak{M}_*$ is separable it follows immediately that X_2 is a Polish space. As any automorphism of $\mathfrak{M}$ has an adjoint in $\mathfrak{M}_*$ (Theorem 2.4.23), we may view the group $\mathrm{Inn}(\mathfrak{M})$ of inner automorphisms of $\mathfrak{M}$ as a subset of $\mathscr{L}(\mathfrak{M}_*)$, the bounded linear operators on $\mathfrak{M}_*$. Equip $\mathscr{L}(\mathfrak{M}_*)$ with the topology of pointwise norm convergence and then the canonical map $f_2; X_2 \mapsto Y_2 \equiv \mathrm{Inn}(\mathfrak{M})$ is continuous, so (i) of condition (2) is satisfied. To establish (ii) of condition (2) consider first the quotient map $f_3; \mathscr{U}(\mathfrak{M}) \mapsto \mathscr{U}(\mathfrak{M})/\mathscr{U}(\mathfrak{Z})$, where $\mathscr{U}(\mathfrak{Z})$ is the group of unitary operators in the center $\mathfrak{Z} = \mathfrak{M} \cap \mathfrak{M}'$. We equip $\mathscr{U}(\mathfrak{M})/\mathscr{U}(\mathfrak{Z})$ with the quotient topology; f_3 is then continuous and open. In particular, $X_1 = \mathscr{U}(\mathfrak{M})/\mathscr{U}(\mathfrak{Z})$ is an analytic Borel space. As $Y_1 \equiv Y_2 \equiv \mathrm{Inn}(\mathfrak{M}) = f_2(\mathscr{U}(\mathfrak{M}))$, the set Y_1 is an analytic Borel space. As the canonical map $f_1; X_1 = \mathscr{U}(\mathfrak{M})/\mathscr{U}(\mathfrak{Z}) \mapsto Y_1 \equiv \mathrm{Inn}(\mathfrak{M})$ is continuous and one-to-one, it follows from (1) that f_1 is a Borel isomorphism. Since f_3 maps open sets into open sets and f_1 maps open sets into Borel sets, $f_2 = f_1 \circ f_3$ maps open sets into Borel sets. Hence f_2 satisfies (i) and (ii) of (2). As $t \mapsto \alpha_t$ is σ-weakly continuous, $t \mapsto \alpha_t{}^* \in \mathscr{L}(\mathfrak{M}_*)$ is weakly, and hence strongly continuous by Corollary 3.1.8. Thus $t \mapsto (\beta_t \alpha_{-t})^* = \alpha^*_{-t}\beta_t{}^*$ is continuous in the topology of pointwise norm convergence in $\mathscr{L}(\mathfrak{M}_*)$. Hence, by statement (2) above, it follows that there exists a Borel map $t \in \mathbb{R} \mapsto W_t' \in \mathscr{U}(\mathfrak{M})$ such that

$$\beta_t \alpha_{-t}(A) = W_t' A W_t'^*, \qquad A \in \mathfrak{M},\ t \in \mathbb{R}.$$

To obtain the estimate on W_t, we repeat the reasoning above with

$$Y_2 = Y_1 = \mathrm{Inn}_\varepsilon(\mathfrak{M}) = \{\alpha; \alpha \in \mathrm{Aut}(\mathfrak{M}),\ \|\alpha - \iota\| \le \varepsilon\},$$
$$X_2 = \mathscr{U}_{\varepsilon'}(\mathfrak{M}) = \{U; U \in \mathscr{U}(\mathfrak{M}),\ \|U - \mathbb{1}\| \le \varepsilon'\},$$

where $\varepsilon' = \{2(1 - \sqrt{1 - \varepsilon^2/4})\}^{1/2}$. Then Y_2 is a closed subset of $\mathrm{Inn}(\mathfrak{M})$, X_2 a closed subset of $\mathscr{U}(\mathfrak{M})$, and by Theorem 3.2.71 and the subsequent remark one has $Y_2 = f_2(X_2)$. Applying statement (2) once more, we find a Borel map $t \in [-\delta, \delta] \mapsto W_t'' \in \mathscr{U}_{\varepsilon'}(\mathfrak{M})$ such that W_t'' implements $\beta_t \alpha_{-t}$ for each t. Defining

$$W_t = \begin{cases} W_t'' & \text{for } |t| \le \delta \\ W_t' & \text{for } |t| > \delta, \end{cases}$$

the proposition follows immediately.

We next aim to modify the map $t \mapsto W_t$ obtained in Proposition 3.2.72 so that after the modification $t \mapsto W_t$ satisfies the one-cocycle relation

$$W_{t+s} = W_t \alpha_t(W_s).$$

We have encountered a similar result earlier in Theorem 2.7.16, and there exist other versions of this theorem, which are also interesting when $\alpha_t = \iota$ for all t (see Notes and Remarks).

Theorem 3.2.73. *Let $\mathfrak{M}$ be a von Neumann algebra with separable predual $\mathfrak{M}_*$, and let α, β be two σ-weakly continuous one-parameter groups of $*$-automorphisms of $\mathfrak{M}$. Assume there exists $\delta > 0$ and $0.47 \simeq \sqrt{71}/18 > \varepsilon \geq 0$ such that*

$$\|\beta_t - \alpha_t\| \leq \varepsilon \quad \text{when } |t| \leq \delta.$$

It follows that there exists a σ-weakly continuous map $t \in \mathbb{R} \mapsto \Gamma_t \in \mathcal{U}(\mathfrak{M})$ such that

$$\begin{aligned} \Gamma_{t+s} &= \Gamma_t \alpha_t(\Gamma_s), \qquad t, s \in \mathbb{R}, \\ \beta_t(A) &= \Gamma_t \alpha_t(A) \Gamma_t^*, \qquad t \in \mathbb{R}, A \in \mathfrak{M}, \\ \|\Gamma_t - \mathbb{1}\| &\leq 10\{2(1 - \sqrt{1 - \varepsilon^2/4})\}^{1/2} \\ &= 5\varepsilon + O(\varepsilon^2) \qquad \text{for } |t| < \delta/4. \end{aligned}$$

PROOF. By Proposition 3.2.72, there exists a Borel map $t \in \mathbb{R} \mapsto W_t \in \mathcal{U}(\mathfrak{M})$ such that

$$\beta_t(A) = W_t \alpha_t(A) W_t^*, \qquad A \in \mathfrak{M}, t \in \mathbb{R},$$

and

$$\|W_t - \mathbb{1}\| \leq \varepsilon'$$

for $|t| \leq \delta$, where $\varepsilon'^2 = 2(1 - \sqrt{1 - \varepsilon^2/4})$. Now, define

$$z(s, t) = W_s \alpha_s(W_t) W_{s+t}^{-1}.$$

Then

$$\begin{aligned} z(s, t) A z(s, t)^* &= (\beta_s \alpha_{-s})(\alpha_s \beta_t \alpha_{-t} \alpha_{-s})(\beta_{s+t} \alpha_{-s-t})^{-1}(A) \\ &= A. \end{aligned}$$

Hence $z(s, t) \in \mathcal{U}(\mathfrak{Z})$, where $\mathfrak{Z} = \mathfrak{M} \cap \mathfrak{M}'$. Also, by a straightforward computation, $(s, t) \in \mathbb{R}^2 \mapsto z(s, t) \in \mathcal{U}(\mathfrak{Z})$ is a two-cocycle, i.e., satisfies the identities

$$\begin{aligned} z(s, 0) &= z(0, t) = \mathbb{1}, \\ z(s, t) z(s + t, u) &= \alpha_s(z(t, u)) z(s, t + u) \end{aligned}$$

for all $s, t, u \in \mathbb{R}$. Note in passing that $t \mapsto W_t$ is a one-cocycle if, and only if, $z(s, t) = \mathbb{1}$ for all s, t. We are going to modify W_t to obtain this condition.

From the definition of z we immediately obtain the estimate

$$\|z(s, t) - \mathbb{1}\| \leq 3\varepsilon' \qquad \text{for } |s| + |t| \leq \delta.$$

Define $t \in \mathbb{R} \mapsto \lambda_t \in \mathscr{U}(\mathfrak{Z})$ inductively by

$$\lambda_0 = \mathbb{1},$$
$$\lambda_{\delta/2\cdot(t+n)} = \lambda_{\delta/2\cdot n} z(\delta/2 \cdot n, \delta/2 \cdot t)$$

for $0 \le t \le 1$, $n \in \mathbb{Z}$. Next define $z'(s, t)$ by

$$z'(s, t) = \lambda_s \alpha_s(\lambda_t) z(s, t) \lambda_{s+t}^{-1}$$

for $s, t \in \mathbb{R}$. It follows that z' is a two-cocycle, and $s, t \mapsto z'(s, t)$ is Borel. Now $z'(\delta n/2, \delta t/2) = \lambda_{\delta n/2} \alpha_{\delta n/2}(\lambda_{\delta t/2}) \lambda_{\delta n/2}^{-1}$, but $\lambda_{\delta t/2} = \mathbb{1}$ for $0 \le t \le 1$. Hence

$$z'(p, t) = \mathbb{1} \quad \text{for } 0 < t \le \delta/2,\ p \in \frac{\delta}{2}\mathbb{Z}.$$

If $0 \le s, t \le \delta/2$ and $s + t \le \delta/2$ then $z'(s, t) = z(s, t)$. If $0 \le s, t \le \delta/2$ but $s + t > \delta/2$ and $\lambda_{s+t} = z(\delta/2, s + t - \delta/2)$ and $z'(s, t) = z(s, t) z(\delta/2, s + t - \delta/2)^{-1}$. Therefore $\|z(s, t) - \mathbb{1}\| \le 3\varepsilon'$ for $0 \le s, t \le \delta/2$ implies

$$\|z'(s, t)\| \le 6\varepsilon' \quad \text{for } 0 \le s, t \le \delta/2.$$

In order to proceed we need the following lemma.

Lemma 3.2.74. (a) *If z is a two-cocycle such that*

$$z(s, t) = \mathbb{1} \quad \textit{for } 0 \le t \le 1$$

then

$$z(s, t) = \mathbb{1} \quad \textit{for } s, t \in \mathbb{R}.$$

(b) *If z is a two-cocycle such that*

$$z(n, t) = \mathbb{1} \quad \textit{for } 0 \le t \le \mathbb{1},\ n \in \mathbb{Z},$$

then for each $t \in \mathbb{R}$, $s \mapsto \alpha_{-s}(z(s, t))$ is periodic with period 1.

PROOF. From the cocycle identity in the form

$$z(n, 1) z(n + 1, t) = \alpha_n(z(1, t)) z(n, 1 + t)$$

we deduce by iteration in both cases covered by the lemma that $z(n, t) = \mathbb{1}$ for $t \ge 0$ and $n \in \mathbb{Z}$. From the cocycle identity in the form

$$z(-n, n) = \alpha_{-n}(z(n, -t)) z(-n, n - t)$$

it follows that $z(n, -t) = \mathbb{1}$ for $t \le n \in \mathbb{Z}$, and in particular $z(n, -1) = \mathbb{1}$ for $n \in \mathbb{Z}$, $n \ge 0$. Hence it follows from the cocycle identity in the form

$$z(n, -1) z(n - 1, t) = \alpha_n(z(-1, t)) z(n, t - 1)$$

by iteration that $z(n, t) = \mathbb{1}$ for $n \in \mathbb{Z}$, $n \ge 0$, $t \in \mathbb{R}$. However, the previous form of the cocycle identity now implies that $z(-n, n - t) = \mathbb{1}$ for $n \in \mathbb{Z}$, $n \ge 0$. Hence $z(n, t) = \mathbb{1}$ for $n \in \mathbb{Z}$, $t \in \mathbb{R}$. Using this argument for $s\mathbb{Z}$ instead of $\mathbb{Z}$, statement (a) follows. To finish the proof of (b) we exploit the cocycle identity in the form

$$z(n, s) z(n + s, t) = \alpha_n(z(s, t)) z(n, s + t)$$

and then use $z(n, s) = z(n, s + t) = \mathbb{1}$ to deduce

$$\alpha_{-n-s}(z(n + s, t)) = \alpha_{-s}(z(s, t)).$$

Hence (b) follows.

Now let us resume the proof of Theorem 3.2.73.

Note that $z'(p, t) = \mathbb{1}$ for $0 \leq t \leq \delta/2$, $p \in \delta/2\mathbb{Z}$ and so statement (b) of Lemma 3.2.74 implies that $s \mapsto \alpha_{-s}(z'(s, t))$ is periodic with period $\delta/2$. Hence the estimate

$$\|z'(s, t) - \mathbb{1}\| \leq 6\varepsilon'$$

is valid for all s when $0 \leq t \leq \delta/2$. Now, let log denote the principal value of the logarithm on the complex plane with a cut along the negative real axis. Since $\varepsilon < \sqrt{71}/18$ it follows that $6\varepsilon' < \sqrt{2}$, and by the estimate on z' we can consistently define

$$y(s, t) = \log(z'(s, t))$$

for $0 \leq t \leq \delta/2$. The cocycle properties of z' then yield the relations

$$y(s, 0) = y(0, t) = 0,$$

$$y(s, t) + y(s + t, u) = \alpha_s(y(t, u)) + y(s, t + u)$$

for $0 \leq t, u, t + u, \leq \varepsilon$. The $\sqrt{2}$-estimate on z' is needed for the last identity. Also, $s, t \mapsto y(s, t)$ is a Borel map. Define a Borel map $c; t \in [0, \delta/2] \mapsto c(t) \in \mathfrak{Z}_{sa}$ by

$$c(t) = -\frac{2}{\delta}\int_0^{\delta/2} ds\, \alpha_{-s}(y(s, t)).$$

From the periodicity of $s \mapsto \alpha_{-s}(y(s, t))$ we have

$$c(t) = -\frac{2}{\delta}\int_u^{\delta/2 + u} ds\, \alpha_{-s}(y(s, t))$$

for all $u \in \mathbb{R}$, and a simple computation, using the relation for y derived above, shows that

$$c(s + t) - \alpha_s(c(t)) - c(s) = y(s, t)$$

when $0 \leq s, t, s + t \leq \delta/2$. Hence, defining $\lambda'; t \in [0, \delta/2] \mapsto \lambda_t' \in \mathcal{U}(\mathfrak{Z})$ by

$$\lambda_t' = \exp(c(t))$$

we find

$$\lambda'_{s+t}\alpha_s(\lambda_t'^{-1})\lambda_s'^{-1} = z'(s, t)$$

for $0 \leq s, t, s + t \leq \delta/2$.

From the estimate $\|z'(s, t) - \mathbb{1}\| \leq 6\varepsilon'$ for $0 \leq s, t \leq \delta/2$ we then obtain by spectral theory the estimates

$$\|y(s, t)\| \leq \arccos(1 - (6\varepsilon')^2/2), \qquad 0 \leq t \leq \delta/2,$$

$$\|c(t)\| \leq \arccos(1 - (6\varepsilon')^2/2), \qquad 0 \leq t \leq \delta/2,$$

and, finally,

$$\|\lambda_t' - \mathbb{1}\| \leq 6\varepsilon'$$

for $0 \leq t \leq \delta/2$. We now extend the map $t \in [0, \delta/2] \mapsto \lambda_t' \in \mathscr{U}(\mathfrak{Z})$ to a Borel map from all of $\mathbb{R}$ into $\mathscr{U}(\mathfrak{Z})$, and define

$$z''(s, t) = \lambda_{s+t}'^{-1} z'(s, t)\lambda_s'\alpha_s(\lambda_t').$$

Then z'' is a two-cocycle, and

$$z''(s, t) = \mathbb{1}$$

for $0 \leq s, t \leq \delta/4$.

Now, replacing $\delta/2$ by $\delta/4$ and ε' by 0, we next define z''' from z'' in the same manner that z' was defined from z. First, define

$$\lambda_0'' = \mathbb{1},$$

$$\lambda''_{\delta(t+n)/4} = \lambda''_{\delta n/4} z''(\delta n/4, \delta t/4)$$

for $0 \leq t \leq 1$, $n \in \mathbb{Z}$, and then

$$z'''(s, t) = \lambda_s''\alpha_s(\lambda_t'')z''(s, t)\lambda_{s+t}''^{-1}.$$

As earlier, we deduce that z''' is a two-cocycle satisfying

$$z'''(p, t) = \mathbb{1} \quad \text{for } 0 \leq t \leq \delta/4,\ p \in \delta/4\mathbb{Z},$$

and

$$z'''(s, t) = \mathbb{1} \quad \text{for } 0 \leq s, t \leq \delta/4.$$

Hence statement (a) of Lemma 3.2.74 implies that $z'''(s, t) = \mathbb{1}$ for all s, t. Defining

$$\lambda_s''' = \lambda_s''\lambda_s'\lambda_s$$

we thus have

$$\begin{aligned} z(s, t) &= \lambda_{s+t}''' z'''(s, t)\lambda_s'''^{-1}\alpha_s(\lambda_t'''^{-1}) \\ &= \lambda_{s+t}'''\lambda_s'''^{-1}\alpha_s(\lambda_t'''^{-1}). \end{aligned}$$

Now if $\Gamma_s = \lambda_s''' W_s$ then Γ_s is a one-cocycle,

$$\Gamma_{s+t} = \Gamma_s\alpha_s(\Gamma_t),$$

and

$$\beta_t(A) = \Gamma_t\alpha_t(A)\Gamma_t^*.$$

But $t \mapsto \Gamma_t$ is Borel, and $\mathfrak{M}_*$ is separable. Therefore $t \mapsto \Gamma_t$ is continuous by the following reasoning. Since $\mathfrak{M}_*$ is separable, $\mathfrak{M}$ has a faithful, normal state ω. If $\mathscr{P}$ is the natural positive cone in the corresponding representation on $\mathfrak{H} = \mathfrak{H}_\omega$ (see Definition 2.5.25) and $\varphi \in \mathfrak{M}_{*+} \mapsto \xi(\varphi) \in \mathscr{P}$ is the associated map defined by Theorem 2.5.31, then it follows from the estimate

$$\|\xi(\varphi_1) - \xi(\varphi_2)\|^2 \leq \|\varphi_1 - \varphi_2\|$$

and the separability of M_{*+} that $\mathscr{P}$ is a separable subset of $\mathfrak{H}$. As $\mathfrak{H} = \mathscr{P} - \mathscr{P} + i(\mathscr{P} - \mathscr{P})$ (Proposition 2.5.26) it follows that $\mathfrak{H}$ is separable. Now if $t \mapsto U_t$ is the canonical unitary implementation of $t \mapsto \alpha_t$ (Corollary 2.5.32) then $t \mapsto V_t = \Gamma_t U_t$ is a unitary representation of $\mathbb{R}$, because of the cocycle property of Γ_t. As $t \mapsto V_t$ is Borel and $\mathfrak{H}$ is separable, it follows by a regularization argument similar to the one used to prove that weak continuity implies strong continuity in Corollary 3.1.8 that $t \mapsto V_t$ is strongly continuous. Hence $t \mapsto \Gamma_t = V_t U_{-t}$ is strongly continuous.

Finally, we derive the estimate on Γ_t by summing up the estimates already derived:

$$\|W_t - \mathbb{1}\| \le \varepsilon', \qquad |t| \le \delta;$$
$$\|\lambda_t - \mathbb{1}\| \le 3\varepsilon', \qquad 0 \le t \le \delta/2;$$
$$\|\lambda_t' - \mathbb{1}\| \le 6\varepsilon', \qquad 0 \le t \le \delta/2;$$
$$\|\lambda_t'' - \mathbb{1}\| = 0, \qquad 0 \le t \le \delta/4.$$

It follows that

$$\|\Gamma_s - \mathbb{1}\| \le \varepsilon' + 3\varepsilon' + 6\varepsilon' = 10\varepsilon'$$

for $|s| \le \delta/4$.

We have now developed the necessary tools to give a sharpened version of the main theorem of approximation theory, Theorem 3.1.36, in the von Neumann case.

Theorem 3.2.75. *Let $\mathfrak{M}$ be a von Neumann algebra with separable predual, and let α, β be two σ-weakly continuous one-parameter groups of *-automorphisms of $\mathfrak{M}$ with generators δ_α and δ_β, respectively. The following conditions are equivalent:*

(1) *there exists ε_1, $0 \le \varepsilon_1 < \sqrt{199}/50 \simeq 0.28$, and $\delta_1 > 0$ such that*

$$\|\alpha_t - \beta_t\| \le \varepsilon_1$$

for $|t| \le \delta_1$,

(2) *there exists ε_2, $0 \le \varepsilon_2 < \sqrt{199}/50$, and $\delta_2 > 0$, an inner automorphism γ of $\mathfrak{M}$, and a bounded derivation δ of $\mathfrak{M}$ such that*

$$\delta_\beta = \gamma(\delta_\alpha + \delta)\gamma^{-1}$$

and

$$\|\alpha_t \circ \gamma \circ \alpha_{-t} - \gamma\| \le \varepsilon_2$$

for $|t| \le \delta_2$.

If these conditions are satisfied then

$$\|\beta_t - \alpha_t\| = \|\alpha_{-t} \circ \gamma \circ \alpha_t - \gamma\| + O(t)$$

and there is a unitary element $W \in \mathfrak{M}$ implementing γ which satisfies the estimate

$$\|W - \mathbb{1}\| \le 10\{2(1 - \sqrt{1 - \varepsilon_1{}^2/4})\}^{1/2}.$$

Thus $\|\gamma - \iota\| \le 10\varepsilon_1 + O(\varepsilon_1{}^2)$.

PROOF. (1) $\Rightarrow$ (2) By Theorem 3.2.73 there exists a strongly continuous map $t \in \mathbb{R} \mapsto \Gamma_t \in \mathcal{U}(\mathfrak{M})$ such that

$$\Gamma_{t+s} = \Gamma_t \alpha_t(\Gamma_s), \qquad s, t \in \mathbb{R},$$
$$\beta_t(A) = \Gamma_t \alpha_t(A)\Gamma_t^*, \qquad A \in \mathfrak{M},\ t \in \mathbb{R},$$
$$\|\Gamma_t - \mathbb{1}\| \le \varepsilon' = 10\{2(1 - \sqrt{1 - \varepsilon_1{}^2/4})\}^{1/2} < \sqrt{2}, \qquad |t| < \delta/4.$$

We may assume that $\mathfrak{M}$ is in a standard representation and hence by Corollary 2.5.32 there exists a unitary implementation $t \mapsto U_t$ of α on $\mathfrak{H}$:

$$\alpha_t(A) = U_t A U_t^*, \qquad t \in \mathbb{R}, A \in \mathfrak{M}.$$

Define $V_t = \Gamma_t U_t$. The cocycle property of Γ_t implies that $t \mapsto V_t$ is a unitary representation of $\mathbb{R}$, and

$$\beta_t(A) = V_t A V_t^*, \quad t \in \mathbb{R}, A \in \mathfrak{M}.$$

We have

$$\|V_t U_{-t} - \mathbb{1}\| = \|\Gamma_t - \mathbb{1}\| \le \varepsilon' < \sqrt{2}$$

for $|t| < \delta/4$, and $\Omega = 4\delta^{-1} \int_0^{\delta/4} dt\, V_t U_{-t} \in \mathfrak{M}$. Hence, by Proposition 3.2.70, there exists a unitary $W \in \mathfrak{M}$ such that

$$\|V_t - W U_t W^*\| = O(t).$$

Defining $\gamma(A) = WAW^*$ and $\hat{\alpha}_t = \gamma\alpha_t\gamma^{-1}$ we then have

$$\|\beta_t - \hat{\alpha}_t\| \le 2\|V_t - WU_t W^*\| = O(t).$$

Hence by Proposition 3.2.69 there exists a bounded derivation δ' of $\mathfrak{M}$ such that

$$\begin{aligned} \delta_\beta &= \delta_{\hat{\alpha}} + \delta' \\ &= \gamma\delta_\alpha\gamma^{-1} + \delta' = \gamma(\delta_\alpha + \delta)\gamma^{-1}, \end{aligned}$$

where $\delta = \gamma^{-1}\delta'\gamma$.

It follows immediately that

$$\alpha_t - \gamma\alpha_t\gamma^{-1} = \exp(t\delta_\alpha) - \exp(t\delta_\beta) + \exp(t\delta_\beta) - \exp(t(\delta_\beta - \delta'))$$

and hence

$$\begin{aligned} \|\alpha_{-t}\gamma\alpha_t - \gamma\| &= \|\alpha_t - \gamma\alpha_t\gamma^{-1}\| \\ &= \|\alpha_t - \beta_t\| + O(t). \end{aligned}$$

Thus the estimate in (2) follows from the estimate in (1), and conversely, the estimate in (1) follows from (2).

The estimate on $\|W - \mathbb{1}\|$ follows from Proposition 3.2.70 and the estimate on $\|\Gamma_t - \mathbb{1}\|$.

We conclude by remarking that the two analogous approximation results (Proposition 3.2.70 and Theorem 3.2.75) can to a certain extent be understood in terms of two special cases. First consider the comparison of the two unitary groups U, V on the Hilbert space $\mathfrak{H}$. If

$$\|U_t - V_t\| = O(t), \qquad t \to 0,$$

then Theorem 3.1.38 implies that the generators of U and V differ by a bounded operator. If, however,

$$\|U_t - V_t\| < \sqrt{2} - \varepsilon_1$$

for all $t \in \mathbb{R}$ then we claim that

$$U_t = WV_tW^*$$

for some unitary W. To prove this one chooses an invariant mean M on $\mathbb{R}$ (see Section 4.3.3 for the definition and discussion of invariant means) and then defines Ω by

$$\Omega = M(UV^{-1}).$$

If $\mathfrak{N}$ is the von Neumann algebra generated by $\{U_t V_{-t}; t \in \mathbb{R}\}$ then it follows that Ω is a well-defined operator in $\mathfrak{N}$ which does not contain zero in its numerical range. Hence the partial isometry W in the polar decomposition $\Omega = W|\Omega|$ is actually unitary. Moreover, the relation

$$U_s U_t V_{-t} = U_{s+t} V_{-s-t} V_s$$

establishes that

$$U_s \Omega = \Omega V_s,$$

and it follows by a simple argument that

$$U_t W = W V_t.$$

A similar result is true for two σ-weakly continuous groups α, β of *-automorphisms of a von Neumann algebra $\mathfrak{M}$. If condition (1) of Theorem 3.2.75 is satisfied for all $t \in \mathbb{R}$, i.e., if $\delta_1 = \infty$, then

$$\beta_t = \gamma \alpha_t \gamma^{-1}$$

for an inner automorphism γ of $\mathfrak{M}$. This is established by combination of the proof of Theorem 3.2.75 with the foregoing discussion of the unitary groups.

Thus in both the foregoing approximation theorems, the boost, i.e., the bounded perturbation, can be understood as arising from $O(t)$ behavior, and the twist, i.e. the automorphism, arises from norm comparability for all t.

Notes and Remarks

Sections 3.1.1 and 3.1.2

The standard reference for semigroup theory is the book by Hille and Phillips [[Hil 1]] which describes the theory to 1956. The following books also contain chapters on the theory and describe aspects which are more recent than 1956: [[But 1]], [[Kat 1]], [[Ree 2]], [[Rie 1]], [[Yos 1]].

In [[Hil 1]] the definition of a semigroup differs from ours insofar that the condition $U_0 = I$ is not imposed and this then leads to an exhaustive investigation of the behavior of U at the origin under various conditions of continuity away from the origin. Nevertheless, the major objects of investigation in [[Hil 1]] are C_0-semigroups as we have defined them. The notion of a $C_0{}^*$-semigroup, or $\sigma(X, X_*)$-continuous semigroup is due to de Leeuw [Lee 1], and is described in [[But 1]]. It should be emphasized that this notion does not coincide with the concept of dual semigroup discussed in [[Hil 1]]. If U is a C_0-semigroup on the Banach space X and one introduces the $C_0{}^*$-semigroup on X^* by duality then U^* is strongly continuous in restriction to a weakly* dense subspace $X_0{}^*$ of X^*. The restriction of U^* to $X_0{}^*$ is then a C_0-semigroup which is defined to be the dual semigroup of U in [[Hil 1]].

The theory of uniformly continuous semigroups is contained in [[Hil 1]] and is attributed to Nagumo [Nagu 1].

The equivalence of weak, and strong, continuity, and differentiability, of C_0-semigroups, Corollary 3.1.8, was first proved by Yosida [Yos 1] who has also developed a theory of equicontinuous semigroups which has much in common with the $\sigma(X, F)$-continuous semigroups which we describe.

The C_0-semigroup version of Theorem 3.1.10 was proved independently by Hille and Yosida in 1948 [[Hil 2]], [Yos 1]. The first exponentiation algorithm in this theorem is due to Yosida and the second to Hille. Lemma 3.1.11, which allows the passage from one algorithm to the other, is much more recent and was first given by Chernoff [Che 1]. The full characterization of generators of C_0-semigroups, i.e., the strongly continuous version of Corollary 3.1.12, was derived almost simultaneously by Feller [Fel 1], Miyadera [Miy 1], and Phillips [Phi 1] in 1953.

The notion of a dissipative operator was introduced by Lumer and Phillips [Lum 1] in 1961. These authors proved the basic properties of these operators,

Propositions 3.1.14 and 3.1.15, and characterized C_0-semigroups of contractions by Theorem 3.1.16. A version of Theorem 3.1.19 is also contained in [Lum 1]: if S is a closed, dissipative operator on a Banach space X and $\|S^n A\|^{1/n} = o(n)$ as $n \to \infty$ and all A in a dense subset of X then S generates a C_0-semigroup of contractions. Another version of this theorem was proved for unitary groups on Hilbert space by Nelson [Nel 1].

The interesting characterization of the domain of the generator of a $C_0{}^*$-semigroup, Proposition 3.1.23, is due to de Leeuw [Lee 1] but is based on earlier results of Butzer [[But 1]]. This proposition is of particular interest in approximation theory where it provides solutions to the so-called saturation problem (see [[But 1]] Chapter 2).

Section 3.1.3

The C_0-version of Theorem 3.1.26 is often called the Trotter–Kato theorem. It was first proved by Trotter [Tro 1] and the proof was subsequently clarified by Kato [Kat 1].

Theorem 3.1.28 is due to Kurtz [Kur 1] although Glimm and Jaffe [Gli 5] independently obtained the same result for the special case of unitary groups on a Hilbert space.

The algorithm contained in Theorem 3.1.30 is due to Chernoff [Che 1] and is a direct generalization of product algorithms first studied by Trotter [Tro 2]. These algorithms, which occur in Corollary 3.1.31, are often referred to as the Trotter product formulae. Chernoff's article [Che 2] discusses their use in the definition of the sum of unbounded operators.

Section 3.1.4

The standard reference for perturbation theory is the book of Kato [[Kat 1]]. Reed and Simon [[Ree 2]], Riesz and Nagy [[Rie 1]], and Hille and Phillips [[Hil 1]] also contain partial discussions.

Theorem 3.1.34 is due to Bratteli and Kishimoto [Bra 11], while the counterexample alluded to in the related remark can be found in Jørgensen [Jør 1].

Section 3.1.5

The comparison of groups in the manner we describe generalizes in one direction certain results of approximation theory, e.g., Proposition 3.1.23. The subject was initiated by the work of Bucholz and Roberts [Buc 1] and a subsequent paper by Robinson. Proposition 3.1.35 and Theorems 3.1.38, 3.1.40 are taken from [Rob 6], but Theorem 3.1.38 includes a refinement due to Johnson [Joh 1]. Theorem 3.1.36 is due to Bratteli, Herman and Robinson [Bra 4].

Section 3.2.1

For the original account of Wigner symmetries and the proof given in Example 3.2.14, see [[Wign 1]]; a more updated account can be found in [Sim 1]. The original algebraic version of Proposition 3.2.2 was proved by N. Jacobson and C. Rickart [Jac 1], while the C^*-version is due to Kadison [Kad 4], [Kad 5]. The inequality in Proposition 3.2.4 is known as Kadison's inequality when A is selfadjoint [Kad 6], and the present version was proved by Stinespring [Sti 1], using ideas which go back to Naimark [[Nai 1]]. Størmer proved in 1963 that the same inequality is true under the assumption that $\varphi(A)$ is normal [Stø 1]. Proposition 3.2.5 is due to B. Russo and H. A. Dye [Rus 1], while the fact that the convex hull of the unitaries contains the open unit sphere was shown by A. G. Robertson [Rober 1]. The short proof used here is due to G. K. Pedersen [[Ped 1]]. An excellent survey of positive maps can be found in [Stø 2]. As for the classification of general positive maps, it is an enormous task to do this even when $\mathfrak{A}$ is the 2×2 matrix algebra and $\mathfrak{B}$ is the 4×4 matrix algebra [Wor 2].

The Jordan decomposition of hermitian functionals in the dual of a C^*-algebra, Proposition 3.2.7, originates with Grothendieck [Gro 1]. Kadison introduced the concept of a full family of states, Definition 3.2.9, and proved Proposition 3.2.10 and Theorem 3.2.11 in [Kad 5]. Our proofs of 3.2.10 and 3.2.11 are different from the original ones. Versions of Corollaries 3.2.12 and 3.2.13 appear in the same paper. An example showing that Corollary 3.2.13 is not true for general von Neumann algebras can be found in [Bra 3].

The fact that all automorphisms of $\mathscr{L}(\mathfrak{H})$ are inner is a special feature of $\mathscr{L}(\mathfrak{H})$ which it shares with no other factor. The automorphisms of the hyperfinite II_1 and II_∞ factor have been classified by A. Connes [Con 5], [Con 6]. He has also constructed a surprising example of a factor which is not antiisomorphic with itself [Con 7]. Note that a von Neumann algebra in standard form is antiisomorphic with its commutant by the antiisomorphism $A \mapsto JA^*J$. Connes' example is a factor which is not isomorphic with its commutant in a standard representation; thus it cannot be antiisomorphic with itself.

The characterization of Jordan automorphisms of von Neumann algebras in standard form as unitary mappings of the natural cone onto itself, Theorem 3.2.15, was found independently by A. Connes [Con 1] and U. Haagerup [Haa 1, 2]. Theorem 3.2.18 has not appeared earlier, but similar ideas, and Lemma 3.2.19, were given by Bratteli and Robinson [Bra 3].

Section 3.2.2

The theory of bounded derivations of C^*- and W^*-algebras was essentially initiated by Kaplansky [Kap 3], who showed, in 1953, that each derivation δ of a type I von Neumann algebra $\mathfrak{M}$ is inner, i.e., $\delta(A) = i[H, A]$ with $H \in \mathfrak{M}$. Nevertheless, no great progress took place until 1965 when work of Kadison [Kad 7] instigated a swift development. Most of the results up to 1970 are described in the book of Sakai [[Sak 1]] and we return to specific references in the notes to Section 3.2.3.

The investigation of unbounded derivations occurred much later and was principally motivated by problems of mathematical physics and in particular the problem of the construction of dynamics in statistical mechanics. An early result in this context was due to Robinson [Rob 7] who, in 1968, used an analytic vector technique to construct a C_0-group of isometries of a UHF algebra from a given generator. The main developments of the subject took place, however, after 1975 and were largely inspired by papers of Sakai [Sak 2] and Bratteli and Robinson [Bra 5].

Proposition 3.2.22, and the idea of applying dissipative operator techniques to derivation theory, are both due to Kishimoto [Kis 1]. The boundedness of an everywhere defined derivation, Corollary 3.2.23, had previously been proved by Sakai [Sak 3], in 1960, by an ingenious calculation which also used a square root technique. The improved result which gives the boundedness of each norm-closed derivation δ with domain $D(\delta)$ invariant under the square root operation was established by Ota [Ota 1] in 1976. Ota's proof relied upon a result of Cuntz [Cun 1]. Cuntz proved that a semisimple Banach *-algebra with identity is a C^*-algebra in an equivalent norm if, and only if, it is closed under the square root operation of positive elements, i.e., selfadjoint elements with nonnegative spectrum. (A Banach *-algebra is called semisimple if it has a faithful *-representation on a Hilbert space.)

Proposition 3.2.24 is adapted from Kadison [Kad 7]. The construction of H_Λ in Example 3.2.25 was first made by Elliott [Ell 2]. The closability criterion, Proposition 3.2.26, is due to Chi [Chi 1] and the analogous implementability criterion, Proposition 3.2.28, to Bratteli and Robinson [Bra 3].

It should be emphasized that there exist norm-densely defined, nonclosable derivations of C^*-algebras. The first examples were given by Bratteli and Robinson [Bra 5] and were based on differentiation on the Cantor subset of the unit interval [0, 1]. This operation defines a derivation of the continuous functions over the set which is nonclosable and can be used to construct examples of nonclosable derivations of UHF algebras. In fact, if $\mathfrak{A}$ is a UHF algebra generated by an increasing sequence of matrix algebras $\mathfrak{A}_n$ one may find a norm-densely defined derivation such that $\mathfrak{A}_n \in D(\delta)$ for each n, $\delta|_{\bigcup_n \mathfrak{A}_n} = 0$, but $\delta \neq 0$. Subsequently, Herman [Her 1] has constructed an extension of the usual differentiation on $C(0, 1)$ which is a nonclosable derivation of $C(0, 1)$. It is unclear whether there exist nonclosable derivations of $\mathscr{LC}(\mathfrak{H})$ (see [Bra 5] or Example 3.2.34 for an analysis of derivations of $\mathscr{LC}(\mathfrak{H})$ and $\mathscr{L}(\mathfrak{H})$).

Functional analysis of the domain of a closed derivation was initiated by Sakai (see Bratteli and Robinson [Bra 5], Powers [Pow 2], and Sakai [Sak 2]). Powers obtained Lemma 3.2.31 concerning the exponential operation and Sakai and Bratteli–Robinson obtained an analogous result for the resolvent operation. It appears, however, most efficient to approach the subject via the modified resolvent as in Proposition 3.2.29. In particular, this immediately yields Corollary 3.2.30 which had previously been obtained by Bratteli and Robinson by a tedious calculation technique [Bra 6]. Theorem 3.2.32 is mentioned in [Bra 6] but it is essentially contained in [Pow 2]. This last reference claimed that if δ is normclosed, $A = A^* \in D(\delta)$, and f is once-continuously differentiable then $f(A) \in D(\delta)$, but McIntosh disproved this result with an explicit counterexample [McI 1]. As A is bounded the question whether $f(A) \in D(\delta)$ depends only on the properties of f on an interval containing the spectrum of A and, in fact, essentially reduces to the behavior of f at the origin. McIntosh's example is a function which behaves like

$$f_\alpha(x) = |x|(\log|\log|x||)^{-\alpha}$$

for $0 \leq \alpha < 1$, in a neighborhood of the origin. McIntosh has also remarked that if g_α has compact support, is twice differentiable away from the origin, and

$$g_\alpha(x) = |x||\log|x||^{-1}(\log|\log|x||)^{-\alpha}$$

in a neighborhood of the origin then

$$\int dp\,|\hat{g}_\alpha(p)||p| \begin{cases} < +\infty & \text{if } \alpha > 1 \\ = +\infty & \text{if } 0 < \alpha \leq 1. \end{cases}$$

Thus $g_\alpha(A) \in D(\delta)$ for $\alpha > 1$ by Theorem 3.2.32. This delineates rather sharply the possible behavior of f.

Examples 3.2.34 and 3.2.35 are drawn from the analysis of derivations of $\mathscr{L}(\mathfrak{H})$ and $\mathscr{LC}(\mathfrak{H})$ given by Bratteli and Robinson [Bra 5]. The method of modifying δ by a bounded derivative so that the perturbed derivation δ^E satisfies $\delta^E(E) = 0$ dates back to Kaplansky [Kap 3]. The conclusion concerning Wigner symmetries which follows from these examples can be stated in a different manner—we have concluded that the cohomology of $\mathbb{R}$ is trivial through *a priori* estimates on unbounded derivations.

Example 3.2.36 is a special case of a result of Elliott [Ell 7], which generalized earlier theorems by Kallman [Kal 1], [Kal 2]. The proof of this result is based on a generalization of the ideas used in Example 3.2.36.

Theorem. *Let $\mathfrak{M}$ be a von Neumann algebra, and let $\{\tau_n\}_{n \geq 1}$ be a sequence of *-automorphisms of $\mathfrak{M}$ converging strongly to a *-automorphism τ, i.e.,*

$$\lim_{n\to\infty} \|\tau_n(A) - \tau(A)\| = 0$$

for all $A \in \mathfrak{M}$. It follows that τ_n converges to τ in norm, i.e.,

$$\lim_{n\to\infty} \sup_{A \in \mathfrak{M}\setminus\{0\}} \|\tau_n(A) - \tau(A)\|/\|A\| = 0.$$

One can use functional analysis on the domain to study closed derivations δ on a UHF algebra. The following theorem was proved for generators by Sakai in his original paper on unbounded derivations [Sak 2] and the extension to general closed derivations was noted in [Bra 5].

Theorem. *Let δ be a closed derivation of a UHF algebra $\mathfrak{A}$. Then there exists an increasing sequence $\{\mathfrak{A}_n\}$ of matrix subalgebras of $\mathfrak{A}$ such that $\bigcup_n \mathfrak{A}_n \subseteq D(\delta)$ and $\bigcup_n \mathfrak{A}_n$ is dense in $\mathfrak{A}$. If δ is a generator, $\bigcup_n \mathfrak{A}_n$ can be chosen to consist of analytic elements for δ.*

This theorem has led several authors to study so-called normal derivations of UHF algebras, i.e., derivations δ with a core of the form $\bigcup_n \mathfrak{A}_n$, but it is still an open problem whether all generators are normal (see in this context Propositions 3.2.52 and 3.2.53).

Another more special consequence of the functional analysis on the domain is that the only closable derivation of the continuous functions on the Cantor set is zero. This is because the projections in the domain of a closable derivation on this algebra must be dense in all projections, and the derivation applied to a projection must be zero as a consequence of abelianness.

There are two recent developments in the theory of unbounded derivations which have not been treated in this text. The first is the classification problem of derivations on abelian C^*-algebras. Here Kurose has obtained a complete classification of all closed derivations on the algebra $C(I)$, where $I = [0, 1]$ is the unit interval. One surprising byproduct of this classification is that any nonzero closed derivation on $C(I)$ has a proper closed extension! See [[Tom 1]] for a review of these developments. The second is the theory of derivations associated with C^*-dynamical systems. Typically one considers a Lie group G acting as a group of *-automorphisms τ on a C^*-algebra $\mathfrak{A}$, and then tries to classify the set of all derivations defined on a class of smooth elements for the action. A typical result is that such a derivation is a linear combination of an approximately inner, or inner, derivation and a linear combination of the generators of the one-parameter subgroups of the action τ with coefficients affiliated to the centre of $\mathfrak{A}$. However, there are examples where the above decomposition fails, and for the moment the theory consists of a large number of special results, and the general pattern is not yet clear. See [[Bra 1]] for a survey of these results.

Section 3.2.3

Spectral theory for abelian groups of automorphisms on operator algebras is old folklore known to many mathematical physicists. For example, regularization by functions with appropriate support properties in momentum space has long been a standard procedure. The form of the theory presented here, however, was developed at a rather late stage by Arveson [Arv 1], Borchers [Bor 4], and Connes [Con 4]. Proposition 3.2.40 can be found in [Con 4], except point (6), which has been noted independently by several authors, the first of whom seem to be Ikunishi and Nakagami [Iku 1]. The Fourier analysis argument alluded to in the proof of (2) $\Rightarrow$ (3) can be found in the book by Rudin [[Rud 3]]. If K is a one-point set the result corresponds to Theorem 2.6.3 in this reference, and the proof readily extends to general K. Proposition 3.2.41 is due to Olesen [Ole 1]. The method of proving the Borchers–Arveson theorem (Theorem 3.2.46) via Propositions 3.2.43 and 3.2.44 is due to Arveson [Arv 1] and Borchers [Bor 4]. We have made some simplifications in this proof. Borchers' original proof of the theorem [Bor 2] made use of the derivation theorem, Corollary 3.2.47, together with a complicated approximation argument. In the case where the unitary group U of Theorem 3.2.46 has a ground state, [Bor 2] gives a very simple argument for the theorem (see Corollary 3.2.60). Borchers' theorem in its general form, as given in [Bor 2], involves representations of $\mathbb{R}^{n+1}$ rather than $\mathbb{R}$, and instead of assuming that the spectrum is positive, he assumes that the spectrum is contained in the positive lightcone. A complete treatment of results of this nature can be found in Chapter 8 of [[Ped 1]]. For conditions ensuring that Lorentz boosts are also in the algebra, see Streater [Str 1].

The derivation theorem, Corollary 3.2.47, is due to Kadison [Kad 7] and Sakai [Sak 4]. The original proof is much shorter, but less intuitive than the one given here. It should be pointed out that the derivation theorem was proved a long time before the mathematical theory of spectra of automorphism groups was worked out. Derivations of C^*-algebras need not be inner, even when the algebra has an identity. Problems in this connection have been studied by several authors, but the most satisfying result in this direction is due to G. Elliott [Ell 3] and C. A. Akemann and G. K. Pedersen [Ake 1]. An extra complication in this case is that the algebra does not necessarily have an identity, and the appropriate notion of innerness is therefore that the derivation δ is implemented by an element H in the so-called multiplier algebra $M(\mathfrak{A})$ of the C^*-algebra $\mathfrak{A}$. Recall that the double dual $\mathfrak{A}^{**}$ of $\mathfrak{A}$ can be viewed as a von Neumann algebra containing $\mathfrak{A}$ as in Section 3.2.1. $M(\mathfrak{A})$ is then defined as the set of $B \in \mathfrak{A}^{**}$ such that $B\mathfrak{A} \subseteq \mathfrak{A}$ and $\mathfrak{A}B \subseteq \mathfrak{A}$. Note that $M(\mathfrak{A})$ is a C^*-algebra such that $\mathfrak{A} \subseteq M(\mathfrak{A}) \subseteq \mathfrak{A}^{**}$, $\mathfrak{A} = M(\mathfrak{A})$ if and only if $\mathfrak{A}$ has an identity. An example is $M(\mathcal{LC}(\mathfrak{H})) = \mathcal{L}(\mathfrak{H})$. As a multiplier clearly defines a derivation of the algebra, the following theorem of Elliott, Akemann and Pedersen is the best possible.

Theorem. *Let $\mathfrak{A}$ be a separable C^*-algebra. Then the following three conditions are equivalent*:

(1) *every bounded derivation of $\mathfrak{A}$ is implemented by a multiplier*;
(2) *every summable central sequence in $\mathfrak{A}$ is trivial*;
(3) *$\mathfrak{A}$ has the form $\mathfrak{A} = \mathfrak{A}_1 \oplus \mathfrak{A}_2$ where $\mathfrak{A}_1$ only has trivial central sequences, and $\mathfrak{A}_2$ is the restricted direct sum of simple C^*-algebras.*

A central sequence $\{B_n\}$ in a C^*-algebra $\mathfrak{B}$ is a uniformly bounded sequence such that $\|[B_n, A]\| \xrightarrow[n=\infty]{} 0$ for all $A \in \mathfrak{B}$. A central sequence is summable if $\sum_n B_n$ converges in the strong operator topology on $\mathfrak{A}^{**}$, and it is trivial if there exists a sequence $\{Z_n\}$ in the center of $M(\mathfrak{B})$ such that $\|(B_n - Z_n)A\| \xrightarrow[n=\infty]{} 0$ for all $A \in \mathfrak{B}$. Earlier it was proved by Sakai [Sak 5], that derivations of simple C^*-algebras $\mathfrak{A}$ are determined by multipliers, without any separability condition on $\mathfrak{A}$.

In the analysis of C^*- and W^*-dynamical systems, Definition 2.7.1, there is another notion of spectrum which differs from the Arveson spectrum and which is more useful. For simplicity, consider a W^*-dynamical system $\{\mathfrak{M}, G, \alpha\}$ where $\mathfrak{M}$ is a factor, and G is a locally compact abelian group. The Γ-spectrum (sometimes called the Connes spectrum) of α is then defined as

$$\Gamma(\alpha) = \bigcap_{E \in \mathfrak{M}^\alpha(\{0\})} \sigma(\alpha|_{\mathfrak{M}_E}).$$

Here E ranges over all nonzero projections in the fixed point algebra $\mathfrak{M}^\alpha(\{0\})$. It is then not hard to show that $\Gamma(\alpha)$ is a closed subgroup of $\hat{G}$. To prove this one uses techniques such as those used to prove spectral properties of ergodic systems in Section 4.3.3 (see Theorem 4.3.33). The following theorem is a generalization of the derivation theorem, Corollary 3.2.47, and is due to Connes [Con 4].

Theorem. *Let G be a locally compact abelian group acting as an automorphism group α on a factor $\mathfrak{M}$, and assume that $\sigma(\alpha)/\Gamma(a)$ is compact. It follows that the complement*

$$\Gamma(\alpha)^\perp = \{t \in G; (\gamma, t) = 1 \text{ for all } \gamma \in \Gamma(\alpha)\}$$

of $\Gamma(\alpha)$ consists of just those $t \in G$ such that α_t is implemented by a unitary element in $\mathfrak{M}^\alpha(\{0\})$.

An immediate corollary is that if α is a single automorphism of $\mathfrak{M}$ such that $\Gamma(\alpha) \neq \mathbb{T}$, then some power α^n, $n \geq 1$, of α is inner.

Properties of $\Gamma(\alpha)$ can give a lot of information about the fixed point algebra $\mathfrak{M}^\alpha(\{0\})$ and the crossed product $W^*(\mathfrak{M}, \alpha)$. Connes [Con 4] has proved the following:

Theorem. *Let G be a locally compact abelian group acting on a factor $\mathfrak{M}$ such that $\Gamma(\alpha)$ is discrete. It follows that $\sigma(\alpha) = \Gamma(\alpha)$ if, and only if, $\mathfrak{M}^\alpha(\{0\})$ is a factor.*

The next theorem is due to Connes and Takesaki [Con 2].

Theorem. *Let G be locally compact abelian group acting on a factor $\mathfrak{M}$. It follows that $W^*(\mathfrak{M}, \alpha)$ is a factor if, and only if, $\Gamma(\alpha) = \hat{G}$.*

The three theorems above have natural generalizations to von Neumann algebras which are not factors, and also to C^*-algebras. In the C^*-algebra setting the natural replacements of factors are not simple C^*-algebras, as one might initially expect, but prime C^*-algebras, i.e., C^*-algebras such that any two nonzero closed two-sided ideals have nonzero intersection. If a C^*-algebra admits a faithful irreducible representation, it is easily seen to be prime; the converse is also true for separable C^*-algebras [Dix 1]. For various other versions of the three theorems above, see [Iku 1], [Ole 2], [Ole 3], [Ole 4], [Kis 2], [Kis 3], [Bra 10]. For an exhaustive treatment of these results, see Chapter 8 of [[Ped 1]].

Section 3.2.4

Theorems 3.2.50 and 3.2.51 stem from various sources. The nonalgebraic aspects are covered by the Banach space theory of Section 3.1. Among the fresh algebraic elements the positivity-preserving properties of the resolvent, condition (C2), was first discussed by Bratteli and Robinson [Bra 6] and the property $\delta(\mathbb{1}) = 0$, condition (A2), was suggested as a replacement of the derivation property by Kishimoto [Kis 1].

The first statement of Proposition 3.2.52 is due to Elliott [Ell 2] and the remaining statements follow from the work of Powers and Sakai on UHF algebras [Pow 3] [Pow 4], or that of Bratteli and Robinson [Bra 5]. Theorem 3.2.53 was proved by Bratteli and Kishimoto [Bra 11]. It is an application of time-dependent perturbation expansion techniques to the construction of dynamics for quantum spin systems.

Section 3.2.5

Spatial derivations arising from invariant states occurred in the work of Bratteli and Robinson [Bra 5], [Bra 6] and Powers and Sakai [Pow 3], [Pow 4] but the general theory outlined in this section was first proposed in [Bra 3]. A generalization of Proposition 3.2.58, in which analytic elements are replaced by quasi-analytic elements, was given by Bratteli, Herman, and Robinson [Bra 7]. Theorem 3.2.59, for Ω separating, was given in [Bra 3]. The discussion of the alternative $H \geq 0$ uses the type of argument associated with the early proofs of the Borchers–Arveson theorem [Bor 2] (see also Kadison's expository article [Kad 8]).

The general version of Theorem 3.2.61 was proved by Bratteli and Haagerup [Bra 9], while the trace state version, which is essentially contained in Lemma 3.2.62, was derived earlier by Bratteli and Robinson [Bra 8]. The last paper was preceded by a paper of Gallavotti and Pulvirenti [Gal 1] in which the theorem was proved for abelian von Neumann algebras. The latter authors were motivated by problems in classical statistical mechanics as outlined in Example 3.2.67, and modeled their proof on the proof of the Tomita–Takesaki theorem. There is indeed an analogy between Lemma 2.5.12 and some results in the proof of Lemma 3.2.62. Lemma 3.2.65 is a special case of results on dual weights derived by Digernes [Dig 1] and Haagerup [Haa 4]. The conclusive example, Example 3.2.68, appears in [Bra 9], where it is even proved that $\mathfrak{M}$ can be taken to be a type I von Neumann algebra.

Section 3.2.6

All the results of this section are taken from the work of Bratteli, Herman, and Robinson [Bra 4]. The basic problem of comparison of neighboring automorphism groups was proposed by Buchholz and Roberts [Buc 1] who obtained a version of the main theorem, Theorem 3.2.75 ,with ε_1, ε_2 replaced by $o(1)$ as $t \to 0$. The line of reasoning based on cohomology and the smoothing of cocycles, Theorem 3.2.73 and Lemma 3.2.74, are due to Buchholz and Roberts.

The use of numerical range techniques to derive the improved version, Proposition 3.2.70, of Theorem 3.1.35 for unitary groups follows a suggestion of Haagerup.

The results on numerical ranges used in the proof can be found in [[Bon 1]], except the result on the sector of the unitary part of a polar decomposition, which was proved by Woronowicz in [Wor 3].

Theorem 3.2.71 is due to Kadison and Ringrose [Kad 9]. For a nice account of this theorem, see Section 8.7 of [[Ped 1]]. Here one can also find a proof of a result of Borchers [Bor 3], which relates the spectral radius $\rho(\alpha - \iota)$ to $\|\alpha - \iota\|$—specifically:

Theorem. *Let α be a *-automorphism of a C^*-algebra $\mathfrak{A}$. If $\rho(\alpha - \iota) < \sqrt{3}$ or if $\|\alpha - \iota\| < 2$, then*

$$\rho(\alpha - \iota) = \|\alpha - \iota\|.$$

The constant $\sqrt{3}$ is the best possible here; Sakai [Sak 6] has produced an automorphism α of a von Neumann algebra $\mathfrak{M}$ such that $\rho(\alpha - \iota) = \sqrt{3}$, but α is outer. In this example, α is the cyclic change of $\mathfrak{M} = \mathfrak{N} \otimes \mathfrak{N} \otimes \mathfrak{N}$, where $\mathfrak{N}$ is a factor not of type I, i.e.,

$$\alpha(A \otimes B \otimes C) = C \otimes A \otimes B$$

Then $\sigma(\alpha)$ is equal to the three cube roots of 1 hence $\rho(\alpha - \iota) = \sqrt{3}$.

The measure theoretic results (1) and (2) used in the proof of Proposition 3.2.72 can be found in [[Arv 1]]. The result (2) is essentially due to Dixmier [Dix 2].

Another version of Theorem 3.2.73, due to Connes [Con 4], is the following:

Theorem. *If $\mathfrak{M}$ is a factor with separable predual $\mathfrak{M}_*$, and $t \in \mathbb{R} \mapsto \alpha_t, \beta_t$ is a pair of σ-weakly continuous representations of $\mathbb{R}$ in the automorphism group of $\mathfrak{M}$ such that $\beta_t\alpha_{-t}$ is inner for each $t \in \mathbb{R}$ then there exists a σ-weakly continuous unitary cocycle $t \mapsto \Gamma_t \in \mathfrak{M}$ connecting α and β.*

A slight generalization of this result can be found in [Han 1]. Another variation of the theorem, where one needs neither that $\mathfrak{M}$ is a factor nor that $\mathfrak{M}_*$ is separable, is that if $\|\alpha_t - \beta_t\| = o(1)$, $t \mapsto 0$, then there exists a norm continuous cocycle $t \mapsto \Gamma_t$ connecting α and β [Buc 1]. A little warning: theorems of this sort are very sensitive to whether the group in question is $\mathbb{R}$ or a cyclic group. The following is an instructive example. Let $\mathfrak{M} = M_2$ be the 2×2 matrices, and $G = \mathbb{Z}_2 \times \mathbb{Z}_2$. Let $a, b \in G$ be the generators of G such that $a^2 = b^2 = e$, $ab = ba$, and let U, V be the matrices

$$U = \begin{pmatrix} 0 & 1 \\ 1 & 0 \end{pmatrix}, \qquad V = \begin{pmatrix} 1 & 0 \\ 0 & -1 \end{pmatrix}.$$

Then

$$UV + VU = 0, \qquad U^2 = \mathbb{1}, \qquad V^2 = \mathbb{1}.$$

It follows that

$$\alpha_a(A) = UAU^*, \qquad \alpha_b(A) = VAV^*$$

defines a representation of G in $\mathrm{Aut}(\mathfrak{M})$ which is pointwise inner. But if there were a unitary representation $t \in G \mapsto U_t$ of G implementing α, then it follows from

$$\alpha_t(U_s) = U_t U_s U_{-t} = U_s$$

that U_s would be contained in the fixed point algebra $\mathfrak{M}^\alpha$ for each $S \in G$. But an explicit computation shows that $\mathfrak{M}^\alpha = \mathbb{C}\mathbb{1}$. Hence no such unitary representation exists.

The use of invariant means to establish the uniform approximation results stated at the end of the section originates in a paper by Fujii, Furuta and Matsumoto [Fuj 1].

Decomposition Theory

4.1. General Theory

4.1.1. Introduction

The aim of decomposition theory is to express a complex structure as a superposition of simpler components. There is no general rule for what is meant by simpler component and this is determined by the particular application. In an algebraic setting it is usual to examine two complementary forms of decomposition, the decomposition of states and the decomposition of representations. In this chapter we principally describe the theory relating to states, but the intimate connection between states and representations allows us to develop and exploit properties of the representations.

Let $\mathfrak{A}$ be a C^*-algebra with identity $\mathbb{1}$. The states $E_{\mathfrak{A}}$, of $\mathfrak{A}$, form a convex subset of the dual $\mathfrak{A}^*$ and $E_{\mathfrak{A}}$ is compact in the weak*, or $\sigma(\mathfrak{A}^*, \mathfrak{A})$-, topology (see Theorem 2.3.15). Generally, we are interested in decomposing a given state ω as a convex combination of states which are extremal points of some closed convex subset K of $E_{\mathfrak{A}}$. The set K might be given directly by some physical requirement, e.g., K could be the set of states which are invariant under some group of *-automorphisms of $\mathfrak{A}$, or K might be given indirectly, e.g., if $\mathfrak{B}$ is an abelian von Neumann subalgebra of the commutant $\pi_\omega(\mathfrak{A})'$ of the representation $(\mathfrak{H}_\omega, \pi_\omega, \Omega_\omega)$ associated with ω then K could be the weak* closure of the set

$$K_{\mathfrak{B}} = \{\omega_T; \omega_T \in E_{\mathfrak{A}}, \omega_T(A) = (T\Omega_\omega, \pi_\omega(A)\Omega_\omega), T \in \mathfrak{B}\}.$$

We will consider various possibilities:

(1) extremal decomposition, i.e., $K = E_{\mathfrak{A}}$ and one attempts to decompose a state into pure states;

(2) central decomposition, i.e., $\mathfrak{B} = \pi_\omega(\mathfrak{A})'' \cap \pi_\omega(\mathfrak{A})'$, the center of $\pi_\omega(\mathfrak{A})''$, and the aim is to express ω as a superposition of factor states;

(3) decomposition at infinity: if ω is a locally normal state of a quasi-local algebra one can introduce the algebra at infinity $\mathfrak{Z}_\omega^\perp \subseteq \pi_\omega(\mathfrak{A})'' \cap \pi_\omega(\mathfrak{A})'$ (see Definition 2.6.4 and Theorem 2.6.5) and then try to decompose ω as a combination of states with trivial algebra at infinity;

(4) ergodic decomposition: if $g \in G \mapsto \tau_g \in \text{Aut}\,\mathfrak{A}$ is a group of *-automorphisms of $\mathfrak{A}$ and K is the set of τ_g-invariant states,

$$K = \{\omega;\, \omega \in E_{\mathfrak{A}},\, \omega(\tau_g(A)) = \omega(A),\, g \in G,\, A \in \mathfrak{A}\},$$

then K is a closed convex subset of $E_{\mathfrak{A}}$ and the decomposition of an invariant state into extremal invariant states is called ergodic decomposition. Various related decompositions naturally occur, decomposition into states invariant under a subgroup, etc.

In each of the above situations one considers a state $\omega \in K$ and attempts to find a measure μ which is supported by the extremal points $\mathscr{E}(K)$, of K, and which decomposes ω in the form

$$\omega(A) = \int d\mu(\omega')\, \omega'(A).$$

This type of investigation involves geometric, measure-theoretic, and algebraic aspects but the various aspects can be separated to a certain extent. Firstly, we ignore the underlying algebraic structure and examine the decomposition of points ω in a convex compact subset K of a locally convex topological Hausdorff vector space. Geometrically, this corresponds to barycentric decomposition of ω into extreme points or, alternatively stated, the examination of measures supported by $\mathscr{E}(K)$ with a fixed center of mass ω. Such decompositions are of greatest interest when the barycentric measure is unique.

The general theory of barycentric decomposition can be divided into two parts. Firstly, one introduces an order relation $\prec$ among the positive measures on K by specifying $\nu \prec \mu$ if, and only if, $\nu(f) \leq \mu(f)$ for all convex continuous real functions on K. If ν, μ have the same barycenter then $\nu \prec \mu$ indicates that the support of μ is further removed from this barycenter, i.e. the support of the larger measure is nearer the boundary of K. Thus, as an initial step toward constructing extremal decompositions, one attempts to find maximal measures with a fixed barycenter ω. Such measures always exist and they can be characterized essentially by geometric properties. Similarly the existence of a unique maximal measure for each $\omega \in K$ is equivalent to a geometric property of K, the set K must be a simplex (see Section 4.1.2). After analysis of maximal measures the second important step is to determine in which sense these measures are supported by the extreme points $\mathscr{E}(K)$. As a first guess it appears reasonable that measurability of $\mathscr{E}(K)$ would be sufficient to ensure that $\mu(\mathscr{E}(K)) = 1$ for all maximal probability measures. One of the mild surprises of the theory is that this is not the case. If $\mathscr{E}(K)$ is a Borel set and the probability measure μ satisfies $\mu(\mathscr{E}(K)) = 1$ then μ is maximal but, conversely, there exist examples for which $\mathscr{E}(K)$ is Borel, μ is maximal, and $\mu(\mathscr{E}(K)) = 0$. Nevertheless, one can show that each maximal probability measure μ is pseudosupported by $\mathscr{E}(K)$ in the sense that $\mu(B) = 1$ for all Baire sets B such that $B \supseteq \mathscr{E}(K)$. In particular, if $\mathscr{E}(K)$ were a Baire set then μ would be supported by $\mathscr{E}(K)$. Unfortunately, $\mathscr{E}(K)$ is a Baire set if, and only if, K is metrizable (see Notes and Remarks) and hence this last criterion is rather special. However, the study of the supports of the probability measures with a fixed barycenter ω can often be reduced to the study of a face F_ω of K and metrizability, etc., of F_ω usually suffice to ensure that μ is supported by $\mathscr{E}(K)$.[1] Thus analysis of the measure-theoretic structure of the faces of K plays a role in the examination of maximal measures.

Next let us sketch the algebraic aspects of decomposition theory which enter when K is a subset of the state space $E_{\mathfrak{A}}$ of the C^*-algebra $\mathfrak{A}$. For motivation it is useful to first examine the decomposition of a given state ω as a finite convex combination of states ω_i,

$$\omega(A) = \sum_{i=1}^{n} \lambda_i \omega_i(A), \qquad A \in \mathfrak{A}.$$

To reexpress this in measure-theoretic language we define the affine functions $\hat{A}$ over $E_{\mathfrak{A}}$ by

$$\hat{A}(\omega) = \omega(A)$$

for all $A \in \mathfrak{A}$ and introduce the Dirac, or point, measure δ_ω. One then has

$$\hat{A}(\omega) = \mu(\hat{A}), \qquad \mu = \sum_{i=1}^{n} \lambda_i \delta_{\omega_i}.$$

Next note that $\lambda_i \omega_i \leq \omega$ and hence

$$\lambda_i \omega_i(A) = (T_i \Omega_\omega, \pi_\omega(A)\Omega_\omega)$$

for some positive T_i in the commutant $\pi_\omega(\mathfrak{A})'$ by Theorem 2.3.19. Thus the decomposition of ω corresponds to a finite decomposition of the identity,

$$\mathbb{1} = \sum_{i=1}^{n} T_i,$$

within the commutant $\pi_\omega(\mathfrak{A})'$. Intuitively, the more general integral decompositions of ω correspond to analogous, but continuous, decompositions of $\mathbb{1}$ in $\pi_\omega(\mathfrak{A})'$ and in this manner the theory is intimately connected with operator decompositions.

The simplest form of finite decomposition occurs when the T_i are mutually orthogonal projections, $T_i = P_i$, in $\pi_\omega(\mathfrak{A})'$. In this case one easily sees that the representation $(\mathfrak{H}_\omega, \pi_\omega)$ is a direct sum of the representations $(\mathfrak{H}_{\omega_i}, \pi_{\omega_i})$. Thus the decomposition is in this sense a decomposition into orthogonal, or independent, components. The family $\{P_i\}$ generates a finite-dimensional subalgebra $\mathfrak{B}$ of $\pi_\omega(\mathfrak{A})'$ and the finite orthogonal decompositions of this type are in one-to-one correspondence with the algebras $\mathfrak{B}$. It is enlightening to

[1] A face F of a compact convex set K is defined to be a subset of K with the property that if $\omega = \sum_{i=1}^{n} \lambda_i \omega_i$ is a finite convex combination of elements in K such that $\omega \in F$ then $\omega_i \in F$ for $i = 1, \ldots, n$. In this book we will not assume that a face is closed subset of K, contrary to the standard usage of this term.

reexpress the decomposition of ω in terms of the projection operator $P = [\mathfrak{B}\Omega_\omega]$. One has

$$P\psi = \sum_{i=1}^{n} \lambda_i^{-1}(P_i\Omega_\omega, \psi)P_i\Omega_\omega$$

and

$$\lambda_i = (\Omega_\omega, P_i\,\Omega_\omega), \qquad \lambda_i\omega_i(A) = (P_i\Omega_\omega, \pi_\omega(A)\Omega_\omega).$$

Using the orthogonality of the P_i and $P_i \in \pi_\omega(\mathfrak{A})'$ one then computes that the associated measure μ satisfies

$$\begin{aligned}
\mu(\hat{A}_1\hat{A}_2\cdots\hat{A}_m) &= \sum_{i=1}^{n} \lambda_i\omega_i(A_1)\omega_i(A_2)\cdots\omega_i(A_m)\\
&= \sum_{i=1}^{n} \lambda_i^{-(m-1)}(\Omega_\omega, \pi_\omega(A_1)P_i\Omega_\omega)\\
&\quad\times (\Omega_\omega, \pi_\omega(A_2)P_i\Omega_\omega)\cdots(\Omega_\omega, \pi_\omega(A_m)P_i\Omega_\omega)\\
&= \sum_{i_1=1}^{n}\cdots\sum_{i_{m-1}=1}^{n} \lambda_{i_1}^{-1}\cdots\lambda_{i_{m-1}}^{-1}(\Omega_\omega, \pi_\omega(A_1)P_{i_1}\,\Omega_\omega)\\
&\quad\times (P_{i_1}\Omega_\omega, \pi_\omega(A_2)P_{i_2}\Omega_\omega)\cdots(P_{i_{m-1}}\Omega_\omega, \pi_\omega(A_m)\Omega_\omega)\\
&= (\Omega_\omega, \pi_\omega(A_1)P\pi_\omega(A_2)P\cdots P\pi_\omega(A_m)\Omega_\omega).
\end{aligned}$$

Among other things this calculation shows that $\{P\pi_\omega(A)P;\ A \in \mathfrak{A}\}$ generates an abelian algebra on $P\mathfrak{H}_\omega$. Moreover, it provides an algorithm for constructing more general measures μ from the projections P such that $P\Omega_\omega = \Omega_\omega$ and $P\pi_\omega(\mathfrak{A})''P$ is abelian. Given such a P one can show that the relations

$$\mu(\hat{A}_1\hat{A}_2\cdots\hat{A}_n) = (\Omega_\omega, \pi_\omega(A_1)P\pi_\omega(A_2)P\cdots P\pi_\omega(A_n)\Omega_\omega) \qquad (*)$$

determine a measure μ with barycenter ω. In fact, this class of measures, the orthogonal measures, is in one-to-one correspondence with such projections and also in one-to-one correspondence with the abelian von Neumann algebras $\mathfrak{B} \subseteq \pi_\omega(\mathfrak{A})'$. The corresponding entities are connected by various rules such as $(*)$ and

$$P = [\mathfrak{B}\Omega_\omega], \qquad \mathfrak{B} = \{\pi_\omega(\mathfrak{A}) \cup P\}', \qquad \text{etc.}$$

It is also of interest that the ordering $\prec$ on the orthogonal measures corresponds to the ordering by inclusion of the related abelian algebras $\mathfrak{B}$ and the ordering by increase of the related projections. In particular, measures which are maximal among the orthogonal measures arise from maximal abelian von Neumann subalgebras $\mathfrak{B}$ of the commutant by spectral decomposition of $\mathfrak{B}$. Under suitably good circumstances, e.g. suitable metrizability properties, the maximal orthogonal measures are also maximal among all measures with the given barycenter.

The algebraic structure also allows one to obtain good measure-theoretic properties even if $\mathfrak{A}$ is not separable and $E_\mathfrak{A}$ nonmetrizable. There are two approaches. One is to look at suitable faces of $E_{\mathfrak{A}}$ with nice separability properties, for example, locally normal states over quasi-local algebras for which the local subalgebras are isomorphic to $\mathscr{L}(\mathfrak{H})$ for some $\mathfrak{H}$. We discuss this point of view throughout Sections 4.2 and 4.3. The second approach relies more strongly on the representation structure underlying the state ω which one is decomposing. Metrizability of $E_{\mathfrak{A}}$ can essentially be replaced by separability of the representation space $\mathfrak{H}_\omega$. This point of view is discussed in Section 4.4. The representation structure also appears indispensable for the discussion of the decomposition at infinity.

4.1.2. Barycentric Decompositions

In this subsection we examine barycentric decompositions of points in a convex compact subset of a locally convex topological vector space. The classic geometric result is the theorem of Carathéodory–Minkowski which establishes that every point of a convex compact subset K of the ν-dimensional Euclidean space $\mathbb{R}^\nu$ can be written as a convex combination of at most $\nu + 1$ extreme points of K. Thus each point is the barycenter of a finite number of point masses distributed over the extreme points of K. The theorem also asserts that the decomposition into extreme points is unique if, and only if, K is a simplex, i.e., if K is affinely isomorphic to the set with projective coordinates $\{(\lambda_1, \lambda_2, \ldots, \lambda_{\nu+1}); \lambda_i \geq 0, \sum \lambda_i = 1\}$. Our purpose is to derive an analogue of this theorem in more general, infinite-dimensional, spaces. First, we introduce some notation which will be used throughout the subsection and recall various generalities.

Let K denote a convex compact subset of a real locally convex topological vector space X and $\mathscr{E}(K)$ the set of extreme points of K. We use $C(K)$ to denote the real continuous functions over K, $S(K)$ the real continuous convex functions, and $A(K)$ the real continuous affine functions, i.e., $S(K) = \{f \in C(K); f(\lambda\omega_1 + (1 - \lambda)\omega_2) \leq \lambda f(\omega_1) + (1 - \lambda)f(\omega_2)$ for all $\omega_1, \omega_2 \in K$ and $0 \leq \lambda \leq 1\}$ and $A(K) = \{f \in C(K); f(\lambda\omega_1 + (1 - \lambda)\omega_2) = \lambda f(\omega_1) + (1 - \lambda)f(\omega_2)$ for all ω_1, $\omega_2 \in K$ and $0 \leq \lambda \leq 1\} = S(K) \cap (-S(K))$. If $f, g \in C(K)$ then $f \geq g$ is understood to mean $f(\omega) - g(\omega) \geq 0$ for all $\omega \in K$.

The set $M_+(K)$ of positive Radon measures over K forms a subset of the dual of $C(K)$ which can be equipped with the weak *-topology, i.e., the $\sigma(C(K)^*, C(K))$-topology. We denote the positive Radon measures with norm one, i.e., the probability measures, by $M_1(K)$.

Next note that as K is compact the Borel sets of K are unambiguously defined as the elements of the σ-algebra $\mathfrak{B}$ generated by the closed, or open, subsets of K. The positive regular Borel measures are then the positive countably additive set functions μ over $\mathfrak{B}$ which satisfy

$$\begin{aligned}\mu(B) &= \sup\{\mu(C); B \supset C, C \text{ closed}\}\\ &= \inf\{\mu(C); B \subset C, C \text{ open}\}.\end{aligned}$$

The Riesz representation theorem establishes a one-to-one correspondence between the positive regular Borel measures and the Radon measures $M_+(K)$ such that $\mu(K) = \|\mu\|$. We use a standard integral notation,

$$\mu(f) = \int d\mu(\omega) f(\omega), \qquad f \in C(K),$$

which implicitly identifies these descriptions.

The Borel sets of K contain all countable unions of closed sets and all countable intersections of open sets. These two types of sets are called F_σ and G_δ respectively. The Baire sets of the compact K are defined as the elements of the σ-algebra $\mathfrak{B}_0$ generated by the closed G_δ-subsets, or the open F_σ-subsets, of K. $\mathfrak{B}_0$ is the smallest σ-algebra of subsets of K such that all the continuous functions on K are measurable. A measure μ_0 on $\mathfrak{B}_0$ is called a Baire measure. A Baire measure is automatically regular, and has a unique extension to a regular Borel measure on $\mathfrak{B}$. Since the restriction of a Borel measure to $\mathfrak{B}_0$ is a Baire measure we have a one-to-one correspondence between Radon measures μ, regular Borel measures $d\mu$, and Baire measures $d\mu_0$ given by

$$\mu(f) = \int d\mu(\omega) f(\omega) = \int d\mu_0(\omega) f(\omega), \qquad f \in C(K).$$

In the sequel, the term measure will be used interchangeably to denote these three concepts.

The support of a measure $\mu \in M_+(K)$ is defined as the smallest closed subset $C \subseteq K$ such that $\mu(C) = \mu(K)$. This subset exists due to the regularity of μ. Analogously, one says that μ is supported by a Borel set $B \subseteq K$ if $\mu(B) = \mu(K)$. Moreover, the measure μ is said to be pseudosupported by an arbitrary set $A \subseteq K$ if $\mu(B) = 0$ for all Baire sets B such that $B \cap A = \emptyset$. Although these latter notions are superficially similar we emphasize that this similarity is rather misleading. There exist examples in which

(1) A is a Borel set,
(2) μ is pseudosupported by A,
(3) $\mu(A) = 0$.

If μ is supported by A we call A a supporting set and a pseudosupporting set is defined in an analogous manner. The stronger notion of a supporting set is of the greatest use but the weaker notion enters naturally in the general theory to follow.

The Dirac, or point, measure with support ω is denoted by δ_ω. Thus $\delta_\omega(f) = f(\omega)$ for all $f \in C(K)$. Each μ with finite support $\{\omega_1, \omega_2, \ldots, \omega_n\}$ is a superposition

$$\mu = \sum_{i=1}^{n} \lambda_i \delta_{\omega_i}$$

of Dirac measures with coefficients $\lambda_i \geq 0$. One has

$$\mu(K) = \|\mu\| = \sum_{i=1}^{n} \lambda_i.$$

Finally, we remark that the measures $M_+(K)$ form a positive cone which has a natural order defined by $\mu \geq \nu$ if $\mu(f) \geq \nu(f)$ for all positive $f \in C(K)$. This cone has an interesting geometric property which will be of subsequent significance. It is a lattice with respect to the order $\geq$. By this we mean that each pair $\mu, \nu \in M_+(K)$ has a least upper bound $\mu \vee \nu$ and a greatest lower bound $\mu \wedge \nu$ in $M_+(K)$. Explicitly, one has

$$\mu \vee \nu = (\mu - \nu)_+ + \nu, \qquad \mu \wedge \nu = \nu - (\mu - \nu)_-,$$

where $(\mu - \nu)_\pm$ denote the positive and negative parts of the signed measure $\mu - \nu$.

After these preliminaries we now introduce the concept of the barycenter $b(\mu)$ of a measure $\mu \in M_1(K)$ by

$$b(\mu) = \int d\mu(\omega)\, \omega,$$

where the integral is understood in the weak sense. The set $M_\omega(K)$ is then defined as the subset of $M_1(K)$ with barycenter ω, i.e.,

$$M_\omega(K) = \{\mu;\, \mu \in M_1(K),\, b(\mu) = \omega\}.$$

The existence of the barycenter of a general measure is not quite evident but is given by the following result which also establishes a useful approximation procedure.

Proposition 4.1.1. *If $\mu \in M_1(K)$ then there exists a unique point $b(\mu) \in K$, the barycenter of μ, such that*

$$f(b(\mu)) = \mu(f) = \int d\mu(\omega') f(\omega')$$

for all $f \in A(K)$.

Moreover, there exists a net $\mu_\alpha \in M_1(K)$ of measures with finite support which is weakly convergent to μ and is such that*

$$b(\mu_\alpha) = b(\mu).$$

PROOF. If μ has finite support, i.e.,

$$\mu = \sum_{i=1}^{n} \lambda_i \delta_{\omega_i}$$

with $\lambda_i \geq 0$ and $\sum \lambda_i = 1$, then $b(\mu)$ exists and is given by

$$b(\mu) = \sum_{i=1}^{n} \lambda_i \omega_i.$$

As K is convex $b(\mu) \in K$. But each $\mu \in M_1(K)$ may be approximated in the weak* topology by a net of measures $\mu_\alpha \in M_1(K)$ with finite support (this is just the Riemann approximation of the integral). The barycenters $b(\mu_\alpha)$ of the μ_α must lie in K and as K is compact there exists a convergent subnet $\{b(\mu_{\alpha'})\}$. Let $b(\mu) \in K$ denote the limit of this subnet; then

$$\begin{aligned} f(b(\mu)) &= \lim_{\alpha'} f(b(\mu_{\alpha'})) \\ &= \lim_{\alpha'} \mu_{\alpha'}(f) = \mu(f) \end{aligned}$$

for all $f \in A(K)$. But $b(\mu)$ is certainly unique because the elements of $A(K)$ separate the points of K by the Hahn–Banach theorem. Thus the barycenter $b(\mu)$ of μ exists.

Next consider all finite partitions, $\mathscr{U} = \{U_i\}_{1 \le i \le n}$, of K in terms of Baire sets U_i. These partitions form a directed set when the order $\mathscr{U} \prec \mathscr{V}$ is taken to mean that each $V_i \in \mathscr{V}$ is a subset of a $U_j \in \mathscr{U}$. Let χ_i denote the characteristic function of U_i and define λ_i, and μ_i, by $\lambda_i = \mu(U_i)$ and $\lambda_i \, d\mu_i = \chi_i \, d\mu$. Let $\omega_i = b(\mu_i)$ and introduce the measure $\mu_{\mathscr{U}} \in M_1(K)$ by

$$\mu_{\mathscr{U}} = \sum_{i=1}^{n} \lambda_i \delta_{\omega_i}.$$

This measure has finite support and

$$b(\mu_{\mathscr{U}}) = \sum_{i=1}^{n} \lambda_i b(\mu_i) = b\left(\sum_{i=1}^{n} \lambda_i \mu_i\right) = b(\mu).$$

Moreover, for $f \in C(K)$ one has

$$\begin{aligned} |\mu_{\mathscr{U}}(f) - \mu(f)| &\le \sum_{i=1}^{n} \lambda_i |f(\omega_i) - \mu_i(f)| \\ &\le \sup_i \sup_{\omega \in U_i} |f(\omega_i) - f(\omega)|. \end{aligned}$$

Finally, for $\varepsilon > 0$ choose a finite family of open sets G_i such that $|f(x) - f(y)| < \varepsilon$ for $x, y \in G_i$ and such that

$$K \subseteq \bigcup_{i=1}^{n} G_i.$$

Defining U_i by

$$U_i = G_i - \bigcup_{j<i} G_j$$

one has

$$|\mu_{\mathscr{U}}(f) - \mu(f)| < \varepsilon.$$

Therefore $\mu_{\mathscr{U}}(f) \to \mu(f)$ for each such f and $\mu_{\mathscr{U}}$ converges weakly* to μ.

Now we turn to the problem of barycentric decomposition, i.e., the problem of associating to each $\omega \in K$ a measure $\mu \in M_\omega(K)$ which is supported by the extremal points $\mathscr{E}(K)$. We follow the program outlined in the introduction and first introduce an order relation on $M_+(K)$.

Definition 4.1.2. The *order relation* $\succ$ is defined on $M_+(K)$ by specifying $\mu \succ \nu$ if, and only if,

$$\mu(f) \geq \nu(f)$$

for all $f \in S(K)$.

It is not immediately evident that the relation $\succ$ is a genuine partial order on $M_+(K)$. Clearly, $\succ$ is reflexive ($\mu \succ \mu$) and transitive ($\mu \succ \nu$, $\nu \succ \rho$ imply $\mu \succ \rho$) but one must prove that it is antisymmetric, i.e., that $\mu \succ \nu$, $\nu \succ \mu$ imply $\mu = \nu$. The following proposition establishes that $\succ$ is a partial ordering and that each $\omega \in K$ is the barycenter of a maximal measure, where maximality of μ means that $\nu \succ \mu$ implies $\nu = \mu$.

Proposition 4.1.3. *The relation $\succ$ on $M_+(K)$ is a partial ordering. If $\mu \succ \nu$ then $\|\mu\| = \|\nu\|$ and if, further, $\|\mu\| = \|\nu\| = 1$ then $b(\mu) = b(\nu)$. Moreover, $\mu \in M_\omega(K)$ if, and only if, $\mu \succ \delta_\omega$.*

Each $\omega \in K$ is the barycenter of a measure $\mu \in M_1(K)$ which is maximal for the order $\succ$.

PROOF. If $\mu \succ \nu$ and $\nu \succ \mu$ then $\mu(f) = \nu(f)$ for all $f \in S(K)$. It follows immediately that $\mu(f - g) = \nu(f - g)$ for all pairs $f, g \in S(K)$, and to conclude that $\mu = \nu$ it suffices to show that $S(K) - S(K)$ is uniformly dense in $C(K)$.

Lemma 4.1.4. *The set $S(K) - S(K)$ is uniformly dense in $C(K)$, i.e., any real continuous function over K may be uniformly approximated by differences of real continuous convex functions.*

PROOF. The Hahn–Banach theorem establishes that the continuous affine functions $A(K)$ separate points of K. Consequently the Stone–Weierstrass theorem implies that the real polynomials of elements of $A(K)$ are uniformly dense in $C(K)$. But every real polynomial whose arguments take values over a compact convex subset of $\mathbb{R}^n$ can be decomposed as the difference of two convex functions over this subset, e.g.,

$$P(x_1, x_2, \ldots, n_n) = \left[P(x_1, x_2, \ldots, x_n) + \lambda \sum_{i=1}^{n} x_i^2\right] - \lambda \sum_{i=1}^{n} x_i^2$$

with λ sufficiently large. Combination of these observations completes the proof.

END OF PROOF OF PROPOSITION 4.1.3. Lemma 4.1.4 implies that $\succ$ is antisymmetric and hence a partial order. Next remark that if $f \in A(K)$ then $f \in S(K)$ and $-f \in S(K)$. Hence $\mu \succ \nu$ implies $\mu(f) = \nu(f)$. In particular,

$$\|\mu\| = \int d\mu(\omega) = \int d\nu(\omega) = \|\nu\|$$

and for $\mu, \nu \in M_1(K)$ one also has

$$b(\mu) = \int d\mu(\omega)\, \omega = \int d\nu(\omega)\, \omega = b(\nu),$$

where the last integrals are interpreted in the weak sense. In particular, $\mu \succ \delta_\omega$ implies $\mu \in M_\omega(K)$. Conversely, if $\mu \in M_\omega(K)$ then Proposition 4.1.1 establishes that μ is the weak* limit of a net $\mu_\alpha \in M_\omega(K)$ of measures with finite support. Thus if $f \in S(K)$ then

$$f(\omega) = f(b(\mu_\alpha)) \leq \mu_\alpha(f)$$

by convexity. Taking the net limit one deduces that $\mu(f) \geq f(\omega)$ or, equivalently, $\mu \succ \delta_\omega$.

Finally, the existence of a maximal $\mu \succ \delta_\omega$ follows from Zorn's lemma if one can show that each set $\{\mu_\alpha\}$ of measures which is totally ordered by the relation $\succ$ has an upper bound. But for $f \in S(K)$ the set $\{\mu_\alpha(f)\}$ is monotonic increasing and, moreover, $\|\mu_\alpha\|$ is independent of α. Therefore $\mu_\alpha(f)$ converges and $\mu_\alpha(g-h)$ converges for all $g, h \in S(K)$. But from Lemma 4.1.4 one concludes that $\{\mu_\alpha\}$ converges in the weak* topology. The limit is clearly a positive Radon measure which majorizes all the μ_α.

The partial characterization of the order $\succ$ given by Proposition 4.1.3 can be greatly extended. If $\nu \in M_1(K)$ is a measure of finite support,

$$\nu = \sum_{i=1}^{n} \lambda_i \delta_{\omega_i},$$

and $\mu \succ \nu$ then one can show that

$$\mu = \sum_{i=1}^{n} \lambda_i \mu_i$$

with $\mu_i \succ \delta_{\omega_i}$. A similar type of comparison result is true for general ν and can be used to completely characterize the order. As we do not at present need such detail we postpone this characterization to Section 4.2, Proposition 4.2.1.

Proposition 4.1.3 establishes the existence of maximal measures, and the geometrical intuition motivating the order relation indicates that the maximal measures should in some sense be supported by the extremal points. In order to localize the support of maximal measures we exploit the geometric properties of convex functions, by use of their upper envelopes.

Definition 4.1.5. If $f \in C(K)$ then its *upper envelope* $\bar{f}$ is defined by

$$\bar{f}(\omega) = \inf\{g(\omega);\ -g \in S(K), g \geq f\}.$$

If $f \in S(K)$ the *associated boundary set* $\partial_f(K)$ is defined by

$$\partial_f(K) = \{\omega;\ \omega \in K, f(\omega) = \bar{f}(\omega)\}.$$

Note that the upper envelope $\bar{f}$ can be discontinuous even although f is continuous. Nevertheless, as $\bar{f}$ is the lower envelope of a family of concave continuous functions it is both concave and upper semicontinuous. The boundary set comprises the points where the convex function f meets its concave upper envelope $\bar{f}$ and should contain the maxima of f. The elementary properties of $\bar{f}$ and $\partial_f(K)$ are summarized as follows.

Proposition 4.1.6. *If $f \in C(K)$ then the upper envelope $\bar{f}$ is concave, upper semicontinuous, and hence measurable. Moreover, if $\mu \in M_+(K)$ then*

$$\mu(\bar{f}) = \inf\{\mu(g);\ -g \in S(K), g \geq f\}.$$

If $f \in S(K)$ the associated boundary set $\partial_f(K)$ is a G_δ and, in particular, a Borel set.

PROOF. The concavity and upper semicontinuity follow by definition. The upper semicontinuity is equivalent to the condition that the sets $\{\omega; \bar{f}(\omega) < a\}$ are open and the Borel measurability of $\bar{f}$ follows immediately.

If $S_f = \{g; -g \in S(K), g \geq f\}$ then for each pair $g_1, g_2 \in S_f$ one has $g_1 \wedge g_2 \in S_f$, where $g_1 \wedge g_2$ is the greatest lower bound of g_1 and g_2. Thus S_f forms a pointwise decreasing net of continuous functions with limit $\bar{f}$. The identification of $\mu(\bar{f})$ then follows from the monotone convergence theorem.

Finally, as f is continuous, $\bar{f} - f$, is upper semicontinuous and the sets

$$S_n = \{\omega; \omega \in K, \bar{f}(\omega) - f(\omega) < 1/n\}$$

are open. But $\bar{f} \geq f$ and hence

$$\partial_f(K) = \bigcap_{n \geq 0} S_n.$$

The following theorem gives the fundamental characterization of maximal measures. Basically, the theorem substantiates the intuition that a maximal measure is supported by the sets for which the convex functions attain their maxima, i.e., the sets where $f = \bar{f}$.

Theorem 4.1.7. *Let K be a convex compact subset of a locally convex Hausdorff space X and let $\mu \in M_+(K)$. The following conditions are equivalent:*

(1) *μ is maximal for the order $\succ$ on $M_+(K)$;*
(2) *μ is supported by every boundary set $\partial_f(K)$, i.e.,*

$$\mu(\partial_f(K)) = \|\mu\|, \qquad f \in S(K);$$

(3) $\mu(\bar{f}) = \mu(f)$ *for all $f \in C(K)$.*

Note that condition (2) is by definition equivalent to $\mu(\bar{f}) = \mu(f)$ for all $f \in S(K)$.

The equivalence of the three conditions follows from a more detailed analysis of the upper envelope of elements of $C(K)$. The necessary information is summarized in the following two lemmas.

Lemma 4.1.8. *If $\bar{f}$ is the upper envelope of $f \in C(K)$ then the set*

$$S = \{(\omega, t); (\omega, t) \in K \times \mathbb{R}, t \leq \bar{f}(\omega)\}$$

is the closed convex hull S' of the set

$$\{(\omega, t); (\omega, t) \in K \times \mathbb{R}, t \leq f(\omega)\}.$$

PROOF. The function $\bar{f}$ is concave, upper semicontinuous, and $\bar{f} \geq f$. Hence S is convex, closed, and $S' \subseteq S$. Next assume there is a $(\omega_0, t_0) \in S$ such that $(\omega_0, t_0) \notin S'$. The Hahn–Banach theorem implies the existence of a continuous linear functional g over $X \times \mathbb{R}$ such that $g(\omega_0, t_0) > 1$ and $g(\omega, t) \leq 1$ for $t \leq f(\omega)$. Now g must have the form $g(\omega, t) = h(\omega) + \lambda t$ where h is a continuous linear functional on X. Thus

$$h(\omega_0) + \lambda t_0 > 1 \quad \text{and} \quad h(\omega) + \lambda f(\omega) \leq 1$$

for all $\omega \in K$. Choosing $\omega = \omega_0$ and subtracting one finds $\lambda(f(\omega_0) - t_0) < 0$. But since $\omega_0 \in K$ and $t_0 > f(\omega_0)$ one must then have $\lambda > 0$. This implies that $\omega \mapsto (1 - h(\omega))/\lambda$ is an affine continuous function which majorizes f and satisfies $t_0 > (1 - h(\omega_0))/\lambda \geq \bar{f}(\omega_0)$. This contradicts the assumption $t_0 \leq \bar{f}(\omega_0)$ and hence $S' = S$.

Lemma 4.1.9. *If $\bar{f}$ is the upper envelope of $f \in C(K)$ then*

$$\bar{f}(\omega) = \sup\{\mu(f);\ \mu \succ \delta_\omega,\ \mu \text{ has finite support}\}.$$

Therefore for each pair $f, g \in C(K)$, one has

$$\overline{(f + g)}(\omega) \leq \bar{f}(\omega) + \bar{g}(\omega).$$

PROOF. If $\mu \succ \delta_\omega$ and μ has finite support then

$$\mu = \sum_{i=1}^{n} \lambda_i \delta_{\omega_i}$$

for some $\lambda_i \geq 0$ such that

$$\sum_{i=1}^{n} \lambda_i = 1, \qquad \sum_{i=1}^{n} \lambda_i \omega_i = \omega.$$

But then

$$\bar{f}(\omega) \geq \sum_{i=1}^{n} \lambda_i \bar{f}(\omega_i) \geq \sum_{i=1}^{n} \lambda_i f(\omega_i) = \mu(f),$$

where the first inequality uses the concavity of $\bar{f}$.

Next, by Lemma 4.1.8 there exist $\lambda_i^\alpha \geq 0$, $\omega_i^\alpha \in K$, $t_i^\alpha \in \mathbb{R}$, and $n_\alpha > 0$ such that

$$\sum_{i=1}^{n_\alpha} \lambda_i^\alpha = 1, \qquad t_i^\alpha \leq f(\omega_i^\alpha)$$

$$\lim_\alpha \sum_{i=1}^{n_\alpha} \lambda_i^\alpha \omega_i^\alpha = \omega, \qquad \lim_\alpha \sum_{i=1}^{n_\alpha} \lambda_i^\alpha t_i^\alpha = \bar{f}(\omega).$$

Define μ_α by

$$\mu_\alpha = \sum_{i=1}^{n_\alpha} \lambda_i^\alpha \delta_{\omega_i^\alpha}$$

and pass to a weakly* convergent subnet $\{\mu_{\alpha'}\}$. If μ denotes the limit of this subnet one has

$$b(\mu) = \lim_{\alpha'} b(\mu_{\alpha'}) = \omega$$

and

$$\mu(f) = \lim_{\alpha'} \mu_{\alpha'}(f)$$

$$\geq \lim_{\alpha'} \sum_{i=1}^{n_{\alpha'}} \lambda_i^{\alpha'} t_i^{\alpha'} = \bar{f}(\omega).$$

But by Proposition 4.1.1 one can then approximate μ in the weak* topology by measures with finite support, barycenter $b(\mu)$, and norm unity. Thus $\bar{f}(\omega)$ is smaller than the supremum of the lemma. The observation of the previous paragraph establishes that it is also larger and hence equality is established.

The second statement of the lemma follows from the first by observing that the supremum of a sum is less than the sum of the suprema.

PROOF OF THEOREM 4.1.7. Let us now return to the proof of Theorem 4.1.7.

(1) $\Rightarrow$ (2) Define $p(g) = \mu(\bar{g})$ for $g \in C(K)$. One has $p(\lambda g) = \lambda p(g)$ for all $\lambda \geq 0$ and $g \in C(K)$. Moreover, by Lemma 4.1.9 one finds $p(g + h) \leq p(g) + p(h)$ for $g, h \in C(K)$. Thus by the Hahn–Banach theorem one can find a linear functional ν over $C(K)$ such that $\nu(f) = \mu(\bar{f})$ for one fixed $f \in S(K)$ and $\nu(g) \leq \mu(\bar{g})$ for $g \in C(K)$. Now if $g \leq 0$ then $\bar{g} \leq 0$ and $\nu(g) \leq \mu(\bar{g}) \leq 0$. Thus $\nu \in M_+(K)$. Moreover, if $-g \in S(K)$ then $g = \bar{g}$ and $\nu(g) \leq \mu(\bar{g}) = \mu(g)$, which implies $\nu \succ \mu$. By the assumed maximality of μ one concludes that $\nu = \mu$ and therefore

$$\mu(f) = \nu(f) = \mu(\bar{f})$$

for the prescribed $f \in S(K)$, i.e., μ is supported by $\partial_f(K)$.

(2) $\Rightarrow$ (3) The map $f \in C(K) \mapsto \mu(\bar{f})$ is subadditive by Lemma 4.1.9. Therefore

$$\mu(\bar{f}) - \mu(\bar{g}) \leq \mu(\overline{f - g}) \leq \mu(\bar{f}) + \mu(\overline{(-g)})$$

for all $f, g \in C(K)$. But $\mu(f) = \mu(\bar{f})$ for $f \in S(K)$ by assumption and $\overline{(-g)} = -g$ for $g \in S(K)$ by definition. Thus

$$\mu(f) - \mu(g) \leq \mu(\overline{f - g}) \leq \mu(f) - \mu(g)$$

for all $f, g \in S(K)$ or, equivalently,

$$\mu(f - g) = \mu(\overline{f - g})$$

for $f, g \in S(K)$. It follows from Lemma 4.1.4 that

$$\mu(h) = \mu(\bar{h})$$

for all $h \in C(K)$.

(3) $\Rightarrow$ (1) Assume $\nu \succ \mu$. If $h \in C(K)$ then by Proposition 4.1.6

$$\begin{aligned} \nu(\bar{h}) &= \inf\{\nu(g);\ -g \in S(K), g \geq h\} \\ &\leq \inf\{\mu(g);\ -g \in S(K), g \geq h\} \\ &= \mu(\bar{h}). \end{aligned}$$

Therefore as $\bar{h} \geq h$

$$\nu(h) \leq \nu(\bar{h}) \leq \mu(\bar{h}) = \mu(h).$$

Thus $\nu(h) \leq \mu(h)$. But replacing h by $-h$, one has $\nu(h) \geq \mu(h)$ and hence $\mu = \nu$, i.e., μ is maximal.

Theorem 4.1.7 establishes that the maximal measures are supported by each of the boundary sets $\partial_f(K)$. We next demonstrate that the set $\mathscr{E}(K)$ of extremal points of K is exactly equal to the intersection of the $\partial_f(K)$. This shows that the maximal measures are supported in some weak sense by $\mathscr{E}(K)$ but some care must be exercised in this form of interpretation because $\mathscr{E}(K)$ is not necessarily measurable. We return to a more detailed discussion of this point after establishing the identification of $\mathscr{E}(K)$.

Proposition 4.1.10. *Let K be a convex compact subset of a locally convex Hausdorff topological vector space. The extremal points $\mathscr{E}(K)$ of K and the boundary sets $\partial_f(K)$ are related by*

$$\mathscr{E}(K) = \bigcap_{f \in S(K)} \partial_f(K).$$

Therefore if $B \subseteq \mathscr{E}(K)$ is a Borel set and $\mu(B) = \|\mu\|$ for some $\mu \in M_+(K)$ then μ is maximal.

PROOF. If $\omega \in K$ but $\omega \notin \mathscr{E}(K)$ then $\omega = (\omega_1 + \omega_2)/2$ for some pair $\omega_1, \omega_2 \in K$ with $\omega_1 \neq \omega_2$. Let f be any affine continuous function with $f(\omega_1) \neq f(\omega_2)$. One has

$$f^2\left(\frac{\omega_1 + \omega_2}{2}\right) = \frac{(f(\omega_1) + f(\omega_2))^2}{4}$$

$$= \frac{f^2(\omega_1) + f^2(\omega_2)}{2} - \frac{(f(\omega_1) - f(\omega_2))^2}{4}.$$

Thus if $-g \in S(K)$ and $g \geq f^2$ then

$$g(\omega) \geq \frac{g(\omega_1) + g(\omega_2)}{2} \geq \frac{f^2(\omega_1) + f^2(\omega_2)}{2} > f^2(\omega).$$

Therefore

$$\overline{f^2}(\omega) \geq \frac{f^2(\omega_1) + f^2(\omega_2)}{2} > f^2(\omega).$$

In particular, $\omega \notin \partial_{f^2}(K)$ and hence

$$\mathscr{E}(K) \supseteq \bigcap_{f \in S(K)} \partial_f(K).$$

Conversely, if $\omega \notin \partial_f(K)$ for some $f \in S(K)$ then $\bar{f}(\omega) > f(\omega)$ and by Lemma 4.1.9 there is a $\mu \in M_1(K)$ with finite support, and barycenter ω such that $\mu(f) > f(\omega)$. Thus if

$$\mu = \sum_{i=1}^{n} \lambda_i \delta_{\omega_i}$$

one has

$$b(\mu) = \sum_{i=1}^{n} \lambda_i \omega_i = \omega$$

and

$$\mu(f) = \sum_{i=1}^{n} \lambda_i f(\omega_i) > f(\omega).$$

It follows from these two conditions that $\omega \notin \mathscr{E}(K)$ and hence

$$\mathscr{E}(K) \subseteq \bigcap_{f \in S(K)} \partial_f(K).$$

These two inclusions establish the desired equality.

The final statement of the theorem follows by noting that if $f \in S(K)$ then $\mu(\partial_f(K)) = \|\mu\|$ and μ is maximal by Theorem 4.1.7.

Each of the boundary sets $\partial_f(K)$ is a Borel set, and even a G_δ, by Proposition 4.1.6, but $\mathscr{E}(K)$ is not necessarily Borel because it is the intersection of a nondenumerable family of $\partial_f(K)$. One can in fact construct examples in which $\mathscr{E}(K)$ is not a Borel set. Thus the sense in which a maximal measure is supported by $\mathscr{E}(K)$ is not immediately evident and it is at this point that the notion of a pseudosupporting set naturally enters.

Theorem 4.1.11. *Let K be a convex compact subset of a locally convex Hausdorff topological vector space. It follows that each maximal $\mu \in M_1(K)$ is pseudosupported by the extremal points $\mathscr{E}(K)$ of K. Therefore $\mu(B) = 1$ for every Baire set B containing $\mathscr{E}(K)$.*

If, moreover, K is metrizable then the Baire sets and the Borel sets coincide, $\mathscr{E}(K)$ is a G_δ-subset, and the following conditions are equivalent:

(1) *$\nu \in M_1(K)$ is maximal;*
(2) *$\nu(\mathscr{E}(K)) = 1$.*

The proof of the first statement relies on two lemmas which are of independent interest. In particular, the first of these establishes a maximum principle which will be reapplied in Chapter 6 in the discussion of variational principles for equilibrium states.

Lemma 4.1.12 (Bauer maximum principle). *Let f be a convex upper semicontinuous function over the compact convex set K. It follows that f attains its maximum at an extremal point of K.*

PROOF. Let

$$\alpha = \sup_{\omega \in K} f(\omega) < +\infty$$

and define

$$F = \{\omega; f(\omega) = \alpha\}.$$

It follows from compactness and semicontinuity that F is closed and nonempty. Moreover, F is stable in the sense that if $\mu \in M_\omega(K)$ and $\omega \in F$ then μ is supported by F. The set of closed subsets of F which are stable in K is directed when ordered with respect to inclusion. Thus by Zorn's lemma and compactness F must contain a minimal nonempty stable subset F_0. Now assume F_0 contains two distinct points ω_1 and ω_2. The Hahn–Banach theorem implies the existence of a continuous linear function g such that $g(\omega_1) > g(\omega_2)$. Let

$$G = \left\{\omega; g(\omega) = \sup_{\omega' \in F_0} g(\omega')\right\} \cap F_0$$

and then G is stable as a subset of F_0 and hence as a subset of K, since F_0 is stable in K. Since $\omega_2 \notin G$ this contradicts the minimality of F_0 and hence $F_0 = \{\omega_0\}$ for some point $\omega_0 \in K$. But because F_0 is stable one must have $\omega_0 \in \mathscr{E}(K)$.

Lemma 4.1.13. *Let μ be a maximal measure and $f_n \in C(K)$ a decreasing sequence of positive functions which converges pointwise to zero on $\mathscr{E}(K)$. It follows that*

$$\lim_{n\to\infty} \mu(f_n) = 0.$$

PROOF. Firstly, we argue that for each n there is a $g_n \in S(K)$ such that $g_n \le f_n$ and

$$\mu(f_n) - \mu(g_n) < \varepsilon.$$

This follows by remarking that $\mu(f_n) = -\mu(\overline{(-f_n)})$, by Theorem 4.1.7, and that

$$\mu(\overline{(-f_n)}) = \inf\{\mu(-g); g \in S(K), g \le f_n\}$$

by Proposition 4.1.6.

Secondly, we claim that the g_n can be chosen to form a decreasing sequence. We argue by induction. Choose g_{n+1} such that $0 \le g_{n+1} \le g_n \wedge f_{n+1}$ and such that

$$\mu(g_n \wedge f_{n+1}) - \mu(g_{n+1}) < \varepsilon + \mu(g_n) - \mu(f_n).$$

This is possible by the preceding argument. Now one has

$$\begin{aligned}\mu(g_n) + \mu(f_{n+1}) &= \mu(g_n \wedge f_{n+1}) + \mu(g_n \vee f_{n+1})\\ &\le \mu(g_n \wedge f_{n+1}) + \mu(f_n)\end{aligned}$$

and hence

$$\mu(f_{n+1}) - \mu(g_{n+1}) < \varepsilon.$$

Thirdly, suppose f_n converges to f and g_n converges to g. It follows that

$$0 \le \mu(f) < \mu(g) + \varepsilon.$$

Now $f = 0$ on $\mathscr{E}(K)$ by assumption and hence $g = 0$ on $\mathscr{E}(K)$. But g is convex and upper semicontinuous and hence by Lemma 4.1.12 one has $g = 0$ on K and

$$0 \le \mu(f) < \varepsilon.$$

Let us return to the proof of the theorem.

PROOF OF THEOREM 4.1.11. As μ is regular and the σ-algebra of Baire sets is generated by the compact G_δ-subsets of K it suffices to prove that $\mu(C) = 0$ for each compact G_δ such that $C \cap \mathscr{E}(K) = \varnothing$. But by Urysohn's lemma there is a bounded sequence of positive functions $f_n \in C(K)$ such that $f_n(\omega) = 1$ for $\omega \in C$ and such that

$$\lim_{n \to \infty} f_n(\omega) = 0$$

for $\omega \notin C$. Therefore

$$0 \le \mu(C) \le \lim_{n \to \infty} \mu(f_n) = 0$$

by Lemma 4.1.13.

Next assume K is metrizable and let C be any closed subset of K. Define C_n by

$$C_n = \left\{\omega;\, d(\omega, \omega') < \frac{1}{n} \quad \text{for some } \omega' \in C\right\},$$

where d is a metric on K. The C_n are open and

$$C = \bigcap_{n \ge 1} C_n,$$

i.e., C is a Baire set. Thus the Baire sets and the Borel sets coincide.

We will prove that $\mathscr{E}(K)$ is a Borel set by establishing the stronger result that $\mathscr{E}(K) = \partial_f(K)$ for a suitable $f \in S(K)$.

As K is metrizable it is automatically separable and replacing the metric d by $d/(1 + d)$, if necessary, one may assume that $d(\omega', \omega'') \le 1$ for all pairs $\omega', \omega'' \in K$. Now let $\{\omega_n\}$ be a dense sequence of points in K and define $f \in C(K)$ by

$$f(\omega) = \sum_{n \ge 1} 2^{-n} d(\omega_n, \omega)^2.$$

It follows that $f \in S(K)$ but for $\omega', \omega'' \in K$ and $\omega' \neq \omega''$ there is an n such that ω_n is closer to ω' than to ω'' and therefore $d(\omega_n, \omega)^2$, and hence f, is strictly convex on the segment (ω', ω''). The first calculation of the proof of Proposition 4.1.10 then gives

$$\bar{f}\left(\frac{\omega' + \omega''}{2}\right) > f\left(\frac{\omega' + \omega''}{2}\right).$$

This establishes that if $\omega \in K$ but $\omega \notin \mathscr{E}(K)$ then $\omega \notin \partial_f(K)$. Therefore

$$\mathscr{E}(K) \supseteq \partial_f(K).$$

But from Proposition 4.1.10 one then concludes that

$$\mathscr{E}(K) = \partial_f(K)$$

and $\mathscr{E}(K)$ is a G_δ-subset by Proposition 4.1.6.

Finally, (1) $\Rightarrow$ (2) by the first part of the theorem and (2) $\Rightarrow$ (1) is contained in Proposition 4.1.10.

Remark. One may obtain a weaker form of the second part of Theorem 4.1.11 for special subsets of K. A Borel subset F is called a stable face of K if $\omega \in F$ and $\mu \succ \delta_\omega$ imply that μ is supported by F. If the stable face F of K is metrizable and separable then the foregoing argument can be used to establish that $\mathscr{E}(F) = \partial_f(K) \cap F = \mathscr{E}(K) \cap F$ for a suitable $f \in S(K)$. One clearly has $\mathscr{E}(F) = \mathscr{E}(K) \cap F$ but if f is constructed as above with the aid of points $\omega_n \in F$ then the same argument shows that $\omega \in F$, but $\omega \notin \mathscr{E}(F)$ implies

$\omega \notin \partial_f(K)$. Therefore $\mathscr{E}(F) \supseteq \partial_f(K) \cap F$ and equality again follows from Proposition 4.1.10. Thus if $\mu \succ \delta_\omega$ for $\omega \in F$ then μ is maximal if, and only if, $\mu(\mathscr{E}(F)) = 1$. Similar considerations on stable faces will occur in Section 4.1.4.

Theorem 4.1.11 together with Proposition 4.1.3 establish that each $\omega \in K$ is the barycenter of a maximal probability measure and the maximal measures are pseudosupported by $\mathscr{E}(K)$. In the best cases, e.g., K metrizable, the maximal measures really are supported by $\mathscr{E}(K)$. We next derive conditions which ensure that each $\omega \in K$ is the barycenter of a unique maximal measure.

In order to formulate the uniqueness conditions it is convenient to assume that K is the base of a convex cone with apex at the origin.[2] This can always be arranged by replacing X by $\mathbb{R} \times X$, identifying K with $\{1\} \times K$, and taking C to be the cone generated by $\{1\} \times K$. Although there are many methods of realizing K as the base of a cone C all such cones are linearly isomorphic because each point of the cone is specified as $\lambda\omega$ with $\lambda \geq 0$ and $\omega \in K$. Thus the affine properties of C depend only on the affine properties of K and are independent of the method of embedding.

There is a natural ordering on the cone C defined by $\xi \geq \eta$ if, and only if, $\xi - \eta \in C$ and C is defined to be a *lattice* if it is a lattice with respect to this order. Moreover, the compact convex set K which forms the base of the cone C is defined to be a *simplex* whenever C is a lattice. Note that this definition of a simplex agrees exactly with the usual definition in finite dimensions. The ν-dimensional simplex $\{(\lambda_1, \lambda_2, \ldots, \lambda_{\nu+1}); \lambda_i \geq 0, \sum \lambda_i = 1\}$ is the base of the $\nu + 1$-dimensional cone $\{(\lambda_1, \lambda_2, \ldots, \lambda_{\nu+1}); \lambda_i \geq 0\}$. A more general example of a simplex is provided by the probability measures $M_1(K)$. This set is the base of the cone $C = M_+(K)$ and we noted at the beginning of the subsection that C is a lattice. A slightly less obvious example is provided by the following.

Proposition 4.1.14. *Let K be a convex compact subset of a locally convex Hausdorff space and define M to be the subset of $M_+(K)$ formed by the measures which are maximal with respect to the order $\succ$. It follows that M is a convex subcone of $M_+(K)$ and every sum, or integral, of measures in M is contained in M. Moreover, if $\mu \in M$ and $\nu \leq \mu$ then $\nu \in M$. Consequently, M is a lattice.*

PROOF. Theorem 4.1.7 established that a measure μ is maximal if, and only if, $\mu(\bar{f} - f) = 0$ for all $f \in S(K)$. It immediately follows that sums, and integrals, of maximal measures are maximal and that the maximal measures form a subcone of $M_+(K)$. But if $f \in S(K)$ then $\bar{f} - f \geq 0$ and $\nu \leq \mu$, together with $\mu(\bar{f} - f) = 0$, implies that $\nu(\bar{f} - f) = 0$, i.e., ν is maximal.

Now if $\mu \wedge \nu$ is the greatest lower bound of $\mu, \nu \in M$ one has $\mu \wedge \nu \leq \mu$ and hence $\mu \wedge \nu \in M$. Furthermore, if $\rho \in M_+(K)$ and $\rho \leq \mu$, $\rho \leq \nu$ then $\rho \in M$ and

[2] A base of a convex cone P with apex at the origin is a convex subset K of P such that $0 \notin K$ and each nonzero $\varphi \in P$ can be written uniquely as $\varphi = \lambda\omega$, where $\lambda > 0$ and $\omega \in K$.

because $\mu \wedge \nu - \rho \leq \mu \wedge \nu \in M$ one has $\mu \wedge \nu - \rho \in M$. Thus $\mu \wedge \nu$ is the greatest lower bound of μ and ν with respect to the natural order of the cone M. But $\mu \vee \nu = \mu + \nu - \mu \wedge \nu \leq \mu + \nu \in M$ and hence $\mu \vee \nu \in M$. Thus M is a lattice.

This proposition is of use in deriving the following characterization of uniqueness for barycentric decompositions by maximal measures.

Theorem 4.1.15. *Let K be a convex compact subset of a locally convex Hausdorff space. The following conditions are equivalent:*

(1) *each $\omega \in K$ is the barycenter of a unique maximal measure;*
(2) *K is a simplex;*
(3) *the upper envelope $\bar{f}$ of each $f \in S(K)$ is affine.*

PROOF. (3) $\Rightarrow$ (1) If $\mu \in M_\omega(K)$ we first show that

$$\bar{f}(\omega) = \mu(\bar{f}).$$

Let $\{\mu_\alpha\}$ be a net of measures in $M_\omega(K)$, each with finite support, such that μ_α converges to μ in the weak* topology. The existence of $\{\mu_\alpha\}$ is assured by Proposition 4.1.1. Now if $g \in -S(K)$ and $g \geq f$ then one has

$$\bar{f}(\omega) = \mu_\alpha(\bar{f}) \leq \mu_\alpha(g) \leq g(\omega),$$

where the first equality uses the affinity of $\bar{f}$. Hence

$$\bar{f}(\omega) \leq \mu(g) \leq g(\omega).$$

Taking the infimum over g and applying Proposition 4.1.6 one finds

$$\bar{f}(\omega) \leq \mu(\bar{f}) = \bar{f}(\omega).$$

Now assume that μ_1, μ_2 are maximal measures with barycenter ω. Thus

$$\mu_1(\bar{f}) = \mu_1(f), \qquad \mu_2(\bar{f}) = \mu_2(f)$$

for all $f \in S(K)$ and it follows that

$$\mu_1(f) = \mu_1(\bar{f}) = \bar{f}(\omega) = \mu_2(\bar{f}) = \mu_2(f).$$

Thus $\mu_1(f) = \mu_2(f)$ for all $f \in S(K)$ or $\mu_1(g - h) = \mu_2(g - h)$ for all $g, h \in S(K)$. Therefore $\mu_1 = \mu_2$ by Lemma 4.1.4 and the maximal measure is unique.

(1) $\Rightarrow$ (2) The maximal measures M form a lattice by Proposition 4.1.14 and so the maximal measures $\mu \in M_1(K)$ form a simplex M_1. But the mapping $\mu \mapsto \omega = b(\mu)$ is a linear isomorphism from M_1 to K. Thus K is a simplex.

(2) $\Rightarrow$ (3) To prove this last implication it is convenient to extend each $f \in C(K)$ to the cone C, with base K, by homogeneity, i.e., one sets $f(\lambda\omega) = \lambda f(\omega)$ for $\omega \in K$, $\lambda \geq 0$. Note that convexity of f then corresponds to subadditivity

$$f(x + y) \leq f(x) + f(y), \qquad x, y \in C,$$

and concavity to superadditivity,

$$f(x + y) \geq f(x) + f(y), \qquad x, y \in C.$$

Now if $f \in S(K)$ then $\bar{f}$ is concave and its extension is automatically superadditive. Thus to deduce that $\bar{f}$ is affine it suffices to prove that its extension is subadditive and hence linear. This proof relies upon the following decomposition property of a lattice.

Lemma 4.1.16. *Let C be a cone which is a lattice with respect to its natural order. Let $x, y \in C$ and assume*

$$x + y = \sum_{i=1}^{n} z_i$$

where $z_1, z_2, \ldots, z_n \in C$. It follows that there exists $x_i, y_i \in C$ such that $x_i + y_i = z_i$ for $i = 1, 2, \ldots, n$, and

$$x = \sum_{i=1}^{n} x_i, \qquad y = \sum_{i=1}^{n} y_i.$$

Proof. Assume to start with that $n = 2$ and define x_i, y_i by

$$x_1 = z_1 \wedge x, \qquad y_1 = z_1 - z_1 \wedge x,$$
$$x_2 = x - z_1 \wedge x, \qquad y_2 = z_2 - x + z_1 \wedge x.$$

It is easy to check all the desired relations with the possible exception of $y_2 \in C$. But this follows by noting that the translate of a least upper bound is the least upper bound of the translates. Explicitly,

$$\begin{aligned} y_2 &= z_1 \wedge x + (z_2 - x) \\ &= (z_1 + (z_2 - x)) \wedge (x + (z_2 - x)) \\ &= y \wedge z_2 \in C. \end{aligned}$$

Now the general case $n > 2$ is completed by an iterative argument. One applies the $n = 2$ case to

$$x + y = z_1 + \sum_{i=2}^{n} z_i$$

to obtain $z_1 = x_1 + y_1$ and

$$\sum_{i=2}^{n} z_i = t + t',$$

where $x_1 + t = x$ and $y_1 + t' = y$. Then one reapplies the special case to

$$t + t' = z_2 + \sum_{i=3}^{n} z_i.$$

After $n - 1$ steps one obtains the general result.

Now let us return to the proof of (2) $\Rightarrow$ (3) in Theorem 4.1.15. By Lemma 4.1.9 one has, for $f \in C(K)$,

$$\bar{f}(x + y) = \sup\left\{\sum_{i=1}^{n} f(z_i);\, z_i \in C,\, \sum_{i=1}^{n} z_i = x + y\right\}$$

and by Lemma 4.1.16

$$\bar{f}(x + y) = \sup\left\{\sum_{i=1}^{n} f(x_i + y_i);\, x_i, y_i \in C,\, \sum_{i=1}^{n} x_i = x,\, \sum_{i=1}^{n} y_i = y\right\}.$$

Specializing to $f \in S(K)$ the convexity gives

$$\begin{aligned}\bar{f}(x + y) &\leq \sup\left\{\sum_{i=1}^{n} f(x_i);\, x_i \in C,\, \sum_{i=1}^{n} x_i = x\right\}\\ &\quad + \sup\left\{\sum_{i=1}^{n} f(y_i);\, y_i \in C,\, \sum_{i=1}^{n} y_i = y\right\}\\ &= \bar{f}(x) + \bar{f}(y).\end{aligned}$$

Thus $\bar{f}$ is both subadditive and superadditive on C, i.e., $\bar{f}$ is additive on C and affine on K.

There are other characterizations of the uniqueness of the barycentric decomposition which are not covered by Theorem 4.1.15 and which are useful in applications. A simple and useful example is given by the following:

Corollary 4.1.17. *Let K be a convex compact subset of a locally convex topological Hausdorff vector space. The following conditions are equivalent:*

(1) *each $\omega \in K$ is the barycenter of a unique maximal measure μ_ω;*
(2) *there exists an affine mapping $\omega \in K \mapsto \nu_\omega \in M_1(K)$ from $\omega \in K$ to a measure ν_ω with barycenter ω.*

If these conditions are satisfied then $\mu_\omega = \nu_\omega$.

PROOF. (1) $\Rightarrow$ (2) If $\omega_1, \omega_2 \in K$ and $0 \leq \lambda \leq 1$ then $\lambda\mu_{\omega_1} + (1 - \lambda)\mu_{\omega_2}$ is maximal because the maximal measures form a cone (Proposition 4.1.14). But

$$b(\lambda\mu_{\omega_1} + (1 - \lambda)\mu_{\omega_2}) = \lambda b(\mu_{\omega_1}) + (1 - \lambda)b(\mu_{\omega_2}) = \lambda\omega_1 + (1 - \lambda)\omega_2.$$

By uniqueness $\lambda\mu_{\omega_1} + (1 - \lambda)\mu_{\omega_2} = \mu_{\lambda\omega_1 + (1-\lambda)\omega_2}$ and the mapping $\omega \mapsto \mu_\omega$ is affine.

(2) $\Rightarrow$ (1) To deduce that ν_ω is a unique maximal measure in $M_\omega(K)$, we have to prove that if $\nu \in M_\omega(K)$ then $\nu \prec \nu_\omega$. First remark that ν may be approximated in the weak* topology by a net of measures with finite support, by Proposition 4.1.1. Now if

$$\nu^\alpha = \sum_{i=1}^{n_\alpha} \lambda_i^\alpha \delta_{\omega_i^\alpha}$$

with $\lambda_i^\alpha \geq 0$, and if $f \in S(K)$, then

$$\begin{aligned}\nu^\alpha(f) &= \sum_{i=1}^{n_\alpha} \lambda_i^\alpha f(\omega_i^\alpha)\\ &\leq \sum_{i=1}^{n_\alpha} \lambda_i^\alpha \nu_{\omega_i^\alpha}(f)\end{aligned}$$

because $\nu_\omega \succ \delta_\omega$, by Proposition 4.1.3. But by affinity of the map $\omega \mapsto \nu_\omega$ one concludes that

$$\nu^\alpha(f) \leq \nu_{\sum_{i=1}^{n_\alpha} \lambda_i^\alpha \omega_i^\alpha}(f) = \nu_\omega(f).$$

Finally, by limiting, $\nu(f) \leq \nu_\omega(f)$, i.e., $\nu \prec \nu_\omega$.

Remark. It should perhaps be emphasized that the problem of deciding the sense in which a maximal measure is supported by $\mathscr{E}(K)$ is not simplified by the assumption that K is a simplex. In fact, it is possible to construct a simplex K with a point ω such that

(1) $\mathscr{E}(K)$ is a Borel subset of K;
(2) the unique maximal measure $\mu_\omega \in M_\omega(K)$ satisfies $\mu_\omega(\mathscr{E}(K)) = 0$.

Even if the unique maximal measure μ_ω is supported by $\mathscr{E}(K)$ this does not mean that there cannot exist other measures $\mu \in M_\omega(K)$ which are pseudo-supported by $\mathscr{E}(K)$ but such that $\mu \neq \mu_\omega$. One can indeed find a simplex K and point ω with the following properties:

(1) $\mathscr{E}(K)$ is a Borel subset of K;
(2) the unique maximal measure $\mu_\omega \in M_\omega(K)$ satisfies $\mu_\omega(\mathscr{E}(K)) = 1$;
(3) there exists a $\mu \in M_\omega(K)$ such that $\mu(\mathscr{E}(K)) = 0$ and $\mu(B) = 0$ for every G_δ-subset B with $B \cap \mathscr{E}(K) = \emptyset$.

The type of decomposition that we have analyzed in this subsection assures, by definition, a representation of the form

$$f(b(\mu)) = \int d\mu(\omega) f(\omega)$$

for all $f \in A(K)$, where $\mu \in M_1(K)$. It is natural to ask to what extent the continuity properties of f may be relaxed without affecting this representation. We conclude this subsection by remarking that the barycentric representation remains valid for affine semicontinuous functions, or for sums and differences of such functions.

Corollary 4.1.18. *Let K be a convex compact subset of a locally convex topological Hausdorff vector space and f an affine upper semicontinuous function over K. If $\mu \in M_1(K)$ has barycenter $b(\mu)$ then*

$$f(b(\mu)) = \int d\mu(\omega)\, f(\omega).$$

PROOF. Firstly, remark that f is equal to its own upper envelope:

$$f(\omega) = \inf\{g(\omega);\ -g \in S(K),\ g \geq f\}.$$

This fact is established by the argument used to prove Lemma 4.1.8. Secondly, by use of the same approximation technique used to prove (3) $\Rightarrow$ (1) in Theorem 4.1.15 one has

$$f(b(\mu)) = \mu_\alpha(f) \leq \mu_\alpha(g) \leq g(b(\mu))$$

for all $g \in -S(K)$ with $g \geq f$. Hence

$$f(b(\mu)) \leq \mu(g) \leq g(b(\mu)).$$

Taking the infimum over g and reapplying Proposition 4.1.6 then gives $f(b(\mu)) = \mu(f)$.

4.1.3. Orthogonal Measures

In the previous subsection we discussed the general theory of barycentric decomposition and in this subsection we consider certain aspects of this theory in application to the state spaces of C^*-algebras. Throughout the subsection $E_{\mathfrak{A}}$ will denote the convex weakly* compact set of states over the C^*-algebra $\mathfrak{A}$ with identity. Most of this section is devoted to the examination of a certain subclass of the set of measures $M_\omega(E_{\mathfrak{A}})$ with barycenter ω. These special measures, the orthogonal measures, were already briefly discussed in the introduction to this section. They are in one-to-one correspondence with the abelian von Neumann subalgebras of the commutant $\pi_\omega(\mathfrak{A})'$ of the cyclic representation associated with ω. Moreover, there is a one-to-one correspondence between these measures and the projections P on $\mathfrak{H}_\omega$ such that $P\Omega_\omega = \Omega_\omega$ and $P\pi_\omega(\mathfrak{A})P$ generates an abelian algebra on $P\mathfrak{H}_\omega$. The first main result of this subsection is the establishment of the correspondences and the second is to show that the correspondences respect the natural orderings of each of the three classes. In particular, maximal abelian subalgebras of $\pi_\omega(\mathfrak{A})'$ give rise to measures which are maximal among the measures $M_\omega(E_{\mathfrak{A}})$. This result will be established in Section 4.2.1. As a preliminary we prove in this section that the measures are maximal among the orthogonal measures with barycenter ω.

Throughout the subsection we adopt the standard algebraic notation introduced in Chapter 2 together with the extra notation introduced in the previous subsection. Moreover, we associate to each element of the C^*-algebra $\mathfrak{A}$ an affine continuous function $\hat{A}$ over the state space $E_{\mathfrak{A}}$ by the definition

$$\hat{A}(\omega) = \omega(A).$$

Note that the Hahn–Banach theorem implies that all affine continuous functions over $E_{\mathfrak{A}}$ have this form.

In order to introduce the notion of an orthogonal measure μ over $E_{\mathfrak{A}}$ it is first necessary to examine the concept of orthogonal states. There are various possibilities of definition which are equated by the following result.

Lemma 4.1.19. *Let ω_1, ω_2 be positive linear functionals over the C^*-algebra $\mathfrak{A}$ and let $\omega = \omega_1 + \omega_2$. The following conditions are equivalent:*

(1) *if ω' is a positive linear functional over $\mathfrak{A}$ satisfying $\omega' \leq \omega_1$ and $\omega' \leq \omega_2$ then $\omega' = 0$;*

(2) *there is a projection $P \in \pi_\omega(\mathfrak{A})'$ such that*

$$\omega_1(A) = (P\Omega_\omega, \pi_\omega(A)\Omega_\omega), \qquad \omega_2(A) = ((\mathbb{1} - P)\Omega_\omega, \pi_\omega(A)\Omega_\omega);$$

(3) *the representation associated with ω is a direct sum of the representations associated with ω_1 and ω_2,*

$$\mathfrak{H}_\omega = \mathfrak{H}_{\omega_1} \oplus \mathfrak{H}_{\omega_2}, \qquad \pi_\omega = \pi_{\omega_1} \oplus \pi_{\omega_2}, \qquad \Omega_\omega = \Omega_{\omega_1} \oplus \Omega_{\omega_2}.$$

PROOF. (1) ⇒ (2) By Theorem 2.3.19 there exists a T such that $0 \leq T \leq \mathbb{1}$, $T \in \pi_\omega(\mathfrak{A})'$, and

$$\omega_1(A) = (T\Omega_\omega, \pi_\omega(A)\Omega_\omega).$$

Introduce $T' = T(\mathbb{1} - T)$; then $0 \leq T' \in \pi_\omega(\mathfrak{A})'$ and the linear functional ω' defined by

$$\omega'(A) = (T(\mathbb{1} - T)\Omega_\omega, \pi_\omega(A)\Omega_\omega)$$

is positive. But $T' \leq T$ and $T' \leq \mathbb{1} - T$. Hence $\omega' \leq \omega_1$, $\omega' \leq \omega_2$, and condition (1) together with cyclicity of Ω_ω imply that $T(\mathbb{1} - T) = 0$, i.e., T is a projection.

(2) ⇒ (1) Theorem 2.3.19 again ensures the existence of a positive $T' \in \pi_\omega(\mathfrak{A})'$ such that

$$\omega'(A) = (T'\Omega_\omega, \pi_\omega(A)\Omega_\omega).$$

Now if $\omega' \leq \omega_1$ and $\omega' \leq \omega_2$ then $T' \leq P$ and $T' \leq \mathbb{1} - P$ But then one has $0 \leq PT'P \leq P(\mathbb{1} - P)P = 0$ and $0 \leq (\mathbb{1} - P)T'(\mathbb{1} - P) \leq (\mathbb{1} - P)P(\mathbb{1} - P) = 0$. Hence $(T')^{1/2}P = 0 = (T')^{1/2}(\mathbb{1} - P)$ and so $(T')^{1/2} = 0$. Thus $\omega' = 0$ and condition (1) is satisfied.

(2) ⇒ (3) If one sets $(\mathfrak{H}_1, \pi_1, \Omega_1) = (P\mathfrak{H}_\omega, P\pi_\omega, P\Omega_\omega)$ then it follows from the uniqueness statement of Theorem 2.3.16 that $(\mathfrak{H}_1, \pi_1, \Omega_1) \simeq (\mathfrak{H}_{\omega_1}, \pi_{\omega_1}, \Omega_{\omega 1})$. Similarly, $(\mathfrak{H}_{\omega_2}, \pi_{\omega_2}, \Omega_{\omega_2}) \simeq ((\mathbb{1} - P)\mathfrak{H}_\omega, (\mathbb{1} - P)\pi_\omega, (\mathbb{1} - P)\Omega_\omega)$.

(3) ⇒ (2) Let P be the orthogonal projector on $\mathfrak{H}_\omega$ with range $\mathfrak{H}_{\omega_1}$. It follows immediately that $P \in \pi_\omega(\mathfrak{A})'$ and ω_1, ω_2 are given in the correct relationship to ω.

Each of the conditions of Lemma 4.1.19 indicates a certain independence of the functionals ω_1 and ω_2 which we take as our definition of orthogonality.

Definition 4.1.20. If ω_1, ω_2 are positive linear functionals over $\mathfrak{A}$ which satisfy any of the three equivalent conditions of Lemma 4.1.19 then they are said to be *orthogonal* and we write $\omega_1 \perp \omega_2$.

If μ is a positive regular Borel measure on $E_\mathfrak{A}$ and for any Borel set $S \subseteq E_\mathfrak{A}$ one has

$$\left(\int_S d\mu(\omega')\omega'\right) \perp \left(\int_{E_\mathfrak{A}\setminus S} d\mu(\omega')\,\omega'\right)$$

then μ is defined to be an *orthogonal measure* on $E_\mathfrak{A}$. The set of orthogonal probability measures on $E_\mathfrak{A}$ with barycenter ω is denoted by $\mathcal{O}_\omega(E_\mathfrak{A})$, or simply $\mathcal{O}_\omega$.

Theorem 2.3.19 established that if two positive linear functionals ω and ω_1, over $\mathfrak{A}$, satisfy $\omega_1 \leq \omega$ then there is a positive $T \in \pi_\omega(\mathfrak{A})'$ such that

$$\omega_1(A) = (T\Omega_\omega, \pi_\omega(A)\Omega_\omega)$$

for all $A \in \mathfrak{A}$. We next establish a continuous analogue of this result which provides the first essential tool in linking the orthogonal measures in $\mathcal{O}_\omega$ with abelian subalgebras of $\pi_\omega(\mathfrak{A})'$.

Lemma 4.1.21. *If $\mu \in M_{\omega}(E_{\mathfrak{A}})$ then there exists a unique linear map*

$$f \in L^{\infty}(\mu) \mapsto \kappa_{\mu}(f) \in \pi_{\omega}(\mathfrak{A})'$$

defined by

$$(\Omega_{\omega}, \kappa_{\mu}(f)\pi_{\omega}(A)\Omega_{\omega}) = \int d\mu(\omega') f(\omega')\omega'(A)$$

which is positive and contractive.

If $L^{\infty}(\mu)$ is equipped with the $\sigma(L^{\infty}(\mu), L^{1}(\mu))$-topology and $\pi_{\omega}(\mathfrak{A})'$ with the weak operator topology then the map $f \mapsto \kappa_{\mu}(f)$ is continuous.

PROOF. For each $f \in L^{\infty}(\mu)$ define the linear functional b_f over $\mathfrak{A}$ by

$$b_f(A) = \int d\mu(\omega') f(\omega')\omega'(A).$$

If $A \geq 0$ then one has

$$|b_f(A)| \leq \|f\|_{\infty} \int d\mu(\omega')\, \omega'(A) = \|f\|_{\infty}\, \omega(A).$$

Moreover, if $f \geq 0$ then b_f is positive. Thus for $f \geq 0$ Theorem 2.3.19 asserts the existence of a positive $\kappa_{\mu}(f) \in \pi_{\omega}(\mathfrak{A})'$ such that

$$b_f(A) = (\Omega_{\omega}, \kappa_{\mu}(f)\pi_{\omega}(A)\Omega_{\omega}), \qquad \|\kappa_{\mu}(f)\| \leq \|f\|_{\infty}.$$

But each $f \in L^{\infty}(\mu)$ is a linear combination of four positive elements of $L^{\infty}(\mu)$ and hence the existence of $\kappa_{\mu}(f)$ follows by superposition. The bound $\|\kappa_{\mu}(f)\| \leq \|f\|_{\infty}$ follows from the general estimate on b_f.

Next note that

$$(\pi_{\omega}(A)\Omega_{\omega}, \kappa_{\mu}(f)\pi_{\omega}(B)\Omega_{\omega}) = \int d\mu(\omega') f(\omega')\omega'(A^*B).$$

Since $\pi_{\omega}(\mathfrak{A})\Omega_{\omega}$ is dense in $\mathfrak{H}_{\omega}$ and $\|\kappa_{\mu}(f)\| \leq \|f\|_{\infty}$ it follows that $f \mapsto (\psi, \kappa_{\mu}(f)\varphi)$ is continuous in the $\sigma(L^{\infty}, L^{1})$- topology on the unit ball of $L^{\infty}(\mu)$ for all $\psi, \varphi \in \mathfrak{H}_{\omega}$. The general continuity property then follows from the Krein–Smulian theorem,

Now Lemma 4.1.19 provides three equivalent ways of describing the orthogonality of two functionals. The second of these is essentially a condition on the map introduced in Lemma 4.1.21 and the following proposition shows that orthogonality of a measure μ is in fact expressible by means of an algebraic condition on κ_{μ}.

Proposition 4.1.22 (Tomita's theorem). *Let μ be a nonnegative regular Borel measure on $E_{\mathfrak{A}}$. The following conditions are equivalent:*

(1) *μ is an orthogonal measure;*
(2) *the map $f \mapsto \kappa_{\mu}(f)$ is a *-isomorphism of $L^{\infty}(\mu)$ into $\pi_{\omega}(\mathfrak{A})'$;*
(3) *the map $f \mapsto \kappa_{\mu}(f)$ is a *-morphism.*

If these conditions are satisfied then

$$\mathfrak{B} = \{\kappa_\mu(f);\ f \in L^\infty(\mu)\}$$

is an abelian von Neumann subalgebra of $\pi_\omega(\mathfrak{A})'$.

PROOF. (1) ⇒ (2) Assume μ orthogonal. By Lemma 4.1.21 the map $f \mapsto \kappa_\mu(f)$ is linear and positive. If f is a projection there exists a Borel set $S \subseteq E_\mathfrak{A}$ such that $f = \chi_S$ is the characteristic function of S. Now by assumption

$$\int_S d\mu(\omega')\,\omega' + \int_{E_\mathfrak{A}\backslash S} d\mu(\omega')\,\omega' = \omega$$

and

$$\int_S d\mu(\omega')\,\omega' \perp \int_{E_\mathfrak{A}\backslash S} d\mu(\omega')\,\omega'.$$

Hence, by Lemma 4.1.19, $\kappa_\mu(f)$ is a projection. If f and g are orthogonal projections then $f \leq \mathbb{1} - g$. Hence $\kappa_\mu(f) \leq \mathbb{1} - \kappa_\mu(g)$ and $\kappa_\mu(f)\kappa_\mu(g) = 0$. Now if f and g are arbitrary projections in $L^\infty(\mu)$, then each of the pairs $\{f(\mathbb{1} - g), fg\}$, $\{fg, (\mathbb{1} - f)g\}$, and $\{f(\mathbb{1} - g), (\mathbb{1} - f)g\}$ is orthogonal. Thus it follows from the decompositions

$$f = fg + f(\mathbb{1} - g) \quad \text{and} \quad g = gf + g(\mathbb{1} - f)$$

that

$$\kappa_\mu(fg) = \kappa_\mu(f)\kappa_\mu(g).$$

Now any elements $f, g \in L^\infty(\mu)$ can be approximated in norm by linear combinations of projections, so by the estimate $\|\kappa_\mu(f)\| \leq \|f\|_\infty$ the relation

$$\kappa_\mu(fg) = \kappa_\mu(f)\kappa_\mu(g)$$

extends to all $f, g \in L^\infty(\mu)$, i.e., κ_μ is a *-morphism. Now we have

$$\begin{aligned}\|\kappa_\mu(f)\Omega_\omega\|^2 &= (\Omega_\omega \kappa_\mu(\bar{f}f)\Omega_\omega)\\ &= \int d\mu(\omega')\,|f(\omega')|^2\end{aligned}$$

for $f \in L^\infty(\mu)$, and this shows that κ_μ is faithful.

(2) ⇒ (3) Trivial.

(3) ⇒ (1) Assume κ_μ is a *-morphism. If S is a Borel set in $E_\mathfrak{A}$ we have that $\kappa_\mu(\chi_S)$ and $\kappa_\mu(\chi_{E_\mathfrak{A}\backslash S})$ are mutually orthogonal projections of sum $\mathbb{1}$; thus by Lemma 4.1.19,

$$\int_S d\mu(\omega')\,\omega' \perp \int_{E_\mathfrak{A}\backslash S} d\mu(\omega')\,\omega'$$

i.e., μ is orthogonal.

Finally, let B_α be a uniformly bounded weakly* convergent net of elements of $\mathfrak{B}$ with limit B. Since κ_μ is an isometry $\kappa_\mu^{-1}(B_\alpha)$ is uniformly bounded in $L^\infty(\mu)$ and hence it has a weakly* convergent subnet with limit f. But by continuity of κ_μ one must have $\kappa_\mu(f) = B$. Hence the unit sphere of $\mathfrak{B}$ is weakly* closed by Theorem 2.4.11.

The foregoing proposition allows us to partially isolate the orthogonal measures $\mathcal{O}_\omega(E_\mathfrak{A})$ among the measures $M_\omega(E_\mathfrak{A})$.

Corollary 4.1.23. *If $\mu \in \mathcal{O}_\omega(E_\mathfrak{A})$ then $\mu \in \mathcal{E}(M_\omega(E_\mathfrak{A}))$.*

PROOF. Assume $\mu \notin \mathcal{E}(M_\omega(E_\mathfrak{A}))$ then $\mu = (\mu_1 + \mu_2)/2$ with $\mu_1, \mu_2 \in M_\omega(E_\mathfrak{A})$ and $\mu_1 \neq \mu_2$. Because $0 < \mu_1 < 2\mu$ one must have $\mu_1 = h\mu$ with $0 < h \in L^\infty(\mu)$ and because $\mu_1 \neq \mu$ one has $\mathbb{1} - h > 0$. But then

$$\begin{aligned}(\Omega_\omega, \kappa_\mu(\mathbb{1} - h)\pi_\omega(A)\Omega_\omega) &= \int d\mu(\omega')(\mathbb{1} - h(\omega'))\hat{A}(\omega') \\ &= \mu(\hat{A}) - \mu_1(\hat{A}) = 0\end{aligned}$$

for all $A \in \mathfrak{A}$ and hence $\kappa_\mu(\mathbb{1} - h) = 0$. This implies, however, that $\mathbb{1} - h = 0$, which is a contradiction.

The measures which are extremal in $M_\omega(E_\mathfrak{A})$ are usually called *simplicial measures.* It is an easy exercise to show that a finite support measure $\mu = \sum \lambda_i \delta_{\omega_i} \in M_\omega(E_\mathfrak{A})$ is simplicial if and only if the states $\omega_1, \omega_2, \ldots, \omega_n$ are affinely independent. A more general characterization of simplicial measures along this line will be derived later (Lemma 4.2.3). Corollary 4.1.23 shows that orthogonal measures are simplicial, but the general relations between the notions of orthogonality, maximality, and simpliciality are rather weak. If there are two distinct maximal measures $\mu_1, \mu_2 \in M_\omega(E_\mathfrak{A})$ then $(\mu_1 + \mu_2)/2$ is maximal in $M_\omega(E_\mathfrak{A})$ by Proposition 4.1.14 but it is not simplicial. If, however, there is a unique maximal $\mu_\omega \in M_\omega(E_\mathfrak{A})$ then it is simplicial because $\nu \in M_\omega(E_\mathfrak{A})$ and $\lambda\nu \leq \mu_\omega$ implies that ν is maximal by another application of Proposition 4.1.14. In Section 4.2.1 we will demonstrate that the uniqueness of the maximal measure μ_ω implies that it is orthogonal and hence simplicial. Next note that if $\omega \notin \mathcal{E}(E_\mathfrak{A})$ then the point measure δ_ω is orthogonal, simplicial, but not maximal. Finally we give an example of a simplicial measure which is maximal but not orthogonal.

EXAMPLE 4.1.24. Let M_2 be the algebra of 2×2 matrices on a two-dimensional Hilbert space $\mathfrak{H}_2$ and choose three unit vectors ψ_1, ψ_2, ψ_3 such that each pair of these vectors is linearly independent. Define states $\omega_1, \omega_2, \omega_3$ on M_2 by $\omega_i(A) = (\psi_i, A\psi_i)$ and a measure μ with barycenter $\omega = (\omega_1 + \omega_2 + \omega_3)/3$ by

$$\mu = \frac{\delta_{\omega_1} + \delta_{\omega_2} + \delta_{\omega_3}}{3}.$$

It follows immediately that μ is not orthogonal but Proposition 4.1.10 implies that it is maximal; it is supported by the pure states $\mathcal{E}(M_2)$ over M_2. Moreover, $\mu \in \mathcal{E}(M_\omega(E_{M_2}))$ because the contrary assumption would imply the existence of a $\mu_1 \in M_\omega(E_{M_2})$ such that $0 < \mu_1 < 2\mu$, which is impossible since the states $\{\omega_1, \omega_2, \omega_3\}$ are affinely independent.

After these preliminaries we now come to the main characterization of orthogonal measures in terms of abelian algebras and projections.

Theorem 4.1.25. *Let $\mathfrak{A}$ be a C^*-algebra with identity and ω a state over $\mathfrak{A}$. There is a one-to-one correspondence between the following three sets:*

(1) *the orthogonal measures $\mu \in M_\omega(E_\mathfrak{A})$;*
(2) *the abelian von Neumann subalgebras $\mathfrak{B} \subseteq \pi_\omega(\mathfrak{A})'$;*
(3) *the orthogonal projections P on $\mathfrak{H}_\omega$ such that*

$$P\Omega_\omega = \Omega_\omega, \qquad P\pi_\omega(\mathfrak{A})P \subseteq \{P\pi_\omega(\mathfrak{A})P\}'.$$

If μ, $\mathfrak{B}$, P are in correspondence one has the following relations:

(1) $\mathfrak{B} = \{\pi_\omega(\mathfrak{A}) \cup P\}'$;
(2) $P = [\mathfrak{B}\Omega_\omega]$;
(3) $\mu(\hat{A}_1\hat{A}_2 \cdots \hat{A}_n) = (\Omega_\omega, \pi_\omega(A_1)P\pi_\omega(A_2)P \cdots P\pi_\omega(A_n)\Omega_\omega)$;
(4) $\mathfrak{B}$ *is *-isomorphic to the range of the map* $f \in L^\infty(\mu) \mapsto \kappa_\mu(f) \in \pi_\omega(\mathfrak{A})'$ *defined by*

$$(\Omega_\omega, \kappa_\mu(f)\pi_\omega(A)\Omega_\omega) = \int d\mu(\omega')\, f(\omega')\hat{A}(\omega')$$

and for $A, B \in \mathfrak{A}$

$$\kappa_\mu(\hat{A})\pi_\omega(B)\Omega_\omega = \pi_\omega(B)P\pi_\omega(A)\Omega_\omega.$$

Remark. In addition to the four relations stated in the theorem there also exists a direct way for constructing μ when $\mathfrak{B}$ is given which will be described in Lemma 4.1.26. Moreover, the correspondence between the $\mathfrak{B}$ and P is valid even if $\mathfrak{A}$ does not have an identity because one can always adjoin an identity.

PROOF. Proposition 4.1.22 associates to each orthogonal measure the abelian von Neumann subalgebra $\mathfrak{B}$ of $\pi_\omega(\mathfrak{A})'$ given by

$$\mathfrak{B} = \{\kappa_\mu(f); f \in L^\infty(\mu)\}.$$

If $P = [\mathfrak{B}\Omega_\omega]$ and $B = B^* \in \mathfrak{B}$ then

$$BP = PBP = (PBP)^* = PB$$

and hence $\mathfrak{B} \subseteq P'$. Moreover, if $f \in L^\infty(\mu)$ and $A \in \mathfrak{A}$ then

$$\begin{aligned}(\Omega_\omega, \kappa_\mu(f)\kappa_\mu(\hat{A})\Omega_\omega) &= (\Omega_\omega, \kappa_\mu(f\hat{A})\Omega_\omega)\\ &= \mu(f\hat{A})\\ &= (\Omega_\omega, \kappa_\mu(f)\pi_\omega(A)\Omega_\omega).\end{aligned}$$

Therefore

$$\kappa_\mu(\hat{A})\Omega_\omega = P\pi_\omega(A)\Omega_\omega$$

and by iteration, using $\kappa_\mu(\hat{A})P = P\kappa_\mu(\hat{A})$, one has

$$\kappa_\mu(\hat{A}_1) \cdots \kappa_\mu(\hat{A}_n)\Omega_\omega = P\pi_\omega(A_1)P \cdots P\pi_\omega(A_n)\Omega_\omega.$$

This gives the relation

$$\begin{aligned}\mu(\hat{A}_1\hat{A}_2\cdots\hat{A}_n) &= (\Omega_\omega, \kappa_\mu(\hat{A}_1\hat{A}_2\cdots\hat{A}_n)\Omega_\omega) \\ &= (\Omega_\omega, \kappa_\mu(\hat{A}_1)\kappa_\mu(\hat{A}_2)\cdots\kappa_\mu(\hat{A}_n)\Omega_\omega) \\ &= (\Omega_\omega, \pi_\omega(A_1)P\pi_\omega(A_2)P\cdots P\pi_\omega(A_n)\Omega_\omega),\end{aligned}$$

i.e., properties (3) and (4) of the theorem are valid with the appropriate choice of P and $\mathfrak{B} \subseteq \{\pi_\omega(\mathfrak{A}) \cup P\}'$.

Next assume an abelian von Neumann subalgebra $\mathfrak{B} \subseteq \pi_\omega(\mathfrak{A})'$ is given. If $\mathfrak{N}$ is a finite-dimensional von Neumann subalgebra of $\mathfrak{B}$ then it is generated by a finite family $P_1, P_2, \ldots, P_n$ of mutually orthogonal projections and one may associate to $\mathfrak{N}$ an orthogonal measure $\mu_\mathfrak{N}$ by the definitions, already used in the introduction,

$$\lambda_i = (\Omega_\omega, P_i\Omega_\omega), \qquad \lambda_i\omega_i(A) = (P_i\Omega_\omega, \pi_\omega(A)\Omega_\omega),$$

$$\mu_\mathfrak{N} = \sum_{i=1}^{n} \lambda_i \delta_{\omega_i}.$$

Note that as Ω_ω is cyclic for $\pi_\omega(\mathfrak{A})$ it is separating for $\pi_\omega(\mathfrak{A})'$ by Proposition 2.5.3 and hence $\lambda_i > 0$. Now the calculation given in the introduction establishes that

$$\mu_\mathfrak{N}(\hat{A}_1\hat{A}_2\cdots\hat{A}_m) = (\Omega_\omega, \pi_\omega(A_1)P_\mathfrak{N}\pi_\omega(A_2)P_\mathfrak{N}\cdots P_\mathfrak{N}\pi_\omega(A_m)\Omega_\omega)$$

for all $A_1, A_2, \ldots, A_m \in \mathfrak{A}$, where $P_\mathfrak{N} = [\mathfrak{N}\Omega_\omega]$. Note that this formula implies that the $P_\mathfrak{N}\pi_\omega(A)P_\mathfrak{N}$ commute. We will establish a similar property for the $P_\mathfrak{B}\pi_\omega(A)P_\mathfrak{B}$ by an approximation argument involving the second statement of the following lemma.

Lemma 4.1.26. *Let $\mathfrak{A}$ be a C^*-algebra with identity, ω a state over $\mathfrak{A}$, $\mathfrak{B}$ an abelian von Neumann subalgebra of $\pi_\omega(\mathfrak{A})'$, and $\mathfrak{M}$, $\mathfrak{N}$ finite-dimensional von Neumann subalgebras of $\mathfrak{B}$. Let $\mu_\mathfrak{M}$, $\mu_\mathfrak{N}$ denote the associated orthogonal measures introduced above. If $\mathfrak{N} \subseteq \mathfrak{M}$ then it follows that $\mu_\mathfrak{N} \prec \mu_\mathfrak{M}$.*

The subalgebras $\mathfrak{N}$ form a directed set when ordered by inclusion and the net $\mu_\mathfrak{N}$ converges in the $\sigma(C(E_\mathfrak{A})^*, C(E_\mathfrak{A}))$-topology to an orthogonal measure $\mu_\mathfrak{B}$ such that

$$\mu_\mathfrak{B}(\hat{A}_1\hat{A}_2\cdots\hat{A}_n) = (\Omega_\omega, \pi_\omega(A_1)P_\mathfrak{B}\pi_\omega(A_2)P_\mathfrak{B}\cdots P_\mathfrak{B}\pi_\omega(A_n)\Omega_\omega) \qquad (*)$$

for all $A_1, A_2, \ldots, A_n \in \mathfrak{A}$, where $P_\mathfrak{B} = [\mathfrak{B}\Omega_\omega]$.

PROOF. The algebras $\mathfrak{N}$ and $\mathfrak{M}$ are generated by finite families $P_1, P_2, \ldots, P_n$ and $Q_1, Q_2, \ldots, Q_m$ of mutually orthogonal projections. If $\mathfrak{N} \subseteq \mathfrak{M}$ then each $P_i \in \mathfrak{N}$ has the form

$$P_i = \sum_{j=m_{i-1}+1}^{m_i} Q_j.$$

Setting $\lambda_i = (\Omega_\omega, P_i\Omega_\omega)$ and $\lambda_j' = (\Omega_\omega, Q_j\Omega_\omega)$ one has

$$\lambda_i = \sum_{j=m_{i-1}+1}^{m_i} \lambda_j'$$

and the discussion of the measures $\mu_{\mathfrak{N}}, \mu_{\mathfrak{M}}$ which precedes the lemma gives the representations

$$\mu_{\mathfrak{N}} = \sum_{i=1}^{n} \lambda_i \delta_{\omega_i}, \qquad \mu_{\mathfrak{M}} = \sum_{j=1}^{m} \lambda_j' \delta_{\omega_j'}$$

where the states ω_i and ω_j' are related by

$$\lambda_i \omega_i = \sum_{j=m_{i-1}+1}^{m_i} \lambda_j' \omega_j'.$$

Thus a simple calculation using convexity and the definition of the order relation $\prec$ establishes that $\mu_{\mathfrak{N}} \prec \mu_{\mathfrak{M}}$.

Next note that

$$\mu_{\mathfrak{N}}(\hat{A}_1 \hat{A}_2 \cdots \hat{A}_n) = (\Omega_\omega, \pi_\omega(A_1) P_{\mathfrak{N}} \pi_\omega(A_2) P_{\mathfrak{N}} \cdots P_{\mathfrak{N}} \pi_\omega(A_n) \Omega_\omega)$$

with $P_{\mathfrak{N}} = [\mathfrak{N}\Omega_\omega]$. But $P_{\mathfrak{N}}$ converges weakly, hence strongly, to $P_{\mathfrak{B}} = [\mathfrak{B}\Omega_\omega]$. Therefore

$$\lim_{\mathfrak{N}} \mu_{\mathfrak{N}}(\hat{A}_1 \hat{A}_2 \cdots \hat{A}_n) = (\Omega_\omega, \pi_\omega(A_1) P_{\mathfrak{B}} \pi_\omega(A_2) P_{\mathfrak{B}} \cdots P_{\mathfrak{B}} \pi_\omega(A_n) \Omega_\omega).$$

But the $\hat{A}_i$ separate points of $E_{\mathfrak{A}}$ and hence the polynomials of the $\hat{A}_i$ are norm dense in $C(E_{\mathfrak{A}})$ by the Stone–Weierstrass theorem. Thus $\mu_{\mathfrak{N}}$ converges in the weak* topology, i.e., the $\sigma(C(E_{\mathfrak{A}})^*, C(E_{\mathfrak{A}}))$-topology, and weak* compactness of $M_\omega(E_{\mathfrak{A}})$ ensures that the limit is a measure $\mu_{\mathfrak{B}} \in M_\omega(E_{\mathfrak{A}})$ which satisfies the required relations $(*)$. But because

$$\mu_{\mathfrak{B}}(\hat{A}_1 \hat{A}_2 \hat{A}_3 \hat{A}_4) = \mu_{\mathfrak{B}}(\hat{A}_1 \hat{A}_3 \hat{A}_2 \hat{A}_4)$$

one concludes that the $P_{\mathfrak{B}} \pi_\omega(A) P_{\mathfrak{B}}$ commute on $P_{\mathfrak{B}} \mathfrak{H}_\omega$.

Note that this establishes a correspondence between $\mathfrak{B} \subseteq \pi_\omega(\mathfrak{A})'$ and a projection $P_{\mathfrak{B}}$ of the type considered in the third set of Theorem 4.1.25. Next we argue that if P is an arbitrary projection of this type then the relations

$$\mu(\hat{A}_1 \hat{A}_2 \cdots \hat{A}_n) = (\Omega_\omega, \pi_\omega(A_1) P \pi_\omega(A_2) P \cdots P \pi_\omega(A_n) \Omega_\omega) \tag{$**$}$$

determine an orthogonal measure. This establishes the third correspondence of the theorem and simultaneously completes the proof of the lemma.

To show that $(**)$ consistently defines a continuous, positive linear functional μ on $C(E_{\mathfrak{A}})$, consider the Gelfand representation $C(K)$ of the abelian C^*-algebra generated by $P\pi_\omega(\mathfrak{A})P$. If $F(z_1, z_2, \ldots, z_n)$ is a polynomial, we have

$$\begin{aligned} |(\Omega_\omega, F(P\pi_\omega(A_1)P, \ldots, P\pi_\omega(A_n)P)\Omega_\omega)| &\le \|F(P\pi_\omega(A_1)P, \ldots, P\pi_\omega(A_n)P)\| \\ &= \sup_{\varphi \in K} |F((P\pi_\omega(A_1)P)(\varphi), \ldots, (P\pi_\omega(A_n)P)(\varphi))| \\ &\le \sup_{\varphi \in E_{\mathfrak{A}}} |F(\hat{A}_1(\varphi), \ldots, \hat{A}_n(\varphi))|. \end{aligned}$$

The last estimate is true since $A \mapsto (P\pi_\omega(A)P)(\varphi)$ is a state on $\mathfrak{A}$ for each $\varphi \in K$. The Stone–Weierstrass theorem now implies the existence of a linear function μ on $C(E_{\mathfrak{A}})$ such that

$$\mu(F(\hat{A}_1, \ldots, \hat{A}_n)) = (\Omega_\omega, F(P\pi_\omega(A_1)P, \ldots, P\pi_\omega(A_n)P)\Omega_\omega)$$

for polynomials F. Since $\mu(\hat{\mathbb{1}}) = (\Omega_\omega, P\Omega_\omega) = 1$ and $\|\mu\| \leq 1$, by the estimate above, μ is a positive Radon measure by Proposition 2.3.11. It remains to prove that μ is orthogonal.

First one remarks that if κ_μ is the map introduced in Lemma 4.1.21 then

$$\begin{aligned}\mu(\hat{A}_1\hat{A}_2) &= (\Omega_\omega, \pi_\omega(A_1)P\pi_\omega(A_2)\Omega_\omega)\\ &= (\Omega_\omega, \pi_\omega(A_1)\kappa_\mu(\hat{A}_2)\Omega_\omega)\end{aligned}$$

and so, by cyclicity,

$$\kappa_\mu(\hat{A})\Omega_\omega = P\pi_\omega(A)\Omega_\omega$$

for all $A \in \mathfrak{A}$. But then one has

$$\begin{aligned}(\Omega_\omega, \pi_\omega(A_1)\kappa_\mu(\hat{A}_2)\kappa_\mu(\hat{A}_3)\Omega_\omega) &= (\Omega_\omega, \pi_\omega(A_2)P\pi_\omega(A_1)P\pi_\omega(A_3)\Omega_\omega)\\ &= \mu(\hat{A}_1\hat{A}_2\hat{A}_3)\\ &= (\Omega_\omega, \pi_\omega(A_1)\kappa_\mu(\hat{A}_2\hat{A}_3)\Omega_\omega).\end{aligned}$$

Hence

$$\kappa_\mu(\hat{A}_2)\kappa_\mu(\hat{A}_3) = \kappa_\mu(\hat{A}_2\hat{A}_3)$$

and

$$\kappa_\mu(\hat{A}_2\hat{A}_3)\Omega_\omega = P\pi_\omega(A_2)P\pi_\omega(A_3)\Omega_\omega$$

for all $A_2, A_3 \in \mathfrak{A}$. Iterating this argument then yields

$$\kappa_\mu(\hat{A}_1\hat{A}_2\cdots\hat{A}_n) = \kappa_\mu(\hat{A}_1)\kappa_\mu(\hat{A}_2)\cdots\kappa_\mu(\hat{A}_n)$$

and consequently for any polynomial $\mathscr{P}$

$$\kappa_\mu(\mathscr{P}(\hat{A}_1, \hat{A}_2, \ldots, \hat{A}_n)) = \mathscr{P}(\kappa_\mu(\hat{A}_1), \kappa_\mu(\hat{A}_2), \ldots, \kappa_\mu(\hat{A}_n)).$$

But $\|\kappa_\mu(f)\| \leq \|f\|_\infty$ for all $f \in L^\infty(\mu)$ and thus another application of the Stone–Weierstrass theorem and the Kaplansky density theorem gives

$$\kappa_\mu(fg) = \kappa_\mu(f)\kappa_\mu(g)$$

for all $f, g \in L^\infty(\mu)$. Thus μ is orthogonal by Proposition 4.1.22.

End of proof of Theorem 4.1.25. Now if μ, $\mathfrak{B}$, P are elements of the three sets considered in the theorem we have defined successive correspondences $\mu \mapsto \mathfrak{B}_\mu$, $\mathfrak{B} \mapsto P_{\mathfrak{B}}$ and $P \mapsto \mu_P$. For consistency we must show that the initial and final measures in this chain of correspondences are identical. But by the argument at the beginning of the proof

$$\mu(\hat{A}_1\hat{A}_2\cdots\hat{A}_n) = (\Omega_\omega, \pi_\omega(A_1)P_{\mathfrak{B}_\mu}\pi_\omega(A_2)P_{\mathfrak{B}_\mu}\cdots P_{\mathfrak{B}_\mu}\pi_\omega(A_n)\Omega_\omega)$$

and from Lemma 4.1.26

$$\mu_{P_{\mathfrak{B}_\mu}}(\hat{A}_1\hat{A}_2\cdots\hat{A}_n) = (\Omega_\omega, \pi_\omega(A_1)P_{\mathfrak{B}_\mu}\pi_\omega(A_2)P_{\mathfrak{B}_\mu}\cdots P_{\mathfrak{B}_\mu}\pi_\omega(A_n)\Omega_\omega).$$

A final application of the Stone–Weierstrass theorem establishes the required equality and places the three sets in one-to-one correspondence.

The only relation in Theorem 4.1.25 which remains unproved is the first. We showed at the beginning of the proof that $\mathfrak{B} \subseteq \{\pi_\omega(\mathfrak{A}) \cup P\}'$. Moreover if $C \in \{\pi_\omega(\mathfrak{A}) \cup P\}'$ then

$$C\Omega_\omega = CP\Omega_\omega = PC\Omega_\omega \in [\mathfrak{B}\Omega_\omega].$$

Hence $C \in \mathfrak{B}$ and $\{\pi_\omega(\mathfrak{A}) \cup P\}' \subseteq \mathfrak{B}$ by the following lemma.

Lemma 4.1.27. *Let $\mathfrak{M}$ be a von Neumann algebra on a Hilbert space $\mathfrak{H}$ with a separating vector $\Omega \in \mathfrak{H}$. Let $\mathfrak{N}$ be an abelian von Neumann subalgebra of $\mathfrak{M}$. It follows that if $A \in \mathfrak{M}$ and $A\Omega \in [\mathfrak{N}\Omega]$ then $A \in \mathfrak{N}$.*

PROOF. If $P = [\mathfrak{N}\Omega]$ then $B \in \mathfrak{N} \mapsto BP$ is a *-isomorphism, and Ω is separating and cyclic for $\mathfrak{N}P$ on $[\mathfrak{N}\Omega]$. By Proposition 2.5.9 there exists a closed operator Q affiliated with $\mathfrak{N}$ such that $\Omega \in D(Q) = D(Q^*)$ and $A\Omega = Q\Omega$. Then, for any $A' \in \mathfrak{M}' \subseteq \mathfrak{N}'$ we have

$$AA'\Omega = A'A\Omega = A'Q\Omega = QA'\Omega,$$

i.e., $A|_{\mathfrak{M}'\Omega} = Q|_{\mathfrak{M}'\Omega}$. As Ω is cyclic for $\mathfrak{M}'$ (Proposition 2.5.3) and Q is closed it follows that Q is bounded and $A = Q$.

Theorem 4.1.25 places the set of orthogonal measures $\mathcal{O}_\omega(E_\mathfrak{A})$ in direct correspondence with the set of abelian von Neumann algebras $\mathfrak{B} \subseteq \pi_\omega(\mathfrak{A})'$ and the set of projections $P = [\mathfrak{B}\Omega_\omega]$. Each of these families has a natural ordering and we next show that the previous correspondences respect these orderings.

Theorem 4.1.28. *Let μ and ν be orthogonal measures on $E_\mathfrak{A}$ with barycenter ω and $\mathfrak{B}_\mu$, P_μ and $\mathfrak{B}_\nu$, P_ν, the abelian subalgebras of $\pi_\omega(\mathfrak{A})'$ and the orthogonal projections given by the correspondences of Theorem 4.1.25. The following conditions are equivalent:*

(1) $\mu \succ \nu$;
(2) $\mathfrak{B}_\mu \supseteq \mathfrak{B}_\nu$;
(3) $P_\mu \geq P_\nu$;
(4) $\mu(|\hat{A} - \mu(\hat{A})|^2) \geq \nu(|\hat{A} - \nu(\hat{A})|^2)$, $A \in \mathfrak{A}$.

PROOF. We will prove (1) ⇒ (4) ⇒ (3) ⇒ (2) ⇒ (1).

(1) ⇒ (4) This is evident because $\mu(\hat{A}) = \nu(\hat{A}) = \hat{A}(\omega)$ and $\hat{A}^*\hat{A}$ is convex.

(4) ⇒ (3) One has

$$\begin{aligned}\mu(\hat{A}^*\hat{A}) &= (\Omega_\omega, \pi_\omega(A^*)P_\mu\pi_\omega(A)\Omega_\omega)\\ &= (\pi_\omega(A)\Omega_\omega, P_\mu\pi_\omega(A)\Omega_\omega).\end{aligned}$$

Now $\mu(\hat{A}^*\hat{A}) \geq \nu(\hat{A}^*\hat{A})$ and hence

$$(\pi_\omega(A)\Omega_\omega, P_\mu\pi_\omega(A)\Omega_\omega) \geq (\pi_\omega(A)\Omega_\omega, P_\nu\pi_\omega(A)\Omega_\omega)$$

for all $A \in \mathfrak{A}$. Since Ω_ω is cyclic for $\pi_\omega(\mathfrak{A})$ it follows that $P_\mu \geq P_\nu$.

(3) ⇒ (2) Since Ω_ω is cyclic for $\pi_\omega(\mathfrak{A})$ it is separating for $\pi_\omega(\mathfrak{A})'$ by Proposition 2.5.3. It follows from Lemma 4.1.27 that

$$\mathfrak{B}_\mu = \{B; B \in \pi_\omega(\mathfrak{A})', B\Omega_\omega \in P_\mu\mathfrak{H}\}$$

and similarly for $\mathfrak{B}_\nu$. Hence $P_\mu \geq P_\nu$ implies $\mathfrak{B}_\mu \supseteq \mathfrak{B}_\nu$.

(2) ⇒ (1) By Lemma 4.1.26 the set of $\mu_\mathfrak{N}$ where $\mathfrak{N}$ ranges over the finite-dimensional subalgebras of $\mathfrak{B}_\mu$ forms an increasing net, with respect to the order $\prec$, with weak* limit μ. Therefore we may choose $\mathfrak{N} \subseteq \mathfrak{B}_\mu$ and $\mathfrak{M} \subseteq \mathfrak{B}_\nu$ such that $\mu_\mathfrak{N} \succ \mu_\mathfrak{M}$ and $\mu_\mathfrak{N} \to \mu$, $\mu_\mathfrak{M} \to \nu$. Thus by limiting one concludes that $\mu \succ \nu$.

Theorem 4.1.28 not only identifies the orderings of the sets considered in Theorem 4.1.25 but it also shows that the ordering $\prec$ on orthogonal measures is equivalent to the ordering of their root mean square deviations Δ_μ, Δ_ν, where

$$\Delta_\mu(A) = \sqrt{\mu(|\hat{A} - \mu(\hat{A})|^2)}.$$

The set Δ_μ gives a natural measure of the distribution of μ about its mean values $\mu(\hat{A}) = \omega(A)$; the larger measure has the larger spread. As we are considering measures μ, ν with a fixed barycenter, $\mu(\hat{A}) = \omega(A) = \nu(\hat{A})$, the ordering $\Delta_\mu(A) \geq \Delta_\nu(A)$, $A \in \mathfrak{A}$, is equivalent to

$$\mu(\hat{A}^*\hat{A}) \geq \nu(\hat{A}^*\hat{A})$$

for all $A \in \mathfrak{A}$. Further, if $A = A_1 + iA_2$ with A_1, A_2 selfadjoint then

$$\mu(\hat{A}^*\hat{A}) = \mu(\hat{A}_1{}^2) + \mu(\hat{A}_2{}^2)$$

and hence the ordering is equivalent to

$$\mu(\hat{A}^2) \geq \nu(\hat{A}^2)$$

for all $A = A^* \in \mathfrak{A}$. As the set $\{\hat{A};\, A = A^* \in \mathfrak{A}\}$ is exactly equal to the set of real affine continuous functions over $E_\mathfrak{A}$ it is natural to ask whether the ordering $\succ$ is always characterized in this last manner. But this is not the case as will be seen in Example 4.1.29. Thus the coincidence of the ordering $\succ$ with the ordering of the root mean square deviations is a special property of orthogonal measures.

EXAMPLE 4.1.29. Let K be a hexagon formed by superposition of two equilateral triangles placed symmetrically in the manner shown in Figure 1. If μ is the probability measure concentrated on the points A, B, C with $\mu(A) = \mu(B) = \mu(C)$ and ν is the probability measure concentrated on D, E, F with $\nu(D) = \nu(E) = \nu(F)$ then μ and ν have the midpoint O of the equilateral triangles as common barycenter. If the distance from the midpoint of the segment BC to the vertex F is sufficiently small then one calculates that $\mu(f^2) \geq \nu(f^2)$ for all $f \in A(K)$. Nevertheless, μ and ν are not comparable in the order $\succ$ because they are both maximal.

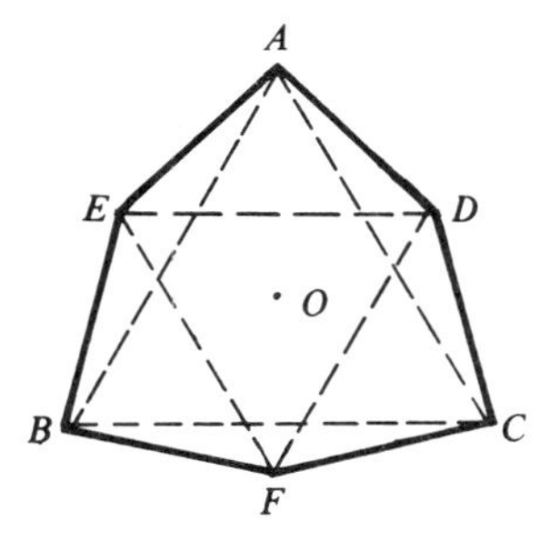

Figure 1

Finally, we remark that Theorem 4.1.28 also has the obvious implication that μ is maximal among the orthogonal measures $\mathcal{O}_\omega(E_\mathfrak{A})$ if, and only if, the corresponding abelian von Neumann algebra $\mathfrak{B} \subseteq \pi_\omega(\mathfrak{A})'$ is maximal abelian. In Section 4.2.1 we will establish that the measures μ corresponding to maximal abelian von Neumann algebras $\mathfrak{B} \subseteq \pi_\omega(\mathfrak{A})'$ are in fact maximal in $M_\omega(E_\mathfrak{A})$.

4.1.4. Borel Structure of States

The discussion, in Section 4.1.2, of barycentric decomposition of points in a convex compact set K illustrated that the measure-theoretic properties of the extremal points $\mathscr{E}(K)$, of K, are of fundamental importance. In this subsection we discuss some of the measure-theoretic properties of the state space $E_\mathfrak{A}$, of a C^*-algebra $\mathfrak{A}$ with identity and its extremal points $\mathscr{E}(E_\mathfrak{A})$. The states $E_\mathfrak{A}$ form a convex weakly* compact subset of the dual $\mathfrak{A}^*$, of $\mathfrak{A}$, and the extremal points of $E_\mathfrak{A}$ are just the pure states over $\mathfrak{A}$.

First remark that if $\mathfrak{A}$ is separable then $E_\mathfrak{A}$ is metrizable. In fact, if $\{A_n\}_{n\geq 1}$ is a uniformly dense sequence in the unit sphere of $\mathfrak{A}$ one can define a metric d on $E_\mathfrak{A}$ by

$$d(\omega_1, \omega_2) = \sum_{n\geq 1} \frac{|(\omega_1 - \omega_2)(A_n)|}{2^n}$$

Since $E_\mathfrak{A}$ is a uniformly bounded subset of $\mathfrak{A}^*$ it follows immediately that the weak* topology on $E_\mathfrak{A}$ is equal to the topology defined by this metric. Consequently, one may conclude from Theorem 4.1.11 that the pure states $\mathscr{E}(E_\mathfrak{A})$ are a G_δ-subset of $E_\mathfrak{A}$.

For a general C^*-algebra $\mathfrak{A}$ the set of pure states may be as pathological as the set of extremal points of a general convex compact set. We first give some examples. The first example illustrates a situation in which the pure states have the advantageous property of being closed.

EXAMPLE 4.1.30. Let $\mathfrak{A}$ be abelian. Then $\mathscr{E}(E_\mathfrak{A})$ is the set of characters on $\mathfrak{A}$ by Proposition 2.3.27. Hence $\mathscr{E}(E_\mathfrak{A})$ is a closed subset of $E_\mathfrak{A}$ and, in particular, the extremal points form a Borel set. Although $\mathscr{E}(E_\mathfrak{A})$ need not be a Baire set, the Riesz representation theorem applied to $\mathfrak{A} = C(\mathscr{E}(E_\mathfrak{A}))$ gives a unique decomposition of any state ω over $\mathfrak{A}$ into pure states

$$\omega(A) = \int d\mu_0(\omega')\, \hat{A}(\omega').$$

Since $\mathscr{E}(E_\mathfrak{A})$ is a Borel set one can identify the unique regular maximal Borel measure μ in $M_\omega(E_\mathfrak{A})$ with the Riesz measure through

$$\mu(B) = \mu_0(B \cap \mathscr{E}(E_\mathfrak{A}))$$

for all Borel sets $B \subseteq E_\mathfrak{A}$.

The foregoing example has a generalization. A C^*-algebra is called n-dimensionally homogeneous if all its irreducible representations are n-dimensional. For example, a C^*-algebra is abelian if, and only if, it is one-dimensionally homogeneous. An n-dimensional homogeneous C^*-algebra is locally of the form $M_n \otimes C(K) = C(M_n, K)$ for compact sets K where M_n is the full matrix algebra of $n \times n$ matrices. The set of pure states of an n-dimensionally homogeneous C^*-algebra $\mathfrak{A}$ is closed in $E_{\mathfrak{A}}$ and, conversely, it is almost true that if $\mathscr{E}(E_{\mathfrak{A}})$ is closed in $E_{\mathfrak{A}}$ then $\mathfrak{A}$ is a finite direct sum of n-dimensionally homogeneous C^*-algebras. The qualification almost is necessary because of examples of the type

$$\mathfrak{A} = \left\{f\,;\, f \in C(M_2; [0, 1]),\, f(0) \in \mathbb{C}\begin{pmatrix} 1 & 0 \\ 0 & 1 \end{pmatrix}\right\}.$$

It is true, however, that if $\mathscr{E}(E_{\mathfrak{A}})$ is compact and the subset of $\mathscr{E}(E_{\mathfrak{A}})$ consisting of characters is open in $\mathscr{E}(E_{\mathfrak{A}})$, then $\mathfrak{A}$ is a finite direct sum of dimensionally homogeneous C^*-algebras (see Notes and Remarks).

The next example gives a further illustration of the fact that the pure states do not have to be closed and, in fact, they can be dense within the state space.

EXAMPLE 4.1.31. Let $\mathfrak{A}$ be a UHF algebra (see Example 2.6.12 and the subsequent remark). Thus $\mathfrak{A}$ has the form

$$\mathfrak{A} = \overline{\bigcup_{n \geq 1} \mathfrak{A}_n},$$

where $\{\mathfrak{A}_n\}_{n \geq 1}$ is an increasing sequence of full-matrix algebras all with the same identity. We claim that the pure states $\mathscr{E}(E_{\mathfrak{A}})$ are weakly* dense in $E_{\mathfrak{A}}$. To prove the density it suffices to show that the restriction of a given state ω to any one of the $\mathfrak{A}_n$ has an extension to a pure state on $\mathfrak{A}$.

If $\mathfrak{A}_n = M_{[n]} =$ the algebra of $[n] \times [n]$ matrices, there exists by Theorem 2.4.21 a positive operator $\rho \in M_{[n]}$ with $\mathrm{Tr}(\rho) = 1$ such that

$$\omega(A) = \mathrm{Tr}(\rho A), \qquad A \in M_{[n]}.$$

Choose $m > n$ so large that $[m] \geq [n]^2$. Since $M_{[m]}$ and $M_{[n]}$ have the same identity $[m]/[n]$ must be an integer and $M_{[m]} = M_{[m]/[n]} \otimes M_{[n]}$. Let $\xi_1, \ldots, \xi_{[n]}$ be an orthonormal basis for $l^2(1, \ldots, [n])$ such that $\rho\xi_i = \lambda_i \xi_i$ and $\eta_1, \ldots, \eta_{[m]/[n]}$ an orthonormal basis for $l^2(1, \ldots, [m]/[n])$. Then

$$M_{[m]} = \mathscr{L}(l^2(1, \ldots, [n])) \otimes \mathscr{L}(l^2(1, \ldots, [m]/[n])).$$

Consider the vector state ω_1 of $M_{[m]}$ defined by the vector

$$\Omega = \sum_{i=1}^{[n]} \lambda_i^{1/2} \xi_i \otimes \eta_i.$$

For $A = A \otimes 1 \in M_{[n]}$ we have

$$\begin{aligned}\omega_1(A) &= \sum_{i=1}^{[n]} \sum_{j=1}^{[n]} \lambda_i^{1/2}\lambda_j^{1/2}(\xi_i \otimes \eta_i, A\xi_j \otimes \eta_j)\\ &= \sum_{i=1}^{[n]} \lambda_i(\xi_i, A\xi_i)\\ &= \mathrm{Tr}(\rho A) = \omega(A).\end{aligned}$$

Hence ω_1 is an extension of ω to $M_{[m]} \subseteq \mathfrak{A}$. But ω_1 is a pure state over $M_{[m]}$ and it then has an extension to a pure state on $\mathfrak{A}$ by Proposition 2.3.24.

The foregoing example is not atypical. In the proof of Proposition 2.6.15 we saw that if a C^*-algebra $\mathfrak{A}$ has a faithful, irreducible representation π on a Hilbert space $\mathfrak{H}$ such that

$$\mathscr{L}\mathscr{C}(\mathfrak{H}) \cap \pi(\mathfrak{A}) = \{0\}$$

then the vector states in this representation are dense in $E_{\mathfrak{A}}$. Hence $\mathscr{E}(E_{\mathfrak{A}})$ is dense in $E_{\mathfrak{A}}$. More generally, $\mathscr{E}(E_{\mathfrak{A}})$ is dense in $E_{\mathfrak{A}}$ if no irreducible representation of $\mathfrak{A}$ contains the compact operators on the representation space. Algebras of this type are called *antiliminal*. The quasi-local algebras occurring in mathematical physics, of which the UHF algebra is an example, are always antiliminal.

After these preliminary examples we next examine a situation in which the subsets of the states have good measurability properties. The assumptions of the following definition are essentially motivated by the structure of the locally normal states over quasi-local algebras which were studied in Section 2.6 and are the basis of the first approach to the measure-theoretic difficulties occurring in the decomposition of states. The second approach to these problems is based upon the weaker assumption of separability of the representation space associated with the state in question. This will be discussed in Section 4.4.

Definition 4.1.32. Let $\mathfrak{C}$ be a C^*-algebra with identity, and F a subset of the state space $E_{\mathfrak{C}}$. F is said to satisfy *separability condition* S if there exists a sequence $\{\mathfrak{C}_n\}_{n\geq 1}$ of sub-C^*-algebras of $\mathfrak{C}$ such that $\bigcup_{n\geq 1} \mathfrak{C}_n$ is dense in $\mathfrak{C}$, and each $\mathfrak{C}_n$ contains a closed, two-sided, separable ideal $\mathfrak{J}_n$ such that

$$F = \{\omega;\, \omega \in E_{\mathfrak{C}},\, \|\omega|_{\mathfrak{J}_n}\| = 1, n \geq 1\}.$$

The simplest application of this definition occurs if $\mathfrak{C}$ is separable and $\mathfrak{J}_n = \mathfrak{C}$ for all $n \geq 1$. In this case the state space $E_{\mathfrak{C}}$ is metrizable and the Borel and Baire structures on $E_{\mathfrak{C}}$ coincide. We intend to study these structures for subsets $F \subseteq E_{\mathfrak{C}}$ satisfying condition S but as a preliminary we need the following lemma.

Lemma 4.1.33. *Let $\mathfrak{A}$ be a C^*-algebra, and $\mathfrak{J} \subseteq \mathfrak{A}$ a closed, two sided ideal in $\mathfrak{A}$. If ω is a state on $\mathfrak{J}$ there exists one and only one state $\tilde{\omega}$ on $\mathfrak{A}$*

which extends ω. Furthermore, if $(\mathfrak{H}_{\tilde{\omega}}, \pi_{\tilde{\omega}}, \Omega_{\tilde{\omega}})$ is the cyclic representation of $\mathfrak{A}$ associated to this state $\tilde{\omega}$, $\pi_{\tilde{\omega}}(\mathfrak{I})$ is strongly dense in $\pi_{\tilde{\omega}}(\mathfrak{A})''$.

PROOF. We may assume that $\mathfrak{A}$ has an identity $\mathbb{1}$. The existence of an extension $\tilde{\omega}$ of ω to $\mathfrak{A}$ follows from Proposition 2.3.24. By Proposition 2.3.11 there exists an approximate identity $\{E_\alpha\}$ for $\mathfrak{I}$ such that

$$\lim_\alpha \omega(E_\alpha^2) = \lim_\alpha \omega(E_\alpha) = 1 = \tilde{\omega}(\mathbb{1}).$$

If $A \in \mathfrak{A}$ it follows that

$$\begin{aligned} |\tilde{\omega}(E_\alpha A - A)| &= |\tilde{\omega}((\mathbb{1} - E_\alpha)A)| \\ &\leq \tilde{\omega}((\mathbb{1} - E_\alpha)^2)^{1/2}\tilde{\omega}(A^*A)^{1/2} \\ &\to 0 \end{aligned}$$

as $\alpha \to \infty$. Hence

$$\tilde{\omega}(A) = \lim_\alpha \tilde{\omega}(E_\alpha A) = \lim_\alpha \omega(E_\alpha A)$$

and this formula shows the uniqueness of the extension $\tilde{\omega}$.

Now by the limit property of $\{E_\alpha\}$ it follows that $\lim \pi_{\tilde{\omega}}(E_\alpha)\Omega_{\tilde{\omega}} = \Omega_{\tilde{\omega}}$, thus $\lim \pi_{\tilde{\omega}}(AE_\alpha)\Omega_{\tilde{\omega}} = \pi_{\tilde{\omega}}(A)\Omega_{\tilde{\omega}}$ for all $A \in \mathfrak{A}$, and hence $\pi_{\tilde{\omega}}(\mathfrak{I})\Omega_{\tilde{\omega}}$ is dense in $\mathfrak{H}_{\tilde{\omega}}$. If $A \in \mathfrak{A}$ and $B \in \mathfrak{I}$ we have

$$\begin{aligned} \pi_{\tilde{\omega}}(AE_\alpha)\pi_{\tilde{\omega}}(B)\Omega_{\tilde{\omega}} &= \pi_{\tilde{\omega}}(A)\pi_{\tilde{\omega}}(E_\alpha B)\Omega_{\tilde{\omega}} \\ &\to \pi_{\tilde{\omega}}(A)\pi_{\tilde{\omega}}(B)\Omega_{\tilde{\omega}} \end{aligned}$$

as $\alpha \to \infty$. It follows that $\pi_{\tilde{\omega}}(AE_\alpha)$ converges strongly to $\pi_{\tilde{\omega}}(A)$ and hence

$$\pi_{\tilde{\omega}}(\mathfrak{I})'' = \pi_{\tilde{\omega}}(\mathfrak{A})''.$$

We are now ready to study the separability condition S.

Proposition 4.1.34. *Let $\mathfrak{C}$ be a C^*-algebra with identity $\mathbb{1}$, and assume that $F \subseteq E_{\mathfrak{C}}$ satisfies the separability condition S. It follows that*

(1) *F is a stable face of $E_{\mathfrak{C}}$;*
(2) *F is a Baire set;*
(3) *the extremal points $\mathscr{E}(F)$ of F form a Baire set and there is a convex continuous function f over $E_{\mathfrak{C}}$ such that*

$$\mathscr{E}(F) = \partial_f(E_{\mathfrak{C}}) \cap F,$$

where $\partial_f(E_{\mathfrak{C}})$ is the boundary set associated with f;
(4) *if $\omega \in F$, the Hilbert space $\mathfrak{H}_\omega$ of the corresponding representation is separable.*

PROOF. We first prove property (2). Let $\{A_{n,k}\}_{k \geq 1}$ be a countable dense subset of the unit ball of the selfadjoint elements of $\mathfrak{I}_n$, $n \geq 1$. The state $\omega \in F$ if, and only if,

$$\sup_{k \geq 1} \omega(A_{n,k}) = 1, \qquad n = 1, 2, \ldots .$$

Therefore

$$F = \bigcap_{n,m\geq 1} V_{n,m}$$

where

$$V_{n,m} = \bigcup_{p\geq 1} V_{n,m,p}$$

and

$$\begin{aligned} V_{n,m,p} &= \bigcap_{q\geq 1}\left\{\omega;\,\omega\in E_{\mathfrak{C}},\,\omega(A_{n,p}) > 1 - \frac{1}{m} - \frac{1}{q}\right\} \\ &= \left\{\omega;\,\omega\in E_{\mathfrak{C}},\,\omega(A_{n,p}) \geq 1 - \frac{1}{m}\right\}. \end{aligned}$$

Thus $V_{n,m,p}$ is closed and it is also a countable intersection of open sets, i.e., $V_{n,m,p}$ is a compact G_δ. Hence F is a Baire set.

Next we prove property (1). The notion of a stable face was introduced in the remark following the proof of Theorem 4.1.11. The set F has this property if, and only if, the conditions $\omega \in F$ and $\mu \succ \delta_\omega$ imply that the measure μ is supported by F. But if $\mu = \mu_1 + \mu_2$ where μ_1 is supported by $V_{n,m}$ and μ_2 by its complement then one has

$$\begin{aligned} \omega(A_{n,k}) &= \mu_1(\hat{A}_{n,k}) + \mu_2(\hat{A}_{n,k}) \\ &\leq (1 - \|\mu_2\|) + \|\mu_2\|(1 - 1/m) \\ &= 1 - \|\mu_2\|/m. \end{aligned}$$

Since $\sup_k \omega(A_{n,k}) = 1$ one must have $\|\mu_2\| = 0$ and it follows that $\mu(F) = 1$.

We now turn to the proof of property (3). Let $\mathbb{R}^{\mathbb{N}^2}$ be the linear space of real double sequences equipped with the topology of pointwise convergence and define a map t; $E_{\mathfrak{C}} \mapsto \mathbb{R}^{\mathbb{N}^2}$ by

$$t(\omega) = \{\hat{A}_{n,k}(\omega)\}_{n,k\geq 1}.$$

The map t is affine and continuous and hence the range $t(E_{\mathfrak{C}})$ of t is a compact convex subset of $[-1, 1]^{\mathbb{N}^2}$. The latter space is metrizable and separable and therefore these properties are shared by $t(E_{\mathfrak{C}})$. One could, for example, equip $t(E_{\mathfrak{C}})$ with the metric

$$d(t(\omega_1), t(\omega_2)) = \sum_{n,k\geq 1} 2^{-k-n}|\hat{A}_{n,k}(\omega_1 - \omega_2)|.$$

Now from Theorem 4.1.11, and the remark after its proof, it follows that $\mathscr{E}(t(E_{\mathfrak{C}}))$ is a Borel set. But metrizability and compactness of $t(E_{\mathfrak{C}})$ ensure that the Borel and Baire sets coincide. Therefore $\mathscr{E}(t(E_{\mathfrak{C}}))$ is a Baire set. Since t is continuous and $E_{\mathfrak{C}}$ is compact one concludes that $t^{-1}(\mathscr{E}(t(E_{\mathfrak{C}})))$ is a Baire set. Next we argue that

$$t^{-1}(\mathscr{E}(t(E_{\mathfrak{C}}))) \cap F = \mathscr{E}(E_{\mathfrak{C}}) \cap F$$

and hence establish that $\mathscr{E}(F) = \mathscr{E}(E_{\mathfrak{C}}) \cap F$ is a Baire set. For this latter equality it is first essential to note that $t(F)$ is a face of $t(E_{\mathfrak{C}})$. But $t(F)$ consists of the double sequences which attain values arbitrarily close to one when k varies, for each n, and the facial property is immediate. It is also important to remark that t restricted to F is faithful. For this, note that $\{A_{n,k}\}_{n,k\geq 1}$ separates the restrictions of the states $\omega \in E_{\mathfrak{A}}$ to $\bigcup_n \mathfrak{J}_n$. But any $\omega \in F$ is uniquely determined by its restriction to $\bigcup_n \mathfrak{J}_n$ by Lemma 4.1.33 and the assumptions in Definition 4.1.32. Hence $t|_F$ is faithful.

Now if $\omega \in \mathscr{E}(E_{\mathfrak{C}}) \cap F$ then ω is extremal in F and $t(\omega)$ is extremal in $t(F)$ because t is faithful in restriction to F. But $t(F)$ is a face of $t(E_{\mathfrak{A}})$ and hence $t(\omega) \in \mathscr{E}(t(E_{\mathfrak{C}}))$, i.e., $\mathscr{E}(E_{\mathfrak{C}}) \cap F \subseteq t^{-1}(\mathscr{E}(t(E_{\mathfrak{C}}))) \cap F$. Conversely, if $\omega \in t^{-1}(\mathscr{E}(t(E_{\mathfrak{C}}))) \cap F$ then $t(\omega) \in \mathscr{E}(t(E_{\mathfrak{C}})) \cap t(F) = \mathscr{E}(t(F))$ since $t(F)$ is a face. But because t is faithful in restriction to F one has $t(\mathscr{E}(F)) = \mathscr{E}(t(F))$ and therefore $t^{-1}(\mathscr{E}(t(E_{\mathfrak{C}}))) \cap F \subseteq \mathscr{E}(E_{\mathfrak{C}}) \cap F$.

Finally, the metrizability and separability of $t(E_{\mathfrak{C}})$ imply the existence of a strictly convex, continuous function g over $t(E_{\mathfrak{C}})$. A constructive procedure for obtaining g is given in the proof of Theorem 4.1.11. Now if f is defined over $E_{\mathfrak{C}}$ by

$$f(\omega) = g(t(\omega)), \qquad \omega \in E_{\mathfrak{C}},$$

then one concludes that f is convex continuous and, moreover, it is strictly convex over F because t is an affine isomorphism on F. Once again this implies that if $\omega \in F$ but $\omega \notin \mathscr{E}(F)$ then $\omega \notin \partial_f(E_{\mathfrak{C}})$, i.e.,

$$\mathscr{E}(F) \supseteq \partial_f(E_{\mathfrak{C}}) \cap F.$$

But equality of the latter sets then follows from Proposition 4.1.10 and the relation

$$\mathscr{E}(F) = \mathscr{E}(E_{\mathfrak{C}}) \cap F.$$

It remains to prove property (4). If $\omega \in F$ then $\pi_\omega(\mathfrak{J}_n)\Omega_\omega$ is dense in $\pi_\omega(\mathfrak{C}_n)\Omega_\omega$ by the last statement in Lemma 4.1.33. As $\bigcup_n \mathfrak{C}_n$ is dense in $\mathfrak{C}$ it follows that $(\bigcup_n \pi_\omega(\mathfrak{J}_n))\Omega_\omega$ is dense in $\mathfrak{H}_\omega$, and hence $\mathfrak{H}_\omega$ is separable.

The simplest application of Proposition 4.1.34 occurs if $\mathfrak{C}$ is separable and $F = E_{\mathfrak{C}}$. The foregoing result is then contained in Theorem 4.1.11 and its proof. Less trivial applications are contained in the following example.

EXAMPLE 4.1.35. If $\mathfrak{M}$ is a von Neumann algebra, let $N_{\mathfrak{M}} = E_{\mathfrak{M}} \cap \mathfrak{M}_*$ be the normal states on $\mathfrak{M}$.

(1) $N_{\mathfrak{M}}$ is a face in $E_{\mathfrak{M}}$.

We must show that if $\varphi \in E_{\mathfrak{M}}$ and $\varphi < \lambda\omega$ for an $\omega \in N_{\mathfrak{M}}$ then $\varphi \in N_{\mathfrak{M}}$. But by Theorem 2.3.19 there exists a $T \in \pi_\omega(\mathfrak{M})_+{}'$ such that $\varphi(A) = (T\Omega_\omega, \pi_\omega(A)T\Omega_\omega)$. As π_ω is a normal representation of $\mathfrak{M}$, (1) follows.

(2) $N_{\mathfrak{M}}$ is dense in $E_{\mathfrak{M}}$.

This follows from Proposition 3.2.10.

(3) $N_{\mathfrak{M}}$ is closed in the norm topology, and $N_{\mathfrak{M}}$ is sequentially complete in the weak* topology.

Since $\mathfrak{M}_*$ is a Banach space, the first assertion is clear. We will not prove the second assertion (see Notes and Remarks).

(4) If $\mathfrak{M}$ is a factor on a separable Hilbert space, the following conditions are equivalent:

(i) $N_{\mathfrak{M}}$ is a stable face in $E_{\mathfrak{M}}$;
(ii) $\mathfrak{M}$ is type I, i.e., $\mathfrak{M} \simeq \mathscr{L}(\mathfrak{H})$ for some Hilbert $\mathfrak{H}$.

If these conditions are satisfied, then $\mathscr{E}(N_{\mathfrak{M}})$ is a Baire subset of $E_{\mathfrak{M}}$. If these conditions are not satisfied, there exists a Borel subset $G \subseteq E_{\mathfrak{M}}$ such that

(a) G supports all maximal measures in $M_+(E_{\mathfrak{M}})$,
(b) $G \cap N_{\mathfrak{M}} = \varnothing$.

PROOF. (ii) ⇒ (i) If $\mathfrak{M} = \mathscr{L}(\mathfrak{H})$, we have that

$$N_{\mathfrak{M}} = \{\omega \in E_{\mathfrak{M}};\ \|\omega|_{\mathscr{L}\mathscr{C}(\mathfrak{H})}\| = 1\}$$

by Proposition 2.6.14. Hence $N_{\mathfrak{M}}$ is a stable face and $\mathscr{E}(N_{\mathfrak{M}})$ is a Baire set, by Proposition 4.1.34. In this case it is easy to see that $\mathscr{E}(N_{\mathfrak{M}})$ is just the vector states on $\mathfrak{M} = \mathscr{L}(\mathfrak{H})$.

(i) ⇒ (ii) Since $\mathfrak{M}$ is a factor on a separable Hilbert space, the unit sphere $\mathfrak{M}_1$ is metrizable in the weak topology, and since $\mathfrak{M}_1$ is compact, the positive part $\mathfrak{M}_{1+}$ of $\mathfrak{M}_1$ contains a countable, dense sequence $\{A_n\}_{n\geq 1}$. But then $\{A_n\}_{n\geq 1}$ separates the points in $N_{\mathfrak{M}}$, and it follows that the function $f \in S(E_{\mathfrak{M}})$ defined by

$$f(\omega) = \sum_{n\geq 1} 2^{-n}\omega(A_n)^2$$

is strictly convex when restricted to $N_{\mathfrak{M}}$. Hence

$$\partial_f(E_{\mathfrak{M}}) \cap N_{\mathfrak{M}} \subseteq \mathscr{E}(N_{\mathfrak{M}}) = \mathscr{E}(E_{\mathfrak{M}}) \cap N_{\mathfrak{M}},$$

where the last equality follows since $N_{\mathfrak{M}}$ is a face in $E_{\mathfrak{M}}$. But if $\mathfrak{M}$ is not a type I factor, $\mathfrak{M}$ has no normal, pure states as a consequence of Proposition 2.4.22 and Theorem 2.4.24. In fact, if ω is a normal, pure state of $\mathfrak{M}$ then $\pi_\omega(\mathfrak{M}) \simeq \mathfrak{M}$, $\pi_\omega(\mathfrak{M}) = \pi_\omega(\mathfrak{M})''$, and $\pi_\omega(\mathfrak{M})'' = \mathscr{L}(\mathfrak{H}_\omega)$; thus $\mathfrak{M} \simeq \mathscr{L}(\mathfrak{H}_\omega)$. Hence $\mathscr{E}(E_{\mathfrak{M}}) \cap N_{\mathfrak{M}} = \emptyset$. Furthermore, $G = \partial_f(E_{\mathfrak{M}})$ supports all maximal probability measures on $E_{\mathfrak{M}}$ by Theorem 4.1.7, and since $G \cap N_{\mathfrak{M}} = \emptyset$, $N_{\mathfrak{M}}$ cannot be a stable face.

From the fourth property, one derives the following.

(5) Let $\mathfrak{A} = \overline{\bigcup_n \mathfrak{A}_n}$ be a quasi-local algebra, where each $\mathfrak{A}_n$ has the form $\mathfrak{A}_n = \mathscr{L}(\mathfrak{H}_n)$, and the $\mathfrak{H}_n$ are separable Hilbert spaces. Then the locally normal states on $\mathfrak{A}$ form a Baire subset of $E_{\mathfrak{A}}$, and a stable face. If each $\mathfrak{A}_n$ is a factor on a separable Hilbert space not of type I, then the set of locally normal states is not a stable face, and is, in fact, contained in a Borel set which has measure zero for all maximal measures.

All the foregoing considerations were aimed at the characterization of properties of the pure states over a C^*-algebra. Such properties are fundamental for the discussion of the barycentric decomposition of a state into pure states. One is interested, however, in other types of decomposition and one might wish to express ω as a superposition of factor states, i.e., states for which the associated von Neumann algebra $\pi_\omega(\mathfrak{A})''$ is a factor. Thus it is also necessary to study properties of the factor states over $\mathfrak{A}$. One method of obtaining information about this latter set is by embedding $\mathfrak{A}$ in a larger C^*-algebra $\mathfrak{C}$ which is chosen such that the pure states over $\mathfrak{C}$ are factor states when restricted to $\mathfrak{A}$. The information obtained concerning pure states on $\mathfrak{C}$ is then translated into information concerning the factor states over $\mathfrak{A}$. In order to exploit this method it is, however, necessary to introduce some additional measure-theoretic concepts.

Let K be a compact Hausdorff space and μ a positive Radon measure on K. A subset $E \subseteq K$ is said to be *μ-negligible* if there exists a Borel set F such that $E \subseteq F$ and $\mu(F) = 0$. Alternatively, a set $E \subseteq K$ is said to be *μ-measurable* if there exists a Borel set F such that $(E \cup F)\backslash(E \cap F)$ is μ-negligible. It is clear that the μ-measurable sets form a σ-algebra and μ can be extended to a measure on this σ-algebra by setting $\mu(E) = \mu(F)$.

Next define a subset of K to be an $F_{\sigma\delta}$ if it is the countable intersection of countable unions of closed sets. A subset $E \subseteq K$ is then defined to be *analytic* if there exists a compact Hausdorff space G, an $F_{\sigma\delta}$-subset $B \subseteq G$, and a continuous map f; $B \mapsto K$ such that $f(B) = E$. The set $\mathscr{A}$ of analytic subsets of K contains the Borel sets and is closed under countable unions and countable intersections. But $\mathscr{A}$ is not closed for the operation of taking complements. In fact if $E \in \mathscr{A}$ and $K \backslash E \in \mathscr{A}$ then E is a Borel set. A useful property of analytic sets is that they are μ-measurable for all regular Borel measures μ on K. Moreover, the corresponding measure of the analytic set E satisfies

$$\begin{aligned}\mu(E) &= \inf\{\mu(U); E \subseteq U, U \text{ is open}\}\\ &= \sup\{\mu(V); V \subseteq E, V \text{ is compact}\}.\end{aligned}$$

Since $\mu(K) < +\infty$ it follows from this that there exists a G_δ-set G and a F_σ-set F such that $F \subseteq E \subseteq G$ and $\mu(G \backslash F) = 0$, i.e., $G \backslash E$ and $E \backslash F$ are μ-negligible.

Finally, if μ is a positive Radon measure and E is μ-measurable we say that μ is *supported* by E if $\mu(E) = \mu(K)$. This extends the previous notion of a supporting set by allowing more general sets than Borel sets.

The above concepts will be used in the context of mappings of states of a C^*-algebra into states of a C^*-subalgebra. The following lemma summarizes some of the most obvious facts.

Lemma 4.1.36. *Let $\mathfrak{A}$ be a C^*-subalgebra of the C^*-algebra $\mathfrak{C}$ and assume that $\mathfrak{A}$ and $\mathfrak{C}$ have a common identity element. Define the restriction map r; $E_{\mathfrak{C}} \mapsto E_{\mathfrak{A}}$ by $(r\omega)(A) = \omega(A)$ for all $A \in \mathfrak{A}$. It follows that if F is a Baire (resp. Borel) subset of $E_{\mathfrak{A}}$ then $r^{-1}(F)$ is a Baire (resp. Borel) subset of $E_{\mathfrak{C}}$. Conversely, if G is a Borel subset of $E_{\mathfrak{C}}$ then $r(G)$ is an analytic subset of $E_{\mathfrak{A}}$.*

Next, let $\tilde{\mu}$ be a positive regular Borel measure on $E_{\mathfrak{C}}$ and define μ on the Borel subsets $F \subseteq E_{\mathfrak{A}}$ by

$$\mu(F) = \tilde{\mu}(r^{-1}(F)).$$

It follows that μ is a regular Borel measure on $E_{\mathfrak{A}}$. If $F \subseteq E_{\mathfrak{A}}$ is μ-measurable then $r^{-1}(F) \subseteq E_{\mathfrak{A}}$ is $\tilde{\mu}$-measurable. Moreover, if G is a Borel subset of $E_{\mathfrak{C}}$ then

$$\mu(r(G)) \geq \tilde{\mu}(G).$$

Hence if $\tilde{\mu}$ is supported by G then μ is supported by $r(G)$ and μ is also supported by an F_σ-subset of $r(G)$.

All statements of the lemma follow from the preceding remarks and the continuity of the map r when $E_{\mathfrak{C}}$ and $E_{\mathfrak{A}}$ are equipped with their respective weak* topologies.

To conclude this subsection we describe a specific form of the restriction mapping considered in Lemma 4.1.36. The relation between the algebras $\mathfrak{A}$ and $\mathfrak{C}$ is arranged so that r provides a connection between the pure states over $\mathfrak{C}$ and the factor states over $\mathfrak{A}$. The following proposition will be of use in the subsequent discussion of extremal decompositions and central decompositions of a state.

Proposition 4.1.37. *Let* $\mathfrak{A}$ *and* $\mathfrak{B}$ *be* C^**-subalgebras of the* C^**-algebra* $\mathfrak{C}$. *Assume that* $\mathfrak{A}$, $\mathfrak{B}$, *and* $\mathfrak{C}$ *have a common identity,* $\mathfrak{A} \subseteq \mathfrak{B}'$, *and* $\mathfrak{A} \cup \mathfrak{B}$ *generates* $\mathfrak{C}$ *as a* C^**-algebra. Define the restriction map* r; $E_{\mathfrak{C}} \mapsto E_{\mathfrak{A}}$ *by* $(r\omega)(A) = \omega(A)$ *for all* $A \in \mathfrak{A}$. *It follows that* r *is weakly*-weakly* continuous and maps* $E_{\mathfrak{C}}$ *onto* $E_{\mathfrak{A}}$. *Moreover,*

$$\mathscr{E}(E_{\mathfrak{A}}) \subseteq r(\mathscr{E}(E_{\mathfrak{C}})) \subseteq F_{\mathfrak{A}},$$

where $F_{\mathfrak{A}}$ *denotes the factor states of* $\mathfrak{A}$. *If* $\mathfrak{B}$ *is abelian then* $\mathscr{E}(E_{\mathfrak{A}}) = r(\mathscr{E}(E_{\mathfrak{C}}))$.

PROOF. The continuity of r is obvious. It follows from Proposition 2.3.24 that r is onto and $\mathscr{E}(E_{\mathfrak{A}}) \subseteq r(\mathscr{E}(E_{\mathfrak{C}}))$.

Let ω be a pure state of $\mathfrak{C}$ and $(\mathfrak{H}_\omega, \pi_\omega, \Omega_\omega)$ the associated cyclic representation. Now $\pi_\omega(\mathfrak{A})$ and $\pi_\omega(\mathfrak{B})$ generate the irreducible set $\pi_\omega(\mathfrak{C})$ and hence

$$\{\pi_\omega(\mathfrak{A}) \cup \pi_\omega(\mathfrak{B})\}'' = \mathscr{L}(\mathfrak{H}_\omega).$$

Let $P \in \pi_\omega(\mathfrak{A})'' \cap \pi_\omega(\mathfrak{A})'$ be a projection. Since $\pi_\omega(\mathfrak{A})'' \subseteq \pi_\omega(\mathfrak{B})'$ it follows that $P \in \pi_\omega(\mathfrak{B})'$ and therefore $P \in \{\pi_\omega(\mathfrak{A}) \cup \pi_\omega(\mathfrak{B})\}' = \pi(\mathfrak{C})' = \mathbb{C}1$. It follows that $\pi_\omega(\mathfrak{A})'' \cap \pi_\omega(\mathfrak{A})' = \mathbb{C}1$ and hence $\pi_\omega(\mathfrak{A})''$ is a factor. But $A \mapsto \pi_\omega(A)[\pi_\omega(\mathfrak{A})\Omega_\omega]$ is the representation of $\mathfrak{A}$ associated to $r\omega$ and hence $r\omega$ is a factor state of $\mathfrak{A}$.

Finally, if $\mathfrak{B}$ is abelian it follows that $\mathfrak{B}$ is contained in the center of $\mathfrak{C}$. Hence for $\omega \in \mathscr{E}(E_{\mathfrak{C}})$ one has

$$\pi_\omega(\mathfrak{B}) \subseteq \pi_\omega(\mathfrak{C}) \cap \pi_\omega(\mathfrak{C})' = \mathbb{C}1.$$

Since $\pi_\omega(\mathfrak{C})$ is the C^*-algebra generated by $\pi_\omega(\mathfrak{A})$ and $\pi_\omega(\mathfrak{B})$ it follows that

$$\pi_\omega(\mathfrak{A}) = \pi_\omega(\mathfrak{C})$$

and consequently $\pi_\omega(\mathfrak{A})$ is irreducible. Thus $r\omega$ is pure and $r(\mathscr{E}(E_{\mathfrak{C}})) \subseteq \mathscr{E}(E_{\mathfrak{A}})$. The reverse containment, however, has already been proved and hence $r(\mathscr{E}(E_{\mathfrak{C}})) = \mathscr{E}(E_{\mathfrak{A}})$.

4.2. Extremal, Central, and Subcentral Decompositions

4.2.1. Extremal Decompositions

In this section we apply the general theory developed in the previous section to some specific decompositions of states of a C^*-algebra $\mathfrak{A}$ with identity. We begin by examining extremal decompositions of a state ω over $\mathfrak{A}$, i.e., decompositions of ω into pure states.

The extremal decompositions of ω will be constructed with the aid of the orthogonal measures of Section 4.1.3. Theorem 4.1.25 associates with each abelian von Neumann subalgebra $\mathfrak{B}$ of $\pi_\omega(\mathfrak{A})'$ an orthogonal measure $\mu_{\mathfrak{B}} \in \mathcal{O}_\omega(E_{\mathfrak{A}})$ and Theorem 4.1.28 asserts that if $\mathfrak{B}$ is maximal abelian then $\mu_{\mathfrak{B}}$ is maximal in $\mathcal{O}_\omega(E_{\mathfrak{A}})$. This latter result is not optimal and our first aim is to prove that $\mu_{\mathfrak{B}}$ is indeed maximal in $M_\omega(E_{\mathfrak{A}})$. Subsequently we deduce that there is a unique maximal probability measure on $E_{\mathfrak{A}}$ with barycenter ω if, and only if, $\pi_\omega(\mathfrak{A})'$ is abelian. The proof of these results involves the comparison of orthogonal measures with general measures and this introduces a number of new difficulties. To overcome these difficulties one needs a deeper understanding of the order relation $\succ$.

The following result gives a characterization of the order relation in the general setting of convex compact sets. It shows that two measures μ, ν are comparable if, and only if, they are comparable component by component.

Proposition 4.2.1 (Cartier–Fell–Meyer theorem). *Let K be a convex compact subset of a locally convex Hausdorff space and μ, $\nu \in M_\omega(K)$ measures with barycenter ω. The following conditions are equivalent*:

(1) $\nu \prec \mu$;
(2) *for every convex combination*

$$\nu = \sum_{i=1}^{n} \lambda_i \nu_i$$

with $\nu_i \in M_{\omega_i}(K)$ there exists a corresponding convex combination

$$\mu = \sum_{i=1}^{n} \lambda_i \mu_i$$

with $\mu_i \in M_{\omega_i}(K)$;
(3) *condition* (2) *is valid and, moreover $\mu_i \succ \nu_i$.*

Proof. (3) $\Rightarrow$ (2) is evident.

(1) $\Rightarrow$ (3) First define a map p from $C(K)^n$ to $\mathbb{R}$ by

$$p(\underline{f}) = \sum_{i=1}^{n} \lambda_i \nu_i(\bar{f}_i),$$

where $\underline{f} = (f_1, f_2, \ldots, f_n)$. It is evident that p is homogeneous but it is also subadditive by Lemma 4.1.9.

Next let Y be the subspace of $C(K)^n$ formed by the vectors $\underline{f} = (f, f, f, \ldots, f)$ and define a linear functional φ on Y by $\varphi(\underline{f}) = \mu(f)$. As $\nu \prec \mu$ one has from Proposition 4.1.6

$$\nu(\bar{f}) \geq \mu(\bar{f}) \geq \mu(f).$$

Therefore

$$\varphi(\underline{f}) \leq \nu(\bar{f}) = \sum_{i=1}^{n} \lambda_i \nu_i(\bar{f}) = p(\underline{f}).$$

It follows from the Hahn–Banach theorem that φ has a linear extension, which we also denote by φ, to $C(K)^n$ such that $\varphi(\underline{f}) \leq p(\underline{f})$. Explicitly, one has

$$-\sum_{i=1}^{n} \lambda_i \nu_i((\overline{-f_i})) \leq \varphi(\underline{f}) \leq \sum_{i=1}^{n} \lambda_i \nu_i(\bar{f}_i). \tag{$*$}$$

Now if $f_i \geq 0$ then $-(\overline{-f_i}) \geq 0$ and one concludes that φ is a positive functional with

$$|\varphi(\underline{f})| \leq \max_{1 \leq i \leq n} \|f_i\|_\infty .$$

Next define μ_k by

$$\lambda_k \mu_k(f) = \varphi(\underline{f_k})$$

where $\underline{f_k}$ is the vector with kth component equal to f and all other components zero. From $(*)$ one deduces that $\mu_k(\mathbb{1}) = 1$, where $\mathbb{1}$ is the identity function and hence μ_k is a probability measure. More generally, one has

$$\mu_k(f) \leq \nu_k(\bar{f}).$$

But if $f \in S(K)$ then $(\overline{-f}) = -f$ and one has

$$\mu_k(f) \geq \nu_k(f).$$

Thus $\mu_k \succ \nu_k$. Finally, choosing $\underline{f} = (f, f, \ldots, f)$ one has

$$\mu(f) = \varphi(\underline{f}) = \sum_{i=1}^{n} \lambda_i \mu_i(f).$$

(2) $\Rightarrow$ (1) Next let $f \in S(K)$ and consider all finite partitions $\mathscr{U} = \{U_i\}_{1 \leq i \leq n}$, of K, in terms of Baire sets U_i. Let χ_i denote the characteristic function of U_i and define λ_i and ν_i by $\lambda_i = \nu(U_i)$ and $\lambda_i \, d\nu_i = \chi_i \, d\nu$. Thus

$$\nu = \sum_{i=1}^{n} \lambda_i \nu_i$$

and by assumption there is a decomposition

$$\mu = \sum_{i=1}^{n} \lambda_i \mu_i$$

such that the μ_i and ν_i have the same barycenter ω_i. Now it is possible by Proposition 4.1.1 to approximate each μ_i by a measure with finite support and barycenter ω_i. Convexity and a limiting argument then give

$$f(\omega_i) \leq \mu_i(f).$$

Moreover,

$$\nu_i(f) \leq f(\omega_i) + \sup_{\omega \in U_i} |f(\omega_i) - f(\omega)|.$$

Therefore

$$\nu(f) \leq \mu(f) + \sup_{1 \leq i \leq n} \sup_{\omega \in U_i} |f(\omega_i) - f(\omega)|.$$

Finally, for $\varepsilon > 0$ one may choose the U_i as in the proof of Proposition 4.1.1 to ensure that

$$\nu(f) \leq \mu(f) + \varepsilon.$$

Therefore $\nu \prec \mu$.

Now we are in a position to prove the main result of this subsection.

Theorem 4.2.2. *Let $\mathfrak{A}$ be a C^*-algebra with identity and ω a state over $\mathfrak{A}$. Let $\mathfrak{B}$ be an abelian von Neumann subalgebra of $\pi_\omega(\mathfrak{A})'$ and $\mu \in \mathcal{O}_\omega(E_{\mathfrak{A}})$ the corresponding orthogonal measure. The following conditions are equivalent:*

(1) *$\mathfrak{B}$ is a maximal abelian subalgebra of $\pi_\omega(\mathfrak{A})'$;*
(2) *μ is maximal among the orthogonal measures $\mathcal{O}_\omega(E_{\mathfrak{A}})$;*
(3) *μ is maximal among the measures $M_\omega(E_{\mathfrak{A}})$.*

PROOF. The equivalence of conditions (1) and (2) is a direct consequence of Theorem 4.1.28, and condition (3) clearly implies condition (2). Therefore it remains to prove that condition (2) implies condition (3).

Assume μ is maximal among the orthogonal measures $\mathcal{O}(E_{\mathfrak{A}})$. Since μ is orthogonal it follows from Theorem 4.1.25 that $\mathfrak{B}$ is *-isomorphic to the range of the map $f \in L^\infty(\mu) \mapsto \kappa_\mu(f) \in \pi_\omega(\mathfrak{A})'$. Therefore, if $P \neq \mathbb{1}$ is a projection in $\mathfrak{B}$, with $P\Omega_\omega \neq 0$, there is an $f \in L^\infty(\mu)$, with $1 > \mu(f) > 0$, such that $P = \kappa_\mu(f)$. Now set $\lambda = \mu(f)$ and define measures μ_1, μ_2 on $E_{\mathfrak{A}}$ by setting

$$\mu_1(g) = \mu(fg)/\lambda,$$

$$\mu_2(g) = \mu((\mathbb{1} - f)g)/(1 - \lambda)$$

for $g \in C(E_{\mathfrak{A}})$. Thus

$$\mu = \lambda\mu_1 + (1 - \lambda)\mu_2$$

and if ω_i denotes the barycenter of μ_i then

$$\omega = \lambda\omega_1 + (1 - \lambda)\omega_2.$$

Next suppose $\nu \in M_\omega(E_{\mathfrak{A}})$ and $\nu \succ \mu$. We will prove that $\nu \prec \mu$. Hence $\nu = \mu$ and μ is maximal among the measures $M_\omega(E_{\mathfrak{A}})$.

First, we remark that by Proposition 4.2.1 there exists a decomposition

$$\nu = \lambda\nu_1 + (1 - \lambda)\nu_2$$

where $\nu_i \in M_{\omega_i}(E_{\mathfrak{A}})$. Now suppose that ρ is a positive measure on $E_{\mathfrak{A}}$ satisfying $\rho \leq \nu$. Since the positive measures form a cone which is a lattice with respect to its natural order (see the discussion prior to Proposition 4.1.1) it follows from Lemma 4.1.16 that there exist measures ρ_1, ρ_2 on $E_{\mathfrak{A}}$ such that

$$\rho = \lambda\rho_1 + (1 - \lambda)\rho_2$$

and $0 \leq \rho_i \leq \nu_i$. Let φ and φ_i denote the barycenters of ρ and ρ_i. Then since $\rho \leq \nu$ one deduces from Theorem 2.3.19 that

$$\varphi(A) = (\Omega_\omega, T\pi_\omega(A)\Omega_\omega),$$

$$\lambda\varphi_1(A) = (\Omega_\omega, T_1\pi_\omega(A)\Omega_\omega),$$

$$(1 - \lambda)\varphi_2(A) = (\Omega_\omega, T_2\pi_\omega(A)\Omega_\omega),$$

for all $A \in \mathfrak{A}$ where T, T_1, and T_2, are positive elements of $\pi_\omega(\mathfrak{A})'$ and $T = T_1 + T_2$. But since $\rho_1 \leq \nu_1$ one has for $A \geq 0$

$$\begin{aligned}(\Omega_\omega, T_1\pi_\omega(A)\Omega_\omega) &= \rho_1(\hat{A}) \\ &\leq \nu_1(\hat{A}) \\ &= \mu_1(\hat{A}) = (\Omega_\omega, P\pi_\omega(A)\Omega_\omega).\end{aligned}$$

Therefore

$$0 \leq T_1 \leq P.$$

Similarly

$$0 \leq T_2 \leq \mathbb{1} - P.$$

It then follows straightforwardly that $T_1 = PT_1P$ and $T_2 = (\mathbb{1} - P)T_2(\mathbb{1} - P)$. In particular both T_1 and T_2 commute with P and since $T = T_1 + T_2$ one concludes that T also commutes with P. But the definition of T is independent of the choice of $P \in \mathfrak{B}$ and hence T commutes with every projection in $\mathfrak{B}$. Since $\mathfrak{B}$ is maximal abelian it follows that $T \in \mathfrak{B}$. Thus we have deduced that if $\nu \succ \mu$ and ρ is a positive measure satisfying $\rho \leq \nu$ then there is a positive element $T \in \mathfrak{B}$ such that

$$\rho(\hat{A}) = (\Omega_\omega, T\pi_\omega(A)\Omega_\omega)$$

for all $A \in \mathfrak{A}$.

Finally, let

$$\nu = \sum_{i=1}^{n} \lambda_i \nu_i$$

be a decomposition of ν in terms of probability measures ν_i and with $\lambda_i > 0$. Then $\rho_i = \lambda_i\nu_i \leq \nu$. Hence there exist positive elements $T_i \in \mathfrak{B}$ such that

$$\rho_i(\hat{A}) = (\Omega_\omega, T_i\pi_\omega(A)\Omega_\omega)$$

for all $A \in \mathfrak{A}$ and

$$\sum_{i=1}^{n} T_i = \mathbb{1}.$$

But since μ is orthogonal there exist, by Theorem 4.1.25, positive $f_i \in L^\infty(\mu)$ such that $T_i = \lambda_i\kappa_\mu(f_i)$. Then if one defines measures μ_i by

$$\mu_i(g) = \mu(f_i g)$$

for $g \in C(E_{\mathfrak{A}})$, it readily follows that each μ_i is a probability measure, the barycenters of μ_i and ν_i are equal, and

$$\mu = \sum_{i=1}^{n} \lambda_i \mu_i.$$

Therefore $\nu \prec \mu$ by Proposition 4.2.1 and the proof of the theorem is complete.

Our next aim is to derive a characterization of uniqueness for a maximal measure with a fixed state as barycenter.

Theorem 4.2.3. *Let $\mathfrak{A}$ be a C^*-algebra with identity, ω a state over $\mathfrak{A}$ and $P = [\pi_\omega(\mathfrak{A})'\Omega_\omega]$. The following conditions are equivalent:*

(1) *there is a unique maximal probability measure μ with barycenter ω;*
(2) *$\pi_\omega(\mathfrak{A})'$ is abelian;*
(3) *$P\pi_\omega(\mathfrak{A})P$ generates an abelian algebra.*

If these conditions are satisfied then μ is the orthogonal measure corresponding to $\pi_\omega(\mathfrak{A})'$.

PROOF. The equivalence of conditions (2) and (3) follows from the considerations of Section 4.1.3. Theorem 4.1.25 establishes that if $\pi_\omega(\mathfrak{A})'$ is abelian then $P\pi_\omega(\mathfrak{A})P$ generates an abelian algebra and conversely. Thus it suffices to prove the equivalence of conditions (1) and (2).

Now suppose condition (2) is false then there exist two distinct maximal abelian subalgebras $\mathfrak{B}_1$ and $\mathfrak{B}_2$ of $\pi_\omega(\mathfrak{A})'$ and the corresponding orthogonal measures μ_1 and μ_2 are distinct maximal measures, with barycenter ω, by Theorem 4.2.2. Thus condition (1) is false and one concludes that (1) $\Rightarrow$ (2).

It remains to prove (2) $\Rightarrow$ (1).

Let μ denote the orthogonal measure associated with the abelian von Neumann algebra $\pi_\omega(\mathfrak{A})'$. Then μ is maximal in $M_\omega(E_{\mathfrak{A}})$ by Theorem 4.2.2. Next let ν be a second maximal measure in $M_\omega(E_{\mathfrak{A}})$. Our aim is to show that μ and ν have a common upper bound $\rho \in M_\omega(E_{\mathfrak{A}})$, and consequently $\rho = \mu$ and $\rho = \nu$ by maximality of μ and ν. Hence $\mu = \nu$ and μ is the unique maximal measure in $M_\omega(E_{\mathfrak{A}})$.

Now μ and ν can be approximated in the weak* topology by measures μ_α and ν_α with finite support by Lemma 4.1.26. Thus it suffices to construct a net $\rho_\alpha \in M_\omega(E_{\mathfrak{A}})$ such that $\mu_\alpha \prec \rho_\alpha$ and $\nu_\alpha \prec \rho_\alpha$ because any weak* limit point ρ of the ρ_α will satisfy $\mu \prec \rho$ and $\nu \prec \rho$. These remarks reduce the problem to finding a common upper bound for two finitely supported measures μ, $\nu \in M_\omega(E_{\mathfrak{A}})$ whenever $\pi_\omega(\mathfrak{A})'$ is abelian.

Let

$$\mu = \sum_{i=1}^{n} \lambda_i \delta_{\omega_i}, \qquad \nu = \sum_{j=1}^{n'} \lambda_j' \delta_{\omega_j'}$$

and define T_i, $Z_j \in \pi_\omega(\mathfrak{A})'$ by

$$T_i = \kappa_\mu(\chi_{\{\omega_i\}}), \qquad Z_j = \kappa_\nu(\chi_{\{\omega_j'\}}),$$

where χ_S denotes the characteristic function of the subset $S \subseteq E_{\mathfrak{A}}$. It follows that

$$\lambda_i \omega_i(A) = (T_i \Omega_\omega, \pi_\omega(A)\Omega_\omega), \qquad \lambda_j' \omega_j'(A) = (Z_j \Omega_\omega, \pi_\omega(A)\Omega_\omega)$$

for all $A \in \mathfrak{A}$, and

$$\sum_{i=1}^{n} T_i = \mathbb{1} = \sum_{j=1}^{n'} Z_j.$$

Since T_i and Z_j are positive and $\pi_\omega(\mathfrak{A})'$ is abelian the products $T_i Z_j$ are positive. Now define λ_{ij} and ω_{ij} by

$$\lambda_{ij} = (T_i Z_j \Omega_\omega, \Omega_\omega),$$
$$\lambda_{ij}\omega_{ij}(A) = (T_i Z_j \Omega_\omega, \pi_\omega(A)\Omega_\omega)$$

and consider the measure

$$\rho = \sum_{i=1}^{n} \sum_{j=1}^{n'} \lambda_{ij} \delta_{\omega_{ij}}.$$

One has

$$\lambda_i \omega_i = \sum_{j=1}^{n'} \lambda_{ij} \omega_{ij}, \qquad \lambda_j' \omega_j' = \sum_{i=1}^{n} \lambda_{ij} \omega_{ij}$$

and a simple application of convexity gives

$$\rho \succ \mu, \qquad \rho \succ \nu.$$

Theorem 4.1.28 characterizes the order relation $\mu \succ \nu$ for orthogonal measures on the state space of a C^*-algebra and Proposition 4.2.1 gives an alternative characterization for general measures on convex compact sets. The next proposition provides an intermediate result in the algebraic setting valid whenever the larger measure μ is simplicial, i.e., $\mu \in \mathscr{E}(M_\omega(E_\mathfrak{A}))$. Noting that orthogonal measures are automatically simplicial, by Corollary 4.1.23, this result can be viewed as a generalization of the equivalence (1)$\Leftrightarrow$(2) in Theorem 4.1.28. It can be used to derive an alternative proof of the last part of Theorem 4.2.3 and will also be useful in the following subsection for the geometric characterization of the orthogonal measure associated with the center of the representation π_ω.

Proposition 4.2.4. *Let $\mathfrak{A}$ be a C^*-algebra with identity, ω a state over $\mathfrak{A}$, and μ, ν two measures in $M_\omega(E_\mathfrak{A})$. Let $L_{1+}^\infty(\mu)$ denote the positive part of the unit ball of $L^\infty(\mu)$ and consider the following two conditions*:

(1) $\mu \succ \nu$;
(2) $\{\kappa_\mu(g);\, g \in L_{1+}^\infty(\mu)\} \supseteq \{\kappa_\nu(f);\, f \in L_{1+}^\infty(\nu)\}$.

It follows that (1) *implies* (2) *and if $\mu \in \mathscr{E}(M_\omega(E_\mathfrak{A}))$ then* (2) *implies* (1). *Moreover the following conditions are equivalent*:

(1′) $\mu \in \mathscr{E}(M_\omega(E_\mathfrak{A}))$;
(2′) *the map $f \in L^\infty(\mu) \mapsto \kappa_\mu(f) \in \pi_\omega(\mathfrak{A})'$ is faithful*;
(3′) *the affine continuous functions over $E_\mathfrak{A}$ are dense in $L^1(\mu)$.*

PROOF. Assume $\mu \succ \nu$ and let $f \in L^{\infty}_{1+}(\nu)$. Define ν_1 and ν_2 by

$$\nu_1(g) = \frac{\nu(fg)}{\nu(f)}, \qquad \nu_2(g) = \frac{\nu((\mathbb{1} - f)g)}{\nu(\mathbb{1} - f)}.$$

It follows that

$$\nu = \lambda\nu_1 + (1 - \lambda)\nu_2$$

with $\lambda = \nu(f)$. Therefore there exists by Proposition 4.2.1 a decomposition of μ,

$$\mu = \lambda\mu_1 + (1 - \lambda)\mu_2,$$

such that the μ_i and ν_i have a common barycenter ω_i. Since $\lambda\mu_i \leq \mu$ there is a $g \in L^{\infty}_{1+}(\mu)$ such that $\lambda\, d\mu_1 = g\, d\mu$. Now one has

$$\begin{aligned}(\Omega_\omega, \kappa_\nu(f)\pi_\omega(A)\Omega_\omega) &= \nu(f\hat{A}) \\ &= \lambda\nu_1(\hat{A}) \\ &= \lambda\mu_1(\hat{A}) \\ &= \mu(g\hat{A}) = (\Omega_\omega, \kappa_\mu(g)\pi_\omega(A)\Omega_\omega)\end{aligned}$$

for all $A \in \mathfrak{A}$. Therefore $\kappa_\nu(f) = \kappa_\mu(g)$, by cyclicity, and condition (2) is valid.

The proof of the converse implication is based upon the alternative characterization of simpliciality provided by condition (2′). Hence we next prove the final statement of the proposition, the equivalence of conditions (1′), (2′), and (3′).

(1′) ⇒ (3′) If condition (3′) is false then there is an $f \in L^\infty(\mu)$ such that $0 \leq f \leq 1$ and $\mu(f\hat{A}) = 0$ for all $A \in \mathfrak{A}$. In particular $\mu(f) = 0$. Define $\mu_{1\pm f}$ by

$$\mu_{1\pm f}(g) = \mu((\mathbb{1} \pm f)g)$$

and note that

$$\mu_{1\pm f}(\hat{A}) = \mu(\hat{A}) = \omega(A).$$

Therefore $\mu_{1\pm f} \in M_\omega(E_{\mathfrak{A}})$ and

$$\mu = \frac{\mu_{1+f} + \mu_{1-f}}{2},$$

i.e., condition (1′) is false.

(3′) ⇒ (2′) If the map $f \mapsto \kappa_\mu(f)$ is not faithful then there is a nonzero f such that $\kappa_\mu(f)$ is zero and then

$$\mu(f\hat{A}) = (\Omega_\omega, \kappa_\mu(f)\pi_\omega(A)\Omega_\omega) = 0$$

for all $A \in \mathfrak{A}$. Thus the affine functions are not dense in $L^1(\mu)$.

(2′) ⇒ (1′) If $\mu \notin \mathscr{E}(M_\omega(E_{\mathfrak{A}}))$ then $\mu = (\mu_1 + \mu_2)/2$ for two distinct measures $\mu_1, \mu_2 \in M_\omega(E_{\mathfrak{A}})$. As $\mu_1 \leq 2\mu$ there is a nonzero $f \in L^\infty(\mu)$ such that $\mu_1(g) = \mu((\mathbb{1} + f)g)$. One then has

$$\begin{aligned}\omega(A) = \mu_1(\hat{A}) &= \mu(\hat{A}) + \mu(f\hat{A}) \\ &= \omega(A) + (\Omega_\omega, \kappa_\mu(f)\pi_\omega(A)\Omega_\omega).\end{aligned}$$

By cyclicity $\kappa_\mu(f) = 0$ and condition (2′) is false.

Now we return to the proof that condition (2) implies condition (1) when $\mu \in \mathscr{E}(M_\omega(E_{\mathfrak{A}}))$.

Let

$$\nu = \sum_{i=1}^{n} \lambda_i \nu_i$$

be a convex decomposition of ν in terms of probability measures ν_i. As $\lambda_i \nu_i \leq \nu$ there are $f_i \in L^\infty_{1+}(\nu)$ such that $\lambda_i \, d\nu_i = f_i \, d\nu$. Clearly,

$$\sum_{i=1}^{n} f_i = \mathbb{1}.$$

Now let $g_i \in L^\infty_{1+}(\mu)$ be such that $\kappa_\mu(g_i) = \kappa_\nu(f_i)$ and define μ_i by $\lambda_i \, d\mu_i = g_i \, d\mu$. Now

$$\begin{aligned}\lambda_i \hat{A}(b(\mu_i)) &= \lambda_i \mu_i(\hat{A}) = \mu(g_i \hat{A}) \\ &= (\Omega_\omega, \kappa_\mu(g_i)\pi_\omega(A)\Omega_\omega) \\ &= (\Omega_\omega, \kappa_\nu(f_i)\pi_\omega(A)\Omega_\omega) \\ &= \nu(f_i \hat{A}) = \lambda_i \nu_i(\hat{A}) = \lambda_i \hat{A}(b(\nu_i)).\end{aligned}$$

Therefore the barycenters $b(\mu_i)$ and $b(\nu_i)$ of μ_i and ν_i, respectively, are equal. But

$$\begin{aligned}\mathbb{1} = \kappa_\nu\left(\sum_{i=1}^{n} f_i\right) &= \sum_{i=1}^{n} \kappa_\nu(f_i) \\ &= \sum_{i=1}^{n} \kappa_\mu(g_i) = \kappa_\mu\left(\sum_{i=1}^{n} g_i\right).\end{aligned}$$

Now $\mu \in \mathscr{E}(M_\omega(E_\mathfrak{A}))$ and hence, by condition (2′), the map κ_μ is faithful. Therefore

$$\sum_{i=1}^{n} g_i = \mathbb{1}$$

and

$$\mu = \sum_{i=1}^{n} \lambda_i \mu_i.$$

One immediately concludes from Proposition 4.2.1 that $\mu \succ \nu$.

We mentioned above that an alternative proof of the implication (2) ⇒ (1) in Theorem 4.2.3 can be constructed with the aid of Proposition 4.2.4. Explicitly if μ is the orthogonal measure corresponding to the abelian algebra $\pi_\omega(\mathfrak{A})'$ then the map $f \in L^\infty(\mu) \mapsto \kappa_\mu(f) \in \pi_\omega(\mathfrak{A})'$ is a *-morphism, and

$$\pi_\omega(\mathfrak{A})' = \{\kappa_\mu(f);\ f \in L^\infty(\mu)\}.$$

As the map is automatically isometric one has

$$(\pi_\omega(\mathfrak{A})')_{1+} = \{\kappa_\mu(f);\ f \in L^\infty_{1+}(\mu)\},$$

where $(\pi_\omega(\mathfrak{A})')_{1+}$ denotes the positive part of the unit ball of $\pi_\omega(\mathfrak{A})'$. Now if $\nu \in M_\infty(E_\mathfrak{A})$ then

$$\{\kappa_\nu(g);\ g \in L^\infty_{1+}(\nu)\} \subseteq (\pi_\omega(\mathfrak{A})')_{1+} = \{\kappa_\mu(f);\ f \in L^\infty_{1+}(\mu)\}.$$

But $\mu \in \mathscr{E}(M_\omega(E_\mathfrak{A}))$ by Corollary 4.1.23 and hence $\nu \prec \mu$ by Proposition 4.2.2. Thus μ is the unique maximal measure in $M_\omega(E_\mathfrak{A})$.

Finally, we can deduce support properties of the orthogonal measures associated with maximal abelian subalgebras of $\pi_\omega(\mathfrak{A})'$ by combining the foregoing results with those of Section 4.1.

Theorem 4.2.5. *Let $\mathfrak{A}$ be a C^*-algebra with identity and ω a state over $\mathfrak{A}$. Let $\mathfrak{B}$ be a maximal abelian subalgebra of $\pi_\omega(\mathfrak{A})'$ and μ the corresponding orthogonal measure over the state space $E_\mathfrak{A}$. It follows that μ is pseudosupported by the pure states $\mathscr{E}(E_\mathfrak{A})$ over $\mathfrak{A}$.*

Moreover, if ω is contained in a face F satisfying the separability condition S *then the extremal points $\mathscr{E}(F)$, of F, form a Baire subset of the pure states over $\mathfrak{A}$, and*

$$\mu(\mathscr{E}(F)) = 1.$$

PROOF. The measure μ is maximal in $M_\omega(E_\mathfrak{A})$ by Theorem 4.2.2. Hence it is pseudosupported by $\mathscr{E}(E_\mathfrak{A})$ by Theorem 4.1.11.

Now assume that $\mathfrak{A}$ contains a sequence of C^*-subalgebras $\{\mathfrak{A}_n\}_{n\geq 1}$ such that $\mathfrak{A} = \overline{\bigcup_n \mathfrak{A}_n}$ and each $\mathfrak{A}_n$ contains a separable ideal $\mathfrak{G}_n$, and assume that

$$\omega \in F = \{\varphi;\, \varphi \in E_\mathfrak{A},\ \|\varphi|_{\mathfrak{G}_n}\| = 1, n = 1, 2, \ldots\}.$$

Then $\mathfrak{H}_\omega$ is separable by Proposition 4.1.34. Hence $\mathscr{L}(\mathfrak{H}_\omega)$ is separable in the weak operator topology. It follows that $\mathfrak{B}$ contains a separable C^*-subalgebra $\mathfrak{B}_0$ such that $\mathfrak{B}_0$ is weakly dense in $\mathfrak{B}$. Let now $\mathfrak{C}$ be the C^*-algebra generated by $\pi_\omega(\mathfrak{A})(=\mathfrak{A})$ and $\mathfrak{B}_0$, and let $\mathfrak{C}_n$ be the C^*-algebra generated by $\mathfrak{B}_0$ and $\mathfrak{A}_n$, and finally, $\mathfrak{J}_n$ the C^*-algebra generated by $\mathfrak{B}_0$ and $\mathfrak{G}_n$. As $\mathfrak{B}_0 \subseteq \mathfrak{A}'$, it follows that $\mathfrak{J}_n$ is a separable ideal in $\mathfrak{C}_n$, and we have that $\bigcup_n \mathfrak{C}_n$ is dense in $\mathfrak{C}$. Define

$$\tilde{F} = \{\tilde{\varphi};\, \tilde{\varphi} \in E_\mathfrak{C},\ \|\tilde{\varphi}|_{\mathfrak{J}_n}\| = 1, n = 1, 2, \ldots\}.$$

Then $\tilde{\omega} \in \tilde{F}$. Since $\mathfrak{C}' = \mathfrak{B}$ is abelian, it follows from Proposition 4.1.34 and Theorem 4.2.4 that $\mathscr{E}(\tilde{F})$ is a Baire set of the form $\partial_f(E_\mathfrak{C}) \cap \tilde{F}$ for some $f \in S(E_\mathfrak{C})$, $\tilde{\mu}(\partial_f(E_\mathfrak{C})) = 1$, $\tilde{\mu}(\tilde{F}) = 1$, and hence $\tilde{\mu}(\mathscr{E}(\tilde{F})) = 1$. Now, $r(\mathscr{E}(\tilde{F})) \subseteq \mathscr{E}(E_\mathfrak{A})$ by Proposition 4.1.37, and $r(\mathscr{E}(\tilde{F}))$ contains an F_σ-set U with $\mu(U) = 1$ by Lemma 4.1.36. It then follows from Proposition 4.1.34 that $\mu(U \cap F) = 1$. But as $U \cap F \subseteq \mathscr{E}(F)$ and $\mathscr{E}(F)$ is a Baire set it follows, finally, that $\mu(\mathscr{E}(F)) = 1$.

We conclude our discussion of extremal decompositions with two examples which illustrate special structures of state spaces. Although Theorem 4.2.4 established criteria for a given state $\omega \in E_\mathfrak{A}$ to be the barycenter of a unique maximal measure, we did not examine criteria which ensure that every $\omega \in E_\mathfrak{A}$ has this property. The general theory of barycentric decompositions, Theorem 4.1.15, establishes that this is equivalent to $E_\mathfrak{A}$ being a simplex and the first example shows that this occurs if, and only if, $\mathfrak{A}$ is abelian.

EXAMPLE 4.2.6. Let $\mathfrak{A}$ be a C^*-algebra; then the following conditions are equivalent:

(1) the state space $E_\mathfrak{A}$ is a simplex;
(2) $\mathfrak{A}$ is abelian;
(3) the positive elements $\mathfrak{A}_+$ of $\mathfrak{A}$ form a lattice.

This equivalence is valid for $\mathfrak{A}$ with, or without, an identity. The proof of the general case is reduced to the special case by adjoining an identity. Next note if $\mathfrak{A}$ is abelian then $\mathfrak{A} = C(X)$, where X is the space of characters of $\mathfrak{A}$ by Theorem 2.1.11(A) in Section 2.3.5, and $E_{\mathfrak{A}}$ is the set of probability measures on $C(X)$. Thus (2) $\Rightarrow$ (1) and (2) $\Rightarrow$ (3). The equivalence of conditions (1) and (2) is then established by the following argument.

(1) $\Rightarrow$ (2) Assume that (2) is false; then there exist two elements $A, B \in \mathfrak{A}$ such that

$$AB - BA = C \neq 0.$$

Moreover, Lemma 2.3.23 asserts the existence of a pure state ω over $\mathfrak{A}$ such that $\omega(C^*C) = \|C\|^2$. The associated irreducible representation $(\mathfrak{H}_\omega, \pi_\omega)$ must be on a space $\mathfrak{H}_\omega$ of dimension greater than one because the contrary assumption would imply $\pi_\omega(C) = 0$. Let ψ_1 and ψ_2 be any two orthogonal unit vectors in $\mathfrak{H}_\omega$ and define $\omega_i(A) = (\psi_i, \pi_\omega(A)\psi_i)$ for $A \in \mathfrak{A}$, $i = 1, 2$. Moreover, introduce

$$\omega_\pm(A) = ((\psi_1 \pm \psi_2), \pi_\omega(A)(\psi_1 \pm \psi_2))/2 \text{ for } A \in \mathfrak{A}.$$

One has

$$\omega_1(A) + \omega_2(A) = \omega_+(A) + \omega_-(A).$$

Hence the two distinct maximal measures $(\delta_{\omega_1} + \delta_{\omega_2})/2$, $(\delta_{\omega_+} + \delta_{\omega_-})/2$ have the same barycenter and $E_{\mathfrak{A}}$ is not a simplex.

The proof of the equivalence is now completed by showing that (3) implies (1).

(3) $\Rightarrow$ (1) Let φ be any hermitian functional over $\mathfrak{A}$ and define $\varphi^{(+)}$ by

$$\varphi^{(+)}(A) = \sup\{\varphi(B); 0 \leq B \leq A\}.$$

The function $\varphi^{(+)}$ is clearly bounded and we next prove that it is linear. For this we argue that if $0 \leq B \leq A_1 + A_2$ with the $A_i \geq 0$ then the lattice property implies the existence of B_1, B_2 such that $B = B_1 + B_2$ and $0 \leq B_1 \leq A_1$, $0 \leq B_2 \leq A_2$. This is established by defining

$$B_1 = A_1 \wedge B, \qquad B_2 = B - B_1.$$

Clearly $0 \leq B_1 \leq A_1$ and $0 \leq B_2$. But $B - A_2 \leq B$ and $B - A_2 \leq A_1$. Therefore one has $B - A_2 \leq A_1 \wedge B$, which is equivalent to $B_2 \leq A_2$. Using this decomposition one then has

$$\begin{aligned}\varphi^{(+)}(A_1 + A_2) &= \sup\{\varphi(B_1 + B_2); 0 \leq B_1 \leq A_1, 0 \leq B_2 \leq A_2\} \\ &= \varphi^{(+)}(A_1) + \varphi^{(+)}(A_2).\end{aligned}$$

Now let ω_1 and ω_2 be positive linear functionals over $\mathfrak{A}$ and define

$$\omega_1 \vee \omega_2 = (\omega_1 - \omega_2)^{(+)} + \omega_2.$$

For $A \geq 0$ and $\varepsilon \geq 0$ one then has a B with $0 \leq B \leq A$ such that

$$(\omega_1 - \omega_2)^{(+)}(A) \leq \omega_1(B) - \omega_2(B) + \varepsilon.$$

Therefore if $\omega \geq \omega_1$ and $\omega \geq \omega_2$ one has

$$\begin{aligned}(\omega_1 \vee \omega_2)(A) &\leq \omega_1(B) + \omega_2(A - B) + \varepsilon \\ &\leq \omega(A) + \varepsilon,\end{aligned}$$

i.e., $\omega_1 \vee \omega_2$ is the least upper bound of ω_1 and ω_2. A similar argument shows that the greatest lower bound of ω_1, ω_2 exists and is given by

$$\omega_1 \wedge \omega_2 = \omega_1 - (\omega_1 - \omega_2)^{(+)}.$$

Hence the positive functionals over $\mathfrak{A}$ form a lattice and $E_{\mathfrak{A}}$ is a simplex.

Note that the $\varphi^{(+)}$ introduced in this proof is just the positive part φ_+ of φ in the Jordan decomposition. The negative part φ_- of φ can be calculated by a similar ansatz and the obvious relation $\varphi_- = (-\varphi)_+$.

EXAMPLE 4.2.7. Let $\mathfrak{A} = M_2$ be the C^*-algebra of all 2×2 complex matrices. It follows from Example 4.2.6 that $E_{\mathfrak{A}}$ is not a simplex, and we will show that $E_{\mathfrak{A}}$ is affinely isomorphic to the unit ball in $\mathbb{R}^3$.

Define the Pauli matrices $\sigma_0, \sigma_1, \sigma_2, \sigma_3$ in $\mathfrak{A}_{sa}$ by

$$\sigma_0 = \begin{pmatrix} 1 & 0 \\ 0 & 1 \end{pmatrix}, \quad \sigma_1 = \begin{pmatrix} 0 & 1 \\ 1 & 0 \end{pmatrix}, \quad \sigma_2 = \begin{pmatrix} 0 & -i \\ i & 0 \end{pmatrix}, \quad \sigma_3 = \begin{pmatrix} 1 & 0 \\ 0 & -1 \end{pmatrix}.$$

Then the real linear span of $\{\sigma_i\}$ is just $\mathfrak{A}_{sa}$. If $\sum_i \alpha_i \sigma_i \in \mathfrak{A}_{sa}$ one computes

$$\mathrm{Tr}\left(\sum_i \alpha_i \sigma_i\right) = 2\alpha_0,$$

$$\det\left(\sum_i \alpha_i \sigma_i\right) = \alpha_0^2 - \alpha_1^2 - \alpha_2^2 - \alpha_3^2.$$

Hence, $\sum_i \alpha_i \sigma_i \in \mathfrak{A}_+$ if, and only if, $\alpha_0 \geq 0, \alpha_0^2 \geq \alpha_1^2 + \alpha_2^2 + \alpha_3^2$, i.e., $\mathfrak{A}_+$ is identified with the positive light cone in the Minkowski space $\mathbb{R}^4$.

Now any state $\omega \in E_{\mathfrak{A}}$ is given by a unique positive matrix $\rho \in \mathfrak{A}_+$ with $\mathrm{Tr}(\rho) = 1$, by $\omega(A) = \mathrm{Tr}(\rho A)$. The map $\omega \mapsto \rho$ is affine. Thus $E_{\mathfrak{A}}$ is affinely isomorphic to

$$\{(\alpha_i) \in \mathbb{R}^4; \alpha_0 = \tfrac{1}{2}, \alpha_1^2 + \alpha_2^2 + \alpha_3^2 \leq \tfrac{1}{4}\},$$

i.e., $E_{\mathfrak{A}}$ is affinely isomorphic to the unit ball in $\mathbb{R}^3$.

Finally, it is easy to construct the natural, positive cone $\mathscr{P}_\tau$ associated with M_2 in the representation defined by $\tau = (\frac{1}{2})\mathrm{Tr}$. In fact, $\{2^{-1/2}\pi_\tau(\sigma_i)\Omega_\tau\}_{0 \leq i \leq 3}$ forms an orthonormal basis for $\mathfrak{H}_\tau$, and since τ is a trace we have $\mathscr{P}_\tau = \pi_\tau(\mathfrak{A}_+)\Omega_\tau$. This gives an identification of $\mathfrak{H}_\tau$ with the Hilbert space $\mathbb{C}^4$ such that

$$\mathscr{P}_\tau = \{(\alpha_i) \in \mathbb{R}^4; \alpha_0 \geq 0, \alpha_1^2 + \alpha_2^2 + \alpha_3^2 \leq \alpha_0^2\},$$

i.e., $\mathscr{P}_\tau$ is just the positive light-cone in $\mathbb{R}^4$.

More generally, it is not hard to show that if $\mathfrak{A}$ is an arbitrary C^*-algebra with identity, and $\omega_1, \omega_2 \in \mathscr{E}(E_{\mathfrak{A}})$, then the face generated by $\{\omega_1, \omega_2\}$ in $E_{\mathfrak{A}}$ is either the line segment between ω_1 and ω_2 (if π_{ω_1} and π_{ω_2} are inequivalent) or is affinely isomorphic to the unit sphere in $\mathbb{R}^3$ (if $\omega_1 \neq \omega_2$ and π_{ω_1} is equivalent to π_{ω_2}). This is an important abstract property of the convex, compact set $E_{\mathfrak{A}}$.

4.2.2. Central and Subcentral Decompositions

In this subsection we examine the decompositions of a state ω which are associated with the von Neumann subalgebras of the center $\mathfrak{Z}_\omega = \pi_\omega(\mathfrak{A})'' \cap \pi_\omega(\mathfrak{A})'$ of the representation $\pi_\omega(\mathfrak{A})$. The orthogonal measure corresponding to $\mathfrak{Z}_\omega$ is usually called the *central measure* and we refer to the measures associated with von Neumann subalgebras of $\mathfrak{Z}_\omega$ as *subcentral measures.* This class of measures is of particular importance in physical applications because the elements of $\pi_\omega(\mathfrak{A})''$ are interpreted as observables of a system in the state ω and the center $\mathfrak{Z}_\omega$ corresponds to the set of invariants of the system. The central measure then gives the probability distribution of the values of these invariants, and the associated decomposition is an expression of ω in terms of states in which the invariants have specific values. Specific subalgebras of $\mathfrak{Z}_\omega$, such as the commutant algebra $\mathfrak{Z}_\omega{}^c$ and the algebra at infinity $\mathfrak{Z}_\omega{}^\perp$ introduced by Definition 2.6.4, might have a particular physical significance and hence the corresponding subcentral decompositions are of interest.

We begin our analysis by showing that the subcentral measures can be characterized by a strengthened condition of orthogonality. To introduce this condition we first recall that two representations π_1 and π_2 of a C^*-algebra $\mathfrak{A}$ are quasi-equivalent if each π_1-normal state is π_2-normal and conversely. As a complement, we say that π_1 and π_2 are *disjoint*, written $\pi_1 \between \pi_2$, if no π_1-normal state is π_2-normal and conversely. Correspondingly, two positive linear functionals ω_1 and ω_2 on $\mathfrak{A}$ are *disjoint*, $\omega_1 \between \omega_2$, if π_{ω_1} and π_{ω_2} are disjoint. If $\mathfrak{A}$ is an abelian C^*-algebra, the notions of quasi-equivalence and disjointness reduce to the notions of equivalence and disjointness of regular Borel measures.

It follows immediately from the definition that $\pi_1 \between \pi_2$ if and only if π_1 and π_2 have no quasi-equivalent subrepresentations, which again is equivalent to π_1 and π_2 having no unitary equivalent subrepresentations (Theorem 2.4.26). For our purposes the following characterization of disjointness of states will be useful.

Lemma 4.2.8. *Let ω_1, ω_2 be positive linear functionals over the C^*-algebras $\mathfrak{A}$, and let $\omega = \omega_1 + \omega_2$. The following conditions are equivalent:*

(1) *ω_1 and ω_2 are disjoint;*
(2) *there is a projection $P \in \pi_\omega(\mathfrak{A})'' \cap \pi_\omega(\mathfrak{A})'$ such that*

$$\omega_1(A) = (\Omega_\omega, P\pi_\omega(A)\Omega_\omega),$$

$$\omega_2(A) = (\Omega_\omega, (\mathbb{1} - P)\pi_\omega(A)\Omega_\omega).$$

In particular, disjointness of ω_1 and ω_2 implies orthogonality.

PROOF. (1) $\Rightarrow$ (2) If ω' is a positive linear functional such that $\omega' \le \omega_1$ and $\omega' \le \omega_2$, then ω' is both π_{ω_1}-normal and π_{ω_2}-normal. Hence $\omega' = 0$. It follows from

Lemma 4.1.19 that $\omega_1 \perp \omega_2$ and, in particular, that there exists a projection $P \in \pi_\omega(\mathfrak{A})'$ with the property that

$$\omega_1(A) = (\Omega_\omega, P\pi_\omega(A)\Omega_\omega),$$

etc. Let $B \in \pi_\omega(\mathfrak{A})'$, $\xi \in P\mathfrak{H}_\omega$ and consider the positive functional

$$\omega'(A) = ((\mathbb{1} - P)B\xi, \pi_\omega(A)(\mathbb{1} - P)B\xi).$$

Then ω' is manifestly π_{ω_2}-normal, and since

$$\omega'(A) \leq \|(\mathbb{1} - P)B\|^2(\xi, \pi_\omega(A)\xi)$$

for $A \geq 0$, ω' is also π_{ω_1}-normal. It follows that $\omega' = 0$ and hence $(\mathbb{1} - P)BP = 0$ for all $B \in \pi_\omega(\mathfrak{A})'$. Thus $P \in \pi_\omega(\mathfrak{A})''$ and, finally, $P \in \pi_\omega(\mathfrak{A})'' \cap \pi_\omega(\mathfrak{A})'$.

(2) $\Rightarrow$ (1) If (2) holds there exists no partial isometry U on $\mathfrak{H}_\omega$ such that $U \in \pi_\omega(\mathfrak{A})'$ and $(\mathbb{1} - P)UP = U$. But this amounts to saying that π_{ω_1} and π_{ω_2} have no unitary equivalent subrepresentations, hence $\omega_1 \between \omega_2$.

The following proposition gives an alternative characterization of the notion of subcentral measure.

Proposition 4.2.9. *Let $\mathfrak{A}$ be a C^*-algebra with identity, ω a state over $\mathfrak{A}$, and let $\mu \in M_\omega(E_\mathfrak{A})$. The following conditions are equivalent:*

(1) *for any Borel set $S \subseteq E_\mathfrak{A}$ one has*

$$\left(\int_S d\mu(\omega')\,\omega'\right) \between \left(\int_{E_\mathfrak{A}\setminus S} d\mu(\omega')\,\omega'\right);$$

(2) *μ is subcentral, i.e., μ is orthogonal, and the associated abelian subalgebra $\kappa_\mu(L^\infty(\mu))$ of $\pi_\omega(\mathfrak{A})'$ is contained in the center $\pi_\omega(\mathfrak{A})'' \cap \pi_\omega(\mathfrak{A})'$ of the representation $\pi_\omega(\mathfrak{A})$.*

PROOF. (1) $\Rightarrow$ (2) It follows from Lemma 4.2.8 that μ is orthogonal and κ_μ maps projections in $L^\infty(\mu)$ into central projections. Since κ_μ is a *-isomorphism by Proposition 4.1.22, $\kappa_\mu(L^\infty(\mu))$ is contained in the center of $\pi_\omega(\mathfrak{A})$.

(2) $\Rightarrow$ (1) If S is a Borel set, then $\kappa_\mu(\chi_S)$ and $\kappa_\mu(\chi_{E_\mathfrak{A}\setminus S})$ are mutually orthogonal projections in the center of $\pi_\omega(\mathfrak{A})$ with sum $\mathbb{1}$; thus

$$\left(\int_S d\mu(\omega')\,\omega'\right) \between \left(\int_{E_\mathfrak{A}\setminus S} d\mu(\omega')\,\omega'\right)$$

by the formula

$$\int_S d\mu(\omega')\,\omega'(A) = (\Omega_\omega, \kappa_\mu(\chi_S)\pi_\omega(A)\Omega_\omega)$$

and Lemma 4.2.8.

It follows from Theorem 4.1.28 that the central measure $\mu_{\mathfrak{Z}_\omega}$ is the smallest measure in $M_\omega(E_\mathfrak{A})$ which maximizes all the subcentral measures. Moreover, Lemma 4.1.26 implies that $\mu_{\mathfrak{Z}_\omega}$ is the weak* limit of the monotone net of subcentral measures of finite support. This latter result can be used to strengthen the geometric characterization of the central measure.

Theorem 4.2.10. *Let $\mathfrak{A}$ be a C^*-algebra with identity and ω a state over $\mathfrak{A}$. It follows that each subcentral measure $\nu \in \mathcal{O}_\omega(E_\mathfrak{A})$ is dominated by all maximal measures $\mu \in M_\omega(E_\mathfrak{A})$.*

Conversely, $\mu_{\mathfrak{Z}_\omega}$ is the largest measure in $M_\omega(E_\mathfrak{A})$ which is dominated by all maximal measures in $M_\omega(E_\mathfrak{A})$.

PROOF. To prove the first statement we argue exactly as in the proof of (2) $\Rightarrow$ (1) in Theorem 4.2.3. One reduces the problem to finding a common upper bound ρ for two measures μ, $\nu \in M_\omega(E_\mathfrak{A})$ of finite support where ν is assumed to be subcentral. But this is achieved by repeating the construction used in the proof of Theorem 4.2.3 and noting that the subcentrality of ν implies $Z_j \in \mathfrak{Z}_\omega$ and hence the products $T_i Z_j$ are positive.

To prove the converse statement we must show that each measure $\mu \in M_\omega(E_\mathfrak{A})$ that is dominated by all the maximal measures in $M_\omega(E_\mathfrak{A})$ is also dominated by the central measure $\mu_{\mathfrak{Z}_\omega}$. Now the maximal orthogonal measures in $\mathcal{O}_\omega(E_\mathfrak{A})$ are maximal in $M_\omega(E_\mathfrak{A})$ by Theorem 4.2.2. Hence $\{\kappa_\mu(f); f \in L^\infty_{1+}(\mu)\}$ is contained in all maximal abelian von Neumann subalgebras of $\pi_\omega(\mathfrak{A})'$, by Theorem 4.2.2 and Proposition 4.2.4. Hence,

$$\{\kappa_\mu(f); f \in L^\infty_{1+}(\mu)\} \subseteq (\mathfrak{Z}_\omega)_{1+},$$

where $(\mathfrak{Z}_\omega)_{1+}$ is the positive part of the unit ball of the center. But

$$(\mathfrak{Z}_\omega)_{1+} = \{\kappa_{\mu_{\mathfrak{Z}_\omega}}(f); f \in L^\infty_{1+}(\mu_{\mathfrak{Z}_\omega})\}$$

and $\mu_{\mathfrak{Z}_\omega} \in \mathscr{E}(M_\omega(E_\mathfrak{A}))$ by Corollary 4.1.23. Therefore

$$\mu \prec \mu_{\mathfrak{Z}_\omega}$$

by Proposition 4.2.4.

To conclude this section we establish the natural result that $\mu_{\mathfrak{Z}_\omega}$ is pseudosupported by the factor states $F_\mathfrak{A}$ and that with some separability assumptions it is actually supported by $F_\mathfrak{A}$.

Theorem 4.2.11. *Let $\mathfrak{A}$ be a C^*-algebra with identity, ω a state over $\mathfrak{A}$, and $\mu_{\mathfrak{Z}_\omega}$ the associated central measure. It follows that $\mu_{\mathfrak{Z}_\omega}$ is pseudosupported by the factor states $F_\mathfrak{A}$ over $\mathfrak{A}$. Moreover, if ω is contained in a face F satisfying the separability condition* S, *then μ is supported by a subset $G \subseteq E_\mathfrak{A}$ such that*

(1) *G is a F_σ-subset of $E_\mathfrak{A}$;*
(2) *$G \subseteq F_\mathfrak{A} \cap F$.*

PROOF. We may assume $\pi_\omega(\mathfrak{A})$ is faithful, because if $\mathfrak{J} = \ker \pi_\omega(\mathfrak{A})$ and ρ; $\mathfrak{A} \mapsto \mathfrak{A}/\mathfrak{J}$ is the quotient map, then φ is a pure state of $\mathfrak{A}/\mathfrak{J}$ if, and only if, $\varphi \circ \rho$ is a pure state of $\mathfrak{A}$, and the states of this form are a weakly*-closed subset of $E_\mathfrak{A}$.

Next we may view $\mathfrak{A}$ as a C^*-subalgebra of the C^*-algebra $\mathfrak{C}$ generated by $\pi_\omega(\mathfrak{A})$ and $\pi_\omega(\mathfrak{A})'$. The commutant of $\mathfrak{C}$ is $\mathfrak{C}' = \mathfrak{Z}_\omega = \pi_\omega(\mathfrak{A})' \cap \pi_\omega(\mathfrak{A})''$. Let $\tilde{\omega}$ be the extension of ω to $\mathfrak{C}$ defined by

$$\tilde{\omega}(C) = (\Omega_\omega, C\Omega_\omega), \qquad C \in \mathfrak{C},$$

and let $\tilde{\mu}$ be the element of $\mathcal{O}_{\tilde{\omega}}(E_{\mathfrak{C}})$ corresponding to $\mathfrak{Z}_{\omega} = \mathfrak{Z}_{\tilde{\omega}}$. If $r;\ E_{\mathfrak{C}} \mapsto E_{\mathfrak{A}}$ is the restriction mapping then $r\tilde{\omega} = \omega$ and $\mu = \tilde{\mu} \circ r^{-1}$. The last identification follows from the relation

$$\tilde{\mu}(\hat{C}_1 \hat{C}_2 \cdots \hat{C}_n) = (\Omega_{\omega}, C_1 P C_2 P \cdots P C_n \Omega_{\omega}),$$

where $P = [\mathfrak{Z}_{\omega}\Omega_{\omega}] = [\mathfrak{Z}_{\tilde{\omega}}\Omega_{\tilde{\omega}}]$.

Now, $\tilde{\mu}$ is a maximal measure in $M_{\tilde{\omega}}(E_{\mathfrak{C}})$, by Theorem 4.2.4, and one concludes from Theorem 4.1.11 that $\tilde{\mu}$ is pseudosupported by the pure states on $\mathfrak{C}$. Next, $r(\mathscr{E}(E_{\mathfrak{C}})) \subseteq F_{\mathfrak{A}}$ by Proposition 4.1.37, and hence μ is pseudosupported by $F_{\mathfrak{A}}$ by Lemma 4.1.36.

Now consider the second statement of the theorem. Let $\{\mathfrak{A}_n\}_{n \geq 1}$ be a sequence of C^*-subalgebras of $\mathfrak{A}$ such that $\mathfrak{A} = \overline{\bigcup_n \mathfrak{A}_n}$, and let $\mathfrak{G}_n$ be a separable ideal in $\mathfrak{A}_n$ such that

$$F = \{\varphi;\ \varphi \in E_{\mathfrak{A}},\ \|\varphi|_{\mathfrak{G}_n}\| = 1, n \geq 1\}.$$

If $\omega \in F$, then $\mathfrak{H}_{\omega}$ is separable by Proposition 4.1.34, and the argument used in the proof of Theorem 4.2.5 implies the existence of a separable C^*-subalgebra $\mathfrak{B}_0 \subseteq \pi_{\omega}(\mathfrak{A})'$ such that $\mathfrak{B}_0$ is weakly dense in $\pi_{\omega}(\mathfrak{A})'$. Now, redefining $\mathfrak{C}$ as the C^*-algebra generated by $\mathfrak{B}_0$ and $\pi_{\omega}(\mathfrak{A})$ and arguing exactly as in the proof of Theorem 4.2.5, we find an F_{σ}-set $U \subseteq F_{\mathfrak{A}}$ such that $\mu(U) = 1$. Since F is a stable face $\mu(F) = 1$, and hence $\mu(F \cap U) = 1$. The regularity of the Borel measure μ then implies the existence of an F_{σ}-subset $G \subseteq F \cap U \subseteq F \cap F_{\mathfrak{A}}$ such that $\mu(G) = 1$.

Remark. In the simplest case covered by Theorem 4.2.11, the case of separable $\mathfrak{A}$, one can prove that the set $F_{\mathfrak{A}}$ of all factor states is a Borel subset of $E_{\mathfrak{A}}$ (see Notes and Remarks).

EXAMPLE 4.2.12. As an application of the foregoing theorem consider a quasi-local algebra constructed as follows. Let I be a countable index set and I_f the directed set of finite subsets of I, where the direction is by inclusion. Associate with each $\alpha \in I$ a separable Hilbert space $\mathfrak{H}_{\alpha}$, with each $\Lambda \in I_f$ the tensor product space

$$\mathfrak{H}_{\Lambda} = \bigotimes_{\alpha \in \Lambda} \mathfrak{H}_{\alpha},$$

and define $\mathfrak{A}_{\Lambda} = \mathscr{L}(\mathfrak{H}_{\Lambda})$. Let $\mathfrak{A}$ be the C^*-algebra generated by the union of the $\mathfrak{A}_{\Lambda}$. The algebra $\mathfrak{A}$ is quasi-local in the sense of Definition 2.6.3 and a state ω over $\mathfrak{A}$ is locally normal in the sense of Definition 2.6.6 if, and only if, $\omega \in F$, where

$$F = \{\omega;\ \omega \in E_{\mathfrak{A}},\ \|\omega|_{\mathscr{LC}(\mathfrak{H}_{\Lambda})}\| = 1, \Lambda \in I_f\}$$

(compare Example 4.1.35(5)). Now consider the algebra at infinity

$$\mathfrak{Z}_{\omega}^{\perp} = \bigcap_{\Lambda \in I_f} \left(\bigcup_{\Lambda' \cap \Lambda = \emptyset} \pi_{\omega}(\mathfrak{A}_{\Lambda'}) \right)''.$$

It follows from Theorem 2.6.10 that $\mathfrak{Z}_{\omega}^{\perp} = \mathfrak{Z}_{\omega}$ and hence the central decomposition coincides with the decomposition at infinity, and conversely. Moreover, Theorems 2.6.10 and 4.2.10 imply that the associated measure $\mu_{\mathfrak{Z}_{\omega}}$ is supported by an F_{σ}-subset of locally normal states with trivial algebra at infinity.

4.3. Invariant States

4.3.1. Ergodic Decompositions

We conclude our description of decomposition theory for states with an examination of states which are invariant under a group of *-automorphisms. Thus in this section we examine both a C^*-algebra $\mathfrak{A}$ with identity and a group G represented as *-automorphisms of $\mathfrak{A}$. We denote the action of G by

$$A \in \mathfrak{A} \mapsto \tau_g(A) \in \mathfrak{A}$$

for all $g \in G$. A state ω over $\mathfrak{A}$ is then said to be *G-invariant* if

$$\omega(A) = \omega(\tau_g(A))$$

for all $g \in G$ and $A \in \mathfrak{A}$. The states $E_{\mathfrak{A}}$ of $\mathfrak{A}$ are a convex weakly* compact subset of the dual $\mathfrak{A}^*$, of $\mathfrak{A}$, and it follows immediately that the G-invariant states form a convex, weakly* closed, hence compact, subset of $E_{\mathfrak{A}}$. We denote this set by $E_{\mathfrak{A}}{}^G$. Our aim is to decompose a state $\omega \in E_{\mathfrak{A}}{}^G$ in terms of the extremal points $\mathscr{E}(E_{\mathfrak{A}}{}^G)$ of $E_{\mathfrak{A}}{}^G$. The extremal G-invariant states $\mathscr{E}(E_{\mathfrak{A}}{}^G)$ are usually called *ergodic states*, or *G-ergodic states*, because of their relative purity, or indecomposability, among the invariant states. The corresponding decomposition is naturally referred to as the *ergodic decomposition* of a state.

Our discussion will be divided into three parts. In this subsection we analyze the existence and uniqueness of the ergodic decomposition. In Sections 4.3.2 and 4.3.3 we examine a variety of characterizations of the ergodic states $\mathscr{E}(E_{\mathfrak{A}}{}^G)$ and in Section 4.3.4 we analyze the decomposition of a G-ergodic state in terms of states with a lower invariance, e.g., invariance under a subgroup $H \subseteq G$. In physical applications the group G represents symmetries of the system and G-invariance of ω reflects the presence of these symmetries in the state ω. The G-ergodic states should correspond to the symmetric pure phases of the system and the decomposition with respect to a subgroup corresponds to an analysis of broken symmetries.

First recall that if ω is G-invariant then by Corollary 2.3.17 there exists a representation of G by unitary operators $U_\omega(G)$ acting on the Hilbert space

$\mathfrak{H}_\omega$ of the cyclic representation $(\mathfrak{H}_\omega, \pi_\omega, \Omega_\omega)$ associated with ω. This representation is uniquely determined by the two requirements.

$$U_\omega(g)\pi_\omega(A)U_\omega(g)^* = \pi_\omega(\tau_g(A))$$

for $A \in \mathfrak{A}$, $g \in G$, and

$$U_\omega(g)\Omega_\omega = \Omega_\omega$$

for $g \in G$. Throughout this section we often use the notation $(\mathfrak{H}_\omega, \pi_\omega, U_\omega, \Omega_\omega)$ to denote this quadruple of space, representations, and cyclic vector.

Next it is useful to define an action τ^* of G on the dual $\mathfrak{A}^*$ by

$$(\tau_g{}^*\varphi)(A) = \varphi(\tau_{g^{-1}}(A))$$

for $g \in G$, $A \in \mathfrak{A}$, and $\varphi \in \mathfrak{A}^*$. We have adopted a convention which ensures that τ^* is a group representation,

$$\begin{aligned}(\tau^*_{g_1g_2}\varphi)(A) &= \varphi(\tau_{(g_1g_2)^{-1}}(A)) \\ &= \varphi(\tau_{g_2^{-1}}\tau_{g_1^{-1}}(A)) = (\tau^*_{g_1}\tau^*_{g_2}\varphi)(A).\end{aligned}$$

We can define an action of G on the algebra $C(E_\mathfrak{A})$ of continuous functions on $E_\mathfrak{A}$ by second transposition. If μ is a Baire measure on $E_\mathfrak{A}$ which is invariant under this action, then we can extend this action to $L^\infty(\mu)$. For simplicity of notation we also denote this latter action by τ. Thus

$$(\tau_g f)(\omega) = f(\tau^*_{g^{-1}}\omega)$$

for $f \in L^\infty(\mu)$ and $\omega \in E_\mathfrak{A}$.

Proposition 4.3.1. *Let $\mathfrak{A}$ be a C^*-algebra with identity, G a group, $g \in G \mapsto \tau_g \in \mathrm{Aut}(\mathfrak{A})$ a representation of G as *-automorphisms of $\mathfrak{A}$, and ω a G-invariant state over $\mathfrak{A}$. There is a one-to-one correspondence between the following:*

(1) *the orthogonal measures μ, over $E_\mathfrak{A}$, with barycenter ω which satisfy the invariance condition*

$$\mu(\tau_g(f_1)f_2) = \mu(f_1 f_2)$$

for all $f_1, f_2 \in L^\infty(\mu)$ and $g \in G$;

(2) *the abelian von Neumann subalgebras $\mathfrak{B}$ of the commutant*

$$\{\pi_\omega(\mathfrak{A}) \cup U_\omega(G)\}';$$

(3) *the orthogonal projections P on $\mathfrak{H}_\omega$ such that*

$$P\Omega_\omega = \Omega_\omega, \qquad U_\omega(g)P = P,$$
$$P\pi_\omega(\mathfrak{A})P \subseteq \{P\pi_\omega(\mathfrak{A})P\}'.$$

PROOF. Theorem 4.1.25 has already established a one-to-one correspondence between orthogonal measures μ, abelian von Neumann subalgebras $\mathfrak{B} \subseteq \pi_\omega(\mathfrak{A})'$, and projections P such that $P\Omega_\omega = \Omega_\omega$, $P\pi_\omega(\mathfrak{A})P \subseteq \{P\pi_\omega(\mathfrak{A})P\}'$. It remains to incorporate the invariance properties and this is easily achieved by use of the explicit relationships, between corresponding elements μ, $\mathfrak{B}$, P, established in Theorem 4.1.25.

First assume μ to satisfy the invariance conditions. Then

$$\begin{aligned}
&(\Omega_\omega, \pi_\omega(A)U_\omega(g)\kappa_\mu(f)U_\omega(g^{-1})\pi_\omega(B)\Omega_\omega)\\
&= (\Omega_\omega, \pi_\omega(\tau_{g^{-1}}(A))\kappa_\mu(f)\pi_\omega(\tau_{g^{-1}}(B))\Omega_\omega)\\
&= \mu(f\tau_{g^{-1}}(\widehat{AB}))\\
&= \mu(f\widehat{AB})\\
&= (\Omega_\omega, \pi_\omega(A)\kappa_\mu(f)\pi_\omega(B)\Omega_\omega)
\end{aligned}$$

for all $A, B \in \mathfrak{A}$, $g \in G$, and $f \in L^\infty(\mu)$. Therefore by cyclicity

$$U_\omega(g)\kappa_\mu(f)U_\omega(g)^* = \kappa_\mu(f).$$

But $\mathfrak{B} = \{\kappa_\mu(f); f \in L^\infty(\mu)\}$ and hence $\mathfrak{B} \subseteq \{\pi_\omega(\mathfrak{A}) \cup U_\omega(G)\}'$.

Next assume $\mathfrak{B} \subseteq \{\pi_\omega(\mathfrak{A}) \cup U_\omega(G)\}'$ and note that $P = [\mathfrak{B}\Omega_\omega]$. It follows immediately that

$$U_\omega(g)P = P.$$

Finally, assume P to be in the third set described in the proposition and note that

$$\begin{aligned}
&\mu(\widehat{\tau_{g_1}(A_1)}\widehat{\tau_{g_2}(A_2)}\cdots\widehat{\tau_{g_n}(A_n)})\\
&= (\Omega_\omega, \pi_\omega(\tau_{g_1}(A_1))P\pi_\omega(\tau_{g_2}(A_2))P\cdots P\pi_\omega(\tau_{g_n}(A_n))\Omega_\omega)\\
&= (\Omega_\omega, \pi_\omega(A_1)P\pi_\omega(A_2)P\cdots P\pi_\omega(A_n)\Omega_\omega)\\
&= \mu(\hat{A}_1\hat{A}_2\cdots\hat{A}_n)
\end{aligned}$$

for all $A_1, A_2, \ldots, A_n \in \mathfrak{A}$ and $g_1, g_2, \ldots, g_n \in G$.

But the Stone–Weierstrass theorem ensures that each $f \in C(E_\mathfrak{A})$ can be uniformly approximated by a polynomial $\mathscr{P}$ in the $\hat{A}_i$ and by isometry $\tau_g f$ can be approximated by $\tau_g\mathscr{P}$ uniformly in $g \in G$. Using two such approximations one concludes that

$$\mu((\tau_g f_1)f_2) = \mu(f_1 f_2)$$

for all $f_1, f_2 \in C(E_\mathfrak{A})$ and $g \in G$.

The ordering of the orthogonal measures described in Proposition 4.3.1 naturally retains all the properties described in Theorem 4.1.28 for the ordering of orthogonal measures. Thus if $\mathfrak{B}_1$, $\mathfrak{B}_2$ are two abelian von Neumann subalgebras of $\{\pi_\omega(\mathfrak{A}) \cup U_\omega(G)\}'$ and $\mu_{\mathfrak{B}_1}, \mu_{\mathfrak{B}_2}$ are the corresponding measures then

$$\mathfrak{B}_1 \subseteq \mathfrak{B}_2 \Leftrightarrow \mu_{\mathfrak{B}_1} \prec \mu_{\mathfrak{B}_2}$$

etc.

The next result concerns the support properties of these special orthogonal measures.

Proposition 4.3.2. *Let $\mathfrak{A}$ be a C^*-algebra with identity, G a group, $g \in G \mapsto \tau_g \in \mathrm{Aut}(\mathfrak{A})$ a representation of G as *-automorphisms of $\mathfrak{A}$, and ω a G-invariant state over $\mathfrak{A}$. If μ is an orthogonal measure with barycenter ω then the following conditions are equivalent:*

(1) $\mu(\tau_g(f_1)f_2) = \mu(f_1 f_2)$ *for all* $f_1, f_2 \in C(E_\mathfrak{A})$ *and* $g \in G$;
(2) *the support μ is contained in the weakly* closed subset $E_\mathfrak{A}{}^G$ formed by the G-invariant states.*

Finally, if μ satisfies these conditions and is maximal then it is pseudo-supported by the ergodic states $\mathscr{E}(E_{\mathfrak{A}}^G)$ and if, moreover, ω is contained in a face F of $E_{\mathfrak{A}}^G$ which satisfies the separability condition S then μ is supported by $\mathscr{E}(E_{\mathfrak{A}}^G)$.

PROOF. Clearly (2) $\Rightarrow$ (1).

(1) $\Rightarrow$ (2) Let $\mathfrak{B} \subseteq \{\pi_\omega(\mathfrak{A}) \cup U_\omega(G)\}'$ be the abelian von Neumann algebra corresponding to μ by the correspondence established in Proposition 4.3.1, and let $\mathfrak{N}$ be a finite-dimensional abelian von Neumann subalgebra of $\mathfrak{B}$. The orthogonal measure $\mu_{\mathfrak{N}}$ corresponding to $\mathfrak{N}$ satisfies the invariance property and has finite support $\{\omega_1, \omega_2, \ldots, \omega_n\}$. But each $\omega_i \in E_{\mathfrak{A}}^G$ because

$$\omega_i(A) = \frac{(P_i\Omega_\omega, \pi_\omega(A)\Omega_\omega)}{(\Omega_\omega, P_i\Omega_\omega)}, \qquad A \in \mathfrak{A},$$

with P_i a projection in $\mathfrak{N}$. Lemma 4.1.26 then establishes that the measures $\mu_{\mathfrak{N}}$ converge in the weak* topology to μ and hence the $\mu_{\mathfrak{N}}$ also converge in the weak* topology on $C(E_{\mathfrak{A}}^G)^*$ to the measure $\nu \in M_\omega(E_{\mathfrak{A}}^G)$ obtained by restricting μ to $E_{\mathfrak{A}}^G$. Thus $\mu(E_{\mathfrak{A}}^G) = \nu(E_{\mathfrak{A}}^G) = 1$, i.e., the support of μ is in $E_{\mathfrak{A}}^G$.

If $\nu_1, \nu_2 \in \mathcal{O}_\omega(E_{\mathfrak{A}})$ are two orthogonal measures satisfying the invariance property then they can be viewed as measures on $E_{\mathfrak{A}}^G$, i.e., $\nu_1, \nu_2 \in M_\omega(E_{\mathfrak{A}}^G)$. We first argue that if $\nu_1 \succ \nu_2$ in $M_\omega(E_{\mathfrak{A}})$ then $\nu_1 \succ \nu_2$ in $M_\omega(E_{\mathfrak{A}}^G)$. For this it suffices to note that each $f \in S(E_{\mathfrak{A}}^G)$ has a convex lower semicontinuous extension $\hat{f}$ to $E_{\mathfrak{A}}$ obtained by setting $\hat{f}(\omega) = +\infty$ for $\omega \in E_{\mathfrak{A}} \backslash E_{\mathfrak{A}}^G$. Thus using the ordering one has

$$\nu_1(f) = \nu_1(\hat{f}) \geq \nu_2(\hat{f}) = \nu_2(f).$$

Consequently, if an orthogonal measure with the invariance property is maximal in $M_\omega(E_{\mathfrak{A}})$ then it is maximal in $M_\omega(E_{\mathfrak{A}}^G)$ and the pseudosupport statement follows from Theorem 4.1.11. Next remark that Proposition 4.1.34 established that

$$\mathscr{E}(F) = \partial_f(E_{\mathfrak{A}}) \cap F$$

for a function $f \in S(E_{\mathfrak{A}})$ which is strictly convex over F. Thus defining $F^G = F \cap E_{\mathfrak{A}}^G$ one concludes that F^G is a face in $E_{\mathfrak{A}}^G$ which satisfies the separability condition S and

$$\mathscr{E}(F^G) = \partial_f(E_{\mathfrak{A}}^G) \cap F^G \subseteq \mathscr{E}(E_{\mathfrak{A}}^G)$$

because f is also strictly convex over $F^G \subseteq F$. But F is stable and μ is supported by $F \cap E_{\mathfrak{A}}^G$ by (2). Further, μ is supported by $\partial_f(E_{\mathfrak{A}}^G)$ by Theorem 4.1.7. Thus μ is supported by $\mathscr{E}(F^G) \subseteq \mathscr{E}(E_{\mathfrak{A}}^G)$.

Next we give a characterization of uniqueness of maximal measures with a fixed barycenter $\omega \in E_{\mathfrak{A}}^G$.

Proposition 4.3.3. *Let $\mathfrak{A}$ be a C^*-algebra with identity, G a group acting as *-automorphisms τ of $\mathfrak{A}$, and ω a G-invariant state over $\mathfrak{A}$. The following conditions are equivalent:*

(1) *$M_\omega(E_{\mathfrak{A}}^G)$ contains a unique maximal measure ν;*
(2) *$\{\pi_\omega(\mathfrak{A}) \cup U_\omega(G)\}'$ is abelian.*

If these conditions are satisfied then ν *is the orthogonal measure corresponding to* $\{\pi_\omega(\mathfrak{A}) \cup U_\omega(G)\}'$.

PROOF. This follows by a repetition of the proof that (1) $\Leftrightarrow$ (2) in Theorem 4.2.3. One now replaces $\pi_\omega(\mathfrak{A})'$ by $\{\pi_\omega(\mathfrak{A}) \cup U_\omega(G)\}'$ and notes that if ν is a G-invariant measure with barycenter ω then $\{\kappa_\nu(f); f \in L^\infty(\nu)\} \subseteq \{\pi_\omega(\mathfrak{A}) \cup U_\omega(G)\}'$. This inclusion follows from the identity

$$\nu(f\tau_{g^{-1}}(\widehat{AB})) = \nu(f\widehat{AB})$$

by the same calculation used at the beginning of the proof of Proposition 4.3.1.

Our next aim is to establish more useful criteria for the uniqueness of a maximal orthogonal measure $\mu \in M_\omega(E_\mathfrak{A}{}^G)$. Note that if ω is in a face F of $E_\mathfrak{A}{}^G$ which satisfies the separability condition S then the unique μ will be supported by the G-ergodic states $\mathscr{E}(E_\mathfrak{A}{}^G)$, by the above proposition. Thus in this case uniqueness of a maximal orthogonal μ on $E_\mathfrak{A}{}^G$ corresponds to uniqueness of ergodic decomposition. In the sequel we consider the uniqueness problem with no further reference to the support properties of μ.

The main new technical tool used in the further development of ergodic decompositions is a result usually referred to as the mean ergodic theorem. We will repeatedly use this result in application to the unitary representation $U_\omega(G)$ of G on $\mathfrak{H}_\omega$. As we are not making any continuity assumptions on $g \mapsto U_\omega(g)$ we need a slightly abstract version of this theorem.

Proposition 4.3.4. (Alaoglu–Birkhoff mean ergodic theorem). *Let* $\mathscr{U}$ *be a family of bounded operators on the Hilbert space* $\mathfrak{H}$ *satisfying*

(1) $\|U\| \le 1$ *for all* $U \in \mathscr{U}$;
(2) $U_1 U_2 \in \mathscr{U}$ *for all* $U_1, U_2 \in \mathscr{U}$.

Furthermore, let E be the orthogonal projection on the subspace of $\mathfrak{H}$ *formed by the vectors invariant under all* $U \in \mathscr{U}$, *i.e.*,

$$E\mathfrak{H} = \{\psi; U\psi = \psi \text{ for all } U \in \mathscr{U}\}.$$

It follows that E is in the strong closure of the convex hull $\mathrm{Co}(\mathscr{U})$ *of* $\mathscr{U}$.

PROOF. First note that if $\psi \in E\mathfrak{H}$ and $U \in \mathscr{U}$ then

$$\|\psi\|^2 = (U\psi, \psi) = (\psi, U^*\psi) \le \|\psi\| \, \|U^*\psi\| \le \|\psi\|^2.$$

Therefore

$$(\psi, U^*\psi) = \|\psi\|^2$$

and

$$\|U^*\psi\| = \|\psi\|.$$

Thus

$$\begin{aligned}\|U^*\psi - \psi\|^2 &= \|U^*\psi\|^2 - (U^*\psi, \psi) - (\psi, U^*\psi) + \|\psi\|^2 \\ &= 0\end{aligned}$$

and one has $U^*\psi = \psi$. Consequently, if $\varphi \in (E\mathfrak{H})^\perp$ then

$$(U\varphi, \psi) = (\varphi, U^*\psi) = (\varphi, \psi) = 0$$

and $U\varphi \in (E\mathfrak{H})^\perp$. Next consider the convex set

$$C_\varphi = \{\chi; \chi = X\varphi, X \in \mathrm{Co}(\mathscr{U})\}.$$

Because each convex closed subset of $\mathfrak{H}$ contains a unique element of minimal norm there must be a minimal element $\chi \in \overline{C_\varphi} \subseteq (E\mathfrak{H})^\perp$. But then for each $U \in \mathscr{U}$ one has

$$\|U\chi\| \leq \|\chi\|$$

and by minimality

$$U\chi = \chi.$$

Thus $\chi = 0$.

Next, if χ is a general element of $\mathfrak{H}$ it has a unique decomposition

$$\chi = \psi + \varphi$$

with $\psi \in E\mathfrak{H}$ and $\varphi \in (E\mathfrak{H})^\perp$. The above argument then gives

$$\inf_{X \in \mathrm{Co}(\mathscr{U})} \|X\chi - \psi\| = \inf_{X \in \mathrm{Co}(\mathscr{U})} \|X\varphi\| = 0.$$

Finally, we may apply the foregoing argument to the family

$$\mathscr{U}_n = \{U \oplus U \oplus \cdots \oplus U; U \in \mathscr{U}\}$$

acting on the direct sum of n copies of $\mathfrak{H}$. Thus for $\varepsilon > 0$ and $\chi_1, \ldots, \chi_n \in \mathfrak{H}$ one may find an $X \in \mathrm{Co}(\mathscr{U})$ such that

$$\sum_{i=1}^{n} \|X\chi_i - E\chi_i\|^2 < \varepsilon.$$

Note that as E is in the strong closure of $\mathrm{Co}(\mathscr{U})$ it follows that one can always choose a net of elements $X_\alpha \in \mathrm{Co}(\mathscr{U})$,

$$X_\alpha = \sum_i \lambda_i^\alpha U_i,$$

such that X_α converges strongly to E. We will apply this result to representations of the group G by unitary operators and then it can be further argued that these approximants can be chosen as averages, or means, over the group. This interpretation as a mean value gives an intuitive explanation of the above limiting process and it is also helpful in understanding several of the limiting processes which occur in the subsequent analysis. Nevertheless, it is not absolutely necessary to the development of ergodic decompositions and we will describe the theory without this extra embellishment (see Notes and Remarks).

EXAMPLE 4.3.5. Let $x \in \mathbb{R}^\nu \mapsto U(x) \in \mathscr{L}(\mathfrak{H})$ be a weakly (strongly) continuous unitary representation of the translation group on the Hilbert space $\mathfrak{H}$ and let E

denote the orthogonal projection on the subspace of $\mathfrak{H}$ formed by the vectors invariant under all $U(x)$. Next let Λ_α be a net of Borel subsets of $\mathbb{R}^\nu$ with the property that

$$\lim_\alpha \frac{|\Lambda_\alpha \Delta(\Lambda_\alpha + y)|}{|\Lambda_\alpha|} = 0$$

for all $y \in \mathbb{R}^\nu$ where $|\Lambda|$ denotes the Lebesgue measure of Λ, and $A \Delta B = (A \cup B) \backslash (A \cap B)$. It follows that

$$\lim_\alpha \left\| \left(\frac{1}{|\Lambda_\alpha|} \int_{\Lambda_\alpha} dx\, U(x) - E \right) \psi \right\| = 0, \qquad \psi \in \mathfrak{H}.$$

To establish this one first remarks that if $\psi \in E\mathfrak{H}$ then the result is trivial. Next take $\psi \in (E\mathfrak{H})^\perp$ and for $y \in \mathbb{R}^\nu$ form $\psi_y = (\mathbb{1} - U(y))\psi$. One then has

$$\begin{aligned} \left\| \left(\frac{1}{|\Lambda_\alpha|} \int_{\Lambda_\alpha} dx\, U(x) - E \right) \psi_y \right\| &= \left\| \frac{1}{|\Lambda_\alpha|} \int_{\Lambda_\alpha \Delta(\Lambda_\alpha + y)} dx\, U(x) \psi \right\| \\ &\leq \frac{\|\psi\| \, |\Lambda_\alpha \Delta(\Lambda_\alpha + y)|}{|\Lambda_\alpha|} \end{aligned}$$

and the result again follows. Finally, if φ is orthogonal to the set

$$\mathfrak{D} = \{\psi_y ; \psi \in (E\mathfrak{H})^\perp, y \in \mathbb{R}^\nu\}$$

then $(1 - U(y))\varphi \in E\mathfrak{H}$ for all $y \in \mathbb{R}^\nu$. But $E(1 - U(y))\varphi = 0$ and hence $(1 - U(y))\varphi = 0$. Thus $\varphi \in E\mathfrak{H}$ and $\mathfrak{D}^\perp = E\mathfrak{H}$. Hence we have established the existence of the limit on a subset of $\mathfrak{H}$ whose linear span is dense and the existence then follows for all $\psi \in \mathfrak{H}$.

The properties which are crucial to the development of ergodic decomposition theory involve approximate commutation of pairs of elements of the C^*-algebra $\mathfrak{A}$ when one element is shifted under the action τ of the automorphism group G. These conditions are usually referred to under the generic heading of *asymptotic abelianness*.

We will study such conditions in the form

$$\inf_{A' \in \operatorname{Co}(\tau_G(A))} |\omega([A', B])| = 0$$

for each pair $A, B \in \mathfrak{A}$ and a certain subclass of states ω. Here we have used $\operatorname{Co}(\tau_G(A))$ to denote the convex hull of $\tau_G(A) = \{\tau_g(A); g \in G\}$. The term asymptotic abelianness is used because in specific applications in which G has a topological structure the condition is usually achieved by taking $A' = \tau_g(A)$ and moving g out of every compact subset of G. We will demonstrate that refinements of this condition can be used to characterize the uniqueness of ergodic decomposition, the subcentrality of these decompositions, and the ergodic structure of the extremal states $\mathscr{E}(E_{\mathfrak{A}}{}^G)$. We emphasize that these criteria are particularly useful because the conditions of asymptotic abelianness are often fulfilled in a strong and easily verifiable form such as norm commutation:

$$\inf_{g \in G} \|[\tau_g(A), B]\| = 0.$$

We begin by immediately introducing the two precise forms of asymptotic abelianness which will be necessary.

Definition 4.3.6. Let $\mathfrak{A}$ be a C^*-algebra with identity, G a group, $g \in G \mapsto \tau_g \in \mathrm{Aut}(\mathfrak{A})$ a representation of G as $*$-automorphisms of $\mathfrak{A}$, and for each $A \in \mathfrak{A}$ let $\mathrm{Co}(\tau_G(A))$ denote the convex hull of $\{\tau_g(A); g \in G\}$. Finally, let ω be a G-invariant state over $\mathfrak{A}$.

The pair $(\mathfrak{A}, \omega)$ is defined to be *G-abelian* if

$$\inf_{A' \in \mathrm{Co}(\tau_G(A))} |\omega'([A', B])| = 0$$

for all $A, B \in \mathfrak{A}$ and all G-invariant vector states ω' of π_ω.

The pair $(\mathfrak{A}, \omega)$ is defined to be *G-central* if

$$\inf_{A' \in \mathrm{Co}(\tau_G(A))} |\omega''([A', B])| = 0$$

for all $A, B \in \mathfrak{A}$ and all states ω'' such that $\omega'' < \lambda\omega'$ for some $\lambda > 0$ and some G-invariant vector state ω' of π_ω.

Note that if $\mathfrak{A}$ is abelian then both these conditions are satisfied for any group of automorphisms G. Further, if $(\mathfrak{A}, \omega)$ is H-abelian, or H-central, where H is a subgroup of G, then $(\mathfrak{A}, \omega)$ is automatically G-abelian or G-central.

Our aim is to show that $(\mathfrak{A}, \omega)$ is G-abelian for a sufficiently large number of G-invariant states ω if, and only if, each such ω is the barycenter of a measure μ, which is unique maximal among the measures such that

$$\mu(\tau_g(f_1)f_2) = \mu(f_1 f_2), \quad f_1, f_2 \in C(E_{\mathfrak{A}}).$$

Subsequently we show that G-centrality characterizes the situation in which the corresponding μ are subcentral.

Proposition 4.3.7. *Let $\mathfrak{A}$ be a C^*-algebra with identity, G a group, and $g \in G \mapsto \tau_g \in \mathrm{Aut}(\mathfrak{A})$ a representation of G as $*$-automorphisms of $\mathfrak{A}$. Next let ω be a G-invariant state over $\mathfrak{A}$ and denote by E_ω the orthogonal projection on the subspace of $\mathfrak{H}_\omega$ formed by the vectors invariant under $U_\omega(G)$.*

Consider the following conditions:

(1) *the pair $(\mathfrak{A}, \omega)$ is G-abelian;*
(2) *$E_\omega \pi_\omega(\mathfrak{A}) E_\omega$ is abelian;*[3]
(3) *$\{\pi_\omega(\mathfrak{A}) \cup U_\omega(G)\}'$ is abelian;*
(4) *there exists a unique maximal measure $\mu \in M_\omega(E_{\mathfrak{A}}{}^G)$.*

It follows that (1) $\Leftrightarrow$ (2) $\Rightarrow$ (3) $\Leftrightarrow$ (4) and if Ω_ω is separating for $\pi_\omega(\mathfrak{A})''$ then (1) $\Leftrightarrow$ (2) $\Leftrightarrow$ (3) $\Leftrightarrow$ (4).

[3] By a slight abuse of language which appears frequently in the sequel, Condition 2 does not mean that $E_\omega \pi_\omega(\mathfrak{A}) E_\omega$ is an algebra, but only that the operators in $E_\omega \pi_\omega(\mathfrak{A}) E_\omega$ commute mutually.

PROOF. (1) ⇒ (2) It suffices to prove (2) for A selfadjoint. Now given $\varepsilon > 0$, $A = A^* \in \mathfrak{A}$, and $\psi \in E_\omega \mathfrak{H}$, there exists by Proposition 4.3.4 a convex combination $S_\lambda(U_\omega)$ of U_ω,

$$S_\lambda(U_\omega) = \sum_{i=1}^{n} \lambda_i U_\omega(g_i),$$

such that

$$\|(S_\lambda(U_\omega) - E_\omega)\pi_\omega(A)\psi\| < \frac{\varepsilon}{2}.$$

Therefore defining $S_\lambda(\tau(A))$ by

$$S_\lambda(\tau(A)) = \sum_{i=1}^{n} \lambda_i \tau_{g_i}(A)$$

and letting $S_\mu(U_\omega)$ denote any other convex combination of U_ω and $S_\mu(A)$ the corresponding convex combination of $\tau_g(A)$ one has

$$\|E_\omega \pi_\omega(A)\psi - \pi_\omega(S_\mu S_\lambda(\tau(A)))\psi\| = \|S_\mu(U_\omega)[E_\omega \pi_\omega(A)\psi - \pi_\omega(S_\lambda(\tau(A)))\psi]\| < \frac{\varepsilon}{2}.$$

Thus

$$\begin{aligned} |(\psi, \pi_\omega(A)E_\omega \pi_\omega(B)\psi) - (\psi, \pi_\omega(B)E_\omega \pi_\omega(A)\psi)| &\le \varepsilon\|B\|\,\|\psi\| \\ &\quad + |(\psi, \pi_\omega([S_\mu S_\lambda(\tau(A)), B])\psi)|. \end{aligned}$$

But the convex combination S_μ is still arbitrary and hence, applying condition (1), one finds

$$(\psi, [E_\omega \pi_\omega(A)E_\omega, E_\omega \pi_\omega(B)E_\omega]\psi) = 0.$$

As this is valid for all $\psi \in E_\omega \mathfrak{H}$ one concludes that condition (2) is valid.

(2) ⇒ (1) If $S_\lambda(\tau(A))$ denotes the above convex combination in $\tau_G(A)$ then condition (2) implies

$$\begin{aligned} |(\psi, \pi_\omega([S_\lambda(\tau(A)), B])\psi)| &\le \|B\|\,\|(S_\lambda(U_\omega) - E_\omega)\pi_\omega(A)\psi\| \\ &\quad + \|B\|\,\|(S_\lambda(U_\omega) - E_\omega)\pi_\omega(A^*)\psi\| \end{aligned}$$

and condition (1) follows from Proposition 4.3.4.

(2) ⇒ (3) First note that as $E_\omega \Omega_\omega = \Omega_\omega$ and $E_\omega \pi_\omega(\mathfrak{A})E_\omega$ is abelian the basic characterization of orthogonal measures, Theorem 4.1.25, places E_ω in correspondence with the abelian von Neumann algebra $\{\pi_\omega(\mathfrak{A}) \cup E_\omega\}'$. But by Proposition 4.3.4 one has $E_\omega \in U_\omega(G)''$. Hence

$$\{\pi_\omega(\mathfrak{A}) \cup E_\omega\}'' \subseteq \{\pi_\omega(\mathfrak{A}) \cup U_\omega(G)\}''$$

and

$$\{\pi_\omega(\mathfrak{A}) \cup U_\omega(G)\}' \subseteq \{\pi_\omega(\mathfrak{A}) \cup E_\omega\}'.$$

Thus $\{\pi_\omega(\mathfrak{A}) \cup U_\omega(G)\}'$ is abelian.

Conditions (3) and (4) are equivalent by Proposition 4.3.3.

Finally, we assume that Ω_ω is separating for $\pi_\omega(\mathfrak{A})''$ and prove that (3) ⇒ (2).

Let $\mathfrak{B}_\omega = \{\pi_\omega(\mathfrak{A}) \cup U_\omega(G)\}'$. One concludes from Theorem 4.1.25 and Proposition 4.3.1 that $\mathfrak{B}_\omega$ is in correspondence with a projection $P = [\mathfrak{B}_\omega \Omega_\omega]$ with the properties $U_\omega(g)P = P$ and $P\pi_\omega(\mathfrak{A})P \subseteq \{P\pi_\omega(\mathfrak{A})P\}'$. In particular, $P \le E_\omega$, and to establish condition (2) it suffices to prove that $P = E_\omega$. But this is equivalent to showing that

Ω_ω is cyclic for $\pi_\omega(\mathfrak{A})' \cap U_\omega(G)'$ in $E_\omega \mathfrak{H}_\omega$. Now as Ω_ω is separating for $\pi_\omega(\mathfrak{A})''$ it is automatically cyclic for $\pi_\omega(\mathfrak{A})'$ by Proposition 2.5.3. Hence the desired conclusion will follow by applying the subsequent result, Proposition 4.3.8, to $\mathfrak{M} = \pi_\omega(\mathfrak{A})'$. In particular, this proposition associates to each $A \in \pi_\omega(\mathfrak{A})'$ an

$$M(A) \in \pi_\omega(\mathfrak{A})' \cap U_\omega(G)'$$

such that

$$M(A)\Omega_\omega = E_\omega A\Omega_\omega.$$

Hence

$$\overline{\{M(A)\Omega_\omega;\, A \in \pi_\omega(\mathfrak{A})'\}} = E_\omega \mathfrak{H}_\omega$$

by cyclicity.

Note that the implication (3) $\Rightarrow$ (2) in Proposition 4.3.7 is not true in general. A simple counterexample is provided by taking G to be the one point group and $\mathfrak{A}$ to be nonabelian. Thus $U_\omega(G)$ is the identity on $\mathfrak{H}_\omega$, condition (3) is equivalent to $\pi_\omega(\mathfrak{A})'$ being abelian, and condition (2) is equivalent to $\pi_\omega(\mathfrak{A})$ being abelian. But this is not the case for all ω. Nevertheless, there are more general situations in which this converse implication is valid and for their analysis we need the following generalization of the mean ergodic theorem to groups of automorphisms of a von Neumann algebra.

Proposition 4.3.8 (Kovacs and Szücs). *Let $\mathfrak{M}$ be a von Neumann algebra on a Hilbert space $\mathfrak{H}$, $g \in G \mapsto U(g) \in \mathcal{L}(\mathfrak{H})$ a unitary representation of the group G on $\mathfrak{H}$ such that $U(g)\mathfrak{M}U(g)^* \subseteq \mathfrak{M}$ for all $g \in G$, and define $\mathfrak{M}^G = \mathfrak{M} \cap U(G)'$. Let E_0 be the projection onto the $U(G)$-invariant vectors on $\mathfrak{H}$.*

If $F_0 \equiv [\mathfrak{M}'E_0] = \mathbb{1}$ then there exists a unique normal G-invariant projection M from $\mathfrak{M}$ onto $\mathfrak{M}^G$. Furthermore, one has

(1) *M is positive and faithful, i.e., if $A \geq 0$ and $M(A) = 0$ then $A = 0$,*
(2) *$\{M(A)\} = \mathfrak{M}^G \cap \overline{\mathrm{Co}\, \tau_G(A)}$ for all $A \in \mathfrak{M}$, where $\overline{\mathrm{Co}\, \tau_G(A)}$ denotes the weakly closed convex hull of $\{\tau_g(A);\, g \in G\}$, $\tau_g(A) = U(g)AU(g)^*$,*
(3) *$M(A)$ is the unique element in $\mathfrak{M}$ such that*

$$M(A)E_0 = E_0 AE_0,$$

(4) *a normal state ω on $\mathfrak{M}$ is G-invariant if, and only if,*

$$\omega(A) = \omega(M(A)), \qquad A \in \mathfrak{M},$$

i.e., $\omega = \omega|_{\mathfrak{M}^G} \circ M$.

If $F_0 \neq \mathbb{1}$ then $F_0 \in \mathfrak{M}^G \cap (\mathfrak{M}^G)'$ and the above results apply with $\mathfrak{M}$ replaced by $F_0\mathfrak{M}F_0$ and $\mathfrak{M}^G$ replaced by $(F_0\mathfrak{M}F_0)^G = \mathfrak{M}^G F_0$.

Proof. By the mean ergodic theorem (Proposition 4.3.4) there is a net $\sum_i \lambda_i^\alpha U(g_i^\alpha)$ in Co $(U(G))$ which converges strongly to E_0. If $A \in \mathfrak{M}$ then

$$\begin{aligned} E_0 AE_0 &= \text{strong} \lim_\alpha \sum_i \lambda_i^\alpha U(g_i^\alpha) AE_0 \\ &= \text{strong} \lim_\alpha \sum_i \lambda_i^\alpha U(g_i^\alpha) AU(g_i^\alpha)^* E_0. \end{aligned}$$

By Proposition 2.4.18, $\mathfrak{M}_1$ is weakly compact, hence the uniformly bounded net $\sum_i \lambda_i^\alpha U(g_i^\alpha)AU(g_i^\alpha)^*$ has a weakly convergent subnet $\sum_i \lambda_i^\beta U(g_i^\beta)AU(g_i^\beta)^*$ converging to $B \in \mathfrak{M}$. Therefore

$$E_0 A E_0 = B E_0.$$

Since $F_0 = \mathbb{1}$, $E_0 \mathfrak{H}$ is cyclic for $\mathfrak{M}'$ and so separating for $\mathfrak{M}$; hence this relation uniquely determines $B \in \mathfrak{M}$, and we can define a map M by $B = M(A)$. The map M is obviously linear and G-invariant. It is positive by the construction of B, and it is faithful since $E_0 \mathfrak{H}$ is separating for $\mathfrak{M}$. Explicitly, if $A \geq 0$ and $M(A) = 0$ then $0 = E_0 A E_0 = (A^{1/2}E_0)^*(A^{1/2}E_0)$. Hence $A^{1/2}E_0 = 0$ and so $A^{1/2} = 0$. Thus $A = 0$. Since

$$\begin{aligned} U(g)BU(g)^{-1}E_0 &= U(g)BE_0 \\ &= U(g)E_0 A E_0 = E_0 A E_0 = BE_0 \end{aligned}$$

we see that $M(\mathfrak{M}) \subseteq \mathfrak{M}^G$. Conversely, if $B \in \mathfrak{M}^G$ then $M(B) = B$ by construction and hence M is a projection onto $\mathfrak{M}^G$. If $B_1 \in \mathfrak{M}^G \cap \overline{\{\mathrm{Co}(\tau_G(A))\}}$ then

$$B_1 E_0 = E_0 B_1 E_0 = E_0 A E_0 = B E_0$$

and hence $B_1 = B$ and

$$\{M(A)\} = \mathfrak{M}^G \cap \overline{\{\mathrm{Co}(\tau_G(A))\}}.$$

If $\xi \in \mathfrak{H}$ is a vector of the form $\xi = B'\xi_0$, where $\xi_0 \in E_0 \mathfrak{H}$ and $B' \in \mathfrak{M}'$, then for $A \geq 0$

$$\begin{aligned} (B'\xi_0, M(A)B'\xi_0) &= (B'M(A)^{1/2}\xi_0, B'M(A)^{1/2}\xi_0) \\ &\leq \|B'\|^2(\xi_0, M(A)\xi_0) \\ &= \|B'\|^2(\xi_0, M(A)E_0\xi_0) \\ &= \|B'\|^2(\xi_0, A\xi_0). \end{aligned}$$

Hence $A \mapsto (B'\xi_0, M(A)B'\xi_0)$ is a normal positive functional. As $F_0 = \mathbb{1}$, these normal functionals span a norm-dense subset of $\mathfrak{M}_*$, and hence M is normal. If ω is a normal G-invariant state on $\mathfrak{M}$ then

$$\begin{aligned} \omega(A) &= \lim_\beta \omega\left(\sum_i \lambda_i^\beta U(g_i^\beta)AU(g_i^\beta)^*\right) \\ &= \omega(M(A)). \end{aligned}$$

Hence $\omega = (\omega|_{\mathfrak{M}^G}) \circ M$, and this establishes a one-to-one correspondence between the normal states on $\mathfrak{M}^G$ and normal G-invariant states on $\mathfrak{M}$.

If $F_0 \neq \mathbb{1}$ then since $U(g)\mathfrak{M}'U(g)^{-1} = \mathfrak{M}'$ for all $g \in G$ and $F_0 = [\mathfrak{M}'E_0]$ we have $U(g)F_0 = F_0 U(g)$. Hence $F_0 \in \mathfrak{M} \cap U(g)' = \mathfrak{M}^G$. Also, if $A \in \mathfrak{M}^G$, then

$$A\mathfrak{M}'E_0 = \mathfrak{M}'AE_0 = \mathfrak{M}'E_0 A,$$

hence $F_0 \in (\mathfrak{M}^G)'$. The last assertion of the proposition then follows by applying the first assertion to the von Neumann algebra $F_0 \mathfrak{M} F_0$ acted upon by the unitary elements $U(g)F_0$ on $F_0 \mathfrak{H}$.

Now we can use this result to obtain the principal characterization of uniqueness of G-invariant maximal measures.

Theorem 4.3.9. *Let $\mathfrak{A}$ be a C^*-algebra with identity, G a group, and $g \in G \mapsto \tau_g \in \mathrm{Aut}(\mathfrak{A})$ a representation of G as $*$-automorphisms of $\mathfrak{A}$. Next* *let ω be a G-invariant state over $\mathfrak{A}$, and E_ω be the orthogonal projection on the subspace of $\mathfrak{H}_\omega$ formed by the vectors invariant under $U_\omega(G)$. Finally, let $N_\omega{}^G$ denote the π_ω-normal G-invariant states over $\mathfrak{A}$. The following conditions are equivalent:*

(1) *the pair $(\mathfrak{A}, \omega)$ is G-abelian;*
(2) *every $\omega' \in N_\omega{}^G$ is the barycenter of a unique maximal measure $\mu' \in M_{\omega'}(E_{\mathfrak{A}}{}^G)$;*
(3) *$\{\pi_{\omega'}(\mathfrak{A}) \cup U_{\omega'}(G)\}'$ is abelian for all $\omega' \in N_\omega{}^G$;*
(4) *$\{E_{\omega'}\pi_{\omega'}(\mathfrak{A})E_{\omega'}\}$ is abelian for all $\omega' \in N_\omega{}^G$.*

Proof. Proposition 4.3.7 has already established the implications $(1) \Leftrightarrow (4) \Rightarrow (2) \Leftrightarrow (3)$ and it remains to prove $(2) \Rightarrow (4)$.

Define $F_\omega = [\pi_\omega(\mathfrak{A})'E_\omega]$ and set $\mathfrak{M} = F_\omega \pi_\omega(\mathfrak{A})''F_\omega$. Next define the action of τ on $\mathfrak{M}$ by

$$\tau_g(A) = U_\omega(g)AU_\omega(g)^*$$

and introduce the corresponding unique G-invariant normal projection $M; \mathfrak{M} \mapsto \mathfrak{M}^G$ of Proposition 4.3.8. Next we extend each $\omega' \in N_\omega{}^G$ to $\pi_\omega(\mathfrak{A})''$ by σ-weak continuity. Now let $\hat{\omega}$ denote the restriction of ω to $\mathfrak{M}$. Since $F_\omega \geq E_\omega$ one has $\omega(F_\omega) = 1$ and $\hat{\omega}$ is a state on $\mathfrak{M}$. Furthermore,

$$N_{\hat{\omega}}{}^G = \{\omega'; \omega' \in N_\omega{}^G, \omega'(F_\omega) = 1\}.$$

Next let $\overline{N_\omega{}^G}$ denote the weak* closure of $N_\omega{}^G$. It follows from Proposition 4.1.14 that the maximal probability measures over $\overline{N_\omega{}^G}$ form a simplex. Hence condition (2) implies, by a linear isomorphism, that $N_\omega{}^G$ is a simplex. But if $\omega_1, \omega_2 \in N_\omega{}^G$ are states such that $(\omega_1 + \omega_2)/2 \in N_{\hat{\omega}}{}^G$, and $\omega_1 \vee \omega_2$ and $\omega_1 \wedge \omega_2$ denote the least upper and greatest lower bounds of ω_1 and ω_2, respectively, then the relations

$$\frac{(\omega_1 + \omega_2)(F_\omega)}{2} = \frac{(\omega_1 \vee \omega_2 + \omega_1 \wedge \omega_2)(F_\omega)}{2} = 1$$

imply that $\omega_1, \omega_2, \omega_1 \vee \omega_2, \omega_1 \wedge \omega_2 \in N_{\hat{\omega}}{}^G$. Thus $N_{\hat{\omega}}{}^G$ is both a face in $N_\omega{}^G$ and a simplex. Now by Proposition 4.3.8 $N_{\hat{\omega}}{}^G = N_{\mathfrak{M}^G}$ through the identification $\omega' = \omega' \circ M$ and, since $N_{\mathfrak{M}^G}$ is a simplex, $\mathfrak{M}^G$ is abelian by Example 4.2.6. Now

$$\begin{aligned} E_\omega \pi_\omega(\mathfrak{A})''E_\omega &= E_\omega F_\omega \pi_\omega(\mathfrak{A})''F_\omega E_\omega \\ &= E_\omega \mathfrak{M} E_\omega \\ &= E_\omega M(\mathfrak{M}) E_\omega \\ &= E_\omega \mathfrak{M}^G E_\omega \end{aligned}$$

and because $E_\omega \in (\mathfrak{M}^G)'$ it follows that $E_\omega \pi_\omega(\mathfrak{A})''E_\omega$ is abelian. Finally, because $N_{\omega'}^G \subseteq N_\omega{}^G$ if $\omega' \in N_\omega{}^G$ the same proof shows that $E_{\omega'}\pi_{\omega'}(\mathfrak{A})''E_{\omega'}$ is abelian for all $\omega' \in N_\omega{}^G$.

One conclusion of this theorem is that the commutation property used to define G-abelianness is, in fact, equivalent to an apparently much stronger and more uniform type of commutation.

Corollary 4.3.10. *Adopt the notation and assumptions of Theorem 4.3.9. The following conditions are equivalent:*

(1) *the pair $(\mathfrak{A}, \omega)$ is G-abelian;*
(2) *for all $A \in \mathfrak{A}$ there exists a net A_α in the convex hull of $\tau_G(A)$ such that*

$$\lim_\alpha \omega'([\tau_g(A_\alpha), B]) = 0$$

for all $\omega' \in N_\omega{}^G$ and all $B \in \mathfrak{A}$, uniformly in $g \in G$.

PROOF. Clearly (2) ⇒ (1) but (1) ⇒ (2) by the estimate used in proving (2) ⇒ (1) in Proposition 4.3.7. Namely, if

$$S_\lambda(\tau(A)) = \sum_{i=1}^n \lambda_i \tau_{g_i}(A)$$

and $\omega'(A) = (\psi, \pi_\omega(A)\psi)$ then one has

$$|(\psi, \pi_\omega([\tau_g S_\lambda(\tau(A)), B])\psi)| \le \|B\| \, \|(S_\lambda(U_\omega) - E_\omega)\pi_\omega(A)\psi\| + \|B\| \, \|(S_\lambda(U_\omega) - E_\omega)\pi_\omega(A^*)\psi\|.$$

Corollary 4.3.11. *Adopt the notation and assumptions of Theorem 4.3.9. The following conditions are equivalent:*

(1) *the pair $(\mathfrak{A}, \omega)$ is G-abelian for all $\omega \in E_{\mathfrak{A}}{}^G$;*
(2) *$E_{\mathfrak{A}}{}^G$ is a simplex;*
(3) *$\{\pi_\omega(A) \cup U_\omega(G)\}'$ is abelian for all $\omega \in E_{\mathfrak{A}}{}^G$;*
(4) *$\{E_\omega \pi_\omega(\mathfrak{A}) E_\omega\}$ is abelian for all $\omega \in E_{\mathfrak{A}}{}^G$.*

This corollary is a global version of Proposition 4.3.7, which involves one $\omega \in E_{\mathfrak{A}}{}^G$, and Theorem 4.3.9, which involves the set of π_ω-normal G-invariant states $N_\omega{}^G$. The global result follows immediately from the local version Theorem 4.3.9 but it is of some interest that the equivalences 1 ⇔ 2 ⇔ 4 may be derived by relatively simple direct arguments. In particular the implication 1 ⇒ 2 is frequently of importance in applications because condition 1 is often easily verifiable and the simplex property, together with some separability, implies that each $\omega \in E_{\mathfrak{A}}{}^G$ has a unique ergodic decomposition.

A direct proof of 2 ⇒ 4 is obtained by a simple variation of the argument used to prove 1 ⇒ 2 in Example 4.2.6, i.e., one assumes that $\{E_\omega \pi_\omega(\mathfrak{A}) E_\omega\}$ is not abelian and constructs two distinct maximal measures in $M_\omega(E_{\mathfrak{A}}{}^G)$.

The proof of 4 ⇒ 1 is contained in the proof of Proposition 4.3.7. It is an elementary computation.

Finally 1 ⇒ 2 may be obtained through use of the mean ergodic theorem and Corollary 4.1.17. The idea of this proof is to first establish that for each $\omega \in E_{\mathfrak{A}}{}^G$ and $A_1, A_2, \ldots, A_n \in \mathfrak{A}$ one has

$$(\Omega_\omega, \pi_\omega(A_1)E_\omega \pi_\omega(A_2)E_\omega \cdots E_\omega \pi_\omega(A_n)\Omega_\omega) = \lim_\alpha \omega(S_{\lambda^\alpha}(\tau(A_1))S_{\lambda^\alpha}(\tau(A_2)) \cdots S_{\lambda^\alpha}(\tau(A_n))),$$

where $S_{\lambda^\alpha}(\tau(A))$, denotes a suitable net of convex combinations of $\tau_G(A)$:

$$S_{\lambda^\alpha}(\tau(A)) = \sum_{i=1}^{n_\alpha} \lambda_i^\alpha \tau_{g_i^\alpha}(A).$$

This is easily arranged by choosing the combination such that the corresponding combination $S_{\lambda^\alpha}(U_\omega)$ of $U_\omega(G)$,

$$S_{\lambda^\alpha}(U_\omega) = \sum_{i=1}^{n_\alpha} \lambda_i^\alpha U_\omega(g_i^\alpha),$$

converges strongly to E_ω on $\mathfrak{H}_\omega$. Next we define a linear functional μ_ω on $C(E_{\mathfrak{A}}^G)$ by

$$\mu_\omega(\hat{A}_1 \cdots \hat{A}_n) = (\Omega_\omega, \pi_\omega(A_1)E_\omega \pi_\omega(A_2)E_\omega \cdots E_\omega \pi_\omega(A_n)\Omega_\omega).$$

G-abelianness implies that $E_\omega \pi_\omega(\mathfrak{A})E_\omega$ is abelian and hence the Stone–Weierstrass theorem and the spectral argument used in the proof of Theorem 4.1.25 establish that μ_ω defines a probability measure on $E_{\mathfrak{A}}^G$ with barycenter ω. Next, if $\omega' \in E_{\mathfrak{A}}^G$ and $\omega' < \lambda\omega$ for some $\lambda > 0$ then

$$\omega'(A) = (C\Omega_\omega, \pi_\omega(A)\Omega_\omega)$$

with $C \in \pi_\omega(\mathfrak{A})' \cap U_\omega(G)'$ by Theorem 2.3.19 and a simple invariance argument. Therefore

$$\begin{aligned} &(C\Omega_\omega, \pi_\omega(A_1)E_\omega \pi_\omega(A_2)E_\omega \cdots E_\omega \pi_\omega(A_n)\Omega_\omega) \\ &= \lim_\alpha \omega'(S_{\lambda^\alpha}(\tau(A_1)) \cdots S_{\lambda^\alpha}(\tau(A_n)) \\ &= (\Omega_{\omega'}, \pi_{\omega'}(A_1)E_{\omega'} \pi_{\omega'}(A_2)E_{\omega'} \cdots E_{\omega'} \pi_{\omega'}(A_n)\Omega_\omega) \\ &= \mu_{\omega'}(\hat{A}_1\hat{A}_2 \cdots \hat{A}_n). \end{aligned}$$

Thus the mapping $\omega \in E_{\mathfrak{A}}^G \mapsto \mu_\omega \in M_\omega(E_{\mathfrak{A}}^G)$ is affine and $E_{\mathfrak{A}}^G$ is a simplex by Corollary 4.1.17. In the next section we will demonstrate that the simplices formed by the invariant states $E_{\mathfrak{A}}^G$ often have the strange geometric characteristic that their extreme points are weakly* dense (see Example 4.3.26).

After this analysis of the implications of G-abelianness we next examine the stronger condition of G-centrality. We begin with a proposition which provides two alternative characterizations of this property.

Proposition 4.3.12. *Let $\mathfrak{A}$ be a C^*-algebra with identity, G a group, $g \in G \mapsto \tau_g \in \mathrm{Aut}(\mathfrak{A})$ a representation of G as *-automorphisms of $\mathfrak{A}$, and ω a G-invariant state over $\mathfrak{A}$. The following conditions are equivalent:*

(1) *the pair $(\mathfrak{A}, \omega)$ is G-central;*
(2) $\{\pi_{\omega'}(\mathfrak{A})'' \cap U_{\omega'}(G)'\}F_{\omega'} = \{\mathfrak{Z}_{\omega'} \cap U_{\omega'}(G)'\}F_{\omega'}$ *for all $\omega' \in N_\omega^G$, where $F_{\omega'} = [\pi_{\omega'}(\mathfrak{A})'E_{\omega'}]$, $E_{\omega'}$ denotes the projection onto the $U_{\omega'}(G)$ invariant vectors in $\mathfrak{H}_{\omega'}$, and $\mathfrak{Z}_{\omega'}$ is the center of $\pi_{\omega'}(\mathfrak{A})''$.*

PROOF. For each $A \in \mathfrak{A}$ we let $M_{\omega'}(A)$ denote the mean value of $F_{\omega'}\pi_{\omega'}(A)F_{\omega'}$ as defined in Proposition 4.3.8. We will demonstrate that both the above conditions are equivalent to the condition

(3) $(\psi, [M_{\omega'}(A), F_{\omega'}\pi_{\omega'}(B)F_{\omega'}]\psi) = 0$

for all $A, B \in \mathfrak{A}$, $\psi \in F_{\omega'}\mathfrak{H}_{\omega'}$, *and* $\omega' \in N_{\omega}{}^{G}$.

(3) ⇒ (1) Let $\omega' \in N_{\omega}{}^{G}$ and let ω'' be a state which is majorized by a multiple of ω'. It follows from Theorem 2.3.19 that ω'' has the form

$$\omega''(A) = (\Omega_{\omega''}, \pi_{\omega'}(A)\Omega_{\omega''})$$

with $\Omega_{\omega''} = C\Omega_{\omega'}$ for some $C = C^* \in \pi_{\omega'}(\mathfrak{A})'$. Now the mean $M_{\omega'}(A)$ is the weak limit of $\pi_{\omega'}(S_{\lambda^\alpha}(\tau(A)))F_{\omega'}$ where $S_{\lambda^\alpha}(\tau(A))$ is a net of convex combinations of $\tau_g(A)$:

$$S_{\lambda^\alpha}(\tau(A)) = \sum_{i=1}^{n_\alpha} \lambda_i{}^\alpha \tau_{g_i{}^\alpha}(A).$$

Referring to the proof of Proposition 4.3.7 we see that the net can be chosen so that the corresponding combination

$$S_{\lambda^\alpha}(U_{\omega'}) = \sum_{i=1}^{n_\alpha} \lambda_i{}^\alpha U_{\omega'}(g_i{}^\alpha)$$

converges strongly to $E_{\omega'}$. Thus one calculates

$$\begin{aligned}
&(\Omega_{\omega''}, [M_{\omega'}(A), F_{\omega'}\pi_{\omega'}(B)F_{\omega'}]\Omega_{\omega''}) \\
&= \lim_\alpha(\Omega_{\omega'}, [\pi_{\omega'}(S_{\lambda^\alpha}(\tau(A)))F_{\omega'}, F_{\omega'}\pi_{\omega'}(B)F_{\omega'}]C^2\Omega_{\omega'}) \\
&= (\Omega_{\omega'}, \pi_{\omega'}(A)E_{\omega'}\pi_{\omega'}(B)C^2\Omega_{\omega'}) - (\Omega_{\omega'}, \pi_{\omega'}(B)E_{\omega'}\pi_{\omega'}(A)C^2\Omega_{\omega'}) \\
&= \lim_\alpha(\Omega_{\omega'}, [\pi_{\omega'}(S_{\lambda^\alpha}(\tau(A)))F_{\omega'}, \pi_{\omega'}(B)]C^2\Omega_{\omega'}) \\
&= (\Omega_{\omega''}, [M_{\omega'}(A), \pi_{\omega'}(B)]\Omega_{\omega''}) \\
&= \lim_\alpha \omega''([S_{\lambda^\alpha}(\tau(A)), B]).
\end{aligned}$$

Therefore (3) ⇒ (1).

(1) ⇒ (3) Let $\omega' \in N_{\omega}{}^{G}$. If ψ is a G-invariant vector in $\mathfrak{H}_{\omega'}$ and $A = A^*$ then for each $\varepsilon > 0$ there exists a convex combination $S_\lambda(\tau(A))$ of $\tau_G(A)$,

$$S_\lambda(\tau(A)) = \sum_{i=1}^{n} \lambda_i \tau_{g_i}(A)$$

such that

$$\|(\pi_{\omega'}(S_\lambda(\tau(A))) - M_{\omega'}(A))\psi\| < \frac{\varepsilon}{6}.$$

Further, we may choose a convex combination $S_\mu(U_{\omega'})$ of $U_{\omega'}(G)$ such that

$$\|(S_\mu(U_{\omega'}) - E_{\omega'})\pi_{\omega'}(S_\lambda(\tau(A)))\psi\| < \frac{\varepsilon}{6}\cdot$$

Now define $S_\mu S_\lambda(\tau(A))$ by successive convex combinations and note that

$$\|(\pi_{\omega'}(S_\mu S_\lambda(\tau(A))) - M_{\omega'}(A))\psi\| < \frac{\varepsilon}{6}.$$

Finally, define a third convex combination $S_\rho S_\mu S_\lambda(\tau(A))$ of $\tau_G(S_\mu S_\lambda(\tau(A)))$ and note that one still has

$$\|(\pi_{\omega'}(S_\rho S_\mu S_\lambda(\tau(A))) - M_{\omega'}(A))\psi\| < \frac{\varepsilon}{6}$$

and

$$\|\pi_{\omega'}(S_\rho S_\mu S_\lambda(\tau(A)))\psi - E_{\omega'}\pi_{\omega'}(A)\psi\| < \frac{\varepsilon}{6}.$$

Now if $C = C^* \in \pi_{\omega'}(\mathfrak{A})'$ and $B = B^* \in \mathfrak{A}$ then one has

$$\begin{aligned}
&|(C\psi, [M_{\omega'}(A), F_{\omega'}\pi_{\omega'}(B)F_{\omega'}]C\psi)| \\
&= 2|\mathrm{Im}(C^2\psi, \pi_{\omega'}(B)F_{\omega'}M_{\omega'}(A)\psi)| \\
&\leq (\varepsilon/3)\|C\|^2\|B\| + 2|\mathrm{Im}(C^2\psi, \pi_{\omega'}(B)F_{\omega'}\pi_{\omega'}(S_\rho S_\mu S_\lambda(\tau(A)))\psi)| \\
&\leq (2\varepsilon/3)\|C\|^2\|B\| + 2|\mathrm{Im}(C^2\psi, \pi_{\omega'}(B)E_{\omega'}\pi_{\omega'}(A)\psi)| \\
&\leq \varepsilon\|C\|^2\|B\| + 2|\mathrm{Im}(C^2\psi, \pi_{\omega'}(B)\pi_{\omega'}(S_\rho S_\mu S_\lambda(\tau(A)))\psi)| \\
&= \varepsilon\|C\|^2\|B\| + |(C\psi, \pi_{\omega'}([S_\rho S_\mu S_\lambda(\tau(A)), B])C\psi)|,
\end{aligned}$$

where we have first used the choice of S_λ, subsequently used the choice of S_μ, and have also exploited the obvious relation $F_{\omega'}E_{\omega'} = E_{\omega'}$. Now we are still at liberty to choose the convex combination S_ρ and hence condition (1) implies

$$(C\psi, [M_{\omega'}(A), F_{\omega'}\pi_{\omega'}(B)F_{\omega'}]C\psi) = 0$$

for A, B selfadjoint elements of $\mathfrak{A}$ and $C = C^* \in \pi_\omega(\mathfrak{A})'$. But condition (3) then follows straightforwardly by density and polarization.

(2) $\Rightarrow$ (3) Again let $\omega' \in N_\omega{}^G$. Condition (2) states that for each $A \in \pi_{\omega'}(\mathfrak{A})$ there is a $C \in \mathfrak{Z}_{\omega'} \cap U_{\omega'}(G)'$ such that

$$M_{\omega'}(A)F_{\omega'} = CF_{\omega'}.$$

Hence

$$[M_{\omega'}(A), F_{\omega'}\pi_{\omega'}(B)F_{\omega'}] = [C, F_{\omega'}\pi_{\omega'}(B)F_{\omega'}] = 0$$

because $F_{\omega'} \in \pi_{\omega'}(\mathfrak{A})''$. Thus condition (3) is valid.

It remains to prove that (3) $\Rightarrow$ (2) and for this the following lemma will be necessary.

Lemma 4.3.13. *If E is an orthogonal projection in a von Neumann algebra $\mathfrak{M}$ and $\mathfrak{M}_E$ is the von Neumann algebra $E\mathfrak{M}E$ on $E\mathfrak{H}$ then it follows that*

$$(\mathfrak{M}_E)' = (\mathfrak{M}')_E.$$

Moreover, if $\mathfrak{Z}$ is the center of $\mathfrak{M}$ then $\mathfrak{Z}_E$ is the center of $\mathfrak{M}_E$.

PROOF. It is evident that $(\mathfrak{M}_E)' \supseteq \mathfrak{M}'E$ and hence $(\mathfrak{M}'E)' \supseteq \mathfrak{M}_E'' = \mathfrak{M}_E$. Conversely, if T acts on $E\mathfrak{H}$ and $T \in (\mathfrak{M}'E)' \subseteq \{E\}'$ then $TE \in \mathfrak{M}$, or $T \in \mathfrak{M}_E$ and $(\mathfrak{M}'E)' \subseteq \mathfrak{M}_E$. Thus $\mathfrak{M}_E = (\mathfrak{M}'E)'$.

Now clearly $\mathfrak{Z}_E$ is contained in the center of $\mathfrak{M}_E$. If, however, $T \in \mathfrak{M}_E \cap \mathfrak{M}_E'$ then $T = T'E$ with $T' \in \mathfrak{M}'$. Let $F = [\mathfrak{M}E]$; then $F \in \mathfrak{Z}$ and we now argue that FT' is in $\mathfrak{Z}_F$. For this note that the mapping $S \in \mathfrak{M}_F' \mapsto SE \in \mathfrak{M}'_E$ is a morphism and because $F = [\mathfrak{M}E]$ it is an isomorphism. Hence as $T'E$ lies in the center of $\mathfrak{M}'_E$ one must have $T'F$ in the center $\mathfrak{Z}_F$ of $\mathfrak{M}_F$. Hence $T'F \in \mathfrak{Z}$ and $T = (T'F)E \in \mathfrak{Z}_E$.

End of Proof of Proposition 4.3.12. If $\omega' \in N_\omega{}^G$ and $\mathfrak{M}^G = \pi_{\omega'}(\mathfrak{A})'' \cap U_{\omega'}(G)'$ then condition (3) states that

$$\mathfrak{M}^G F_{\omega'} = F_{\omega'}\mathfrak{M}^G F_{\omega'} \subseteq (F_{\omega'}\pi_{\omega'}(\mathfrak{A})''F_{\omega'})'.$$

But $F_{\omega'} \in \pi_{\omega'}(\mathfrak{A})''$ and hence Lemma 4.3.13 implies that

$$(F_{\omega'}\pi_{\omega'}(\mathfrak{A})''F_{\omega'})' = F_{\omega'}\pi_{\omega'}(\mathfrak{A})'F_{\omega'} = \pi_{\omega'}(\mathfrak{A})'F_{\omega'}.$$

Combining these relations gives

$$\mathfrak{M}^G F_{\omega'} \subseteq \pi_{\omega'}(\mathfrak{A})'F_{\omega'}.$$

But one then has

$$\begin{aligned}\mathfrak{M}^G F_{\omega'} &\subseteq (\pi_{\omega'}(\mathfrak{A})'' \cap U_{\omega'}(G)' \cap \pi_{\omega'}(\mathfrak{A})')F_{\omega'} \\ &= (\mathfrak{Z}_{\omega'} \cap U_{\omega'}(G)')F_{\omega'}.\end{aligned}$$

The reverse inclusion is, however, obvious because $\mathfrak{Z}_{\omega'} \cap U_{\omega'}(G)' \subseteq \mathfrak{M}^G$ and hence condition (2) is valid.

Next we come to the second principal result concerning ergodic decompositions.

Theorem 4.3.14. *Let $\mathfrak{A}$ be a C^*-algebra with identity, G a group, and $g \in G \mapsto \tau_g \in \mathrm{Aut}(\mathfrak{A})$ a representation of G as $*$-automorphisms of $\mathfrak{A}$. Next, let ω be a G-invariant state over $\mathfrak{A}$ and E_ω the orthogonal projection on the subspace of $\mathfrak{H}_\omega$ formed by the vectors invariant under $U_\omega(G)$. Finally let $N_\omega{}^G$ denote the π_ω-normal G-invariant states over $\mathfrak{A}$. The following conditions are equivalent:*

(1) *the pair $(\mathfrak{A}, \omega)$ is G-central;*
(2) *every $\omega' \in N_\omega{}^G$ is the barycenter of a unique maximal measure $\mu' \in M_{\omega'}(E_{\mathfrak{A}}{}^G)$, and this measure is subcentral;*
(3) *$\{\pi_{\omega'}(\mathfrak{A}) \cup U_{\omega'}(G)\}' = \mathfrak{Z}_{\omega'} \cap U_{\omega'}(G)'$, for all $\omega' \in N_\omega{}^G$, where $\mathfrak{Z}_{\omega'}$ is the center of $\pi_{\omega'}(\mathfrak{A})''$;*
(4) *$\{E_{\omega'}\pi_{\omega'}(\mathfrak{A})E_{\omega'}\}$ is abelian and*

$$\{\pi_{\omega'}(\mathfrak{A}) \cup U_{\omega'}(G)\}'' \cap \{\pi_{\omega'}(\mathfrak{A}) \cup U_{\omega'}(G)\}' = \mathfrak{Z}_{\omega'} \cap U_{\omega'}(G)'$$

for all $\omega' \in N_\omega{}^G$, where $\mathfrak{Z}_{\omega'}$ is the center of $\pi_{\omega'}(\mathfrak{A})''$.

Proof. First consider conditions (2) and (3). If (2) is valid then μ' is the orthogonal measure corresponding to $\{\pi_{\omega'}(\mathfrak{A}) \cup U_{\omega'}(G)\}'$ by Proposition 4.3.3 and by the definition of a subcentral measure

$$\{\pi_{\omega'}(\mathfrak{A}) \cup U_{\omega'}(G)\}' \subseteq \mathfrak{Z}_{\omega'} \cap U_{\omega'}(G)'.$$

As the reverse inclusion is trivial condition (3) is valid. Conversely, if (3) is valid then $\{\pi_{\omega'}(\mathfrak{A}) \cup U_{\omega'}(G)\}'$ is abelian and condition (2) follows from Proposition 4.3.3.

Next we prove that (1) $\Rightarrow$ (3) but for this we need the following lemma.

Lemma 4.3.15. *Let $\mathfrak{M}$ be a abelian von Neumann algebra on a Hilbert space $\mathfrak{H}$ with a cyclic vector Ω. It follows that $\mathfrak{M}$ is maximal abelian, i.e.,*

$$\mathfrak{M}' = \mathfrak{M}.$$

PROOF. Since $\mathfrak{M} \subseteq \mathfrak{M}'$ the vector Ω is separating for $\mathfrak{M}$ by Proposition 2.5.3. If Δ and J are the modular operator and modular conjugation associated with $(\mathfrak{M}, \Omega)$ then one has $\Delta = \mathbb{1}$, $J = S$ because $\|SA\Omega\|^2 = \|A^*\Omega\|^2 = \|A\Omega\|^2$.

But $JAJB\Omega = JAB^*\Omega = BA^*\Omega = A^*B\Omega$ in this case. Hence $JAJ = A^*$. Since $J\mathfrak{M}J = \mathfrak{M}'$ by Theorem 2.5.14, it follows that $\mathfrak{M} = \mathfrak{M}^* = \mathfrak{M}'$.

END OF PROOF OF THEOREM 4.3.14. (1) $\Rightarrow$ (3) Note by Theorem 4.3.9 that condition (1) implies that $E_{\omega'}\pi_{\omega'}(\mathfrak{A})''E_{\omega'}$ is abelian. Moreover, if

$$\mathfrak{M} = \{\pi_{\omega'}(\mathfrak{A}) \cup U_{\omega'}(G)\}'',$$

$$E_{\omega'}\pi_{\omega'}(\mathfrak{A})''E_{\omega'} = E_{\omega'}\mathfrak{M}E_{\omega'}.$$

Also $E_{\omega'} \in U_{\omega'}(G)'' \subseteq \mathfrak{M}$ by the mean ergodic theorem. But $\Omega_{\omega'}$ is cyclic for $E_{\omega'}\pi_{\omega'}(\mathfrak{A})E_{\omega'}$ in $E_{\omega'}\mathfrak{H}_{\omega'}$, and therefore by successive applications of Lemmas 4.3.15 and 4.3.13 one has

$$\begin{aligned} E_{\omega'}\pi_{\omega'}(\mathfrak{A})''E_{\omega'} &= (E_{\omega'}\mathfrak{M}E_{\omega'})'E_{\omega'} \\ &= E_{\omega'}\mathfrak{M}'E_{\omega'}. \end{aligned}$$

But then one has

$$\begin{aligned} \{\pi_{\omega'}(\mathfrak{A})' \cap U_{\omega'}(G)'\}E_{\omega'} &= \{\pi_{\omega'}(\mathfrak{A})'' \cap U_{\omega'}(G)'\}E_{\omega'} \\ &= \{\mathfrak{Z}_{\omega'} \cap U_{\omega'}(G)'\}E_{\omega'}, \end{aligned}$$

where the second equality follows from condition (2) of Proposition 4.3.12. Finally, one notes that $E_{\omega'}\mathfrak{H}_{\omega'}$ is cyclic for $\pi_{\omega'}(\mathfrak{A})''$ and hence

$$\pi_{\omega'}(\mathfrak{A})' \cap U_{\omega'}(G)' = \mathfrak{Z}_{\omega'} \cap U_{\omega'}(G)'.$$

(3) $\Rightarrow$ (4) As $\mathfrak{Z}_{\omega'}$ is abelian one automatically has $\{\pi_{\omega'}(\mathfrak{A}) \cup U_{\omega'}(G)\}'$ abelian and then $E_{\omega'}\pi_{\omega'}(\mathfrak{A})E_{\omega'}$ is abelian by Theorem 4.3.9. But again defining

$$\mathfrak{M} = \{\pi_{\omega'}(\mathfrak{A}) \cup U_{\omega'}(G)\}''$$

one has

$$\begin{aligned} \mathfrak{Z}_{\omega'} \cap U_{\omega'}(G)' &\subseteq \mathfrak{M} \cap \mathfrak{M}' \\ &\subseteq \mathfrak{M}' = \mathfrak{Z}_{\omega'} \cap U_{\omega'}(G)', \end{aligned}$$

where the first two inclusions are trivial and the third follows from condition (3).

(4) $\Rightarrow$ (3) Once again remark that

$$E_{\omega'}\pi_{\omega'}(\mathfrak{A})''E_{\omega'} = E_{\omega'}\mathfrak{M}E_{\omega'}$$

and $E_{\omega'} \in U_{\omega'}(G)'' \subseteq \mathfrak{M}$, where $\mathfrak{M} = \{\pi_{\omega'}(\mathfrak{A}) \cup U_{\omega'}(G)\}''$. Hence because $E_{\omega'}\pi_{\omega'}(\mathfrak{A})''E_{\omega}$ is abelian the argument used in the proof of (1) $\Rightarrow$ (3) gives

$$E_{\omega'}\mathfrak{M}E_{\omega'} = E_{\omega'}\mathfrak{M}'E_{\omega'}.$$

But then applying the second statement of Lemma 4.3.13 one has

$$E_{\omega'}\mathfrak{M}'E_{\omega'} = E_{\omega'}(\mathfrak{M} \cap \mathfrak{M}')E_{\omega'}.$$

Therefore by condition (4)

$$\{\pi_{\omega'}(\mathfrak{A}) \cup U_{\omega'}(G)\}'E_{\omega'} = \{\mathfrak{Z}_{\omega'} \cap U_{\omega'}(G)'\}E_{\omega'}.$$

Finally, as $E_{\omega'}\mathfrak{H}_{\omega'}$ is cyclic for $\pi_{\omega}(\mathfrak{A})''$ condition (3) follows.

(3) $\Rightarrow$ (1) By Proposition 4.3.12 it suffices to prove that

$$\{\pi_{\omega'}(\mathfrak{A})'' \cap U_{\omega'}(G)'\}F_{\omega} = \{\mathfrak{Z}_{\omega'} \cap U_{\omega'}(G)'\}F_{\omega'}.$$

For this consider the map κ_μ associated with the orthogonal measure μ corresponding to $\{\pi_{\omega'}(\mathfrak{A}) \cup U_{\omega'}(G)\}'$ and the map M constructed in Proposition 4.3.8. One has for $A \in \mathfrak{A}$

$$M(A)E_{\omega'} = E_{\omega'}\pi_{\omega'}(A)E_{\omega'} = \kappa_\mu(\hat{A})E_{\omega'}.$$

Now the range of κ_μ is contained in $\{\pi_{\omega'}(\mathfrak{A}) \cup U_{\omega'}(G)\}'$ and this latter set is contained in $\mathfrak{Z}_{\omega'}$ by condition (3). Therefore multiplying the last relation by $\pi_{\omega'}(\mathfrak{A})'$ one finds

$$\begin{aligned} M(A)F_{\omega'} &= \kappa_\mu(\hat{A})F_{\omega'} \\ &\subseteq \{\mathfrak{Z}_{\omega'} \cap U_{\omega'}(G)'\}F_{\omega'}. \end{aligned}$$

But the range of M on $\pi_{\omega}(\mathfrak{A})''$ is equal to $\{\pi_{\omega'}(\mathfrak{A})'' \cap U_{\omega'}(G)'\}F_{\omega'}$ and hence

$$\begin{aligned} \{\mathfrak{Z}_{\omega'} \cap U_{\omega'}(G)'\}F_{\omega'} &\subseteq \{\pi_{\omega'}(\mathfrak{A})'' \cap U_{\omega'}(G)'\}F_{\omega'} \\ &\subseteq \{\mathfrak{Z}_{\omega'} \cap U_{\omega'}(G)'\}F_{\omega'} \end{aligned}$$

and the proof is complete.

As with G-abelianness the property of G-centrality is actually equivalent to a seemingly much stronger commutation property.

Corollary 4.3.16. *Adopt the assumptions of Theorem* 4.3.14. *The following conditions are equivalent*:

(1) *the pair* $(\mathfrak{A}, \omega)$ *is G-central*;
(2) *for all* $A \in \mathfrak{A}$ *there exists a net* A_α *in the convex hull of* $\{\tau_g(A); g \in G\}$ *such that*

$$\lim_\alpha \omega''([\tau_g(A_\alpha), B]) = 0$$

for all ω'' *such that* $\omega'' < \lambda\omega'$ *for some* $\lambda > 0$ *and* $\omega' \in N_\omega{}^G$, *and for all* $B \in \mathfrak{A}$, *uniformly for* $g \in G$.

Condition (2) clearly implies condition (1) but the converse follows from rewriting the estimates used to prove Proposition 4.3.12. One first chooses a net of convex combinations $S_{\lambda^\alpha}(\tau(A))$ of $\tau_G(A)$ such that $\pi_\omega(S_{\lambda^\alpha}(\tau(A)))E_\omega$ converges strongly to $M_\omega(A)E_\omega$. Subsequently, one chooses a second net S_{μ^β} of convex combinations of U_ω such that $S_{\mu^\beta}(U_\omega)$ converges strongly to E_ω. Then defining $A_{\alpha,\beta}$ by

$$A_{\alpha,\beta} = S_{\mu^\beta}S_{\lambda^\alpha}(\tau(A))$$

one checks from the estimates used in the proof of Proposition 4.3.12 that this double net converges with the correct uniformity in g.

4.3.2. Ergodic States

In the previous subsection we analyzed the decomposition of G-invariant states in terms of maximal orthogonal measures on the space $E_{\mathfrak{A}}{}^G$. This theory is a direct generalization of the decomposition theory for states over an abelian C^*-algebra $\mathfrak{A}$; the abelian property of $\mathfrak{A}$ is replaced by G-abelianness, or G-centrality, and the states $E_{\mathfrak{A}}$ are replaced by the G-invariant states $E_{\mathfrak{A}}{}^G$. Thus the G-ergodic states $\mathscr{E}(E_{\mathfrak{A}}{}^G)$ correspond to the pure states $\mathscr{E}(E_{\mathfrak{A}})$ of the abelian algebra and it is to be expected that they will have similar properties. In this subsection we analyze this analogy for general G and in the subsequent subsection we examine further details for locally compact abelian groups which act continuously. Throughout both subsections we assume that G acts as a group of $*$-automorphisms τ of $\mathfrak{A}$ and it will be irrelevant whether or not $\mathfrak{A}$ possesses an identity.

The pure states of an abelian C^*-algebra have several distinctive features. Firstly, they generate irreducible representations, which are automatically one-dimensional, and as a consequence they factorize, i.e.,

$$\omega(AB) = \omega(A)\omega(B)$$

for all $A, B \in \mathfrak{A}$, by Corollary 2.3.21. Each of these properties has an analogue for the G-ergodic states but the group G and its representation $U_\omega(G)$ intervene in a crucial manner. If E_ω denotes the orthogonal projection on the subspace of $\mathfrak{H}_\omega$ formed by the vectors invariant under $U_\omega(G)$ then the following chart lists equivalent properties for a state ω over an abelian algebra and the analogous properties for a G-abelian pair $(\mathfrak{A}, \omega)$ (we subsequently show that the latter properties are also equivalent).

Abelian	G-Abelian
$\mathfrak{H}_\omega$ one-dimensional	$E_\omega \mathfrak{H}_\omega$ one-dimensional
$\omega \in \mathscr{E}(E_{\mathfrak{A}})$	$\omega \in \mathscr{E}(E_{\mathfrak{A}}{}^G)$
$\pi_\omega(\mathfrak{A})$ irreducible	$\{\pi(\mathfrak{A}) \cup U_\omega(G)\}$ irreducible
$\omega(AB) = \omega(A)\omega(B)$	$\inf_{A' \in \mathrm{Co}\, \tau_G(A)} \lvert\omega(A'B) - \omega(A)\omega(B)\rvert = 0$

In the last entry $\mathrm{Co}\, \tau_G(A)$ again denotes the convex hull of the set

$$\{\tau_g(A); g \in G\}.$$

The approximate factorization property which occurs in this case is often called a cluster property, or a mixing property. In the sequel we also examine more general cluster properties of the type

$$\inf_{B' \in \mathrm{Co}\, \tau_G(B)} \lvert\omega(AB'C) - \omega(AC)\omega(B)\rvert = 0.$$

The cluster property with two elements is equivalent to a spectral property of $U_\omega(G)$, the property that E_ω has rank one, and under quite general circumstances stronger spectral restrictions can be deduced from three element cluster properties. This will be illustrated in Section 4.3.3.

Several of the equivalences indicated in the above chart are independent of the assumption that the pair $(\mathfrak{A}, \omega)$ is G-abelian. We begin with a result which characterizes the concepts of irreducibility and ergodicity.

Theorem 4.3.17. *Take $\omega \in E_{\mathfrak{A}}{}^G$. Let E_ω denote the orthogonal projection on the subspace of $\mathfrak{H}_\omega$ formed by the vectors invariant under $U_\omega(G)$ and let $\mathbb{1}_\omega$ denote the identity on $\mathfrak{H}_\omega$. Consider the following conditions:*

(1) *E_ω has rank one;*
(2) *ω is G-ergodic, i.e., $\omega \in \mathscr{E}(E_{\mathfrak{A}}{}^G)$;*
(3) *$\{\pi_\omega(\mathfrak{A}) \cup U_\omega(G)\}$ is irreducible on $\mathfrak{H}_\omega$;*
(4) *$\{\mathfrak{Z}_\omega \cap U_\omega(G)'\} = \{\mathbb{C}\mathbb{1}_\omega\}$,*

where

$$\mathfrak{Z}_\omega = \pi_\omega(\mathfrak{A})'' \cap \pi_\omega(\mathfrak{A})'.$$

It follows that (1) $\Rightarrow$ (2) $\Leftrightarrow$ (3) $\Rightarrow$ (4).

If, moreover, the pair $(\mathfrak{A}, \omega)$ is G-abelian (2) $\Rightarrow$ (1) *and if $(\mathfrak{A}, \omega)$ is G-central then all the conditions are equivalent.*

PROOF. (2) $\Leftrightarrow$ (3) From Theorem 2.3.19 and G-invariance one sees that (2) is false if, and only if, there exists a nonzero selfadjoint $T \in \{\pi_\omega(\mathfrak{A}) \cup U_\omega(G)\}'$ which is not proportional to $\mathbb{1}_\omega$. But by Proposition 2.3.8 the latter condition is equivalent to the falseness of (3). Thus conditions (2) and (3) are simultaneously false and hence simultaneously true.

(1) $\Rightarrow$ (3) By assumption E_ω is the projection onto Ω_ω. As this vector is cyclic for $\pi_\omega(\mathfrak{A})$ it follows that $\{\pi_\omega(\mathfrak{A}) \cup E_\omega\}$ is irreducible. The mean ergodic theorem, Proposition 4.3.4, implies, however, that $\{\pi_\omega(\mathfrak{A}) \cup E_\omega\} \subseteq \{\pi_\omega(\mathfrak{A}) \cup U_\omega(G)\}''$ and hence this latter set is irreducible.

(3) $\Rightarrow$ (4) Condition (3) is equivalent to

$$\{\pi_\omega(\mathfrak{A})' \cap U_\omega(G)'\} = \{\mathbb{C}\mathbb{1}_\omega\}.$$

But one has

$$\{\mathfrak{Z}_\omega \cap U_\omega(G)'\} \subseteq \{\pi_\omega(\mathfrak{A})' \cap U_\omega(G)'\}$$

and hence condition (4) is valid.

Next assume that the pair $(\mathfrak{A}, \omega)$ is G-abelian. If $\{\pi_\omega(\mathfrak{A}) \cup U_\omega(G)\}$ is irreducible on $\mathfrak{H}_\omega$ then $E_\omega\{\pi_\omega(\mathfrak{A}) \cup U_\omega(G)\}E_\omega$ is irreducible on $E_\omega\mathfrak{H}_\omega$. But

$$E_\omega\{\pi_\omega(\mathfrak{A}) \cup U_\omega(G)\}E_\omega = E_\omega\pi_\omega(\mathfrak{A})E_\omega$$

and this latter set is abelian by Proposition 4.3.7. Thus E_ω must have rank one, i.e., (3) $\Rightarrow$ (1).

Finally, if $(\mathfrak{A}, \omega)$ is G-central then

$$\{\pi_\omega(\mathfrak{A}) \cup U_\omega(G)\}' = \{\mathfrak{Z}_\omega \cap U_\omega(G)'\}$$

by Theorem 4.3.14 and hence (4) $\Rightarrow$ (3).

Note that if $\mathfrak{A} = \mathscr{L}(\mathfrak{H})$ for some Hilbert space $\mathfrak{H}$, ω is a vector state, and G is the one-point group represented by the identity automorphism then condition (2) is true but condition (1) is false whenever the dimension of $\mathfrak{H}$ is greater than one. Thus G-abelianness is necessary for the implication (2) $\Rightarrow$ (1). The next example shows that G-centrality is necessary for the implication (4) $\Rightarrow$ (3).

EXAMPLE 4.3.18. Let $\mathfrak{A} = M_2$ be the algebra of complex 2×2 matrices $\{A_{ij}\}$, G the group of diagonal unitary elements in $\mathfrak{A}$, and $\tau_g(A) = gAg^{-1}$. The G-invariant states of $\mathfrak{A}$ are given by

$$\omega_\lambda(\{A_{ij}\}) = \lambda A_{11} + (1 - \lambda)A_{22}$$

for $0 \leq \lambda \leq 1$, and there are two distinct G-ergodic states ω_0 and ω_1. One has $E_{\omega_\lambda} = [\pi_{\omega_\lambda}(D)\Omega_{\omega_\lambda}]$, where D is the set of diagonal matrices and hence $(\mathfrak{A}, \omega_\lambda)$ is G-abelian. But if $\lambda \neq 0$ or 1 then ω_λ is not G-ergodic. Nevertheless $\mathfrak{Z}_{\omega_\lambda} \cap U_{\omega_\lambda}(G)' = \{\mathbb{C}1_{\omega_\lambda}\}$. Thus the implication (4) $\Rightarrow$ (3) in Theorem 4.3.17 is false and consequently $(\mathfrak{A}, \omega_\lambda)$ is not G-central. Note that for all λ different from 0 or 1 the vector Ω_{ω_λ} is separating for $\pi_{\omega_\lambda}(\mathfrak{A})$. This will be of subsequent interest.

Let $\omega \in E_{\mathfrak{A}}{}^G$. Condition (4) of Theorem 4.3.17,

$$\{\mathfrak{Z}_\omega \cap U_\omega(G)'\} = \{\mathbb{C}1_\omega\},$$

is sometimes called *central ergodicity* and the corresponding states are said to be *centrally ergodic*. The equivalence and disjointness properties of these states are of some interest and the next theorem generalizes Proposition 2.4.27 to these states.

Theorem 4.3.19. *If $\omega_1, \omega_2 \in E_{\mathfrak{A}}{}^G$ are centrally ergodic then ω_1 and ω_2 are either quasi-equivalent or disjoint, and they are quasi-equivalent if, and only if, $(\omega_1 + \omega_2)/2$ is centrally ergodic. Furthermore, the following two conditions are equivalent:*

(1) *all centrally ergodic states are ergodic, i.e., if $\omega \in E_{\mathfrak{A}}{}^G$ and*

$$\mathfrak{Z}_\omega \cap U_\omega(G)' = \mathbb{C}1_\omega$$

then

$$\pi_\omega(\mathfrak{A})' \cap U_\omega(G)' = \mathbb{C}1_\omega;$$

(2) *if ω_1 and ω_2 are centrally ergodic, then either $\omega_1 = \omega_2$ or $\omega_1 \between \omega_2$.*

In particular, if $\{\mathfrak{A}, \omega\}$ is G-central for all $\omega \in E_{\mathfrak{A}}{}^G$ then all centrally ergodic states are ergodic, and any pair ω_1, ω_2 of centrally ergodic states are either equal or disjoint.

PROOF. Assume that $\omega_1, \omega_2 \in E_{\mathfrak{A}}{}^G$ are centrally ergodic, with associated representations $(\mathfrak{H}_i, \pi_i, U_i, \Omega_i)$. Define

$$\mathfrak{H} = \mathfrak{H}_1 \oplus \mathfrak{H}_2, \qquad \pi = \pi_1 \oplus \pi_2, \qquad U = U_1 \oplus U_2,$$

$$\mathfrak{Z} = \pi(\mathfrak{A})'' \cap \pi(\mathfrak{A})', \qquad \mathfrak{Z}^G = \mathfrak{Z} \cap U(G)',$$

$$P_1 = \mathbb{1}_{\mathfrak{H}_1} \oplus 0, \qquad P_2 = 0 \oplus \mathbb{1}_{\mathfrak{H}_2},$$

$$E = [\pi(\mathfrak{A})(\Omega_1 \oplus \Omega_2)], \qquad P = [\pi(\mathfrak{A})' P_1 \, \mathfrak{H}].$$

Since $P_1 \in \pi(\mathfrak{A})'$, we have $P \in \pi(\mathfrak{A})'' \cap \pi(\mathfrak{A})'$. Moreover, since $P_1 \in U(G)'$ and $U(g)\pi(\mathfrak{A})'U(g)^* = \pi(\mathfrak{A})'$ for all $g \in G$ it follows that $P \in U(G)'$, and we conclude that $P \in \mathfrak{Z}^G$.

Clearly, $PP_1 = P_1 P = P_1$. Define $Q = PP_2 = P_2 P$ and then $Q \in \pi(\mathfrak{A})'$. Thus $P_2 Q$, viewed as an operator on $\mathfrak{H}_2$, is contained in $\pi_2(\mathfrak{A})'$. Also $Q \in U(G)'$ and hence $P_2 Q \in U(G)'$. But since $P \in \pi(\mathfrak{A})''$ there exists a net $\{A_\alpha\}$ in $\mathfrak{A}$ such that $\pi(A_\alpha) \to P$ strongly, and then $\pi(A_\alpha)P_2 = \pi_2(A_\alpha) \to QP_2$. Hence $QP_2 \in \pi_2(\mathfrak{A})''$, and finally, $P_2 Q \in \pi_2(\mathfrak{A})'' \cap \pi_2(\mathfrak{A})' \cap U_2(G)' = \mathfrak{Z}_2{}^G = \mathbb{C}\mathbb{1}_{\mathfrak{H}_2}$, where the last equality uses the central ergodicity of ω_2.

In conclusion, if $P = [\pi(\mathfrak{A})' P_1 \, \mathfrak{H}]$ there are two possibilities:

$$\text{(a)} \quad P = P_1; \qquad \text{(b)} \quad P = P_1 + P_2 = \mathbb{1}_{\mathfrak{H}}.$$

If (a) holds then $P_1 \in \pi(\mathfrak{A})''$. Thus if U is any partial isometry in $\pi(\mathfrak{A})'$ then $UP_1 = P_1 U$. Therefore π_1 and π_2 have no unitarily equivalent subrepresentations, and π_1 and π_2 are disjoint by the remarks preceding Lemma 4.2.8. If $\omega = (\omega_1 + \omega_2)/2$ then $\pi_\omega(A) = \pi(A)E$, and $P_1 E$ identifies with a nontrivial projection in $\mathfrak{Z}_\omega{}^G$. Hence ω is not centrally ergodic.

If (b) holds, then $\pi(\mathfrak{A})' P_1 \mathfrak{H}$ is dense in $\mathfrak{H}$, and it follows by polar decomposition that if Q_0 is any nonzero projection in $\pi_2(\mathfrak{A})'$ there exists a nonzero partial isometry $U \in \pi(\mathfrak{A})'$ such that $UU^* \leq Q_0$, $U^*U \leq P_1$. Hence $Q_0 \pi_2$ contains a subrepresentation which is unitarily equivalent with a subrepresentation of π_1. Interchanging the roles of π_1 and π_2, it follows that any subrepresentation of π_1 contains a representation which is unitarily equivalent to a subrepresentation of π_2, and hence π_1 and π_2 are quasi-equivalent by Theorem 2.4.26. In this case $\mathfrak{Z}^G = \mathbb{C}P = \mathbb{C}\mathbb{1}_{\mathfrak{H}}$ and $\omega = (\omega_1 + \omega_2)/2$ is centrally ergodic by similar reasoning to case (a).

We now turn to the proof of the equivalence of the statements (1) and (2) in the theorem.

(1) $\Rightarrow$ (2) Let ω_1 and ω_2 be two nondisjoint centrally ergodic states in $E_{\mathfrak{A}}{}^G$. Then ω_1 and ω_2 are quasi-equivalent and $(\omega_1 + \omega_2)/2$ is centrally ergodic by the first part of the theorem. But then $(\omega_1 + \omega_2)/2 \in \mathscr{E}(E_{\mathfrak{A}}{}^G)$ by assumption (1), and hence $\omega_1 = \omega_2 = (\omega_1 + \omega_2)/2$.

(2) $\Rightarrow$ (1) Let ω be a centrally ergodic state and let $\omega_1, \omega_2 \in E_{\mathfrak{A}}{}^G$ be two states such that $\omega = (\omega_1 + \omega_2)/2$. Since $\omega_i \leq 2\omega$ there exists a vector $\Omega_i \in \mathfrak{H}_\omega$ such that

$$\omega_i(A) = (\Omega_i, \pi_\omega(A)\Omega_i), \qquad A \in \mathfrak{A},$$

by Theorem 2.3.19. Hence π_{ω_i} is unitarily equivalent to the subrepresentation of π_ω determined by the projection $P_i = [\pi_\omega(\mathfrak{A})\Omega_i] \in \pi_\omega(\mathfrak{A})'$. But then $\mathfrak{Z}_{\omega_i} = P_i \mathfrak{Z}_\omega$ by Lemma 4.3.13 and so $\mathfrak{Z}^G_{\omega_i} = P_i \mathfrak{Z}_\omega{}^G = \mathbb{C}P_i$. Hence ω_1 and ω_2 are centrally ergodic and since $\omega = (\omega_1 + \omega_2)/2$ is centrally ergodic it follows by the first part of the theorem that ω_1 and ω_2 are quasi-equivalent. But then $\omega_1 = \omega_2$ by assumption (2), and hence ω is extremal in $E_{\mathfrak{A}}{}^G$, i.e., ω is ergodic.

The last statements of the theorem now follow from the implication (4) $\Rightarrow$ (2) of Theorem 4.3.17 and the implication (1) $\Rightarrow$ (2) of the present theorem.

Let us now return to the examination of criteria for ergodicity. The next result shows that G-abelianness and G-centrality in Theorem 4.3.17 can to a certain extent be replaced by a separation property of Ω_ω.

Theorem 4.3.20. *Let $\omega \in E_{\mathfrak{A}}{}^G$. Let E_ω denote the orthogonal projection on the subspace of $\mathfrak{H}_\omega$ formed by the vectors invariant under $U_\omega(G)$ and let $\mathbb{1}_\omega$ denote the identity on $\mathfrak{H}_\omega$.*

Consider the following conditions:

(1) $\{\pi_\omega(\mathfrak{A})'' \cap U_\omega(G)'\} = \{\mathbb{C}\mathbb{1}_\omega\}$;
(2) *E_ω has rank one;*
(3) *ω is G-ergodic, i.e., $\omega \in \mathscr{E}(E_{\mathfrak{A}}{}^G)$;*
(4) *$\{\pi_\omega(\mathfrak{A}) \cup U_\omega(G)\}$ is irreducible on $\mathfrak{H}_\omega$.*

It follows that (1) $\Rightarrow$ (2) $\Rightarrow$ (3) $\Leftrightarrow$ (4).

Moreover, (1) *implies that Ω_ω is separating for $\pi_\omega(\mathfrak{A})''$. Conversely, if Ω_ω is separating for $\pi_\omega(\mathfrak{A})''$ then all four conditions are equivalent.*

PROOF. First assume condition (1) is true and define P_ω by

$$P_\omega = [\pi_\omega(\mathfrak{A})'\Omega_\omega].$$

It follows that

$$P_\omega \in \{\pi_\omega(\mathfrak{A})'' \cap U_\omega(G)'\}.$$

Thus $P_\omega = \mathbb{1}_\omega$ by assumption. Hence Ω_ω is cyclic for $\pi_\omega(\mathfrak{A})'$ and separating for $\pi_\omega(\mathfrak{A})''$ by Proposition 2.5.3.

(1) $\Rightarrow$ (2) By Proposition 4.3.8 there exists a unique normal G-invariant projection M; $\pi_\omega(\mathfrak{A})'' \mapsto \pi_\omega(\mathfrak{A})'' \cap U_\omega(G)'$ such that $M(A)E_\omega = E_\omega A E_\omega$. Thus if $M(\mathfrak{A}) = \mathbb{C}\mathbb{1}_\omega$ then E_ω has rank one.

The implications (2) $\Rightarrow$ (3) $\Leftrightarrow$ (4) follow from Theorem 4.3.17.

Finally, assume that Ω_ω is separating for $\pi_\omega(\mathfrak{A})''$ and let Δ, J be the modular operator and modular conjugation associated with the pair $\{\pi_\omega(\mathfrak{A})'', \Omega_\omega\}$. Since

$$\begin{aligned} J\Delta^{1/2}U_\omega(g)A\Omega_\omega &= S\tau_g(A)\Omega_\omega \\ &= \tau_g(A^*)\Omega_\omega = U_\omega(g)J\Delta^{1/2}A\Omega_\omega \end{aligned}$$

for $A \in \pi_\omega(\mathfrak{A})''$ and $\pi_\omega(\mathfrak{A})''\Omega_\omega$ is a core for $\Delta^{1/2}$ it follows that $J\Delta^{1/2}U_\omega(g) = U_\omega(g)J\Delta^{1/2}$. By the uniqueness of the polar decomposition one then concludes that $JU_\omega(g) = U_\omega(g)J$. But $\pi_\omega(\mathfrak{A})'' = J\pi_\omega(\mathfrak{A})'J$ and hence

$$\{\pi_\omega(\mathfrak{A})'' \cap U_\omega(G)'\} = J\{\pi_\omega(\mathfrak{A})' \cap U_\omega(G)'\}J.$$

Thus (4) $\Rightarrow$ (1).

Theorem 4.3.20 is similar to Theorem 4.3.17 but contains the extra condition $\{\pi_\omega(\mathfrak{A})'' \cap U_\omega(G)'\} = \{\mathbb{C}\mathbb{1}_\omega\}$ and omits the condition of central ergodicity, i.e., $\{\mathfrak{Z}_\omega \cap U_\omega(G)'\} = \{\mathbb{C}\mathbb{1}_\omega\}$. We have, however, already remarked in Example 4.3.18 that this latter condition does not imply ergodicity of ω even if Ω_ω is separating for $\pi_\omega(\mathfrak{A})''$. The next example complements this by showing that ergodicity of ω does not necessarily imply $\{\pi_\omega(\mathfrak{A})'' \cap U_\omega(G)'\} = \mathbb{C}\mathbb{1}_\omega$ even when $\{\mathfrak{A}, \omega\}$ is G-central.

EXAMPLE 4.3.21. Let $\mathfrak{A} = \mathscr{L}(\mathfrak{H})$ for some Hilbert space of dimension greater than one and let U be any unitary operator with a unique eigenvector Ω corresponding to the eigenvalue zero. Let $G = \mathbb{Z}$ and $\tau_n(A) = U^n A U^{-n}$. The state ω defined by

$$\omega(A) = (\Omega, A\Omega)$$

is G-invariant and E_ω is the rank one projection with range Ω. Therefore ω is G-ergodic. It also follows that ω is the only π_ω-normal G-invariant state. Hence $(\mathfrak{A}, \omega)$ is G-central because $\pi_\omega(\mathfrak{A})' = \mathfrak{Z}_\omega = \mathbb{C}\mathbb{1}_\omega$. Nevertheless, there are nontrivial $A \in \mathfrak{A}$ which commute with U, for example U itself, and hence condition (1) of Theorem 4.3.20 is false.

Next we examine characterizations of ergodicity in terms of cluster properties. The criteria provided by the following results are not direct criteria for ergodicity but give conditions for the projection E_ω on the invariant vectors to have rank one. Thus Theorem 4.3.17 and Theorem 4.3.20 can then be applied to obtain ergodicity criteria. The following result is essentially an algebraic rephrasing of the mean ergodic theorem.

Theorem 4.3.22. *Let $\omega \in E_{\mathfrak{A}}{}^G$ and let E_ω denote the orthogonal projection on the subspace of $\mathfrak{H}_\omega$ formed by the vectors invariant under $U_\omega(G)$. Further, for each $A \in \mathfrak{A}$ let $\mathrm{Co}\, \tau_G(A)$ denote the convex hull of $\{\tau_g(A); g \in G\}$. The following conditions are equivalent:*

(1) *E_ω has rank one;*
(2) *$\inf_{B' \in \mathrm{Co}\,\tau_G(B)} |\omega(AB') - \omega(A)\omega(B)| = 0$ for all $A, B \in \mathfrak{A}$;*
(3) *For each $B \in \mathfrak{A}$ there exists a net $\{B_\alpha\} \subseteq \mathrm{Co}\, \tau_G(B)$ such that*

$$\lim_\alpha |\omega(A\tau_g(B_\alpha)) - \omega(A)\omega(B)| = 0$$

for all $A \in \mathfrak{A}$ uniformly for $g \in G$.

PROOF. (1) ⇒ (3) By the mean ergodic theorem, Proposition 4.3.4, there exists a net of convex combinations $S_{\lambda^\alpha}(U_\omega)$ of $U_\omega(g)$,

$$S_{\lambda^\alpha}(U_\omega) = \sum_{i=1}^{n_\alpha} \lambda_i{}^\alpha U_\omega(g_i{}^\alpha),$$

such that $U_\omega(g)S_{\lambda^\alpha}(U_\omega)$ converges strongly on $\mathfrak{H}_\omega$, uniformly in g, to the rank one projection E_ω on Ω_ω. Hence if $B_\alpha = S_{\lambda^\alpha}(\tau(B))$, where

$$S_{\lambda^\alpha}(\tau(B)) = \sum_{i=1}^{n_\alpha} \lambda_i{}^\alpha \tau_{g_i{}^\alpha}(B)$$

one has

$$\lim_\alpha |\omega(A\tau_g(B_\alpha)) - \omega(A)\omega(B)| = 0$$

uniformly in g.

(3) $\Rightarrow$ (2) This is evident.

(2) $\Rightarrow$ (1) Given $\varepsilon > 0$ the mean ergodic theorem implies that for each $B \in \mathfrak{A}$ one may choose a $B' \in \text{Co}\,\tau_G(B)$ such that

$$\|(\pi_\omega(B') - E_\omega \pi_\omega(B))\Omega_\omega\| < \varepsilon.$$

Thus

$$\begin{aligned}
&|(\Omega_\omega, \pi_\omega(A)E_\omega \pi_\omega(B)\Omega_\omega) - (\Omega_\omega, \pi_\omega(A)\Omega_\omega)(\Omega_\omega \pi_\omega(B)\Omega_\omega)| \\
&= |(\Omega_\omega, \pi_\omega(A')E_\omega \pi_\omega(B)\Omega_\omega) - (\Omega_\omega, \pi_\omega(A)\Omega_\omega)(\Omega_\omega, \pi_\omega(B')\Omega_\omega)| \\
&\leq \varepsilon\|A\| + |\omega(A'B') - \omega(A)\omega(B')|
\end{aligned}$$

for all $A' \in \text{Co}\,\tau_G(\text{A})$. Hence condition (2) implies

$$(\Omega_\omega, \pi_\omega(A)E_\omega \pi_\omega(B)\Omega_\omega) = (\Omega_\omega, \pi_\omega(A)\Omega_\omega)(\Omega_\omega, \pi_\omega(B)\Omega_\omega)$$

and by cyclicity E_ω is the rank one projection onto Ω_ω.

The next theorem establishes that the rank one property of E_ω can also be characterized by a three-element cluster property, if Ω_ω is separating for $\pi_\omega(\mathfrak{A})''$.

Theorem 4.3.23. *Adopt the assumption and notations of Theorem* 4.3.22. *Consider the following conditions:*

(1) *E_ω has rank one;*
(2) $\inf_{B' \in \text{Co}\,\tau_G(B)} |\omega(AB'C) - \omega(AC)\omega(B)| = 0$ *for all $A, B, C \in \mathfrak{A}$;*
(3) *for each $B \in \mathfrak{A}$ there exists a net $\{B_\alpha\} \subseteq \text{Co}\,\tau_G(B)$ such that*

$$\lim_\alpha |\omega(A\tau_g(B_\alpha)C) - \omega(AC)\omega(B)| = 0$$

for all $A, C \in \mathfrak{A}$, uniformly for $g \in G$.

It follows that (1) $\Leftarrow$ (2) $\Leftarrow$ (3) *and if Ω_ω is separating for $\pi_\omega(\mathfrak{A})''$ then all three conditions are equivalent.*

PROOF. (3) $\Rightarrow$ (2) This is evident.

(2) $\Rightarrow$ (1) If $\mathfrak{A}$ possesses an identity $\mathbb{1}$ then setting $C = \mathbb{1}$ one finds that condition (2) implies condition (2) of Theorem 4.3.22. But this latter condition is equivalent to condition (1). If $\mathfrak{A}$ does not possess an identity we choose for $\varepsilon > 0$ a C such that

$$\|\pi_\omega(C)\Omega_\omega - \Omega_\omega\| < \varepsilon.$$

Then it follows that

$$|\omega(AB') - \omega(A)\omega(B)| \leq 2\varepsilon\|A\|\,\|B\| + |\omega(AB'C) - \omega(AC)\omega(B)|.$$

The desired conclusion follows again from Theorem 4.3.22.

Finally, assume Ω_ω is separating for $\pi_\omega(\mathfrak{A})''$ and consider the implication (1) $\Rightarrow$ (3).

If E_ω has rank one, then Theorem 4.3.22 establishes the existence of a net $\{B_\alpha\} \subseteq \text{Co}\,\tau_G(B)$ such that $\pi_\omega(\tau_g(B_\alpha))\Omega_\omega$ converges weakly to $\omega(B)\Omega_\omega$. If, however, Ω_ω is separating for $\pi_\omega(\mathfrak{A})''$ then it is cyclic for $\pi_\omega(\mathfrak{A})'$ and it follows that $\pi_\omega(\tau_g(B_\alpha))$ converges weakly to $\omega(B)\mathbb{1}_\omega$. But this is equivalent to condition (3).

Note that Example 4.3.21 illustrates a situation in which Ω_ω is the unique $U_\omega(G)$-invariant vector in $\mathfrak{H}_\omega$ and hence the cluster properties of Theorem 4.3.22 are valid. Nevertheless, conditions (2) and (3) of Theorem 4.3.23 are not valid in this example because $\mathfrak{A}$ contains many nontrivial elements which are invariant under the action τ of the group G. Thus the conditions of Theorem 4.3.23 are not generally equivalent. Nevertheless, there are a variety of circumstances which assure equivalence. For example, the separation property for Ω_ω which was assumed in Theorem 4.3.23 can be replaced by a form of asymptotic abelianness. If, in particular, the G-invariant state ω satisfies

$$\inf_{B' \in \operatorname{Co} \tau_G(B)} |\omega(A[B', C])| = 0$$

for all $A, B, C \in \mathfrak{A}$ then it follows from the uniformity established by condition (3) of Theorem 4.3.22 that the two-point cluster property

$$\inf_{B' \in \operatorname{Co} \tau_G(B)} |\omega(AB') - \omega(A)\omega(B)| = 0, \qquad A, B \in \mathfrak{A},$$

is equivalent to the three-point cluster property

$$\inf_{B' \in \operatorname{Co} \tau_G(B)} |\omega(AB'C) - \omega(AC)\omega(B)| = 0.$$

The foregoing results describe the general connections between cluster properties, ergodicity, and spectral properties, i.e., uniqueness of Ω_ω. Without further assumptions on the structure of G, or the continuity of its action, it is difficult to elaborate the theory more fully and results are fragmentary. Thus in the next subsection we specialize to the important case of locally compact abelian groups with continuous action, but we first conclude the general discussion with a few remarks concerning the directions in which the theory has developed and three illustrative examples.

First notice that under suitable algebraic assumptions G-ergodicity of ω is equivalent to the existence of a net $\{B_\alpha\} \subseteq \operatorname{Co} \tau_G(B)$ such that

$$\lim_\alpha |\omega(AB_\alpha) - \omega(A)\omega(B)| = 0$$

for all $A, B \in \mathfrak{A}$. This expression of clustering in terms of convex combinations can be interpreted as an abstract method of stating that the functions $g \in G \mapsto \omega(A\tau_g(B)) \in \mathbb{C}$ have mean value $\omega(A)\omega(B)$. There are various methods of emphasizing this notion of mean, or average, value. For general G one may examine the subset of bounded functions over G such that the norm closed convex hulls of their left and their right translates contain constant functions, i.e., the *almost periodic functions* (see Section 4.3.4). For such functions the constant is unique and has the obvious interpretation as a mean value. The mean ergodic theorem states that the matrix elements of every unitary representation of G have mean values in this sense. Alternatively, one can define an *invariant mean* as a state over the C^*-algebra $C_b(G)$ of bounded continuous functions over G which is invariant under left, or right, translations. Even if a mean of this type exists it is not necessarily unique but it will coincide with the foregoing mean over the almost periodic functions. With an utter disregard for etymology, the groups with invariant states in this sense are called *amenable*. Not every group is amenable but for the amenable groups the mean values can be constructed by various explicit methods and the clustering criteria are reexpressed by each of these methods, e.g., if $G = \mathbb{R}^\nu$ and $U_\omega(\mathbb{R}^\nu)$ is continuous then the above form of clustering is equivalent to

$$\lim_\alpha \frac{1}{|\Lambda_\alpha|} \int_{\Lambda^\alpha} dx\, \omega(A\tau_x(B)) = \omega(A)\omega(B),$$

where the sets Λ_α satisfy the conditions described in Example 4.3.5 and the equivalence follows from this example. In the next section we will use mean values for locally compact abelian groups.

Next we remark that there are various other stronger cluster properties which can be used to classify stronger notions of ergodicity. For example, one could examine those ω for which zero is in the convex hull of the clustering functions, i.e., the ω such that

$$0 \in \mathrm{Co}\{|\omega(A\tau_g(B)) - \omega(A)\omega(B)|;\, g \in G\}$$

for all $A, B \in \mathfrak{A}$. Note that this would be impossible if $U_\omega(G)$ has a non-invariant eigenvector and thus this type of clustering implies a stronger spectral property. One can also partially hierarchize the G-ergodic states in terms of spectral properties (see Examples 4.3.28 and 4.3.34). Furthermore, it is of interest to examine the states for which there exists a net $g^\alpha \in G$ such that

$$\lim_\alpha |\omega(A\tau_{g^\alpha}(B)) - \omega(A)\omega(B)| = 0$$

for all $A, B \in \mathfrak{A}$. This type of cluster property occurs in classical ergodic theory and the following example shows that it can arise in the general noncommutative theory for quite unexpected reasons.

EXAMPLE 4.3.24 (Strong mixing). Let G be a group acting as $*$-automorphisms τ of a C^*-algebra $\mathfrak{A}$ and assume the existence of a net $g^\alpha \in G$ such that the asymptotic abelianness condition

$$\lim_\alpha \|[A, \tau_{g^\alpha}(B)]\| = 0$$

is satisfied for all $A, B \in \mathfrak{A}$. If ω is a factor state over $\mathfrak{A}$, i.e., if $\pi_\omega(\mathfrak{A})''$ is a factor, then it follows that the cluster property

$$\lim_\alpha |\omega(A\tau_{g^\alpha}(B)) - \omega(A)\omega(\tau_{g^\alpha}(B))| = 0$$

is satisfied for all $A, B \in \mathfrak{A}$. To deduce this we first remark that as $\pi_\omega(\mathfrak{A})''$ is a factor then the C^*-algebra $\mathfrak{B}$, generated by $\pi_\omega(\mathfrak{A}) \cup \pi_\omega(\mathfrak{A})'$, on $\mathfrak{H}_\omega$, is irreducible. Next note that the vector $\eta_A = \pi_\omega(A)\Omega_\omega - \omega(A)\Omega_\omega$ is orthogonal to Ω_ω. Therefore there is an hermitian $T \in \mathscr{L}(\mathfrak{H}_\omega)$ such that

$$T\eta_A = 0, \qquad T\Omega_\omega = \Omega_\omega.$$

It follows from the Kadison transitivity theorem, cited in the notes and remarks to Section 2.4.2, that T may be chosen in $\mathfrak{B}$. Thus, introducing C_1 and C_2 by

$$C_1 = T(\pi_\omega(A) - \omega(A)\mathbb{1}_\omega), \quad C_2 = (\mathbb{1}_\omega - T)(\pi_\omega(A) - \omega(A)\mathbb{1}_\omega),$$

one has $C_1\Omega_\omega = 0$, $C_2{}^*\Omega_\omega = 0$ and $\pi_\omega(A) = \omega(A)\mathbb{1}_\omega + C_1 + C_2$. Therefore

$$\omega(A\tau_{g^\alpha}(B)) - \omega(A)\omega(\tau_{g^\alpha}(B)) = (\Omega_\omega, [C_1, \pi_\omega(\tau_{g^\alpha}(B))]\Omega_\omega).$$

But $C_1 \in \mathfrak{B}$ and hence for each $\varepsilon > 0$ there is a finite family of $A_i \in \mathfrak{A}$, $i = 1, 2, \ldots, n$, and $B_i \in \pi_\omega(\mathfrak{A})'$ such that

$$\left\| C_1 - \sum_{i=1}^{n} \pi_\omega(A_i)B_i \right\| < \frac{\varepsilon}{\|B\|}.$$

Thus

$$|\omega(A\tau_{g^\alpha}(B)) - \omega(A)\omega(\tau_{g^\alpha}(B))| < \varepsilon + \sum_{i=1}^{n} \|B_i\| \, \|[A_i, \tau_{g^\alpha}(B)]\|.$$

The cluster property follows immediately.

Note that in the foregoing derivation G-invariance of ω was not necessary. If one assumes that ω is G-invariant and that there exists a sequence of nets $\{g_i{}^\alpha\}_{i\geq 1} \subseteq G$ such that

$$\lim_\alpha \|[A, \tau_{(g_i{}^\alpha)^{-1}g_j{}^\alpha}(B)]\| = 0$$

for all $i \neq j$, and all $A, B \in \mathfrak{A}$, then one can derive the multiple clustering property

$$\lim_\alpha |\omega(\tau_{g_1{}^\alpha}(A_1)\tau_{g_2{}^\alpha}(A_2) \cdots \tau_{g_n{}^\alpha}(A_n)) - \omega(A_1)\omega(A_2) \cdots \omega(A_n)| = 0$$

for all $A_1, A_2, \ldots, A_n \in \mathfrak{A}$. For this it suffices to apply the foregoing reasoning $n - 1$ times with the successive choices

$$A = A_k, \qquad B = \tau_{(g_k{}^\alpha)^{-1}g_{k+1}^\alpha}(A_{k+1}) \cdots \tau_{(g_k{}^\alpha)^{-1}g_n{}^\alpha}(A_n), \qquad k = 1, 2, \ldots, n - 1.$$

If $G = \mathbb{R}^\nu$ one often encounters situations for which

$$\lim_{|x|\to\infty} \|[A, \tau_x(B)]\| = 0$$

for all $A, B \in \mathfrak{A}$, and in this case one concludes that each $\mathbb{R}^\nu$-invariant factor state ω satisfies the property

$$\lim |\omega(\tau_{x_1}(A_1)\tau_{x_2}(A_2) \cdots \tau_{x_n}(A_n)) - \omega(A_1)\omega(A_2) \cdots \omega(A_n)| = 0$$

for all $A_1, \ldots, A_n \in \mathfrak{A}$, and all $n \geq 2$. The limit is understood to mean that $|x_i - x_j|$ tends to infinity for all pairs $i \neq j$. This last cluster property, with $\nu = 1$, was originally introduced in classical ergodic theory under the name of strong mixing of all orders. The property for $n = 2$ is simply called strong mixing.

The uniform asymptotic abelianness condition

$$\lim_\alpha \|[A, \tau_{g^\alpha}(B)]\| = 0$$

played a crucial role in this last example. The uniformity allows one to deduce mixing for every factor state ω. This condition is of further interest because it clearly implies G-abelianness and G-centrality for every G-invariant state. There are various other forms of asymptotic abelianness which are stronger than G-abelianness but weaker than the above uniform condition. Although none of these other properties appear to have the same type of generic characteristic as G-abelianness and G-centrality (by this we mean the structure described in Theorems 4.3.9 and 4.3.14), they do occur naturally in examples and are of interest for particular problems. For example, the pair $(\mathfrak{A}, \omega)$ is usually defined to be weakly asymptotically abelian if there is a net $g^\alpha \in G$ such that

$$\lim_\alpha (\varphi, \pi_\omega([A, \tau_{g^\alpha}(B)])\psi) = 0$$

for all $\varphi, \psi \in \mathfrak{H}_\omega$ and $A, B \in \mathfrak{A}$. Weak asymptotic abelianness in mean is defined by replacing the pointwise limit by a mean value. Example 4.3.24 immediately yields an application of these notions. If $(\mathfrak{A}, \omega)$ is weakly asymptotically abelian and ω is a factor state then ω is strongly mixing and if the asymptotic abelianness is in mean then the mixing is in mean. The proof of both these statements is identical to the proof given in Example 4.3.24. Note, however, that the proof of mixing to all orders depends crucially on uniformity of the asymptotic abelian condition.

If ω is a faithful factor state, i.e., if $\pi_\omega(\mathfrak{A})''$ is a factor and Ω_ω is separating for $\pi_\omega(\mathfrak{A})''$ then weak asymptotic abelianness and strong clustering (mixing) are actually equivalent. The above discussion shows that weak asymptotic abelianness implies clustering but to establish the converse we remark that the clustering condition

$$\lim_\alpha |\omega(A\tau_{g^\alpha}(B)) - \omega(A)\omega(\tau_{g^\alpha}(B))| = 0$$

for all $A, B \in \mathfrak{A}$ is equivalent to

$$\lim_\alpha |(\psi, (\pi_\omega(\tau_{g^\alpha}(B)) - \omega(\tau_{g^\alpha}(B))\mathbb{1}_\omega)\Omega_\omega)| = 0$$

for all $B \in \mathfrak{A}$, $\psi \in \mathfrak{H}_\omega$. But Ω_ω is cyclic for $\pi_\omega(\mathfrak{A})'$ by Proposition 2.5.3, and hence this latter property is equivalent to

$$\lim_\alpha |(\psi, (\pi_\omega(\tau_{g^\alpha}(B)) - \omega(\tau_{g^\alpha}(B))\mathbb{1}_\omega)\varphi)| = 0$$

for all $\varphi, \psi \in \mathfrak{H}_\omega$. Weak asymptotic abelianness follows immediately. Once again this result does not require G-invariance of ω but if ω is G-invariant then the weak asymptotic abelianness becomes equivalent to strong clustering (mixing) of all orders.

At this point it is worth mentioning another interplay between topologies and types of limits. If $(\mathfrak{A}, \omega)$ is weakly asymptotically abelian for all $\omega \in E_\mathfrak{A}$ then it follows from the Hahn–Banach theorem that

$$\inf_{B' \in \mathrm{Co}\,\tau_G(B)} \|[A, B']\| = 0$$

for all $A, B \in \mathfrak{A}$, i.e., one has a uniform asymptotic abelianness in mean. This is a special case of a general theorem due to Mazur, which establishes that if a sequence of elements converges weakly on a Banach space then some convex combination of the sequence converges uniformly.

Next, we emphasize that the ergodicity criteria can sometimes be strengthened by examination of the structure of the particular group G under consideration. In fact, this type of generalization has been less exploited but it is of interest in applications to mathematical physics where special symmetry groups such as the Poincaré group, or the Euclidean group, have particular significance. We will illustrate this point with an example involving the Euclidean group $\mathbb{E}^\nu$. This group consists of the translations $\mathbb{R}^\nu$ and the orthogonal group $\mathbb{O}^\nu$ of rotations acting on $\mathbb{R}^\nu$. We assume that the groups act continuously on $\mathfrak{A}$, that ω is $\mathbb{E}^\nu$-invariant and $\mathbb{R}^\nu$-ergodic. Moreover, we assume that Ω_ω is the unique (up to a phase factor) $U_\omega(\mathbb{R}^\nu)$ invariant vector in $\mathfrak{H}_\omega$. It then follows from Example 4.3.5 that ω satisfies the cluster property

$$\lim_{L\to\infty} \frac{1}{(2L)^\nu} \int_{|x_1|\leq L, \ldots, |x_\nu|\leq L} dx\, \omega(A\tau_x(B)) = \omega(A)\omega(B)$$

for all $A, B \in \mathfrak{A}$. The invariance of ω under rotations allows one to conclude, however, that the stronger cluster property

$$\lim_{|x|\to\infty} \omega(A\tau_x(B)) = \omega(A)\omega(B)$$

for all $A, B \in \mathfrak{A}$ is also valid. The idea behind this derivation is a smoothing with respect to rotations, a method which is described in the following example.

EXAMPLE 4.3.25. Let $U; (x, R) \mapsto U(x, R)$ be a strongly continuous unitary representation of the Euclidean group $\mathbb{E}^\nu$ with $\nu \geq 2$. If E is the projection on the subspace of $\mathfrak{H}$ formed by the vectors invariant under $U(\mathbb{R}^\nu, 1)$ then one can establish that

$$\lim_{|x|\to\infty} (\varphi, U(x, 1)\psi) = (\varphi, E\psi)$$

for all $\varphi, \psi \in \mathfrak{H}$. The proof is based upon a regularization procedure which involves smoothing over rotations. Let $\mathcal{N}$ be a neighborhood of the identity in the rotation group such that

$$\|\psi\|\,\|U(0, R)\varphi - \varphi\| < \frac{\varepsilon}{2}, \qquad \|\varphi\|\,\|U(0, R)\psi - \psi\| < \frac{\varepsilon}{2},$$

for all $R \in \mathcal{N}$. Thus if f is a positive C^∞-function with support in $\mathcal{N}$, and integral unity, then

$$\left| \int dR\, f(R)(\varphi, U(0, R)^* U(x, 1) U(0, R)\psi) - (\varphi, U(x, 1)\psi) \right| < \varepsilon$$

and consequently,

$$|(\varphi, U(x, 1)\psi) - (\varphi, E\psi)| \leq \left| \int d(\varphi, E(p)\psi) \int dR\, f(R) e^{i(Rp)x} \right| + \varepsilon,$$

where E denotes the spectral measure of U except that the point $\{0\}$ is given zero weight. It thus suffices to argue that there is an f of the above type which satisfies

$$\lim_{|x|\to\infty} \int dR\, f(R)e^{i(Rp)x} = 0$$

uniformly for p in compact subsets of $\mathbb{R}^\nu\setminus\{0\}$. This is not difficult but we omit the details.

Finally, we examine a geometric property of the set of G-ergodic states which occurs in many applications. Corollary 4.3.11 established that $E_{\mathfrak{A}}{}^G$ is a simplex whenever $(\mathfrak{A}, \omega)$ is G-abelian for each $\omega \in E_{\mathfrak{A}}{}^G$. The next example demonstrates that the set of extreme points $\mathscr{E}(E_{\mathfrak{A}}{}^G)$ of such simplices can be dense. This density property is reminiscent of the density of pure states established in Example 4.1.31 but is more interesting in the simplex situation. The extra interest arises because one can demonstrate that there exists a unique (up to affine isomorphisms) simplex such that the set of extreme points is dense (see Notes and Remarks). Thus the various spaces of G-invariant states described in the following example are all affinely isomorphic.

EXAMPLE 4.3.26. Let $\mathfrak{A}$ denote a UHF algebra as introduced in Examples 2.6.12, 3.2.25, and 4.1.31 but choose the index set I to be equal to $\mathbb{Z}^\nu$. For $x \in \mathbb{Z}^\nu$ let $\mathfrak{H}_x$ denote the underlying Hilbert space (see Example 2.6.12) and assume the dimension of $\mathfrak{H}_x$ is independent of x. Next for each $a \in \mathbb{Z}^\nu$ choose a unitary mapping $V_x(a); \mathfrak{H}_x \mapsto \mathfrak{H}_{x+a}$ such that $V_x(0)$ is the identity in $\mathfrak{H}_x$ and $V_x(a_1 + a_2) = V_{x+a_2}(a_1)V_x(a_2)$. Furthermore, for each finite subset $\Lambda \subset \mathbb{Z}^\nu$ define $V_\Lambda(a)$ by

$$V_\Lambda(a) = \bigotimes_{x\in\Lambda} V_x(a).$$

Thus one has $V_{a+\Lambda}(-a) = V_\Lambda(a)^*$. One can now introduce an action τ of $\mathbb{Z}^\nu$ as *-automorphisms of $\mathfrak{A}$ by defining

$$\tau_a(A) = V_\Lambda(a)A\,V_{\Lambda+a}(-a)$$

for each $A \in \mathfrak{A}_\Lambda$. In this way τ is consistently defined on the union of the $\mathfrak{A}_\Lambda$, $\Lambda \subset \mathbb{Z}^\nu$, as an isometric *-isomorphism and can be extended by continuity to $\mathfrak{A}$. Note that if $A \in \mathfrak{A}_\Lambda$ then $\tau_x(A) \in \mathfrak{A}_{\Lambda+x}$. Now we claim that $\mathscr{E}(E_{\mathfrak{A}}^{\mathbb{Z}^\nu})$ is dense in $E_{\mathfrak{A}}^{\mathbb{Z}^\nu}$. To establish this we first let Λ_L denote the cube

$$\Lambda_L = \{x; x = \{x_1, x_2, \ldots, x_\nu\}, |x_\nu| \le L\}$$

and remark that if $\omega \in E_{\mathfrak{A}}{}^G$ then ω in restriction to $\mathfrak{A}_{\Lambda_L}$ is determined by a density matrix ρ_{Λ_L} in the form

$$\omega(A) = \mathrm{Tr}_{\mathfrak{H}_{\Lambda_L}}(\rho_{\Lambda_L}A), \qquad A \in \mathfrak{A}_{\Lambda_L}.$$

Moreover, if $A \in \mathfrak{A}_{\Lambda_0}$ then

$$\omega(A) = \mathrm{Tr}_{\mathfrak{H}_{\Lambda_L}}(\rho_{\Lambda_L}\tau_x(A))$$

for all $x \in \mathbb{Z}^\nu$ such that $\Lambda_0 + x \subset \Lambda_L$. Next we build a periodic approximation ω_L, of ω, by the following procedure. Let $\{\Lambda_L{}^n\}_{n\ge1}$ denote a sequence of mutually disjoint translates of Λ_L whose union is equal to $\mathbb{Z}^\nu$. For each finite $\Lambda \subset \mathbb{Z}^\nu$ there will exist a finite subset I_Λ of the positive integers such that

$$\Lambda \subseteq \bigcup_{n \in I_\Lambda} \Lambda_L{}^n$$

and we define ω_L on $\mathfrak{A}_\Lambda$ by setting

$$\omega_L(A) = \mathrm{Tr}\left(\left(\bigotimes_{n \in I_\Lambda} \rho_{\Lambda_L{}^n}\right) A\right)$$

(the $\rho_{\Lambda_L{}^n}$ are the density matrices determined by ω). Thus ω_L is specified on the union of the $\mathfrak{A}_\Lambda$ and it extends by continuity to a state over $\mathfrak{A}$. Next we define a $\mathbb{Z}^\nu$-invariant state $\tilde{\omega}_L$ by

$$\tilde{\omega}_L(A) = |\Lambda_L|^{-1} \sum_{x \in \Lambda_L} \omega_L(\tau_x(A)),$$

where $|\Lambda_L| = (2L + 1)^\nu$ indicates the number of points in Λ_L. Now it easily follows that if $A \in \mathfrak{A}_\Lambda$ then

$$\begin{aligned} |\omega(A) - \tilde{\omega}_L(A)| &\le |\Lambda_L|^{-1} \sum_{\substack{x \in \Lambda_L \\ x + \Lambda \not\subset \Lambda_L}} |\omega_L(\tau_x(A)) - \omega(A)| \\ &\le 2\|A\| \, |\Lambda_L|^{-1} \sum_{\substack{x \in \Lambda_L \\ x + \Lambda \not\subset \Lambda_L}} 1 \\ &\xrightarrow[L = \infty]{} 0 \end{aligned}$$

and one concludes that $\tilde{\omega}_L$ converges in the weak* topology to ω. Finally, we argue that $\tilde{\omega}_L \in \mathscr{E}(E_{\mathfrak{A}}{}^{\mathbb{Z}^\nu})$. For this we first remark that $\mathfrak{A}$ is asymptotically abelian in the strong sense:

$$\lim_{|x| \to \infty} \|[A, \tau_x(B)]\| = 0.$$

This property follows easily for $A, B \in \mathfrak{A}_\Lambda$ from the product structure of $\mathfrak{A}$ and for general A, B by norm continuity. Now it follows from Theorem 4.3.17 that $\tilde{\omega}_L \in \mathscr{E}(E_{\mathfrak{A}}{}^{\mathbb{Z}^\nu})$ if, and only if, it satisfies the cluster property given by condition (2) of Theorem 4.3.22. But if $A, B \in \mathfrak{A}_\Lambda$ and B' is defined by

$$B' = |\Lambda_{L'}|^{-1} \sum_{y \in \Lambda_{L'}} \tau_y(B)$$

then

$$\begin{aligned} \lim_{L' \to \infty} \tilde{\omega}_L(AB') &= \lim_{L' \to \infty} |\Lambda_L|^{-1} |\Lambda_{L'}|^{-1} \sum_{\substack{x \in \Lambda_L \\ y \in \Lambda_{L'}}} \omega_L(\tau_x(A)\tau_{x+y}(B)) \\ &= |\Lambda_L|^{-1} \sum_{x \in \Lambda_L} \omega_L(\tau_x(A))\tilde{\omega}_L(B) \\ &= \tilde{\omega}_L(A)\tilde{\omega}_L(B). \end{aligned}$$

The cluster property for general A, B follows by continuity and we conclude that $\tilde{\omega}_L \in \mathscr{E}(E_{\mathfrak{A}}{}^{\mathbb{Z}^\nu})$.

Finally, $E_{\mathfrak{A}}{}^{\mathbb{Z}^\nu}$ is a simplex by asymptotic abelianness and Corollary 4.3.11.

4.3.3. Locally Compact Abelian Groups

In this subsection we continue the analysis of G-invariant and G-ergodic states ω over a C^*-algebra $\mathfrak{A}$ but with a specialization for G and its action τ. Firstly, we assume that G is a locally compact abelian group and, secondly, we assume strong continuity of the unitary representation $U_\omega(G)$ of G generated by ω. The second assumption is automatically fulfilled if the action τ of G on $\mathfrak{A}$ is strongly or, equivalently, weakly continuous. We adopt the notation already used in Sections 2.7.1 and 3.2.3 for G. In particular, $\hat{G}$ denotes the dual group, or character group, of G and dP_ω the projection-valued measure associated with $U_\omega(G)$ by the SNAG theorem, i.e., the spectral decomposition of $U_\omega(G)$ is given by

$$U_\omega(t) = \int_{\hat{G}} dP_\omega(\gamma)\overline{(\gamma, t)}.$$

The aim of this subsection is the analysis of the spectrum $\sigma(U_\omega)$, of $U_\omega(G)$, and the spectrum $\sigma(\hat{\tau})$ of the group of automorphisms $\hat{\tau}$, of $\pi_\omega(\mathfrak{A})''$, obtained by the canonical extension

$$\hat{\tau}_t(A) = U_\omega(t)AU_\omega(t)^*$$

for $A \in \pi_\omega(\mathfrak{A})''$. Subsequently, we examine the spectra of $\hat{\tau}$ acting on $\pi_\omega(\mathfrak{A})'$. The formal definition of these spectra are given by Definition 3.2.37 and $\sigma(U_\omega)$ is exactly the support of P_ω. Note that if $f \in L^1(G)$ and $\hat{\tau}_f(A)$, $U_\omega(f)$ denote the regularized operators

$$\hat{\tau}_f(A) = \int dt\; f(t)\hat{\tau}_t(A), \qquad U_\omega(f) = \int dt\; f(t)U_\omega(t),$$

then one has

$$U_\omega(f)A\Omega_\omega = \hat{\tau}_f(A)\Omega_\omega$$

for all $A \in \pi_\omega(\mathfrak{A})''$. But if $\operatorname{supp}\hat{f} \cap \sigma(\hat{\tau}) = \emptyset$ then $\hat{\tau}_f(A) = 0$ for all $A \in \pi_\omega(\mathfrak{A})''$ and by cyclicity one must have $U_\omega(f) = 0$. This shows that $\sigma(U_\omega) \subseteq \sigma(\hat{\tau})$; but the converse is generally false as can be illustrated by taking $G = \mathbb{R}$ and arranging that $\sigma(U_\omega) \subseteq [0, \infty\rangle$. Nevertheless, we will examine situations for which the two spectra are identical.

It is also of interest to examine the *point spectra* $\sigma_p(U_\omega)$ and $\sigma_p(\hat{\tau})$ of $U_\omega(G)$ and $\hat{\tau}$. The point spectrum of $U_\omega(G)$ is directly defined by

$$\sigma_p(U_\omega) = \{\gamma;\, \gamma \in \hat{G},\, P_\omega(\{\gamma\}) \neq 0\},$$

i.e., $\gamma \in \sigma_p(U_\omega)$ if, and only if, there is a nonzero eigenvector $\psi_\gamma \in \mathfrak{H}_\omega$ such that

$$U_\omega(t)\psi_\gamma = \overline{(\gamma, t)}\psi_\gamma$$

for all $t \in G$. Similarly, $\sigma_p(\hat{\tau})$ is defined as the set of characters γ such that $\mathfrak{M}^{\hat{\tau}}(\{\gamma\}) \neq 0$, i.e., the set of γ for which there exists a nonzero $A_\gamma \in \mathfrak{M} \equiv \pi_\omega(\mathfrak{A})''$ satisfying the eigenvalue equation

$$\hat{\tau}_t(A_\gamma) = \overline{(\gamma, t)}A_\gamma$$

for all $t \in G$. We will find conditions under which the sets $\sigma_p(U_\omega)$ and $\sigma_p(\hat{\tau})$ are related.

Throughout the rest of the subsection we adopt the above notation and assumptions and for simplicity we use the notation

$$P[\gamma] = P(\{\gamma\}).$$

We generally use additive notation for the group operations but for brevity we occasionally use multiplicative notation, i.e., $\gamma_1\gamma_2$ in place of $\gamma_1 + \gamma_2$ and γ^{-1} in place of $-\gamma$. Finally, we associate with $\sigma_p(U_\omega)$ the projection $\hat{P}_\omega$ defined by

$$\hat{P}_\omega = \sum_{\gamma \in \hat{G}} P_\omega[\gamma].$$

We refer to $\hat{P}_\omega \mathfrak{H}_\omega$ as the subspace of $U_\omega(G)$-almost periodic vectors. The motivation for this nomenclature will be clarified in Section 4.3.4.

The first and most complete spectral result is obtained for G-ergodic faithful states.

Theorem 4.3.27. *Let $\omega \in E_{\mathfrak{A}}{}^G$. Assume that G is locally compact abelian, the corresponding unitary representation $U_\omega(G)$ is strongly continuous, and let dP_ω denote the projection valued spectral measure of $U_\omega(G)$.*

If Ω_ω is separating for $\pi_\omega(\mathfrak{A})''$ the following statements are true:

(1) *$\sigma(U_\omega) = \sigma(\hat{\tau})$, $\sigma_p(U_\omega) = \sigma_p(\hat{\tau})$ and both sets are symmetric, i.e., closed under the inverse operation $\gamma \to -\gamma$ in $\hat{G}$;*
(2) *if ω is G-ergodic then for each $\gamma \in \sigma_p(U_\omega) = \sigma_p(\hat{\tau})$ there exists a unit vector $\psi_\gamma \in \mathfrak{H}_\omega$ and a unitary $V_\gamma \in \pi_\omega(\mathfrak{A})''$ such that*

$$U_\omega(t)\psi_\gamma = \overline{(\gamma, t)}\psi_\gamma, \quad \hat{\tau}_t(V_\gamma) = \overline{(\gamma, t)}V_\gamma$$

and ψ_γ, V_γ are both unique up to a phase;
(3) *if ω is G-ergodic then the two sets $\sigma(U_\omega)$ and $\sigma_p(U_\omega)$ are both subgroups of $\hat{G}$;*
(4) *if ω is G-ergodic and the annihilator H_ω of $\sigma_p(U_\omega)$ is the closed subgroup of G defined by*

$$H_\omega = \{t; t \in G, (\gamma, t) = 1 \text{ for all } \gamma \in \sigma_p(U_\omega)\}$$

then

$$\begin{aligned}\{V_\gamma; \gamma \in \sigma_p(\hat{\tau})\}'' &= \pi_\omega(\mathfrak{A})'' \cap \hat{P}_\omega' \\ &\subseteq \pi_\omega(\mathfrak{A})'' \cap E_\omega(H_\omega)' = \pi_\omega(\mathfrak{A})'' \cap U_\omega(H_\omega)'\end{aligned}$$

where $\hat{P}_\omega$ is the projection onto the subspace of $U_\omega(G)$-almost periodic vectors and $E_\omega(H_\omega)$ is the projection onto the subspace of $U_\omega(H_\omega)$-invariant vectors. Equality of the four sets occurs if, and only if, $E_\omega(H_\omega) = \hat{P}_\omega$ and, in particular, if $\sigma_p(U_\omega)$ is countable and closed.

Proof. As Ω_ω is separating for $\pi_\omega(\mathfrak{A})''$ it is cyclic for $\pi_\omega(\mathfrak{A})'$ by Proposition 2.5.3. This will be used throughout the proof.

(1) If $\hat{\tau}_f$ and $U_\omega(f)$ denote the regularization of $\hat{\tau}$ and U_ω with $f \in L^1(G)$ then it follows from the relation

$$U_\omega(f)A\Omega_\omega = \hat{\tau}_f(A)\Omega_\omega$$

and the separating character of Ω_ω that $U_\omega(f) = 0$ if, and only if, $\hat{\tau}_f = 0$. Hence $\sigma(U_\omega) = \sigma(\hat{\tau})$ by Definition 3.2.37.

Next note that if $A_\gamma \in \pi_\omega(\mathfrak{A})''$ satisfies the eigenvalue equation

$$\hat{\tau}_t(A_\gamma) = \overline{(\gamma, t)}A_\gamma$$

then $\psi_\gamma = A_\gamma\Omega_\omega$ satisfies

$$U_\omega(t)\psi_\gamma = \hat{\tau}_t(A_\gamma)\Omega_\omega = \overline{(\gamma, t)}\psi_\gamma.$$

Therefore $\sigma_p(\hat{\tau}) \subseteq \sigma_p(U_\omega)$. Conversely, assume $\gamma \in \sigma_p(U_\omega)$ and choose $A \in \pi_\omega(\mathfrak{A})''$ such that

$$P_\omega[\gamma]A\Omega_\omega \neq 0.$$

Now $P_\omega[\gamma]$ is the subspace of $\mathfrak{H}_\omega$ formed by the vectors invariant under the unitary representation $t \mapsto (\gamma, t)U_\omega(t)$ of G. Therefore Proposition 4.3.4 implies the existence of a net

$$S_{\lambda^\alpha}(\gamma U_\omega) = \sum_{i=1}^{n_\alpha} \lambda_i^\alpha(\gamma, t_i^\alpha)U_\omega(t_i^\alpha)$$

of convex combinations of γU_ω such that $S_{\lambda^\alpha}(\gamma U_\omega)$ converges strongly to $P_\omega[\gamma]$. Consider the corresponding net

$$S_{\lambda^\alpha}(\gamma\hat{\tau}(A)) = \sum_{i=1}^{n_\alpha} \lambda_i^\alpha(\gamma, t_i^\alpha)\hat{\tau}_{t_i^\alpha}(A)$$

of convex combinations of $\gamma\hat{\tau}(A)$. One has $\|S_{\lambda^\alpha}(\gamma\hat{\tau}(A))\| \leq \|A\|$ and $S_{\lambda^\alpha}(\gamma\hat{\tau}(A))$ converges strongly on the dense subspace $\pi_\omega(\mathfrak{A})'\Omega_\omega$. Hence $S_{\lambda^\alpha}(\gamma\hat{\tau}(A))$ converges strongly to a bounded operator A_γ. Clearly, one has $A_\gamma \in \pi_\omega(\mathfrak{A})''$, $\|A_\gamma\| \leq \|A\|$, and

$$A_\gamma C\Omega_\omega = CP_\omega[\gamma]A\Omega_\omega$$

for all $C \in \pi_\omega(\mathfrak{A})'$. The special choice $C = \mathbb{1}_\omega$ gives $A_\gamma \neq 0$ and it then follows that

$$\begin{aligned}\hat{\tau}_t(A_\gamma)C\Omega_\omega &= U_\omega(t)A_\gamma(U_\omega(t)^*CU_\omega(t))\Omega_\omega \\ &= CU_\omega(t)P_\omega[\gamma]A\Omega_\omega \\ &= \overline{(\gamma, t)}A_\gamma C\Omega_\omega\end{aligned}$$

for all $C \in \pi_\omega(\mathfrak{A})'$. Thus

$$\hat{\tau}_t(A_\gamma) = \overline{(\gamma, t)}A_\gamma$$

and $\sigma_p(U_\omega) \subseteq \sigma_p(\hat{\tau})$. This together with the previous inclusion establishes equality of the point spectra.

Symmetry of $\sigma(\hat{\tau})$ follows from Lemma 3.2.42 and symmetry of $\sigma_p(\hat{\tau})$ follows because

$$\begin{aligned}\hat{\tau}_t(A_\gamma{}^*) &= \hat{\tau}_t(A_\gamma)^* \\ &= (\gamma, t)A_\gamma{}^* = \overline{(\gamma^{-1}, t)}A_\gamma{}^*.\end{aligned}$$

(2) We assume ω is G-ergodic.

If A_γ denotes the eigenelement of $\hat{\tau}$ constructed above then $A_\gamma{}^*A_\gamma \in \pi_\omega(\mathfrak{A})'' \cap U_\omega(G)'$. Thus one concludes from Theorem 4.3.20 that $A_\gamma{}^*A_\gamma$ is a nonzero multiple λ_A of the identity $\mathbb{1}_\omega$. Consequently, $|A_\gamma| = \lambda_A^{1/2}\mathbb{1}_\omega \neq 0$ and if $A_\gamma = V_\gamma|A_\gamma|$ denotes the polar decomposition of A_γ then V_γ is unitary. It also immediately follows that

$$\hat{\tau}_t(V_\gamma) = \frac{\hat{\tau}_t(A_\gamma)}{\lambda_A^{1/2}} = (\overline{\gamma, t})V_\gamma$$

and V_γ is also an eigenelement of $\hat{\tau}$. Now if $W_\gamma{}^* \in \pi_\omega(\mathfrak{A})''$ is any other unitary eigenelement one has

$$\hat{\tau}_t(W_\gamma{}^*V_\gamma) = \hat{\tau}_t(W_\gamma)^*\hat{\tau}_t(V_\gamma) = W_\gamma{}^*V_\gamma .$$

Thus $W_\gamma{}^*V_\gamma \in \pi_\omega(\mathfrak{A})'' \cap U_\omega(G)'$ and another application of Theorem 4.3.20 establishes that $W_\gamma{}^*V_\gamma$ is a multiple of the identity, i.e., W_γ differs from V_γ by at most a phase.

Next note that $\{A_\gamma\Omega_\omega;\, A \in \pi_\omega(\mathfrak{A})''\}$ is certainly dense in $P_\omega[\gamma]\mathfrak{H}_\omega$ because

$$A_\gamma\Omega_\omega = P_\omega[\gamma]A\Omega_\omega .$$

Hence $P_\omega[\gamma]\mathfrak{H}_\omega$ must consist of multiples of $V_\gamma\Omega_\omega$, i.e., $P_\omega[\gamma]$ has rank one and the corresponding eigenelement of $U_\omega(G)$ is unique up to a phase.

(3) If $\gamma_1, \gamma_2 \in \sigma_{\mathrm{p}}(\hat{\tau})$ and $V_{\gamma_1}, V_{\gamma_2}$ are the corresponding unitary eigenelements then $V_{\gamma_1}V_{\gamma_2}$ is unitary and satisfies the eigenvalue equation

$$\hat{\tau}_t(V_{\gamma_1}V_{\gamma_2}) = (\overline{\gamma_1, t})(\overline{\gamma_2, t})V_{\gamma_1}V_{\gamma_2} = (\overline{\gamma_1\gamma_2, t})V_{\gamma_1}V_{\gamma_2}.$$

Thus $\gamma_1 + \gamma_2 (= \gamma_1\gamma_2) \in \sigma_{\mathrm{p}}(\hat{\tau})$ and this additivity property combined with the previously proved symmetry establishes that $\sigma_{\mathrm{p}}(\hat{\tau})(= \sigma_{\mathrm{p}}(U_\omega))$ is a subgroup of $\hat{G}$.

Finally, assume $\gamma_1, \gamma_2 \in \sigma(U_\omega)$ and let N be an arbitrary neighborhood of $\gamma_1 + \gamma_2$. If M is a neighborhood of the identity in $\hat{G}$ such that $\gamma_1 + \gamma_2 + M + M \subseteq N$ then the equivalence (1) $\Leftrightarrow$ (2) of Proposition 3.2.40 implies the existence of functions $f_1, f_2 \in L^1(G)$ such that $\operatorname{supp} \hat{f}_i \subseteq \gamma_i + M$ and $U_\omega(f_i) \neq 0$. Hence there exist $A_1, A_2 \in \mathfrak{A}$ such that the corresponding regularizations $\hat{\tau}_{f_i}(A_i)$ satisfy

$$\hat{\tau}_{f_i}(A_i)\Omega_\omega \neq 0, \qquad i = 1, 2.$$

Thus $\sigma_{\hat{\tau}}(\hat{\tau}_{f_i}(A_i)) \subseteq \gamma_i + M$, where we now use the terminology of spectral subspaces introduced in Section 3.2.3. Moreover, one concludes from Lemmas 3.2.38 and 3.2.42 that $\sigma_{\hat{\tau}}(\hat{\tau}_t(\hat{\tau}_{f_1}(A_1))\hat{\tau}_{f_2}(A_2)) \subseteq \gamma_1 + \gamma_2 + M + M$ for all $t \in G$. Therefore if we can show that

$$\hat{\tau}_t(\hat{\tau}_{f_1}(A_1))\hat{\tau}_{f_2}(A_2)\Omega_\omega \neq 0$$

for some $t \in G$ then we can conclude from Proposition 3.2.40 that $\gamma_1 + \gamma_2 \in \sigma(U_\omega)$. But

$$\|\hat{\tau}_t(\hat{\tau}_{f_1}(A_1))\hat{\tau}_{f_2}(A_2)\Omega_\omega\|^2 = \omega(\hat{\tau}_{f_2}(A_2)^*\hat{\tau}_t(\hat{\tau}_{f_1}(A_1)^*\hat{\tau}_{f_1}(A_1))\hat{\tau}_{f_2}(A_2))$$

and as ω is G-ergodic

$$\inf_{A \in \mathrm{Co}\,\tau_G(\hat{\tau}_{f_1}(A_1)^*\hat{\tau}_{f_1}(A_1))} |\omega(\hat{\tau}_{f_2}(A_2)^*A\hat{\tau}_{f_2}(A_2)) - \omega(\hat{\tau}_{f_2}(A_2{}^*)\hat{\tau}_{f_2}(A_2)) \cdot \omega(\hat{\tau}_{f_1}(A_1{}^*)\hat{\tau}_{f_1}(A_1))| = 0$$

by Theorem 4.3.23. Thus the desired conclusion is valid and $\sigma(U_\omega)\,(= \sigma(\hat{\tau}))$ is additive. The additivity together with the symmetry established in part (1) then prove that $\sigma(U_\omega)$ is a subgroup of $\hat{G}$.

(4) As $(\gamma, t_1 + t_2) = (\gamma, t_1)(\gamma, t_2)$ and $(\gamma, -t) = \overline{(\gamma, t)}$ it follows that H_ω is a subgroup of G. But H_ω is closed because $t \mapsto (\gamma, t)$ is continuous. The definition of H_ω and the eigenvalue equation for V_γ imply that $V_\gamma \in U_\omega(H_\omega)'$ and hence

$$\{V_\gamma;\, \gamma \in \sigma_p(\hat{\tau})\}'' \subseteq \pi_\omega(\mathfrak{A})'' \cap U_\omega(H_\omega)'.$$

But if V_γ and ψ_{γ_1} denote the eigenelements of $\hat{\tau}$ and $U_\omega(G)$ then

$$U_\omega(t)V_\gamma\psi_{\gamma_1} = \overline{(\gamma\gamma_1, t)}V_\gamma\psi_{\gamma_1}$$

and hence

$$V_\gamma P_\omega[\gamma_1] = P_\omega[\gamma\gamma_1]V_\gamma P_\omega[\gamma_1].$$

Now taking the conjugate of this relation and using $V_\gamma^* = V_{\gamma^{-1}}$ one obtains

$$P_\omega[\gamma_1]V_\gamma = P_\omega[\gamma_1]V_\gamma P_\omega[\gamma^{-1}\gamma_1].$$

Therefore

$$\begin{aligned}\hat{P}_\omega V_\gamma &= \sum_{\gamma_1} P_\omega[\gamma_1]V_\gamma P_\omega[\gamma^{-1}\gamma_1] \\ &= \sum_{\gamma_1} P_\omega[\gamma\gamma_1]V_\gamma P_\omega[\gamma_1] = V_\gamma\hat{P}_\omega\end{aligned}$$

and hence

$$\{V_\gamma;\, \gamma \in \sigma_p(\hat{\tau})\}'' \subseteq \pi_\omega(\mathfrak{A})'' \cap \hat{P}_\omega'.$$

Conversely, if $A \in \pi_\omega(\mathfrak{A})'' \cap \hat{P}_\omega'$ one has

$$\hat{P}_\omega A\Omega_\omega = A\Omega_\omega.$$

Consequently,

$$A\Omega_\omega = \sum_\gamma A_\gamma\Omega_\omega,$$

where $A_\gamma \in \pi_\omega(\mathfrak{A})''$ are the eigenelements constructed in the proof of part (2). More generally,

$$AC\Omega_\omega = \sum_\gamma A_\gamma C\Omega_\omega$$

for all $C \in \pi_\omega(\mathfrak{A})'$. Consequently, if $X \in \{V_\gamma;\, \gamma \in \sigma_p(\hat{\tau})\}'$ then

$$\begin{aligned}(C_1\Omega_\omega, [A, X]C_2\Omega_\omega) &= \sum_\gamma \{(A_\gamma^*C_1\Omega_\omega, XC_2\Omega_\omega) - (C_1\Omega_\omega, XA_\gamma C_2\Omega_\omega)\} \\ &= 0\end{aligned}$$

for all $C_1, C_2 \in \pi_\omega(\mathfrak{A})'$, because $[A_\gamma, X] = 0$. Thus $A \in \{V_\gamma;\, \gamma \in \sigma_p(\hat{\tau})\}''$ and

$$\{V_\gamma;\, \gamma \in \sigma_p(\hat{\tau})\}'' = \pi_\omega(\mathfrak{A})'' \cap \hat{P}_\omega' \subseteq \pi_\omega(\mathfrak{A})'' \cap U_\omega(H_\omega)'.$$

But it follows from the mean ergodic theorem that $E_\omega(H_\omega) \in U_\omega(H_\omega)''$ and hence

$$\pi_\omega(\mathfrak{A})'' \cap E_\omega(H_\omega)' \supseteq \pi_\omega(\mathfrak{A})'' \cap U_\omega(H_\omega)'.$$

Thus if $A \in \pi_\omega(\mathfrak{A})'' \cap E_\omega(H_\omega)'$ and $t \in H_\omega$ then

$$\begin{aligned}U_\omega(t)AU_\omega(-t)\Omega_\omega &= U_\omega(t)A\Omega_\omega \\ &= E_\omega(H_\omega)A\Omega_\omega \\ &= A\Omega_\omega\end{aligned}$$

and since Ω_ω is separating $A \in \pi_\omega(\mathfrak{A})'' \cap U_\omega(H_\omega)'$, i.e.,

$$\pi_\omega(\mathfrak{A})'' \cap E_\omega(H_\omega)' = \pi_\omega(\mathfrak{A})'' \cap U_\omega(H_\omega)'.$$

If $E_\omega(H_\omega) = \hat{P}_\omega$ then one has equality between all four sets and conversely equality of the sets gives

$$\begin{aligned}\hat{P}_\omega &= [\{V_\gamma; \gamma \in \sigma_{\mathrm{p}}(\hat{\tau})\}''\Omega_\omega] \\ &= [\{\pi_\omega(\mathfrak{A})'' \cap U_\omega(H_\omega)'\}\Omega_\omega] = E_\omega(H_\omega).\end{aligned}$$

Finally, assume $\sigma_{\mathrm{p}}(U_\omega)$ is closed; then the dual $\hat{H}_\omega$ of H_ω is given by $\hat{H}_\omega = \hat{G}/\sigma_{\mathrm{p}}(U_\omega)$. Let $\varphi;\ \gamma \in \hat{G} \mapsto \gamma + \sigma_{\mathrm{p}}(U_\omega) \in \hat{H}_\omega$ denote the natural homomorphism of $\hat{G}$ onto $\hat{H}_\omega$ then each Borel subset of $\hat{H}_\omega$ is the image under φ of a Borel subset of $\hat{G}$. Now if for $\psi \in \mathfrak{H}_\omega$ one introduces a spectral measure μ_ψ on $\hat{G}$ by

$$(\psi, U_\omega(t)\psi) = \int_{\hat{G}} d\mu_\psi(\gamma)\, \overline{(\gamma, t)}$$

for all $t \in G$, and a second spectral measure ν_ψ on $\hat{H}_\omega$ by restriction, i.e.,

$$(\psi, U_\omega(s)\psi) = \int_{\hat{H}_\omega} d\nu_\psi(\gamma)\, \overline{(\gamma, s)}$$

for all $s \in \hat{H}_\omega$, one then has

$$\nu_\psi(S) = \mu_\psi(\varphi^{-1}(S))$$

for each Borel set $S \subseteq \hat{H}_\omega$. In particular,

$$\nu_\psi(\{0\}) = \mu_\psi(\{\sigma_{\mathrm{p}}(U_\omega)\}).$$

Now assume that $\sigma_{\mathrm{p}}(U_\omega)$ is countable and consider an increasing sequence of finite subsets S_n with characteristic functions χ_{S_n} such that $\bigcup_n S_n = \sigma_{\mathrm{p}}(U_\omega)$. To each such subset corresponds a projection operator $P_n \leq \hat{P}_\omega$ and

$$\mu_\psi(S_n) = \mu_\psi(\chi_{S_n}) = (\psi, P_n\psi).$$

Now the P_n converge weakly to $\hat{P}_\omega$ and the χ_{S_n} are a uniformly bounded sequence which converges pointwise to the characteristic function of $\sigma_{\mathrm{p}}(U_\omega)$. Thus by the Lebesgue dominated convergence theorem

$$\begin{aligned}\mu_\psi(\{\sigma_{\mathrm{p}}(U_\omega)\}) &= \lim_{n\to\infty} \mu_\psi(\chi_{S_n}) \\ &= \lim_{n\to\infty} (\psi, P_n\psi) = (\psi, \hat{P}_\omega\psi).\end{aligned}$$

Therefore

$$\begin{aligned}(\psi, E_\omega(H_\omega)\psi) &= \nu_\psi(\{0\}) \\ &= \mu_\psi(\{\sigma_{\mathrm{p}}(U_\omega)\}) = (\psi, \hat{P}_\omega\psi)\end{aligned}$$

and consequently $E_\omega(H_\omega) = \hat{P}_\omega$.

Remark. This last equality is a result concerning unitary representations U of G which is independent of the algebraic structure. If the point spectrum $\sigma_{\mathrm{p}}(U)$ of U is a closed, countable, subgroup of $\hat{G}$ and H denotes its annihilator then one has the corresponding equality, but the same argument also establishes that the restriction $U|_H$ of U to H has no point spectrum other than at zero.

The structure of the point spectrum exhibited by the foregoing theorem is of interest in the subsequent discussion of broken symmetries. Note that if $\omega \in \mathscr{E}(E_{\mathfrak{A}}{}^G)$ and if $U_\omega(G)$ has eigenvectors which are not multiples of Ω_ω then ω is not ergodic for the stabilizer group H_ω. This follows from Theorem 4.3.27 because

$$\pi_\omega(\mathfrak{A})'' \cap U_\omega(H_\omega)' \supseteq \{V_\gamma; \gamma \in \sigma_{\mathrm{p}}(\hat{\tau})\}'' \neq \mathbb{C}1_\omega.$$

If $\sigma_{\mathrm{p}}(U_\omega)$ is countable and closed then the general algebraic structure is simplified by the equality $E_\omega(H_\omega) = \hat{P}_\omega$. In particular, one has the identification

$$\{V_\gamma; \gamma \in \sigma_{\mathrm{p}}(U_\omega)\}'' = \pi_\omega(\mathfrak{A})'' \cap U_\omega(H_\omega)'.$$

This is of interest because the algebra on the left has a tendency to be abelian. The uniqueness of the V_γ implies that $\gamma \mapsto V_\gamma$ is a representation up to a phase factor of the abelian group $\sigma_{\mathrm{p}}(U_\omega)$. Thus if the factors can be chosen so that it is a representation then the V_γ commute. This is obviously the case if $\sigma_{\mathrm{p}}(U_\omega)$ is a cyclic group but it is possible under more general circumstances.[4] These algebraic simplifications will be of significance in the subsequent discussion of decomposition theory for G-ergodic states with nontrivial point spectrum $\sigma_{\mathrm{p}}(U_\omega)$. We return to these questions in Section 4.3.4 but we immediately illustrate how the foregoing result limits the spectral possibilities.

EXAMPLE 4.3.28. Let $G = \mathbb{R}$ and consider a G-ergodic state ω for which U_ω is continuous and Ω_ω is separating for $\pi_\omega(\mathfrak{A})''$. The spectrum $\sigma(U_\omega)$ $(=\sigma(\hat{\tau}))$ is then a closed subgroup of $\hat{\mathbb{R}}$ $(=\mathbb{R})$ containing $\{0\}$. There are three possibilities:

(1) $\sigma(U_\omega) = \sigma_{\mathrm{p}}(U_\omega) = \{0\}$;
(2) $\sigma(U_\omega) = \sigma_{\mathrm{p}}(U_\omega)$ is isomorphic to $\mathbb{Z}$;
(3) $\sigma(U_\omega) = \mathbb{R}$.

In the second case $\pi_\omega(\mathfrak{A})''$ is abelian by the remark preceding the example. In the third case there are three possibilities for $\sigma_{\mathrm{p}}(U_\omega)$:

(3_1) $\sigma_{\mathrm{p}}(U_\omega) = \{0\}$;
(3_2) $\sigma_{\mathrm{p}}(U_\omega)$ is isomorphic to $\mathbb{Z}$;
(3_3) $\sigma_{\mathrm{p}}(U_\omega)$ is a dense subgroup of $\mathbb{R}$.

In the first situation $H_\omega = \mathbb{R}$, in the second H_ω is isomorphic to $\mathbb{Z}$, and in the third H_ω is trivial.

Note that in the second case $\pi_\omega(\mathfrak{A})'' \cap U_\omega(H_\omega)'$ is abelian by the remark preceding the example.

[4] See a more detailed discussion of this point in the Notes and Remarks to Section 3.2.6.

If $G = \mathbb{R}^\nu$ and $\sigma(U_\omega) = \mathbb{R}^\nu$ there are many more possibilities for $\sigma_p(U_\omega)$, e.g., one could have

$$\sigma_p(U_\omega) = \mathbb{Z}^{\nu_1} \times \{0\} \quad \text{for } \nu_1 \leq \nu.$$

The properties of the point spectra of G-invariant states described in Theorem 4.3.27 can be derived by an independent argument which has interest in its own right because it characterizes $\sigma_p(U_\omega)$, etc., through an algebraic property of the projection $\hat{P}_\omega$ onto the almost periodic vectors. Under the assumptions of the theorem one can prove that

$$\hat{P}_\omega \pi_\omega(\mathfrak{A})'' \hat{P}_\omega = (\hat{P}_\omega \pi_\omega(\mathfrak{A})' \hat{P}_\omega)'$$

and this condition is in turn equivalent to

$$P_\omega[\gamma_1] A P_\omega[\gamma\gamma_1] B P_\omega[\gamma_2] = P_\omega[\gamma_1] B P_\omega[\gamma_2 \gamma^{-1}] A P_\omega[\gamma_2]$$

for all $A \in \pi_\omega(\mathfrak{A})''$, $B \in \pi_\omega(\mathfrak{A})'$, and $\gamma_1, \gamma_2, \gamma \in \hat{G}$. This latter condition implies the spectral properties by a series of arguments which we will exploit below for states where Ω_ω is not separating for $\pi_\omega(\mathfrak{A})''$.

The experience of decomposition theory for G-invariant states as described in the previous sections indicates that the same general ergodic structure pervades either if Ω_ω is separating for $\pi_\omega(\mathfrak{A})''$ or if $\mathfrak{A}$ satisfies a weak form of asymptotic abelianness. This is to a large extent the case for the point spectrum and the appropriate commutation property is a generalized form of G-abelianness.

Definition 4.3.29. Let $\mathfrak{A}$ be a C^*-algebra, G a locally compact abelian group represented as *-automorphisms τ of $\mathfrak{A}$, and ω a G-invariant state over $\mathfrak{A}$ such that the corresponding unitary representation U_ω of G is strongly continuous. The pair $(\mathfrak{A}, \omega)$ is defined to be G_Γ-*abelian* if

$$\inf_{A' \in \mathrm{Co}\, \gamma\tau(A)} |(\varphi_1, \pi_\omega([A', B])\varphi_2)| = 0$$

for all $A, B \in \mathfrak{A}$, $\gamma \in \hat{G}$, and $\varphi_1, \varphi_2 \in \hat{P}_\omega \mathfrak{H}_\omega$, where $\mathrm{Co}\, \gamma\tau(A)$ denotes the convex hull of $\{(\gamma, t)\tau_t(A); t \in G\}$.

In the subsequent subsection we will demonstrate that G_Γ-abelianness characterizes uniqueness of maximal measures $\mu \in M_\omega(E_\mathfrak{A})$ which are supported by an appropriate class of periodic, or almost periodic, states. Thus G_Γ-abelianness shares some of the features characteristic of G-abelianness. In the same way that G-abelianness can be characterized by commutativity of $\{E_\omega \pi_\omega(\mathfrak{A})'' E_\omega\}$, Proposition 4.3.7, G_Γ-abelianness can be characterized by commutativity of $\{\hat{P}_\omega \pi_\omega(\mathfrak{A})'' \hat{P}_\omega\}$. Moreover, this property can be characterized by use of an invariant mean. Such means were briefly mentioned in Section 4.3.2 in the context of general groups and it is relatively simple to establish their existence if G is a locally compact abelian group.

First consider the C^*-algebra $C_b(G)$ of bounded continuous functions, over G, equipped with the usual supremum norm, etc. The group G acts as a group of *-automorphisms τ of $C_b(G)$ by $(\tau_t f)(t') = f(tt')$. By transposition G acts on the dual $C_b(G)^*$ of $C_b(G)$ and leaves the convex set of states invariant. Since G is abelian the Markov–Kakutani fixed point theorem guarantees the existence of G-invariant states over $C_b(G)$ and it is these states which are called invariant means. One can similarly define invariant means over any C^*-subalgebra of $C_b(G)$ which is left invariant by τ. Invariant means can often be used to replace the nets of convex combinations that we have previously used. To illustrate this consider a strongly continuous unitary representation U of G on the Hilbert space $\mathfrak{H}$. Application of an invariant mean M to the functions

$$t \in G \mapsto (\varphi, U(t)\psi) \in C_b(G)$$

defines a bounded linear operator E on $\mathfrak{H}$ such that

$$M((\varphi, U\psi)) = (\varphi, E\psi).$$

The mean ergodic theorem, Proposition 4.3.4, implies that there exists a net $\sum_i \lambda_i^\alpha U(t_i^\alpha)$ in $\mathrm{Co}(U(G))$ such that $\sum_i \lambda_i^\alpha U(t_i^\alpha) \to E_0$ strongly, where E_0 is the orthogonal projection onto the U-invariant vectors. But this implies that the functions $t \mapsto \sum_i \lambda_i^\alpha(\varphi, U(t)U(t_i)\psi)$ tend to $(\varphi, E_0\psi)$ uniformly in t, and hence the invariance of M implies

$$\begin{aligned}(\varphi, E\psi) &= M((\varphi, U\psi)) \\ &= M\left(\sum_i \lambda_i^\alpha(\varphi, UU(t_i)\psi\right) \\ &= (\varphi, E_0\psi).\end{aligned}$$

Thus E is the orthogonal projection on the subspace of $\mathfrak{H}$ spanned by the U-invariant vectors. This yields another version of the mean ergodic theorem for the representation U; each invariant mean satisfies $M(U) = E$. This statement can be used to rephrase the definition of G-abelianness. The pair $(\mathfrak{A}, \omega)$ is G-abelian if, and only if,

$$M(\omega'([\tau(A), B])) = 0$$

for all $A, B \in \mathfrak{A}$, and all G-invariant vector states ω' of $\pi_\omega(\mathfrak{A})$. A similar rephrasing of G_Γ-abelianness is given by the following:

Proposition 4.3.30. *The following conditions are equivalent*:

(1) *the pair* $(\mathfrak{A}, \omega)$ *is* G_Γ*-abelian*;
(2) $M((\varphi_1, \pi_\omega([\gamma^{-1}\tau(A), B])\varphi_2) = 0$ *for all* $A, B \in \mathfrak{A}$, $\gamma \in \hat{G}$, *and* $\varphi_1, \varphi_2 \in \hat{P}_\omega\mathfrak{H}_\omega$ *for some invariant mean* M;
(3) $\{\hat{P}_\omega\pi_\omega(\mathfrak{A})''\hat{P}_\omega\}$ *is abelian*;
(4) $P_\omega[\gamma_1]AP_\omega[\gamma\gamma_1]BP_\omega[\gamma_2] = P_\omega[\gamma_1]BP_\omega[\gamma_2\gamma^{-1}]AP_\omega[\gamma_2]$ *for all* $A, B \in \pi_\omega(\mathfrak{A})''$ *and all* $\gamma, \gamma_1, \gamma_2 \in \hat{G}$.

Proof. We begin by proving (1) $\Leftrightarrow$ (4) and subsequently establish that (3) $\Leftrightarrow$ (4) and (2) $\Leftrightarrow$ (4).

(1) $\Leftrightarrow$ (4) The range of $P_\omega[\gamma]$ is the subspace of $\mathfrak{H}_\omega$ invariant under the unitary representation γU_ω of G. Therefore, given $\varepsilon > 0$ and ψ_1 an element of a finite subset $\mathscr{K}$ of $\mathfrak{H}_\omega$, one can find a convex combination S_λ of $\gamma\gamma_1 U_\omega$,

$$S_\lambda(\gamma\gamma_1 U_\omega) = \sum_{i=1}^{n} \lambda_i(\gamma\gamma_1, t_i)U_\omega(t_i)$$

such that

$$\|(S_\lambda(\gamma\gamma_1 U_\omega) - P_\omega[\gamma\gamma_1])\psi_1\| < \varepsilon/2.$$

This is again an application of Proposition 4.3.4. Similarly, if $\psi_2 \in \mathscr{K}$ then a second application of this proposition to the unitary representation $\gamma\gamma_1 U_\omega \oplus \gamma^{-1}\gamma_2 U_\omega$ on $\mathfrak{H}_\omega \oplus \mathfrak{H}_\omega$ establishes that S_λ can be chosen such that

$$\|(S_\lambda(\gamma^{-1}\gamma_2 U_\omega) - P_\omega[\gamma^{-1}\gamma_2])\psi_2\| < \varepsilon/2.$$

Next note that if S_μ denotes any other convex combination then

$$\|(S_\mu(\gamma\gamma_1 U_\omega)S_\lambda(\gamma\gamma_1 U_\omega) - P_\omega[\gamma_1\gamma])\psi_1\| < \varepsilon/2,$$
$$\|(S_\mu(\gamma^{-1}\gamma_2 U_\omega)S_\lambda(\gamma^{-1}\gamma_2 U_\omega) - P_\omega[\gamma^{-1}\gamma_2]\psi_2\| < \varepsilon/2.$$

This follows because of invariance and boundedness, e.g.,

$$S_\mu(\gamma\gamma_1 U_\omega)P_\omega[\gamma_1\gamma] = P_\omega[\gamma_1\gamma]$$

and $\|S_\mu(\gamma\gamma_1 U_\omega)\| \leq 1$. Now if $A \in \pi_\omega(\mathfrak{A})''$, $\varphi_i \in P_\omega[\gamma_i]\mathfrak{H}_\omega$, and $S_\mu S_\lambda(\gamma^{-1}\hat{\tau}(A)) \in \mathrm{Co}\,\gamma^{-1}\hat{\tau}(A)$ is defined by

$$S_\mu S_\lambda(\gamma^{-1}\hat{\tau}(A)) = \sum_{i,j} \lambda_i \mu_j(\gamma^{-1}, t_i t_j)\hat{\tau}_{t_i t_j}(A)$$

one then has

$$S_\mu S_\lambda(\gamma^{-1}\hat{\tau}(A))^*\varphi_1 = S_\mu(\gamma\gamma_1 U_\omega)S_\lambda(\gamma\gamma_1 U_\omega)A^*\varphi_1,$$
$$S_\mu S_\lambda(\gamma^{-1}\hat{\tau}(A))\varphi_2 = S_\mu(\gamma^{-1}\gamma_2 U_\omega)S_\lambda(\gamma^{-1}\gamma_2 U_\omega)A\varphi_2.$$

One concludes that if $\psi_1 = A^*\varphi_1$, $\psi_2 = A\varphi_2$ then

$$|(\varphi_1, (AP_\omega[\gamma_1\gamma]B - BP_\omega[\gamma_2\gamma^{-1}]A)\varphi_2) - (\varphi_1, [S_\mu S_\lambda(\gamma^{-1}\hat{\tau}(A)), B]\varphi_2)| < \varepsilon.$$

But the convex combination S_μ is still arbitrary and hence (1) $\Rightarrow$ (4) and (4) $\Rightarrow$ (1) for φ_i of the special form $P_\omega[\gamma_i]\chi_i$. Condition (1) follows by approximating $\varphi_i \in \hat{P}_\omega\mathfrak{H}_\omega$ by finite superpositions of vectors of the form $P_\omega[\gamma_i]\chi_i$.

(3) $\Leftrightarrow$ (4) Condition (3) follows from condition (4) by summation over $\gamma, \gamma_1, \gamma_2 \in \hat{G}$. Conversely, if $\varphi_i = P_\omega[\gamma_i]\chi_i$ one has

$$(\varphi_1, \hat{\tau}_t(A)\hat{P}_\omega B\varphi_2) = (\varphi_1, B\hat{P}_\omega\hat{\tau}_t(A)\varphi_2)$$

for all $A, B \in \pi_\omega(\mathfrak{A})''$, and the decompositions

$$\hat{P}_\omega = \sum_{\gamma \in \hat{G}} P_\omega[\gamma\gamma_1] = \sum_{\gamma \in \hat{G}} P_\omega[\gamma^{-1}\gamma_2]$$

give

$$\sum_{\gamma \in \hat{G}} \overline{(\gamma, t)}(\varphi_1, AP_\omega[\gamma_1\gamma]B\varphi_2) = \sum_{\gamma \in \hat{G}} \overline{(\gamma, t)}(\varphi_1, BP_\omega[\gamma^{-1}\gamma_2]A\varphi_2).$$

Condition (4) follows from orthogonality of the characters γ, i.e., by Fourier transformation.

(2) $\Leftrightarrow$ (4) Let $\varphi_i = P_\omega[\gamma_i]\chi_i$ and then one has

$$\begin{aligned} &M((\varphi_1, \pi_\omega([\gamma^{-1}\hat{\tau}(A), B])\varphi_2)) \\ &\quad = M((\varphi_1, (\pi_\omega(A)(\gamma\gamma_1)^{-1}U_\omega^{-1}\pi_\omega(B) - \pi_\omega(B)\gamma_2\gamma^{-1}U_\omega\pi_\omega(A))\varphi_2)) \\ &\quad = (\varphi_1, (\pi_\omega(A)P_\omega[\gamma\gamma_1]\pi_\omega(B) - \pi_\omega(B)P_\omega[\gamma_2\gamma^{-1}]\pi_\omega(A))\varphi_2), \end{aligned}$$

where we have used the fact that $P_\omega[\gamma\gamma_1]$ is the projection onto the subspace invariant under the unitary representation $\gamma\gamma_1 U_\omega$ and hence $M(\gamma\gamma_1 U_\omega) = P_\omega[\gamma\gamma_1]$. This formula establishes the desired equivalence.

This proposition now allows us to derive conclusions for the point spectra similar to those of Theorem 4.3.27.

Theorem 4.3.31. *Let $\omega \in E_{\mathfrak{A}}{}^G$, where G is a locally compact abelian group, and assume that the pair $(\mathfrak{A}, \omega)$ is G_Γ-abelian. The following statements are true:*

(1) *$\sigma_{\mathrm{p}}(U_\omega) \subseteq \sigma_{\mathrm{p}}(\hat{\tau})$ and both sets are symmetric;*

(2) *if ω is G-ergodic then for each $\gamma \in \sigma_{\mathrm{p}}(U_\omega)$ there exists a unit vector $\psi_\gamma \in \mathfrak{H}_\omega$ and a unitary element $V_\gamma \in \pi_\omega(\mathfrak{A})'$ such that*

$$U_\omega(t)\psi_\gamma = (\overline{\gamma, t})\psi_\gamma, \qquad \hat{\tau}_t(V_\gamma) = (\overline{\gamma, t})V_\gamma$$

and ψ_γ, V_γ are both unique up to a phase;

(3) *if ω is G-ergodic then $\sigma_{\mathrm{p}}(U_\omega)$ is a subgroup of G. If H_ω is the annihilator of $\sigma_{\mathrm{p}}(U_\omega)$ then*

$$\begin{aligned} \{V_\gamma; \gamma \in \sigma_{\mathrm{p}}(U_\omega)\}'' &= \pi_\omega(\mathfrak{A})' \cap \hat{P}_\omega{}' \\ &\subseteq \pi_\omega(\mathfrak{A})' \cap E_\omega(H_\omega)' = \pi_\omega(\mathfrak{A})' \cap U_\omega(H_\omega)', \end{aligned}$$

where $\hat{P}_\omega$ is the projection onto the subspace spanned by the $U_\omega(G)$-almost periodic vectors and $E_\omega(H_\omega)$ is the projection on the subspace of $U_\omega(H_\omega)$-invariant vectors. The four sets are equal if, and only if, $E_\omega(H_\omega) = \hat{P}_\omega$ and, in particular, if $\sigma_{\mathrm{p}}(U_\omega)$ is countable and closed;

(4) *if $\hat{P}_\omega$ is the identity then $\pi_\omega(\mathfrak{A})''$ is maximal abelian.*

PROOF. (1) Condition (4) of Proposition 4.3.30 implies that

$$(\Omega_\omega, AP_\omega[\gamma]B\Omega_\omega) = (\Omega_\omega, BP_\omega[\gamma^{-1}]A\Omega_\omega)$$

for all $A, B \in \pi_\omega(\mathfrak{A})''$. Thus $\gamma \in \sigma_p(U_\omega)$ implies $\gamma^{-1} \in \sigma_{\mathrm{p}}(U_\omega)$ and hence $\sigma_{\mathrm{p}}(U_\omega)$ is symmetric. Moreover, if

$$\hat{\tau}_t(A) = (\overline{\gamma, t})A$$

then

$$\hat{\tau}_t(A^*) = (\gamma, t)A^* = (\overline{-\gamma, t})A^*$$

and so $\sigma_{\mathrm{p}}(\hat{\tau})$ is symmetric. Next let $\gamma \in \sigma_{\mathrm{p}}(U_\omega)$ and $A \in \pi_\omega(\mathfrak{A})''$ be such that

$$P_\omega[\gamma]A\Omega_\omega \neq 0$$

If M is an invariant mean over $C_b(G)$ then one can define a bounded operator A_γ by

$$(\varphi, A_\gamma \psi) = M((\varphi, \gamma\hat{\tau}(A)\psi))$$

and because

$$\begin{aligned}(\varphi, A_\gamma \Omega_\omega) &= M((\varphi, \gamma U_\omega A\Omega_\omega)) \\ &= (\varphi, P_\omega[\gamma]A\Omega_\omega)\end{aligned}$$

one concludes that $A_\gamma \neq 0$. It easily follows that

$$\hat{\tau}_t(A_\gamma) = \overline{(\gamma, t)}A_\gamma$$

and hence $\sigma_p(U_\omega) \subseteq \sigma_p(\hat{\tau})$.

(2) and (3) Let $\gamma \in \sigma_p(U_\omega)$, and $\omega \in \mathscr{E}(E_{\mathfrak{A}}{}^G)$. The set $\{\pi_\omega(\mathfrak{A}) \cup U_\omega(G)\}$ is irreducible, on $\mathfrak{H}_\omega$, by Theorem 4.3.17, and hence $P_\omega[\gamma]\{\pi_\omega(\mathfrak{A}) \cup U_\omega(G)\}P_\omega[\gamma]$ is irreducible on $P_\omega[\gamma]\mathfrak{H}_\omega$. But

$$P_\omega[\gamma]\{\pi_\omega(\mathfrak{A}) \cup U_\omega(G)\}P_\omega[\gamma] = P_\omega[\gamma]\pi_\omega(\mathfrak{A})P_\omega[\gamma]$$

and, since $(\mathfrak{A}, \omega)$ is G-abelian, the right hand set is abelian (see Proposition 4.3.30).

Therefore $P_\omega[\gamma]$ must have rank one and this establishes the existence, and uniqueness up to a phase, of ψ_γ.

Next choose $A \in \pi_\omega(\mathfrak{A})''$ such that $P_\omega[\gamma]A\Omega_\omega \neq 0$. If B is any other element of $\pi_\omega(\mathfrak{A})''$ one then has

$$\begin{aligned}\|BP_\omega[\gamma]A\Omega_\omega\|^2 &= (\Omega_\omega, A^*P_\omega[\gamma]B^*BP_\omega[\gamma]A\Omega_\omega) \\ &= (\Omega_\omega, A^*P_\omega[\gamma]AP_\omega[0]B^*B\Omega_\omega)\end{aligned}$$

by condition (4) of Proposition 4.3.30 (0 is the identity in $\hat{G}$ in the additive notation.) But by the previous paragraph $P_\omega[0]$ has rank one and $\Omega_\omega \in P_\omega[0]\mathfrak{H}_\omega$. Thus

$$\|BP_\omega[\gamma]A\Omega_\omega\|^2 = \|B\Omega_\omega\|^2\|P_\omega[\gamma]A\Omega_\omega\|^2 \leq \|A\|^2\|B\Omega_\omega\|^2$$

Hence once can define a bounded operator A_γ by

$$A_\gamma B\Omega_\omega = BP_\omega[\gamma]A\Omega_\omega$$

and then extending by continuity. One has $\|A_\gamma\| \leq \|A\|$ and one easily computes that

$$\hat{\tau}_t(A_\gamma) = \overline{(\gamma, t)}A_\gamma.$$

Moreover, $A_\gamma \in \pi_\omega(\mathfrak{A})'$ because if $B, C \in \pi_\omega(\mathfrak{A})''$ then

$$CA_\gamma B\Omega_\omega = CBP_\omega[\gamma]A\Omega_\omega = A_\gamma CB\Omega_\omega.$$

Now $A_\gamma{}^*A_\gamma \in \pi_\omega(\mathfrak{A})' \cap U_\omega(G)'$ and hence if ω is G-ergodic $A_\gamma{}^*A_\gamma$ is a nonzero multiple of the identity. The unitary element $V_\gamma \in \pi_\omega(\mathfrak{A})'$ occurring in the polar decomposition of A_γ is then an eigenelement of $\hat{\tau}$ and the proof of (2) and the subgroup property of $\sigma_p(U_\omega)$ are identical to the proofs of similar properties in Theorem 4.3.27. Next, note that $\hat{P}_\omega\pi_\omega(\mathfrak{A})''\hat{P}_\omega$ is abelian and $\hat{P}_\omega\Omega_\omega = \Omega_\omega$. Thus $\{\pi_\omega(\mathfrak{A}) \cup \hat{P}_\omega\}'$ is abelian by the general correspondence theory for orthogonal measures Theorem 4.1.25. But one also has $[\{V_\gamma; \gamma \in \sigma_p(U_\omega)\}''\Omega_\omega] = \hat{P}_\omega$ and hence, by the same theorem,

$$\{V_\gamma; \gamma \in \sigma_p(U_\omega)\}'' = \{\pi_\omega(\mathfrak{A}) \cup \hat{P}_\omega\}'.$$

But the proof that

$$\{V_\gamma; \gamma \in \sigma_p(U_\omega)\}'' \subseteq \pi_\omega(\mathfrak{A})' \cap U_\omega(H_\omega)' = \pi_\omega(\mathfrak{A})' \cap E_\omega(H_\omega)'$$

is identical to the proof of the similar properties proved in statement (4) of Theorem 4.3.27. One interchanges $\pi_\omega(\mathfrak{A})''$ and $\pi_\omega(\mathfrak{A})'$ in the argument. Finally, $\hat{P}_\omega = E_\omega(H_\omega)$ implies equality of the four sets but if the sets are equal then

$$\begin{aligned}\hat{P}_\omega &= [\{\pi_\omega(A) \cup \hat{P}_\omega\}'\Omega_\omega] \\ &= [\{\pi_\omega(\mathfrak{A}) \cup E_\omega(H_\omega)\}'\Omega_\omega] = E_\omega(H_\omega)\end{aligned}$$

by the correspondence of Theorem 4.1.25. The equality $\hat{P} = E_\omega(H_\omega)$ follows whenever $\sigma_p(U_\omega)$ is countable and closed by the same argument used to prove the equality in Theorem 4.3.27.

(4) If $\hat{P}_\omega = \mathbb{1}_\omega$ then condition (3) of Proposition 4.3.30 together with Lemma 4.3.15 imply that $\pi_\omega(\mathfrak{A})''$ is maximal abelian.

Corollary 4.3.32. *Let $\omega \in \mathscr{E}(E_{\mathfrak{A}}{}^G)$, where G is a locally compact abelian group and $\mathfrak{A}$ has an identity. Assume that the point spectrum $\sigma_p(U_\omega)$ of U_ω is a countable, closed, subgroup of $\hat{G}$, and let H_ω denote its annihilator. The following conditions are equivalent:*

(1) *the pair $(\mathfrak{A}, \omega)$ is G_Γ-abelian;*
(2) *the pair $(\mathfrak{A}, \omega)$ is H_ω-abelian;*
(3) *for each H_ω-invariant π_ω-normal state ω' there exists a unique maximal measure $\mu' \in M_{\omega'}(E_{\mathfrak{A}}^{H_\omega})$.*

If, moreover, $\sigma_p(U_\omega)$ is cyclic then these conditions are satisfied.

PROOF. It follows from the third statement of Theorem 4.3.31 that

$$\{V_\gamma; \gamma \in \sigma_p(U_\omega)\}'' = \pi_\omega(\mathfrak{A})' \cap \hat{P}_\omega{}' = \pi_\omega(\mathfrak{A})' \cap U_\omega(H_\omega)'$$

but $(\mathfrak{A}, \omega)$ is G_Γ-abelian if, and only if, the second algebra is abelian (Proposition 4.3.30) and $(\mathfrak{A}, \omega)$ is H_ω-abelian if, and only if, the third algebra is abelian. Thus (1) $\Leftrightarrow$ (2). The equivalence (2) $\Leftrightarrow$ (3) follows from Theorem 4.3.9 applied to H_ω. Finally, if $\sigma_p(U_\omega)$ is cyclic then the uniqueness of the V_γ ensures that $\{V_\gamma; \gamma \in \sigma_p(U_\omega)\}''$ is abelian and hence all three conditions are satisfied.

We remark in passing that the final statement of the corollary is not true for general $\sigma_p(U_\omega)$.

After this discussion of the point spectrum we next analyze the complete spectrum with the assumption of a clustering condition which is stronger than ergodicity. Note that we cannot generally expect the spectra $\sigma(U_\omega)$ and $\sigma(\hat{\tau})$ to be equal nor can we expect $\sigma(U_\omega)$ to be symmetric because for $G = \mathbb{R}$ there are many counterexamples. Nevertheless, additivity of $\sigma(U_\omega)$ is possible.

Theorem 4.3.33. *Let $\omega \in E_{\mathfrak{A}}{}^G$, where G is a locally compact abelian group, and assume that the corresponding unitary representation $U_\omega(G)$, of G, is strongly continuous. Further assume that*

$$\inf_{A' \in \mathrm{Co}\,\hat{\tau}_G(A)} |\omega(BA'C) - \omega(A)\omega(BC)| = 0$$

for all $A, B, C \in \mathfrak{A}$. It follows that the spectrum $\sigma(U_\omega)$ of $U_\omega(G)$ is additive.

PROOF. The proof is identical to the proof of additivity given in Theorem 4.3.27(3). For $\gamma_1, \gamma_2 \in \sigma(U_\omega)$ one forms regularized elements $\hat{\tau}_{f_1}(A_1)$, $\hat{\tau}_{f_2}(A_2)$ with spectra in neighborhoods containing γ_1 and γ_2 such that $\hat{\tau}_{f_i}(A_i)\Omega_\omega \neq 0$, $i = 1, 2$. One then uses the cluster property to prove that $\hat{\tau}_t(\hat{\tau}_{f_1}(A_1))\hat{\tau}_{f_2}(A_2)\Omega_\omega$ is a nonzero element of $\mathfrak{H}_\omega$ for some $t \in G$ and this element has spectrum in a neighborhood containing $\gamma_1 + \gamma_2$.

Remark. If in Theorem 4.3.33, G is the product of two subgroups G_1 and G_2, then the additivity of the spectra of $U_\omega|_{G_1}$ and $U_\omega|_{G_2}$ follows from the three-point cluster property for G_2 alone. It is also of interest that this cluster property gives additivity of the spectrum of $U_\omega|_{G_1}$ even if G_2 is not locally compact abelian.

Note that the cluster condition in Theorem 4.3.33 implies the ergodicity of ω by Theorems 4.3.17 and 4.3.23. Moreover, if Ω_ω is separating for $\pi_\omega(\mathfrak{A})''$ then the cluster property is equivalent to ergodicity by Theorems 4.3.20 and 4.3.23. Further note that if the cluster property is slightly more uniform, e.g., if

$$\inf_{A' \in \mathrm{Co}\,\gamma\hat{\tau}(A)} |\omega(B\hat{\tau}_t(A')C) - \omega(A)\omega(BC)| = 0$$

for all $A, B, C \in \mathfrak{A}$ uniformly for $t \in G$, then one easily deduces that

$$\inf_{A_1' \in \mathrm{Co}\,\gamma\hat{\tau}(A_1)} |\omega(B[A_1', A_2]C)| = 0$$

for all $A_1, A_2, C \in \mathfrak{A}$. Therefore the pair $(\mathfrak{A}, \omega)$ is G_Γ-abelian and the results of Theorem 4.3.31 concerning the point spectrum apply. A particular example of how this can occur is the following.

EXAMPLE 4.3.34. Let $G = \mathbb{R}$ and ω an $\mathbb{R}$-ergodic state such that $t \mapsto U_\omega(t)$ is strongly continuous and

$$\lim_{T \to \infty} \frac{1}{2T} \int_{-T}^{T} dt\, |\omega(B[\tau_t(A_1), A_2]C)| = 0.$$

Then $(\mathfrak{A}, \omega)$ is G_Γ-abelian, and Theorem 4.3.17 and Example 4.3.5 imply that

$$\lim_{T \to \infty} \frac{1}{2T} \int_{-T}^{T} dt\, \omega(B\tau_t(A)C) = \omega(A)\omega(BC).$$

Hence Theorems 4.3.31 and 4.3.33 imply that $\sigma_p(U_\omega)$ is a closed subgroup of $\mathbb{R}$ and $\sigma(U_\omega)$ is a closed, additive subset of $\mathbb{R}$. One has first the possibility

(1) $\sigma(U_\omega) \subseteq [0, \infty\rangle$ or $\sigma(U_\omega) \subseteq \langle -\infty, 0]$ and $\sigma_p(U_\omega) = \{0\}$.

There remain two possibilities: either all points in $\sigma(U_\omega)$ are isolated, or not. In the first case one has that $\sigma(U_\omega) = \sigma_p(U_\omega)$ and hence

(2) $\sigma(U_\omega) = \{0\}$ or $\sigma(U_\omega)$ is isomorphic to $\mathbb{Z}$, and in both cases $\pi_\omega(\mathfrak{A})''$ is maximal abelian.

In the remaining case there exist $a, b > 0$ such that $a, -b \in \sigma(U_\omega)$ and one of the points, for example a, is not isolated in $\sigma(U_\omega)$. (The case that 0 is not isolated can be treated by a small variation of the argument.) By additivity one has that

$$ma - nb \in \sigma(U_\omega)$$

for all $m, n \in \mathbb{Z}_+$. If a and b are not commensurate it follows that $\sigma(U_\omega)$ is dense in $\mathbb{R}$, and hence $\sigma(U_\omega) = \mathbb{R}$. If a and b are commensurate, there exist positive $m, n \in \mathbb{Z}$ such that $ma = nb \equiv c$. But then $\pm c \in \sigma(U_\omega)$ and $c = ma$ is not isolated. It follows that 0 is not isolated, and hence $\sigma(U_\omega)$ contains at least one of the half intervals $\pm[0, \infty\rangle$. By additivity $\sigma(U_\omega) = \mathbb{R}$, and hence the last possibility is

(3) $\sigma(U_\omega) = \mathbb{R}$, and either $\sigma_p(U_\omega) = \{0\}$ or $\sigma_p(U_\omega)$ is isomorphic to $\mathbb{Z}$ or $\sigma_p(U_\omega)$ is a dense subgroup of $\mathbb{R}$.

Note that in case (1) it follows from the Borchers–Arveson theorem (Theorem 3.2.46) that $U_\omega(\mathbb{R}) \subseteq \pi_\omega(\mathfrak{A})''$. Hence ergodicity, which is equivalent to

$$\pi_\omega(\mathfrak{A})' \cap U_\omega(\mathbb{R})' = \mathbb{C}1_\omega.$$

implies $\pi_\omega(\mathfrak{A})' = \mathbb{C}1_\omega$ and hence $\pi_\omega(\mathfrak{A})'' = \mathscr{L}(\mathfrak{H}_\omega)$. More generally, one can prove that $\pi_\omega(\mathfrak{A})''$ is a von Neumann algebra of type III if, and only if, ω_{Ω_ω} is not a trace when restricted to $\pi_\omega(\mathfrak{A})'$. (See Notes and Remarks.)

We conclude this subsection with a method for isolating the point spectrum of a unitary group and a criterion for $\sigma_p(U_\omega)$ to consist of the single point $\{0\}$. We begin by remarking that the mean ergodic theorem can be rephrased to give the following rather different looking result.

Lemma 4.3.35. *Let U be a strongly continuous unitary representation of the locally compact abelian group G acting on a Hilbert space $\mathfrak{H}$ and let M denote an invariant mean over $C_b(G)$. It follows that*

$$M(|(\varphi, U\psi)|^2) = \sum_{\gamma \in \sigma_p(U)} |(\varphi, P[\gamma]\psi)|^2$$

for all $\varphi, \psi \in \mathfrak{H}$, where P denotes the spectral projector, and σ_p the point spectrum of U.

PROOF. Let $\bar{\mathfrak{H}}$ denote the conjugate space to $\mathfrak{H}$ (see footnote on page 70) and define $\hat{\mathfrak{H}} = \mathfrak{H} \otimes \bar{\mathfrak{H}}$. To each $\varphi \in \mathfrak{H}$ we then associate $\hat{\varphi} = \varphi \otimes \bar{\varphi} \in \hat{\mathfrak{H}}$. Next introduce $\bar{U}$ on $\bar{\mathfrak{H}}$ by $\bar{U}(t)\bar{\varphi} = \overline{U(t)^*\varphi}$ and define $\hat{U} = U \otimes \bar{U}^*$. One then has

$$\begin{aligned}(\hat{\varphi}, \hat{U}(t)\hat{\psi}) &= (\varphi, U(t)\psi)(\bar{\varphi}, \overline{U(t)^*\bar{\psi}}) \\ &= (\varphi, U(t)\psi)(\bar{\varphi}, \overline{U(t)\psi}) = |(\varphi, U(t)\psi)|^2.\end{aligned}$$

Moreover,

$$\hat{U}(t)\varphi \otimes \bar{\psi} = \int dP(\gamma_1)\varphi \otimes \int \overline{dP(\gamma_2)}\, \bar{\psi}\, \overline{(\gamma_1\gamma_2^{-1}, t)}$$

and hence the projection valued measure P associated with U satisfies

$$d\hat{P}(\gamma) = \int dP(\gamma_1) \otimes \overline{dP}(\gamma^{-1}\gamma_1).$$

Using the fact that the convolution of a discrete measure with a continuous measure is continuous one easily concludes that the projection $\hat{E}$ on the subspace of $\hat{\mathfrak{H}}$ formed by the $\hat{U}$ invariant vectors is

$$\hat{E} = \sum_{\gamma \in \sigma_p(U)} P[\gamma] \otimes \bar{P}[\gamma].$$

Therefore

$$\begin{aligned}(\hat{\varphi}, \hat{E}\hat{\psi}) &= \sum_{\gamma \in \sigma_p(U)} (\varphi, P[\gamma]\psi)(\bar{\varphi}, \overline{P[\gamma]}\bar{\psi}) \\ &= \sum_{\gamma \in \sigma_p(U)} |(\varphi, P[\gamma]\psi)|^2.\end{aligned}$$

Thus the statement of the lemma is equivalent to the mean ergodic theorem for $\hat{U}$, i.e., $M(\hat{U}) = \hat{E}$.

The existence of an invariant mean on $C_b(G)$ allows one to reformulate some of the criteria for ergodicity given in the previous subsection. For example, one easily concludes that if $\omega \in E_{\mathfrak{A}}{}^G$ and E_ω is the orthogonal projection on the subspace of $\mathfrak{H}_\omega$ formed by the vectors invariant under $U_\omega(G)$ then E_ω is rank one if, and only if,

$$M(\omega(A\tau(B)) - \omega(A)\omega(B)) = 0$$

for all $A, B \in \mathfrak{A}$ and some invariant mean M. Furthermore, if Ω_ω is separating for $\pi_\omega(\mathfrak{A})''$ then these conditions are equivalent to

$$M(\omega(A\tau(B)C) - \omega(AC)\omega(B)) = 0$$

for all $A, B, C \in \mathfrak{A}$ and some M.

These last criteria generalize as follows.

Proposition 4.3.36. *Let $\omega \in E_{\mathfrak{A}}{}^G$, where G is a locally compact abelian group and assume that the corresponding unitary representation U_ω of G is strongly continuous. Consider the following conditions:*

(1) *Ω_ω is the unique (up to a factor) eigenvector of $U_\omega(G)$;*
(2) *$M(|\omega(A\tau(B)) - \omega(A)\omega(B))|) = 0$ for all $A, B \in \mathfrak{A}$ and some invariant mean M;*
(3) *$M(|\omega(A\tau(B)C) - \omega(AC)\omega(B)|) = 0$ for all $A, B, C \in \mathfrak{A}$, and some invariant mean M.*

One has (1) $\Leftrightarrow$ (2) $\Leftarrow$ (3) and if Ω_ω is separating for $\pi_\omega(\mathfrak{A})''$ then all three conditions are equivalent.

Proof. First note that as M is a state over $C_b(G)$ it satisfies the Cauchy–Schwarz inequality, Lemma 2.3.10, and hence

$$M(|f|)^2 \leq M(|f|^2) \leq M(|f|)\|f\|_\infty$$

for all $f \in C_b(G)$. Therefore condition (2) is equivalent to

$$M(|\omega(A\tau(B)) - \omega(A)\omega(B)|^2) = 0$$

for all $A, B \in \mathfrak{A}$.

Next introduce φ_A and ψ_B by

$$\varphi_A = \pi_\omega(A)^*\Omega_\omega - \omega(A)\Omega_\omega, \qquad \psi_B = \pi_\omega(B)\Omega_\omega - \omega(B)\Omega_\omega$$

and remark that

$$|(\varphi_A, U_\omega(t)\psi_B)|^2 = |\omega(A\tau_t(B)) - \omega(A)\omega(B)|^2.$$

Thus by Lemma 4.3.35 one has

$$M(|\omega(A\tau(B)) - \omega(A)\omega(B)|^2) = \sum_{\gamma \in \sigma_p(U_\omega)} |(\varphi_A, P_\omega[\gamma]\psi_B)|^2$$

and condition (2) is equivalent to each term on the right-hand side being zero. It follows immediately that this is equivalent to condition (1).

The implication (3) $\Rightarrow$ (2) follows by choosing C to be the identity, if $\mathfrak{A}$ possesses an identity, or by approximating Ω_ω by $\pi_\omega(C)\Omega_\omega$ in the general case. The converse implication, for Ω_ω separating, follows by remarking that (3) is equivalent to

$$M(|(\varphi, \pi_\omega(\tau(B))\psi) - (\varphi, \psi)\omega(B)|) = 0$$

for all $\varphi, \psi \in \mathfrak{H}_\omega$ and $B \in \mathfrak{A}$. But as Ω_ω is cyclic for $\pi_\omega(\mathfrak{A})'$ by assumption, this is equivalent to

$$M(|(\varphi, C\pi_\omega(\tau(B))\Omega_\omega) - (\varphi, C\Omega_\omega)\omega(B)|) = 0$$

for all $\varphi \in \mathfrak{H}_\omega$, $C \in \pi_\omega(\mathfrak{A})'$, and $B \in \mathfrak{A}$. But this last condition is implied by the condition

$$M(|(\varphi, \pi_\omega(\tau(B))\Omega_\omega) - (\varphi, \Omega_\omega)\omega(B)|) = 0$$

for all $\varphi \in \mathfrak{H}_\omega$ and $B \in \mathfrak{A}$, and this is equivalent to condition (2).

The second implication of Proposition 4.3.36 first arose in classical ergodic theory, i.e., the case of abelian $\mathfrak{A}$ and $G = \mathbb{R}$, and it is usually called *weak mixing.*

4.3.4. Broken Symmetry

Throughout this section we concentrated on the problem of decomposing a given G-invariant state into G-ergodic states. In this subsection we examine decompositions of a G-ergodic state into states with a lower symmetry.

Given a G-ergodic state ω there are various ways of decomposing it into states which are not G-invariant. For example, if $\pi_\omega(\mathfrak{A})''$ is not a factor then one could consider the central decomposition of ω discussed in Section 4.2.2. There are several possibilities for the states occurring in this decomposition.

(1) The states could retain invariance under a subgroup H of G. This occurs if $\mathfrak{Z}_\omega \subseteq U_\omega(H)'$. Note that in this case if the pair $(\mathfrak{A}, \omega)$ is H-central then $\mathfrak{Z}_\omega = \{\pi_\omega(\mathfrak{A}) \cup U_\omega(H)\}'$, by Theorem 4.3.14, and the central and H-ergodic decompositions coincide. Thus from the point of view of the central decomposition H would be the natural symmetry group of ω.

(2) The states could lose all invariance but nevertheless retain some traces of symmetry such as almost periodicity. This behavior arises if the action of G on $\mathfrak{Z}_\omega$ is multiperiodic in an appropriate sense.

(3) The states could simply lose all form of invariance and symmetry.

The first two situations in which a residual symmetry remains are of particular interest and it is this type of problem which we intend to discuss. Such decompositions are related to the phenomena of spontaneous symmetry breaking which occur in physics.

First we examine the decomposition of G-ergodic states into states invariant under a normal subgroup H and establish conditions under which the decomposition corresponds to an averaging over the quotient group G/H. Subsequently, we specialize to locally compact abelian groups and examine decompositions into almost periodic states. The existence of nontrivial decompositions and the choice of a natural symmetry group is then closely related to the properties of the point spectrum discussed in the previous section.

We begin with an existence result for decomposition with respect to a normal subgroup. This result is of interest because it does not require any particular assumption on the structure of $\mathfrak{A}$ other than the strong continuity of the action of the group.

Theorem 4.3.37. *Let G be a topological group and H a closed normal subgroup of G such that the quotient group G/H is compact. Let $d\dot{g}$ be the normalized Haar measure on G/H. Assume that G acts as a strongly continuous group of *-automorphisms τ of a C^*-algebra $\mathfrak{A}$ with identity and take $\omega \in \mathscr{E}(E_{\mathfrak{A}}{}^G)$. It follows that there exists a probability measure μ_ω, with barycenter ω, supported by a closed subset of $\mathscr{E}(E_{\mathfrak{A}}{}^H)$, and hence maximal for the order $\succ$. Furthermore, there exists an $\tilde{\omega} \in \mathscr{E}(E_{\mathfrak{A}}{}^H)$ such that*

$$\mu_\omega(f) = \int_{G/H} d\dot{g}\, f(\tau_g{}^*\tilde{\omega})$$

for each $f \in C(E_{\mathfrak{A}}{}^H)$ and, in particular,

$$\omega(A) = \int_{G/H} d\dot{g}\, \tilde{\omega}(\tau_g(A))$$

for all $A \in \mathfrak{A}$. If the pair $(\mathfrak{A}, \omega)$ is H-abelian then μ_ω is the unique maximal measure on $E_{\mathfrak{A}}{}^H$ with barycenter ω and coincides with the orthogonal measure corresponding to $\{\pi_\omega(\mathfrak{A}) \cup U_\omega(H)\}'$.

Proof. First note that if $A \in \mathfrak{A}$ and $\omega \in E_{\mathfrak{A}}{}^H$ then $\hat{A}(\tau_g{}^*\omega)$ is a bounded continuous function on G which is invariant under translations by H. Therefore it defines a bounded continuous function over G/H. Next we introduce the average of $\hat{A}$ over the group G/H by

$$\langle \hat{A} \rangle(\omega) = \int_{G/H} d\dot{g}\, \hat{A}(\tau_g{}^*\omega) = \int_{G/H} d\dot{g}\, \omega(\tau_{g^{-1}}(A))$$

and prove that $\langle \hat{A} \rangle$ is continuous over $E_{\mathfrak{A}}{}^H$.

Let $\dot{g}$ denote the image of g under the quotient map p of G onto G/H. Thus $\omega(\tau_g(A)) = \omega(\tau_{\dot{g}}(A))$. Now if $\varepsilon > 0$ and $g \in G$ then there is an open neighborhood $N(g)$ of g such that

$$\|\tau_g(A) - \tau_{g'}(A)\| < \varepsilon/4$$

for all $g' \in N(g)$. This uses the strong continuity of τ. But as p is open $\dot{N}(\dot{g}) = p(N(g))$ is an open neighborhood of $\dot{g} \in G/H$, and because for any $\dot{g}' \in \dot{N}(\dot{g})$ there exists a $g' \in p^{-1}(\dot{g}') \cap N(g)$ one must have

$$|\omega(\tau_{\dot{g}}(A)) - \omega(\tau_{\dot{g}'}(A))| < \frac{\varepsilon}{4}$$

for all $\dot{g}' \in \dot{N}(\dot{g})$. Next let $N(g_i)$, $i = 1, 2, \ldots, n$, be a finite family of such neighborhoods chosen such that the corresponding $\dot{N}(\dot{g}_i)$ cover the compact space G/H and consider an $\omega' \in E_{\mathfrak{A}}{}^H$ with the property

$$|\omega(\tau_{\dot{g}_i}(A)) - \omega'(\tau_{\dot{g}_i}(A))| < \frac{\varepsilon}{2}$$

for all $i = 1, 2, \ldots, n$. Since each $\dot{g} \in G/H$ must lie in some $\dot{N}(\dot{g}_i)$ one then has

$$\begin{aligned} |\omega(\tau_{\dot{g}}(A)) - \omega'(\tau_{\dot{g}}(A))| &\leq |\omega(\tau_{\dot{g}}(A)) - \omega(\tau_{\dot{g}_i}(A))| \\ &\quad + |\omega(\tau_{\dot{g}_i}(A)) - \omega'(\tau_{\dot{g}_i}(A))| + |\omega'(\tau_{\dot{g}_i}(A)) - \omega'(\tau_{\dot{g}}(A))| < \varepsilon \end{aligned}$$

for all $\dot{g} \in G/H$. Therefore $\omega'(\tau_{\dot{g}}(A))$ converges to $\omega(\tau_{\dot{g}}(A))$, uniformly in $\dot{g} \in G/H$, as ω' converges in the weak* topology to ω. Thus $\omega \in E_{\mathfrak{A}}{}^H \mapsto \langle \hat{A} \rangle(\omega)$ is continuous.

Next define the average of $\omega \in E_{\mathfrak{A}}{}^H$ by

$$\langle \omega \rangle(A) = \int_{G/H} d\dot{g}\; \omega(\tau_{\dot{g}}(A)) = \langle \hat{A} \rangle(\omega).$$

Clearly, $\langle \omega \rangle \in E_{\mathfrak{A}}{}^G$ and $\langle \omega \rangle = \omega$ for all $\omega \in E_{\mathfrak{A}}{}^G$. Now for a fixed $\omega \in \mathscr{E}(E_{\mathfrak{A}}{}^G)$ define the set K_ω by

$$K_\omega = \{\omega';\, \omega' \in E_{\mathfrak{A}}{}^H,\, \langle \omega' \rangle(A) = \omega(A) \quad \text{for all } A \in \mathfrak{A}\}.$$

As $\langle \hat{A} \rangle$ is continuous the sets

$$\{\omega';\, \omega' \in E_{\mathfrak{A}}{}^H,\, \langle \omega' \rangle(A) = \omega(A)\}$$

are closed and hence K_ω is a closed subset of the compact set $E_{\mathfrak{A}}{}^H$. But K_ω is convex and nonempty because $\omega \in K_\omega$, and hence it has extreme points.

We next argue that $\mathscr{E}(K_\omega) \subseteq \mathscr{E}(E_{\mathfrak{A}}{}^H)$. Assume, conversely, that $\tilde{\omega} \in \mathscr{E}(K_\omega)$ but

$$\tilde{\omega} = \lambda \tilde{\omega}_1 + (1 - \lambda)\tilde{\omega}_2$$

for $\tilde{\omega}_1, \tilde{\omega}_2 \in E_{\mathfrak{A}}{}^H$ distinct from $\tilde{\omega}$ and $0 < \lambda < 1$. As $\tilde{\omega} \in \mathscr{E}(K_\omega)$ one must have $\tilde{\omega}_1 \notin K_\omega$ or $\tilde{\omega}_2 \notin K_\omega$. Suppose $\tilde{\omega}_1 \notin K_\omega$; then $\langle \tilde{\omega}_1 \rangle \neq \omega$. But

$$\omega = \langle \tilde{\omega} \rangle = \lambda \langle \tilde{\omega}_1 \rangle + (1 - \lambda) \langle \tilde{\omega}_2 \rangle,$$

which contradicts the assumption $\omega \in \mathscr{E}(E_{\mathfrak{A}}{}^G)$. Therefore $\tilde{\omega} \in \mathscr{E}(E_{\mathfrak{A}}{}^H)$.

Now if $f \in C(E_{\mathfrak{A}}{}^H)$ then $f(\tau_g{}^*\tilde{\omega})$, with $\tilde{\omega} \in \mathscr{E}(K_\omega)$, is a continuous function over G which is invariant under right translations by H and one can define

$$\mu_\omega(f) = \int_{G/H} d\dot{g}\; f(\tau_g{}^*\tilde{\omega}).$$

It follows that μ_ω is a positive functional over $C(E_{\mathfrak{A}}{}^H)$ and hence a positive Radon measure. Now the orbit $\mathcal{O}_{\tilde{\omega}} = \{\tau_g{}^*\tilde{\omega};\, \dot{g} \in G/H\}$ of $\tilde{\omega}$ is closed because G/H is compact. Moreover, if $\dot{g} \in G/H$ then $\tau_{\dot{g}}{}^*$ is an affine isomorphism of $E_{\mathfrak{A}}{}^H$ onto $E_{\mathfrak{A}}{}^H$ and hence maps $\mathscr{E}(E_{\mathfrak{A}}{}^H)$ onto $\mathscr{E}(E_{\mathfrak{A}}{}^H)$. Thus $\mathcal{O}_{\tilde{\omega}}$ is a closed subset of $\mathscr{E}(E_{\mathfrak{A}}{}^H)$. But $\mathcal{O}_{\tilde{\omega}}$ is a supporting set of μ_ω and μ_ω is maximal by Proposition 4.1.10. But

$$\mu_\omega(\hat{A}) = \langle\hat{A}\rangle(\tilde{\omega}) = \langle\tilde{\omega}\rangle(A) = \omega(A)$$

for each $A \in \mathfrak{A}$ and one concludes that μ_ω has barycenter ω.

Finally, if the pair $(\mathfrak{A}, \omega)$ is H-abelian then $\{\pi_\omega(\mathfrak{A}) \cup U_\omega(H)\}'$ is abelian, by Proposition 4.3.7, and μ_ω is the corresponding orthogonal measure by the fourth statement of Proposition 4.3.7.

The final statement of this theorem is of the greatest interest because it describes a situation in which the decomposition into H-ergodic states exists in the appropriate form and is unique. Note that the $\tilde{\omega}$ occurring in the decomposition is also unique in the sense that any other $\tilde{\omega} \in \mathscr{E}(E_{\mathfrak{A}}{}^H)$ which gives a similar decomposition must lie in the orbit $\mathcal{O}_\omega$ of $\tilde{\omega}$ under τ^*. We next give another version of this theorem which requires less continuity of the action of G but some separability. In contrast to the above proof one exploits the H-abelianness of $(\mathfrak{A}, \omega)$ to establish the existence of the desired decomposition. Note that this theorem applies to a σ-weakly continuous group of *-automorphisms of a von Neumann algebra.

Theorem 4.3.38. *Let G be a topological group and H a closed normal subgroup of G such that the quotient group G/H is compact. Let $d\dot{g}$ be the normalized Haar measure on G/H. Assume that G acts as a group of *-automorphisms τ of a C^*-algebra $\mathfrak{A}$ with identity. Take $\omega \in \mathscr{E}(E_{\mathfrak{A}}{}^G)$ and assume that the pair $(\mathfrak{A}, \omega)$ is H-abelian and that the unitary representation U_ω of G is weakly continuous.*

If ω is in a face F of $E_{\mathfrak{A}}{}^H$ which satisfies the separability condition S then it follows that there is an $\tilde{\omega} \in \mathscr{E}(E_{\mathfrak{A}}{}^H) \cap F$ such that $g \in G/H \mapsto \tau_g{}^*\tilde{\omega}$ is $d\dot{g}$-measurable and

$$\omega(A) = \int_{G/H} d\dot{g}\ \tilde{\omega}(\tau_{\dot{g}}(A))$$

for all $A \in \mathfrak{A}$. This decomposition coincides with that given by the unique maximal measure $\mu \in M_\omega(E_{\mathfrak{A}}{}^H)$, i.e., the orthogonal measure corresponding to $\{\pi_\omega(\mathfrak{A}) \cup U_\omega(H)\}'$, and one has

$$\mu(f) = \int_{G/H} d\dot{g}\ f(\tau_g{}^*\tilde{\omega})$$

for each $f \in C(E_{\mathfrak{A}}{}^H)$.

PROOF. As $(\mathfrak{A}, \omega)$ is H-abelian the commutant $\{\pi_\omega(\mathfrak{A}) \cup U_\omega(H)\}'$ is abelian by Proposition 4.3.7. Hence Proposition 4.3.3 implies that $M_\omega(E_{\mathfrak{A}}{}^H)$ contains a unique maximal measure μ and this measure is the orthogonal measure corresponding to $\{\pi_\omega(\mathfrak{A}) \cup U_\omega(H)\}'$. Now if $f \in C(E_{\mathfrak{A}}{}^H)$ then the action of τ on f is defined by

$$(\tau_g f)(\omega') = f(\tau_g{}^*\omega')$$

but problems arise because this action is not necessarily continuous. We circumvent these problems by exploiting properties of μ.

First note that if $A, B \in \mathfrak{A}$ and $f \in C(E_{\mathfrak{A}}{}^H)$ then one has

$$\kappa_\mu(\tau_g f) = U_\omega(g)\kappa_\mu(f)U_\omega(g)^{-1}$$

by a simple calculation using the invariance of μ, i.e.,

$$\begin{aligned}(\Omega_\omega, \pi_\omega(A)\kappa_\mu(\tau_g f)\pi_\omega(B)\Omega_\omega) &= \mu(\widehat{AB}\tau_g f) \\ &= \mu(\widehat{\tau_{g^{-1}}(AB)} f) \\ &= (\Omega_\omega, \pi_\omega(A)U_\omega(g)\kappa_\mu(f)U_\omega(g)^{-1}\pi_\omega(B)\Omega_\omega).\end{aligned}$$

(The G-invariance of μ follows from uniqueness.) Next let E_ω denote the projection onto the subspace of $\mathfrak{H}_\omega$ spanned by the $U_\omega(H)$-invariant vectors. The basic correspondences for orthogonal measures give

$$E_\omega \mathfrak{H}_\omega = [\{\kappa_\mu(f); f \in C(E_{\mathfrak{A}}{}^H)\}''\Omega_\omega]$$

and, moreover,

$$\begin{aligned}\mu(\tau_g(f)f_n) &= (\Omega_\omega, \kappa_\mu(\tau_g(f)f_n)\Omega_\omega) \\ &= (\Omega_\omega, \kappa_\mu(f)U_\omega(g)^*\kappa_\mu(f_n)\Omega_\omega) \\ &= (\Omega_\omega, \kappa_\mu(f)U_\omega(\dot{g})^*E_\omega\kappa_\mu(f_n)\Omega_\omega)\end{aligned}$$

for all $f, f_n \in C(E_{\mathfrak{A}}{}^H)$, where $\dot{g}$ denotes the image of g in the quotient group G/H. Now note that

$$\int_{G/H} d\dot{g}\, U_\omega(\dot{g})E_\omega = E_\omega(G)$$

defines a projection onto a subspace of $U_\omega(G)$-invariant states. But the pair $(\mathfrak{A}, \omega)$ is H-abelian, hence G-abelian, and $\omega \in \mathscr{E}(E_{\mathfrak{A}}{}^G)$. Thus it follows from Theorem 4.3.17 that $E_\omega(G)$ has rank one, i.e., $E_\omega(G)$ is the projection onto Ω_ω. Therefore we conclude that

$$g \in G \mapsto \mu((\tau_g f)f_n) = \mu((\tau_{\dot{g}} f)f_n)$$

is a continuous function over G/H, for all $f, f_n \in C(E_{\mathfrak{A}}{}^H)$, and

$$\int_{G/H} d\dot{g}\, \mu((\tau_{\dot{g}} f)f_n) = \mu(f)\mu(f_n).$$

Now F is a stable face by Proposition 4.1.34. It follows that μ is supported by F. Moreover, μ is supported by $\mathscr{E}(E_{\mathfrak{A}}{}^H)$ by Proposition 4.3.2. Therefore μ is supported by $F \cap \mathscr{E}(E_{\mathfrak{A}}{}^H)$ and there must exist an $\tilde{\omega} \in F \cap \mathscr{E}(E_{\mathfrak{A}}{}^H)$ which is contained in the support of μ. Next let $\{A_{n,m}\}_{m \geq 1}$ be a countable dense subset of the selfadjoint elements of the two-sided ideals $\mathfrak{I}_n$, $n \geq 1$, used in the definition of the separability condition S, Definition 4.1.32. We introduce a sequence of weak* neighborhoods $N_p(\tilde{\omega})$, $p \geq 1$, of $\tilde{\omega}$ by

$$N_p(\tilde{\omega}) = \{\omega';\, \omega' \in E_{\mathfrak{A}}{}^H, |(\tilde{\omega} - \omega')(A_{n,m})| < 1/p \quad \text{for } 1 \leq n, m \leq p\}.$$

It then follows that

$$\{\tilde{\omega}\} = \bigcap_{p \geq 1} N_p(\tilde{\omega}) \cap F.$$

Now if $f_p \in C(E_{\mathfrak{A}}{}^H)$, $p \geq 2$, are chosen such that $0 \leq f_p \leq 1$, $f_p(\omega') = 1$ for $\omega' \in N_p(\tilde{\omega})$, and $f_p(\omega') = 0$ for $\omega' \in E_{\mathfrak{A}}{}^H \backslash N_{p-1}(\tilde{\omega})$ then one has

$$\left| \frac{\mu((\tau_{\dot{g}} f) f_p)}{\mu(f_p)} \right| \leq \|f\|$$

and

$$\lim_{p \to \infty} \left| \frac{\mu((\tau_{\dot{g}} f) f_p)}{\mu(f_p)} - (\tau_{\dot{g}} f)(\tilde{\omega}) \right| = 0$$

for $f \in C(E_{\mathfrak{A}}{}^H)$, and for each $\dot{g} \in G/H$. Hence the Lebesgue dominated convergence theorem implies that $\dot{g} \in G/H \mapsto (\tau_{\dot{g}} f)(\tilde{\omega})$ is $d\dot{g}$-measurable and

$$\begin{aligned} \int d\dot{g}\, (\tau_{\dot{g}} f)(\tilde{\omega}) &= \lim_{p \to \infty} \int_{G/H} d\dot{g}\, \mu((\tau_{\dot{g}} f) f_p)/\mu(f_p) \\ &= \mu(f). \end{aligned}$$

In particular,

$$\omega(A) = \mu(\hat{A}) = \int_{G/H} d\dot{g}\, \omega(\tau_{\dot{g}}(A)).$$

Although the foregoing theorems give existence results for decomposition of a G-ergodic state with respect to a normal subgroup H they do not indicate any natural choice of H. If one specializes to locally compact abelian groups G then the choice of H is directly related to properties of the point spectrum $\sigma_{\mathrm{p}}(U_\omega)$ of U_ω. Recall that if $\omega \in \mathscr{E}(E_{\mathfrak{A}}{}^G)$ then the point spectrum of U_ω is a subgroup of $\hat{G}$ if Ω_ω is separating for $\pi_\omega(\mathfrak{A})''$ by Theorem 4.3.27, or if the pair $(\mathfrak{A}, \omega)$ is G_Γ-abelian by Theorem 4.3.31. If H_ω is the annihilator of $\sigma_{\mathrm{p}}(U_\omega)$ then all eigenvectors of $U_\omega(G)$ will be invariant under $U_\omega(H_\omega)$. This indicates that ω is not H_ω-ergodic and there should exist nontrivial decompositions with respect to H_ω, or any subgroup of H_ω. Now note that G/H_ω is the dual of $\sigma_{\mathrm{p}}(U_\omega)$ and G/H_ω is compact if, and only if, $\sigma_{\mathrm{p}}(U_\omega)$ is discrete. Thus one has the following:

Corollary 4.3.39. *Let $\omega \in \mathscr{E}(E_{\mathfrak{A}}{}^G)$, where G is a locally compact abelian group and $\mathfrak{A}$ has an identity. Assume that the point spectrum $\sigma_{\mathrm{p}}(U_\omega)$ of $U_\omega(G)$ is a discrete subgroup of $\hat{G}$ and let H_ω denote its annihilator. If the pair $(\mathfrak{A}, \omega)$ is H_ω-abelian and if ω is in a face F of $E_{\mathfrak{A}}^{H_\omega}$ which satisfies the separability condition S then there exists an $\tilde{\omega} \in \mathscr{E}(E_{\mathfrak{A}}^{H_\omega}) \cap F$ such that $\dot{t} \in G/H_\omega \mapsto \tau_{\dot{t}}{}^*\tilde{\omega}$ is measurable with respect to the Haar measure $d\dot{t}$ on G/H_ω and*

$$\omega(A) = \int_{G/H_\omega} d\dot{t}\, \tilde{\omega}(\tau_{\dot{t}}(A))$$

for all $A \in \mathfrak{A}$.

This result is a direct transcription of Theorem 4.3.38 but in this special setting one can deduce more concerning $\tilde{\omega}$.

The above comments on the points spectrum indicate that the annihilator group is the natural symmetry group of the system. But experience with physical examples indicates that the characterization of the natural symmetry group is related to an improvement of cluster properties for the constituent states occurring in the ergodic decompositions. In this respect it is of interest that under a hypothesis similar to those of the corollary one can indeed show that $\tilde{\omega}$ may be chosen to be weakly mixing for H_ω, i.e.,

$$M(|\tilde{\omega}(A\tau(B)) - \tilde{\omega}(A)\tilde{\omega}(B)|) = 0$$

for all $A, B \in \mathfrak{A}$, where M is any invariant mean over $C_b(H_\omega)$. The proof of this statement relies, however, upon some representation theory, and will be given in the following section, Theorem 4.4.12. It appears reasonable to conjecture that this strengthening of cluster properties is the correct quantitative characterization of the natural symmetry group but as results of this type are scarce the position is unclear.

Decompositions with respect to normal subgroups possess a number of characteristic features. As the set of states occurring in the decomposition is the orbit $\{\tau_{\dot{g}}{}^*\tilde{\omega};\, \dot{g} \in G/H\}$ of one fixed state under the action of the quotient group G/H, these states simultaneously satisfy various types of algebraic property, e.g., $(\mathfrak{A}, \tau_{\dot{g}}{}^*\tilde{\omega})$ is H-abelian, or H-central, if, and only if, $(\mathfrak{A}, \tilde{\omega})$ is H-abelian, or H-central. Thus if $(\mathfrak{A}, \tilde{\omega})$ is H-central then it follows by a slight modification of the last statement of Theorem 4.3.19 that distinct states $\tau_{\dot{g}}{}^*\tilde{\omega}$ of the orbit generate disjoint representations of $\mathfrak{A}$. Nevertheless, there is a unitary equivalence relation between the associated unitary representations of H which one establishes as follows.

For simplicity let $(\mathfrak{H}_{\dot{g}}, \pi_{\dot{g}}, U_{\dot{g}}, \Omega_{\dot{g}})$ denote the representation associated with $\tau_{\dot{g}}{}^*\tilde{\omega}$ and $(\mathfrak{H}, \pi, U, \Omega)$ the representation associated with $\tilde{\omega}$. We define a map $V_{\dot{g}}$ from $\mathfrak{H}_{\dot{g}}$ to $\mathfrak{H}$ by

$$V_{\dot{g}}\pi_{\dot{g}}(A)\Omega_{\dot{g}} = \pi(\tau_{\dot{g}}(A))\Omega.$$

One then has

$$\begin{aligned} \|V_{\dot{g}}\pi_{\dot{g}}(A)\Omega_{\dot{g}}\|^2 &= \tilde{\omega}(\tau_{\dot{g}}(A^*A)) \\ &= \|\pi_{\dot{g}}(A)\Omega_{\dot{g}}\|^2 \end{aligned}$$

and hence $V_{\dot{g}}$ extends to a bounded isometric map. But $V_{\dot{g}}$ is invertible because

$$V_{\dot{g}}^{-1}\pi(A)\Omega = \pi_{\dot{g}}(\tau_{\dot{g}^{-1}}(A))\Omega_{\dot{g}}$$

defines a bounded isometric inverse. Therefore $V_{\dot{g}}$ is a unitary map from $\mathfrak{H}_{\dot{g}}$ to $\mathfrak{H}$. Now one calculates

$$\begin{aligned} (V_{\dot{g}}\pi_{\dot{g}}(A)\Omega_{\dot{g}}, U(h)V_{\dot{g}}\pi_{\dot{g}}(B)\Omega_{\dot{g}}) &= (\pi(\tau_{\dot{g}}(A))\Omega, \pi(\tau_h\tau_{\dot{g}}(B))\Omega) \\ &= (\pi(\tau_{\dot{g}}(A))\Omega, \pi(\tau_{\dot{g}}\tau_{\dot{g}^{-1}h\dot{g}}(B))\Omega) \\ &= (V_{\dot{g}}\pi_{\dot{g}}(A)\Omega_{\dot{g}}, V_{\dot{g}}\pi_{\dot{g}}(\tau_{\dot{g}^{-1}h\dot{g}}(B))\Omega) \\ &= (\pi_{\dot{g}}(A)\Omega_{\dot{g}}, U_{\dot{g}}(\dot{g}^{-1}h\dot{g})\pi_{\dot{g}}(B)\Omega) \end{aligned}$$

for all $h \in H$, and hence

$$V_{\dot{g}}^* U(h) V_{\dot{g}} = U_{\dot{g}}(\dot{g}^{-1} h \dot{g}).$$

Thus the unitary representation $U_{\dot{g}}(H)$ is unitarily equivalent to the unitary representation $U(\dot{g} H \dot{g}^{-1})$. In particular, if G is abelian then $U_{\dot{g}}(H)$ and $U(H)$ are unitarily equivalent. More generally, the equivalence is between the family of representations $U_{\dot{g}}(\dot{g}^{-1} H \dot{g})$ which incorporate the correct orientation of the state within the quotient group.

We next examine almost periodic decompositions of G-ergodic states for locally compact abelian groups G. First we need to review a number of basic facts concerning almost periodic functions.

Let $L^\infty(G)$ denote the bounded functions over G equipped with the supremum norm,

$$\|f\|_\infty = \sup_{t \in G} |f(t)|,$$

and $C_b(G)$ the C^*-algebra of bounded continuous functions over G. The *trigonometric functions* $T(G)$ are defined as the Banach *-subalgebra of $C_b(G)$ formed by finite linear combinations of characters of G and the C^*-algebra $A(G)$ of *almost periodic functions* as the uniform closure of $T(G)$.

The foregoing definition states that a function f is almost periodic if, and only if, it is the uniform limit of a family of finite linear combinations of characters of G. There are other possible characterizations and the original definition of Bohr is in terms of approximate periods. If τ denotes the action of G on the bounded functions, i.e., $(\tau_t f)(t') = f(t + t')$ then $t \in G$ is called an *ε-period* of f if, and only if,

$$\|\tau_t f - f\|_\infty < \varepsilon.$$

With this definition the following conditions are equivalent (see Notes and Remarks):

(1) $f \in A(G)$;
(2) f is continuous and for each $\varepsilon > 0$ there is a finite subset $B \subseteq G$ such that $A_\varepsilon + B = G$, where A_ε denotes the set of ε-periods of f;
(3) $f \in L^\infty(G)$ and the uniform closure of the orbit $\{\tau_t f\,;\, t \in G\}$ is compact.

The equivalence of conditions (2) and (3) can be clarified if one remarks that a set in a complete metric space has a compact closure if, and only if, it is totally bounded. Thus condition (3) is equivalent to the statement that for each $\varepsilon > 0$ there exists a finite number of elements $t_1, t_2, \ldots, t_n \in G$ such that

$$\inf_{1 \leq i \leq n} \|\tau_t f - \tau_{t_i} f\|_\infty < \varepsilon$$

for all $t \in G$. This is equivalent to the statement that for each $t \in G$ there is a $t' \in B = \{t_1, \ldots, t_n\}$ such that $t - t'$ is an ε-period of f.

Suppose now that G acts as a group of *-automorphisms τ of a C^*-algebra $\mathfrak{A}$ with identity. It is natural to define a state ω over $\mathfrak{A}$ to be *almost periodic* if the functions $t \in G \mapsto \omega(\tau_t(A))$ are in $A(G)$ for all $A \in \mathfrak{A}$. If $E_{\mathfrak{A}}^{A(G)}$ denotes the corresponding set of states then $E_{\mathfrak{A}}^{A(G)}$ is convex but not necessarily weakly* closed. This causes difficulties if one attempts to develop the decomposition theory of almost periodic states. Nevertheless, one has the following result for orthogonal measures.

Proposition 4.3.40. *Let $\mathfrak{A}$ be a C^*-algebra with identity, G a locally compact abelian group acting as *-automorphisms τ of $\mathfrak{A}$, and ω a G-invariant state over $\mathfrak{A}$. Assume that U_ω is strongly continuous and let $\hat{P}_\omega$ denote the projection on the subspace of $U_\omega(G)$-almost periodic vectors. There is a one-to-one correspondence between the following:*

(1) *the orthogonal measures μ over $E_{\mathfrak{A}}$ with barycenter ω which are such that*

$$t \in G \mapsto \mu(\tau_t(f_1)f_2)$$

is almost periodic for all $f_1, f_2 \in C(E_{\mathfrak{A}})$;

(2) *the abelian von Neumann subalgebras $\mathfrak{B}$ of the commutant*

$$\{\pi_\omega(\mathfrak{A}) \cup \hat{P}_\omega\}';$$

(3) *the orthogonal projections P on $\mathfrak{H}_\omega$ such that*

$$P\Omega_\omega = \Omega_\omega, \qquad P \leq \hat{P}_\omega$$
$$P\pi_\omega(\mathfrak{A})P \subseteq \{P\pi_\omega(\mathfrak{A})P\}'.$$

Remark. This proposition is in absolute analogy with the corresponding result for invariant states, Proposition 4.3.1. Note that in condition (2) of this latter proposition one has the identification

$$\{\pi_\omega(\mathfrak{A}) \cup U_\omega(G)\}' = \{\pi_\omega(\mathfrak{A}) \cup E_\omega\}',$$

where E_ω is the projection on the subspace of $U_\omega(G)$-invariant vectors (for this equality see the conclusion of the proof of (4) in Theorem 4.3.27). Moreover the property $U_\omega(g)P = P$ in condition (3) of Proposition 4.3.1 can be reexpressed as $P \leq E_\omega$.

Proof. We again exploit the general correspondences between orthogonal measures μ, abelian von Neumann subalgebras $\mathfrak{B} \subseteq \pi_\omega(\mathfrak{A})'$, and projections P such that $P\Omega_\omega = \Omega_\omega$ and $P\pi_\omega(\mathfrak{A})P \subseteq \{P\pi_\omega(\mathfrak{A})P\}'$, established by Theorem 4.1.25. We concentrate on incorporating the conditions of almost periodicity.

(1) $\Rightarrow$ (2) First assume μ to satisfy the almost periodic condition; then

$$t \in G \mapsto (\Omega_\omega, \pi_\omega(A)U_\omega(t)\kappa_\mu(f)\Omega_\omega) = \mu(\tau_{-t}(\hat{A})f)$$

is almost periodic for each $A \in \mathfrak{A}$ and $f \in C(E_{\mathfrak{A}})$. By cyclicity, one concludes that $[\kappa_\mu(f)\Omega_\omega] \leq \hat{P}_\omega$. Next remark that

$$\begin{aligned} \kappa_\mu(f)U_\omega(t)\pi_\omega(B)\Omega_\omega &= \pi_\omega(\tau_t(B))\kappa_\mu(f)\Omega_\omega \\ &= \pi_\omega(\tau_t(B))\hat{P}_\omega\kappa_\mu(f)\Omega_\omega \\ &= \sum_{\gamma \in \hat{G}} (\gamma, t)U_\omega(t)\pi_\omega(B)P_\omega[\gamma]\kappa_\mu(f)\Omega_\omega. \end{aligned}$$

Therefore as

$$(\pi_\omega(A)\Omega_\omega, \kappa_\mu(f)\hat{P}_\omega\pi_\omega(B)\Omega_\omega) = \sum_{\gamma' \in \hat{G}} M(\gamma'(\pi_\omega(A)\Omega_\omega, \kappa_\mu(f)U_\omega\pi_\omega(B)\Omega_\omega))$$

one has

$$\begin{aligned}(\pi_\omega(A)\Omega_\omega, \kappa_\mu(f)\hat{P}_\omega\pi_\omega(B)\Omega_\omega) &= \sum_{\gamma', \gamma \in \hat{G}} M(\gamma'\gamma(\pi_\omega(A)\Omega_\omega, U_\omega\pi_\omega(B)P_\omega[\gamma]\kappa_\mu(f)\Omega_\omega)) \\ &= \sum_{\gamma', \gamma \in \hat{G}} (\pi_\omega(A)\Omega_\omega, P_\omega[\gamma'\gamma]\pi_\omega(B)P_\omega[\gamma]\kappa_\mu(f)\Omega_\omega) \\ &= (\pi_\omega(A)\Omega_\omega, \hat{P}_\omega\pi_\omega(B)\hat{P}_\omega\kappa_\mu(f)\Omega_\omega) \\ &= (\pi_\omega(A)\Omega_\omega, \hat{P}_\omega\kappa_\mu(f)\pi_\omega(B)\Omega_\omega).\end{aligned}$$

This means that $\kappa_\mu(f) \in \{\pi_\omega(\mathfrak{A}) \cup \hat{P}_\omega\}'$. But $\mathfrak{B} = \{\kappa_\mu(f); f \in C(E_\mathfrak{A})\}''$ and hence $\mathfrak{B} \subseteq \{\pi_\omega(\mathfrak{A}) \cup \hat{P}_\omega\}'$. Thus we have established that (1) $\Rightarrow$ (2). But as $P = [\mathfrak{B}\Omega_\omega]$ it immediately follows that $P \leq \hat{P}_\omega$, i.e., (2) $\Rightarrow$ (3). It remains to prove that (3) $\Rightarrow$ (1). But

$$\begin{aligned}\mu(\tau_t(f_1)f_2) &= (\Omega_\omega, \kappa_\mu(f_1)U_\omega(t)^{-1}\kappa_\mu(f_2)\Omega_\omega) \\ &= (\Omega_\omega, \kappa_\mu(f_1)U_\omega(t)^{-1}\hat{P}_\omega\kappa_\mu(f_2)\Omega_\omega)\end{aligned}$$

by the basic structure theorem for orthogonal measures, Theorem 4.1.25, and thus this function is almost periodic for all $f_1, f_2 \in C(E_\mathfrak{A})$.

The next result characterizes the basic properties of the measures introduced in Proposition 4.3.40.

Proposition 4.3.41. *Let $\mathfrak{A}$ be a C^*-algebra with identity, G a locally compact abelian group acting as *-automorphisms τ of $\mathfrak{A}$, ω a G-invariant state, and assume that U_ω is strongly continuous. If $\mu \in \mathcal{O}_\omega(E_\mathfrak{A})$ and*

$$t \in G \mapsto \mu(\tau_t(f_1)f_2)$$

is almost periodic for all $f_1, f_2 \in C(E_\mathfrak{A})$ then the support of μ is contained in the weak-closure $\bar{E}_\mathfrak{A}^{A(G)}$ of the convex set $E_\mathfrak{A}^{A(G)}$ of almost periodic states. If, moreover, μ is maximal then it is pseudosupported by $\mathcal{E}(\bar{E}_\mathfrak{A}^{A(G)})$ and if ω is contained in a face F which satisfies the separability condition S then μ is supported by $\mathcal{E}(\bar{E}_\mathfrak{A}^{A(G)})$.*

If $\{\pi_\omega(\mathfrak{A}) \cup \hat{P}_\omega\}'$ is abelian then $M_\omega(\bar{E}_\mathfrak{A}^{A(G)})$ contains a unique maximal measure. Conversely, if $M_\omega(\bar{E}_\mathfrak{A}^{A(G)})$ contains a unique maximal measure such that $\mu(\tau(f_1)f_2) \in A(G)$ for all $f_1, f_2 \in C(E_\mathfrak{A})$ then $\{\pi_\omega(\mathfrak{A}) \cup \hat{P}_\omega\}'$ is abelian. In both cases μ is the orthogonal measure corresponding to $\{\pi_\omega(\mathfrak{A}) \cup \hat{P}_\omega\}'$.

PROOF. The proof is similar to that of Propositions 4.3.2 and 4.3.3 with invariance replaced by almost periodicity. The first statement follows by simple repetition of the arguments used to establish Proposition 4.3.2 but the second statement needs a slight modification of the previous proofs.

First consider a measure $\nu \in M_\omega(\bar{E}_\mathfrak{A}^{A(G)})$ with finite support in $E_\mathfrak{A}^{A(G)}$ then

$$\{\kappa_\nu(f); f \in C(E_\mathfrak{A})\} \subseteq \{\pi_\omega(\mathfrak{A}) \cup \hat{P}_\omega\}'$$

by the same calculation used to prove Proposition 4.3.40. For this it is important that $\nu(\tau(f_1)f_2) \in A(G)$ but this follows from the assumed support property of ν. Now if $\{\pi_\omega(\mathfrak{A}) \cup \hat{P}_\omega\}'$ is abelian then it determines an orthogonal measure μ over $E_\mathfrak{A}$, by Proposition 4.3.40, and this measure satisfies the almost periodic condition $\mu(\tau(f_1)f_2) \in A(G)$. Hence repetition of the argument used to prove (2) $\Rightarrow$ (1) in Theorem 4.2.3 establishes that $\mu \succ \nu$. But the measures in $M_\omega(\bar{E}_\mathfrak{A}^{A(G)})$ with finite support in $E_\mathfrak{A}^{A(G)}$ are weakly* dense and hence $\mu \succ \nu$ for all $\nu \in M_\omega(\bar{E}_\mathfrak{A}^{A(G)})$, i.e., μ is maximal. If, conversely, there is a unique maximal measure μ which satisfies the almost periodic condition then

$$\{\kappa_\mu(f); f \in C(E_\mathfrak{A})\} \subseteq \{\pi_\omega(\mathfrak{A}) \cup \hat{P}_\omega\}'$$

by Proposition 4.3.40. But the argument used to prove (1) $\Rightarrow$ (2) in Theorem 4.2.3 then establishes that these two algebras are equal and abelian.

Finally, we examine the decomposition of a G-ergodic state into almost periodic states in the case that the pair $(\mathfrak{A}, \omega)$ is G_Γ-abelian. This is equivalent to $\hat{P}_\omega \pi_\omega(\mathfrak{A})\hat{P}_\omega$ being abelian, by Proposition 4.3.30, and the basic correspondences of orthogonal measures then imply that $\{\pi_\omega(\mathfrak{A}) \cup \hat{P}_\omega\}'$ is abelian. Thus there is a unique maximal measure $\mu \in M_\omega(\bar{E}_\mathfrak{A}^{A(G)})$, by Proposition 4.3.41. It is tempting to interpret the decomposition of ω defined by μ as a decomposition of ω into almost periodic states but the difficulty with this interpretation is that μ is not necessarily supported by $E_\mathfrak{A}^{A(G)}$ but only by its weak* closure. Nevertheless, we first show that μ is equivalent to an ergodic measure m on a compact Hausdorff space M on which G acts in an almost periodic manner. Ergodicity in this context means that every m-measurable G-invariant subset of M has measure zero, or measure one. It is this property which provides an analogy with the previous subgroup decompositions, in which the orbit under the group action of each point in the support of μ had measure one.

Proposition 4.3.42. *Let $\omega \in E_\mathfrak{A}{}^G$, where $\mathfrak{A}$ has an identity, and let μ be an orthogonal G-invariant measure with barycenter ω, i.e., $\mu(\tau_g(f)) = \mu(f)$ for all $f \in C(E_\mathfrak{A})$ and $g \in G$. There exists a compact Hausdorff space M, an action $\hat{\tau}^*$ of G on M, a G-invariant Baire measure m on M, and a measure-preserving *-isomorphism κ of $L^\infty(E_\mathfrak{A}; \mu)$ onto $L^\infty(M; m)$ which commutes with the action of G.*

If the pair $(\mathfrak{A}, \omega)$ is G-abelian and $\omega \in \mathscr{E}(E_\mathfrak{A}{}^G)$ then m is ergodic for the action $\hat{\tau}^*$.

If, moreover, G is locally compact abelian, the pair $(\mathfrak{A}, \omega)$ is G_Γ-abelian, and μ is the orthogonal measure corresponding to the projector $\hat{P}_\omega$ onto the $U_\omega(G)$-almost periodic vectors, then M can be chosen such that the action $\hat{\tau}^*$ is almost periodic, i.e., the orbit $\{\hat{\tau}_t{}^*\varphi; t \in G\}$ of each $\varphi \in C(M)$ has compact closure in $C(M)$.

PROOF. Let $\mathfrak{B} \subseteq \pi_\omega(\mathfrak{A})'$ be the abelian von Neumann subalgebra of the commutant corresponding to μ. The G-invariance of μ ensures that

$$U_\omega(g)\mathfrak{B}U_\omega(g)^{-1} \subseteq \mathfrak{B}$$

for all $g \in G$ and hence one can define an automorphism group $\hat{\tau}$ of $\mathfrak{B}$ by

$$\hat{\tau}_g(B) = U_\omega(g)BU_\omega(g)^{-1}.$$

Next let m be the G-invariant state over $\mathfrak{B}$ defined by

$$m(B) = (\Omega_\omega, B\Omega_\omega)$$

for $B \in \mathfrak{B}$. Then m identifies with a probability measure on the spectrum M of $\mathfrak{B}$ with support equal to the spectrum since Ω_ω is separating for $\mathfrak{B}$, and $\mathfrak{B}$ can be identified with $L^\infty(M; m)$. The action $\hat{\tau}$ of G on $\mathfrak{B}$ defines an action $\hat{\tau}^*$ of the group on M by transposition. The *-isomorphism κ of $L^\infty(E_\mathfrak{A}; \mu)$ onto $L^\infty(M; m)$ is then introduced by

$$f \in L^\infty(E_\mathfrak{A}; \mu) \mapsto \kappa(f) = \kappa_\mu(f) \in \mathfrak{B} = L^\infty(M; m),$$

where κ_μ is the linear map associated with μ by Lemma 4.1.21. Note that

$$\begin{aligned}\kappa(\tau_g f) &= \kappa_\mu(\tau_g f)\\ &= U_\omega(g)\kappa_\mu(f)U_\omega(g)^{-1} = \hat{\tau}_g{}^*\kappa(f)\end{aligned}$$

where the covariant transformation law for κ_μ follows from the G-invariance of μ by the calculation used in the proof of Theorem 4.3.38. Furthermore,

$$m(\kappa(f)) = (\Omega_\omega, \kappa_\mu(f)\Omega_\omega) = \mu(f).$$

Next assume that the pair $(\mathfrak{A}, \omega)$ is G-abelian. The condition $\omega \in \mathscr{E}(E_\mathfrak{A}{}^G)$ is then equivalent to Ω_ω being the unique, up to a factor, $U_\omega(G)$- invariant vector in $\mathfrak{H}_\omega$, by Theorem 4.3.17. Thus if $P \in \mathfrak{B}$ is a projection onto a G-invariant M-measurable set $\hat{\tau}_g(P) = P$ and hence

$$U_\omega(g)P\Omega_\omega = P\Omega_\omega = \lambda\Omega_\omega$$

for all $g \in G$ and some $\lambda \in \mathbb{C}$. But Ω_ω is separating for $\mathfrak{B}$, because $\mathfrak{B} \subseteq \pi_\omega(\mathfrak{A})'$, and therefore $P = \mathbb{1}$, or 0. Thus m is ergodic for the action of G.

Finally, if $\omega \in \mathscr{E}(E_\mathfrak{A}{}^G)$ and $(\mathfrak{A}, \omega)$ is G_Γ-abelian then the unitary eigenelements $V_\gamma \in \pi_\omega(\mathfrak{A})'$ of the automorphism group $\hat{\tau}$ satisfy

$$\{V_\gamma; \gamma \in \sigma_{\rm p}(U_\omega)\}'' = \{\pi_\omega(\mathfrak{A}) \cup \hat{P}_\omega\}'$$

by Theorem 4.3.31. Let $\mathfrak{B}_0$ be the C^*-algebra generated by $\{V_\gamma; \gamma \in \sigma_{\rm p}(U_\omega)\}$ and then $\mathfrak{B} = \mathfrak{B}_0'' = \{\pi_\omega(\mathfrak{A}) \cup \hat{P}_\omega\}'$ is the von Neumann algebra associated with the orthogonal measure determined by $\hat{P}_\omega$. Thus M may be replaced by the spectrum of $\mathfrak{B}_0$ in the above construction so $\mathfrak{B}_0 = C(M) \subseteq L^\infty(M; m) = \mathfrak{B}$. As each element $B \in \mathfrak{B}_0$ is the uniform limit of polynomials of the eigenelements of $\hat{\tau}$ it follows that $\hat{\tau}$ acts in a strongly continuous manner on $\mathfrak{B}_0$ and the uniform closure of the orbit $\{\tau_t(B); t \in G\}$ is compact. The latter property is a consequence of the arguments which give the alternative characterizations of almost periodicity mentioned above.

Remark. If in the first statement of the proposition one assumes that $[\mathfrak{B}\Omega_\omega]$ is separable, where $\mathfrak{B} \subseteq \pi_\omega(\mathfrak{A})'$ is the von Neumann algebra associated with μ, then one can deduce more concerning the *-isomorphism κ. It follows by a general result of von Neumann's (see Notes and Remarks) that there exist subsets $E_\mu \subseteq E_{\mathfrak{A}}$ and $M_m \subseteq M$ with μ and m measure one, and a Borel isomorphism η from E_μ to M_m such that

$$f(\eta\omega) = (\kappa^{-1}f)(\omega)$$

for all $\omega \in E_\mu$ and $f \in L^\infty(M; m)$.

The final statement of Proposition 4.3.42 can be improved if one has more information concerning the support of the orthogonal measure μ associated with $\hat{P}_\omega$. Our next aim is to show that if the support of μ is contained in $E_{\mathfrak{A}}^{A(G)}$ then the corresponding decomposition of ω can often be expressed as an average of a state $\tilde{\omega} \in \mathscr{E}(E_{\mathfrak{A}}^{A(G)})$ over its orbit. To deduce this result and to create an analogy with the previous normal subgroup decompositions it is useful to have a fourth characterization of almost periodic functions in terms of a compactification of G.

If G is a locally compact abelian group let $\hat{G}_{\mathrm{d}}$ denote the dual of G equipped with the discrete topology. The dual $\bar{G}$ of $\hat{G}_{\mathrm{d}}$ is a compact group by Pontryagin duality. Now let α be the map of G into $\bar{G}$ defined by

$$(\gamma, t) = (\alpha t, \gamma)$$

for all $f \in G$, and $\gamma \in \hat{G}_{\mathrm{d}}$. One can show that α is a continuous isomorphism of G onto a dense subgroup $\alpha(G)$ of $\bar{G}$. The group $\bar{G}$ is usually called the *Bohr compactification* of G. It can be used to characterize the almost periodic functions $A(G)$ by the following criterion.

Let f be a bounded function over G; then the following conditions are equivalent:

(1) $f \in A(G)$;
(2) there is an $h \in C(\bar{G})$ such that

$$f(t) = h(\alpha t)$$

for all $t \in G$.

Note that as $\alpha(G)$ is dense in $\bar{G}$, this characterization of $A(G)$ establishes a *-isomorphism $\alpha; f \in C(\bar{G}) \mapsto \alpha f \in A(G)$ such that

$$(\alpha f)(t) = f(\alpha t)$$

for all $t \in G$.

Next remark that G has an action τ on $A(G)$ given by $(\tau_t f)(t') = f(t + t')$ and the existence of a unique Haar measure $d\bar{t}$ on $\bar{G}$ ensures the existence of a unique invariant mean M on $A(G)$. One has the relation

$$M(\alpha f) = \int_{\bar{G}} d\bar{t}\; f(\bar{t})$$

for all $f \in C(\bar{G})$. The existence and uniqueness of the mean M can also be established from the other characterizations of $A(G)$. For example, M can be described as a linear functional over $A(G)$ satisfying $M(\gamma) = 0$ for each character distinct from the identity ι and $M(\iota) = 1$. To see this let M' be any invariant mean over $A(G)$. If $\gamma \in \hat{G}$ then

$$(\tau_t \gamma, t') = (\gamma, tt') = (\gamma, t)(\gamma, t')$$

and consequently

$$M'(\tau_t \gamma) = M'(\gamma) = (\gamma, t)M'(\gamma)$$

for all $t \in G$. Hence $M'(\gamma) = 0$ for all γ except the identity. This establishes that the invariant mean M is uniquely determined on $T(G)$ and hence, by continuity, uniquely determined on $A(G)$. A third characterization of M is given in terms of the orbits $\{\tau_t f; t \in G\}$. For each $f \in A(G)$ one can demonstrate that the closed convex hull of the associated orbit contains a unique constant $M(f)$ and the constants define the invariant mean.

Now let us return to the characterization of almost periodic decompositions.

Proposition 4.3.43. *Let $\omega \in \mathscr{E}(E_{\mathfrak{A}}{}^G)$ where G is a locally compact abelian group and $\mathfrak{A}$ has an identity. Assume that the pair $(\mathfrak{A}, \omega)$ is G_Γ-abelian, let $H_\omega \subseteq G$ denote the annihilator of the point spectrum $\sigma_p(U_\omega)$ of U_ω, and let μ denote the orthogonal measure corresponding to the projection $\hat{P}_\omega$ onto the $U_\omega(G)$-almost periodic vectors. Further assume that*

(1) *ω lies in a face F which satisfies the separability condition* S;
(2) *the support of μ is contained in $E_{\mathfrak{A}}^{A(G)}$.*

It follows that there exists an $\tilde{\omega} \in \mathscr{E}(E_{\mathfrak{A}}^{A(G)}) \cap E_{\mathfrak{A}}^{H_\omega}$ such that

$$f(\omega) = M(f(\tau^* \tilde{\omega}))$$

for all $f \in C(\bar{E}_{\mathfrak{A}}^{A(G)})$, where M denotes the unique invariant mean on $A(G/H_\omega)$. In particular,

$$\omega(A) = M(\tilde{\omega}(\tau(A)))$$

for all $A \in \mathfrak{A}$.

Proof. As the pair $(\mathfrak{A}, \omega)$ is G_Γ-abelian $\hat{P}_\omega \pi_\omega(\mathfrak{A})\hat{P}_\omega$ is abelian by Proposition 4.3.30 and $\{\pi_\omega(\mathfrak{A}) \cup \hat{P}_\omega\}'$ is abelian by the basic correspondences for orthogonal measures. It then follows from Proposition 4.3.41 that μ is supported by $\mathscr{E}(\bar{E}_{\mathfrak{A}}^{A(G)})$. Let S_μ denote the support of μ. It follows from Theorem 4.3.31 that $\{\pi_\omega(\mathfrak{A}) \cup \hat{P}_\omega\}' \subseteq \{\pi_\omega(\mathfrak{A}) \cup E_\omega(H_\omega)\}'$. Consequently, applying Proposition 4.3.2 with $G = H_\omega$ one concludes that $S_\mu \subseteq E_{\mathfrak{A}}^{H_\omega}$. But by assumption $S_\mu \subseteq E_{\mathfrak{A}}^{A(G)}$ and hence

$$\begin{aligned} S_\mu &\subseteq \mathscr{E}(\bar{E}_{\mathfrak{A}}^{A(G)}) \cap E_{\mathfrak{A}}^{A(G)} \cap E_{\mathfrak{A}}^{H_\omega} \\ &\subseteq \mathscr{E}(E_{\mathfrak{A}}^{A(G)}) \cap E_{\mathfrak{A}}^{H_\omega}. \end{aligned}$$

Finally, μ is supported by F, and hence supported by $F \cap \mathscr{E}(E_{\mathfrak{A}}^{A(G)}) \cap E_{\mathfrak{A}}^{H_\omega}$.

Next note that G-invariance of μ implies that if $\tilde{\omega} \in S_\mu$ then $\tau_t{}^*\tilde{\omega} \in S_\mu$ for all $t \in G$, and hence τ defines an action on $C(S_\mu)$. Moreover, each almost periodic function over G is in correspondence with a continuous function over the Bohr compactification

$\bar{G}$ of G. Thus one can define an action τ^* of $\bar{G}$ on each $\omega \in S_\mu \subseteq E_{\mathfrak{A}}^{A(G)}$ and, by transposition, an action $\bar{\tau}$ on $f \in C(S_\mu)$. The resulting function $\bar{t} \in \bar{G} \mapsto \bar{\tau}_{\bar{t}} f$ is continuous. But if $f, f_n \in C(S_\mu)$ then

$$\begin{aligned}\mu(\tau_t(f)f_n) &= (\Omega_\omega, \kappa_\mu(f)U_\omega(t)^{-1}\kappa_\mu(f_n)\Omega_\omega) \\ &= (\Omega_\omega, \kappa_\mu(f)U_\omega(t)^{-1}\hat{P}_\omega\kappa_\mu(f_n)\Omega_\omega)\end{aligned}$$

by the basic structure theorem for orthogonal measures (Theorem 4.1.25) and the covariant transformation law for κ_μ derived in the proof of Theorem 4.3.38. Thus

$$\mu(\tau_t(f)f_n) = \sum_{\gamma \in \sigma_p(U_\omega)} (\gamma, t)(\Omega_\omega, \kappa_\mu(f)P_\omega[\gamma]\kappa_\mu(f_n)\Omega_\omega).$$

In particular, $\mu(\tau(f)f_n) \in A(G/H_\omega)$ and

$$M(\mu(\tau(f)f_n)) = \mu(f)\mu(f_n)$$

because $M(\gamma) = 0$ for all $\gamma \in \hat{G}$, except the identity ι, and $P_\omega[\iota]$ is the projection on Ω_ω by Theorem 4.3.31(2). Expressed in terms of $\overline{G/H_\omega}$ this last relation then gives

$$\mu(f) = \mu(f_n)^{-1} \int_{\overline{G/H_\omega}} d\bar{t}\; \mu(\bar{\tau}_{\bar{t}}(f)f_n),$$

where $d\bar{t}$ is again the Haar measure. Now one can repeat the argument based on the Lebesgue dominated convergence theorem used in the proof of Theorem 4.3.38, to deduce that

$$\mu(f) = \int_{\overline{G/H_\omega}} d\bar{t}\; f(\bar{\tau}_{\bar{t}}^*\tilde{\omega})$$

for any $\tilde{\omega} \in S_\mu \cap \mathscr{E}(E_{\mathfrak{A}}^{A(G)}) \cap \mathscr{E}(E_{\mathfrak{A}}^{H_\omega})$.

Note that $E_{\mathfrak{A}}^{A(G)} \cap E_{\mathfrak{A}}^{H_\omega} = E_{\mathfrak{A}}^{A(G/H_\omega)}$ and hence the decomposition is really into almost periodic states over the quotient group G/H_ω.

Proposition 4.3.43 is unsatisfactory because it requires the assumption that μ has support in $E_{\mathfrak{A}}^{A(G)}$. To illustrate the nature of this assumption consider the case that $\sigma_p(U_\omega)$ is discrete, and hence G/H_ω is compact. Thus G/H_ω is equal to its own compactification, $A(G/H_\omega) = C(G/H_\omega)$, and the orbits $\dot{t} \in G/H_\omega \mapsto \tilde{\omega}(\tau_{\dot{t}}(A))$ are automatically continuous for each $\tilde{\omega} \in E_{\mathfrak{A}}^{A(G/H_\omega)}$ and each $A \in \mathfrak{A}$. Thus Proposition 4.3.43 gives a decomposition of $\omega \in \mathscr{E}(E_{\mathfrak{A}}{}^G)$ of the same type as Corollary 4.3.39,

$$\omega(A) = \int_{G/H_\omega} d\dot{t}\; \tilde{\omega}(\tau_{\dot{t}}(A)),$$

but the assumed support property of μ introduces a continuity of $\tau_{\dot{t}}^*\tilde{\omega}$ which is not in general present. Of course if τ acts in a strongly continuous manner on $\mathfrak{A}$ then the special case of Proposition 4.3.43 and Corollary 4.3.39 coincide. To avoid the support assumption for μ and nevertheless obtain a decomposition of ω as the mean of another state, with ergodic properties, over its orbit it appears necessary to broaden the notion of almost periodic state. One should find some analogue of measurability of the orbit, i.e., the correct notion of almost periodicity should be related to measurability of the orbit with respect to the Haar measure on $\bar{G}$.

Finally, we note that it is not clear what cluster properties characterize the $\tilde{\omega} \in \mathscr{E}(\mathfrak{A}^{A(G)})$ but if ω and $\tilde{\omega}$ are related as in Proposition 4.3.43, it is not difficult to demonstrate that

$$M(\tilde{\omega}(A\tau(B))) = \tilde{\omega}(A)\omega(B).$$

4.4. Spatial Decomposition

Hitherto in this chapter we have been interested in decomposing a given state ω on an operator algebra $\mathfrak{A}$ into other states. The general scheme of these decompositions, developed in section 4.1.3, was to choose an abelian von Neumann subalgebra $\mathfrak{B} \subseteq \pi_\omega(\mathfrak{A})'$, and decompose ω "over the spectrum of $\mathfrak{B}$." If $\mathfrak{B}$ is finite-dimensional, i.e., $\mathfrak{B}$ is spanned by a finite sequence P_1, $P_2, \ldots, P_n$ of mutually orthogonal projections with $\sum_{i=1} P_i = \mathbb{1}$, then the decomposition of ω takes the form

$$\omega = \sum_{i=1}^{n} \lambda_i \omega_i,$$

where $\lambda_i \omega_i(A) = (\Omega_\omega, P_i \pi_\omega(A)\Omega_\omega)$. Closely associated with this decomposition of ω one also has a spatial decomposition of the representation

$$(\mathfrak{H}_\omega, \pi_\omega, \Omega_\omega),$$

i.e., defining

$$\mathfrak{H}_i = P_i \mathfrak{H}_\omega,$$

$$\pi_i = P_i \pi_\omega,$$

$$\Omega_i = \frac{P_i \Omega_\omega}{\|P_i \Omega_\omega\|},$$

then $(\mathfrak{H}_i, \pi_i, \omega_i)$ identifies with $(\mathfrak{H}_{\omega_i}, \pi_{\omega_i}, \Omega_{\omega_i})$ and one has

$$\mathfrak{H}_\omega = \bigoplus_{i=1}^{n} \mathfrak{H}_{\omega_i},$$

$$\pi_\omega = \bigoplus_{i=1}^{n} \pi_{\omega_i}.$$

This decomposition of π_ω is called the *spatial decomposition* of π_ω defined by $\mathfrak{B}$, and this subsection is devoted to a generalization of this concept to general abelian von Neumann subalgebras $\mathfrak{B} \subseteq \pi_\omega(\mathfrak{A})'$. Of course the decomposition in the general case will take the form of a direct integral rather than a direct sum, and we first have to develop a theory of direct integrals of Hilbert spaces. It turns out that a satisfactory theory of this sort can only be developed when both the measure space and the Hilbert space have suitable separability properties. As this theory has been treated extensively in most of the standard textbooks on operator algebras, we principally content ourselves with a review of the results. Most of the problems are of a measure-theoretic nature, and the proofs have the flavor of the proof of Proposition 4.1.34.

The examination of decomposition theory via the decomposition of representations has several advantages over the foregoing approach, which concentrated almost completely on states. For example to obtain a "good" support properties of the maximal orthogonal measures μ_ω occurring in the extremal, center, and ergodic decompositions we consistently assumed that the state ω was in a face F satisfying the separability condition S. Although this assumption is natural in applications of decomposition theory to quasi-local algebras it appears slightly artificial in the general setting. Representation theory allows us to conclude that the measures μ_ω have good support properties under the weaker, and more natural, assumption that the representation space $\mathfrak{H}_\omega$ is separable. Furthermore, one can exploit the decomposition of representations to obtain results concerning the decomposition with respect to the algebra at infinity and to prove that the decomposition of an ergodic state with respect to a normal subgroup leads to improved cluster properties.

4.4.1. General Theory

First recall that a finite positive measure μ on a measure space Z is called a *standard measure* if there exists a μ-negligible subset $E \subseteq Z$ such that $Z \backslash E$ is a standard measure space, i.e., the Borel structure on $Z \backslash E$ is the Borel structure defined by a Polish space (a complete, separable, metric space). Note in particular that if $E_{\mathfrak{A}}$ is the state space of a C^*-algebra with unit, $F \subseteq E_{\mathfrak{A}}$ is a face satisfying the separability condition S, $\omega \in F$, and μ is a Baire measure with ω as barycenter, then μ is supported by F by Proposition 4.1.34 and one deduces easily that μ is a standard measure. Now assume that μ is a standard measure on Z, and furthermore that there is a Hilbert space $\mathfrak{H}(z)$, associated to each $z \in Z$. Then, under circumstances we shall describe, one can define a direct integral Hilbert space

$$\mathfrak{H} = \int_Z^\oplus d\mu(z)\, \mathfrak{H}(z).$$

There are two equivalent ways of doing this. One is rather short and concrete but a bit artificial from a fundamental viewpoint, and the other is longer but nevertheless more intuitive and applicable. We will describe both methods.

Definition 4.4.1A. Let $\mathfrak{H}_1 \subseteq \mathfrak{H}_2 \subseteq \cdots \subseteq \mathfrak{H}_{\aleph_0}$ be a sequence of Hilbert spaces chosen once and for all, with $\mathfrak{H}_n$ having dimension n. Let $\{Z_n\}$ be a partition of Z into measurable subsets, and define $\mathfrak{H}(z) = \mathfrak{H}_n$ for $z \in Z_n$. Then

$$\mathfrak{H} = \int_Z^{\oplus} d\mu(z)\, \mathfrak{H}(z)$$

is defined as the set of all functions f defined on Z such that

(i) $f(z) \in \mathfrak{H}_n \subseteq \mathfrak{H}_{\aleph_0}$ for $z \in Z_n$,
(ii) $z \mapsto f(z) \in \mathfrak{H}_{\aleph_0}$ is μ-measurable, i.e., $z \mapsto (\xi, f(z))$ is μ-measurable for all $\xi \in \mathfrak{H}_{\aleph_0}$,
(iii) $\int_Z d\mu(z)\, \|f(z)\|^2 < \infty$.

(Note that (ii) and separability of $\mathfrak{H}_{\aleph_0}$ imply that $z \mapsto \|f(z)\|^2$ is measurable.)

The linear operations on $\mathfrak{H}$ are defined in the obvious manner, and the inner product by

$$(f, g) = \int_Z d\mu(z)(f(z), g(z)).$$

The space $\mathfrak{H}$ is called the *direct integral of the Hilbert spaces* $\mathfrak{H}(z)$.

For the other definition we need a condition on the collection $\mathfrak{H}(z)$ of Hilbert spaces, which is trivially fulfilled in Definition 4.4.1A.

Definition 4.4.1B. Let $\mathscr{F}$ be the space of functions on Z such that $\xi(z) \in \mathfrak{H}(z)$ for each $z \in Z$. $\{\mathfrak{H}(z); z \in Z\}$ is called a *measurable family* if there exists a sequence $\{\xi_n\}$ of functions in $\mathscr{F}$ such that

(1) $z \mapsto (\xi_n(z), \xi_m(z))$ is μ-measurable for all n, m,
(2) $\{\xi_n(z); n = 1, 2, \ldots\}$ is dense in $\mathfrak{H}(z)$ for each $z \in Z$.

If $\{\mathfrak{H}(z)\}$ is a measurable family and V is a linear subspace of $\mathscr{F}$ then V is said to be *measurable* if it satisfies:

(1) for $\xi, \eta \in V$, $z \mapsto (\xi(z), \eta(z))$ is measurable;
(2) there exists a countable family $\{\xi_n; n = 1, 2, \ldots\}$ in V such that $\{\xi_n(z); n = 1, 2, \ldots\}$ is dense in $\mathfrak{H}(z)$ for each $z \in Z$.

Now, it follows by Zorn's Lemma that any subspace satisfying (1) and (2) is contained in a maximal subspace $V \subseteq \mathscr{F}$ satisfying (1) and (2).

Let $\mathfrak{H}$ be the set of elements ξ in such a maximal V satisfying

$$\int d\mu(z) \|\xi(z)\|^2 < \infty.$$

Then $\mathfrak{H}$ is a Hilbert space if the inner product is defined by

$$(\xi, \eta) = \int d\mu(z)\, (\xi(z), \eta(z)).$$

We write

$$\mathfrak{H} = \int_Z^{\oplus} d\mu(z)\, \mathfrak{H}(z)$$

and call $\mathfrak{H}$ the *direct integral of the family* $\{\mathfrak{H}(z)\}$.

The equivalence of Definitions 4.4.1A and 4.4.1B, and the independence of the latter from the particular maximal measurable subspace V employed, follow from the next structure theorem, which is proved by a method reminiscent of the Gram–Schmidt orthogonalization procedure.

Theorem 4.4.2. *Let μ be a standard measure on Z, $\{\mathfrak{H}(z);\, z \in Z\}$ a measurable family of Hilbert spaces, $\mathfrak{H}_1 \subseteq \mathfrak{H}_2 \subseteq \cdots \subseteq \mathfrak{H}_{\aleph_0}$ an increasing sequence of Hilbert spaces such that* $\dim(\mathfrak{H}_n) = n$, *and define*

$$Z_n = \{z \in Z;\, \dim(\mathfrak{H}(z)) = n\}.$$

Then Z_n is a partition of Z into measurable subsets.

If V is a maximal measurable subset of $\mathscr{F}$, and $\mathfrak{H}$ the associated Hilbert space, then there exists a unitary operator

$$U;\, \mathfrak{H} \mapsto \bigoplus_{n=1}^{\aleph_0} (L^2(Z_n, d\mu) \otimes \mathfrak{H}_n)$$

of the following form: for each $z \in Z_n \subseteq Z$ there exists a unitary operator $U(z);\, \mathfrak{H}(z) \mapsto \mathfrak{H}_n$, and if $\xi \in \mathfrak{H}$ then

$$(U\xi)(z) = U(z)\xi(z).$$

Here we view $\bigoplus_{n=1}^{\aleph_0} (L^2(Z_n, d\mu) \otimes \mathfrak{H}_n)$ as functions η from Z into $\mathfrak{H}_{\aleph_0}$ such that $\eta(z) \in \mathfrak{H}_n$ for $z \in Z_n$.

It follows immediately from this theorem, that if V_0, V_1 are two maximal measurable subspaces of $\mathscr{F}$, and $\mathfrak{H}_0, \mathfrak{H}_1$ are the associated Hilbert spaces, then there exists for each $z \in Z$ a unitary $V(z)$ on $\mathfrak{H}(z)$ such that

$$(V\xi)(z) = V(z)\xi(z), \qquad \xi \in \mathfrak{H}_0$$

defines a unitary operator $V;\, \mathfrak{H}_0 \mapsto \mathfrak{H}_1$. Hence

$$\mathfrak{H}_0 = \int_Z^{\oplus} d\mu(z)\, \mathfrak{H}(z)$$

is uniquely defined up to this type of unitary equivalence.

Since μ is a standard measure, it follows that

$$\int_Z^{\oplus} d\mu(z)\, \mathfrak{H}(z)$$

is separable. It is also useful to note that if

$$V \subseteq \int_Z^{\oplus} d\mu(z)\, \mathfrak{H}(z)$$

is a subspace such that

(1) $\{\xi(z); \xi \in V\}$ is dense in $\mathfrak{H}(z)$ for each $z \in Z$ and
(2) if $\xi \in V$ and $f \in L^\infty(Z; d\mu)$ then $(f\xi)(z) = f(z)\xi(z)$ defines an element on V,

then V is dense in $\int_Z^\oplus d\mu(z)\, \mathfrak{H}(z)$ (and this can of course be used to define the direct integral without invoking maximal measurable subspaces). The proof is simple. If $\xi \in V^\perp$ then

$$\int d\mu(z)\, f(z)(\xi(z), \eta(z)) = 0$$

for all $f \in L^\infty(Z; d\mu)$ and $\eta \in V$. Hence $(\xi(z), \eta(z)) = 0$ for z μ-almost everywhere and so $\xi(z) = 0$ for z μ-almost everywhere, i.e., $\xi = 0$. We will encounter similar results later, e.g., Theorem 4.4.5.

We next define the decomposable and diagonalizable operators on the direct integral space

$$\mathfrak{H} = \int_Z^\oplus d\mu(z)\, \mathfrak{H}(z)$$

defined by the maximal measurable subspace V. For each $z \in Z$ let $T(z) \in \mathscr{L}(\mathfrak{H}(z))$. $z \mapsto T(z)$ is called *a measurable family of operators* if $(z \mapsto T(z)\xi(z)) \in V$ for each $\xi \in V$. If this is the case then $z \mapsto \|T(z)\|$ is measurable. If this function is essentially bounded then $\xi \in \mathfrak{H}$ implies that $T\xi \in \mathfrak{H}$, where the latter vector is defined by

$$(T\xi)(z) = T(z)\xi(z).$$

The mapping $\xi \mapsto T\xi$ defines a bounded operator T on $\mathfrak{H}$. Any operator of this form is called *decomposable* and is denoted by

$$T = \int_Z^\oplus d\mu(z)\, T(z).$$

It is clear that if T and S are decomposable operators then $T + S$, TS, and T^* are decomposable, and the following relations hold:

$$T + S = \int_Z^\oplus d\mu(z)\, (T(z) + S(z));$$

$$TS = \int_Z^\oplus d\mu(z)\, (T(z)S(z));$$

$$T^* = \int_Z^\oplus d\mu(z)\, T(z)^*;$$

$$\|T\| = \text{ess sup}\{\|T(z)\|; z \in Z\}.$$

If T is a decomposable operator and in addition $T(z)$ is a scalar operator on $\mathfrak{H}(z)$ for each z then T is called a *diagonalizable operator*. Writing

$$\int_Z^\oplus d\mu(z)\, \mathfrak{H}(z) = \bigoplus_{n=1}^{\aleph_0} (L^2(Z_n, d\mu) \otimes \mathfrak{H}_n)$$

one deduces that the decomposable operators identify with the von Neumann algebra

$$\bigoplus_{n=1}^{\aleph_0} (L^\infty(Z_n, d\mu) \otimes \mathscr{L}(\mathfrak{H}_n))$$

and the diagonalizable operators identify with the abelian von Neumann subalgebra

$$\bigoplus_{n=1}^{\aleph_0} (L^\infty(Z_n, d\mu) \otimes \mathbb{1}_{\mathfrak{H}_n}).$$

The latter algebra is just the center of the former, and we show next that this is the general form of an abelian von Neumann algebra on a separable Hilbert space.

Theorem 4.4.3. *Let $\mathfrak{Z}$ be an abelian von Neumann algebra on a separable Hilbert space $\mathfrak{H}$. It follows that there exists a standard measure μ on a measure space Z, a measurable family $z \mapsto \mathfrak{H}(z)$ of Hilbert spaces, and a unitary map*

$$U; \mathfrak{H} \mapsto \int_Z^\oplus d\mu(z)\, \mathfrak{H}(z)$$

such that $U\mathfrak{Z}U^*$ is just the algebra of diagonalizable operators on

$$\int_Z^\oplus d\mu(z)\, \mathfrak{H}(z).$$

We have already proved this theorem in the case that $\mathfrak{Z}$ has a cyclic and separating vector in the introductory remarks to Section 2.5. In this case all the $\mathfrak{H}(z)$ are one-dimensional. Using the structure theorem for isomorphisms between von Neumann algebras deduced in Theorem 2.4.26, the proof of Theorem 4.4.3 is mainly a technical exercise.

We next mention some results on the form of von Neumann algebras $\mathfrak{M}$ consisting of decomposable operators and containing the diagonalizable operators $\mathfrak{Z}$, i.e., $\mathfrak{Z} \subseteq \mathfrak{M} \subseteq \mathfrak{Z}'$.

Definition 4.4.4. Let μ be a standard measure on Z, $z \mapsto \mathfrak{H}(z)$ a measurable family of Hilbert spaces over Z, with direct integral

$$\mathfrak{H} = \int_Z^\oplus d\mu(z)\, \mathfrak{H}(z).$$

For each z, let $\mathfrak{M}(z)$ be a von Neumann algebra on $\mathfrak{H}(z)$. The family

$$\{\mathfrak{M}(z); z \in Z\}$$

is called a *measurable family of von Neumann algebras* if there exists a sequence

$$A_n = \int_Z^{\oplus} d\mu(z)\, A_n(z)$$

of decomposable operators on $\mathfrak{H}$ such that $\mathfrak{M}(z)$ is the von Neumann algebra generated by $\{A_n(z);\, n = 1, 2, \ldots\}$ for almost every z. In this situation, the von Neumann algebra $\mathfrak{M} = (\{A_n;\, n = 1, 2, \ldots\} \cup \mathfrak{Z})''$ generated by

$$\{A_n;\, n = 1, 2, \ldots\}$$

and the diagonalizable operators is called the *direct integral* of the family $\{\mathfrak{M}(z)\}$ and is written

$$\mathfrak{M} = \int_Z^{\oplus} d\mu(z)\, \mathfrak{M}(z).$$

The justification for this terminology is provided by the following theorem, which shows that the sequence $\{A_n\}$ only plays a spurious role in the definition of $\mathfrak{M}$. The proof makes use of the bicommutant theorem and the fact that $\mathfrak{Z}$ is contained in the center of $\mathfrak{M}$.

Theorem 4.4.5. *Let $\mathfrak{M}$ and $\mathfrak{N}$ be two direct integral von Neumann algebras. It follows that $\mathfrak{M} \subseteq \mathfrak{N}$ if, and only if, $\mathfrak{M}(z) \subseteq \mathfrak{N}(z)$ for almost all $z \in Z$.*

In particular, $\mathfrak{M}$ is uniquely determined by the measurable family $\{\mathfrak{M}(z)\}$, i.e., an operator A is contained in $\mathfrak{M}$ if, and only if, A is decomposable with $A(z) \in \mathfrak{M}(z)$ for almost all $z \in Z$.

As any von Neumann algebra on a separable Hilbert space is separable in the σ-weak topology, it follows from Definition 4.4.4 that a von Neumann algebra $\mathfrak{M}$ on

$$\mathfrak{H} = \int_Z^{\oplus} d\mu(z)\, \mathfrak{H}(z)$$

is a direct integral if and only if

$$\mathfrak{Z} \subseteq \mathfrak{M} \subseteq \mathfrak{Z}'.$$

Hence Theorem 4.4.5 gives a characterization of these von Neumann algebras. The next proposition shows, not surprisingly, that the direct integral operation commutes with the operations of countable intersections and formation of the commutant among these von Neumann algebras.

Proposition 4.4.6. *Let μ be a standard measure and*

$$\mathfrak{H} = \int_Z^{\oplus} d\mu(z)\, \mathfrak{H}(z)$$

a direct integral Hilbert space

(a) *If*

$$\mathfrak{M} = \int_Z^{\oplus} d\mu(z)\, \mathfrak{M}(z)$$

is a direct integral von Neumann algebra on $\mathfrak{H}$ *then* $\mathfrak{M}'$ *also has this property and*

$$\mathfrak{M}' = \int_Z^{\oplus} d\mu(z)\, \mathfrak{M}(z)'.$$

(b) *If*

$$\mathfrak{M}_n = \int_Z^{\oplus} d\mu(z)\, \mathfrak{M}_n(z)$$

is a sequence of direct integral von Neumann algebras then the intersection of the $\mathfrak{M}_n$ *is a direct integral von Neumann algebra and*

$$\bigcap_n \mathfrak{M}_n = \int_Z^{\oplus} d\mu(z) \left(\bigcap_n \mathfrak{M}_n(z)\right).$$

One obtains an interesting corollary of the last result by letting

$$\mathfrak{H} = \int_Z^{\oplus} d\mu(z)\, \mathfrak{H}(z)$$

be the decomposition of $\mathfrak{H}$ corresponding to the center $\mathfrak{Z} = \mathfrak{M} \cap \mathfrak{M}'$ by Theorem 4.4.3. Using (a) and putting $\mathfrak{M}_1 = \mathfrak{M}$ and $\mathfrak{M}_2 = \mathfrak{M}'$, in the last proposition we obtain

$$\mathfrak{Z} = \int_Z^{\oplus} d\mu(z) (\mathfrak{M}(z) \cap \mathfrak{M}(z)').$$

But as $\mathfrak{Z}$ is just the diagonalizable operators on $\mathfrak{H}$, we see that

$$\mathfrak{M}(z) \cap \mathfrak{M}(z)' = \mathbb{C}1_{\mathfrak{H}(z)}$$

for almost all z, i.e., $\mathfrak{M}(z)$ is a factor for almost all z. This reduces the classification problem of von Neumann algebras with separable predual to the two problems of classifying standard measures, and of classifying factors with separable predual. But any standard measure space Z is either countable, with all subsets measurable, or isomorphic to the measure space consisting of the unit interval [0, 1] equipped with its usual Borel structure. Thus the hardest part of the problem is the classification of factors, which we briefly sketched in Section 2.7.3.

We next turn to the most important topic of this subsection, the decomposition of representations. Let

$$\mathfrak{H} = \int_Z^{\oplus} d\mu(z)\, \mathfrak{H}(z)$$

be a direct integral Hilbert space and let π be a nondegenerate representation of a C^*-algebra $\mathfrak{A}$ on $\mathfrak{H}$ such that each operator $\pi(A)$ is decomposable. Using standard separability arguments it is not hard to show that for μ-almost all $z \in Z$ there exists a representation $\pi(z)$ of $\mathfrak{A}$ on $\mathfrak{H}(z)$ such that $z \mapsto \pi(z)(A)$ is measurable and

$$\pi(A) = \int_Z^{\oplus} d\mu(z)\, \pi(z)(A)$$

for all $A \in \mathfrak{A}$. One then of course says that $z \mapsto \pi(z)$ is measurable and writes

$$\pi = \int_Z^{\oplus} d\mu(z)\, \pi(z).$$

The next theorem is an immediate consequence of the abovementioned fact and Theorem 4.4.3.

Theorem 4.4.7. *Let $\mathfrak{A}$ be a C^*-algebra, π a nondegenerate representation of $\mathfrak{A}$ on a separable Hilbert space $\mathfrak{H}$, and $\mathfrak{B}$ an abelian von Neumann subalgebra of $\pi(\mathfrak{A})'$. It follows that there exists a standard measure space Z, a positive, bounded, measure μ on Z, a measurable family $z \mapsto \mathfrak{H}(z)$ of Hilbert spaces on Z, a measurable family $z \mapsto \pi(z)$ of representations of $\mathfrak{A}$ on $\mathfrak{H}(z)$, and a unitary map*

$$U; \mathfrak{H} \mapsto \int_Z^{\oplus} d\mu(z)\, \mathfrak{H}(z)$$

such that $U\mathfrak{B}U^*$ is just the set of diagonalizable operators on

$$\int_Z^{\oplus} d\mu(z)\, \mathfrak{H}(z)$$

and

$$U\pi(A)U^* = \int_Z^{\oplus} d\mu(z)\pi(z)(A)$$

for all $A \in \mathfrak{A}$.

This is the fundamental theorem for the spatial decomposition of representations, and in the general setting there are immediately two natural choices of $\mathfrak{B}$:

(1) $\mathfrak{B}$ is maximal abelian in $\pi(\mathfrak{A})'$ (extremal decomposition);
(2) $\mathfrak{B} = \pi(\mathfrak{A})' \cap \pi(\mathfrak{A})''$ (factor decomposition).

The next corollary is an analogue of Theorem 4.2.5 and Theorem 4.2.11 for spatial decompositions. In fact, it follows from Theorem 4.4.9 in the next subsection that these two theorems are consequences of the corollary in the case where ω is contained in a face F satisfying the separability condition S. This gives a proof of these theorems without the use of barycentric decomposition theory, and in fact generalizes the support conclusion in these theorems to the case where ω generates a representation on a separable Hilbert space.

Corollary 4.4.8. *Let $\mathfrak{A}$ be a C^*-algebra, π a nondegenerate representation of $\mathfrak{A}$ on a separable Hilbert space $\mathfrak{H}$, $\mathfrak{B}$ an abelian von Neumann subalgebra of $\pi(\mathfrak{A})'$, and*

$$\pi = \int_Z^{\oplus} d\mu(z)\, \pi(z)$$

the decomposition of π corresponding to $\mathfrak{B}$.

(a) *the following two statements are equivalent:*

(1) *$\pi(z)$ is irreducible for μ-almost all $z \in Z$;*
(2) *$\mathfrak{B}$ is maximal abelian in $\pi(\mathfrak{A})'$.*

(b) *if $\mathfrak{B} = \pi(\mathfrak{A})' \cap \pi(\mathfrak{A})''$ then the $\pi(z)$ are factor representations for μ-almost all $z \in Z$, and in this case*

$$\pi(\mathfrak{A})'' = \int_Z^{\oplus} d\mu(z)\, (\pi(z)(\mathfrak{A}))''.$$

PROOF. (a1) $\Rightarrow$ (a2) Assume that the $\pi(z)$ are irreducible for μ-almost all z and let $A \in \mathfrak{B}' \cap \pi(\mathfrak{A})'$. As $A \in \mathfrak{B}'$ it follows that A is decomposable,

$$A = \int_Z^{\oplus} d\mu(z)\, A(z).$$

But for any $B \in \mathfrak{A}$ we have

$$\pi(B)A = A\pi(B).$$

Therefore

$$\pi(z)(B)A(z) = A(z)\pi(z)(B)$$

for μ-almost all z. As $\pi(\mathfrak{A})''$ is σ-weakly separable it follows that $A(z) \in \pi(z)(\mathfrak{A})' = \mathbb{C}1_{\mathfrak{H}(z)}$ for μ-almost all z. Hence A is diagonalizable and $A \in \mathfrak{B}$.

(a2) $\Rightarrow$ (a1) If $\mathfrak{B}$ is maximal abelian in $\pi(\mathfrak{A})'$, then $\mathfrak{B}'$ is the von Neumann algebra generated by $\pi(\mathfrak{A})$ and $\mathfrak{B}$. Definition 4.4.4 then implies that

$$\mathfrak{B}' = \int_Z^{\oplus} d\mu(z)(\pi(z)(\mathfrak{A}))''$$

and hence, by Proposition 4.4.6(a)

$$\mathfrak{B} = \int_Z^{\oplus} d\mu(z)(\pi(z)(\mathfrak{A}))'.$$

But since $\mathfrak{B}$ consists of just the diagonalizable operators on $\mathfrak{H}$, it follows that $(\pi(z)(\mathfrak{A}))' = \mathbb{C}1_{\mathfrak{H}(z)}$ for μ-almost all z.

(b) This is an immediate consequence of Proposition 4.4.6, (a) and (b), and Definition 4.4.4.

4.4.2. Spatial Decomposition and Decomposition of States

In this subsection we connect the theory of spatial decomposition developed in 4.4.1 with the decomposition theory for states developed earlier. We also use this connection to prove two results on support properties of the orthogonal measures which we could not obtain earlier.

Let $\mathfrak{A}$ be a C^*-algebra with identity, and let ω be a state over $\mathfrak{A}$ which is contained in a face $F \subseteq E_{\mathfrak{A}}$ satisfying the separability condition S, Definition 4.1.32. If $\mu \in M_\omega(E_{\mathfrak{A}})$ is a Baire measure with barycenter ω, it follows from Proposition 4.1.34 that μ may be viewed as a measure on F. Hence μ is a standard measure on F. For $\omega' \in F$ define $\mathfrak{H}(\omega') = \mathfrak{H}_{\omega'}$ as the representation Hilbert space associated with ω'. We shall show that $\omega' \in F \mapsto \mathfrak{H}(\omega')$ is a measurable family of Hilbert spaces in the sense of Definition 4.4.1B. To this end, let $\{A_{n,k}\}_{n,k \geq 1}$ be the sequence of elements of $\mathfrak{A}$ used in characterizing F in the proof of Proposition 4.1.34, with the minor difference that we now assume the $\{A_{n,k}\}_{k \geq 1}$ are dense in the whole ideal $\mathfrak{I}_n$ for each n. Lemma 4.1.33 then implies that $\{\pi_{\omega'}(A_{n,k})\}_{n,k \geq 1}$ is a strongly dense sequence in $\pi_{\omega'}(\mathfrak{A})''$, for each $\omega' \in F$, and in particular $\{\pi_{\omega'}(A_{n,k})\Omega_{\omega'}\}_{n,k \geq 1}$ is dense in $\mathfrak{H}(\omega')$ for each $\omega' \in F$. As

$$\begin{aligned}\omega' \in F \mapsto (\pi_{\omega'}(A_{m,l})\Omega_{\omega'}, \pi_{\omega'}(A_{n,k})\Omega_{\omega'}) \\ = \omega'(A_{m,l}^* A_{n,k})\end{aligned}$$

is continuous, hence measurable, $\omega' \in F \mapsto \mathfrak{H}(\omega')$ is a measurable family of Hilbert spaces and we may form the direct integral

$$\mathfrak{H}_\mu = \int_F^\oplus d\mu(\omega')\, \mathfrak{H}(\omega').$$

For each $A \in \mathfrak{A}$, $\omega' \mapsto \pi_{\omega'}(A)$ is a measurable family of operators on $\mathfrak{H}$, and we may form the direct integral representation

$$\pi_\mu = \int_F^\oplus d\mu(\omega')\, \pi_{\omega'}.$$

Let $\Omega_\mu \in \mathfrak{H}_\mu$ be the unit vector defined by

$$\Omega_\mu = \int_F^\oplus d\mu(\omega')\, \Omega_{\omega'}.$$

Then

$$\begin{aligned}(\Omega_\mu, \pi_\mu(A)\Omega_\mu) &= \int_F d\mu(\omega')\, (\Omega_{\omega'}, \pi_{\omega'}(A)\Omega_{\omega'}) \\ &= \int_F d\mu(\omega')\, \omega'(A) \\ &= \omega(A) \\ &= (\Omega_\omega, \pi_\omega(A)\Omega_\omega)\end{aligned}$$

for all $A \in \mathfrak{A}$. Let

$$E_\mu = [\pi_\mu(\mathfrak{A})\Omega_\mu] \in \pi_\mu(\mathfrak{A})'.$$

It follows from Theorem 2.3.16 that the isometry U from $\mathfrak{H}_\omega$ into $\mathfrak{H}_\mu$ with range $E_\mu \mathfrak{H}_\mu$ defined by

$$U\pi_\omega(A)\Omega_\omega = \pi_\mu(A)\Omega_\mu$$

establishes a unitary equivalence between the representations π_ω and $E_\mu \pi_\mu$. The remarkable fact is now that $E_\mu = \mathbb{1}_{\mathfrak{H}_\mu}$ if, and only if, μ is an orthogonal measure on $E_\mathfrak{A}$. Among other things this provides the structural connection between spatial decompositions and decomposition of states, i.e., the connection between Theorem 4.1.25 and Theorem 4.4.7.

Theorem 4.4.9 (Effros). *Let $\mathfrak{A}$ be a C^*-algebra with identity, $F \subseteq E_\mathfrak{A}$ a face satisfying the separability condition* S, *μ a Baire probability measure on $E_\mathfrak{A}$ with barycenter $\omega \in F$. Let*

$$\pi_\mu = \int_F^\oplus d\mu(\omega')\, \pi_{\omega'}$$

be the direct integral representation of $\mathfrak{A}$ on

$$\mathfrak{H}_\mu = \int_F^\oplus d\mu(\omega')\, \mathfrak{H}_{\omega'}$$

described before the theorem, and let E_μ be the range projection of the isometry U; $\mathfrak{H}_\omega \mapsto \mathfrak{H}_\mu$ which establishes the canonical unitary equivalence between π_ω and $E_\mu \pi_\mu$. The following conditions are equivalent:

(1) $E_\mu = \mathbb{1}_{\mathfrak{H}_\mu}$;
(2) *μ is an orthogonal measure.*

If these conditions are satisfied then

$$\pi_\mu = \int_F^\oplus d\mu(\omega')\, \pi_{\omega'}$$

is the direct integral decomposition of π_ω with respect to the abelian von Neumann subalgebra $\mathfrak{B}_\mu \subseteq \pi_\omega(\mathfrak{A})'$ corresponding to the orthogonal measure μ.

PROOF. (1) $\Rightarrow$ (2) With the notation established before the theorem, condition (1) is equivalent to the cyclicity of Ω_μ for π_μ or the unitarity of U. We want to show that the map

$$f \in L^\infty(\mu) \mapsto \kappa_\mu(f) \in \pi_\omega(\mathfrak{A})'$$

defined in Lemma 4.1.21, is a *-morphism because it then follows from Tomita's theorem, Proposition 4.1.22, that μ is orthogonal.

Let $f \mapsto T(f)$ be the natural *-isomorphism between $L^\infty(\mu)$ and the diagonalizable operators on

$$\mathfrak{H}_\mu = \int_F^\oplus d\mu(\omega')\, \mathfrak{H}_{\omega'},$$

i.e., for $f \in L^\infty(\mu)$

$$T(f) = \int_F^\oplus d\mu(\omega')\, f(\omega')\mathbb{1}_{\mathfrak{H}_{\omega'}}.$$

Now one has

$$\begin{aligned}
(\pi_\mu(A)\Omega_\mu,\, U\kappa_\mu(f)U^*\pi_\mu(B)\Omega_\mu) &= (\pi_\omega(A)\Omega_\omega,\, \kappa_\mu(f)\pi_\omega(B)\Omega_\omega) \\
&= (\Omega_\omega,\, \kappa_\mu(f)\pi_\omega(A^*B)\Omega_\omega) \\
&= \int_F d\mu(\omega')\, f(\omega')\omega'(A^*B) \\
&= \int_F d\mu(\omega')\, (\pi_{\omega'}(A)\Omega_{\omega'},\, f(\omega')\mathbb{1}_{\mathfrak{H}_{\omega'}}\pi_{\omega'}(B)\Omega_{\omega'}) \\
&= (\pi_\mu(A)\Omega_\mu,\, T(f)\pi_\mu(B)\Omega_\mu).
\end{aligned}$$

It follows that $\kappa_\mu(f) = U^*T(f)U$, and so $f \mapsto \kappa_\mu(f)$ is a *-morphism. Note also that $U\mathfrak{B}_\mu U^*$ is just the set of diagonalizable operators on $\mathfrak{H}_\mu$, and

$$\pi_\omega = \int_F^\oplus d\mu(\omega')\, \pi_{\omega'}$$

is just the spatial decomposition of π_ω corresponding to $\mathfrak{B}_\mu \subseteq \pi_\omega(\mathfrak{A})'$ in this case. This proves the last statement of the theorem.

(2) $\Rightarrow$ (1) First note that the linear span of

$$\{T(f)\pi_\mu(A_{n,k})\Omega_\mu;\, f \in L^\infty(\mu);\, n, k \geq 1\}$$

is dense in $\mathfrak{H}_\mu = \int_F^\oplus d\mu(\omega')\, \mathfrak{H}_{\omega'}$ by the remark after Theorem 4.4.2. Thus to prove that Ω_μ is cyclic for π_μ we must show that for any $\varepsilon > 0$, $f \in L^\infty(\mu)$, and $A \in \mathfrak{A}$ there exists a $B \in \mathfrak{A}$ such that

$$\|\pi_\mu(B)\Omega_\mu - T(f)\pi_\mu(A)\Omega_\mu\| < \varepsilon.$$

But

$$\begin{aligned}
\|\pi_\mu(B)\Omega_\mu - T(f)\pi_\mu(A)\Omega_\mu\|^2 &= (\Omega_\mu, \pi_\mu(B^*B)\Omega_\mu) - (\Omega_\mu, T(f)\pi_\mu(B^*A)\Omega_\mu) \\
&\quad - (\Omega_\mu, T(\bar{f})\pi_\mu(A^*B)\Omega_\mu) + (\Omega_\mu, T(\bar{f}f)\pi_\mu(A^*A)\Omega_\mu) \\
&= \int_F d\mu(\omega')\, \omega'(B^*B) - \int_F d\mu(\omega')\, f(\omega')\omega'(B^*A) \\
&\quad - \int_F d\mu(\omega')\, \overline{f(\omega')}\omega'(A^*B) + \int_F d\mu(\omega')\, |f(\omega')|^2\omega'(A^*A) \\
&= (\Omega_\omega, \pi_\omega(B^*B)\Omega_\omega) - (\Omega_\omega, \kappa_\mu(f)\pi_\omega(B^*A)\Omega_\omega) \\
&\quad - (\Omega_\omega, \kappa_\mu(\bar{f})\pi_\omega(A^*B)\Omega_\omega) + (\Omega_\omega, \kappa_\mu(\bar{f}f)\pi_\omega(A^*A)\Omega_\omega).
\end{aligned}$$

If μ is orthogonal then

$$\kappa_\mu(\bar{f}f) = \kappa_\mu(f)^*\kappa_\mu(f)$$

by Tomita's theorem, and hence one obtains

$$\|\pi_\mu(B)\Omega_\mu - T(f)\pi_\mu(A)\Omega_\mu\|^2 = \|\pi_\omega(B)\Omega_\omega - \kappa_\mu(f)\pi_\omega(A)\Omega_\omega\|^2.$$

But as Ω_ω is cyclic for π_ω, the last expression can be made less than ε^2 by an appropriate choice of B.

Remark. Even if μ is not an orthogonal measure it is clear from the last part of the proof that the span of the subspaces $T(f)E_\mu \mathfrak{H}_\mu$, $f \in L^\infty(\mu)$, are dense in $\mathfrak{H}_\mu$. As $T(f) \in \pi_\mu(\mathfrak{A})'$ it follows that π_ω and π_μ are quasi-equivalent for all measures $\mu \in M_\omega(E_{\mathfrak{A}})$ (see Definition 2.4.25).

If $\mathfrak{A}$ is a C^*-algebra with identity, ω a state of $\mathfrak{A}$ such that $\mathfrak{H}_\omega$ is separable and $\mu \in \mathcal{O}_\omega(E_{\mathfrak{A}})$ then μ is isomorphic to a standard measure by the following reasoning: the abelian subalgebra $\mathfrak{B}$ of $\pi_\omega(\mathfrak{A})'$ corresponding to μ is isomorphic to $L^\infty(\mu)$ by Proposition 4.1.22. But as $\mathfrak{B}$ acts on the separable Hilbert space $\mathfrak{H}_\omega$, it follows that $L^\infty(\mu)$ contains a σ-weakly dense separable C^*-subalgebra $\mathfrak{C}$. If $\mathfrak{C} = C(X)$ is the Gelfand representation, of $\mathfrak{C}$, X is a second countable compact Hausdorff space, and hence metrizable. Now μ defines a normal state on $L^\infty(\mu)$, and hence a state on $C(X)$ by restriction. This state is represented by a regular Borel measure μ_0 on X, and the inclusion of $C(X)$ in $L^\infty(\mu)$ defines a *-isomorphism of $L^\infty(X; d\mu_0)$ onto $L^\infty(E_{\mathfrak{A}}; d\mu)$. Identifying measurable subsets modulo null sets with projections, this *-isomorphism preserves measure, and hence μ is *-isomorphic with the standard measure μ_0 in this sense. Hence, repeating the procedure in the introduction to this subsection and the method in the proof of Theorem 4.4.9, we may make the identification

$$\pi_\omega = \int_{E_{\mathfrak{A}}}^{\oplus} d\mu(\omega')\, \pi_{\omega'}.$$

We then obtain the following partial generalizations of Theorems 4.2.5 and 4.2.11 and Proposition 4.3.2.

Theorem 4.4.10. *Let $\mathfrak{A}$ be a C^*-algebra with identity, and ω a state over $\mathfrak{A}$ such that the corresponding representation Hilbert space $\mathfrak{H}_\omega$ is separable. Let $\mathfrak{B}$ be an abelian von Neumann subalgebra of $\pi_\omega(\mathfrak{A})'$, and $\mu \in \mathcal{O}_\omega(E_{\mathfrak{A}})$ the corresponding orthogonal measure.*

(1) *if $\mathfrak{B}$ is maximal abelian in $\pi_\omega(\mathfrak{A})'$ then there exists a μ-measurable subset $B \subseteq E_{\mathfrak{A}}$ such that B consists of pure states and $\mu(B) = 1$;*
(2) *if $\mathfrak{B} = \pi_\omega(\mathfrak{A})' \cap \pi_\omega(\mathfrak{A})''$, then there exists a μ-measurable subset $B \subseteq E_{\mathfrak{A}}$ such that B consists of factor states and $\mu(B) = 1$;*
(3) *if G is a group, $g \in G \mapsto \tau_g \in \mathrm{Aut}(\mathfrak{A})$ a representation of G as *-automorphisms of $\mathfrak{A}$, $\omega \in E_{\mathfrak{A}}{}^G$, and $\mathfrak{B}$ is maximal abelian in*

$$\{\pi_\omega(\mathfrak{A}) \cup U_\omega(G)\}'$$

then there exists a μ-measurable subset $B \subseteq E_{\mathfrak{A}}$ consisting of G-ergodic states such that $\mu(B) = 1$.

PROOF. (1) and (2) follow immediately from Corollary 4.4.8. As for (3) note that the support of μ is contained in $E_{\mathfrak{A}}{}^G$ by Proposition 4.3.2, and we may write

$$\pi_\omega = \int_{E_{\mathfrak{A}}{}^G}^{\oplus} d\mu(\omega')\, \pi_{\omega'},$$

$$\Omega_\omega = \int_{E_{\mathfrak{A}}{}^G}^{\oplus} d\mu(\omega')\Omega_{\omega'}.$$

But then

$$\begin{aligned} U_\omega(g)\pi_\omega(A)\Omega_\omega &= \pi_\omega(\tau_g(A))\Omega_\omega \\ &= \int_{E_{\mathfrak{A}}{}^G}^{\oplus} d\mu(\omega')\pi_{\omega'}(\tau_g(A))\Omega_{\omega'} \\ &= \int_{E_{\mathfrak{A}}{}^G}^{\oplus} d\mu(\omega')\, U_{\omega'}(g)\pi_{\omega'}(A)\Omega_{\omega'} \end{aligned}$$

for all $A \in \mathfrak{A}$ and hence the operators $U_\omega(g)$ are decomposable, with

$$U_\omega(g) = \int_{E_{\mathfrak{A}}{}^G}^{\oplus} d\mu(\omega')\, U_{\omega'}(g).$$

But as $\mathfrak{B}$ is maximal abelian in $\{\pi_\omega(\mathfrak{A}) \cup U_\omega(G)\}'$, it follows that the set of decomposable operators

$$\mathfrak{B}' = \int_{E_{\mathfrak{A}}{}^G}^{\oplus} d\mu(\omega')\, \mathscr{L}(\mathfrak{H}_{\omega'})$$

is the von Neumann algebra generated by $\mathfrak{B}$ itself and the operators

$$\pi_\omega(A) = \int_{E_{\mathfrak{A}}{}^G}^{\oplus} d\mu(\omega')\pi_{\omega'}(A), \qquad A \in \mathfrak{A},$$

$$U_\omega(g) = \int_{E_{\mathfrak{A}}{}^G}^{\oplus} d\mu(\omega')U_{\omega'}(g), \qquad g \in G.$$

It follows from Theorem 4.4.5 that $\{\pi_{\omega'}(\mathfrak{A}) \cup U_{\omega'}(G)\}'' = \mathscr{L}(\mathfrak{H}_{\omega'})$ for μ-almost all ω', but the latter condition is equivalent to the ergodicity of ω' by Theorem 4.3.17.

We next consider decompositions at infinity. Let

$$(\mathfrak{A}, \{\mathfrak{A}_\alpha\}_{\alpha \in I})$$

be a quasi-local algebra, as introduced in Definition 2.6.3. Let ω be a state over $\mathfrak{A}$, and

$$\mathfrak{Z}_\omega{}^\perp = \bigcap_{\alpha \in L} \left(\bigcup_{\beta \perp \alpha} \pi_\omega(\mathfrak{A}_\beta) \right)''$$

the corresponding algebra at infinity (Definition 2.6.4). In Theorem 2.6.5 we proved that $\mathfrak{Z}_\omega{}^\perp$ is a subalgebra of the center $\mathfrak{Z}_\omega$ of the representation π_ω, and gave a characterization of states with trivial algebra at infinity. We now prove, under suitable separability assumptions, that the orthogonal measure

μ corresponding to the decomposition of ω at infinity, i.e., the decomposition defined by $\mathfrak{Z}_\omega{}^\perp$, is supported by a measurable subset of states with trivial algebra at infinity.

Theorem 4.4.11. *Let*

$$(\mathfrak{A}, \{\mathfrak{A}_\alpha\}_{\alpha \in I})$$

be a quasilocal algebra, ω a state of $\mathfrak{A}$, and assume:

(1) *there exists an increasing sequence $\{\alpha_n\}_{n \geq 1}$, in I such that for any $\alpha \in I$ there exists an n with $\alpha < \alpha_n$;*
(2) *each $\mathfrak{A}_{\alpha_n}$ contains a separable closed twosided ideal $\mathfrak{J}_n$ such that $\|\omega|_{\mathfrak{J}_n}\| = 1$.*

It follows that the orthogonal measure μ corresponding to the algebra at infinity $\mathfrak{Z}_\omega{}^\perp$ is supported by a measurable set consisting of states with trivial algebra at infinity.

PROOF. Since

$$\mathfrak{Z}_{\omega'}^\perp = \bigcap_{n \geq 1} \left(\bigcup_{\beta \perp \alpha_n} \pi_{\omega'}(\mathfrak{A}_\beta) \right)''$$

for any $\omega' \in E_\mathfrak{A}$, by (1) of Definition 2.6.3, we might as well assume that $I = \{1, 2, \ldots\}$ and set $\mathfrak{A}_n = \mathfrak{A}_{\alpha_n}$. By assumption (2), the state ω is contained in a face $F \subseteq E_\mathfrak{A}$ satisfying the separability condition S (Definition 4.1.32). Hence Theorem 4.4.9 implies that the direct integral decomposition of π_ω corresponding to $\mathfrak{Z}_\omega{}^\perp$ has the form

$$\pi_\omega = \int_F^\oplus d\mu(\omega')\, \pi_{\omega'}.$$

Now, for any $\omega' \in F$ let

$$\mathfrak{B}_{\omega', n} = \left\{ \bigcup_{\beta \perp \alpha_n} \pi_{\omega'}(\mathfrak{A}_\beta) \right\}''.$$

It follows that

$$\mathfrak{Z}_{\omega'}^\perp = \bigcap_n \mathfrak{B}_{\omega', n}.$$

As $\mathfrak{Z}_\omega{}^\perp \subseteq \mathfrak{B}_{\omega, n}$ for each n, it then follows from Definition 4.4.4 that

$$\mathfrak{B}_{\omega, n} = \int_F^\oplus d\mu(\omega')\, \mathfrak{B}_{\omega', n}.$$

Proposition 4.4.6(b) now implies

$$\begin{aligned} \mathfrak{Z}_\omega{}^\perp &= \bigcap_n \mathfrak{B}_{\omega, n} \\ &= \int_F^\oplus d\mu(\omega') \left(\bigcap_n \mathfrak{B}_{\omega', n} \right) \\ &= \int_F^\oplus d\mu(\omega')\, \mathfrak{Z}_{\omega'}^\perp. \end{aligned}$$

But as $\mathfrak{Z}_\omega^\perp$ are just the diagonalizable operators in the decomposition

$$\pi_\omega = \int_F^\oplus d\mu(\omega')\, \pi_{\omega'},$$

it follows that $\mathfrak{Z}_{\omega'}^\perp = \mathbb{C}1_{\mathfrak{H}_{\omega'}}$ for μ-almost all $\omega' \in F$.

To conclude this discussion we use spatial decomposition theory to improve a previous result, Corollary 4.3.39, on ergodic decomposition with respect to a subgroup. The following result describes a situation in which an ergodic decomposition is in fact a decomposition into weakly mixing states, i.e., states with a higher degree of ergodicity than would be expected.

Theorem 4.4.12. *Let $\omega \in \mathscr{E}(E_{\mathfrak{A}}^G)$, where G is a second countable locally compact abelian group acting strongly continuously on a C^*-algebra $\mathfrak{A}$ with identity. Assume that the point spectrum $\sigma_p(U_\omega)$ of U_ω is a discrete subgroup of $\hat{G}$ and let H_ω denote its annihilator. Assume that $(\mathfrak{A}, \omega)$ is G_Γ-abelian, that ω is in a face F satisfying the separability condition S, and let $\tilde{\omega} \in \mathscr{E}(E_{\mathfrak{A}}^{H_\omega}) \cap F$ be a state such that $\dot{t} \in G/H_\omega \mapsto \tau_t^*\tilde{\omega}$ is measurable with respect to the Haar measure $d\dot{t}$ on G/H_ω and*

$$\omega(A) = \int_{G/H_\omega} d\dot{t}\, \tilde{\omega}(\tau_t(A))$$

for all $A \in \mathfrak{A}$. It follows that $\tilde{\omega}$ is weakly mixing for H_ω, i.e.,

$$M(|\tilde{\omega}(A\tau(B)) - \tilde{\omega}(A)\tilde{\omega}(B)|) = 0$$

for all $A, B \in \mathfrak{A}$ and any invariant mean M over $C_b(H_\omega)$.

PROOF. First note that as $\sigma_p(U_\omega)$ is discrete it is automatically closed. Now as G is second countable it follows from the definition of the topology on the dual $\hat{G}$ that $\hat{G}$ is second countable. Thus as $\sigma_p(U_\omega) \subseteq \hat{G}$ is discrete it must also be countable. Thus it follows from Corollary 4.3.32 that the pair $(\mathfrak{A}, \omega)$ is H_ω-abelian.

Now the existence of $\tilde{\omega}$ follows from Corollary 4.3.39. It also follows from the proof of this corollary that

$$f \in C(E_{\mathfrak{A}}) \mapsto \mu(f) = \int_{G/H_\omega} d\dot{t}\, f(\tau_t^*\tilde{\omega})$$

defines an orthogonal measure μ on $E_{\mathfrak{A}}$. But as G/H_ω is second countable μ is standard and it follows from Theorem 4.4.9 that the spatial decomposition of π_ω corresponding to μ exists and can be written as

$$\mathfrak{H}_\omega = \int_{G/H_\omega}^\oplus d\dot{t}\, \mathfrak{H}_{\tau_t^*\tilde{\omega}},$$

$$\pi_\omega = \int_{G/H_\omega}^\oplus d\dot{t}\, \pi_{\tau_t^*\tilde{\omega}}.$$

Then the vectors of the form

$$\int_{G/H_\omega}^{\oplus} d\dot{t}\, f(t)\pi_{\tau_t{}^*\tilde{\omega}}(A)\Omega_{\tau_t{}^*\tilde{\omega}},$$

where $A \in \mathfrak{A}$ and $f \in L^\infty(G/H)$ are dense in $\mathfrak{H}_\omega$, by the remark after Theorem 4.4.2.

Next, as G is a second countable topological group, G is metrizable (see Notes and Remarks), and as G is locally compact, it is complete in this metric. Thus G is a Polish topological space. As the quotient map $G \mapsto G/H_\omega$ is continuous and open, it follows from the measure theoretic result (2) used in the proof Proposition 3.2.72 that there exists a Borel map η; $G/H_\omega \mapsto G$ such that $\dot{\eta}(\dot{t}) = \dot{t}$. We now define an isometry U; $\mathfrak{H}_{\tilde{\omega}} \mapsto \mathfrak{H}_\omega = \int_{G/H}^{\oplus} d\dot{t}\, \mathfrak{H}_{\tau_t{}^*\tilde{\omega}}$ as follows. For $A \in \mathfrak{A}$ set

$$U\pi_{\tilde{\omega}}(A)\Omega_{\tilde{\omega}} = \int_{G/H}^{\oplus} d\dot{t}\, \pi_{\tau_t{}^*\tilde{\omega}}(\tau_{\eta(\dot{t})^{-1}}(A))\Omega_{\tau_t{}^*\tilde{\omega}}.$$

This actually defines a vector in

$$\mathfrak{H}_\omega = \int_{G/H}^{\oplus} d\dot{t}\, \mathfrak{H}_{\tau_t{}^*\tilde{\omega}}$$

because if $f \in L^\infty(G/H)$ and $B \in \mathfrak{A}$ then

$$\begin{aligned} \dot{t} \mapsto (f(\dot{t})\pi_{\tau_t{}^*\tilde{\omega}}(B)\Omega_{\tau_t{}^*\tilde{\omega}},\ &\pi_{\tau_t{}^*\tilde{\omega}}(\tau_{\eta(\dot{t})^{-1}}(A))\Omega_{\tau_t{}^*\tilde{\omega}}) \\ &= \overline{f(\dot{t})}\tilde{\omega}(\tau_{\eta(\dot{t})}(B^*)A) \end{aligned}$$

is a Borel map due to the strong continuity of τ. Hence $U\pi_{\tilde{\omega}}(A)\Omega_{\tilde{\omega}} \in \mathfrak{H}_\omega$ by Definition 4.4.1B.

Next, note that

$$\begin{aligned} \|U\pi_{\tilde{\omega}}(A)\Omega_{\tilde{\omega}}\|^2 &= \int_{G/H} d\dot{t}\, \tau_t{}^*\tilde{\omega}(\tau_{\eta(\dot{t})^{-1}}(A^*A)) \\ &= \int_{G/H} d\dot{t}\, \tilde{\omega}(A^*A) \\ &= \tilde{\omega}(A^*A) = \|\pi_{\tilde{\omega}}(A)\Omega_{\tilde{\omega}}\|^2. \end{aligned}$$

Hence U is a well-defined isometry from $\mathfrak{H}_{\tilde{\omega}}$ into $\mathfrak{H}_\omega$. Also, for $t \in H_\omega$ we have

$$\begin{aligned} UU_{\tilde{\omega}}(s)\pi_{\tilde{\omega}}(A)\Omega_{\tilde{\omega}} &= U\pi_{\tilde{\omega}}(\tau_s(A))\Omega_{\tilde{\omega}} \\ &= \int_{G/H}^{\oplus} d\dot{t}\, \pi_{\tau_t{}^*\tilde{\omega}}(\tau_{\eta(\dot{t})^{-1}s}(A))\Omega_{\tau_t{}^*\tilde{\omega}} \\ &= \int_{G/H}^{\oplus} d\dot{t}\, U_{\tau_t{}^*\tilde{\omega}}(s)\pi_{\tau_t{}^*\tilde{\omega}}(\tau_{\eta(\dot{t})^{-1}}(A))\Omega_{\tau_t{}^*\tilde{\omega}} \\ &= U_\omega(s)U\pi_{\tilde{\omega}}(A)\Omega_{\tilde{\omega}}, \end{aligned}$$

where the last identity comes from the identification

$$\Omega_\omega = \int_{G/H}^{\oplus} d\dot{t}\, \Omega_{\tau_t{}^*\tilde{\omega}},$$

which implies

$$U_{\omega}(s) = \int_{G/H}^{\oplus} d\dot{t} \; U_{\tau_t{}^*\tilde{\omega}}(s).$$

It follows that

$$UU_{\tilde{\omega}}(t) = U_{\omega}(t)U, \quad t \in H_{\omega},$$

i.e., $t \mapsto U_{\tilde{\omega}}(t)$ is unitarily equivalent with a subrepresentation of $t \mapsto U_{\omega}(t)$. But $\sigma_{\mathrm{p}}(U_{\omega})$ is closed and countable by the argument given at the beginning of the proof. Thus it follows from the remark after the proof of Theorem 4.3.27 that U_{ω}, restricted to H_{ω}, has no point spectrum except at zero. As $U_{\tilde{\omega}}$ is unitarily equivalent to a subrepresentation of $U_{\omega}|_{H_{\omega}}$ it follows that $U_{\tilde{\omega}}$ has no point spectrum except at zero. But $\tilde{\omega}$ is H_{ω}-ergodic and hence it follows from Proposition 4.3.36 that $\tilde{\omega}$ is weakly H_{ω}- mixing, i.e.,

$$M(|\tilde{\omega}(A\tau(B)) - \tilde{\omega}(A)\tilde{\omega}(B)|) = 0$$

for all $A, B \in \mathfrak{A}$, and all invariant means M on $C_{\mathrm{b}}(H_{\omega})$.

Notes and Remarks

Section 4.1.2

The modern theory of barycentric decomposition was initiated by the work of Choquet. In 1956 he established that every point ω in a metrizable compact convex set is the barycenter of a probability measure μ_ω supported by the G_δ-set of extreme points $\mathscr{E}(K)$. In the same period he introduced the general notion of a simplex and proved that each $\omega \in K$ is the barycenter of a unique maximal measure if, and only if, K is a simplex. These versions of Theorems 4.1.11 and 4.1.15 laid the foundations for subsequent development. (See [Cho 1], [Cho 2], [Cho 3], [Cho 4].)

The next important contribution to the development of the subject was due to Bishop and de Leeuw [Bis 1]. These authors were the first to introduce an order relation on the positive measures $M_+(K)$. Their order relation $\gg$ differed from the relation $\succ$ that we have used and was defined by

$$\mu \gg \nu$$

if, and only if,

$$\mu(f^2) \geq \nu(f^2)$$

for all $f \in A(K)$. These two relations are in general distinct (see Example 4.1.29 which is due to Skau [Ska 1]). The relation $\succ$ was subsequently introduced by Choquet [Cho 5]. The existence of an order relation allowed the use of Zorn's lemma to establish the existence of maximal measures with a given barycenter (Proposition 4.1.3) and to discuss decompositions for nonmetrizable K. In particular, Bishop and de Leeuw demonstrated that the maximal measures in $M_1(K)$ are pseudo supported by $\mathscr{E}(K)$ and constructed examples to show that $\mathscr{E}(K)$ can be arbitrarily bad, from the point of view of measure theory, if K is nonmetrizable. Mokobodski later constructed the example, mentioned in the remark preceding Corollary 4.1.18, of a simplex K such that $\mathscr{E}(K)$ is a Borel subset of K but $\mu_\omega(\mathscr{E}(K)) = 0$ [Mok 1]. In 1972 MacGibbon established the rather surprising result that if $\mathscr{E}(K)$ is a Baire set then K is automatically metrizable. In fact, for metrizability it suffices that $\mathscr{E}(K)$ be an analytic set [MacG 1].

The first general review of the developments of decomposition theory in the period 1956–1963 was given by Choquet and Meyer [Cho 6]. This article included many new features and refinements of the theory and remained for several years a principal source of access to the theory. Subsequently there have appeared many excellent descriptions of the subject, e.g., [[Alf 1]], [[Cho 1]], [Lan 1], [[Phe 1]].

Section 4.1.3

Algebraic decomposition theory is more recent and expository articles are scarcer. Chapter 3 of Sakai's book [[Sak 1]] contains a partial description of the theory and our discussion has been largely influenced by the article of Skau [Ska 1].

The first author to examine decomposition theory in terms of measures on the state space $E_{\mathfrak{A}}$ of a C^*-algebra $\mathfrak{A}$ was Segal in 1951 [Seg 2] but the general theory was developed much later. There were two basic sources of motivation: firstly, the introduction by Tomita, in 1956, of the concept of orthogonal measure [Tomi 2] and secondly, the analysis of invariant states by Kastler, Robinson and Ruelle, in 1966 (see notes to Section 4.3.1). In particular, Ruelle was the first to apply the Choquet theory to the decomposition of states.

Tomita proved the one-to-one correspondence between the orthogonal measures in $\mathcal{O}_\omega(E_{\mathfrak{A}})$ and abelian von Neumann subalgebras of $\pi_\omega(\mathfrak{A})'$, i.e., the correspondence between the first and second sets in Theorem 4.1.25. The full version of Theorem 4.1.25 combines contributions from various sources and it is not clear to whom it should be most rightly attributed. The correspondence between the second and third sets occurs explicitly in the work of Ruelle [Rue 1] and implicitly in the article of Skau [Ska 1]. These two references also contain versions of the second basic structure theorem for orthogonal measures, Theorem 4.1.28.

Condition (4) of Theorem 4.1.28 is equivalent to $\mu \gg \nu$, where $\gg$ is the Bishop-de Leeuw order relation defined in the notes to Section 4.1.2. Thus for orthogonal measures the two order relations $\gg$ and $\succ$ coincide. This equivalence was first remarked by Skau [Ska 1]. Lemma 4.1.27, which is crucial to our proofs of the structure theorems, was given implicitly by Ruelle [Rue 1] and explicitly by Skau [Ska 2].

Section 4.1.4

The results on n-dimensionally homogeneous C^*-algebras mentioned prior to Example 4.1.31 were proved independently by Fell [Fell 1] and Takesaki and Tomiyama [Tak 4]. These papers also contain an analysis of the global structure of n-dimensionally homogeneous C^*-algebras and contrary to the von Neumann result these are not generally of the form $M_n \otimes C_0(X)$, where

X is a locally compact Hausdorff space. There exists, for example, a 3-dimensionally homogeneous C^*-algebra over the 6-sphere with no projections except 0 and 1.

The separability condition S was first introduced by Ruelle [Rue 2], who had applications to locally normal states on quasi-local algebras in mind. A generalization of the condition occurs in [Rue 1]. Although Ruelle did not emphasize the geometric and measure-theoretic properties of the set F of states satisfying the separability condition S the proof of Proposition 4.1.34 is essentially contained in [Rue 1] and [Rue 2].

The fact that the set N_m of normal states on a von Neumann algebra $\mathfrak{M}$ is sequentially complete was proved by Akemann [Ake 2].

Details concerning analytic sets and associated concepts are given in [[Cho 1]]. In particular, we have used the capacity theorem, Theorem 9.7 of this reference, for the measurability properties of analytic sets.

The use of the restriction map in the decomposition theory of states is very old and dates back at least to Tomita's earlier work [Tomi 3].

Section 4.2.1

The Cartier–Fell–Meyer theorem, Proposition 4.2.1, appeared in [Car 2] in the context of general compact convex sets and was a crucial element in Ruelle's and Skau's proofs of the structure theorems for orthogonal measures. This result can be considered as the natural generalization of the ordering properties of finite sequences of real numbers described much earlier by Hardy, Littlewood, and Polya (see [[Hard 1, p. 45]]).

The equivalence of conditions (1) and (2) in Theorem 4.2.2 was a longstanding open question which was finally resolved by Henrichs [Hen 1]. Theorem 4.2.3 and a version of Theorem 4.2.5 are due to Skau [Ska 1]. Proposition 4.2.4 also occurs in [Ska 1] but the equivalence $(1') \Leftrightarrow (3')$ was derived by Douglas [Dou 1] for a general convex compact set K.

Example 4.2.6 is essentially due to Sherman [She 1] and Fukamiya, Misonou, and Takeda [Fuk 2].

Section 4.2.2

The central decomposition of states was analyzed by Sakai in 1965 [Sak 1]. His definition of central measure differs from the definition that we have adopted but it is straightforward to show that the two concepts coincide. Sakai's version of the central decomposition is described at length in [[Sak 1]] and this reference also contains a proof that the factor states $F_{\mathfrak{A}}$ over a norm separable C^*-algebra $\mathfrak{A}$ form a Borel subset of $E_{\mathfrak{A}}$.

The first analysis of central decomposition with Choquet theory was given by Wils [Wils 1], who subsequently announced the geometric characterization of the central measure given by Theorem 4.2.10 [Wils 2]. Wils also extended the notion of a central measure to an arbitrary compact convex set in a locally convex Hausdorff space. Details of this generalization are given in [[Alf 1]] and [Wils 3].

Sections 4.3.1 and 4.3.2

The analysis of G-invariant states originated with Segal, who introduced the terminology G-ergodic and characterized ergodicity by irreducibility of $\{\pi_\omega(\mathfrak{A}) \cup U_\omega(G)\}$ in 1951 [Seg 3]. The main developments of the theory came much later, however, and were principally inspired by problems of mathematical physics.

In the early 1960s various authors studied problems of irreducibility of relativistic quantum fields and this study essentially incorporated the examination of $\mathbb{R}$-invariant states ω such that $\sigma(U_\omega) \subseteq [0, \infty\rangle$. It was realized that this spectral condition and some commutativity implied $U_\omega(\mathbb{R}) \subseteq \pi_\omega(\mathfrak{A})''$ and hence ergodicity and purity of ω coincide (Example 4.3.34). Moreover, ergodicity was characterized by "uniqueness of the vacuum," i.e., uniqueness up to a factor of the $U_\omega(G)$-invariant vector. This latter property then indicated that the decomposition of ω into pure states was determined by a "diagonalization" of the operators

$$E_\omega \pi_\omega(\mathfrak{A})'' E_\omega$$

and various vestigial forms of decomposition with respect to the orthogonal measure corresponding to E_ω appeared (see for example, [Ara 5], [Bor 1], [Ree 1], [Rue 3]).

During the same period analysis of nonrelativistic field theoretic models indicated that G-invariant states ω with $\{\pi_\omega(\mathfrak{A}) \cup U_\omega(G)\}$ irreducible played a principal role in the description of thermodynamic equilibrium but the notion of G-ergodicity had been lost (see, for example, [Ara 8], [Ara 9], [Haag 3], [Rob 3]). Finally in 1966 Doplicher, Kastler and Robinson [Dop 1] gave a version of Theorems 4.3.17 and 4.3.22 characterizing ergodic states for $\mathbb{R}^\nu$. (The notion of asymptotic abelianness due to Robinson first appeared in print in this last reference.) Subsequently, Kastler and Robinson [Kas 1] and Ruelle [Rue 2] gave independent accounts of ergodic decomposition. These papers contained most of the elements necessary for the complete theory as we have described it, e.g., Choquet theory, asymptotic abelianness, spectral analysis, subgroup decomposition, etc., and triggered an avalanche of other works on the subject.

Ruelle's construction of the ergodic decomposition was inspired by earlier considerations of states of classical statistical mechanics which provide an intuitive explanation for the structure of the orthogonal measure [Rue 4]. Envisage for simplicity a C^*-algebra $\mathfrak{A}$ generated by one element A and its translates $\tau_x(A)$, $x \in \mathbb{R}^3$, under the group $\mathbb{R}^3$. Physically, one could interpret this as describing a system with one observable A, e.g., the number of particles at the origin, and $\tau_x(A)$ corresponds to the observable at the point x. Now one can associate with A and each subsystem $\Lambda \subseteq \mathbb{R}^3$ a macroscopic observable

$$A_\Lambda = \frac{1}{|\Lambda|} \int_\Lambda dx\, \tau_x(A)$$

e.g., A_Λ would correspond to the particle density in Λ. The distribution of values of these macroscopic observables in a state ω is then dictated by the set of moments $\omega((A_\Lambda)^n)$. But if ω is $\mathbb{R}^3$-invariant, i.e., if the system is homogeneous in space, then the size independent distribution is determined by the moments for infinite Λ, i.e., by

$$\begin{aligned}\lim_{\Lambda \to \infty} \omega((A_\Lambda)^n) &= \lim_{\Lambda \to \infty} \frac{1}{|\Lambda|^n} \int_{\Lambda^n} dx_1 \cdots dx_n\, (\Omega_\omega, \pi_\omega(A) U_\omega(x_2 - x_1) \pi_\omega(A) \cdots \\ &\qquad \cdots U_\omega(x_n - x_{n-1}) \pi_\omega(A) \Omega_\omega \\ &= (\Omega_\omega, \pi_\omega(A) E_\omega \pi_\omega(A) E_\omega \cdots E_\omega \pi_\omega(A) \Omega_\omega).\end{aligned}$$

Thus the value of the orthogonal measure μ, corresponding to E_ω, on A^n represents the nth moment of the distribution of the values of the macroscopic observable. The state ω has a precise value for this observable if, and only if, $\mu(\hat{A}^n) = \mu(\hat{A})^n$ and this corresponds exactly to the $\mathbb{R}^3$-ergodicity of ω. Thus this model motivates the characterization of pure thermodynamic phases by ergodic states.

The notion of G-abelianness was introduced by Lanford and Ruelle [Lan 3], who actually defined it by requiring $E_\omega \pi_\omega(\mathfrak{A}) E_\omega$ to be abelian for all $\omega \in E_{\mathfrak{A}}{}^G$. They then proved that this definition coincided with ours for all $\omega \in E_{\mathfrak{A}}{}^G$, i.e., they established the equivalence of conditions (1) and (2) of Proposition 4.3.7. It was first mentioned in [Kas 1] that G-abelianness was sufficient to assure a unique ergodic decomposition for locally compact abelian G and the general result was given in [Lan 3]. The fact that uniqueness of barycentric decomposition for all normal invariant states implies G-abelianness (Theorem 4.3.9) first appeared in the seminar notes of Dang Ngoc and Guichardet [Dan 1], [[Gui 2]]. The notion of G-centrality was introduced by Doplicher, Kastler and Størmer [Dop 2] and Theorem 4.3.14 also first appeared in [Dan 1], [[Gui 2]]. The abstract version of the mean ergodic theorem, Proposition 4.3.4, appeared in [Ala 1] and the algebraic version, Proposition 4.3.8, in [Kov 1].

Much of the earlier analysis of ergodic states used some form of mean value over the group G to express the appropriate conditions of asymptotic abelianness, etc., and this restricted the results to amenable groups. A locally compact topological group G is defined to be amenable if there exists a state M over the C^*-algebra $C_b(G)$ of bounded continuous functions over G which is invariant under right translations. Not all groups are amenable but this class contains the compact groups, locally compact abelian groups, and locally compact soluble groups. Moreover, each closed subgroup of an amenable group is amenable and the quotient of an amenable group G by a closed normal subgroup H is amenable. Conversely, if G/H and H are amenable then G is amenable. Noncompact semisimple Lie groups are not amenable nor is the free group on two generators. There are a host of equivalent conditions which characterize amenability and, for example, a locally compact group G is amenable if, and only if, for each compact $K \subseteq G$ there is a net of Borel sets $U_\alpha \subseteq G$ with $\mu(U_\alpha) < \infty$ and

$$\lim_\alpha \mu(gU_\alpha \Delta U_\alpha)/\mu(U_\alpha) = 0$$

for all $g \in K$, where μ is some fixed Haar measure. The net U_α can then be used to give an explicit form for the invariant mean

$$M(f) = \lim_\alpha \frac{1}{\mu(U_\alpha)} \int_{U_\alpha} d\mu(g)\, f(g).$$

This is a generalization of the type of mean used in Example 4.3.5 (see [[Gre 1]].

An alternative approach to the theory for general G is to introduce a mean on a subspace of the bounded functions $B(G)$ over G. Following Godement [God 1], Doplicher and Kastler [Dop 3] considered the set of $f \in B(G)$ such that the closed convex hulls of their left, and right, translates both contain a constant. This constant is then unique and defines the mean $M(f)$ of f. This set contains the functions of positive type over G and hence the matrix elements of unitary representations of G have means. The difficulty with this approach is that it is not evident whether functions such as

$$g \in G \mapsto \omega([\tau_g(A), B])$$

have a mean and this has to be added as a hypothesis.

Besides G-abelianness and G-centrality there are various stronger notions of commutation which have been discussed and hierarchized in [Dop 2]. Examples 4.3.18 and 4.3.21 are taken from this article.

An aspect of ergodic decomposition theory which we have not covered is the theory of decomposition of positive linear functionals over *-algebras of unbounded operators, for example Wightman functionals over a so-called Borchers algebra. This theory has pathologies with no analogues in the C^*-theory, i.e., there exists infinite-dimensional irreducible *-representations of abelian *-algebras, and analogously there may exist extremal Wightman functionals which are not clustering. These problems have been analyzed by Borchers and Yngvason in a series of papers, [Bor 5], [Yng 1].

A version of Theorem 4.3.19 first occurred in [Kas 1], an elaboration was given by Størmer [Stø 3], and the complete result was derived by Nagel [Nag 1].

The fact that a separating property of Ω_ω can replace asymptotic abelianness for the characterization of ergodic states was pointed out by Jadczyk [Jad 1], who gave a version of Theorems 4.3.20 and 4.3.23.

Example 4.3.24 is based on an idea occurring in [Kas 1]. A discussion and interpretation of mixing properties in classical ergodic theory can be found in Arnold and Avez [[Arn 1]]. The theorem of Mazur quoted in the discussion of asymptotic abelianness can be found in the book of Yosida [[Yos 1]]. It is an immediate consequence of the fact that if K is a closed, convex subset of a Banach space X, and $x \in X \backslash K$ is a point, then there exists a continuous affine map φ on X such that $\varphi(x) > 0$ and $\varphi(K) \subseteq \langle -\infty, 0]$.

Example 4.3.25 is taken from [Rob 4].

Example 4.3.26 is due to Ruelle. The first example of a metrizable simplex whose extreme points are dense was constructed by Poulsen [Pou 1] in 1961, and Lindenstrauss, Olsen and Sternfeld showed in 1976 that this simplex was unique up to an affine homeomorphism [Lin 1]. These authors prove also that the Poulsen simplex S is homogeneous in the sense that any affine homeomorphism between two faces F_1 and F_2 of S can be extended to an affine automorphism of S. Moreover, they prove that S is universal in the sense that any metrizable simplex can be realized as a closed face of S. The Poulsen simplex is in fact characterized by homogeneity and universality among the metrizable simplices. The other extreme among simplices K are those for which $\mathscr{E}(K)$ is a closed set. These are the probability measures on compact Hausdorff spaces and are usually called Bauer simplices. (See, for example [[Alf 1]].)

Sections 4.3.3 and 4.3.4

Theorem 4.3.27 originated in [Jad 1] and Theorem 4.3.31 comes from [Kas 1]. This latter reference also contains the notion of G_Γ-abelianness. The additivity of the total spectrum under the conditions of Theorem 4.3.33 is much older. The idea of the proof dates back at least to 1961 [Wig 1].

The condition for a factor to be type III given in Example 4.3.34 is due to Størmer [Stø 4] but the first result of this type was due to Hugenholtz [Hug 1].

Theorem 4.3.37 is taken from [Rob 4]. It is essentially due to Ginibre. The first result of this type occurred in [Kas 1] for locally compact abelian groups. Theorem 4.3.38 is due to Robinson [Rob 5].

The equivalence of the various characterizations of almost periodic functions that we have used can be found in [[Dix 2]]. In fact, the theory generalizes to nonabelian groups if one replaces the characters by coefficients of irreducible finite-dimensional unitary representations. In this context the results on the point spectrum of a G-ergodic state have been partially generalized by Doplicher and Kastler [Dop 3].

Proposition 4.3.42 can be found in [Rue 1]. This reference also gives an elaboration of the last statement of the proposition and in particular a decomposition of $\omega \in \mathscr{E}(E_{\mathfrak{A}}{}^G)$ of the form

$$\omega(A) = \int_{\bar{G}} d\bar{g}\, (\kappa \hat{A})(\tau_{\bar{g}}{}^* x)$$

for some $x \in M$. The action of $\bar{G}$ on M is in fact transitive, hence this decomposition is independent of the choice of $x \in M$. Moreover, if one adopts the separability assumptions of the remark following Proposition 4.3.42 one can write $(\kappa(\hat{A})(\tau_g{}^*x) = \hat{A}(\eta^{-1}\tau_g{}^*x)$, where η is the Borel isomorphism from $E_\mu \subseteq E_{\mathfrak{A}}$ to $M_m \subseteq M$ which determines κ. It is then tempting to use the fact that κ commutes with the action of G to write $\eta^{-1}\tau_{\bar{g}}{}^*x = \eta^{-1}\tau_g{}^*\eta\tilde{\omega} = \tau_{\bar{g}}{}^*\tilde{\omega}$ for some $\tilde{\omega} \in E_\mu$. Unfortunately it appears difficult to justify this conclusion except in the trivial case that G is a finite group.

The result of von Neumann quoted in the remark following Proposition 4.3.42 can be found in [[Dix 1, Appendix IV]].

Section 4.4.1

The theory of spatial decomposition goes back to von Neumann [Neu 3], and has not undergone essential changes since. We have followed the approaches in [[Dix 1]], [[Dix 2]], [[Sak 1]], [[Sch 1]]. The classification of standard measure spaces mentioned after Proposition 4.4.6 was proved by Mackey in [Mac 1].

Section 4.4.2

Theorem 4.4.9 and the subsequent remark are due to Effros [Eff 3]. The decomposition at infinity, Theorem 4.4.11, was studied by Ruelle in [Rue 1], while the weak mixing property in Theorem 4.4.12 was proved by Kastler and Robinson [Kas 1]. A proof of the metrizability of a second countable locally compact group can be found in [[Hew 1]], Theorem 8.3.

References

Books and Monographs

[[Alf 1]] Alfsen, E.
Compact Convex Sets and Boundary Integrals. Springer-Verlag, Berlin–Heidelberg–New York (1971).

[[Arn 1]] Arnold, V. I., and A. Avez.
Problèmes ergodiques de la mecanique classique. Gauthier–Villars, Paris (1967).

[[Arv 1]] Arveson, W.
An Invitation to C-Algebras.* Springer-Verlag, Berlin–Heidelberg–New York (1976).

[[Bon 1]] Bonsall, F. F., and J. Duncan.
Numerical Ranges of Operators on Normed Spaces and Elements of Normed Algebras, Cambridge University Press, Cambridge (1971).

[[Bou 1]] Bourbaki, N.
Espaces vectoriels topologiques. Hermann, Paris (1955, 1966).

[[Bra 1]] Bratteli, O.
Derivations, Dissipations, and Group Actions on C-Algebras.* Lecture Notes in Mathematics 1229. Springer-Verlag, New York (to be published 1986).

[[But 1]] Butzer, P. L., and H. Berens.
Semi-Groups of Operators and Approximation Theory. Springer-Verlag, Berlin–Heidelberg–New York (1967).

[[Cho 1]] Choquet G.
Lectures on Analysis (J. Marsden, T. Lance, and S. Gelbart, eds.). Benjamin, New York (1969).

[[Dav 1]] Davies, E. B.
Quantum Theory of Open Systems. Academic Press, New York (1976).

[[Dix 1]] Dixmier, J.
Les algèbres d'opérateurs dans l'espace Hilbertien (Algèbres de von Neumann), 2nd ed. Gauthier–Villars, Paris (1969).

[[Dix 2]] Dixmier, J.
Les C-algèbres et leurs représentations*. Gauthier–Villars, Paris (1964).

[[Dun 1]] Dunford, N. and J. T. Schwartz. *Linear Operators*, Vol. I–III. Wiley-Interscience, New York–London–Sydney–Toronto (1971).

[[Eva 1]] Evans, D. E., and J. T. Lewis.
Dilations of irreversible evolutions in algebraic quantum theories. Comm. Dublin Inst. for Advanced Studies, Series A, **24** (1977).

[[Gre 1]] Greenleaf, F. P.
Invariant Means on Topological Groups. van Nostrand–Reinhold, New York–Toronto–London–Melbourne (1969).

[[Gui 1]] Guichardet, A.
Special topics in topological algebras. Gordon & Breach, New York (1968).

[[Gui 2]] Guichardet, A.
Systèmes dynamiques non commutatifs. Societé Mathematique de France Asterisque 13–14 (1974).

[[Hard 1]] Hardy, G. H., J. E. Littlewood, and G. Polya.
Inequalities, 2nd ed. Cambridge University Press, Cambridge (1952).

[[Hew 1]] Hewitt, E., and K. A. Ross.
Abstract Harmonic Analysis I. Springer-Verlag, Berlin–Heidelberg–New York (1963).

[[Hil 1]] Hille, E., and R. S. Phillips.
Functional Analysis and Semi-Groups, rev. ed. American Mathematical Society Colloq. Publ. Vol. 31, Providence, R. I. (1957).

[[Hil 2]] Hille, E.
Functional Analysis and Semi-Groups. American Mathematical Society, New York (1948).

[[Kat 1]] Kato, T.
Perturbation Theory for Linear Operators, 2nd ed. Springer-Verlag, Berlin–Heidelberg–New York (1976).

[[Köt 1]] Köthe, G.
Topologische Lineare Räume, 2nd ed. Springer-Verlag, Berlin–Heidelberg–New York (1966).

[[Mac 1]] Mackey, G. W.
The Theory of Unitary Group Representations. University of Chicago Press, Chicago–London (1976).

[[Nai 1]] Naimark, M. A.
Normed Algebras. Wolters–Noordhoff, Groningen (1972).

[[Neu 1]] von Neumann, J.
Collected Works. Pergamon Press, New York (1961).

[[Neu 2]] von Neumann, J.
Mathematical Foundations of Quantum Mechanics. Princeton University Press, Princeton (1955).

[[Ped 1]] Pedersen, G. K.
Introduction to C-Algebra Theory.* Academic Press, New York–San Francisco–London (1979).

[[Phe 1]] Phelps, R. R.
Lectures on Choquet's Theorem. Van Nostrand–Reinhold, New York–Toronto–London–Melbourne (1966).

[[Ree 1]] Reed, M., and B. Simon.
Methods of Modern Mathematical Physics. Vol. I: Functional Analysis. Academic Press, New York–San Francisco–London (1973).

[[Ree 2]] Reed, M., and B. Simon.
Methods of Modern Mathematical Physics. Vol. II: Fourier Analysis, Self-Adjointness. Academic Press, New York–San Francisco–London (1975).

[[Rie 1]] Riesz, F., and B. Sz-Nagy.
Leçons d'analyse fonctionnelle, 6th ed. Gauthier–Villars, Paris (1972).

[[Rud 1]] Rudin, W.
Real and Complex Analysis, 2nd ed. McGraw–Hill, New York (1974).

[[Rud 2]] Rudin, W.
Functional Analysis. McGraw–Hill, New York (1973).

[[Rud 3]] Rudin, W.
Fourier Analysis on Groups. Interscience, New York (1962).

[[Sak 1]] Sakai, S.
C-Algebras and W*-Algebras.* Springer-Verlag, Berlin–Heidelberg–New York (1971).

[[Sch 1]] Schwartz, J.
W-Algebras*, Gordon & Breach, New York (1967).

[[Sto 1]] Stone, M. H.
Linear Transformations in Hilbert Spaces and Their Applications to Analysis. American Mathematical Society Colloq. Publ. Vol. 15, New York (1932).

[[Str 1]] Streit, L.
Quantum Dynamics Models and Mathematics. Springer-Verlag, Berlin–Heidelberg–New York (1976).

[[Tak 1]] Takesaki, M.
Theory of Operator Algebras I. Springer-Verlag, New York–Heidelberg–Berlin (1979).

[[Tit 1]] Titchmarsh, E. C.
The Theory of Functions, 2nd ed. Oxford University Press, Oxford (1939).

[[Tom 1]] Tomiyama, J.
The Theory of Closed Derivations in the Algebra of Continuous Functions on the Unit Interval. Lecture Notes, Tsing Hua Univ. (1983).

[[Wae 1]] van der Waerden, B. L.
Sources of Quantum Mechanics. North–Holland, Amsterdam (1967).

[[Wign 1]] Wigner, E. P.
Gruppentheorie und Ihre Anwendung auf die Quantenmechanik der Atmospektren. Friedr. Vieweg, Braunschweig (1931).

[[Yos 1]] Yosida, K.
Functional Analysis, 2nd ed. Springer-Verlag. Berlin–Heidelberg–New York (1968).

Articles

[Ake 1] Akemann, C. A., and G. K. Pedersen.
Central sequences and inner derivations of separable C^*-algebras, *Amer. J. Math.* **101** (1979), 1047–1061.

[Ake 2] Akemann, C. A.
The dual space of an operator algebra. *Trans. Amer. Math. Soc.* **126** (1967), 286–302.

[Ala 1] Alaoglu, L., and G. Birkhoff.
General ergodic theorems, *Ann. Math.* **41** (1940), 293–309.

[Ara 1] Araki, H., and G. Elliott.
On the definition of C^*-algebras, *Pub. R.I.M.S., Kyoto Univ.* **9** (1973), 93–112.

[Ara 2] Araki, H.
Some properties of modular conjugation operator of a von Neumann algebra and a non-commutative Radon–Nikodym theorem with a chain rule, *Pac. J. Math.* **50** (1974), 309–354.

[Ara 3] Araki, H.
Introduction to relative Hamiltonian and relative entropy, preprint, Marseille, p. 782 (1975).

[Ara 4] Araki, H., and A. Kishimoto.
On clustering property, *Rep. Math. Phys.* **10** (1976), 275–281.

[Ara 5] Araki, H.
On representations of the canonical commutation relations, *Commun. Math. Phys.* **20** (1971), 9–25.

[Ara 6] Araki, H., and E. J. Woods.
A classification of factors, *Pub. R.I.M.S., Kyoto Univ.* **4** (1968), 51–130.

[Ara 7] Araki, H.
On the algebra of all local observables, *Prog. Theor. Phys.* **32** (1964), 844–854.

[Ara 8] Araki, H., and E. J. Woods.
Representations of the canonical commutation relations describing a non-relativistic infinite free Bose gas, *J. Math. Phys.* **4** (1963), 637–662.

[Ara 9] Araki, H., and W. Wyss.
Representations of canonical anticommutation relations, *Helv. Phys. Acta* **37** (1964), 136–159.

[Arv 1] Arveson, W.
On groups of automorphisms of operator algebras, *J. Func. Anal.* **15** (1974), 217–243.

[Bis 1] Bishop, E., and K. de Leeuw.
The representation of linear functionals by measures on sets of extreme points, *Ann. Inst. Fourier, Grenoble* **9** (1959), 305–331.

[Bor 1] Borchers, H. J.
On the structure of the algebra of field operators, *Il Nuovo Cimento* **24** (1962), 214–236.

[Bor 2] Borchers, H. J.
Energy and momentum as observables in quantum field theory, *Commun. Math. Phys.* **2** (1966), 49–54.

[Bor 3] Borchers, H. J.
Characterization of inner *-automorphisms of W^*-algebras, *Pub. R.I.M.S., Kyoto Univ.* **10** (1974), 11–49.

[Bor 4] Borchers, H. J.
Über Ableitungen von C^*-algebren, *Nachr. Göttingen Akad.* **2** (1973).

[Bor 5] Borchers, H. J., and J. Yngvason.
On the algebra of field operators. The weak commutant and integral decomposition of states, *Commun. Math. Phys.* **42** (1975), 231–252. [See also *ibid.* **43** (1975), 255–271; *ibid.* **47** (1976), 197–213.]

[Bra 1] Bratteli, O.
Inductive limits of finite-dimensional C^*-algebras, *Trans. Amer. Math. Soc.* **171** (1972), 195–234.

[Bra 2] Bratteli, O.
The center of approximately finite-dimensional C^*-algebras, *J. Func. Anal.* **21** (1976), 195–201.

[Bra 3] Bratteli, O., and D. W. Robinson.
Unbounded derivations of von Neumann algebras, *Ann. Inst. H. Poincaré* **25** (A) (1976), 139–164.

[Bra 4] Bratteli, O., R. H. Herman, and D. W. Robinson.
Perturbations of flows on Banach spaces and operator algebras, *Commun. Math. Phys.* **59** (1978), 167–196.

[Bra 5] Bratteli, O., and D. W. Robinson.
Unbounded derivations of C^*-algebras, *Commun. Math. Phys.* **42** (1975), 253–268.

[Bra 6] Bratteli, O., and D. W. Robinson.
Unbounded derivations of C^*-algebras II, *Commun. Math. Phys.* **46** (1976), 11–30.

[Bra 7] Bratteli, O., R., Herman, and D. W. Robinson.
Quasi-analytic vectors and derivations of operator algebras, *Math. Scand.* **39** (1976), 371–381.

[Bra 8] Bratteli, O., and D. W. Robinson.
Unbounded derivations and invariant trace states, *Commun. Math. Phys.* **46** (1976), 31–35.

[Bra 9] Bratteli, O., and U. Haagerup.
Unbounded derivations and invariant states, *Commun. Math. Phys.* **59** (1978), 79–95.

[Bra 10] Bratteli, O.
Crossed products of UHF algebras by product type actions, *Duke Math. J.* **46** (1979), 1–23.

[Bra 11] Bratteli, O., and A. Kishimoto.
Generation of semi groups and two-dimensional quantum lattice systems, *J. Func. Anal.* **35** (1980), 344–368.

[Buc 1] Bucholz, D., and J. E. Roberts.
Bounded perturbations of dynamics, *Comm. Math. Phys.* **49** (1976), 161–177.

[Car 1] Cartier, P., and J. Dixmier.
Vecteurs analytiques dans les representations des groupes de Lie, *Amer. J. Math.* **80** (1958), 131–145.

[Car 2] Cartier, P., J. M. G. Fell, and P. A. Meyer.
Comparison des mesures portées par un ensemble convex compact, *Bull. Soc. Math. France* **92** (1964), 435–445.

[Che 1] Chernoff, P.
Note on product formulas for operator semi-groups, *J. Func. Anal.* **2** (1968), 238–242.

[Che 2] Chernoff, P.
Semi-group product formulas and addition of unbounded operators, *Bull. Amer. Math. Soc.* **76** (1970), 395.

[Chi 1] Chi, D. P.
Derivations in C^*-algebras, thesis, Univ. of Pennsylvania (1976).

[Cho 1] Choquet, G.
Existence des representations integrales au moyen des points extremaux dans les cones convexes, *C. R. Acad. Sci. Paris* **243** (1956), 699–702.

[Cho 2] Choquet, G.
Existence des representations dans les cones convexes, *C. R. Acad. Sci.* Paris **243** (1956), 736–737.

[Cho 3] Choquet, G.
Unicité des representations integrales au moyens des points extremaux dans les cones convexes reticulés, *C. R. Acad. Sci. Paris* **243** (1956), 555–557.

[Cho 4] Choquet, G.
Existence unicité des representations integrales au moyen des points extrémaux dans les cones convexes, *Seminaire Bourbaki* **139** (1956).

[Cho 5] Choquet, G.
Le théorème de representation intégrales dans les ensembles convexes compacts, *Ann. Inst. Fourier, Grenoble* **10** (1960), 333–344.

[Cho 6] Choquet, G., and P. A. Meyer.
Existence et unicité des representations intégrales dans les convexes compacts quelquonque, *Ann. Inst. Fourier, Grenoble* **13** (1963), 139–154.

[Choi 1] Choi, M. D., and E. Effros.
Nuclear C^*-algebras and injectivity, the general case, *Indiana Univ. Math. J.* **26** (1977), 443–446.

[Com 1] Combes, F.
Poids associé à une algebre Hilbertienne à gauche, *Comp. Math.* **23** (1971), 49–77.

[Con 1] Connes, A.
Characterization des espaces vectoriels ordonnés sous-jacents aux algèbres de von Neumann, *Ann. Inst. Fourier, Grenoble* **24** (1974), 121–155.

[Con 2] Connes, A., and M. Takesaki.
The flow of weights on factors of type III, *Tohoku Math. J.* **29** (1977), 473–575.

[Con 3] Connes, A.
On hyperfinite factors of type III and Kriegers factors, *J. Func. Anal.* **18** (1975), 318–327.

[Con 4] Connes, A.
Une classification des facteurs de type III, *Ann. Sci. Ecole Norm. Sup.* **6** (1973), 133–252.

[Con 5] Connes, A.
Classification of injective factors, *Ann. Math.* **104** (1976), 73–115.

[Con 6] Connes, A.
Outer conjugacy classes of automorphisms of factors, *Ann. Sci. Ecole Norm. Sup.* **8** (1975), 383–420.

[Con 7] Connes, A.
A factor not anti-isomorphic to itself, *Ann. Math.* **101** (1975), 536–554.

[Con 8] Connes, A.
Factors of type III_1, property L_λ' and closure of inner automorphisms, *J. Operator Theory* **14** (1985), 189–211.

[Cun 1] Cuntz, J.
Locally C^*-equivalent algebras, *J. Func. Anal.* **23** (1976), 95–106.

[Dae 1] van Daele, A.
A new approach to the Tomita–Takesaki theory of generalized Hilbert algebras, *J. Func. Anal.* **15** (1974), 378–393.

[Dae 2] van Daele, A., and M. Rieffel.
A bounded operator approach to Tomita–Takesaki theory. *Pac. J. Math.* **69** (1977), 187–221.

[Dan 1] Dang Ngoc, N., and F. Ledrappier.
Les systèmes dynamiques simpliciaux, *C. R. Acad. Sci. Paris* **277** (1973), 777–779.

[Dig 1] Digernes, T.
Duality for weights on covariant systems and its applications, thesis UCLA (1975).

[Dix 1] Dixmier, J.
Sur les C^*-algebras, *Bull. Soc. Math. France* **88** (1960), 95–112.

[Dix 2] Dixmier, J.
Dual et quasidual d'une algèbre de Banach involutive, *Trans. Amer. Math. Soc.* **104** (1962), 278–283.

[Dop 1] Doplicher, S., D. Kastler, and D. W. Robinson.
Covariance algebras in field theory and statistical mechanics, *Commun. Math. Phys.* **3** (1966), 1–28.

[Dop 2] Doplicher, S., D. Kastler, and E. Størmer.
Invariant states and asymptotic abelianness, *J. Func. Anal.* **3** (1969), 419–434.

[Dop 3] Doplicher, S., and D. Kastler.
Ergodic states in a non-commutative ergodic theory, *Commun. Math. Phys.* **7** (1968), 1–20.

[Dou 1] Douglas, R. C.
On extremal measures and subspace density, *Michigan Math. J.* **11** (1964), 644–652.

[Dun 1] Dunford, N.
Uniformity in linear spaces, *Trans. Amer. Math. Soc.* **44** (1938), 305–356.

[Eff 1] Effros, E. G., and F. Hahn.
Locally compact transformation groups and C^*-algebras, *Mem. Amer. Math. Soc.* **75** (1967), 1–92.

[Eff 2] Effros, E. G., and C. Lance.
Tensor products of operator algebras, *Adv. Math.* **25** (1977), 1–34.

[Eff 3] Effros, E. G.
On the representations of C^*-algebras, thesis, Harvard Univ. (1961).

[Ell 1] Elliott, G.
On the classification of inductive limits of sequences of semi-simple finite-dimensional algebras, *Algebra* **38** (1976), 29–44.

[Ell 2] Elliott, G.
Derivations of matroid C^*-algebras, *Invent. Math.* **9** (1970), 253–269.

[Ell 3] Elliott, G.
Some C^*-algebras with outer derivations III, *Ann. Math.* **106** (1977), 121–143.

[Ell 4] Elliott, G. A., and E. J. Woods.
The equivalence of various definitions for a properly infinite von Neumann algebra to be approximately finite dimensional, *Proc. Amer. Math. Soc.* **10** (1976), 175–178.

[Ell 5] Elliott, G. A.
On approximately finite dimensional von Neumann algebras I and II, *Math. Scand.* **39** (1976), 91–101; *Can. Math. Bull.*

[Ell 6] Elliott, G. A.
The Mackey–Borel structure on the spectrum of an approximately finite dimensional C^*-algebra, *Trans. Amer. Math. Soc.* **233** (1977).

[Ell 7] Elliott, G. A.
Convergence of automorphisms in certain C^*-algebras, *J. Func. Anal.* **11** (1972), 204–206.

[Fel 1] Feller, W.
On the generation of unbounded semi-groups of bounded linear operators, *Ann. Math.* **58** (1953), 166–174.

[Fell 1] Fell, J. M. G.
The structure of algebras of operator fields, *Acta Math.* **106** (1961), 233–280.

[Fuj 1] Fujii, M. T. Furuta, and K. Matsumoto.
Equivalence of operator representations of semi-groups, *Math. Jap.* **20** (1975), 253–256.

[Fuk 1] Fukamiya, M.
On a theorem of Gelfand and Naimark and the B^*-algebra, *Kumamoto J. Sci.* **1** (1952), 17–22.

[Fuk 2] Fukamiya, M., Y. Misonou, and Z. Takeda.
On order and commutativity of B^*-algebras, *Tohoku Math. J.* **6** (1954), 89–93.

[Gal 1] Gallavotti, G., and M. Pulvirenti.
Classical KMS condition and Tomita–Takesaki theory, *Commun. Math. Phys.* **46** (1976), 1–9.

[Gel 1] Gelfand, I. M., and M. A. Naimark.
On the imbedding of normed rings into the ring of operators in Hilbert space, *Mat. Sb.* **12** (1943), 197–213.

[Gel 2] Gelfand, I. M.
Normierte Ringe, *Mat. Sb.* **9** (1941), 3–24.

[Gli 1] Glimm, J., and R. V. Kadison.
Unitary operators in C^*-algebras, *Pac. J. Math.* **10** (1960), 547–556.

[Gli 2] Glimm, J.
On a certain class of operator algebras, *Trans. Amer. Math. Soc.* **95** (1960), 318–340.

[Gli 3] Glimm, J.
Families of induced representations, *Pac. J. Math.* **12** (1962), 885–911.

[Gli 4] Glimm, J.
Type I C^*-algebras, *Ann. Math.* **73** (1961), 572–612.

[Gli 5] Glimm, J., and A. Jaffe.
Singular perturbations of self-adjoint operators, *Comm. Pure Appl. Math.* (1969), 401–414.

[God 1] Godement, R.
Les fonctions de type positif et la théorie des groupes, *Trans. Amer. Math. Soc.* **63** (1948), 1.

[Gro 1] Grothendieck, A.
Un résultat sur le dual d'une C^*-algebre, *J. Math. Pures Appl.* **36** (1957), 97–108.

[Gui 1] Guichardet, A.
Products tensoriels infinis et representations des relations d'anticommutations, *Ann. Sci. Ecole Norm. Sup.* **83** (1966), 1–52.

[Haa 1] Haagerup, U.
The standard form of von Neumann algebras, *Math. Scand.* **37** (1975), 271–283.

[Haa 2] Haagerup, U.
The standard form of von Neumann algebras, preprint Copenhagen, n°15 (1973).

[Haa 3] Haagerup, U.
Normal weights on von Neumann algebras, *J. Func. Anal.* **19** (1975), 302–317.

[Haa 4] Haagerup, U.
On the dual weights for von Neumann algebras I (to appear).

[Haa 5] Haagerup, U
Connes' bicentralizer problem and uniqueness of the injective factor of type III_1, *Acta. Math.*, to appear.

[Haag 1] Haag, R.
Discussion des "axiomes" et des propriétés asymptotiques d'une théorie des champs locale avec particules composées, in *Les Problèmes mathematiques de la thèorie quantique des champs.* Pub. CNRS Vol. 85 (1959).

[Haag 2] Haag, R., R. V. Kadison, and D. Kastler.
Nets of C^*-algebras and classification of states, *Commun. Math. Phys.* **16** (1970), 81–104.

[Haag 3] Haag, R.
The mathematical structure of the Bardeen–Cooper–Schrieffer model, *Il Nuovo Cimento* **25** (1962), 287–299.

[Han 1] Hansen, F., and D. Olesen.
Perturbations of centre-fixing dynamical systems, *Math. Scand.* **41** (1977), 295–307.

[Har 1] Harish-Chandra,
Representations of a semi-simple Lie group on a Banach space I, *Trans. Amer. Math. Soc.* **75** (1953), 185–243.

[Heg 1] Hegerfeldt, G. C.
On canonical commutation relations and infinite dimensional measures, *J. Math. Phys.* **13** (1972), 45–50.

[Hen 1] Henrichs, R. W.
On decomposition theory for unitary representations of locally compact groups, *J. Func. Anal.* **31** (1979) 101–114.

[Her 1] Herman, R. H.
Private communication.

[Hug 1] Hugenholtz, N. M.
On the factor type of equilibrium states in quantum statistical mechanics, *Commun. Math. Phys.* **6** (1967), 189–193.

[Iku 1] Ikunishi, A., and Y. Nakagami.
On invariants $G(\sigma)$ and $\Gamma(\sigma)$ for an automorphism group of a von Neumann algebra, *Pub. R.I.M.S., Kyoto Univ.* **12** (1976), 1–30.

[Jac 1] Jacobson, N., and C. Rickart.
Jordan homomorphisms of rings, *Trans. Amer. Math. Soc.* **69** (1950), 479–502.

[Jad 1] Jadczyk, A. Z.
On some groups of automorphisms of von Neumann algebras with cyclic and separating vectors, *Commun. Math. Phys.* **13** (1969), 142–153.

[Joh 1] Johnson, B. E.
Perturbations of semigroups, *Quart. J. Math. Oxford*, 2nd Ser. **36** (1985) 315–324.

[Jor 1] Jordan, P., and E. Wigner.
Über der Paulische Aquivalenz-Verbot, *Z. Physik* **47** (1928), 631–

[Jør 1] Jørgensen, P. E. T.
Approximately reducing subspaces for unbounded linear operators, *J. Func. Anal.* **23** (1976), 392–414.

[Kad 1] Kadison, R. V.
Lectures on operator algebras, in *Applications of mathematics to problems in theoretical physics*, in (F. Lurcat, ed.). Gordon & Breach, New York (1967).

[Kad 2] Kadison, R. V.
Irreducible operator algebras, *Proc. Nat. Acad. Sci. USA* **43** (1957), 273–276.

[Kad 3] Kadison, R. V.
Operator algebras with a faithful weakly closed representation, *Ann. Math.* **64** (1956), 175–181.

[Kad 4] Kadison, R. V.
Isometries of operator algebras, *Ann. Math.* **54** (1951), 325–338.

[Kad 5] Kadison, R. V.
Transformation of states in operator theory and dynamics, *Topology* **3** (1965), 177–198.

[Kad 6] Kadison, R. V.
A generalized Schwarz inequality and algebraic invariants for operator algebras, *Ann. Math.* **56** (1952), 494–503.

[Kad 7] Kadison, R. V.
Derivations of operator algebras, *Ann. Math.* **83** (1966), 280–293.

[Kad 8] Kadison, R. V.
Some analytic methods in the theory of operator algebras, in *Lecture Notes in Mathematics*, Vol. 140. Springer-Verlag, Berlin–Heidelberg–New York (1970).

[Kad 9] Kadison, R. V., and J. R. Ringrose.
Derivations and automorphisms of operator algebras, *Commun. Math. Phys.* **4** (1967), 32–63.

[Kal 1] Kallman, R. R.
Unitary groups and automorphisms of operator algebras, *Amer. J. Math.* **91** (1969), 785–806.

[Kal 2] Kallman, R. R.
One-parameter groups of *-automorphisms of II_1 von Neumann algebras, *Proc. Amer. Math. Soc.* **24** (1970), 336–340.

[Kap 1] Kaplansky, I.
Cited by J. A. Schultz in a review of [Fuk 1], *Math. Rev.* **14** (1953), 884.

[Kap 2] Kaplansky, I.
A theorem on rings of operators, *Pac. J. Math.* **1** (1951), 227–232.

[Kap 3] Kaplansky, I.
Modules over operator algebras, *Amer. J. Math.* **75** (1953), 839–859.

[Kas 1] Kastler D., and D. W. Robinson.
Invariant states in statistical mechanics, *Commun. Math. Phys.* **3** (1966), 151–180.

[Kat 1] Kato, T.
Remarks on pseudo-resolvents and infinitesimal generators of semi-groups, *Proc. Jap. Acad.* **35** (1959), 467–468.

[Kel 1] Kelley, J. L., and R. L. Vaught.
The positive cone in Banach algebras, *Trans. Amer. Math. Soc.* **74** (1953), 44–55.

[Kis 1] Kishimoto, A.
Dissipations and derivations, *Commun. Math. Phys.* **47** (1976), 25–32.

[Kis 2] Kishimoto, A., and H. Takai.
On the invariant $\Gamma(\alpha)$ in C^*-dynamical systems, *Tohoku Math. J.* **30** (1978), 83–94.

[Kis 3] Kishimoto, A., and H. Takai.
Some remarks on C^*-dynamical systems with a compact abelian group, *Pub. R.I.M.S., Kyoto Univ.* **14** (1978), 383–397.

[Kov 1] Kovács, I., and J. Szücs.
Ergodic type theorems in von Neumann algebras, *Acta Sci. Math.* **27** (1966), 233–246.

[Kre 1] Krein, M., and D. Milman.
On extreme points of regularly convex sets, *Stud. Math.* **9** (1940), 133–138.

[Kri 1] Krieger, W.
On ergodic flows and the isomorphism of factors, *Math. Ann.* **223** (1976), 19–70.

[Kur 1] Kurtz, T. G.
Extensions of Trotter's operator semigroup approximation theorems, *J. Func. Anal.* **3** (1969), 354–375.

[Lan 1] Lanford, O. E.
Selected topics in functional analysis, in *The Proceedings of the 1970 Les Houches Summer School* (R. Stora, ed.). Gordon & Breach, New York (1971).

[Lan 2] Lanford, O. E., and D. Ruelle.
Observables at infinity and states with short range correlations, *Commun. Math. Phys.* **13** (1969), 194–215.

[Lan 3] Lanford, O. E., and D. Ruelle.
Integral representations of invariant states on B^*-algebra, *J. Math. Phys.* **8** (1967), 1460–1463.

[Lanc 1] Lance, C.
Tensor products of C^*-algebras in *C^*-Algebras and Their Applications to Statistical Mechanics and Quantum Field Theory*. Editrice Compositori, Bologna (1975).

[Land 1] Landstad, M. B.
Duality theory for covariant systems, *Trans. Amer. Math. Soc.* **248** (1979), 223–267.

[Lee 1] de Leeuw, K.
On the adjoint semi-group and some problems in the theory of approximation, *Math. Z.* **73** (1960), 219–234.

[Lin 1] Lindenstrauss, J., G. Olsen, and Y. Sternfeld.
The Poulsen simplex, *Ann. Inst. Fourier, Grenoble* **28** (1978), 91–114.

[Lum 1] Lumer, G., and R. S. Phillips.
Dissipative operators in a Banach space, *Pac. J. Math.* **11** (1961), 679–689.

[Mac 1] Mackey, G. W.
Borel structure in groups and their duals, *Trans. Amer. Math. Soc.* **85** (1975), 134–165.

[MacG 1] MacGibbon, B.
A criterion for the metrizability of a compact convex set in terms of the set of extreme points, *J. Func. Anal.* **11** (1972), 385–392.

[Mar 1] Maréchal, O.
Topologie et structure borélienne sur l'ensemble des algèbres de von Neumann, *C. R. Acad. Sci. Paris* **276** (1973), 847–850.

[McI 1] McIntosh, A.
Functions and derivations of C^*-algebras, *J. Func. Anal.* **30** (1978), 264–275.

[Miy 1] Miyadera, I.
Generation of a strongly continuous semi-group of operators, *Tohoku Math. J.* **4** (1952), 109–114.

[Mok 1] Mokobodski, G.
Balayage défini par un cône convexe de fonctions numériques sur un space compact, *C. R. Acad. Sci. Paris* **254** (1962), 803–805.

[Mur 1] Murray, F. J., and J. von Neumann.
On rings of operators, *Ann. Math.* **37** (1936), 116–229; On rings of operators II, *Trans. Amer. Math. Soc.* **41** (1937), 208–248; On rings of operators IV, *Ann. Math.* **44** (1943), 716–808.

[Nag 1] Nagel, B.
Some results in non-commutative ergodic theory, *Commun. Math. Phys.* **26** (1972), 247–258.

[Nagu 1] Nagumo, M.
Einige analytische Untersuchungen in linearen metrischen Ringen, *Jap. J. Math.* **13** (1936), 61–80.

[Nak 1] Nakagami, Y.
Dual action on a von Neumann algebra and Takesakis duality theorem for a locally compact group, *Publ. R.I.M.S., Kyoto Univ.* **12** (1977), 727–775.

[Nel 1] Nelson, E.
Analytic vectors, *Ann. Math.* **70** (1969), 572–615.

[Neu 1] Von Neumann, J.
Zur algebra der funktional operationen und theorie der normalen operatoren, *Math. Ann.* **102** (1929), 370–427.

[Neu 2] Von Neumann, J.
On rings of operators III, *Ann. Math.* **41** (1940), 94–161.

[Neu 3] Von Neumann, J.
On rings of operators, reduction theory, *Ann. Math.* **50** (1949), 401–485.

[Oga 1] Ogasawara, T.
A theorem on operator algebras, *J. Sci. Hiroshima Univ.* **18** (1955), 307–309.

[Ole 1] Olesen, D.
On norm continuity and compactness of spectrum, *Math. Scand.* **35** (1974), 223–236.

[Ole 2] Olesen, D.
Inner *-automorphisms of simple C^*-algebras, *Commun. Math. Phys.* **44** (1975), 175–190.

[Ole 3] Olesen, D., G. K. Pedersen, and E. Størmer, with appendix by G. A. Elliott.
Compact abelian groups of automorphisms of simple C^*-algebras, *Invent. Math.* **39** (1977), 55–64.

[Ole 4] Olesen, D., and G. K. Pedersen.
Applications of the Connes spectrum to C^*-dynamical systems, *J. Func. Anal.* **30** (1978), 179–197.

[Ôta 1] Ôta, S.
Certain operator algebras induced by *-derivations in C^*-algebras on an infinite inner product space, *J. Func. Anal.* **30** (1978), 238–244.

[Ped 1] Pedersen, G. K., and M. Takesaki.
The Radon–Nikodym theorem for von Neumann algebras, *Acta Math.* **130** (1973), 53–87.

[Ped 2] Pedersen, G. K.
Isomorphisms of U.H.F. algebras, *J. Func. Anal.* **30** (1978), 1–16.

[Phi 1] Phillips, R. S.
Perturbation theory for semi-groups of linear operators, *Trans. Amer. Math. Soc.* **74** (1953), 199–221.

[Pou 1] Poulsen, E. T.
A simplex with dense extreme points, *Ann. Inst. Fourier, Grenoble* **11** (1961), 83–87.

[Pow 1] Powers, R.
Representations of uniformly hyperfinite algebras and their associated von Neumann rings, *Ann. Math.* **86** (1967), 138–171.

[Pow 2] Powers, R.
A remark on the domain of an unbounded derivation of a C^*-algebra, *J. Func. Anal.* **18** (1975), 85–95.

[Pow 3] Powers, R., and S. Sakai.
Existence of ground states and KMS states for approximately inner dynamics, *Commun. Math. Phys.* **39** (1975), 273–288.

[Pow 4] Powers, R., and S. Sakai.
Unbounded derivations in operator algebras, *J. Func. Anal.* **19** (1975), 81–95.

[Ree 1] Reeh, H., and S. Schlieder.
Über den Zerfall der Feldoperatoralgebra im Falle einer Vakuumentartung, *Il Nuovo Cimento* **26** (1962), 32–42.

[Robe 1] Roberts, J. E.
Cross products of von Neumann algebras by group duals, *Symposia Math.* **20** (1976), 335–363.

[Rober 1] Robertson, A. G.
A note on the unit ball in C^*-algebras, *Bull. London Math. Soc.* **6** (1974), 333–335.

[Rob 1] Robinson, D. W.
A characterization of clustering states, *Commun. Math. Phys.* **41** (1975), 79–88.

[Rob 2] Robinson, D. W.
Normal and locally normal states, *Commun. Math. Phys.* **19** (1970), 219–234.

[Rob 3] Robinson, D. W.
The ground state of the Bose gas, *Commun. Math. Phys.* **1** (1965), 159–174.

[Rob 4] Robinson, D. W., and D. Ruelle.
Extremal invariant states, *Ann. Inst. Henri Poincaré* **6** (1967), 299–310.

[Rob 5] Robinson, D. W. Unpublished.

[Rob 6] Robinson, D. W.
The approximation of flows, *J. Func. Anal.* **24** (1977), 280–290.

[Rob 7] Robinson, D. W.
Statistical mechanics of quantum spin systems II, *Commun. Math. Phys.* **7** (1968), 337–348.

[Rue 1] Ruelle, D.
Integral representation of states on a C^*-algebra, *J. Func. Anal.* **6** (1970), 116–151.

[Rue 2] Ruelle, D.
States of physical systems, *Commun. Math. Phys.* **3** (1966), 133–150.

[Rue 3] Ruelle, D.
On the asymptotic condition in quantum field theory, *Helv. Phys. Acta* **35** (1962), 147–163.

[Rue 4] Ruelle, D.
Correlation functionals, *J. Math. Phys.* **6** (1965), 201–220.

[Rus 1] Russo, B., and H. A. Dye.
A note on unitary operators in C^*-algebras, *Duke Math. J.* **33** (1966), 413–416.

[Sak 1] Sakai, S.
On the central decomposition for positive functionals on C^*-algebras, *Trans. Amer. Math. Soc.* **118** (1965), 406–419.

[Sak 2] Sakai, S.
On one parameter groups of *-automorphisms on operator algebras and the corresponding unbounded derivations, *Amer. J. Math.* **98** (1976), 427–440.

[Sak 3] Sakai, S.
On a conjecture of Kaplansky, *Tohoku Math. J.* **12** (1960), 31–33.

[Sak 4] Sakai, S.
Derivations of W^*-algebras, *Ann. Math.* **83** (1966), 273–279.

[Sak 5] Sakai, S.
Derivations of simple C^*-algebras, *J. Func. Anal.* **2** (1968), 202–206.

[Sak 6] Sakai, S.
Automorphisms and tensor products of operator algebras, *Amer. J. Math.* **97** (1975), 889–896.

[Seg 1] Segal, I. E.
Irreducible representations of operator algebras, *Bull. Amer. Math. Soc.* **61** (1947), 69–105.

[Seg 2] Segal, I. E.
Decomposition of operator algebras I, *Mem. Amer. Math. Soc.* **9** (1951), 1–67.

[Seg 3] Segal, I. E.
A class of operator algebras which are determined by groups, *Duke Math. J.* **18** (1951), 221–265.

[She 1] Sherman, S.
Order in operator algebras, *Amer. J. Math.* **73** (1951), 227–232.

[Sim 1] Simon, B.
Quantum dynamics, from automorphism to Hamiltonian, in *Studies in Mathematical Physics*, (Lieb, Simon, Wightman, eds). Princeton Univ. Press, Princeton (1976).

[Ska 1] Skau, C. F.
Orthogonal measures on the state space of a C^*-algebra, in *Algebras in Analysis* (J. H. Williamson, ed.). Academic Press, New York–San Francisco–London (1975).

[Ska 2] Skau, C. F.
Finite subalgebras of a von Neumann algebra, *J. Func. Anal.* **25** (1977), 211–235.

[Sti 1] Stinespring, W. F.
Positive functions on C^*-algebras, *Proc. Amer. Math. Soc.* **6** (1955), 211–216.

[Stø 1] Størmer, E.
Positive linear maps of operator algebras, *Acta Math.* **110** (1963), 233–278.

[Stø 2] Størmer, E.
Positive linear maps of C^*-algebras, in *Lecture Notes in Physics* Vol. **29** Springer-Verlag, Berlin–Heidelberg–New York (1974).

[Stø 3] Størmer, E.
Symmetric states of infinite tensor products of C^*-algebras, *J. Func. Anal.* **3** (1969), 48–68.

[Stø 4] Størmer, E.
Automorphisms and invariant states of operator algebras, *Acta Math.* **127** (1971), 1–9.

[Str 1] Streater, R. F.
Lorentz invariance of the Wightman functions in $P(\phi)_2$, *Commun. Math. Phys.* **26** (1972), 109–120.

[Tak 1] Takesaki, M.
Covariant representations of C^*-algebras and their locally compact automorphism groups, *Acta Math.* **119** (1967), 272–303.

[Tak 2] Takesaki, M.
Duality for crossed products and the structure of von Neumann algebras of type III, *Acta Math.* **131** (1973), 249–310.

[Tak 3] Takesaki, M.
Tomita's theory of modular Hilbert-algebras and its application, in *Lecture Notes in Mathematics* Vol. 128. Springer-Verlag, Berlin–Heidelberg–New York (1970).

[Tak 4] Takesaki, M., and J. Tomiyama.
Applications of fibre bundles to a certain class of C^*-algebras, *Tohoku Math. J.* **13** (1961), 498–523.

[Taka 1] Takai, H.
The quasi orbit space of continuous C^*-dynamical systems, *Trans. Amer. Math. Soc.* **216** (1976), 105–113.

[Tom 1] Tomiyama, J.
On the projection of norm 1 in W^*-algebras I, II, III, *Proc. Jap. Acad. Sci.* **33** (1957), 608–612; *Tohoku Math. J.* **10** (1958), 204–209; *ibid.* **11** (1959), 125–129.

[Tomi 1] Tomita, M.
Quasi-standard von Neumann algebras, unpublished.

[Tomi 2] Tomita, M.
Harmonic analysis on locally compact groups, *Math. J. Okayama Univ.* **5** (1956), 133–193.

[Tomi 3] Tomita, M.
Spectral theory of operator algebras I, *Math. J. Okayama Univ.* **9** (1959), 63–98.

[Tro 1] Trotter, H.
Approximation of semigroups of operators, *Pac. J. Math.* **8** (1959), 887–919.

[Tro 2] Trotter, H.
On the product of semigroups of operators, *Proc. Amer. Math. Soc.* **10** (1959), 545–551.

[Wig 1] Wightman, A. S.
Theoretical physics, Inst. Atomic Energy Authority Vienna (1963).

[Wig 2] Wightman, A. S.
Hilbert's sixth problem: Mathematical treatment of the axioms of physics, in *Mathematical developments arising from Hilbert problems*, F. E. Browder ed., Proc. Sympos. Pure Math. **28** (1976).

[Wil 1] Wils, W.
Désintegration centrale des formes positives sur les C^*algèbres, *C. R. Acad. Sci. Paris* **267** (1968), 810–812.

[Wil 2] Wils, W.
Désintegration centrale dans une partie convexe compacte d'un espace localement convexe, *C. R. Acad. Sci. Paris* **269** (1969), 702–704.

[Wil 3] Wils, W.
The ideal center of partially ordered vector spaces, *Acta Math.* **127** (1971), 41–77.

[Wor 1] Woronowicz, S. L.
Unpublished notes.

[Wor 2] Woronowicz, S. L.
Nonextendible positive maps, *Commun. Math. Phys.* **51** (1976), 243–282.

[Wor 3] Woronowicz, S. L.
A remark on the polar decomposition of m-sectorial operators. *Lett. Math. Phys.* **1** (1977), 429–433.

[Yng 1] Yngvason, J.
On the decomposition of Wightman functionals in the Euclidean framework. *Rep. Math. Phys.* **13** (1978), 101–115.

[Yos 1] Yosida, K.
On the differentiability and the representation of one-parameter semigroups of linear operators, *J. Math. Soc. Japan* **1** (1948), 15–21.

[Zel 1] Zeller-Meier, G.
Products croisés d'une C^*-algèbre par un groupe d'automorphismes, *J. Math. Pures et Appl.* **47** (1968), 101–239.

List of Symbols

Standard Notation

$\mathfrak{A}, \mathfrak{B}, \mathfrak{C}$	C^*-algebras
$\mathfrak{M}, \mathfrak{N}, \mathfrak{Z}$	von Neumann algebras
$\mathfrak{I}$	Ideals in C^* and von Neumann algebras
$A, B, C,$	elements in C^* and von Neumann algebras
ω, φ	states
$\mathfrak{H}$	a Hilbert space
π	a^*-morphism
Ω, ξ, η, ψ	vectors in Hilbert space
P, E, F	orthogonal projections
G	a group
Λ	a bounded region in an Euclidean space or on a lattice
μ, ν	measures

Special Symbols

$\|A\|$	Absolute value of A	Before Proposition 2.2.10
$\\|A\\|$	C^*-norm of A	Before Definition 2.1.1
A_+, A_-	Positive and negative part of A	Proposition 2.2.10
A^*	Adjoint of A	Before Definition 2.1.1
A^{-1}	Inverse of A	Before Definition 2.2.1

Symbol	Description	Reference
$\hat{A}$	Gelfand transform, or affine functional defined by A	Before Theorem 2.1.11(B) and beginning of Section 4.1.3
$A(K)$	The real affine continuous functions on the compact convex set K	Beginning of Section 4.1.2
$\tilde{\mathfrak{A}}$	Unit extension of $\mathfrak{A}$	Definition 2.1.6
$\mathfrak{A}_+$	The positive part of $\mathfrak{A}$	Definition 2.2.7
$\bigotimes_{i=1}^{n} \mathfrak{A}_i, \bigotimes_{\alpha} \mathfrak{A}_\alpha$	Finite resp. infinite tensor products of C^*-algebras	Section 2.7.2
$(\mathfrak{A}, G, \alpha)$	C^*-dynamical system	Definition 2.7.1
$\mathfrak{A} \bigotimes_\alpha G$	C^*-cross product	Definition 2.7.2
$\hat{\alpha}_\gamma$	Dual action of α_t	Before Theorem 2.7.4
$B_{\mathfrak{A}}$	The positive linear functionals of $\mathfrak{A}$ with norm less than or equal to one	Theorem 2.3.15
$b(\mu)$	Barycenter of μ	Proposition 4.1.1
$C_0(X)$	The continuous complex functions on the locally compact Hausforff space X vanishing at infinity	Example 2.1.4
$C(X)$	The continuous complex functions on the compact Hausforff space X	Example 2.1.4
$C^*(\mathfrak{A}, \alpha)$	C^*-crossed product	Definition 2.7.2
C_0	weakly continuous	Definition 3.1.2
$C_0{}^*$	weakly* continuous	Definition 3.1.2
$\mathbb{C}$	The complex numbers	
$Co(\tau_G(A))$	The convex hull of $\{\tau_g(A); g \in G\}$	Definition 4.3.6
$D(S)$	Domain of the linear operator S	
Δ	Modular operator	Definition 2.5.10
Δ_{ξ_1, ξ_2}	relative modular operator	Lemma 2.5.34
$(D\psi : D\varphi)_t$	Radon–Nikodym cocycle	Theorem 2.7.16
δ_ω	Unit point measure on ω	Before Proposition 4.1.1
$\partial_f(K)$	Boundary set of f	Definition 4.1.5
$\Delta_\mu(A)$	Root mean square deviation	Before Example 4.1.29
$E_{\mathfrak{A}}$	State space of $\mathfrak{A}$	Definition 2.3.14
$E_{\mathfrak{A}}^G$	The G-invariant states on $\mathfrak{A}$	Beginning of Section 4.3.1
E_ω	Projection onto $U_\omega(G)$-invariant vectors	Proposition 4.3.7
$\mathscr{E}(K)$	The extremal points of the convex set K	Beginning of Section 4.1.2
F	$F = J\Delta^{-1/2}$	Definition 2.5.10
$\bar{f}(\omega)$	Upper envelope of f	Definition 4.1.5
$F_{\mathfrak{A}}$	The set of factor states on $\mathfrak{A}$	Before Theorem 4.2.10
F_ω	$F_\omega = [\pi_\omega(\mathfrak{A})' E_\omega]$	Proposition 4.3.8
$\Gamma(\mathfrak{M})$	Invariant Γ of $\mathfrak{M}$	Theorem 2.7.23
$G(S)$	Graph of the linear operator S	Before Lemma 3.1.27
$\Gamma(\alpha)$	Γ-spectrum of α	Notes and Remarks to Section 3.2.3
$\hat{G}$	The dual group of the abelian locally compact group G	Beginning of Section 3.2.3
(γ, t)	$\gamma \in \hat{G}$ acting on $t \in G$	Beginning of Section 3.2.3
$\hbar$	Planck's constant	Introduction
$(\mathfrak{H}, \pi)$	Representation on $\mathfrak{H}$	Definition 2.3.2

$(\mathfrak{H}, \pi, \Omega)$	Cyclic representation on $\mathfrak{H}$ with cyclic vector Ω	Definition 2.3.5
$(\mathfrak{H}_\omega, \pi_\omega, \Omega_\omega)$	Cyclic representation defined by the state ω	Definition 2.3.18
$\mathfrak{H}_\omega$	Hilbert space in cyclic representation defined by ω	Definition 2.3.18
$\bigotimes_{i=1}^{n} \mathfrak{H}_i$	Finite tensor product of Hilbert spaces	Section 2.7.2
$\bigotimes_{\alpha}^{\Omega_\alpha} \mathfrak{H}_\alpha$	Infinite tensor product of Hilbert spaces	Section 2.7.2
$\mathbb{1}$	Identity of an operator algebra	After Example 2.1.4
ι	Identity automorphism of an operator algebra	
I	Identity on a Banach space	
$\mathfrak{I}_Y^U$		Definition 3.2.37
J	Modular conjugation	Definition 2.5.10
$j(A)$	$j(A) = JAJ$	Before Definition 2.5.25
J_{ξ_1, ξ_2}	Relative modular conjugation	Lemma 2.5.35
$\mathscr{K}(\mathfrak{A}, G)$	Continuous functions $G \mapsto \mathfrak{A}$ with compact support	Definition 2.7.2
$\kappa_\mu(f)$		Lemma 4.1.21
$\mathscr{L}(X)$	The bounded operators on the Banach space X	Example 2.1.2
$\mathscr{LC}(\mathfrak{H})$	The compact operators on $\mathfrak{H}$	Example 2.1.3
$\mathscr{L}_*(\mathfrak{H})$	The predual of $\mathscr{L}(\mathfrak{H})$	Definition 2.4.4
$L^p(\mu)$, $L^p(X, \mu)$	L^p-spaces of complex functions, $1 \leq p \leq +\infty$	Beginning of Section 2.5
$L^p(X, G)$	L^p-functions from the locally compact group G into the Banach space X	Definitions 2.7.2 and 2.7.3
$\|\Lambda\|$	Area of Λ, or number of points in Λ	
$\mathfrak{M}'$, $\mathfrak{M}''$	Commutant, resp. bicommutant of set $\mathfrak{M}$	Beginning of Section 2.4
$[\mathfrak{M}\mathfrak{K}]$	The closed linear span of $\mathfrak{M}$, $\mathfrak{K}$, or the projection into this	Definition 2.4.10
$\mathfrak{M}_*$	The predual of $\mathfrak{M}$	Definition 2.4.16
$(\mathfrak{M}, G, \alpha)$	W^*-dynamical system	Definition 2.7.3
$\mathfrak{M} \otimes_\alpha G$	W^*-crossed product	Definition 2.7.3
$\bigotimes_{i=1}^{n} \mathfrak{M}_i$	Finite W^*-tensor product	Section 2.7.2
$\bigotimes_{\alpha}^{\Omega_\alpha} \mathfrak{M}_\alpha$	Infinite W^*-tensor product	Section 2.7.2
$M_+(K)$	The positive Radon measures on the convex compact set K	Beginning of Section 4.7.2
$M_1(K)$	The probability measures on K	Beginning of Section 4.7.2
$M_\omega(K)$	The probability measures with barycenter ω	Before Proposition 4.1.1
$\mu \vee \nu$	Least upper bound	Before Proposition 4.1.1
$\mu \wedge \nu$	Greatest lower bound	Before Proposition 4.1.1
$\mu \succ \nu$	Choquet ordering	Definition 4.1.2

$\mu \gg \nu$	Bishop–de Leeuw ordering	Notes and Remarks to Section 4.1.2
$M(A)$	Mean of A	Proposition 4.3.8
$\mathfrak{M}_E$	$\mathfrak{M}_E = E\mathfrak{M}E$	Lemma 4.3.13
$N_{\mathfrak{A}}$	The normal states on the C^*-algebra $\mathfrak{A}$ on a Hilbert space	Proposition 2.6.15
$N_{\omega}{}^G$	The G-invariant, ω-normal states on $\mathfrak{A}$	Theorem 4.3.9
ω_{Ω}	Vector state associated to the vector Ω	After Def. 2.3.9
Ω_{ω}	Cyclic vector associated to the state ω	Definition 2.3.18
$o(t)$	Smaller order than t	Beginning of Section 3.1.5
$O(t)$	Order t	Beginning of Section 3.1.5
$\omega_1 \perp \omega_2$	Orthogonal states	Definition 4.1.20
$\mathcal{O}_{\omega}(E_{\mathfrak{A}}), \mathcal{O}_{\omega}$	The orthogonal probability measures on $E_{\mathfrak{A}}$ with barycenter ω	Definition 4.1.20
$\omega_1 \between \omega_2$	Disjoint states	Beginning of Section 4.2.2
$\pi_1 \simeq \pi_2$	Unitarily equivalent representations	After Proposition 2.3.8
$\pi_1 \approx \pi_2$	Quasi-equivalent representations	Definition 2.4.24
$\pi_1 \between \pi_2$	Disjoint representations	Beginning of Section 4.2.2
π_{ω}	The *-morphism in the cyclic representation associated to ω	Definition 2.3.18
$\mathscr{P}, \mathscr{P}_{\Omega}$	Natural positive cone	Definition 2.5.25
$\hat{P}_{\omega}$	Projection onto U_{ω}-almost periodic vectors	Theorem 4.3.27
$r_{\mathfrak{A}}(A)$	Resolvent set of $A \in \mathfrak{A}$	Definitions 2.2.1 and 2.2.4
$\rho(A)$	Spectral radius of A	Proposition 2.2.2
$R(S)$	Range of the linear operator S	
$\mathbb{R}$	The real numbers	
$\sigma_{\mathfrak{A}}(A)$	Spectrum of $A \in \mathfrak{A}$	Definitions 2.2.1 and 2.2.4
$\sigma(\mathfrak{A})$	Spectrum of the abelian C^*-algebra $\mathfrak{A}$	Definition 2.3.25
S	$S = J\Delta^{1/2}$	Definition 2.5.10
$\sigma_t{}^{\omega}$	Modular automorphism group associated to ω	Definition 2.5.15
$\sigma(X, F)$	Topology on the linear space X defined by the subspace F of the dual	Before Definition 2.5.17
S_{ξ_1, ξ_2}	$S_{\xi_1, \xi_2} = J_{\xi_1, \xi_2}\Delta^{1/2}_{\xi_1, \xi_2}$	Lemma 2.5.33
$S(\mathfrak{M})$	Invariant S of $\mathfrak{M}$	Definition 2.7.22
$\sigma_U(Y)$	U-spectrum of Y	Definition 2.3.37
$\sigma(U)$	Spectrum of U	Definition 2.3.37
$S(K)$	The real continuous convex functions on the compact convex set K	Beginning of Section 4.1.2
$S_{\lambda}(U), S_{\lambda}(\tau(A))$	Convex combinations of $U(g)$'s resp. $\tau_g(A)$'s	Proof of Proposition 4.3.7
$\sigma_p(U), \sigma_p(\tau)$	Point spectra of U resp. τ	Beginning of Section 4.3.3
$\int_Z^{\oplus} d\mu(z)\xi(z)$	Direct integral of vectors	Definition 4.4.1
$\int_Z^{\oplus} d\mu(z)\mathfrak{H}(z)$	Direct integral of Hilbert spaces	Definition 4.4.1
$\int_Z^{\oplus} d\mu(z)T(z)$	Direct integral of operators	Before Theorem 4.4.3

$\int_Z^{\oplus} d\mu(z)\mathfrak{M}(z)$	Direct integral of von Neumann algebras	Definition 4.4.4
$\int_Z^{\oplus} d\mu(z)\pi(z)$	Direct integral of representations	Before Theorem 4.4.7
$\mathscr{T}(\mathfrak{H})$	The trace class operators on	Proposition 2.4.3
$\tau(X, F)$	Mackey topology	Before Definition 3.1.2
U_ω	Unitary representation associated to the G-invariant state	Corollary 2.3.17
$U_\mu(A)$	$U_\mu(A) = \int d\mu(t)\, U_t(A)$	Proposition 3.1.4
$\mathscr{U}(A, \varepsilon)$	Neighborhood of A	Before Definition 2.1.1
$W^*(\mathfrak{M}, \alpha)$	W^*-crossed product	Definition 2.7.3
X_*	Predual of X	Before Definition 2.5.17
$X^U(E), X_0{}^U(E)$	Spectral subspaces corresponding to $E \subseteq \hat{G}$	Definition 3.2.37
$\mathfrak{Z}(\mathfrak{M})$	Center of $\mathfrak{M}$	Definition 2.4.8
$\mathfrak{Z}_\omega$	Center of $\pi_\omega(\mathfrak{A})''$	Before Definition 2.6.4
$\mathfrak{Z}_\omega{}^c$	Commutant algebra	Definition 2.6.4
$\mathfrak{Z}_\omega{}^\perp$	Algebra at infinity	Definition 2.6.4

Index